Antiproton 1984

Antiproton 1984

Proceedings of the VII European Symposium on Antiproton Interactions held in Durham, 9–13 July 1984

Edited by M R Pennington

Institute of Physics Conference Series Number 73

Adam Hilger Ltd, Bristol and Boston

CODEN IPHSAC 73 1–530 (1985)

British Library Cataloguing in Publication Data

European Symposium on Antiproton Interactions
(7th: 1984: Durham)
Antiproton 1984: proceedings of the VII
European Symposium on Antiproton Interactions
held in Durham, 9–13 July 1984.—(Conference
series/Institute of Physics, ISSN 0305–2346; 73)
1. Antiprotons
I. Title II. Pennington, M. R. III. Series
539.7′212 QC793.5.P722

ISBN 0-85498-164-0
ISSN 0305-2346

Organising Committee

A Astbury (TRIUMF, Vancouver, Canada), P Dalpiaz (Ferrara, Italy), J R Fry (Liverpool, UK), U Gastaldi (CERN, Geneva, Switzerland), R Klapisch (CERN, Geneva, Switzerland), M R Pennington (Durham, UK), J G Rushbrooke (Cambridge, UK) and G A Smith (Pennsylvania, USA)

Honorary Editor

M R Pennington

Published on behalf of The Institute of Physics by Adam Hilger Ltd
Techno House, Redcliffe Way, Bristol BS1 6NX, England
PO Box 230, Accord, MA 02018, USA

Printed in Great Britain by J W Arrowsmith Ltd, Bristol

Preface

The last two years have been one of the most exciting periods in particle physics and antiproton interactions have been at the heart of this. The discovery of the intermediate vector bosons of the weak interaction at the CERN $\bar{p}p$ Collider, together with the first physics results from the low energy antiproton ring (LEAR), herald the start of a new age of discovery in high energy physics. The stretching of our knowledge (whether into new regimes of higher energies, or to earlier times in the history of the Universe, or, with much higher statistics, results on low energy processes) confronts our current understanding, testing our theories to their limits, hopefully revealing new and deeper insights into the nature of matter.

This conference, supported by travel funds from the UK Science and Engineering Research Council and held in the beautiful cathedral city of Durham, fruitfully brought together those working in all areas of research triggered by antiproton interactions. These Proceedings cover all these aspects from filling in fine details of the canvas that makes up particle physics and its interface with the physics of the nucleus to broad sweeps into the totally unknown. We hope all those who consult these Proceedings will capture some of the interest in this challenging subject.

These Proceedings are dedicated to the memory of Claude Broll, who, together with his family, participated in this conference and who was so tragically killed in an accident the week after.

M R Pennington
Durham
September 1984

Contents

v Preface

xi List of participants

Section 1: Status report on particle physics

1–16 High energy physics: where are we now?
R L Jaffe

17 Discussion on Section 1

Section 2: Collider physics

19–35 Latest results from the UA1 experiment on W and Z^0 and missing transverse energy events
UA1 collaboration; presented by J Tuominiemi

37–50 Large mass electron–neutrino pairs in UA2
UA2 collaboration; presented by S Loucatos

51–68 Theoretical implications of recent collider results
R D Peccei

69–80 Jets at ISR, FNAL and SPS
T Åkesson

81–97 Latest results on high p_T jets in UA2
UA2 collaboration; presented by L Fayard

99–116 Jet phenomenology
B R Webber

117–118 Discussion on Section 2

Section 3: Low energy antiproton reactions

119–130 LEAR—Machine and experimental areas: experience and future plans after one year of operation
M Chanel, P Lefèvre, D Möhl and D J Simon; presented by M Chanel

131–135 First observation of *K* x-rays of antiprotonic hydrogen atoms
S Ahmad et al; presented by R Landua

137–142 X-rays from antiprotonic hydrogen, deuterium and helium
J D Davies et al; presented by J D Davies

143–145 Study of $\bar{p}p \rightarrow e^+e^-$ reaction at rest
G Bardin et al; presented by P Dalpiaz

147–155 Results from PS 172: $\bar{p}p$ total cross-sections and spin effects in $\bar{p}p \rightarrow K^+K^-$, $\pi^+\pi^-$, $\bar{p}p$ above 200 MeV/c
C I Beard et al; presented by S Dalla Torre-Colautti and L Linssen

157–164 Measurement of $\bar{p}p$ cross-sections at low $\bar{p}$ momenta
W Bruckner et al; presented by T Walcher

165–174 Annihilation mechanisms
I S Shapiro

175–180 Antiprotonic atom spectroscopy at LEAR (PS 176)
H Koch et al; presented by H Koch

181–186 Antiprotonic molybdenum: resonant coupling of atomic and nuclear states and the impact of annihilation on the nucleus (PS 186)
T von Egidy et al; presented by T von Egidy

187–194 PS 184: a study of $\bar{p}$–nucleus interaction with a high resolution magnetic spectrometer
D Garreta et al; presented by D Garreta

195–200 PS 187—a good statistics study of antiproton interactions with nuclei: preliminary results
J W Sunier et al; presented by J W Sunier

201–203 Investigation of the $\bar{p}p \rightarrow \bar{\Lambda}\Lambda$ reaction near threshold at LEAR
P D Barnes et al; presented by B E Bonner

205–212 Low energy antiproton interaction studies at Brookhaven
M Sakitt et al; presented by M Sakitt

213–225 What antiproton physics can tell us about nuclear physics
A M Green

227–233 Discussion on Section 3

Section 4: The very small to the very large

235–242 Antimatter in the universe
N C Rana and A W Wolfendale; presented by A W Wolfendale

243–250 Particle physics and cosmology
P J E Peebles

251–258 Interaction of $\bar{p}$ with ^{4}He and Ne nuclei at 49 and 180 MeV
F Balestra et al; presented by G Bendiscioli

259–262 Measurement of the antineutron mass
M Cresti et al; presented by M Cresti

263–278 Do protons decay?
P J Litchfield

279–280 Discussion on Section 4

Section 5: Hadron spectroscopy

281–286 Current issues in meson spectroscopy
G Karl

287–296 p̄p annihilation at rest from atomic p-states
S Ahmad et al; presented by C Amsler

297–304 Investigations on baryonium and other rare p̄p annihilation modes using high resolution π^0 spectrometers (PS 182)
L Adiels et al; presented by P Pavlopoulos

305–313 A search for structure in the charged meson spectra from proton–antiproton annihilations at rest
A Angelopoulos et al; presented by D Schultz

315–321 In search of p̄N bound states
D Bridges et al; presented by T Kalogeropoulos

323–328 Preliminary results from antineutron–proton annihilation cross-section measurements from 100–500 MeV/*c* at the AGS
D Lowenstein et al; presented by L Pinsky

329–338 Antiproton results from KEK
K Nakamura and T Tanimori

339–347 Formation of charmonium states in antiproton–proton annihilation
C Baglin et al; presented by M Macri

349–354 QCD prediction for charmonium production in p̄p collisions
E L Berger, P H Damgaard and K Tsokos; presented by P H Damgaard

355–374 Experimental observations of isolated large transverse momentum leptons with associated jets in experiment UA1
UA1 collaboration; presented by D DiBitonto

375–394 How to search for new particles at p̄p colliders
V Barger

395–401 Discussion on Section 5

Section 6: Physics of soft processes†

403–409 Proton–proton and antiproton–proton elastic scattering at ISR energies in the t interval $0.05 < |t| \leq 3.0$ GeV^2
F Rimondi

411–419 Odderon effects at the Collider and beyond
B Nicolescu

† No manuscript was received of the report on results from UA4 by P L Braccini (Pisa/Perugia).

421–432 Results from the UA5 experiment at the CERN p̄p Collider
UA5 collaboration; presented by C Geich-Gimbel

433–446 Proton–proton and proton–antiproton elastic scattering at high energy
A Donnachie

447–462 The Lund string model
B Andersson

463–464 Discussion on Section 6

Section 7: The future

465–469 Future plans for the UA5 experiment—p̄p interactions at 900 GeV CM energy
UA5 collaboration; presented by D R Ward

471–488 Physics from future colliders
G L Kane

489 Discussion on Section 7

Section 8: Poster sessions

491–494 Relativistic impulse approximation for antiproton scattering
A T M Aerts

495–498 Low energy nucleon–antinucleon interactions and a hybrid model
F Ando

499–502 A model of antiproton annihilation in nuclei implying two nucleons
J Cugnon and J Vandermeulen; presented by J Vandermeulen

503–506 Parametrization of low energy p̄p scattering data
W Derks, P H Timmers and J J de Swart; presented by J J de Swart

507–510 The antiprotonic x-ray cascade in Mo isotopes (PS 186)
F J Hartmann et al; presented by F J Hartmann

511–514 Calculation of parton fragmentation functions from jet calculus: gluon applications
K E Lassila and A Ng; presented by K E Lassila

515–518 Antineutron production from the p̄–carbon interaction at 590 MeV/*c*
K Nakamura et al; presented by K Nakamura and T Tanimori

519–522 Model analysis of the p̄p annihilation process at low energies
S Saito, Y Hattori, J H Kim and T Yamagata; presented by T Yamagata

523–526 Measurement of the real-to-imaginary ratios of the p̄p and p̄n forward amplitudes between 367 and 764 MeV/*c*
T Takeda et al; presented by K Nakamura

527–530 Author Index

Participants

Aerts A T M	CERN, Geneva, Switzerland
Åkesson T	CERN, Geneva, Switzerland
Allan A	Durham, UK
Amsler C	Zurich, Switzerland
Andersson B	Lund, Sweden
Ando F	Shinshu Univ., Matsumoto, Japan
Aslanides E	Strasbourg, France
Au K L	Durham, UK
Backenstoss G	Basel, Switzerland
Barger V	Madison, Wisconsin, USA
Barnes V	Purdue Univ., Lafayette, USA
Bassompierre G	LAPP, Annecy, France
Bendiscioli G	Pavia, Italy
Biswas N N	Notre Dame, USA
Bonner B E	CERN, Switz./Los Alamos, USA
Braccini P L	Perugia/INFN, Pisa, Italy
Bressani T	Torino, Italy
Broll C	LAPP, Annecy, France
Brown L M	Northwestern Univ., Evanston, USA
Calabrese R	Ferrara, Italy
Capon G	CERN, Geneva, Switzerland
Carter M	Durham, UK
Cester-Regge R	Torino, Italy
Chanel M	CERN, Switzerland
Cooper M	Basel, Switz./Los Alamos, USA
Cresti M	Padova, Italy
Dalla Torre S	Trieste, Italy
Dallman D	IHEP, Vienna, Austria
Dalpiaz P	Ferrara, Italy
Damgaard P H	Nordita, Copenhagen, Denmark
Davies J D	Birmingham, UK
De Swart J J	Nijmegen, Netherlands
DiBitonto D	CERN, Geneva, Switzerland
Donnachie A	Manchester, UK
Duboc J	LPNHE, Paris VI, France
Dumbrajs O	KfK, Karlsruhe, Fed. Germany
von Egidy T	Tech. Univ. München, Fed. Germany
Fayard L	LAL, Orsay, France
Franklin M	Lawrence Berkeley Laboratory, USA
Fry J R	Liverpool, UK
Galster S	KfK, Karlsruhe, Fed. Germany
Garreta D	CEN, Saclay, France

Gastaldi U	CERN, Geneva, Switzerland
Geesaman D	Argonne National Lab., USA
Geich-Gimbel C	Bonn, Fed. Germany
Glover E W N	Durham, UK
Grayson S L	Durham, UK
Green A M	Helsinki, Finland
Guaraldo C	INFN, Frascati, Italy
Haidenbauer J	Graz, Austria
Hamann N	Freiburg, Fed. Germany
Hartmann J	Tech. Univ. München, Fed. Germany
Hertzog D W	CERN, Switz./Carnegie-Mellon, USA
Jabiol M A	CEN, Saclay, France
Jaffe R L	MIT, Cambridge, USA
Jeannet E	Neuchatel, Switzerland
Kalogeropoulos T E	Syracuse, USA
Kane G L	Michigan, USA
Karl G	Oxford, UK/Guelph, Canada
Kenney V P	Notre Dame, USA
Klapisch R	CERN, Geneva, Switzerland
Koch H	KfK, Karlsruhe, Fed. Germany
Laberrigue-Frolow J	LPNHE, Univ. P & M Curie, Paris, France
Laloum C M	Collège de France, Paris, France
Landshoff P V	Cambridge, UK
Landua R	CERN, Geneva, Switzerland
Lassila K E	Ames, Iowa, USA
Lazarus D M	Brookhaven Nat. Lab., Upton, USA
Limentani S	Padova, Italy
Linssen L	NIKHEF, Amsterdam, Netherlands
Litchfield P J	Rutherford Appleton Lab., Chilton, UK
Loucatos S	DPhPE, CEN, Saclay, France
Lowenstein D I	Brookhaven Nat. Lab., Upton, USA
Lundby A	CERN, Geneva, Switzerland
Macri M	Genova, Italy
Maeshima K	CERN, Geneva, Switzerland
Mandelkern M A	Univ. California, Irvine, USA
Martin A D	Durham, UK
Mayes B W	Houston, USA
Miller D M	Northwestern Univ., Evanston, USA
Mitaroff W	IHEP, Vienna, Austria
Monnand E	CEN, Grenoble, France
Mutchler G S	Rice Univ., Houston, USA
Nakamura K E	KEK, Japan
Nicolescu B	IPN, Orsay, France
Oades G C	Aarhus, Denmark
Pavlopoulos P	CERN, Geneva, Switzerland
Peacock A W	Durham, UK

Peccei R D	MPI, Munich, Fed. Germany
Peebles P J E	Princeton, USA
Pennington M R	Durham, UK
Phillips G C	Rice Univ., Houston, USA
Pinsky L S	Houston, USA
Plessas W	Graz, Austria
Ransome R D	MPI, Heidelberg, Fed. Germany
Repond J	Basel, Switzerland
Rimondi F	Bologna, Italy
Robinson D K	Case West. Res. Univ., Cleveland, USA
Rochester G A	Durham, UK
Rouhani S	Durham, UK
Sakitt M	Brookhaven Nat. Lab., Upton, USA
Sartori G	Padova, Italy
Schultz D	Univ. California, Irvine, USA
Shapiro I S (written contribution only)	Lebedev Phys. Inst., Moscow, USSR
Shibata T A	MPI, Heidelberg, Fed. Germany
Shubert R	Cal. State Univ., Fullerton, USA
Smith G A	Penn State Univ., USA
Speiers N A	Durham, UK
Straumann U	CERN, Geneva, Switzerland
Sunier J W	Los Alamos, USA
Sutton C	*New Scientist*, London, UK
Tanimori T	Tokyo, Japan
Tanimoto M	Ehime Univ., Matsuyama, Japan
Tauscher F G L	Basel, Switzerland
Tecchio L	Torino, Italy
Tuominiemi J	Helsinki, Finland
Vandermeulen J	Liège, Belgium
Vrana J	Collège de France, Paris, France
Walcher T	MPI, Heidelberg, Fed. Germany
Ward D R	Cambridge, UK
Webb J N	Durham, UK
Webber B R	Cambridge, UK
Whalley M R	Durham, UK
Williams M C	Basel, Switzerland
Wolf A	CERN, Switz./Karlsruhe, Fed. Germany
Wolfe D	New Mexico, USA
Wolfendale A W	Durham, UK
Worrall A D	Durham, UK
Yamagata T	Tokyo Metropolitan Univ., Japan

Inst. Phys. Conf. Ser. No. 73: Section 1
Paper presented at VII Eur. Symp. Antiproton Interactions, Durham 1984

High energy physics: where are we now?

R. L. Jaffe*

Center for Theoretical Physics
Laboratory for Nuclear Science and Department of Physics
Massachusetts Institute of Technology
Cambridge, Massachusetts 02139

These were two spectacular years for nucleon anti-nucleon physics. They have brought us the discovery of the Z^0 and W[1,2], the demonstration of a clean jet signal at the $Sp\bar{p}S$ collider[3], perhaps the discovery of the t-quark[4], the turn on of LEAR[5], and the beginning of serious study of the physics potential of supercolliders. In addition the UA1 and UA2 collaborations have reported a number of unexpected and intriguing events (see Table I) which have taken theorists by surprise and have raised high expectations for the next collider run. Undoubtedly, some classes of events will go away and others will have rather prosaic explanations, but it seems to me that on balance there is rather strong evidence for new physics at $\sim$100 GeV in hadron-hadron colliders. So perhaps the answer to the question posed by the organizers in the ambitious title of this talk should be: "For a change, we don't know!"

A talk with this title given anytime in the past decade would probably enumerate the tests and successes of the standard $[SU(3)\times SU(2)_L\times U(1)]$ model. Instead I want to concentrate on some of its problems. In general, it is strong on symmetry and weak on dynamics. The dynamical problems fall into two categories:

1. The origins of symmetry breakdown. [What determines masses and angles?]

2. The dynamics of confining gauge theory. [How does quantum chromodynamics explain hadron dynamics?]

Despite a decade of work and many clever ideas on both subjects, little is understood of either. Given my own interests, I will concentrate on the phenomenological aspects of these problems. First, I will succumb to the irresistable temptation to discuss the anomalous collider events. Second, I will discuss two topics in QCD at low energies: glueballs and strange matter. Finally, I will examine some exciting possibilities for low energy theory at the highest energy colliders, namely to model a strongly interacting Higg's sector.

*Supported in part by funds provided by the U.S. Department of Energy (DOE) under contract DE-AC02-76ER03069.

1. Anomalous Collider Events

This past June a workshop on theoretical physics at supercollider energies took place at Berkeley. One working group examined the anomalous events recently reported at the $Sp\bar{p}S$ and at PETRA. The results are given in Tables I and II [6]. Table I summarizes the events themselves, giving standard model background sources and rates along with my own judgement [≡(signal/background)×new physics potential] of the level of excitement they engender. Table II summarizes some of the theoretical models which have been proposed over the past few months to explain these events. We believe Table I to be reasonably complete, but Table II suffers many omissions and new models appear almost daily. A more complete version will appear in Ref. 6. Time doesn't permit me to discuss either the events or the models in much detail. [*]

It is apparent from Table I that the rates for some classes of peculiar events far exceed background estimates. Certain spectacular events, in particular, defy standard model explanation. For example, monojet A reported by UA1 consists of a 80^{+20}_{-13} GeV muon accompanied by 25 GeV of neutral energy; in addition there is an unbalanced P_T of 66±8 GeV. Monojet B and the $ej\not{p}_T$ events B and C from UA2 are only slightly less spectacular. Even models specifically designed to produce a $j\not{p}_T$ or $ej\not{p}_T$ signature in general have a hard time generating such isolated spectacular events.

TABLE I

$Sp\bar{p}S$ ANOMALIES [6]

EVENT	#	GROUP	REF.	BACKGROUND: SOURCE	BACKGROUND: COMMENTS	FEATURES	RATING
$j\not{p}_T$ $jj\not{p}_T$ $>2j\not{p}_T$	6 0 1	UA1	7,8	$q\bar{q}\to Z^0 g$, $Z^0\to\nu\bar{\nu}$	too low by ∿20 for $j\not{p}_T$, even lower for $nj\not{p}_T$	jets have low invariant mass and low multiplicity	***
$ej(s)\not{p}_T$	4	UA2	9	$q\bar{q}\to Wg$	at least two events (B&C) survive	hard e separated from j(s)	***
$\gamma\not{p}_T$	2	UA1	7,8	$W\to\not{e}\nu$	1 event may be missed electron	E_γ=53, 54 GeV	**
$e^+e^-\gamma$ $\mu^+\mu^-\gamma$	2 1	UA1,2 UA1	10,11	QED	depends on how bremsstrahlung is calculated	$m(\ell^+\ell^-\gamma)\sim M_Z$; EM shower isolated from lepton pair	**
$j(s)Z^0$, $Z^0\to\ell^+\ell^-$	5	UA1	8	?	anomalous activity associated with Z^0 but not W	more jets, larger E_T, larger n_{ch} than expected or seen in W production	*
$\mu^+\mu^-j(s)$ $\mu^\pm\mu^\pm j(s)$	7 3	UA1	8	heavy flavor decays		6 GeV<m(μμ)<22 GeV many associated K^0, Λ	*

Notation:

- j(s): hadron jet(s), perhaps including charged leptons
- $\not{p}_T$: imbalance in momentum transverse to beam
- γ: large EM shower with no associated charged track

*For further discussion of models see the talk by R. Peccei in these proceedings.

TABLE II

MODELS FOR NEW PHYSICS AT $Sp\bar{p}S$ [6]

MODEL	REF.	$j\not{p}_T$	$ej\not{p}_T$	$\gamma\not{p}_T$	$\ell\bar{\ell}\gamma$	jZ^0	$\mu\mu j$	COMMENTS
$Z\to\ell^*\bar{\ell}$ ↳ $\ell\gamma$	12,13	-	-	✓	☼■	-	-	$m_{\ell^*} \sim 80$ GeV Fails Dalitz plot
$Z^0\to x\gamma$ ↳ $\ell\bar{\ell}$	14-18	-	-	✓	☼■	-	-	Fails Dalitz plot
$R\to\ell\bar{\ell}\gamma$	19	-	-	✓	☼✓	-	-	Attempts to explain Dalitz plot
$Z\leftrightarrow(Q\bar{Q})$	20	-	-	✓	☼■	-	-	Fails Dalitz plot. Requires several fine tunings.
Composite Z,W	21-24	-	-	✓	☼■	-	-	Fails Dalitz plot.
WW bound state	25	-	-	-	☼■	-	-	Fails Dalitz plot, production rate too low
Higgs	26	☼■	☼✓	☼✓	-	✓	-	No source of monojets, width ∿ 200 GeV
ν_4	27	*	-	☼✓	*	-	-	*These signals exist but would not look like observed elements. Requires fine tuning.
ODOR	28	☼✓*	-	✓†	-	☼✓	-	*Not event A †Rate too low
SUSY $gg\to\tilde{g}\tilde{g}$ $\to\tilde{q}\tilde{\bar{q}}$ $\to\tilde{g}\tilde{\omega}$	29-33	★*	✓	✓†	-	-	-	*Event A unlikely. †Requires hadronic jet with $n_{ch}=0$.

Notation:

- ☼ Primary motivation for model
- ✓ Qualitative explanation
- ★ Quantitative explanation
- \- No explanation
- ■ Model fails to explain what it claims to.

The $\ell^+\ell^-\gamma$ events deserve separate mention. Kinematically the events resemble bremsstrahlung (i.e., one lepton photon pair has a small invariant mass) but the rate is too large. A new estimate by UA2 [34] which integrates over all $O(\alpha)$ configurations less likely than those observed gives a probability

$$\frac{P(UA1:e^+e^-\gamma) + P(UA2:e^+e^-\gamma)}{P(12\ Z\to e^+e^-\ \text{events})} \sim \frac{1}{25}$$

which is small but perhaps not too small.

Theoretical attempts to ascribe the $\ell^+\ell^-\gamma$ events to new physical mechanisms fail to account for their kinematic structure. This is best seen in the Dalitz plot [19,35] of Fig. 1. x_H and x_L are related to lepton energies E_L and E_H in the $\ell^+\ell^-\gamma$ rest frame:

$$x_H = 1 - \frac{2E_L}{m(\ell_1\ell_2\gamma)} = \frac{m^2{}_H(\ell\gamma)}{m^2(\ell_1\ell_2\gamma)}$$

and likewise for x_L. m_H (m_L) is the higher (lower) of the two lepton-photon invariant masses. By convention $x_H > x_L$.

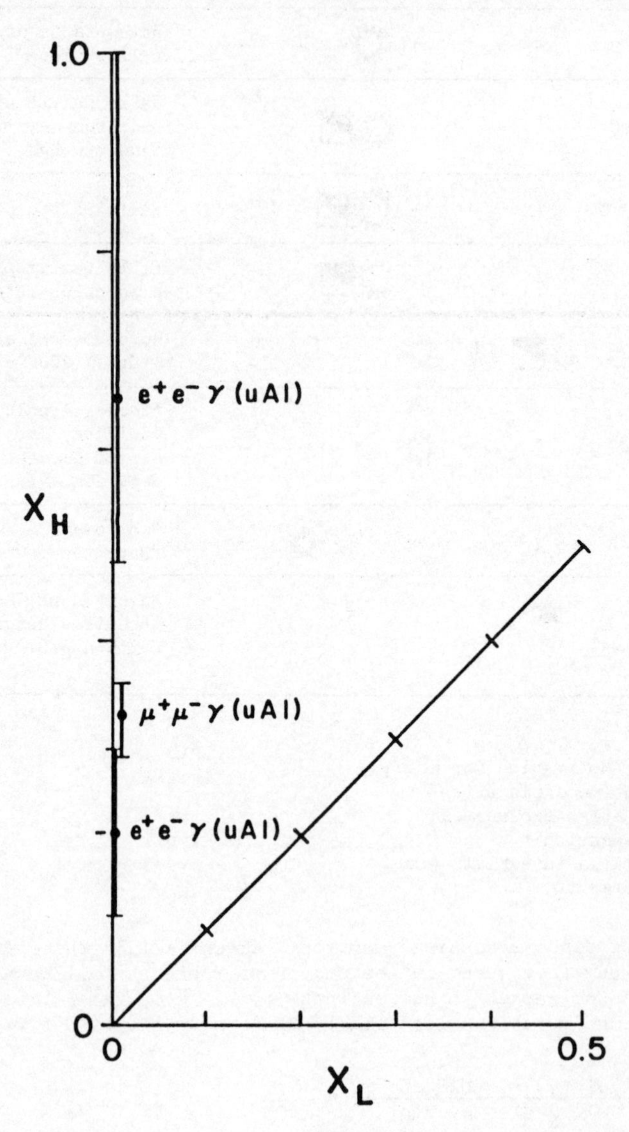

Figure 1

The three events cluster very close to one edge of the Dalitz plot, reflecting the smallness of $m_L(\ell\gamma)$ which enters quadratically into E_L. Such a Dalitz plot is grossly inconsistent with everything proposed by theorists except a low mass excited lepton ℓ^* ($m_{\ell^*} \lesssim 10$ GeV), ($Z\to\ell^*\bar{\ell}$ followed by $\ell^*\to\ell\gamma$) and, of course, bremsstrahlung. A light ℓ^* is very problematic: it would have been seen at PETRA and PEP barring bizarre behavior by its form factor. Also it is hard to reconcile a light e* with the absence of an electron form factor out to very large Q^2. The Dalitz plots for a set of theoretical models are shown in Fig. 2 [taken from Ref. [35]]. Even models which specifically attempt to explain the Dalitz plots are unable to account for the observed clustering [19]. To accept these models one must believe the clustering of the experimental data evident in Fig. 1 to be a statistical fluctuation. This seems less likely to me than the bremsstrahlung explanation.

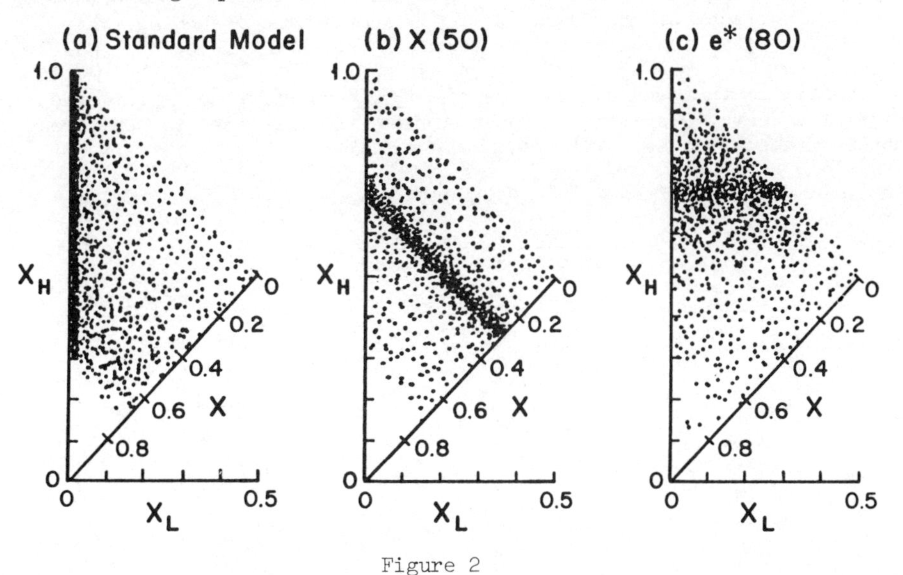

Figure 2

The anomalous jet events have not spawned as many models and most of those have serious flaws if the data are taken at face value. The models fall into two classes: 1. those which introduce new massive neutral particles to account for the $\not{p}_T$ signal (SUSY, sequential neutral heavy lepton (ν_4), mirror neutrinos), and 2. those which introduce new, strongly coupled particles which decay often into the bosons (Z^0, W, γ and perhaps Higgs scalar) of the standard model. Supersymmetry, an example of the first class, does fairly well but cannot easily account for UA1 monojet event A or the $\gamma\not{p}_T$ event(s). Since supersymmetry scenarios will be discussed elsewhere at this conference I will instead describe a not particularly successful (but not ruled out) but rather instructive model due to Glashow. [28] He supposes the existence of another (confining) gauge interaction which he calls "odor". So the gauge group of nature is enlarged to $G_{odor}\times SU(3)\times SU(2)_L\times U(1)$. Odor is not technicolor: it does not provide weak interaction symmetry breaking. It is not necessary, but neither are the anomalous jet events, except perhaps to optimistic supersymmetrists.

Glashow takes $\Lambda_{odor} \sim 1$ GeV and $M_Q \sim 65$ GeV* where M_Q is the lightest

*Actually $M_Q \sim 75$ GeV fits the data better.

"quark" carrying odor (Q also carries color and electroweak quantum numbers). Λ and M_Q are chosen so odor has a typical "onium" spectrum beginning at $2M_Q \sim 260$ GeV where continuum production begins. The model has the following good features:

1. Since odor-quarks carry color they can be copiously produced via $gg \to Q\bar{Q}$. Since they carry $SU(2)_L \times U(1)$ their annihilation decays can involve electroweak gauge bosons and Higgs scalars: $Q\bar{Q} \to Z^0\gamma$, $Q\bar{Q} \to Z^0 H$ etc.

2. Since M_Q is so large (and odor cannot be screened by ordinary color) only $Q\bar{Q}$ "onium" states are produced at the collider. These decay entirely by annihilation.

3. Since Λ_{odor} is small $Q\bar{Q} \to GGG$ (G is an "odor gluon") does not swamp other annihilation decays. In particular the branching ratio for decay into electroweak bosons is not negligible.

The scenario is that UA1 and UA2 are seeing the electroweak annihilation decays of a heavy $Q\bar{Q}$ system. A rough estimate of the rates for different channels (taken from Ref. [6]) is given in Table III:

TABLE III

DECAYS OF ODORONIUM

Normalized to ~5 monojets

1^- DECAY TO	EXPECTED EVENTS	POSSIBLE SIGNATURE
ggg	45	jjj bump at m=150 GeV
GGG	45	—
G^2g^3	1/2	
G^3g^2	1/2	$j(s)\not{p}_T$
$ZH, Z\to\nu\bar{\nu}$	5	$j\not{p}_T$
$ZGG, Z\to\ell^+\ell^-$	1	$\ell^+\ell^-\not{p}_T$
$Zgg, Z\to\ell^+\ell^-$	1	$\ell^+\ell^-j(s)$
$ZGG, Z\to q\bar{q}$	10	$jj\not{p}_T$
$\gamma H \quad H\to b\bar{b}$	5	γj
γgg	5	γjj

0^- DECAY TO	EXPECTED EVENTS	POSSIBLE SIGNATURE
gg	15	jj bump at m=150 GeV
GG	15	-
g^2G^2	1/7	$jj\not{p}_T$
$Z\gamma, Z\to\ell^+\ell^-$	1/20	$\ell^+\ell^-\gamma$ with $m(\ell^+\ell^-\gamma)$=150 GeV
$HGG, H\to b\bar{b}$	1	$j\not{p}_T$
$Z\gamma, Z\to\nu\bar{\nu}$	1/3	$\gamma\not{p}_T$

Note that the monojet signal requires a light Higgs. If $M_{Higgs}<2m_b$ the low multiplicity of some monojets can be understood via the $\tau\bar{\tau}$ decay of the Higgs.[*] Other predictions of Glashow's scheme are the existence of stable odor baryons and antiodor baryons with a relic abundance of $n_{odorbaryon}/n_{baryon} \sim 10^{-7}$, and a lower rate of $\gamma \not{p}_T$ than observed.[**]

It appears to be easier to invent explanations of the anomalous jet events than $\ell^+\ell^-\gamma$ events but no explanations naturally explain all event categories, topologies and rates. With further data some classes of events are sure to disappear while others will be modified and new ones may appear, so even the correct explanation would probably not appear especially attractive at this time.

2. QCD at Low Energies

The dynamics of confining gauge theories are still only poorly understood. Consider the following questions: 1. What are the qualitative features of the spectrum of QCD in the absence of quarks?, 2. What is the ground state of a collection of A baryons? It is remarkable that such basic questions cannot be definitively answered yet. Both have attracted considerable attention over the past year. The first is, of course, the question of glueballs. The second relates to a speculation by Witten that quark matter rather than nuclei may be the ground state of QCD with fixed baryon number.

a. Glueballs

It has been recognized for nearly a decade [36,37,38] that the spectrum of hadrons should contain gluonic excitations. Finding them has been another matter. Experimentally little has changed since last summer's reviews by Chanowitz [39] and Close.[40] The prime candidate is still the $\iota(1440)$ [41] followed by the $\theta(1700)$ [42] and somewhat less convincingly by the $\xi(2220)$ [43] and a set of resonances seen in the $\phi\phi$ channel.[44] In addition it has been argued strongly that there is no 0^{++} glueball below $2M_K \sim 1$ GeV [45] and that the $\pi\pi$ spectrum in $p\bar{p} \to p\bar{p}(\pi\pi)_{central}$ can be accounted for in terms of conventional quark model resonances.[46]

Models for glueballs abound. There are bag models, non-relativistic gluon models, duality-OZI models, predictions of QCD sum rules and of lattice Monte Carlo calculations. In a review at the Vanderbilt conference Sharpe[47] compared the models (see Fig. 3) and found them in rough agreement, perhaps as good as might be expected given the naivety of the models. [See Ref. 47 for further references on individual models.] A more cynical reading of Fig. 3 might be as follows: agreement in the 0^{-+} and 2^{++} channels is prejudiced by the observation of the $\iota(0^{-+}?)$ and the $\theta(2^{++}??)$; most of the 0^{++} mass predictions wander below the 1 GeV limit of [45] and in the 1^{-+} channel, potentially the most interesting because of its exotic J^{PC}, there is little agreement.

---

*A light Higg's decaying appreciably to $\tau\bar{\tau}$ is a source for low multiplicity and low invariant mass monojets irrespective of the production mechanism.
**See also the comment of G. Kane for another possible problem.

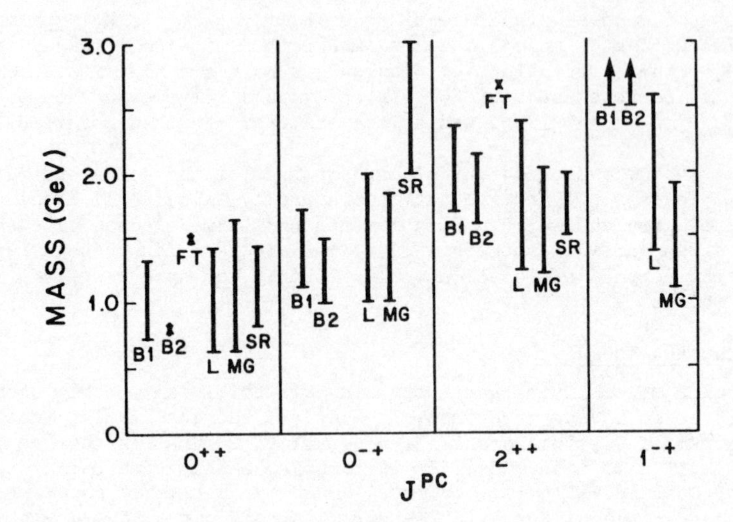

Fig. 3: Glueball mass predictions: B1 and B2-bag models; FT-fluxtube model; L-lattice Monte Carlo calculations; MG-massive gluon model; SR-QCD sum rules.

The difficulty in identifying glueballs experimentally makes it all the more important to reach some consensus on the theoretical expectations. To that end I would like to study the glueball spectrum using a minimal model [48] which is an elaboration of an approach originally suggested by Bjorken.[49] The model is based on two assumptions

1. The lightest states in a field theory are those created from the vacuum by the simplest (i.e. lowest dimension) local gauge invariant composite operators.
2. Among the states selected by (1), the lightest are those which can still be coupled to the vacuum when the fundamental (quark and gluon) fields are replaced by a single mode in some mean field theory.

Rather than defend these assumptions, let me illustrate their predictions and limitations in the $Q\bar{Q}$ meson sector of QCD. The lowest dimension meson interpolating fields have dimension 3. They are listed in Table IV with flavor and color indices suppressed.

TABLE IV
$Q\bar{Q}$ MESONS

OPERATOR	J^{PC}	SINGLE MODE
$\bar{q}\gamma_5 q$	0^{-+}	✓
$\bar{q}\gamma_\mu q$	1^{--}	✓
$\bar{q}q$	0^{++}	×
$\bar{q}\gamma_\mu\gamma_5 q$	0^{-+}	✓
	1^{++}	×
$\bar{q}\sigma_{\mu\nu} q$	1^{--}	✓
	1^{+-}	×

Also listed are the J^{PC} quantum numbers of all single particle states which can be created by each operator from the vacuum. [Note, the operator $\bar{q}\gamma_\mu q$ does not create a 0^{+-} state because of the constraint of current conservation.] So much for step (1). Now replace the field q by a single mode of definite, but arbitrary energy, angular momentum and parity in a frame where one imagines constructing a Hamiltonian mean field theory for QCD. A simple calculation [48] shows that only the 0^{-+} and 1^{--} states can be created from the vacuum by the resulting composite operators. From this we conclude that the lightest mesons have 0^{-+} and 1^{--} quantum numbers. The others, 0^{++}, 1^{+-} and 1^{++}, form part but not necessarily all of the spectrum of low mass meson excitations. In fact the 2^{++} is missing. The advantage of this approach is that it is simple and stresses features common to most if not all sensible models (e.g., non-relativistic quark models, bag models, QCD sum rules etc.). The disadvantages are that it gives information only about the lightest states, and it does not resolve fine and hyperfine structure (e.g., the 0^{-+} and 1^{--} are indistinguishable).

The analogous calculation for glueballs begins with dimension 4 operators constructed from the gluon field strength $G_{\mu\nu}$. The operators and the quantum numbers of the states they create are given in Table V.

TABLE V

LIGHT GLUEBALLS

OPERATOR	J^{PC}	SINGLE MODE?
$S=\mathrm{Tr}G_{\mu\nu}G^{\mu\nu}$	0^{++}	✓
$P=\mathrm{Tr}\tilde{G}_{\mu\nu}G^{\mu\nu}$	0^{-+}	×
$T_{\alpha\beta}=\mathrm{Tr}G_{\alpha\lambda}G^{\lambda}{}_{\beta}-\frac{1}{4}g_{\alpha\beta}S$	2^{++}	✓
$L_{\mu\nu\rho\sigma}=\mathrm{Tr}G_{\mu\nu}G_{\rho\sigma}-(g_{\mu\rho}T_{\nu\sigma}+g_{\nu\sigma}T_{\mu\rho}-g_{\mu\sigma}T_{\nu\rho}-g_{\nu\rho}T_{\mu\sigma})+\frac{1}{12}(g_{\mu\rho}g_{\nu\sigma}-g_{\mu\sigma}g_{\nu\rho})S+\frac{1}{24}\,\varepsilon_{\mu\nu\rho\sigma}P$	2^{-+}	×

The exotic $J^{PC}=1^{-+}$ quantum numbers are notably absent from the Table. Naively the traceless, symmetric tensor $T_{\alpha\beta}$ can create the 1^{-+} state:

$$\langle p,\varepsilon|T_{\alpha\beta}|0\rangle = f_{1^{-+}}(p_\alpha\varepsilon_\beta+p_\beta\varepsilon_\alpha) \tag{3.1}$$

with $p\cdot\varepsilon=0$. However in a QCD without quarks $T_{\alpha\beta}$ is (up to a term $\propto g_{\alpha\beta}S$ irrelevant to this argument) the conserved stress tensor $\partial^\alpha T_{\alpha\beta}=0$, so

$$\langle p,\varepsilon|\partial^\alpha T_{\alpha\beta}|0\rangle = \varepsilon_\alpha M^2 f_{1^{-+}} = 0. \tag{3.2}$$

Either $M^2=0$, corresponding to spontaneous breakdown of translation invariance and a 1^{-+} Goldstone boson; or $f_{1^{-+}} = 0$, corresponding to no 1^{-+} state in the spectrum of this operator. Thus there should be no very light 1^{-+} glueball in the pure glue sector. The appearance of a light 1^{-+} state in the lattice calculations [50] (see Fig. 3) is probably an artifact of breaking translation invariance when the lattice spacing is finite. A careful study of the dependence of the 1^{-+} mass on the lattice spacing should reveal that it does not scale according to the renormalization group and disappears from the light glueball spectrum in the continuum limit. The light 1^{-+} glueball found in non-relativistic gluon models (see Fig. 3) is undoubtedly an artifact of breaking color gauge invariance by giving the gluon a mass. When quarks are included the 1^{-+} state can no longer be excluded. Using $\partial_\mu T^{\mu\nu}=0$ the operator which creates it can be replaced by $\bar{q}\gamma_\mu G^{\mu\nu}q$ which identifies the state as a "hybrid" meson rather than a glueball.

Higher dimension glueball interpolating fields yield more states. Dimension 5 operators like $G_a^{\mu\nu}(D^\alpha G^{\rho\sigma})_a$ are constrained by the equations of motion $D^\alpha G_\alpha{}^\rho=0$ and yield only 1^{++} and 3^{++} states. Dimension 6 operators of the forms $f_{abc}G_aG_bG_c$ and $d_{abc}G_aG_bG_c$ give the states listed in Table VI.[*] Here the mode analysis is more complex.

TABLE VI

DIMENSION 6 GLUEBALL OPERATORS

OPERATOR	J^{PC}
$f_{abc}G_aG_bG_c$	$0^{\pm+}$, $1^{\pm+}$, $2^{\pm+}$
$d_{abc}G_aG_bG_c$	$1^{\pm-}$, $2^{\pm-}$, $3^{\pm-}$

It is necessary to assume a definite angular momentum (j) and parity (π) for the lowest mode. For $j^\pi = 1^+$ [**] only 0^{++}, 1^{+-} and 3^{+-} survive. For $j^\pi = 1^-$ [49,51] only 0^{-+}, 1^{--} and 3^{--} survive. In any event the spectrum is still rather sparse.

The "predictions" of our minimal model are summarized in Fig. 4. I conclude that the lightest glueballs are likely to be 0^{++} and 2^{++} with 0^{-+}, 1^{++}, 2^{-+} and 3^{++} among the lowest excitations. There is no sign of a glueball with exoctic J^{PC} among the lightest states. Fine and hyperfine structure and mixing of the 0^{++} with the condensate in the QCD vacuum could modify these conclusions somewhat. In any event the "agreement" among models regarding 0^{++}, 2^{++} and 0^{-+} glueball masses appears to be a generic feature of models. The 2^{-+} should probably be added to this short list, whereas the predictions of light 1^{-+} exotic glueballs seems to me unreliable.

b. Strange Matter

Nuclei appear stable. Nevertheless Witten [52] has suggested that quark matter rather than a nucleus is the ground state of a fixed number of baryons. Surprisingly this possibility is not ruled out by experiment or theory. It can be tested by searching for stable quark matter with properties very different from ordinary matter.[53]

The crux of Witten's argument is that quark matter in equilibrium with the weak interactions ("strange matter") contains many strange quarks. Consider, for example, a free Fermi gas of quarks with an energy per baryon of order M_N ($\sim$1 GeV). Suppose initially only (massless) u and d quarks are present. Because of the exclusion principle all energy levels up to $\sim M_N/3$ are filled. If the mass of the strange quark is below this, then it is energetically favorable for u and d quarks to convert to s quarks via the weak interactions ($u\to s+e+\bar{\nu}$, $u+d\to s+u$). In equilibrium the strangeness per baryon is of order unity (it is exactly one if $m_s=0$). Thus,

$$\left(\frac{E}{A}\right)_{\substack{\text{non-strange}\\ \text{quark matter}}} > \left(\frac{E}{A}\right)_{\substack{\text{strange}\\ \text{matter}}}$$

---

*Other dimension 6 operators of the form GDDG have yet to be studied.
**Predicted by bag models.

All this was well known some time ago.[54]

The apparent stability of nuclei requires that quark matter with the <u>same flavor composition as nuclei</u>, <u>i.e</u>. only u and d quarks, have an energy per baryon greater than nuclei. Otherwise nuclei would decay by strong interactions into lumps of quark matter. Nuclei do not decay into strange matter because the transition is very high order in G_F making the rate negligibly small. Thus, for example, iron may be unstable against decay to a lump of strange matter with A=56 and S≈40, but the transition is $\sim G_F^{80}$! Witten's scenario requires

$$\left(\frac{E}{A}\right)_{\substack{\text{non-strange}\\ \text{quark matter}}} > \left(\frac{E}{A}\right)_{\text{nuclei}} > \left(\frac{E}{A}\right)_{\substack{\text{strange}\\ \text{matter}}}$$

Whether this occurs in QCD is a matter of detailed calculation and depends on parameters like α_s and m_s. I will return to this point below.

The question arises: Why isn't the universe made of strange matter? Witten's answer is: Perhaps it is. He suggests that the missing, non-luminous matter required to close the universe is in the form of stable, macroscopic ($r\sim1$ cm.) "nuggets" of very dense strange matter which are relics of the confining phase transition which occured when the temperature was ~100 MeV.

If that transition was first order and if the chiral symmetry breaking phase transition did not occur earlier, then it is quite likely that immediately after the transition most of the baryon number in the universe was in the form of strange matter nuggets.[52] As the universe cooled the baryon number evaporated from the nuggets, and it is uncertain whether they survived to temperatures low enough that evaporation ceases. If they evaporate away they fail to account for the dark matter. Nevertheless, strange matter may have been created in supernovae or the interior of neutron stars, and if it is stable it should be present in the earth and as a component in cosmic rays.

As a guide to searches for strange matter, Farhi and I have studied its properties at zero temperature and external pressure.[53] Our results are encouraging. We find a reasonably large window in the parameter space in which Witten's scenario holds. Unfortunately the parameters (B - the bag constant[*], m_s - the strange quark mass, and α_c both renormalized at $\sim M_N/3$) are too poorly known to decide whether strange matter or nuclei are stable.

Choosing parameter values for which strange matter is stable, it is possible to paint a rather detailed picture of its properties. In particular it has a low hadronic charge to baryon number (Z/A) ratio. For $m_s=0$, u=d=s so Z=0. Even as $m_s\to M_N/3$, Z/A remains very small. Small droplets of strange matter ($A\lesssim10^7$), "strangelets", probably have a large dynamically generated surface tension $\sigma\cong(70\text{ MeV})^3\cong9\text{ MeV/fm}^2$ produced by the confinement of the quark wavefunctions. At low baryon number the surface tension destabilizes strangelets so there is some A_{min} below which strange matter is not stable. One consequence of this is that light nuclei should not decay into strangelets by a low order (hence fast) weak interaction. On the other hand Z/A

*The bag constant enters as the energy density of the deconfined phase compared to the true (confining) vacuum at zero temperature.

is so small that Coulomb effects in general do not destabilize large strangelets. Thus, in contrast to ordinary nuclei, the periodic table of strangelets does not end at some maximum A but instead connects smoothly to bulk systems (Witten's "nuggets") and thence to quark stars.

If strange matter is stable it should be present at some abundance in terrestrial material. Its chemistry is diverse since it exists with a wide range of charges, but its physical characteristics: large A hence large mass and small Z/A, are very distinctive. It may be found in heavy metal ores and concentrated by ultracentrafugation or insolubility. Relatively clean substances exposed for long periods to a possible cosmic flux may be rich in strange matter. Examples are moon rocks and south polar ice cores. Given a sample thought to be enriched in strange matter, sub-Coulomb-barrier activation by uranium ions provides a means of identifying it.[55] Suppose a sample is irradiated with uranium ions with energy well below the Coulomb barrier.[*] In the absence of strangelets little happens. Because of their low Z/A, the coulomb barrier of strangelets is typically much lower than that of ordinary matter. A uranium ion incident on a strangelet would be largely absorbed liberating roughly 238 ΔE (ΔE is the binding energy per nucleon of strange matter) or as much as several GeV of energy in the form of pions, kaons, baryons and photons. The detection of these subbarrier "explosions" would be an unmistakable signal for strange matter.

Very recently, De Rujula and Glashow [56] have considered methods of detecting a flux of strange matter in cosmic radiation. Assuming a local density of dark matter of $\sim 10^{-24}/\mathrm{cm}^3$ and characteristic galactic velocities $\sim$250 km/sec they calculate an annual earth infall of $\sim 10^9$ gm. The consequences of this flux depend entirely on the size of the incident lumps or "nuclearites" as they are called in Ref. 56. Nuclearites with $R \gtrsim 10^{-4}$ cm punch through the earth. Only those with $R \lesssim 200$ Å stop in the earth's crust. DeRujula and Glashow suggest a variety of ways of looking for them ranging from anomalous epilinear earthquakes ($R \gtrsim 10^{-4}$ cm) and otherwise inexplicable linear fossils ($R \gtrsim 10^{-5}$ cm) down to events in the IMB proton decay detector ($R \gtrsim 10^{-10}$ cm) and etched trails in ancient mica ($R \gtrsim 3\times 10^{-9}$ cm). It should be possible to put useful limits on a terrestrial or cosmic abundance of strange matter in the near future.

3. Low Energy Theory at the High Energy Frontier

There has been considerable excitement recently concerning a speculation made almost 25 years ago by Skyrme [57], and revitalized by Witten [58] and by Balachandran et al. [59] The idea is that baryons may be described, at least qualitatively, as solitons in an effective field theory of mesons. Skyrme pointed out that the minimal non-linear sigma model, introduced by Gell-Mann and Levy as an effective field theory from which to derive the consequences of chiral symmetry for pions, possesses classical soliton solutions which are stable provided an additional term is added to the sigma model Lagrangian. In the non-linear sigma model the three isospin components of the pion field are gathered together into a unitary (2×2) matrix U(x),

$$U(x) = \exp\, i\vec{\tau}\cdot\vec{\pi}(x)\ , \quad U^{+}U=UU^{+}=1 \tag{3.1}$$

*The reason for using uranium is that the Coulomb barrier for uranium incident on ordinary matter is nearly independent of the composition of the target.

where $\vec{\tau}$ are the Pauli matrices. Skyrme's Lagrangian was

$$L = \frac{1}{16} f_\pi^2 \, Tr(R_\mu R^\mu) + \frac{1}{32e^2} Tr[R_\mu,R_\nu]^2 \qquad (3.2)$$

where $R_\mu = U^+\partial_\mu U$ and f_π is the pion decay constant. The first term is the minimal one which generates soft pion dynamics. The second term, which vanishes in the soft pion limit, stabilizes the soliton:

$$U_0(x) = \exp i\vec{\tau}\cdot\hat{x}F(r), \qquad (3.3)$$

where F(r) is determined numerically to minimize the energy. The size and energy of the soliton are proportional to $1/ef_\pi$ so as $e\to\infty$ it disappears. Skyrme speculated that his soliton is a baryon. Finkelstein and Rubenstein [60] showed it could be quantized as a fermion or a boson. Recently Witten [58] used the current algebra anomalies required by the underlying quark gauge theory [61] to connect the statistics of the soliton with the number of colors in QCD. Adkins, Nappi and he [62] showed that the ground state of the quantized soliton has I=J=1/2 and the first excitation has I=J=3/2.

What has all this to do with the high energy frontier? In the standard model as the Higgs mass becomes large the Higgs sector becomes strongly interacting [63,64] and is described at low energies ($<<M_H$) by a non-linear sigma model. This is easy to see if we rewrite the usual complex Higg's doublet (ϕ^0,ϕ^-) as a two by two matrix:

$$M = \sigma + i\vec{\tau}\cdot\vec{\Phi} \qquad (3.4)$$

with

$$\begin{aligned} i\phi^- &= \Phi_1 + i\Phi_2 \\ i\phi^0 &= \Phi_3 + i\sigma \end{aligned} \qquad (3.5)$$

then the Higgs potential can be written

$$V_{HIGGS} = \frac{\lambda}{4}\left(\frac{1}{2}Tr(M^+M)-v^2\right)^2 \qquad (3.6)$$

$$= \frac{M^2_{HIGG}}{8v^2}\left(\sigma^2 + \vec{\Phi}^2 - v^2\right)^2 \qquad (3.7)$$

where v, the vacuum expectation value of the Higgs field, is fixed at $\sim$250 GeV by the W mass. Clearly $M_{HIGGS}>>v$ requires $\lambda>>1$, so the Higgs sector becomes strongly interacting and difficult to interpret. Naively only solutions satisfying

$$\sigma^2 + \vec{\Phi}^2 \equiv \det M = v^2 \qquad (3.8)$$

would be important at low energies. Explicit calculations in perturbation theory by Appelquist and Bernard [65] bear this out. If one defines M=vU then (3.4) and (3.8) require $U^+U=1$ so the Higg's sector at low energies reduces to the non-linear sigma model. Of course the three pions of the non-linear model are eaten by the gauge bosons giving them mass and longitudinal components. The chiral low energy theorems of the sigma model reproduce the tree graph predictions of the standard model for the interactions of longitudinally polarized W's and Z's.

In the limit $M_H/v>>1$ of the standard model loops with circulating heavy Higgs particles induce further non-renormalizable couplings among the quanta of the non-linear sigma model.[65] These interactions are scaled by inverse powers of M_H and include Skyrme terms among others. These calculations should be taken with a grain of salt since the strong couplings which develop as $M_H/v \to \infty$ are most likely a signal of a new dynamics in the Higg's sector. The correct effective Lagrangian beyond the chiral limit should be determined by this dynamics not by extrapolation of the linear model. QCD illustrates the point: suppose pions were the only known hadrons. The correct effective Lagrangian for pions beyond the chiral limit (ρ,ω resonances, etc.) is not given by sending $m_{sigma} \to \infty$ in a linear sigma model, instead it is determined by quarks and gluons.

Nevertheless the $M_H \to \infty$ limit of the standard model suggests that if the Higg's scalar is not found at low mass new strong interactions await us in the TeV range. It is possible that the new dynamics generates Skyrme-like terms which stabilize the solitons of the non-linear sigma model.[66] D'Hoker and Farhi [67] have argued that these stable particles are the technibaryons of an underlying technicolor dynamics in the same fashion that Skyrme's solitons are the baryons of QCD. The prospect that the dynamics of weak symmetry violation is some sort of recapitulation of the strong interactions is exciting and has begun to generate speculation on phenomenology in the TeV domain.[68]

An immediate problem concerns how to study these interactions experimentally. It is well known that most low energy phenomena are "protected", i.e., decoupled from a strongly interacting Higg's sector. The natural place to look is in the scattering of longitudinal W's and Z's [64] in analogy to $\pi\pi$ scattering in QCD. Recently Cahn and Dawson [69] have pointed out that this is not a hopeless undertaking. At very high energies, $\sqrt{\hat{s}} \gg M_W$ ($\hat{s}$ is the squared center of mass energy of colliding partons), quarks are accompanied by a cloud of almost-real W's and Z's in analogy to the virtual photon approximation of QED. The flux of longitudinal W's and Z's is substantial, though smaller than the transverse flux by a factor of $\log \hat{s}/M_W^2$. Furthermore interesting processes, like the formation of a scalar resonance (coupled to $W_\mu W^\mu S$), select out longitudinal W's. The cross sections are small (less than a picobarn) but not beyond the reach of supercolliders provided a good signature for W and Z production can be devised.

The conclusion of this is a prescription for an exciting future at supercolliders: 1. Hope the Higg's sector is not discovered at masses well below 1 TeV, and 2. Keep studying the low energy behavior of strongly interacting (gauge) field theory.

REFERENCES

1. UA1 Collaboration, G. Arnison et al., Phys. Lett. 122B (1983) 103; G. Arnison et al., Phys. Lett. 126B (1983) 398.
2. UA2 Collaboration, G. Banner et al., Phys. Lett. 129B (1983) 130; P. Bagnaia et al., Phys. Lett. 129B (1983) 130.
3. UA1 Collaboration, G. Arnison et al., Phys. Lett. 132B (1983) 214; UA2 Collaboration, P. Bagnaia et al., CERN preprint CERN-EP/84-12.
4. D. DiBitonto, presentation to this conference.
5. See numerous presentations at this conference.
6. L. Hall, R. L. Jaffe and J. L. Rosner, in preparation, to appear in Proceedings of the 1984 Snowmass PSSC Study.
7. UA1 Collaboration, G. Arnison et al., CERN preprint EP/84-42 (1984).
8. C. Rubbia, Talk at the Fourth Workshop on $\bar{p}p$ Collider Physics, Bern, March 1984.

9. UA2 Collaboration, P. Bagnaia et al., CERN preprint EP/84-40 (1982); A. Roussarie, Talk at the Fourth Workshop on $\bar{p}p$ Collider physics, Bern, March 1984.
10. UA1 Collaboration, G. Arnison et al., Phys. Lett. 135B (1984) 250.
11. UA2 Collaboration, P. Bagnaia et al., CERN preprint EP/84-39 (1984).
12. K. Enqvist and J. Maalanyi, Phys. Lett. 135B (1984) 329.
13. N. Cabibbo, L. Maiani and Y. Srivastava, Phys. Lett. 139B (1984) 459.
14. F. M. Renard, Phys. Lett. 126B (1983) 59.
15. U. Baur, H. Fritzsch and H. Faissner, Phys. Lett. 135B (1984) 313.
16. R. Peccei, Phys. Lett. 136B (1984) 121.
17. F. M. Renard, Phys. Lett. 139B (1984) 449.
18. W. Hollik, F. Schrempp and B. Schrempp, Phys. Lett. 140B (1984) 424.
19. R. Holdom, Stanford University preprint ITP-765 (1984).
20. S. King and S. Sharpe, Harvard University Preprint HUTP-84/A026 (1984).
21. F. M. Renard, Phys. Lett. 116B (1982) 264.
22. G. Gounaris, R. Kögerler and D. Schildknecht, Phys. Lett. 137B (1984) 261.
23. R. W. Robinett, University of Massachusetts Preprint UMHEP-197 (1984); P. Collins and N. Speirs, Durham University Preprint DTP/84/10 (1984).
24. M. Leurer, H. Haravi and R. Barbieri, Weizmann Institute Preprint, WIS-84/7/Ph (1984).
25. M. Veltman, Phys. Lett. 139B (1984) 310.
26. H. Georgi and S. L. Glashow, Harvard University Preprint HUTP-84/A019.
27. L. M. Krauss, Harvard University Preprint HUTP-84/A021; see also V. Barger, W. Y. Keung, and R. Phillips, Phys. Lett. 141B (1984) 126.
28. S. L. Glashow, Harvard University Preprint HUTP-84/A037 (1984).
29. E. Reya and D. P. Roy, University of Dortmund Preprint DO-TH 84/03 (1984).
30. J. Ellis and H. Kowalski, CERN Preprint TH-3843 CERN (1984); DESY Preprint, DESY 84-045 (1984).
31. V. Barger, K. Hagiwara, J. Woodside and W.-Y. Keung, University of Wisconsin Preprint MAD/PH/184 (1984).
32. L. J. Hall and M. Suzuki, Nucl. Phys. B231, 419 (1984).
33. H. E. Haber and G. Kane, University of California Santa Cruz Preprint UCSC-TH-169-84 (1984).
34. UA2 Collaboration, P. Bagnaia et al. CERN Preprint (1984).
35. V. Barger, H. Baer and K. Hagiwara, Wisconsin Preprint MAD/PH/176 (1984).
36. P. G. O. Freund and Y. Nambu, Phys. Rev. Lett. 34, 1646 (1975).
37. H. Fritzsch and P. Minkowski, Nuovo Cimento 30A (1975) 393.
38. R. L. Jaffe and K. Johnson, Phys. Lett. 60B (1976) 201.
39. M. S. Chanowitz, Proceedings of the XIVth International Conference on Multiparticle Dynamics, 1983.
40. F. E. Close, Proceedings of HEP 83 (Rutherford Appelton Laboratory, Didcot, 1983), 361.
41. D. Scharre et al., Phys. Lett. 97B (1980) 329.
42. C. Edwards et al., Phys. Rev. Lett. 48, 458 (1982).
43. K. F. Einsweiler et al. in Proceedings of HEP 83 (Rutherford Appelton Laboratory, Didcot, 1983).
44. A. Etkin et al., Phys. Rev. Lett. 49, 1620 (1982); see also S. J. Lindenbaum, Proceedings of HEP 83 (Rutherford Appelton Laboratory, Didcot, 1983), 351.
45. S. R. Sharpe, R. L. Jaffe and M. R. Pennington, Harvard University preprint HUTP-84/A017 to be published in Phys. Rev. D.
46. D. Morgan and M. R. Pennington, Phys. Lett. 137B (1984) 411.
47. S. Sharpe, Invited Talk at the Symposium on High Energy e^+e^- Interactions, 6th International Conference, Vanderbilt University,

April 1984, Harvard University preprint HUTP-84/A029.
48. R. L. Jaffe, K. Johnson and Z. Ryzak, to be published.
49. J. D. Bjorken, Proceedings of the 1979 SLAC Summer Institute on Particle Physics.
50. K. Ishikawa et al., Z. Phys. C21, 167 (1983).
51. D. Robson, Nucl. Phys. B130 (1977) 328.
52. E. Witten, Phys. Rev. D30, 272 (1984).
53. E. Farhi and R. L. Jaffe, MIT preprint MIT-CTP-1160.
54. S. A. Chin and A. K. Kerman, Phys. Rev. Lett. 43, 1292 (1979).
55. R. Cahn, E. Farhi and R. L. Jaffe, work in progress.
56. A. DeRujula and S. L. Glashow, Harvard University preprint HUTP-84/A039.
57. T. H. R. Skyrme, Proc. Roy. Soc. London, A260, 127 (1961), Nucl. Phys. 31, 556 (1962).
58. E. Witten, Nucl. Phys. B223 (1983) 422, ibid. 433.
59. A. P. Balachandran, V. P. Nair, S. G. Rajeev and A. Stern, Phys. Rev. Lett. 49, 1124 (1982), Phys. Rev. D27, 1153 (1983).
60. D. Finkelstein and J. Rubenstein, J. Math. Phys. 9, 1762 (1968).
61. J. Wess and B. Zumino, Phys. Lett. 37B (1971) 95.
62. G. S. Adkins, C. R. Nappi and E. Witten, Nucl. Phys. B228 (1983) 552.
63. M. Veltman, Phys. Lett. 70B (1977) 253.
64. B. W. Lee, C. Quigg and H. B. Thacker, Phys. Rev. Lett. 38, 883 (1977).
65. T. Appelquist and C. Bernard, Phys. Rev. D22, 200 (1980).
66. J. M. Gibson and H. C. Tze, Nucl. Phys. B183 (1981) 524; J. M. Gibson, Virginia Polytechnic Institute Preprint VPI-HEP-83/1 (1983).
67. E. D'Hoker and E. Farhi, MIT Preprint MIT-CTP-1132 (1984).
68. For discussion and references see M. K. Gaillard, in Proceedings of the 1984 Snowmass PSSC Study.
69. R. Cahn and S. Dawson, Phys. Lett. 136B (1984) 196, S. Dawson, LBL Preprint LBL-17497 (1984), see also G. Kane, University of Michigan Preprint UM TH 83-25 (1984).

Discussion on Section 1

H. Koch (KFK, Univ. Karlsruhe): Are there theoretical predictions for the decay of glueballs?

Speaker, R.L. Jaffe: Yes, there are, but they are very model dependent regarding both overall width and branching ratios. The problem is compounded by mixing between glueballs and ordinary $q\bar{q}$ mesons. It depends sensitively on the spectrum of nearby meson resonances and this affects decays significantly.

M. Cooper (Basel/Los Alamos): Why, if the universe is made of 99% strangelets, have astronomers not observed any effects of this strange matter other than in the virial theorems for galaxies?

Speaker, R.L. Jaffe: Strange matter is very dense, $\rho \sim 10^{15}$gm/cm^3. So a sparse distribution of lumps of strange matter (e.g. one 1 cc lump in a cube with dimension of the order of the solar system) suffices to close the universe without affecting astronomical observations.

G. Karl (Guelph/Oxford): Wouldn't the existence of strange matter affect the detailed dynamics of neutron stars, if they contain quark matter?

Speaker, R.L. Jaffe: It has always been realized that if a neutron star contains quark matter, then it would contain many strange quarks. If strange matter is stable it increases the likelihood that neutron stars are really quark stars. In a paper several years ago Fechner and Joss argued that the phenomenology of quark stars closely resembles that of neutron stars.

G.L. Kane (Michigan): I want to point out that Glashow's model is probably excluded because it predicts about 5 energetic e^+e^- pairs and 5 energetic $\mu^+\mu^-$ pairs, which would be very easy to observe and have not been seen. This result is due to Maiani and myself.

Latest results from the UA1 experiment on W and Z^0 and missing transverse energy events

J. Tuominiemi

Department of High Energy Physics, University of Helsinki,
Siltavuorenpenger 20 C, SF-00170 Helsinki 17, Finland

Abstract. Latest results of the analysis of the 1982-83 data of the UA1 experiment, related to production of the intermediate vector bosons, are reviewed. These include a complete study of the Z^o decaying into two muons revealing one event with an additional hard photon, observation of events with very large missing transverse energy associated with emission of a jet or photon(s) and comparison of jet activity in the W, Z^o and multijet events. Indications of new phenomena are discussed.

1. Introduction

The intermediate vector bosons (IVB's) $W^{\pm}$ and Z^o were discovered in the UA1 and UA2 experiments at the CERN $p\bar{p}$ collider in autumn 1982 and spring 1983. The results on the properties of the $W^{\pm}$ decaying into the electron [1] and muon channel [2] as well as Z^o decaying into the electron channel [3] have been published and reviewed in many conferences earlier [4] and will not be discussed here. Instead we will here review some results obtained in a further analysis of the 1982-83 data of the UA1 experiment that are related to the intermediate vector bosons.

First the results on Z^o decaying into two muons are discussed. One event of this type was already reported in the first publication on the Z^o [3]. In chapter 2 we give results on the complete study of this process.

In the UA1 experiment the charged IVB can be detected in two essentially independent ways. In the first method the W decays are searched for by looking for high transverse momentum electrons emitted in the $p\bar{p}$-collision. In the second method we look for processes with large missing transverse momentum i.e. for processes where a hard neutrino could be emitted. About 90 % of the W's found by the electron identification were observed with the missing E_T method. The succesful application of this method to the IVB search lead us to make a complete inclusive study of all large missing transverse energy events. This resulted in seven events, which are yet defying a "standard" explanation and hint to new phenomena in the energy domain above 100 GeV. These events are discussed in chapter 3. A possible explanation of the events is the existence of a new high mass state which would decay into IVB's and jets. To explore this possibility we have made a study of those IVB events where also jets are observed. The results of this study are shown in chapter 4.

The UA1 detector has been widely covered in several articles and conference talks, and we discuss it only briefly in the following. For details of the parts of the detector and their performance we refer to these articles [5].

2. Observation of $Z^o \to \mu^+\mu^-$

In the UA1 experiment the muons are detected in about 75 % of the full solid angle [2]. A muon emerging from the $p\bar{p}$ interaction first traverses the central track detector (CD), the electromagnetic calorimeter and the magnetized hadron calorimeter. After 60 cm of additional iron shielding (except in the forward region where the iron has been added only for the runs in 1984-85) it enters the muon chamber system which consists of 2 × 4 drift chamber planes. The momenta of the muons are determined with the CD by their deflection in the central dipole field of 0.7 T. The momentum uncertainty due to the measurement error of each point in the CD is $\Delta p/p \sim 0.5\ \% \times p$ [GeV/c]. The track position and angle measurements in the muon chambers permit a second, essentially independent measurement of the momentum. For muons with $p_T > 20$ GeV/c the precision of this measurement is comparable to that in the CD.

Finally the muon momenta can be determined by relying on overall transverse momentum conservation. For $Z^o \to \mu^+\mu^-$ events, where no neutrino is emitted, the transverse momentum of the recoil hadronic debris must be equal to the transverse momentum of the $(\mu^+\mu^-)$ system. The muon momenta and the transverse energy flow are adjusted in a fit to obtain an over-all balanced event.

Jets are defined using the UA1 jet algorithm [6] applied to energy vectors defined from calorimeter cells. A correction is applied to the measured energy (∿ + 25 %) and momentum (∿ + 20 %) of each jet, as a function of the pseudorapidity η, azimuth ϕ and transverse energy E_T of the jet, on the basis of test beam data and Monte Carlo studies. The jet momentum resolution is typically 20 % for a calorimeter jet of E_T = 15 GeV, and improves with increasing jet transverse energy. In this study [7] we have considered all jets with $E_T \geq 7.5$ GeV and pseudorapidity less than 2.5, keeping in mind, however, that the jet finding procedure and jet energy measurement become less reliable when E_T is smaller than 15 GeV.

During the 1983 data-taking period about 2.5×10^6 events were recorded on tape for an integrated luminosity of 108 nb^{-1}. Of these about 1.0×10^6 events were muon triggers. All the events were passed through a fast filter program which selected muon candidates with $p_T > 3$ GeV/c or $p > 6$ GeV/c [2]. This filter program selected 7.2×10^4 events.

The 17326 events from the fast filter program which contained a muon candidate with $p_T > 5$ GeV/c were fully reconstructed, and then passed through an automatic selection program which applied stricter CD track quality and CD-muon chamber matching requirements than were used at the filter stage. This yielded a reduced sample of 6781 events.

A dimuon selection was made requiring two $p_T > 5$ GeV/c muon tracks in the event, each satisfying the criteria used in the inclusive muon selection. This gave 26 candidate dimuon events which were scanned on a high resolution graphics display facility. Of these events, 7 were identified as cosmic rays and 6 as leakage of hadron showers through cracks in the calorimeters, leaving 13 candidates after scanning.

In addition to the dimuon selection described above, all events with a dimuon trigger and at least one reconstructed CD track with $p_T > 7$ GeV/c were examined. Two extra dimuon events were found in this way. In one event both CD tracks were very short and failed the track length requirements of the filter. In the other event one muon passed very close to the edge of the chamber and was seen in only one projection. Hence the final dimuon sample (both muons $p_T > 5$ GeV/c) contains 15 events.

We have considered a number of possible backgrounds to the muon sample. High p_T charged hadrons can fake muons either by penetrating the calorimeters and additional iron shielding without interaction, or by leakage of the hadronic shower. These backgrounds have been measured in a test beam and are found to be $\leq 10^{-4}$ per incident hadron, after requiring matching between CD and muon chamber tracks. The leakage-induced background is then negligible. The dominant background process is pion and kaon decay. The probability for a pion (kaon) to decay before reaching the calorimeter is $\sim 0.02/p_T$ ($\sim 0.11/p_T$). This background has been evaluated from events with a single high p_T muon candidate by calculating the probability for decay of another high p_T particle (pion or kaon) in each event. The background is found to be $< 10^{-3}$ events. The absence of like-sign $\mu\mu$ pairs of high $\mu\mu$-masses confirms that the background is small.

The background contribution due to heavy flavour jets, with semileptonic decays has been estimated from events with a large-p_T lepton accompanied bya recoil jet ($< 10^{-3}$). The Drell-Yan continuum [8] yields a background of 0.1 events for masses greater than 60 GeV/c^2. Production of W^+W^- pairs was found to be completely negligible.

The mass of the dimuon system of the 15 events found is shown in fig. 1. Ten events cluster at low mass values whereas five events are found to have mass values around 90 GeV/c^2 and are interpreted as Z^o decays in the following.

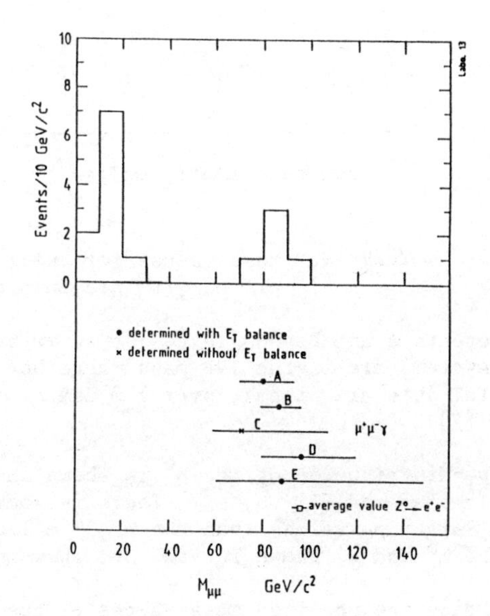

Fig. 1. Invariant mass of the $\mu^+\mu^-$ system. The bottom part of the figure shows the mass values with their errors for the 5 Z^o candidates A to E.

Including the muon in the calculation of the vector energy, all events are consistent with having no missing transverse energy. It is therefore justified to apply the method of transverse momentum balance, as described above. It also means that the transverse momentum, $p_T^{\mu\mu}$ of the dimuon system is balanced by the transverse energy flow of the "rest of the event".

Three events show a topology different from what was naively expected for Z^o-production. Two events, D and E show small $p_T^{\mu\mu}$ and have no jet activity.

In event C a single neutral electromagnetic deposition of E = 28.3 GeV, consistent in shower development with one or several γ (π^0) was found near to one of the muons. Otherwise the event is very "quiet", the scalar E_T of the "rest of the event" is 15.6 GeV, smaller than the average activity in minimum bias events. Two similar events were found in the electron decay channel by the UA1 and UA2 Collaborations respectively [3a,9,10]. These events were interpreted as $Z^0 \to e^+e^-\gamma$. For both electron events it is necessary to include the γ to reconstruct a mass that is consistent with the Z^0-mass. In addition, the p_T of the $e^+e^-\gamma$ system is significantly smaller than the p_T of the e^+e^- system. The same tendency is seen in the $\mu^+\mu^-\gamma$-event, but the measurement errors are much larger. The mass values and transverse momenta are listed in table 1. As for the electron events, the γ is close to one of the leptons. Both the angle $\alpha_o(\mu^+\gamma) = 7.9^o \pm 0.1$ and the fraction $k_o = 0.35 \pm 0.04$ of the energy carried by the γ are too large to be accounted for by bremsstrahlung. For muons the probability of external bremstrahlung is negligible. The probability for internal bremstrahlung is calculated to be $P(\alpha > \alpha_o, k > k_o) = 0.007$. Hence less than 0.04 $Z \to \mu^+\mu^-$ events are expected to show a bremsstrahlung as observed in event C.

Combination	Mass GeV/c^2	p_T GeV/c
$\mu^+\mu^-\gamma$	88.4^{+46}_{-15}	5.7 ± 2.0 [1)]
$\mu^+\mu^-$	70.9^{+37}_{-12}	10.4 ± 2.4 [1)]
$\mu^+\gamma$	5.0 ± 0.4	$30.^{+3}_{-2}$
$\mu^-\gamma$	52.5^{+28}_{-9}	$31.^{+43}_{-14}$

1) From calorimetry only

Table 1. Mass values for the $Z^o \to \mu\mu\gamma$ event C. The parameters of the γ/π^o are $E_T = 10.5 \pm 0.9$ GeV, $\eta = -1.7$, $\phi = -152^o$.

The events A and B show large jet activity (fig. 2). The masses of the jet systems are in the IVB mass range but the masses of the muons and the central jets are higher, over 120 GeV/c^2 (for detailed information, see ref. [7]).

The p_T-distribution of the Z^o is shown in fig. 3a together with the data on $Z^o \to e^+e^-$ and $W^\pm \to \nu\mu$, $e\nu$. There is some indication that the Z^o events have larger p_T values than the W. The longitudinal momentum distributions of the Z^o and W, shown in fig. 3b, show no differences.

According the observed mass values of the four $\mu^+\mu^-$ events and the $\mu^+\mu^-\gamma$ event we obtain

$$m_{\mu\mu} = 85.8^{+7.0}_{-5.4} \text{ GeV}/c^2$$

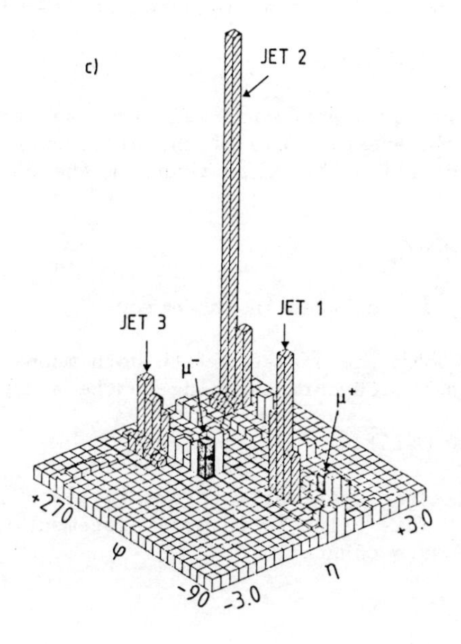

Fig. 2. Transverse energy distribution as a function of rapidity (η) and azimuthal angle (φ) for event B.

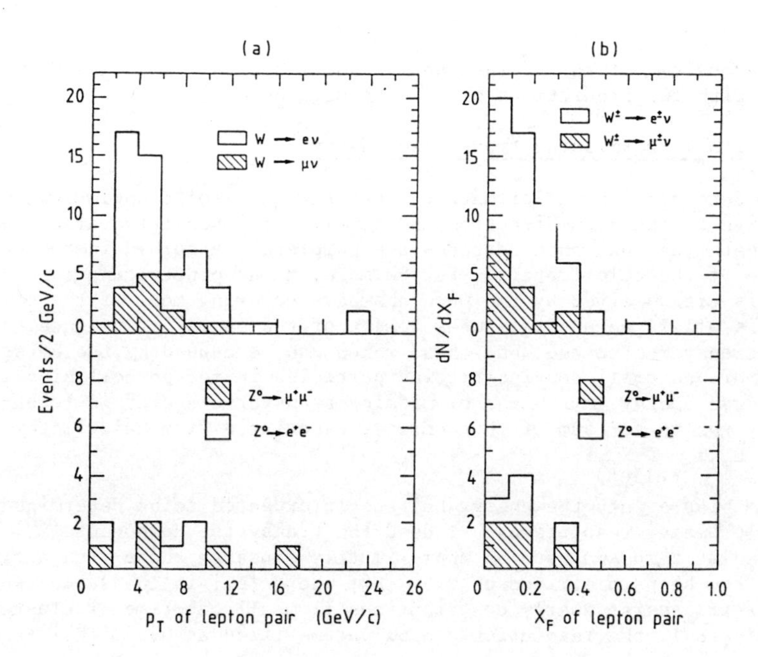

Fig. 3. (a) Transverse momentum distributions of the lepton pairs originating from $W \to \ell^+\nu$ and $Z^0 \to \ell^+\ell^-$ decays, (b) Feynman x-distributions of these lepton pairs.

consistent with the mass value of the Z^o as measured for $Z^o \rightarrow e^+e^-$

$$m_{ee} = 95.6 \pm 1.4 \pm 2.9 \text{ GeV}/c^2$$

Here the first error accounts for the statistical error and the second for the uncertainty in the overall energy scale of the electromagnetic calorimeters. The average value of all 9 Z^o events found in the UA1 experiment is

$$m_{Z^o} = 93.9 \pm 2.9 \text{ GeV}/c^2$$

where systematic uncertainty is included in the error.

The cross section obtained from the 4 events with both muons having track lengths longer than 40 cm in the CD corresponding to the acceptance 0.37 is

$$(\sigma \cdot B)_{\mu\mu} = 100 \pm 50 \ (\pm 15) \text{ pb}$$

where the last error includes the systematic uncertainties both in acceptance and in luminosity. This value is in good agreement with the cross section measured in the electron channel,

$$(\sigma \cdot B)_{ee} = 41 \pm 21 \ (\pm 7) \text{ pb}$$

The average value from all Z^o events is then

$$(\sigma \cdot B)_{\ell\ell} = 58 \pm 21 \ (\pm 9) \text{ pb}.$$

The measured values of the Z^o mass and production cross sections are well consistent with the predictions of the Standard Model [11].

3. Events with large missing transverse energy

In the UA1 detector the calorimeters cover the full solid angle down to 0.2^o relative to the beam direction and havelittle insensitive area. Both electromagnetic and hadronic cascades are completely absorbed in the sensitive volume of the calorimeters. Furthermore, muons penetrating the calorimeters are observed by the muon chambers covering most of the solid angle. This allows an accurate measurement of the energy misbalance in the plane transverse to the beam axis, which can be caused by the emission of one or more neutral, non-interacting particles in the $p\bar{p}$ collision. The transverse energy imbalance or missing transverse energy is reconstructed by the vector sum of the energies associated with calorimeter clusters, $\vec{E}_T^M = - \sum_{cells} \vec{E}_T, i$.

For minimum bias events the energy balance is expected to be determined by overall calorimeter resolutions. Indeed the transverse components E_x^M, E_y^M are centered at zero and have an approximately Gaussian shape with a r.m.s. width that can be parametrized as $0.5 \sqrt{|E_T|}$ where $|E_T|$ is the scalar sum (in GeV) of transverse energy contributions from all calorimeter clusters. For $\sqrt{(E_x^M)^2 + (E_y^M)^2}$ the resolution can be parametrized as $0.65 \sqrt{|E_T|}$ (see fig. 4).

The following results of an inclusive study of large missing transverse energy events come from an event sample corresponding to an integrated luminosity $\int L\, dt = 113 \text{ nb}^{-1}$ (they have been published in ref. [12]). In this run no dedicated trigger was provided by the requirement of the missing energy alone. We therefore used the sample of events recorded with the following signatures: (i) a jet of $E_T > 25$ GeV, (ii) an electromagnetic cluster of $E_T > 10$ GeV, (iii) a muon of $p_T > 5$ GeV/c, or (iv) a scalar transverse energy $|E_T| > 60$ GeV in the region $|\eta| < 1.5$. The initial selection of events started from a sample of 2.5×10^6 events, out of which 1.5×10^6 are calorimeter triggers. A first selection consists of the following requirements: (1) $|\Delta E_M| > 15$ GeV; (2) $|E| < 700$ GeV to remove multiple interactions; (3) technical cuts to remove reconstruction errors in the forward electromagnetic calorimetry (bouchons); (4) removal of cosmic ray and beam halo events, in which most of the energy comes from the outer calorimeter segments.

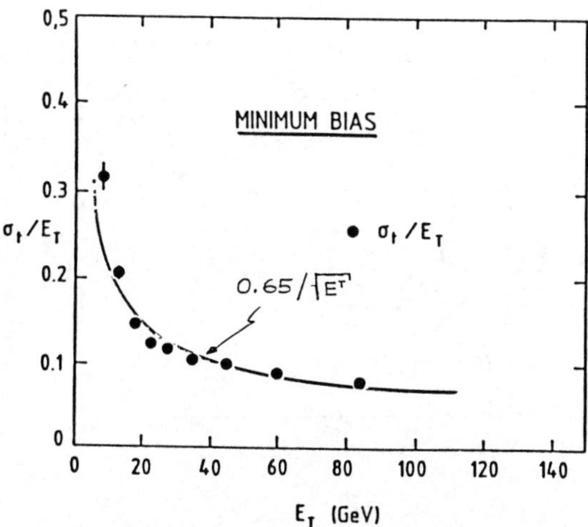

Fig. 4. The r.m.s. width of the missing transverse energy as a function of total scalar transverse energy in minimum bias events.

This first selection gave 29962 events which were fully reconstructed by standard UA1 programs, and E_T^M was recalculated. A more refined selection was then performed: (1) $|E_T^M| > 4\sigma$ with $\sigma = 0.7 * \sqrt{|E_T|}$; (2) E_T^M must not point to within ± 20 degrees of the vertical. The last cut is necessary because of the reduced efficiency of calorimetry in that region. After this second selection 1159 events were left.

Events were then scanned by physicists on the interactive graphic facility and in most of the cases the value of E_T^M was found to be faked by (1) cosmic rays; (2) beam halo particles; or (3) reconstruction problems. A total of 77 events were declared genuine. They are plotted in fig. 5a. There are fifty identified $W^\pm \to e^\pm + \nu$ events [1] which are then removed from the sample leaving twenty-seven additional events (fig. 5b).

Their topology separates them into;

- (i) 2 events with a single, isolated neutral electromagnetic cluster
- (ii) 17 events with a single-jet
- (iii) 5 events with two jets
- (iv) 3 events with more than two jets

With an additional cut on $E_T^M > 40$ GeV we are left with

- 2 events with a neutral electromagnetic cluster (π^0/γ) and large E_T^M back-to-back in azimuth ("photon events"),
- 5 events with a single jet and large E_T^M back-to-back in azimuth ("monojet events")
- 1 event with more than two jets and large E_T^M.

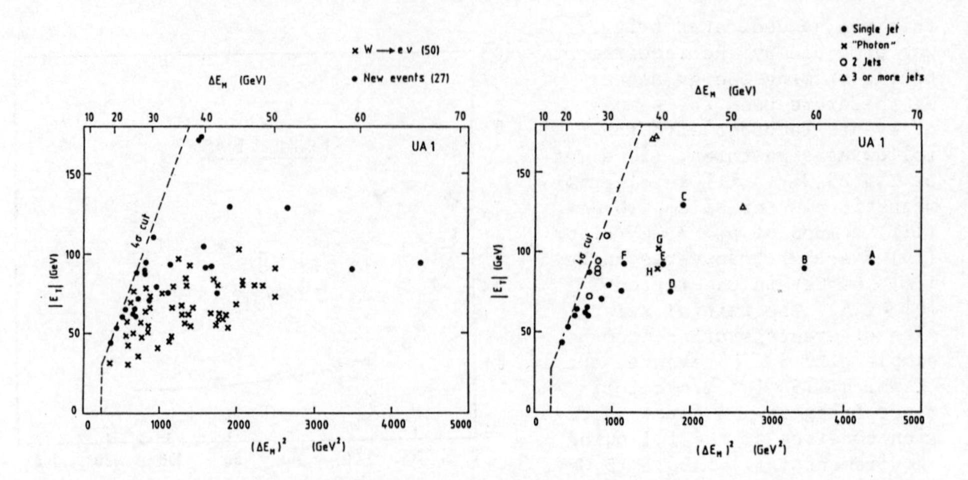

Fig. 5a) Scatter plot of missing transverse energy squared vs total scalar transverse energy for events which have E_T^M in excess of 4 standard deviations.
5b) Same as 5a with $W \rightarrow e\nu$ events removed.

The background to the photon events from $W \rightarrow e\nu$, from cosmic rays and from a pile-up of several neutrals is estimated to be less than 0.01 events [12]. The main background to the monojets comes from QCD jet production where all but one jet is missed due to fluctuations in the jet energy measurement. This background however decreases exponentially with E_T^M and is found to be negligible above E_T^M = 40 GeV (see fig. 6). Part of the monojet events

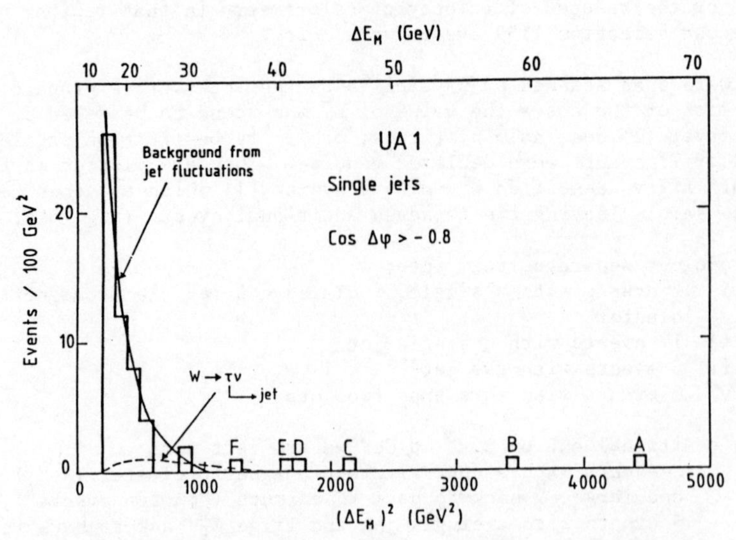

Fig. 6. Distribution of missing transverse energy for events with $\cos\Delta\varphi > -0.8$ where $\Delta\varphi$ is the azimuthal angle between the E_T^M vector and the jet. Background estimates from jet events and $W \rightarrow \tau\nu$ are also shown.

can come from the decay $W \rightarrow \tau\nu_\tau$ where τ is decaying hadronically into a jet. Estimates for the upper limit of this contribution vary between zero and two events without full simulation of trigger and efficiencies [12, 13]. The same conclusions apply to the contribution of a heavy sequential lepton of mass > 20 GeV/c^2. Finally no appreciable contribution is predicted from QCD jet production in association with a Z^o decaying into neutrinos. Our preliminary calculations indicate that the multijet event is consistent with the background due to fluctuations in the energy measurements of the jets.

Finally we show in fig. 7 the E_T of the photon or the jet versus the missing transverse energy for the seven events selected. Several possible interpretations can be presented for these events:

(i) production of prompt neutrinos
(ii) production of a heavy particle decaying into $Z^o + \gamma$ or Z^o + jet with an invisible decay $Z^o \rightarrow \nu\bar{\nu}$
(iii) production of supersymmetric particles
(iv) other, less conventional processes

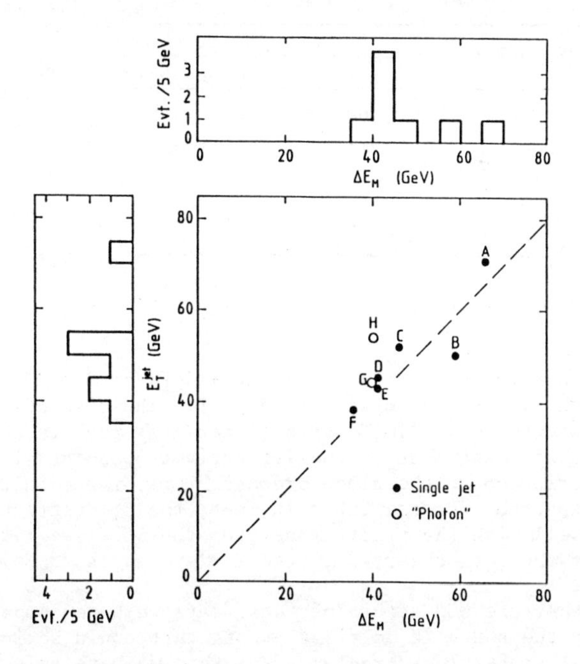

Fig. 7. Scatter plot of missing E_T versus jet or photon(s) E_T for the photon and monojet events.

For a detailed discussion of these interpretation we refer to Peccei's talk in these proceedings. Should interpretation (ii) be correct, the following processes should also be seen

$X \to Z^o\gamma$, $\hookrightarrow \ell\bar{\ell}$ $X \to Z^o$ + jets, $\hookrightarrow \ell\ell$ $\ell = e, \nu$

$X \to W\gamma$, $\hookrightarrow \ell\nu$ $X \to W$ + jets, $\hookrightarrow \ell\nu$ $X \to$ jets

The small branching ratios of decays $Z^o \to e^+e^-$ and $Z^o \to \mu^+\mu^-$ (estimated to 3 %) may of course be the reason that these processes with a jet or photon associated have not been seen with the present statistics. It is in any case of great interest to study more closely the W and Z^o events which have jet activity.

3. Hadronic jets associated with W and Z^o

The jets have been reconstructed in the W and Z^o events with the UA1 jet finding algorithm as described in chapter 2. The rates of occurrence of jets in the IVB events are given in table 2.

IVB decay channel	No. events all	with jets
$W \to e\nu$	68	21
$W \to \mu\nu$	14	4
$Z^o \to e^+e^-$	4	3
$Z^o \to \mu^+\mu^-$	5	3

Table 2. Number of IVB events with jets.

All jets have $E_T > 5$ GeV since the efficiency of the jet algorithm goes to zero below that value.

In order to look for the processes $X \to W$ + jets or $X \to Z^o$ + jets we plot the mass of the IVB + jets system in fig. 8. The resolution is $\sim 0.2 \times$ [mass (W + jet) - 80] GeV/c^2. No significant mass peak is observed. Hence it seems that at least most of the jet activity observed in the IVB events is better attributed to the gluon bremsstrahlung of the initial state partons as expected from the QCD. In fact, the E_T distribution of these jets agrees well with the predictions from the QCD (see ref. [1b]). The highest E_T of the jets observed in the IVB events is 23 GeV.

To check further the QCD nature of these jets we have compared them with those jets in the hadronic multijet events that could be interpreted as due to initial state bremsstrahlung. For this we have selected a sample of multijet events in the following way. First we require the jet algorithm to reconstruct at least two jets. The system of the two highest E_T jets is then required to have an invariant mass in the region 70 - 90 GeV/c^2.

Finally, the angle θ between the direction of the two highest E_T jets and the average beam direction in the rest frame of the jet pair has to satisfy $|\cos\theta| < 0.2$. By selecting angles close to 90° we enhance the separation between initial and final state bremsstrahlung. With these cuts we are

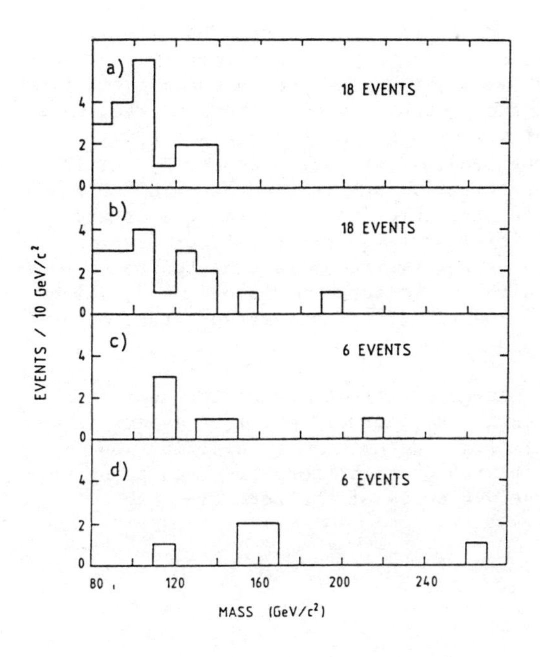

Fig. 8a) Mass of (W + jet). For the 4 events having ≥ 1 hadronic jets the highest E_T jet has been used.
b) Mass of (W + jets). All reconstructed jets used.
c) Mass of (Z^o + jet). Only highest E_T jet used.
d) Mass of (Z^o + jets). All reconstructed jets used.

left with 387 hadronic jet events from which 163 have at least one additional jet to the two highest E_T jets. These additional jets are compared with the jets in the IVB events. In table 3 the rate of occurrence of the additional jets produced in the hadronic multijet events are compared with the rate of occurrence of jets produced in association with an IVB [14]. These numbers are not corrected for any acceptance effects, in particular

Event type	Number of events: With jet activity	Number of events: Without jet activity	Total	Fraction with jet activity (f)
Jet events	163	224	387	0.42 ± 0.04
W events	25	57	82	0.30 ± 0.07
Z^o events	6	3	9	0.64 ± 0.14
Relative rates:	f_{IVB}/f_{Jet} = 0.69 ± 0.18 (acceptance corrected)			

Table 3. Rate of occurrence of jets in the IVB events and "additional" jets in the hadronic multijet events.

we do not know the detection and reconstruction efficiency of our jet algorithm for the low transverse energy jets that dominate the sample. This efficiency depends on the details of hadronization process of the soft parton and is therefore difficult to estimate because of the large uncertainties in understanding these processes. However, these uncertainties may largely cancel out in the ratios of the jet rates of the hadronic events and IVB events. From table 3 we can see that the rate of occurrence of jets in the IVB events is slightly smaller than that of jets in the hadronic events. Naively, we might expect the ratio of these rates to be 4/9, which is the ratio of the probability of a quark to radiate a gluon (IVB events) to the probability of a gluon to radiate a gluon (hadronic jet events, assuming the leading jets are produced by initial state gluons only). In a more quantitative analysis one should include such details as differences in the structure functions for quarks and gluons and the contribution of initial state quarks to jet production. After these corrections the ratio of the two rates is expected to be between 4/9 and 1, in agreement with the measured ratio 0.7 ± 0.2, which has been corrected for those experimental acceptance effects that do not cancel in the ratio.

In fig. 9 we show the transverse energy distributions of the jets in the IVB events and the additional jets in the hadronic multijet events. The distributions agree very well. In fig. 10 the angular distributions of the same jets are shown. Again the two distributions are seen to be very similar. They also fit very closely a curve of the form $(1 - \cos\theta^*)^{-1}$ typical of bremsstrahlung processes.

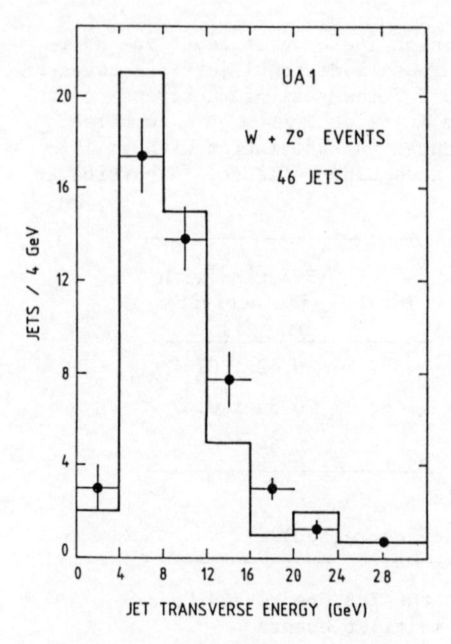

Fig. 9. The transverse energy distribution of the jets in the IVB events (histogram) compared with the corresponding distribution of the additional jets in the hadronic multijet events (data points), normalized to the number jets in the IVB events.

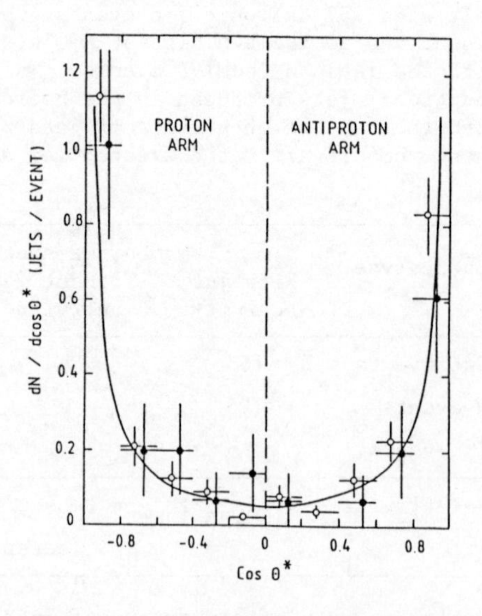

Fig. 10. The angular distribution of jets in the IVB events (closed circles) compared with the corresponding angular distribution for additional jets in the multijet events (open circles). The curve shows a $(1 - |\cos\theta^*|)^{-1}$ dependence normalized to the IVB data in the interval $|\cos\theta^*| < 0.95$.

The apparent suppression of final state bremsstrahlung ($\cos\theta^* \approx 0$) is related to the fairly wide cone around the leading jets used in our jet algorithm.

These results show that the jets produced in association with the IVB's have similar properties to a sample of jets that are believed to arise from gluon bremsstrahlung processes. No significant evidence is seen for the processes $X \to W +$ jets or $X \to Z^0 +$ jets.

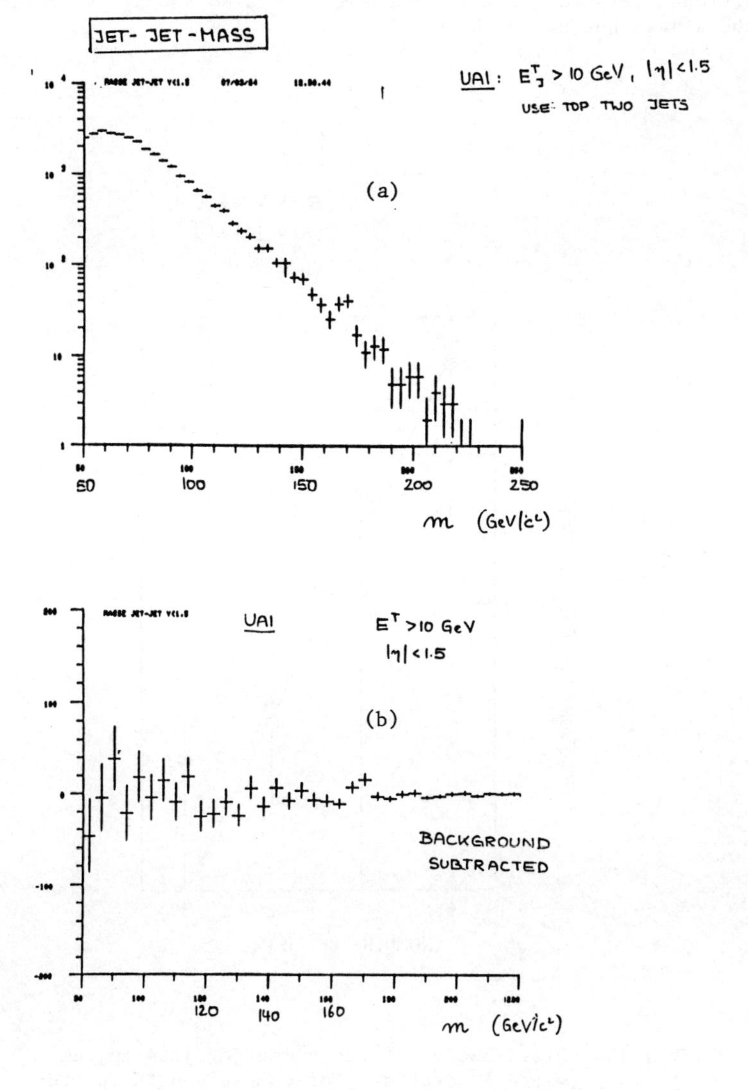

Fig. 11. (a) Mass of two jets in the hadronic jet events
(b) Same as (a) with an exponential background $\exp[Am + Bm^2 + C]$ subtracted.

To complete the search for the new state X we have studied also the invariant mass distribution of two jets in the hadronic jet events. This is shown in fig. 11. The estimated resolution in the two jet mass is ~ 16 %. No significant structure is yet observed.

As a final remark on the IVB + jet(s) events we note that the Z^o events seem to have somewhat more jet activity than the W events, as is already suggested by tables 2 and 3. In fact, 5 Z^o events have at least two jets whereas one would expect only 0.66 such events as there are 5 such events in the W sample. The distributions of the number of jets in the W and Z^o events are shown in fig. 12. The corresponding distribution of the number of additional jets in the multijet sample is also shown. This agrees well with the W data but not with the Z^o data.

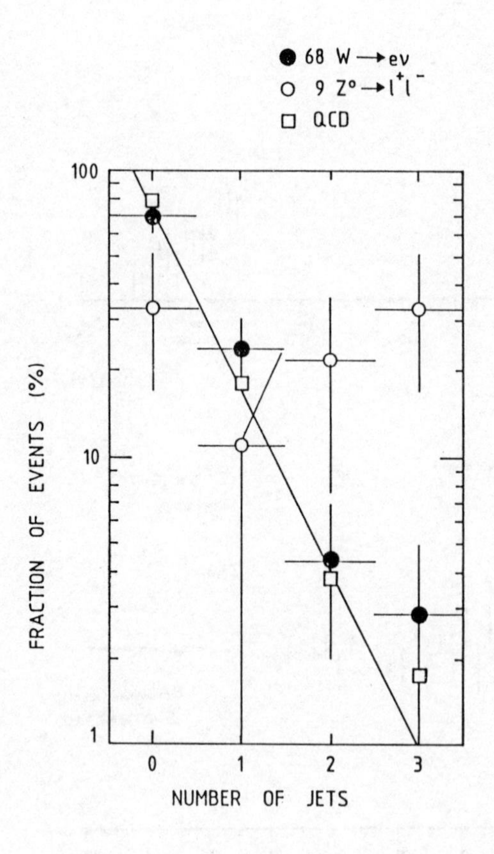

Fig. 12. Distribution of the number of jets in the W and Z^o events. For comparison the corresponding distribution of the number of additional jets in the hadronic multijet events is shown (QCD).

4. Conclusions

From this analysis of the 1982-83 data of the UA1 experiment we list the following conclusions.

We have observed the Z^o decay into $\mu^+\mu^-$ channel with the mass and rate consistent with the measurements for $Z^o \to e^+e^-$ and the predictions of the Standard Model. In addition, one of the $Z^o \to \mu^+\mu^-$ events is observed to have an associated hard photon, similarly to the two events of the type $Z^o \to e^+e^-\gamma$ reported earlier by the UA1 and UA2 experiments.

In the inclusive study of large missing transverse energy events we have observed (i) two events with a high E_T photon(s) and large missing transverse energy and (ii) five events with one high E_T jet and large missing transverse energy, for which no conventional explanation has been found. Several new explanations have already been tried on these events with varying degree of success. We refer to Peccei's talk in these proceedings for a review of these attempts. We mention here only two possibilities. If the events were due to the production of a gluino pair, the rate would give us, according to a recent calculation [15], a lower limit of 40 GeV/c^2 for the gluino mass. Another possible mechanism would be the decay of an object heavier than the IVB's to a Z^o and jets or photons with the subsequent decay of the Z^o to a neutrino pair. One would then expect also events with $W \to e\nu$, $\mu\nu$ and $Z^o \to e^+e^-$, $\mu^+\mu^-$ associated with high transverse energy jets. Our results indicate, however, that the events of this type observed so far are consistent with initial state gluon bremsstrahlung as expected according from the QCD. No significant structure in the jet-jet mass is yet seen either.

Finally we have observed indications of a larger jet activity in the Z^o events when compared to the W events.

The intriguing results presented here are yet all based on a rather small numbers of events. We therefore wait with great excitement for the new information on these interesting, possibly entirely new phenomena to be collected in the next, higher luminosity $p\bar{p}$ collider runs at CERN in 1984 and 1985.

References

1. a) UA1 Collaboration, G. Arnison et al., Phys. Lett. 122B (1983) 103
   b) UA1 Collaboration, G. Arnison et al., Phys. Lett. 129B (1983) 273
   c) UA2 Collaboration, M. Banner et al., Phys. Lett. 122B (1983) 476

2. UA1 Collaboration, G. Arnison et al., Phys. Lett. 134B (1984) 469

3. a) UA1 Collaboration, G. Arnison et al., Phys. Lett. 126B (1983) 398
   b) UA2 Collaboration, P. Bagnaia et al., Phys. Lett. 129B (1983) 130

4. For reviews of the experimental results on the IVB's, see e.g.
   C. Rubbia, "The Physics of the Proton-Antiproton Collider", invited talk at the International Europhysics Conference on High Energy Physics, Brighton, 1983, CERN preprint EP/83-168 (1983)
   E. Radermacher, CERN EP/84-41 (1984), to be published in Progress in Particle and Nuclear Physics
   H. Wahl, Proceedings of the XV Symposium on Multiparticle Dynamics, Lund (1984)

5. M. Barranco Luque et al., Nucl. Instrum. Methods 176 (1980) 75
   M. Calvetti et al., Nucl. Instrum. Methods 176 (1980) 255
   M. Calvetti et al., IEEE Trans. Nucl. Sci. NS-30 (1983) 71
   M. Corden et al., Phys. Scr. 25 (1982) 5 and 11
   K. Eggert et al., Nucl. Instrum. Methods 176 (1980) 217
   The UA1 Collaboration is preparing a comprehensive report on the detector to be published in Nucl. Instrum. Methods

6. UA1 Collaboration, G. Arnison et al., Phys. Lett. 132B (1983) 214
   The corrections to jet energies and errors are discussed in "Correction tables and error estimates for jet energies and momenta", M. Della Negra et al., UA1 Technical Note UA1/TN/84-43 (unpublished).

7. UA1 Collaboration, G. Arnison et al., CERN EP/84-100 (1984) submitted to Phys. Lett. B

8. F.E. Paige and S.D. Protopopescu, ISAJET program, BNL 29777 (1981)
   B. Humpert, Phys. Lett. 85B (1979) 293

9. UA1 Collaboration, G. Arnison et al., Phys. Lett. 135B (1984) 250

10. UA2 Collaboration, P. Bagnaia et al., Phys. Lett. 129B (1983) 130

11. S.L. Glashow, Nucl. Phys. 22 (1961) 579
    S. Weinberg, Phys. Rev. Lett. 19 (1967) 1264
    A. Salam, Proc. 8th Nobel Symposium, Aspenäsgarden, 1968 (Almqvist and Wiksell, Stockholm, 1968), p. 367
    For recent calculations including higher order effects, see:
    W.J. Marciano and Z. Parsa, Proc. 1982 DPF Summer Study on Elementary Particle Physics and Future Facilities, Snowmass, 1982 (AIP, New York, 1983), p. 155

12. UA1 Collaboration, G. Arnison et al., Phys. Lett. 139B (1984) 115

13. P. Aurenche, R. Kinnunen, LAPP-TH-108 (1984)

14. S. Geer, "Hadronic jet activity associated with the intermediate vector boson at the CERN SPS collider, UA1 collaboration", Proceedings of the XIXth Recontre de Moriond, Elementary Particle Physics Meeting on New Particle Production at High Energies, La Plagne, March 1984 (to be published).

15. J. Ellis and H. Kowalski, CERN TH-3843 (1984) and private communication.

Inst. Phys. Conf. Ser. No. 73: Section 2
Paper presented at VII Eur. Symp. Antiproton Interactions, Durham 1984

Large mass electron–neutrino pairs in UA2

The UA2 collaboration

presented by S. Loucatos

CEN-Saclay, DPhPE, 91191 Gif-sur-Yvette Cedex, France.

Abstract The production of electrons with very high transverse momentum has been studied in the UA2 experiment at the CERN $P\bar{P}$ collider ($\sqrt{s}$ = 540 GeV). From a sample of events containing an electron candidate with P_T > 15 GeV/c, we extract a clear signal resulting from the production of the charged intermediate vector boson $W^{\pm}$, which subsequently decays into an electron and a neutrino. We also observe events in which the electron-"neutrino" pair is produced in association with hard jets ; their interpretation in terms of W production via known processes looks unlikely.

1. Introduction

We report here the results from a search for electrons with P_T > 15 GeV/c in events with missing transverse energy observed by the UA2 Collaboration at the CERN $p\bar{p}$ collider ($\sqrt{s}$ = 540 GeV) from a data sample corresponding to a total integrated luminosity of 131 nb^{-1} collected during the 1982 and 1983 periods of collider operation.

These electrons are expected to be produced in the processes :

$$p + \bar{p} \rightarrow W^{\pm} + \text{anything}, \quad W^{\pm} \hookrightarrow e^{\pm} + \nu\ (\nu) \quad \text{and}$$

$$p + \bar{p} \rightarrow Z^{0} + \text{anything}, \quad Z^{0} \hookrightarrow e^{+}e^{-} \text{ or } e^{+}e^{-}\gamma$$

according to the electroweak theory [Glashow, Weinberg, Salam], (for a review see [Ellis]).

The results presented here have already appeared in publications [Banner, Bagnaia a, b, c].

2. Apparatus

The UA2 detector has been described in detail elsewhere [Mansoulié, Beer, Borer, Conta,]. We briefly recall its main features.

Apart for two narrow cones along the beams, the detector provides full azimuthal coverage in three distinct regions of polar angles : $40° < \theta < 140°$, the central region, and $20° < \theta < 40°$, $140° < \theta$ 160°, the forward regions.

In the centre of the detector a set of coaxial cylindrical drift and proportional chambers detect the charged particles produced in the collision and measure the position of the event vertex.

An array of 480 calorimeter cells, each cell covering a similar domain of longitudinal phase-space (15° of azimuth and $\simeq$ 0.2 units of rapidity), measures electron energies E(e) to a good accuracy. Each cell is segmented longitudinally to provide electron-hadron separation. While hadron showers are usually contained in the 4.5 absorption lengths of the central calorimeter, providing a measurement of jet energies, they only deposit a fraction of their energy in the forward calorimeters which are $\simeq$ 1.0 absorption length thick. These forward regions are equipped with magnetic spectrometers which provide additional rejection power against hadrons and converted photons when searching for electrons. In addition they measure the momenta of charged jet fragments, the energy of π^0's being measured in the calorimeter cells.

In both the central and forward regions electron identification is significantly improved by preshower counters which accurately measure the match between the observed incident track and the developing shower.

3. Data Analysis

Trigger thresholds were applied to linear sums of signals from matrices of 2 x 2 calorimeter cells. A signal was generated whenever the linear sum from at least one such matrix exceeded an 8 GeV threshold in transverse energy.

Approximately 7 x 10^5 triggers were recorded during the 1982 and 1983 runs, corresponding to an integrated luminosity L = 131 nb^{-1}.

High-p_T electrons are identified by requiring that the event satisfies the following main criteria :

a) the presence of a localised cluster of energy deposition in the first compartment of the calorimeters, with only a small energy leakage in the hadronic compartment.

b) the presence of a reconstructed charged particle track which points to the energy cluster. The pattern of energy deposition must agree with that expected from an isolated electron incident along the track direction.

c) the presence of a hit in the preshower counter, with an associated pulse height larger than that of a minimum ionising particle (m.i.p.).The distance of the hit from the track must be consistent with the space resolution of the counter itself. Both these features are distinctive of the early shower developed in the converter by a high-energy electron.

In practice, because the central and forward detectors are not identical, these criteria are applied in different ways in the two regions. In order to remove electrons from photon conversion in the detector material, we require that the track coordinates are found in at least one the two innermost chambers of the vertex detector (C_1 or C_2). Furthermore, in the forward detectors we reject the event if we observe another particle of opposite charge sign produced with an angular separation smaller than 30 mr from the electron candidate.

In the forward region we require in addition that the charged particle momentum $\vec{p}$, as measured in the spectrometer, and the energy deposition in the calorimeter agree within errors.

The overall efficiency of the electron identification criteria is 0.76 ± 0.05 in the central region and 0.80 ± 0.05 in th forward regions.

The present analysis is concerned only with electrons having $p_T^e > 15$ GeV/c (approximately 8 x 10^4 triggers). After applying the aforementioned cuts this sample is reduced to 225 events containing a total of 227 electron candidates, of which 200 are observed in the central detector and 27 in the forward ones.

4. Topology of the Events Containing an Electron Candidate

The sample of 225 events still contains, in addition to genuine electrons, fake electrons resulting from misidentified high-p_T hadrons or jets of hadrons.

In the case of genuine electrons, because of lepton number conservation the event must contain either another electron of opposite charge (e.g. $Z^0 \rightarrow e^+e^-$) or a neutrino (e.g. $W \rightarrow e\nu$), which are also emitted, in general, with high p_T.

If the electron candidate is not genuine, but results from a misidentified high-p_T hadron or jet of hadrons, we expect another jet of high-p_T hadrons to be present in the same event at an approximately opposite azimuth. Such a configuration is typical of events containing high-p_T hadronic jets [Bagnaia d, e].

In order to examine the topology of these events, we search for high-p_T jets by using the fine segmentation of the calorimeters to group into clusters all adjacent cells containing an energy deposition in excess of 400 MeV. This procedure to identify high-p_T particles or jets of particles has been described in detail elsewhere [Beer, Fayard, contribution to this conference and Bagnaia d, e].

In the forward detectors, however, the calorimeter thickness is equivalent to 1 absorption length only and hadronic showers are not fully contained. In that case we use the calorimeters to measure the energy of electromagnetic showers, and the magnetic spectrometers to measure charged particle momenta.

To each cluster we associate a momentum $\vec{p}_{jet}$ with magnitude equal to the cluster energy and directed from the event vertex to the cluster centroid. To select preferentially high-p_T particle resulting from the same hard collision which produced the electron candidate, we define as a jet any cluster with a transverse mementum in excess of 3 GeV/c.

We find that 45 events contain no additional high-p_T cluster, the electron candidate being the only high-p_T particle observed in the detector. The p_T distribution of the electron candidates in these events is shown in Fig. 1a. Such events either contain a neutrino, as in the case of $W \rightarrow e\nu$ decays, or contain other high-p_T particles which have escaped detection because they were emitted outside the detector acceptance. Two-jet events, with one of the jets misidentified as an electron and the other undetected, are expected to contribute to the latter category.

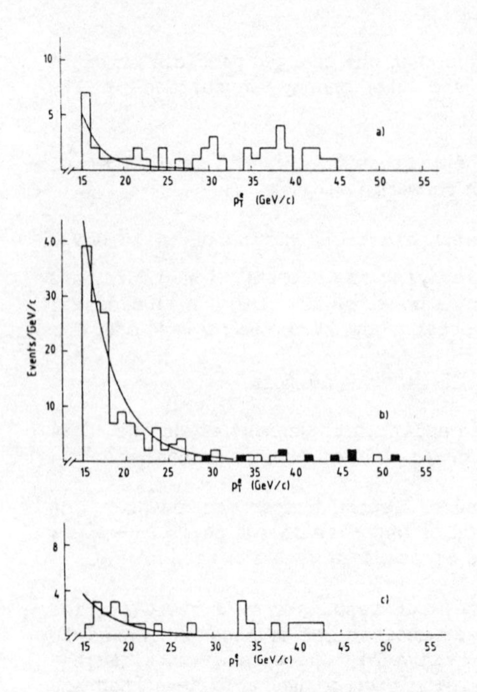

Fig.1 (a)Transverse momentum distribution of the electron candidates in events with no additional jets. (b)Transverse momentum distribution of the electron candidates in events with additional jets having ρ_{opp}>0.2 (see Eq.1). The dark area corresponds to the eight Z_0 events for which only the electron with the higher p_T^e is plotted. (c)Transverse momentum distribution of the electron candidates in events with additional jets having ρ_{opp}<0.2. The curves represent the background estimates as discussed in Sect. 5.

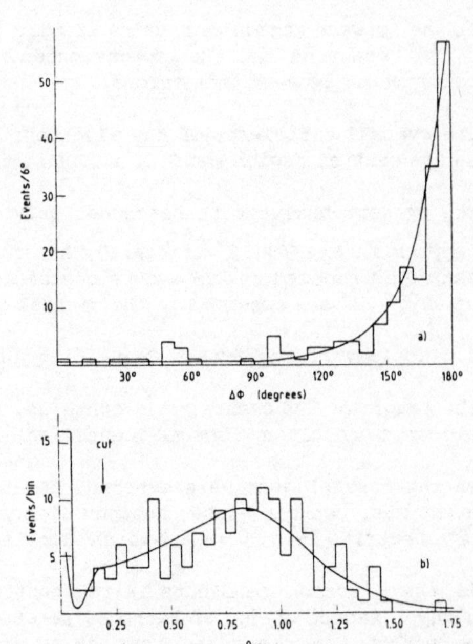

Fig.2 (a)Distribution of the azimuthal separation $\Delta\phi$ between the momentum vector of the electron candidate and that of the associated jet having the highest p_T. (b)Distribution of ρ_{opp} (see Eq.1) for events containing an electron candidate and at least one associated jet. The curves represent the background estimates as discussed in Sect. 5.

The remaining 180 events contain at least one additional high-p_T cluster. For these events, Fig. 2a shows the distribution of the azimuthal separation $\Delta\phi$ between the momentum vector of the electron candidate, $\vec{p}_e$, and that of the jet with the highest transverse momentum. The peak at $\Delta\phi = 180°$ corresponds to configurations in which the electron candidate is approximately back-to-back in azimuth to a high-p_T jet. We expect these events to be contaminated by two-jet events in which one of the two jets has been misidentified as an electron.

In order to reduce the two-jet background in this sample we consider for each event all the jets whose momenta $\vec{p}_{jet}$ are separated in azimuth from $\vec{p}_e$ by an angle $\Delta\phi > 120°$, and we use these jets to define the quantity

$$\rho_{opp} = -\vec{p}_T^{\,e} \cdot \Sigma\vec{p}_{T,jet}/|\vec{p}_T^{\,e}|^2 \qquad (1)$$

where $\vec{p}_T^{\,e}$ and $\vec{p}_{T,jet}$ are two-dimensional vectors obtained projecting $\vec{p}_e$ and $\vec{p}_{jet}$ on a plane perpendicular to the beams.

The distribution of ρ_{opp} for the 180 events is shown in Fig. 2b. We reject all events which have $\rho_{opp} > 0.2$ (a total of 156 events). The p_T^e distribution for these events is shown in Fig. 1b. This sample contains the eight $Z^0 \to e^+e^-$ (or $e^+e^-\gamma$) events reported in [Bagnaia a], for which the higher p_T^e value is plotted in Fig. 1b, and 148 events consisting of an electron candidate with jet activity at opposite azimuthal angles.

The p_T^e distribution of the remaining 24 events with $\rho_{opp} < 0.2$ is shown in Fig. 1c. Of these, 16 events have $\rho_{opp} = 0$, which corresponds to no jet activity wharsoever in an azimuthal sector ± 60° wide opposite to the electron direction.

5. Background Estimate to the Electron Spectra

The results of the previous section suggest that the sample of 148 events with jet activity at azimuthal angles opposite to that of the electron candidates consists mostly of two-jet events in which one of the two jets has been misidentified as an electron.

In the following analysis these 148 events are taken as a pure background sample. We note that the rate of occurrence of these events is lower than that of inclusive jet production [Fayard, Bagnaia e] in the same p_T region by a factor of $\sim 3.6 \times 10^4$ for the central detector and $\sim 2.6 \times 10^5$ for the forward ones.

Backgrounds to the two events samples of Fig. 1a and 1c, which consist of electron candidates without and with jets, respectively, are estimated as follows. We consider a sample of events recorded using the trigger described in section 3, in which, however, no energy cluster with $p_T > 15$ GeV/c passes the electron cuts. In each of these events we take as an electron the energy cluster with $p_T > 15$ GeV/c (if any), having the smallest hadronic energy leakage. These fake electrons are then subdivided into three samples, called A, B and C, having respectively the topological configurations of Figs. 1a, b and c. The background contributions to the 69 electron candidates in the two non back-to back topologies (45 with no additional jets and 24 with jets having $\rho_{opp} < 0.2$) are then estimated directly from the p_T distributions of samples A and C, multiplied by a factor equal to the ratio of the total number of electron candidates with $\rho_{opp} > 0.2$ (148 events) to the total number of events in sample B. These background estimates are shown as smooth curves in Fig. 1a and 1c.

6. The $W \to e\nu$ Event Sample

The two event samples of Fig. 1a and 1c are expected to contain electrons from $W \to e\nu$ decay. The combined p_T^e distribution of these 69 electron candidates is shown in Fig. 3 together with the background estimate, which amounts to 21.4 ± 1.3 events in total. There is a clear accumulation of events near $p_T^e \simeq 40$ GeV/c, which is distinctive of the Jacobian peak expected for $W \to e\nu$ decay. For $p_T^e > 25$ GeV the distribution of Fig. 3 contains 37 events with an estimated background of 1.5 ± 0.1 events.

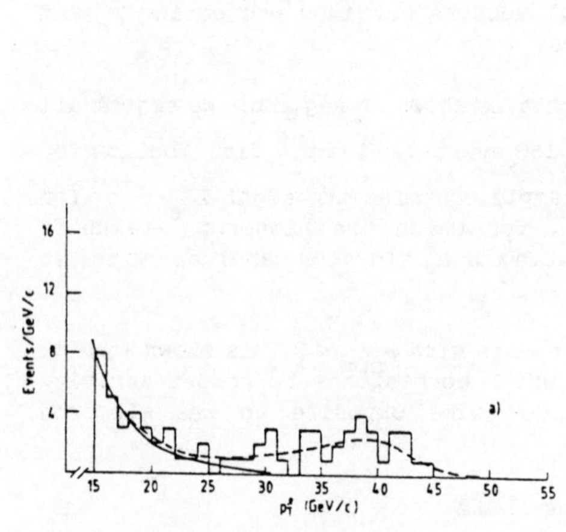

Fig. 3. Transverse momentum distribution of the electron candidate in the $W \to e\nu$ event sample. Full curve: background estimates as discussed in Sect. 5. Dashed curve : the sum of all expected contributions as discussed in [Bagnaia 84a].

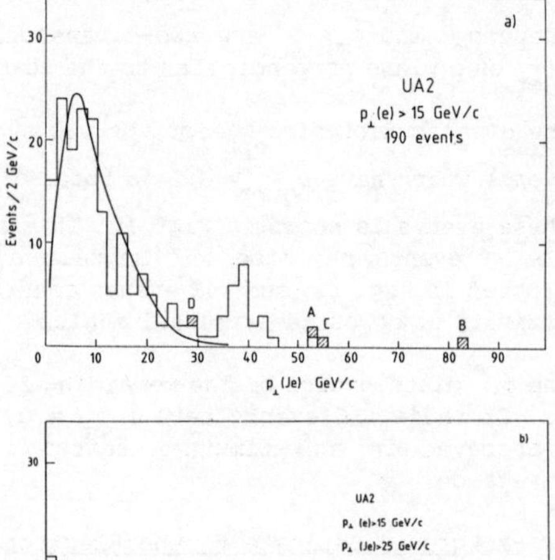

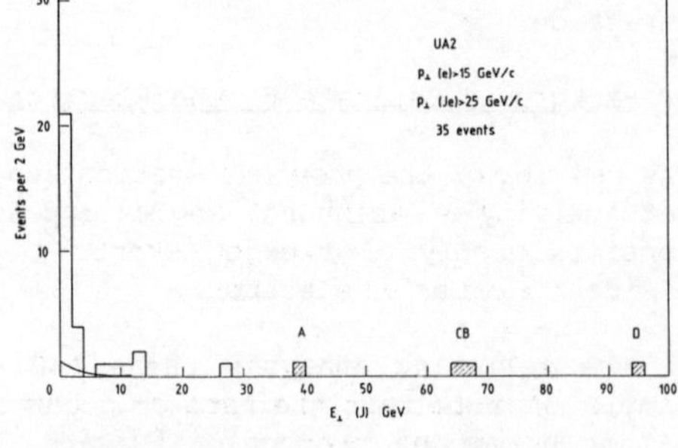

Fig.4 (a)Transverse momentum distribution of the system of electron and jets in the initial sample of 190 events. (b)Transverse energy distribution of the system of jets in the sample of 35 events having $p_\perp(Je)>25$ GeV/c. The event at $E_\perp(J)\simeq 26$ GeV is the largest transverse momentum W mentioned in Sect. 6. The lines correspond to the calculated background contaminations. The four events having $p_\perp(Je)>25$ GeV/c, $E_\perp(J)>30$ GeV, are cross-hatched.

The distributions of the variables used to define the electron identification criteria, such as the ratio E_{had}/E_{cl} (or E_{leak}/E_{em} in the forward regions), the distance between the track and preshower signal, and the charge mesured in the preshower counter are consistent with an electron behaviour.

The presence of a neutrino in $W \to e\nu$ events can be detected by the resulting transverse momentum imbalance. For each events we compute

$$\vec{p}_T^{\,\nu} = \vec{p}_T^{\,W} - \vec{p}_T^{\,e} \qquad (4)$$

where all vectors are projected onto a plane normal to the beams. The W transverse momentum $\vec{p}_T^{\,W}$ can be determined from the momenta of all other particles, or jets of particles, observed in the event in addition to the electron candidate

$$\vec{p}_T^{\,W} = -(\Sigma\, \vec{p}_{T,jet}' + \xi\, \vec{p}_T^{\,sp}) \qquad (5)$$

where the sum extends over all observed jets, $\vec{p}_T^{\,sp}$ is the total transverse momentum carried by the system of all other particles not belonging to jets, and ξ is a correction factor which takes into account the incomplete detection of the rest of the event ($\xi = 1$ for an ideal detector) determined using the eight Z^0 events previously reported [Bagnaia a]. The mean value of p_T^W is $< p_T^W > = 6.9 \pm 1.0$ GeV/c. There are 5 events with $p_T^W > 12$ GeV/c, the highest p_T^W value being 29.6 GeV/c. QCD predictions [Altarelli], are consistent with the observed distribution.

7. <u>Determination of the W Mass</u>

A value of the W mass is extracted from the 37 $W \to e\nu$ candidates with $p_T^e > 25$ GeV/c by comparing their two-dimensional distribution $d^2n/dp_T^e d\theta_e$, where θ_e is the measured electron polar angle, to that expected from $W \to e\nu$ decay. A Monte Carlo program is used to generate the distribution $d^2n/dp_T^e d\theta_e$ for different values of the W mass M_W. The W longitudinal momentum distribution is obtained using the quark structure functions of the proton with scaling violations, as given by [Gluck et al.]. A p_T^W distribution is generated using the shape given in [Altarelli], where we allow the average value $< p_T^W >$ to vary in order to take into account uncertainties of QCD predictions. The decay angular distribution is described by the standard V-A coupling and the detector response is taken into account. A fixed value of the W width, $\Gamma_W = 2.7$ GeV/c^2, is used.

The best fit to the experimental distribution $d^2n/dp_T^e d\theta_e$ gives $M_W = 83.1 \pm 1.8$ GeV/c^2, where the error includes an uncertainty of $\pm$ 1 GeV/c^2 which results from the effect of varying $< p_T^W >$ between 4 and 10 GeV/c in the fit. This uncertainty is added in quadrature to the statistical error.

In addition to the quoted error, we must take into account the systematic uncertainty in the calorimeter energy calibration, which amounts to an average value of ± 1.5 % with a cell-to-cell uncertainty having a r.m.s. of 2.2 %. The latter reduces to $2.2/\sqrt{37} \simeq 0.4$ % for our event sample.

To summarise we quote

$$M_W = 83.1 \pm 1.9 \text{ (stat.)} \pm 1.3 \text{ (syst.) GeV/c}^2 \qquad (6)$$

where we choose to keep the two errors separate because most of the systematic errors cancel when comparing the experimental values of M_W and M_Z.

8. Cross Sections. Z^0 Mass and Width

Results on cross sections for the process $p + \bar{p} \to W^{\pm}$ + anything followed by the decay $W \to e\nu$, as well and for the process $p + \bar{p} \to Z^0$ anything followed by $Z^0 \to e^+e^-$ are shown in Table I, together with the corresponding results of the UA1 experiment [Arnison] and theoretical predictions of [Altarelli et al.]. In the same table, results on the masses of the IVB's and weak interaction parameters computed from the observed masses are shown, compared to theoretical predictions of the standard model. Δr is a radiative correction as defined in [Marciano].

TABLE I. Comparison of the UA1 and UA2 results with theoretical expectations. Values given assume 100% correlation in the m_W and m_Z systematic uncertainties.

	UA1	UA2	Standard Model with $\sin^2\theta_W = 0.217 \pm 0.014$
m_W (GeV)	$80.9 \pm 1.5 \pm 2.4$	$83.1 \pm 1.9 + 1.3$	$83.0^{+2.9}_{-2.7}$
m_Z (GeV)	$95.6 \pm 1.5 \pm 2.9$	$92.7 \pm 1.7 + 1.4$	$93.8^{+2.4}_{-2.2}$
$m_Z - m_W$ (GeV)	$14.7 \pm 2.1 \pm 0.4$	$9.6 \pm 2.5 + 0.2$	10.8 ± 0.5
$\sin^2\theta_W = 1 - m_W^2/m_Z^2$ (a)	0.284 ± 0.035	0.196 ± 0.035	0.217 ± 0.014
Δr (a)	$0.252 \pm 0.072 \pm 0.045$	$-0.025 \pm 0.170 + 0.030$	0.0696 ± 0.0020
$\sin^2\theta_W = \left(\frac{38.65\ \text{GeV}}{m_W}\right)^2$ (b)	$0.228 \pm 0.008 \pm 0.014$	$0.216 \pm 0.010 + 0.007$	0.217 ± 0.014
ρ (b)	$0.928 \pm 0.038 \pm 0.016$	$1.020 \pm 0.050 + 0.010$	1
$(\sigma.B)^{W^{\pm} \to e^{\pm}\nu}$	$530 \pm 80 \pm 90$ pb	$530 \pm 100 \pm 100$ pb	370^{+110}_{-60} pb
$(\sigma.B)^{Z^\circ \to e^+e^-}$	$71 + 24 \pm 13$ pb	$110 \pm 40 \pm 20$ pb	42^{+12}_{-6} pb

a) $\sin^2\theta_W$ and Δr are calculated from UA1 and UA2 data under the hypothesis that $\rho = 1$.

b) $\sin^2\theta_W$ and ρ are calculated from data using the theoretical value of Δr (column 3).

The Z^0 width is obtained by two methods. The method based on the observed mass dispersion of the $Z^0 \rightarrow e^+e^-$ events gives $\Gamma_Z < 6.5 \pm 0.6$ GeV/c^2 (90 % C.L.) while the method based on the observed number of $Z^0 \rightarrow e^+e^-$ events gives $\Gamma_Z < 2.6 \pm 0.3$ GeV/c^2 (90 % C.L.). Within the standard model, these upper limits can be related to the number of additional light neutrinos ΔN_ν. The first method gives $\Delta N_\nu < 22 \pm 3$ ($\Delta N_\nu < 41 \pm 4$) at the 90 % (95 %) confidence level, assuming $\sin^2\theta_W = 0.22$ and a value of the t-quark mass $m_t > M_Z/2$ (we choose this value because it gives the highest upper limit for ΔN_ν).

The second method gives $\Delta N_\nu \leq 0$ ($\Delta N\nu < 2$) at the 90 % (95 %) confidence level.

9. <u>The decay $Z^0 \rightarrow e^+e^-\gamma$</u>

We have reported [Bagnaia a] $Z^0 \rightarrow e^+e^-\gamma$ event containing a photon with an energy k = 24 GeV and an 11 GeV electron clearly separated by an angle ω_{lab} = 31°, excluding, therefore, internal bremsstrahlung.

Using the calculation of radiative corrections by [Albert et al., Berends et al.] we estimate the probabilty that in a sample of 8 events $Z^0 \rightarrow e^+e^-$ we observe a radiative decay containing a resolved γ. Different methods give probabilities between 5 % and 19 %.

10. <u>Search for the Decay $W \rightarrow e\nu\gamma$</u>

Given the interest in unexpected decay modes of the Z^0 we have also looked for events compatible with the decay $W \rightarrow e\nu\gamma$ in the full data sample. In the central detector, we search for events containing an electron candidate with $p_T^e > 8$ GeV/c which passes the electron identification cuts described in section 3, and an additional photon with a momentum k in excess of 8 GeV/c. A photon is defined as an energy cluster which satisfies the same criteria on size and hadronic leakage as an electron, but has no charged particle track pointing to it. If also the photon cluster is seen in the central detector, we require an angular separation $\omega > 30°$ between the cluster centroids in order to resolve the $e\gamma$ pair.

Six events satisfy these conditions. However, none of them survives the additional requirement of a transverse momentum imbalance compatible with the presence of a neutrino having $p_T^\nu > 10$ GeV/c. A Monte Carlo estimate of the number of $W \rightarrow e\nu\gamma$ decays which are expected to satisfy all of these requirements as a result of radiative corrections gives 0.1 event.

In the forward detectors we can identify $e\gamma$ pairs with very small opening angles by releasing the condition that the electron momentum p and the energy E agree within the measuring errors. In this case, however, we limit our search to electron transverse momenta in excess of 20 GeV/c (as measured in the calorimeter), for which we expect a background contribution of 0.2 events. We find one event which contains a cluster of transverse energy E_T = 41 GeV (E = 86.2 GeV) and a track of 3.6 GeV/c momentum pointing to the cluster and satisfying all other electron identification criteria. Large missing p_T is detected in this event as expected in the case of an associated neutrino. The estimated background for $p_T^e > 35$ GeV/c is 0.01 events. Since the opening angle of this $e\gamma$ pair is compatible with zero, we also consider the effect of external bremsstrahlung and we obtain a probability of 0.5 % per $W \rightarrow e\nu$ decay, or 4.5 % to observe one such event in the sample of 9 $W \rightarrow e\nu$ and $e\nu\gamma$ candidates with $p_T^e > 20$ GeV/c detected in the forward detectors.

11. Observation of Electrons Produced in Association with Hard Jets and Large Missing Transverse Momentum

In the preceding analysis we discarded events in which a significant amount of transverse energy was detected at opposite azimuth to the electron candidate. This procedure efficiently rejected two-jet events in which one of the jets was misidentified as an electron. Other events, however containing a genuine electron in the final state, may have also been rejected by this cut.

We will now proceed to a systematic search for events containing an electron-"neutrino" pair in the final state, whatever the transverse energy at opposite azimuth to the electron may be. The loss of rejection power against background, resulting from having relaxed this constraint, is compensated by the requirement that both the electron and the "neutrino" have large transverse moment.

Our detector does not distinguish between a neutrino and other possible non-interacting particles, such as the photino postulated by supersymmetry theories [Fayet]. In the remainder of this letter the word "neutrino" must therefore be understood in a broad sense.

We discard from the sample of 225 events obtained in section 3 events collected in the 1982 period, during which the azimuthal coverage of the UA2 detector was incomplete [Mansoulié]. We also discard 10 events in which the electron candidate is observed near the interface between the central region and one of the forward regions and is associated with nearby calorimeter energy in each of the two regions.

Together, the initial sample is reduced to 190 events, and the corresponding integrated luminosity to 116 nb^{-1}.

For each event we evaluate the momentum p, energy E and mass m of various sets of particles such as the system of all jets (J) or the system of electron and jets, (Je). The quantity $p_\perp(Je)$ measures the missing transverse momentum in the set of all detected particles resulting in clusters having $E_\perp > 3$ GeV. In a typical event, the softer particles carry together only a small transverse momentum and, if no large transverse momentum particle has escaped detection, the observation of a large $p_\perp(Je)$ reveals the presence of a neutrino (ν) with $p_\perp(\nu) \simeq p_\perp(Je)$.

The distribution of $p_\perp(Je)$ is shown in Fig. 4a. The background evaluation follows the method described in section 5 and measures the probability that a multijet event contains both a misidentified electron and an undetected jet (or jets) escaping the UA2 acceptance. Its distribution is shown as a curve superimposed on the data of Fig. 4a. The background contribution amounts to 10.0 ± 0.9 events for $p_\perp(Je) > 20$ GeV/c, 3.4 ± 0.3 events for $p_\perp(Je) > 25$ GeV/c and 1.2 ± 0.1 events for $p_\perp(Je) > 30$ GeV/c.

In the remainder of this talk we restrict the analysis to the sample of 35 events which have $p_\perp(Je) > 25$ GeV/c, and which are therefore candidates for containing a large transverse momentum neutrino in the final state.

12. Data Analysis and Background Evaluation

From the results of the previous analysis we expect our sample to contain a number of events in which the observed electron carries most of the detected transverse energy, and is therefore accompanied by no jet or by jets having small transverse energies. The distribution of the sum of the jet transverse energies, $E_T(J)$, confirms this expectation (Fig. 4b). The 31 events having $E_\perp(J) < 30$ GeV all belong to the samples of Figs. 1a and 1c where they were interpreted in terms of $W \rightarrow e\nu$ decays. We do not comment further on these events. Instead we consider the four events having $E_\perp(J) > 30$ GeV. The distribution of the events in the $p_\perp(Je)$, $E_\perp(J)$ plane is shown in Fig. 5 for each of the signal and background samples. The background contamination expected in the region $p_\perp(Je) > 25$ GeV/c, $E_\perp(J) > 30$ GeV is $\simeq 0.45 \pm 0.04$ events. In each of the four signal events the electron candidate is detected in the central region of the UA2 detector. Their transverse momentum configurations are illustrated in Fig. 6.

For each event, in which $p_\perp(Je)$ and $E_\perp(J)$ take values p_0 and E_0 respectively, we evaluate the expected background contaminations B_1 and B_2 corresponding to the following configurations :

$$B_1 \; : \; p_\perp(Je) > p_0, \; E_\perp(J) > 30 \text{ GeV},$$

$$B_2 \; : \; p_\perp(Je) > 25 \text{ GeV/c}, \; E_\perp(J) > E_0$$

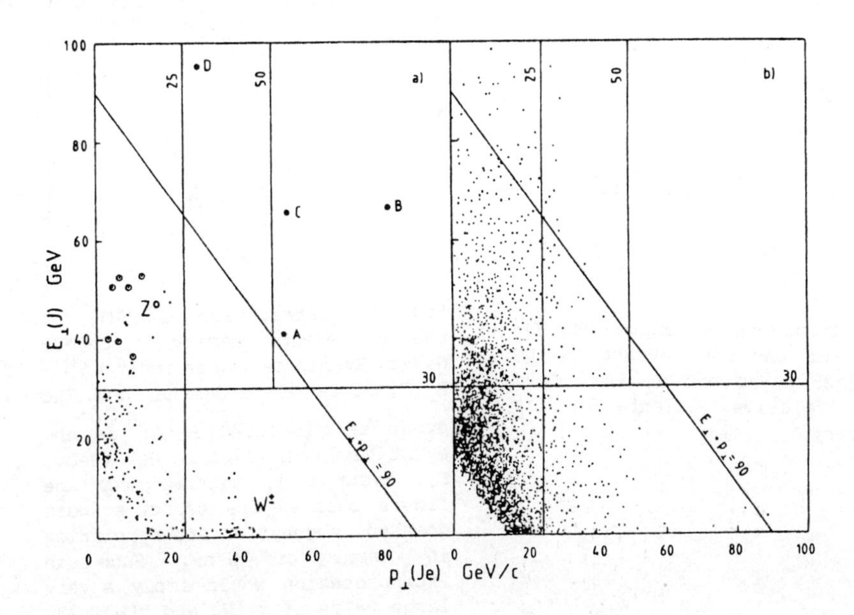

Fig.5 (a)Distribution of the 190 events of the initial sample in the $p_\perp(Je)$, $E_\perp(J)$ plane. Seven Z^0 events are circled (the eighth one was collected during the 1982 period, see [Bagnaia]. The W region is indicated. (b)Distribution of the background sample in the $p_\perp(Je)$, $E_\perp(J)$ plane. A reduction factor of 141 must be applied to infer from this sample the background contamination to the sample of (a). Lines of constant $p_\perp(Je)$=25 and 50 GeV/c, $E_\perp(J)$=30 GeV and $p_\perp(Je)+E_\perp(J)$=90 GeV are indicated.

In each of the four events, either B_1 or B_2 is always less than 0.02. It is remarkable that there is no background event in the region $p_\perp(Je) > 50$ GeV/c, $E_\perp(J) > 30$ GeV, which contains events A to C.

The distribution of variables used for electron identification is consistent with that of the electrons in the $W \rightarrow e\nu$ sample. As far as neutrino identification is concerned, we have checked that missing p_T is not due to tracks hitting passive parts of the detector, muons, cosmic rays or beam-gas background.

Large transverse momentum particles may have escaped detection because they were produced at small angle to the beam line. We considered for each event the possibility that a jet having the same transverse momentum as the neutrino candidate be produced at $\theta = 15°$ (or $165°$). Under such an assumption we can calculate a lower limit for the invariant mass of the system of all large transverse momentum particles produced in the event, including the small angle undetected jet. In the case of events A to C they correspond to impossible or very unlikely kinematical configurations.

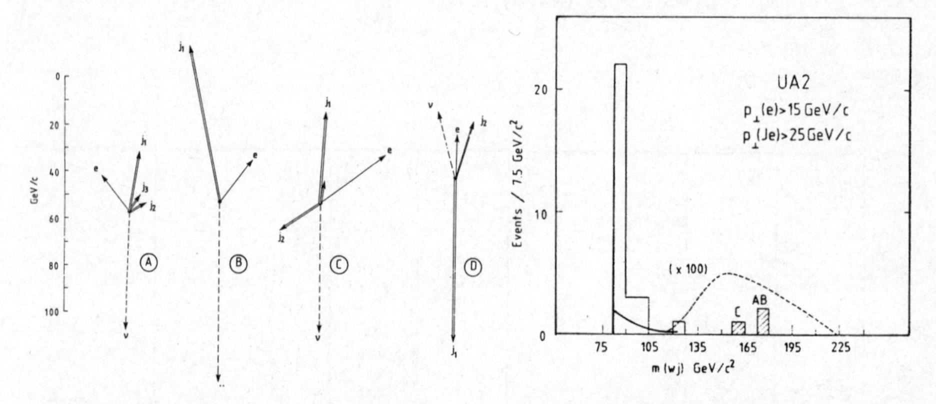

Fig.6 Transverse momentum configuration of the four events having $p_\perp(Je)>25$ GeV/c and $E_\perp(J)>30$ GeV. Their relative orientation is arbitrary.

Fig. 7. Distribution of m(WJ) for the 35 events having $p_\perp(Je)>25$ GeV/c. Events having in addition $E_\perp(J)>30$ GeV are cross-hatched. The event at $m(WJ) \simeq 120$ GeV/c^2 is the event having $E_\perp(J) \simeq 26$ GeV (Sect. 6). Event D is omitted from the figure because its narrow $e\nu$ pair does not suggest an interpretation in terms of $W \rightarrow e\nu$. Such an interpretation would imply a very large value of $p_\perp(W)$ and $m(WJ) \simeq 230$ GeV/c^2. The smooth line corresponds to the calculated background. The dashed line is the background distribution (multiplied by 100) for events having $E_\perp(J)>30$ GeV.

More generally an event appearing at (p_0, E_0) in the $p_\perp(Je)$, $E_\perp(J)$ plane (Fig. 5) may be interpreted as containing an undetected jet which would account, at least in part, for the measured value of p_0. If this jet had been detected, the same event would have appeared at (p'_0, E'_0) with $p'_0 < p_0$ and $p'_0 + E'_0 \geq p_0 + E_0$. But the only events having $p_\perp(Je) + E_\perp(J) > 90$ GeV are precisely the four events of Fig. 6. The absence of other events in this region, even with low values of $p_\perp(Je)$, excludes such an interpretation.

13. Event Interpretation

In this section we study possible sources for events A to D, under the assumption that they contain a genuine $e\nu$ pair.

While in three events (A to C) the $e\nu$ pair has a large azimythal opening ($\Delta\phi > 120°$), event D contains a narrow $e\nu$ pair ($\Delta\phi \simeq 17°$). It consists of a large transverse momentum jet ($p_\perp(j_1) \simeq 70$ GeV/c) emitted at opposite azimuth to the electron-neutrino pair and to a smaller transverse momentum jet ($p_\perp(j_2) \simeq 25$ GeV/c). The invariant mass of the $(e\nu j_2)$ system depends upon the unknown value of $p_L(\nu)$. It takes its minimum value, 23 ± 6 GeV/c^2, when the neutrino has the same rapidity as the (ej_2) system. In this case $m(e\nu j_1 j_2) = 145 \pm 15$ GeV/c^2. The configuration of this event suggests an interpretation in terms of a quark-antiquark pair, one member of which decays semileptonically. However, because of the absence of other events with similar topologies and because of the similarity of its configuration with that of a two-jet event, we prefer to defer such an interpretation until other events of the same kind have been observed.

In the three other events (A to C) the transverse mass of the $e\nu$ pair ranges between 56 GeV/c^2 and 82 GeV/c^2, suggesting an interpretation in terms of a $W \to e\nu$ decay, the W boson being the only known particle with a large enough mass. It is indeed possible to adjust the unknown value of $p_L(\nu)$ to obtain $m(e\nu) = m(W)$ and to describe the events in terms of associated W-jet(s) production. There are in general two solutions to this problem, associated with different values of $p_L(WJ) \equiv x_F(WJ)\sqrt{s}/2$ and of $m(WJ)$. The distribution of $m(WJ)$ for the solution giving the smaller value of $|x_F(WJ)|$ is shown in Fig. 7 for all events having $p_\perp(Je) > 25$ GeV/c. While the 31 events having $E_\perp(J) < 30$ GeV cluster in the neighbourhood of $m(WJ) = m(W)$, the three events populate the region $160 \leq m(WJ) \leq 180$ GeV/c^2, which might suggest an interpretation in terms of a heavy object decaying into a W boson and a system J of other particles. However, the significance of this observation is weakened by the fact that t_L background events have a similar $m(WJ)$ distribution (Fig. 7), implying that the clustering in mass might simply result from kinematics. In addition such an interpretation should account for the fact that in events A and B, J consists essentially of a single jet, while in event C it consists of a large mass pair of jets.

We remark that a reasonable upper limit to the rate of occurence of events containing a $W \to e\nu$ produced via known processes in association with a system J of large transverse momentum jets can be obtained from the relation :

$$\sigma(p\bar{p} \to W + J + \ldots) / \sigma(p\bar{p} \to W + \ldots) \simeq$$

$$\simeq \sigma(p\bar{p} \to j_1 j_2 + J + \ldots) / \sigma(p\bar{p} \to j_1 j_2 + \ldots)$$

where $j_1 j_2$ is a pair of jets having the same configuration as the $e\nu$ pair ascribed to a W decay.

The results include the different acceptances of the UA2 apparatus to $e\nu$ pairs and to jet pairs. The numbers of W-jets events expected are 0.4, 1.2×10^{-2} and 7×10^{-3} respectively for events A, B and C. They indicate that events B and C, if they indeed contain a genuine $W \rightarrow e\nu$, are difficult to understand in terms of associated W-jets(s) production via known processes. Event A, on the other hand, has an unlikely high value of x_F(WJ), corresponding to a W longitudinal momentum of nearly 150 GeV/c.

More generally we note that if the $e\nu$ pair in events A to C is coupled to the initial state via a real or virtual W, as expected in standard theories, other events in which the $e\nu$ pair is replaced by a jet pair should occur as well, at a rate at least 6 times as high (for the two lighter quark families). However, this factor is reduced by acceptance effects and a search for such events is beyond the scope of the present lecture.

12. References

Albert D., Marciano W.J., Wyler D., Parsa Z., 1980 Nucl. Phys. B166, 46
Altarelli G., Ellis R.K., Greco M. and Martinelli G., 1984 TH 3851-CERN
Arnison G. et al., (UA1 collaboration) 1983 Phys. Lett. 129B, 273
Bagnaia P. et al., (UA2 collaboration) 1983 a Phys. Lett. 129B, 130
Bagnaia P. et al., 1984 b Z. Phys. C - Particles and Fields 24, 1
Bagnaia P. et al., 1984 c Phys. Lett 139B, 105
Bagnaia P. et al., 1983 d Z. Phys. C - Particles and Fields 20, 117
Bagnaia P. et al., 1984 e Phys. Lett. 138B, 430
Banner M. et al., (UA2 collaboration) 1983 Phys. Lett. 122B, 476
Beer A. et al., 1983 Nucl. Instrum. Methods 224, 360
Berends F.A., Kleiss R., Hard Photon effects in $W^{\pm}$ and Z^0 decay, University of Leiden, The Netherlands, Nov. 1983
Borer K. et al., CERN-EP/83-177 (1983) Nucl. Instrum. Methods (to be published)
Conta C. et al., 1984 Nucl. Instrum. Methods 224, 65
Ellis J. et al., 1982 Ann. Rev. Nucl. Part. Sci. 32, 443
Fayard L., 1984 Contribution to this Conference
Fayet P. and Ferrara, 1977 Phys. Rep. 32C, 249 and references therein
Glashow S.L. 1961 Nucl. Phys. 22 579
Gluck M. et al., 1982 Z. Phys. C - Particles and Fields 13, 119
Mansoulié B., Proc. of the 3rd Moriond Workshop on $p\bar{p}$ Physics (1983), p. 609, éditions Frontières 1983
Marciano W.J. Electroweak interactions. Proc. of the 1983 International Symposium on Lepton and Photon Interactions at High Energy, Cornell, August 1983 p. 80 (review)
Salam A. 1968 Proceedings of the 8th Nobel Symposium, Aspenasgarden p. 367 Stockholm : Almqvist and Wiksell
Weinberg S. 1967 Phys. Rev. Lett. 19, 1264

Inst. Phys. Conf. Ser. No. 73: Section 2
Paper presented at VII Eur. Symp. Antiproton Interactions, Durham 1984

Theoretical implications of recent collider results

R.D. Peccei*

Max-Planck-Institut für Physik & Astrophysik, Munich (Fed.Rep. Germany)

Abstract: After discussing the comparison of the properties of the W and Z bosons found at the CERN collider with what is expected in the standard model, I critically overview various theoretical speculations concerning some recently reported exotic events, like radiative Z decays, monojets, hot photons, and jet activity. No overwhelmingly favored theoretical explanation appears to spring forth for all the existing exotica.

* Permanent Address after Sept 15, 1984, Theory Group, DESY, Hamburg, Fed.Rep. Germany

1. Standard Model Comparisons

The discovery of the $W^{\pm}$ bosons (Arnison et al., 1983a; Banner et al., 1983) and of the Z^{0} boson (Arnison et al., 1983b; Bagnaia et al., 1983) at the CERN Sp$\bar{p}$S collider with masses in the ranges predicted by the Glashow-Salam-Weinberg model (Glashow 1961; Salam 1968; Weinberg 1967) is one of the outstanding successes of this theory. Because radiative corrections alter significantly the lowest order mass formulas, the precise prediction of the weak boson masses, in terms of parameters measured in low energy weak processes, requires the careful evaluation of these corrections. Although it is possible to define a variety of renormalized $\sin\theta_W$ - differing from each other by terms of $O(\alpha)$ - it has proven particularly convenient to adopt a renormalization framework where the renormalized $\sin\theta_W$ is defined directly in terms of the weak boson masses (Sirlin 1980):

$$\sin^2\theta_W = 1 - \left(\frac{M_W}{M_Z}\right)^2 \qquad (1)$$

After radiative corrections are computed (Marciano and Sirlin 1980, 1981; Llewellyn Smith and Wheater 1981) the lowest order best fit value extracted from deep inelastic ν_μ scattering and polarized eD scattering by Kim et al. (Kim et al., 1980):

$$\sin^2\theta_W{}_{\text{lowest order}} = 0.229 \pm 0.009 \qquad (2)$$

gets substantially altered. For the definition of $\sin\theta_W$ adopted in Eq(1), Marciano and Sirlin and Llewellyn Smith and Wheater find

$$\sin^2\theta_W = \begin{matrix} 0.217 \pm 0.014 \\ 0.218 \pm 0.020 \end{matrix} \qquad \begin{matrix} (3a) \\ (3b) \end{matrix}$$

where the upper (lower) value above applies for ν_μ scattering (eD scattering).

With the definition of $\sin\theta_W$ of Eq (1) the masses of M_W and M_Z follow from a formula which, to logarithmical approximation, is identical to the lowest order formula, except that α is replaced by $\alpha(M_W)$ - the fine structure constant evaluated at the W mass scale - and the Fermi constant is that extracted from, radiatively corrected, μ-decay. To be more precise, the W and Z masses are given by the formula (Marciano 1984a):

$$M_W = \left[\frac{\pi\alpha}{\sqrt{2}\, G_\mu \sin^2\theta_W (1-\Delta r)}\right]^{1/2} ; \quad M_Z = \frac{M_W}{\cos\theta_W} \qquad (4)$$

Here α is the fine structure constant, $\alpha^{-1} = 137.035963$. and G_μ is the Fermi constant extracted from the, radiatively corrected, muon lifetime (Sirlin 1984)

$$\frac{1}{\tau_\mu} = \frac{G_\mu^2 m_\mu^5}{192\pi^3}\left(1 - \frac{8 m_e^2}{m_\mu^2}\right)\left(1 + \frac{3 m_\mu^2}{5 M_W^2}\right)\left[1 + \frac{\alpha}{2\pi}\left(\frac{25}{4} - \pi^2\right)\left(1 + \frac{2\alpha}{3\pi}\ln\frac{m_\mu}{m_e}\right)\right] \qquad (5)$$

The quantity Δr is the complete $O(\alpha)$ radiative correction, whose value (Marciano 1984a), for m_t = 40 GeV and $m_H = m_Z$, is rather substantial:

$$\Delta r = 0.0696 \pm 0.0020 \qquad (6)$$

However, as indicated above, the main effect in Δr can be accounted for by a shift of α to $\alpha(M_W)$, this corresponding to a Δr of approximately 0.073.

Numerically, Eq (4) leads to the predictions

$$M_W = \frac{(38.65 \pm 0.04)}{\sin\theta_W}\ \text{GeV} \quad ; \quad M_Z = \frac{(77.30 \pm 0.08)}{\sin 2\theta_W}\ \text{GeV} \tag{7}$$

Using Eq (3a) this leads to the theoretical values

$$M_W = 83.0 \begin{smallmatrix} +\ 2.9 \\ -\ 2.7 \end{smallmatrix}\ \text{GeV} \quad ; \quad M_Z = 93.8 \begin{smallmatrix} +\ 2.4 \\ -\ 2.2 \end{smallmatrix}\ \text{GeV} \tag{8}$$

which are in excellent agreement with the weighted average of the UA1 and UA2 data

$$\overline{M}_W = 82.1 \pm 1.7\ \text{GeV} \quad ; \quad \overline{M}_Z = 93.0 \pm 1.7\ \text{GeV} \tag{9}$$

Three remarks are in order:

(1) The radiative effects are indeed large, something that has been emphasized already for a number of years (Antonelli, Consoli and Corbo 1980; Veltman 1980; K. Aoki et al. 1981; Bardin, Christova and Fedorenko 1982). Using Eq (4), with $\Delta r = 0$ and the lowest order best value for $\sin\theta_W$ of Eq (2), gives $M_W \approx 78.2$ GeV, $M_Z \approx 89$ GeV. The data clearly seems to favor somewhat larger values than these. The approximate 5 GeV shift in M_W and M_Z is due to a 6% downward shift in $\sin^2\theta_W$ and a 7% decrease due to the $(1-\Delta r)$ factor.

(2) The statistics at the collider are still rough, and there are UA1/UA2 disparities. For example, the W mass determined by UA1 for the $e\nu$ mode is $M_W = 80.9 \pm 1.5 \pm 2.4$ GeV, while UA2 gives the value $M_W = 83.1 \pm 1.9 \pm 1.3$ GeV. After the forthcoming collider run, it is likely, however, that what will prevent an accurate test of higher order corrections will be the error on $\sin\theta_W$, from low energy experiments. Of course, one can eliminate θ_W from Eq (7) and obtain an interesting constraint between the W and Z masses (Consoli, Lo Presti, Maiani 1983; Hioki 1982; Sirlin 1984)

$$M_Z = M_W \left[1 - \left(\frac{(38.65 \pm 0.04)}{M_W(\text{GeV})} \right)^2 \right]^{-1/2} \tag{10}$$

which directly tests the radiative corrections. The lowest order numerical factor in Eq (10) would be 37.3 GeV instead of 38.65 GeV .

(3) One can use Eq (7) and the average UA1 and UA2 W and Z masses to extract a value for $\sin\theta_W$. Using the weighted average of both determinations gives

$$\sin^2\theta_W = 0.222 \pm 0.007 \tag{11}$$

The collider value for the Weinberg angle is seen to be in excellent agreement from that extracted from low energy experiments, Eq (3).

Although the mass of the Z^0 at the collider appears standard, three out of the 12-16 Z^0 decays into lepton pairs seen are accompanied by a rather hard photon. These "anomalous" radiative decays have generated an enormous amount of interest and, in the next section, I will discuss some of the theoretical speculations on their origin. Here, however, I first examine whether these events are really that extraordinary. The total decay rate, to $O(\alpha)$, for $Z^0 \to \ell\bar{\ell}\gamma$ (Albert et al., 1980).

$$\Gamma(Z^0 \to \ell\bar{\ell}\gamma) = \Gamma(Z^0 \to \ell\bar{\ell}) \left[1 + \frac{3\alpha}{4\pi} \right] , \tag{12}$$

is clearly small. However, this rate is not a useful judge of the bremsstrahlung expectation. What is observed physically is the ratio

$$R(Z \to e\bar{e}\gamma\,;\, w,\delta) = \frac{\Gamma(Z \to e\bar{e}\gamma\,;\, E_\gamma > w;\ \text{coll. angle} > \delta)}{\Gamma(Z \to e\bar{e}(\gamma);\, E_\gamma < w;\ \text{coll. angle} < \delta)} \tag{13}$$

in which one imposes certain cuts on the minimum photon energy, w, and collinearity angle, δ, between the photon and the lepton, which one can measure. The ratio R goes up by making the cuts on w and δ more and more stringent.

A careful reanalysis of R has been performed recently by a number of people (Berends and Kleiss 1983; Passarino 1983; Marciano 1984a; Caffo, Gatto and Remiddi 1984; Fleischer and Jegerlehner 1984). Fleischer and Jegerlehner, in particular, present results in which no small angle or soft photon approximations are made. These approximations can alter the results non negligibly. For instance, for $\delta = 5^\circ$, $\epsilon = w/M_Z = 0.1$, which are reasonable experimental values, the exact R is R = 0.034, while R = 0.023 is the small angle, $\omega \ll 1$, approximation. As can be seen from Figs 1 and 2 below, taken from Fleischer and Jegerlehner, R is very sensitive to cuts in δ and w. However, for sensible cuts, as those indicated above, R $\simeq$ 2-3% is a reasonable expectation. I should note incidentally, that similar calculations for W radiative decay give ratios R ($W \to e\nu\gamma$) two to three times smaller than R ($Z \to e\bar{e}\gamma$), for similar cuts.

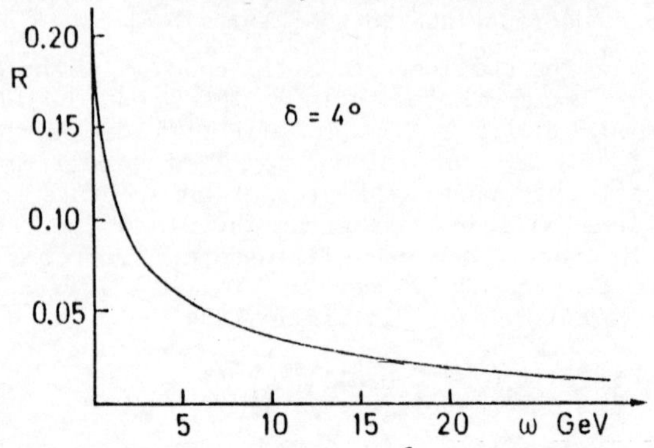

Fig 1: Dependence of R on w for fixed δ, from Fleischer and Jegerlehner 1984

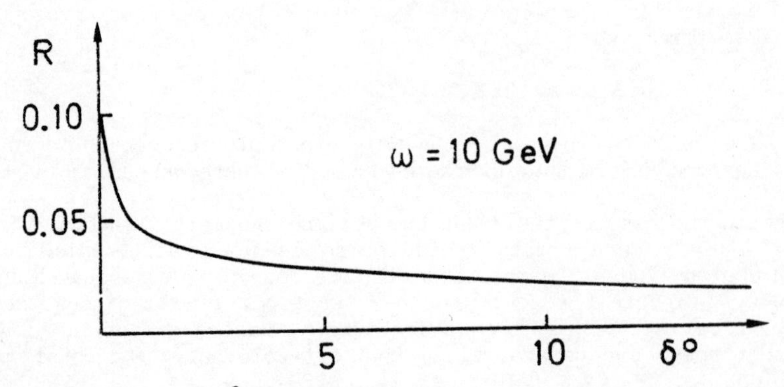

Fig 2: Dependence of R on δ for fixed w, from Fleischer and Jegerlehner 1984

2. Speculations Concerning Radiative Z^0 Decays

Given the above analysis the crucial question is whether the 3 radiative Z^0 decays observed by the UA1 and UA2 collaborations (Arnison et al., 1983b, 1983a; Bagnaia et al., 1983) are new physics or just a statistical fluctuation. Clearly one really needs more statistics to tell what is going on. Nevertheless, a number of brave (foolish?) theorists have taken the nominal UA1 - UA2 radiative rate

$$R_{UA1-UA2} \simeq 0.20 - 0.25 \qquad (14)$$

seriously, and have ventured a myriad suggestions for the origin of this very high radiative rate. These speculations can be broadly classified into four classes:

(i) Models which involve new scalar excitations (Baur, Fritzsch and Faissner 1984, Peccei 1984a, Renard 1984)
(ii) Models involving new excited leptons (Cabibbo, Maiani and Srivastava 1984, Enquist and Maalampi 1984, Renard 1984)
(iii) Models introducing new states degenerate with the Z^0 (Marciano 1984b; Matsuda and Matsuoka 1984; Holdom 1984)
(iv) Models with enhanced $Z\ell\bar{\ell}\gamma$ effective couplings (Gounaris, Kögeler and Schildknecht 1984, Tomozawa 1984, Duncan and Veltman 1984, Renard 1984, Barroso et al., 1984).

All the suggestions above to "explain" an $R \simeq 20\%$ suffer themselves from some problems. Except for models of type (iii), the kinematics of the observed $\ell\bar{\ell}\gamma$ events appear to be different than those of the suggested solutions (Renard 1984; Barger, Baer and Hagiwara 1984). Models involving new degenerate states and models which invoke some enhanced $Z\ell\bar{\ell}\gamma$ coupling have difficulties in producing enough of the new states or in justifying the desired enhancement. Finally, some models, particularly those of type (i), run into conflict with g-2 bounds (del Aguila, Mendez and Pascual 1984; Suzuki 1984; Drell and Parke 1984).

The kinematical configuration - but not the rate - of the three radiative decays is much like one would expect from bremsstrahlung. This is shown graphically in Fig 3. Because all three events have one of the e-γ invariant masses which is small, they necessarily sit at one edge of the Dalitz plot. Sequential decays of the type $Z^0 \to \bar{\ell}\ell^*$; $\ell^* \to \ell\gamma$ or $Z^0 \to S\gamma$; $S \to \ell\bar{\ell}$ would give lines of constant x_H or x, respectively. An enhanced $Z\ell\bar{\ell}\gamma$ coupling also would have no particular reason for populating the small x_L region.

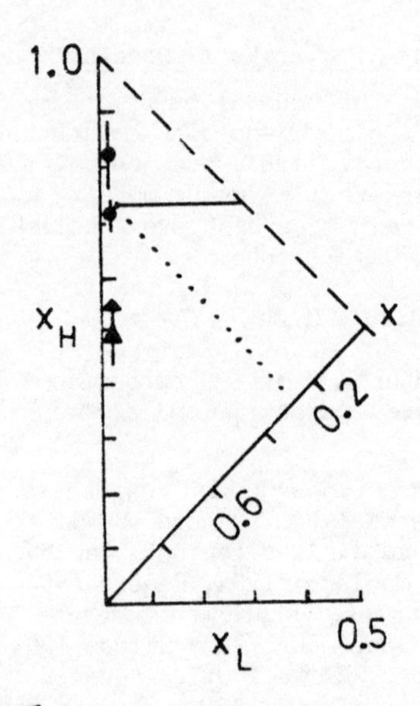

Fig 3: Distribution of $\ell\bar{\ell}\ell$ events in the Dalitz plot. Here

$$x_H = m^2_{e\gamma}/M^2_Z \quad ; \quad x_L = m'^2_{e\gamma}/M^2_Z \quad ; \quad x = m^2_{e\bar{e}}/M^2_Z$$

with $m_{e\gamma} > m'_{e\gamma}$. The solid line corresponds to the expected distribution for an excited lepton interpretation, the dotted line is the distribution expected in the case of a sequential decay to a scalar.

This kinematical pattern, however, can be reproduced if the radiative decays are due to the presence of a scalar or pseudoscalar state, roughly degenerate with the Z^o, which decays dominantly via the graph of Fig 4.

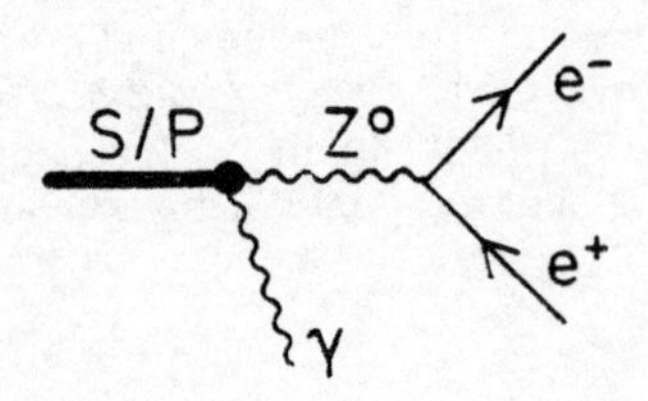

Fig 4: Graph which is supposed to dominate the S/P radiative decays

As emphasized by Matsuda and Matsuoka, helicity conservation tends to allign one of the emitted leptons along the direction of the photon. Further, the Z^o propagator in Fig 4, if $M_{Z^o} \simeq M_{S/P}$, will force the $e\bar{e}$ pair to have a mass peak around 50 to 60 GeV. These are the characteristics of the events seen at the collider.

It is worthwhile to examine, therefore, the scenario (iii) with a bit more care, since, after all, it is the only scenario which can reproduce the

the kinematical characteristics of the collider events. In doing so one uncovers a variety of problems, which also tend to make this proposition rather suspect. I see four principal problems and a query:

(1) The $S/P\ Z^0\gamma$ vertex in Fig 4 is much bigger than what one could expect, if the S/P state was some onia. The coupling that Marciano (Marciano 1984b) needs, to obtain a sufficient number of $\ell\bar{\ell}\gamma$ decays, is about a factor of 10 bigger than that estimated by Guberina et al. (Guberina et al., 1980) for an onia - given the already very big production cross section for S/P (see below) assumed! Such a large coupling can only arise if the S/P <u>and</u> the Z^0 are composed of the same constituents.

(2) The vertex $S/P\gamma\gamma$ must be very small. If not, it is more likely that the S/P decay via 2γ, than radiatively. Further, the peaking in the $\ell\bar{\ell}$ mass would tend to be erased in the radiative decay itself. It seems difficult to ignore this problem entirely, by postulating that this coupling is indeed small. This is because, in a composite Z^0 scenario, required by point (1) above, vector dominance relates the $S/P\gamma\gamma$ and $S/P Z^0\gamma$ vertices. To wit, if g are the effective coupling constants, one has:

$$g_{S/P\gamma\gamma} = g_{S/P Z\gamma} \sin\theta_W \simeq \frac{1}{2} g_{S/P Z\gamma} \tag{15}$$

(3) If Fig (4) dominates, then one predicts that

$$\frac{\Gamma(S/P \to \nu\bar{\nu}\gamma)}{\Gamma(S/P \to \ell\bar{\ell}\gamma)} \simeq 6 \tag{16}$$

Two events of the type energetic photon plus missing energy ("hot photon" events) have been reported by the UA1 Collaboration (Arnison et al., 1984b), which perhaps could be associated with the decay $S/P \to \nu\bar{\nu}\gamma$. However, the transverse masses reported for these events seem too large $[m_T = 93 \pm 5$ GeV, 84 ± 6 GeV$]$ and this interpretation can be questioned. At any rate, it is clear that more events leading to missing energy plus hard photons are needed, if there exists really an S/P state. The statistics are marginal, but against this interpretation.

(4) The production of S/P is probably the most challenging problem. In the scenario of Marciano and of Matsuda and Matsuoka this must occur via gluon gluon fusion (Fig 5a, below), while for the model of Holdom it occurs, in associated production with a quark, by quark-gluon fusion (Fig 5b, below)

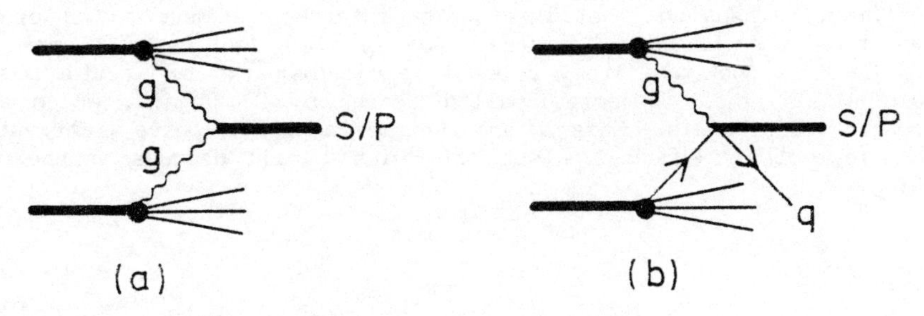

Fig 5: (a)Production of S/P state in the models of Marciano 1984b and Matsuda and Matsuoka 1984
(b)Production of this state in the model of Holdom 1984

To obtain a production rate which suffices, the decay $\Gamma(S/P \to 2g)$ must be of order of at least a GeV. This is $10^3 - 10^4$ times what one would expect for a Higgs meson or a Technipion and it implies that the production rate comes close to saturating the unitarity-bound (see section 3 for more details on this point). Holdom's scenario runs into similar problems. The dimension 7 gqqS/P vertex is scaled by a parameter Λ_c, typifying the scale of compositeness. Only by taking $\Lambda_c \simeq 100$ GeV - a dubious choice - does one get enough S/P production.

(5) The most nagging query, however, remains: why is this state here at all? Sec. 3, in a different vein, provides a partial, but probably unsatisfactory, answer.

I shall not discuss at all the possibility that the radiative Z^0 decays are due to the deexcitation of the Z^0 to a scalar state, or states, of mass around 50 GeV, as this is covered in some detail already in my Bern report (Peccei 1984b). I will, however, make some comments on the excited lepton scenario. This scenario is actually the most straight forward one, were it not for the kinematical problem mentioned above. The scenario does have a theoretical puzzle in that, if the compositeness scale Λ_c is really greater than a TeV (Eichten, Lane and Peskin 1983), the rather "low" mass of the excited leptons needed is unexplained.

Excited electrons and muons give contributions to (g-2). However, as long as the effective lepton - excited lepton - gauge boson vertex is chiral (proportional to $1 - \gamma_5$) then the (g-2) contribution is quadratically dependent on the lepton masses:

$$\delta(g-2) \simeq \alpha \left(\frac{m_e}{M_{e^*}}\right)^2 \qquad (17)$$

Hence, rather low masses for excited leptons are allowed (Renard 1982) and M_μ^*, $M_e^* \sim$ 70-80 GeV do not run into any (g-2) problems. If such states exist, then a ratio $R \simeq 0.2$ is rather easily achieved by postulating a magnetic transition between the ordinary leptons, the excited leptons and the weak gauge bosons, whose scale is set by $1/M_e^*$ and not by $1/\Lambda_c$ (Cabibbo, Maiani and Srivastava 1984). This transition can naturally be chiral by having the excited leptons couple, for example, only to the SU(2) doublet lepton fields.

Cabibbo, Maiani and Srivastava constructed specific SU(2) x U(1) models for the excited leptons in which these states had, alternatively, I_W = 0, 1/2, 1 and 3/2. For each of these models, specific predictions follow for other weak radiative decays like $Z \to \nu\bar{\nu}\gamma$ or $W \to e\nu\gamma$. With more data, these predictions can eventually be tested and (if correct!) a successful model identified. Perhaps the most interesting inference, however, that can be drawn from the possible existence of excited leptons with mass of O(100 GeV) is that also very probably excited quarks of similar masses should exist. These objects, called starks by De Rujula, Maiani and Petronzio (De Rujula, Maiani and Petronzio 1984), have interesting phenomelogical consequences - some of which I will discuss in the next section.

3. Spring Exotica

At the $p\bar{p}$ meeting in March of this year at Bern the UA1 and UA2 Collaborations unveiled a variety of spectacular events (Rubbia 1984, Schacher 1984; Roussaire 1984. More details are contained in Arnison et al., 1984b; Bagnaia et al., 1984). These events contain large transverse momenta and/or large transverse masses and include:

Monojets (Jet or Jets plus missing energy)
Hot Photons (Hard photon plus missing energy)
"W" + Jets (Jet(s) plus lepton plus missing energy)
Di-Jet shoulder at M $\gtrsim$ 150 GeV

One of the peculiarites of the observations in that different phenomena were detected by UA1 and UA2, with UA1 being responsible for the monojets and the hot photons and UA2 for the rest.

I begin my discussion of these exotic phenomena with three general observations and then I'll enter into some selected discussion of the avalanche of theoretical papers which they have generated. (There are certainly more theoretical explanations than exotic events!):

(1) It is unlikely that all the collider exotica is some not well understood background or that they are all instrumental glitches. However, backgrounds do exist. For instance, as shown in Fig 6, ordinary QCD processes can give rise to W's plus a jet. However, the events reported by UA2 have enormous E_T (40-70 GeV), and are thus unlikely to be of pure QCD origin. It is clearly important in the Fall run of the collider to get more experimental information on the tails of ordinary processes, to see whether the "new" phenomena really stand out or not.

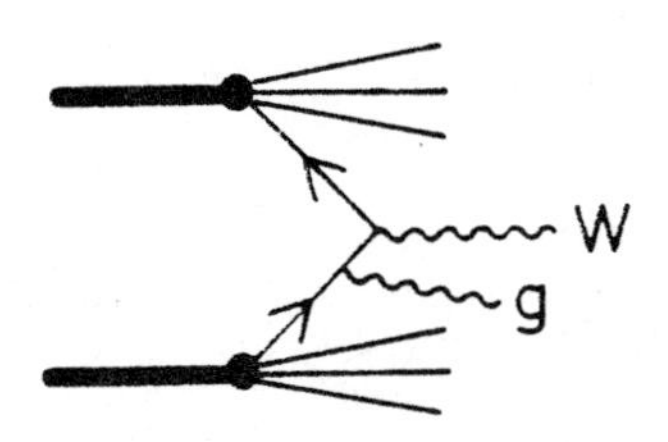

Fig 6: Background to W + Jet exotica

(2) The energy scale associated with the UA1 and UA2 events appear to be connected with phenomena at a scale higher than the Z^o. For instance, the transverse mass (which in general underestimates the real mass) of the jet + missing energy system in the UA1 Monojets ranges from 85 $\pm$ 12 GeV to 130 $\pm$ 16 GeV. Thus the connection of these new phenomena with the anomalous Z^o radiative decays is not so clear. The "hot photon" events could be $Z^o \rightarrow \nu\bar{\nu}\gamma$ or $S/P \rightarrow \nu\bar{\nu}\gamma$ decays and thus be related to the radiative decays. But, at any rate, they would be new physics!

(3) Although, as mentioned above, the theoretical speculations to "explain" the exotic events abound, independent of the underlying physics, the "mechanics" of the models to generate the exotica is essentially the same. Basically, one needs a sizable production for some "heavy object" ($\sigma_{H.O.} \simeq$ 1-10 nb). This "heavy object" then decays with a 5-10% probability into a Jet plus some electroweak boson. More sophisticated models have also a tiny ($\sim$ 1%) electroweak boson plus electroweak boson decay rate.

Broadly speaking the speculations for the Spring Exotica can be divided into four different classes. These are:

(i) Speculations based on Supersymmetry (Ellis and Kowalski, 1984a and 1984b; Haber and Kane 1984a; Reya and Roy 1984; Barger, Hagiwara, Woodside and Keung 1984; Allan, Glover and Martin 1984)
(ii) Speculations involving Stark production (De Rujula, Maiani and Petronzio 1984; Pancheri and Srivastava 1984; Kühn and Zerwas 1984)
(iii) Speculations involving a low mass scale technicolor scenario - the, so called, Fat Higgs scenario (Georgi and Glashow 1984; Düsedau, Lüst and Zeppenfeld 1984; Peccei, 1984c)
(iv) Speculations involving composite colored Vector Bosons (Fritzsch 1984; Baur and Streng 1984; Gounaris and Nicolaidis 1984)

I will discuss (i) in some detail, make some remarks on (ii) and discuss briefly the physics (and improbable assumptions) of (iii), which has some overlap with (iv). I will postpone the discussion of the even wilder speculations of Odoronia (Glashow 1984) to section 4.

In the last few years there has been intense theoretical speculation on the possibility that all particles may have supersymmetric partners. (This subject is reviewed in Ellis 1984; Haber and Kane 1984b) These, so called, spartners have the same quantum number properties as their partners but differ by 1/2 a unit in spin. If Supersymmetry (Susy) were exact, the spartners would be mass degenerate with the ordinary partners. Thus, clearly, one must assume that susy is broken, so that the absence of positive evidence for spartners is no fatal hindrance. Although the breaking of susy is largely arbitrary, there is some theoretical support for supposing that the lightest susy partner is the spin 1/2 photino. Further, it is quite natural to assume that, for all flavors, the squarks and sleptons have common mass $\tilde{m}$, with $\tilde{m} \gtrsim 20$ GeV to avoid conflict with e^+e^- experiments (Yamada 1983). The mass of the susy partners of the gluons - the gluinos - is largely uncertain, and one can entretain the possibilities that $\tilde{m} < m_{\tilde{g}}$ or vice versa.

If squarks ($\tilde{q}$) or gluinos ($\tilde{g}$) were sufficiently light, they would be abundantly produced at the $Sp\bar{p}S$ collider by ordinary QCD processes, like $q\bar{q} \rightarrow \tilde{g}\tilde{g}$, etc. (Kane and Leveille 1982; Harrison and Llewellyn Smith 1983). Events with Jets plus missing energy would be a natural by-product of squark or gluinos production, with subsequent decays into photinos ($\tilde{q} \rightarrow q\tilde{\gamma}$; $\tilde{g} \rightarrow q\bar{q}\tilde{\gamma}$). The photino, being neutral, would give rise to the missing energy. In fact, if the $\tilde{q}$ or $\tilde{g}$ are too light, one risks to produce too much exotica.

The result of various recent independent calculations can be summarized as follows:

(1) Depending on whether $\tilde{m} < m_{\tilde{g}}$ or $m_{\tilde{g}} < \tilde{m}$, the monojet UA1 data admits an interpretation as being due to the production of squarks with masses in the range $\tilde{m} \simeq 30$-40 GeV (Ellis and Kowalski 1984b; Barger, Hagiwara and Keung 1984; Allan, Glover and Martin 1984) or, alternatively, as being due to gluino production with $m_{\tilde{g}} \simeq 35$-40 GeV (Ellis and Kowalski 1984a and 1984b; Reya and Roy 1984)
(2) Although ideally $\tilde{g}\tilde{g}$ production, followed by $\tilde{g} \rightarrow q\bar{q}\tilde{\gamma}$ decay (or $\tilde{q}\tilde{q}$ production, followed by $\tilde{q} \rightarrow q\tilde{\gamma}$ decay) should give rise to 4 Jets plus missing energy (2 Jets plus missing energy), these Jets are largely coalesced by the experimental algorithm used by the UA1 Collaboration to extract Jets. Fig 7 and Fig 8 show this phenomena, as calculated explicitly by Ellis and Kowalski 1984b.

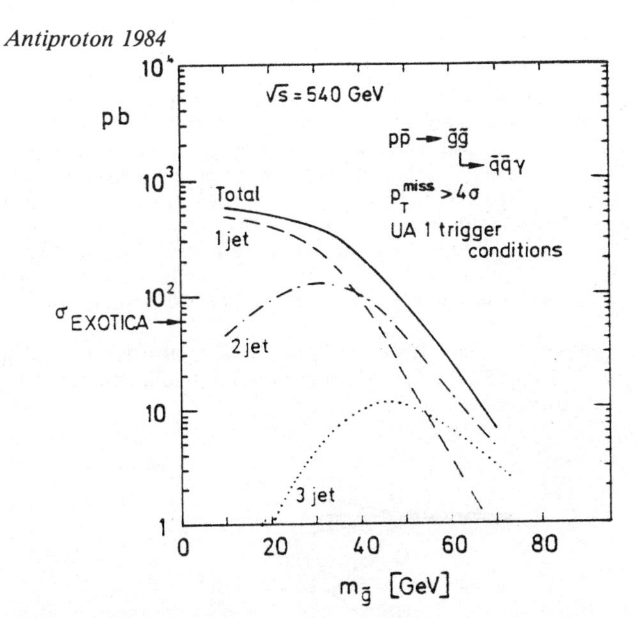

Fig 7: Jet decomposition for gluino production, with UA1 trigger conditions

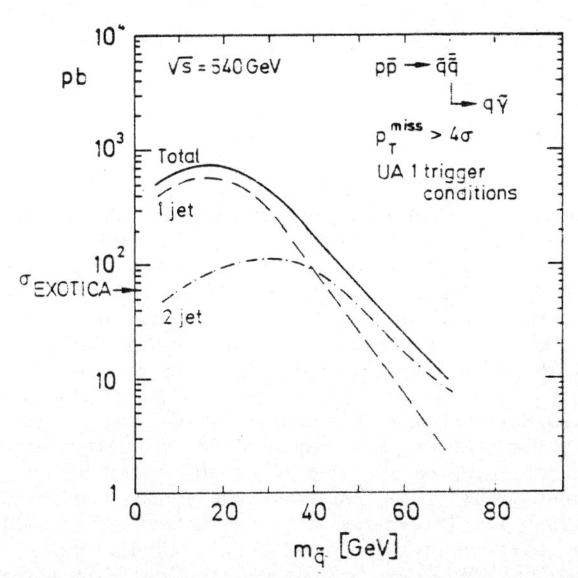

Fig 8: Jet decomposition for squark production, with UA1 trigger conditions

It is obvious from these figures that if $m_{\tilde{g}}$ or $\tilde{m}$ is too light, then too much exotica would ensue.

(3) Although both squark and gluino production, with $\tilde{m}$ and $m_{\tilde{g}}$ in the ranges given in (1), produce the experimentally observed monojet cross section, the squark interpretation is slightly favored. The Jets in the UA1 data are extremely "thin", containing just a few particles, which make them unlikely to be by-products of multijet coalescence, as is the case for gluinos. Further, there appears to be a long

$(p_\perp)_{missing}$ tail which fits better the harder $p_\perp$ produced from the two-body $\tilde{q} \to q\tilde{\gamma}$ decay. Clearly more data, unbiased if at all possible by Jet-algorithms, is needed to decide the issue. However, the "thinness" of the observed jets is already a worry - also for squarks.

(4) Not all exotica is explained by susy. For instance, the "hot photons" are a problem. Haber and Kane 1984a, suggest that the "W" plus Jet events seen by the UA2 Collaboration might be due to $\tilde{w}\tilde{g}$ (or $\tilde{w}\tilde{q}$) production, for sufficently light Wino. In these cases the missing energy would be due to both photino and sneutrino production.

Excited quarks (starks), if they exist, may provide an alternative explanation of the Spring Exotica. Stárks can be produced by quark gluon fusion, as shown schematically in Fig 9.

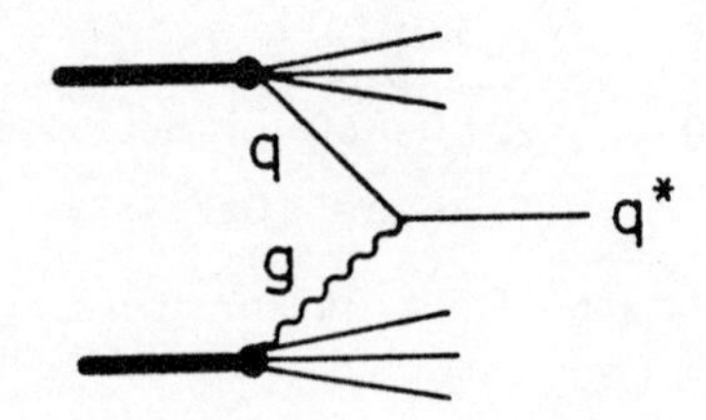

Fig 9: Stark production by quark gluon fusion

Using the same kind of magnetic coupling (scaled by $1/M^*$), as employed for the excited leptons, to parametrize the qq*g vertex, De Rujula, Maiani and Petronzio 1984 find abundant q* production at the collider, up to massses of order of 150 GeV. The decay of the starks back into the quark gluon channel dominates, giving rise to two jet events. However, these two jet events may not stand out against the normal QCD background. Of course, the UA2 shoulder in the 2 jet invariant mass around 150 GeV can temptily be interpreted as stark production, but this is probably too rash.

Starks of $M^* > 120$ Gev can have a sizable ($\sim$10%) decay rate into quarks plus W, Z or γ. These decays could account for the UA1 monojets, through the sequential process $q^* \to qZ$; $Z \to \nu\bar{\nu}$. They are, of course, natural candidates for the UA2 "W" plus Jet events. Although the absolute rate of expected exotica is a little dependent on parameters, the relative rate of various processes is bounded (Kühn and Zerwas 1984). These authors find that, in the simplest SU(2) x U(1) doublet model for the starks, one expects for $M^* = 150$ GeV

$$\frac{N(\gamma + \text{Jet})}{N(W \to e\nu + \text{Jet})} \geq 1.6 \qquad (18a)$$

$$\frac{N(\gamma + \text{Jet})}{N(Z \to \nu\bar{\nu} + \text{Jet})} \geq 2.8 \qquad (18b)$$

Interpreting the 3 UA2 events of "W" plus Jet and the UA1 monojet events (event A does not fit this pattern, because of the presence of a muon in the jet) as being due to stark decay, clearly begins to be dangerous, because of the apparent non existence of correspondingly many γ plus Jet events. Of course, the above bounds can be vitiated if the starks have more complicated weak isospin assignments (Pancheri and Srivastava 1984). At any rate, for the stark interpretation to hold, events with large transverse mass of a photon-jet system must eventually begin to appear in the Fall collider run.

The last scenario for the Spring exotica which I want to discuss here is the one in which the spectacular collider events are attributed to the presence of some low mass remnant of a technicolor theory. this "Fat Higgs" scenario is already discussed in some detail in Peccei 1984c, and so I shall be brief here. In a usual technicolor scenario (Susskind 1979; Weinberg 1979) one supposes that the interactions among the technipions - which give mass to the W bosons, and eventually produce a strong interaction among them - are just like those among pions, but scaled up by a factor $(\sqrt{2}G_F)^{-1/2}/f_\pi \simeq 2500$. The existence of an attractive, I = 0, s-wave partial wave and a broad σ enhancement in π-π scattering has its counterpart in a broad 0^+ resonance at $M \simeq 1.5$ TeV, with perhaps a 1 TeV width, in the technicolor theory (The Fat Higgs). The ρ meson of π-π scattering, similarly implies the presence of a Technirho state at $M \simeq 1.7$ TeV, etc. Clearly, these states, even if they existed, would have no implication for the present collider experiments.

To make technicolor relevant for the Spring exotica requires that one assume that the technipion interactions, although having the same threshold behaviour as those of pions, become stronger much sooner. This is shown schematically in Fig 10, where the behaviour of the π-π A_0^0 partial wave amplitude, naively extrapolated from threshold, is contrasted with what one would want to happen in the technicolor theory.

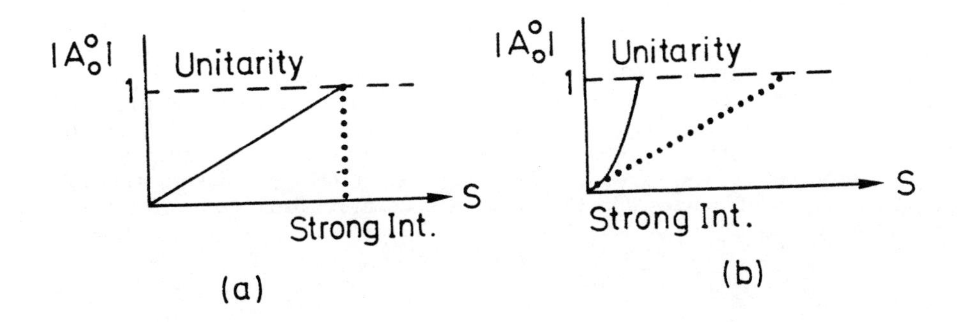

Fig 10: (a)Threshold s-wave π-π scattering rises linearly with s until it saturates unitarity
(b)Desired behaviour of the technipion scattering amplitude

The bold assumption that this might indeed happen, and be the root cause for the collider peculiar events, has been put forth recently by Georgi and Glashow 1984. However, before one can really ascribe the exotica to the production of a low mass Fat Higgs, one needs a mechanism for producing this object. Gluon fusion, as shown in Fig 11, seems to be the only possibility. Unfortunately, as already observed when discussing the production of the S/P state - which is really very akin to the Fat Higgs - one needs unnaturally large gluon couplings to generate enough exotica.

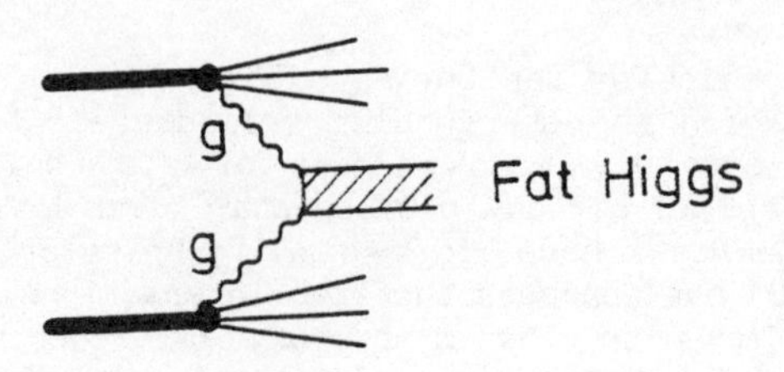

Fig 11: Production of Fat Higgs by gluon fusion

Georgi and Glashow, to produce enough exotica, assumed that the production cross section of an 150 GeV Fat Higgs at the collider could amount to 10 nb. In fact, (Düsedau, Lüst and Zeppenfeld 1984) unitarity restricts this production cross section to perhaps only a tenth of a nanobarn. So it is very unlikely that this could be the mechanism for generating all the exotica, unless the decay of the Fat Higgs into channels containing a jet plus an electroweak boson is comparable to the two gluon channels. This latter type of possibility has many features in common to the colored vector boson scenario (Fritzsch 1984; Baur and Streng 1984).

4. Jet Activity

The latest bit of apparent non standard behaviour, which was discussed in this meeting by Tuominiemi (Tuominiemi, 1984), is a tendency of the Z^o events, of the UA1 Collaboration, to have an unusual amount of jets associated with them. Such jet activity does not seem to be so prevalent in the UA2 Z^o events. Qualitatively, if this observation is not just some statistical fluctuation, this could point to the Z^o being composite of constituents which have color. The jets could then be associated with gluon radiation off these constituents. A schematic sketch of this idea is presented in Fig 12

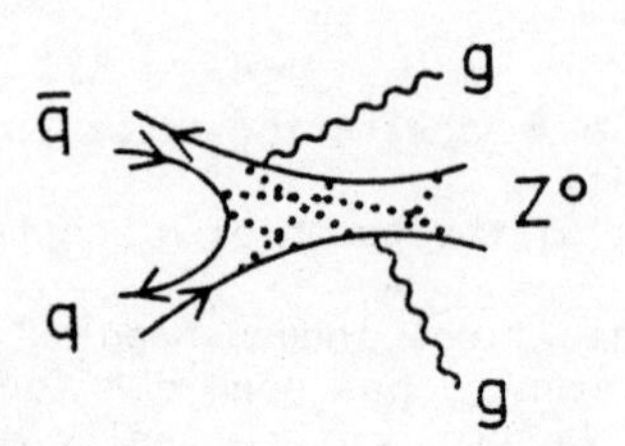

Fig 12: Origin of jet activity in a picture in which both quarks and the Z^o are composite of colored constituents

Quantitatively, however, it is difficult to implement such a picture without, at the same time, spoiling other desirable properties.

Glashow (Glashow 1984) has proposed recently a clever, if a bit outlandish, idea to explain the jet activity, by inventing yet another new strong interaction force "odor". Not only does odor provide a way for the Z^o's to be produced with additional jets, but it also gives one another mechanism for generating the monojet events of UA1 (Odor does not lead to "W" plus Jet events). The basic idea of Glashow is that there are two sources for Z^o's, one directly from $q\bar{q}$ production and the second arising from the decay of Odoronia to a Z^o plus a light Higgs. Z^o's coming from this second source would often be accompanied by Jets from the Higgs decay into Jet-Jet or $\tau\bar{\tau}$. These accompanying jets would be characteristically "thin jets". If the Odoronia produced Z^o's decayed into neutrinos, as they would do a good fraction of the time, then the Higgs initiated jets would be a natural source for the "thin" monojets seen by UA1.

Glashow's mechanism works only because Odoronia - which is the quarkonia of new quarks carrying the new quantum number odor - is much more likely to be produced that a similar (heavy) $q\bar{q}$ bound state. Basically, higher Odoronia bound states cannot dissociate into states containing an odor quark and an ordinary quark, and so must all cascade down to the lowest Odoronia state. This is why the production cross section is big. The J=1 Odoronia ground state has itself almost an equal probability of decaying into three gluons or into HZ^o and/or $H\gamma$, since the Higgs coupling to the heavy odor quarks ($m_{q_o} \simeq$ 60-70 GeV?) is large (see Fig 13). So this unusual suggestion can really produce enough exotic events. However, as Kane and Maiani (Kane and Maiani 1984) point out, the Glashow scheme may be too much of a good thing. Beside the decay channels shown in Fig 13, Odoronia can have also sizable decays into other channels.

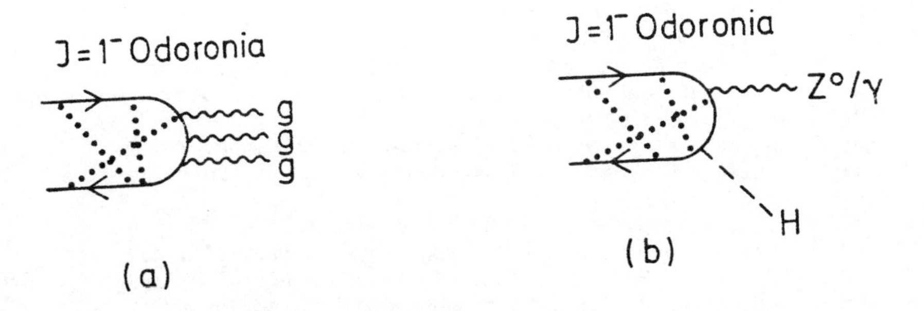

Fig 13: Odoronia decays in the Glashow scheme

In particular, Kane and Maiani estimate that for every Odoronia induced monojet there should exist both an e^+e^- and a $\mu^+\mu^-$ direct Odoronia decay. These, as yet unseen high mass dilepton events, are a powerful constraint on the whole scheme.

5. Concluding Remarks

The Sp$\bar{p}$S collider has confirmed important predictions of the standard model, concerning the existence of W and Z bosons with specifically predicted masses. At the same time, in the last year, experiments at the collider have produced a set of tantalyzing events, which appear to be difficult to explain conventionally. The prudent attitude would be to wait until the Fall run of the collider is completed, to see better which of the various phenomena are confirmed and which are rendered moot. Maybe then it will become clear if there is indeed some new physics. My own feeling is that it is unlikely that all the phenomena observed will quietly fade away. There is some new physics here, although it is difficult to clearly identify its sources. Indeed, the impression that emerges from looking at all the various theoretical explanations is that nothing really fits all phenomena terribly well. So before a really coherent picture emerges it will be necessary that some of the exotica should fade away and some become even more prevalent. The results of the Fall run at the Sp$\bar{p}$S collider are awaited with baited breath.

References

del Aguila F, Mendez A, Pascual R, 1984, Univ. of Barcelona preprint UAB-FT-105
Albert D, Marciano W, Wyler D and Parsa Z, 1980, Nucl. Phs. B166, 460
Allan A R, Glover E W N and Martin A D, 1984, Univ. of Durham, preprint DTP /84/20
Antonelli F, Consoli M and Corbo G, 1980, Phys. Lett. 91B, 90
Aoki K et al, 1981, Prog. Theor. Phys. 65, 1001
Arnison G et al., 1983a, Phys. Lett. 122B, 103
Arnison G et al., 1983b, Phys. Lett. 126B, 398
Arnison G et al., 1984a, Phys. Lett. 135B, 250
Arnison G et al., 1984b, Phys. Lett. 139B, 115
Bagnaia P et al., 1983, Phys. Lett. 129B, 130
Bagnaia P et al., 1984, Phys. Lett. 139B, 105
Banner M et al., 1983, 122B, 476
Bardin D, Christova P and Fedorenko O, 1982, Nucl. Phys. B197, 1
Barger V, Baer H and Hagiwara K, 1984, Univ. of Wisconsin preprint MAD/PH/176
Barger V, Hagiwara K, Keung W Y, 1984, Phys. Lett. 145B, 147
Barger V, Hagiwara K, Woodside J and Keung W Y, 1984, Phys. Rev. Lett. 53, 641
Barroso A et al., 1984, Univ. of Sussex preprint
Baur U, Fritzsch H and Faissner H, 1984, Phys. Lett. B135, 313
Baur U, Streng K H, 1984, Max-Planck-Institute preprint MPI/PAE/PTh 50/84
Berends F and Kleiss R, 1983, Leiden University preprint
Cabibbo N, Maiani L, Srivastava Y, 1984, Phys. Lett. 139B, 459
Caffo M, Gatto R and Remiddi E, 1984, Univ. of Geneva preprint, UGVA-DPT 1984/01-145
Consoli M, Lo Presti S, Maiani L, 1983, Nucl. Phys. B223, 474
De Rujula A, Maiani L, Petronzio R, 1984, Phys. Lett. 140B, 253
Drell S D and Parke S J, 1984, SLAC preprint, SLAC-PUB 3308
Düsedau D, Lüst D, Zeppenfeld D, 1984, Max-Planck-Institute preprint, MPI/PAE/PTh 55/84
Duncan J and Veltman M, 1984, Phys. Lett. 139B, 310
Eichten E, Lane K and Peskin M, 1983, Phys. Rev. Lett. 50, 811
Ellis J, 1984, in Proceedings of the Munich Advanced Study Institute, Quarks Leptons and Beyond, Plenum Press to be published

Ellis J and Kowalski H, 1984a, Phys. Lett. 142B, 441
Ellis J and Kowalski H, 1984 b, DESY preprint, DESY 84-045
Enquist K and Maalampi J, 1984, Phys. Lett. 135B, 329
Fleischer J and Jegerlehner F, 1984, Bielefeld University preprint, Bi-TP 1984/6
Fritzsch H, 1984, in Proceedings of the Workshop on Feasibility of Hadron Colliders in the LEP tunnel, Lausanne-CERN, March 1984
Georgi H and Glashow S L, 1984, Phys. Lett. B143, 155
Glashow S L, 1961, Nucl. Phys. 22, 579
Glashow S L, 1984, Phys. Lett. B143, 130
Gounaris G, Kögeler R and Schildknecht D, 1984, Phys. Lett.137B, 261
Gounaris G and Nicolaidis A, 1984, Phys. Lett. B to be published
Guberina B et al., 1980, Nucl. Phys. B174, 317
Haber H and Kane G, 1984a Phys. Lett. 142B, 212
Haber H and Kane G, 1984b, Phys. Reports to be published
Harrison P R and Llewellyn Smith C, 1983, Nucl. Phys. B213, 223; E B223, 542
Hioki Z, 1982, Prog. Theor. Phys. 68, 2134
Holdom B, 1984, Phys. Lett. 143B, 241
Kane G and Leveille J P, 1982, Phys. Lett. 112B, 227
Kane G and Maiani L, 1984, Univ. of Rome preprint
Kim J et at., 1980, Rev. Mod. Phys. 53, 211
Kühn J and Zerwas P, 1984, Univ. of Aachen preprint PITHA 84/14
Llewelyn Smith C and Wheater J, 1981, Phys. Lett. 105B, 486
Marciano W, 1984a, in Proceedings of the Fourth Topical Conference on Proton-Antiproton Collider Physics, Bern, March 1984
Marciano W, 1984b, Phys. Rev. Lett. 53, 975
Marciano W and Sirlin A, 1980, Phys. Rev. D22, 2695
Marciano W and Sirlin A, 1981, Nucl. Phys. B189, 442
Matsuda M and Matsuoka T, 1984, Phys. Lett. 144B, 443
Pancheri G and Srivastava Y, 1984, Frascati preprint LNF - 84(10) P
Passarino G, 1983, Phys. Lett. 130B, 115
Peccei R D, 1984a, Phys. Lett. B136, 121
Peccei R D, 1984b, in Proceedings of the Fourth Topical Conference on Proton-Antiproton Collider Physics, Bern, March 1984
Peccei R D, 1984c, in Proceedings of the International Conference on Neutrino Physics and Astrophysics, Nordkirchen, June 1984
Renard F M, 1982, Phys. Lett. 116B, 264
Renard F M, 1984, Phys. Lett. 139B, 449
Reya E and Roy D P, 1984, Phys. Rev. Lett. 53, 881
Roussaire A, 1984, in Proceedings of the Fourth Topical Conference on Proton-Antiproton Collider Physics, Bern, March 1984
Rubbia C, 1984, in Proceedings of the Fourth Topical Conference on Proton-Antiproton Collider Physics, Bern, March 1984
Salam A, 1968, in Elementary Particle Theory: Relativistic Groups and Analyticity, ed. by N. Svartholm, Almquist and Wiksell, Stockholm
Schacher J, 1984, in Proceedings of the Fourth Topical Conference on Proton-Antiproton Collider Physics, Bern, March 1984
Sirlin A, 1980, Phys. Rev. D22, 971
Sirlin A, 1984, Phys. Rev. D29, 89
Susskind L, 1979, Phys. Rev.D20, 2619
Suzuki M, 1984, Phys. Lett. 143B, 237
Tomozawa Y, 1984, Univ. of Michigan preprint UM-TH 84-2
Tuominiemi J, 1984, these proceedings
Veltman M, 1980, Phys. Lett. 91B, 95
Weinberg S, 1967, Phys. Rev. Lett. 19, 1264
Weinberg S, 1979, Phys. Rev. D19, 1277

Yamada S, 1983, Proceedings of the Lepton-Photon Conference at Cornell, Ithaca, August 1983

Jets at ISR, FNAL and SPS

T. Åkesson

CERN, Geneva, Switzerland

Abstract. Experimental jet physics at the CERN Intersecting Storage Rings (ISR) and Super Proton Synchrotron (SPS) and at the Fermi National Accelerator Laboratory (FNAL) are reviewed. Only calorimeter-triggered experiments are presented. Results from the ISR are shown on jet fragmentation, E_T flow, and charge distributions. A di-jet cross-section measured at FNAL is also shown.

1. Introduction

It is now two years since the first evidence on jet dominance was presented at the 21st International Conference on High-Energy Physics in Paris (1982) by the UA2 and AFS Collaborations. Jets in hadron collisions have since then changed from being a topical subject to being the major background source at the p$\bar{p}$ Collider. In this talk I will show some measurements on jets made at lower centre-of-mass energies and rediscuss some properties of jet dominance. I will start with the ISR results of the CERN-Michigan State-Oxford-Rockefeller (CMOR) Collaboration and of the Axial Field Spectrometer (AFS) Collaboration [Brookhaven-Cambridge-CERN-Niels Bohr Institute-London (QMC)-Lund-Pennsylvania-Pittsburgh-Rutherford-Tel Aviv Collaboration]. The CMOR results can be studied in more detail in Angelis et al. (1984) and the AFS results in Åkesson et al. (1983, 1984).

2. Jet Dominance

The CMOR experiment (Fig. 1) measures the transverse energy with a combination of shower counters and lead-glass blocks. Since this equipment ideally measures energy only from particles giving electromagnetic showers, mainly π^0's, CMOR make their analysis as a function of total neutral transverse energy E_T^0. Figure 2 shows their distribution of neutral transverse energy. The data correspond to an integrated luminosity of 2.7×10^{37} cm^{-2}. The distribution is not corrected for effects from the apparatus. At the AFS (see Fig. 3) the energy measurement is made with a combined electromagnetic and hadronic calorimeter and hence the total transverse energy E_T is measured. The AFS calorimeter is a sandwich of uranium, copper, and scintillator planes, and has a unique energy resolution of hadronic showers. Figure 4 shows the distributions of total transverse energy measured by the AFS at three different centre-of-mass energies, 30, 45, and 63 GeV, corresponding to the integrated luminosities of 9.9×10^{34}, 6.8×10^{35}, 3×10^{36} cm^{-2}. These spectra are also uncorrected for apparatus effects. It should be noted that the 63 GeV measurement spans 10 orders of magnitude in cross-section.

It has been known for quite some time that in the low-E_T region the events look very isotropic, while in the high-E_T region the events are completely dominated by jet production. This has been discussed many times before and I just want to point out some details on the variation of event shape with transverse energy. Figure 5 shows the circularity (sphericity) distributions in different bins of E_T. We will concentrate on the 63 GeV centre-of-mass energy data. If we measure the event

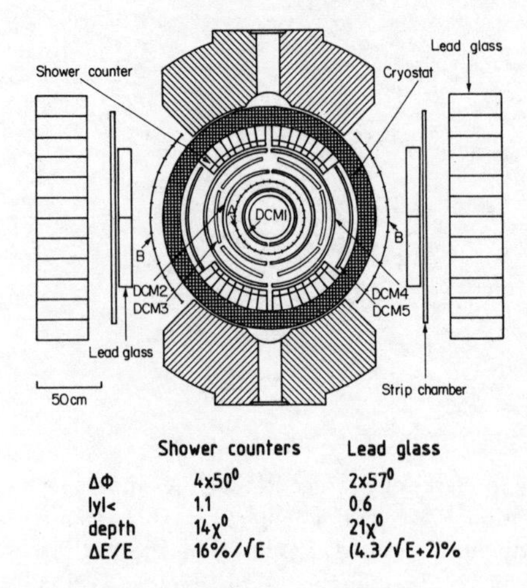

Fig. 1 The CMOR experimental set-up

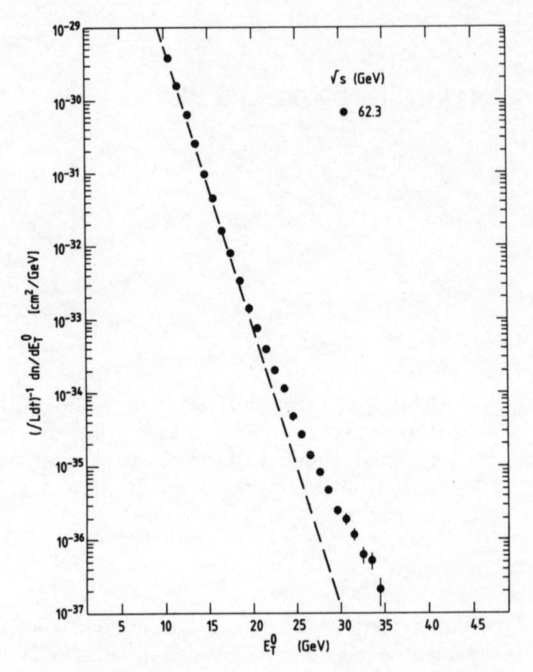

Fig. 2 Uncorrected spectrum for neutral transverse energy measured by the CMOR Collaboration at $\sqrt{s} = 63$ GeV.

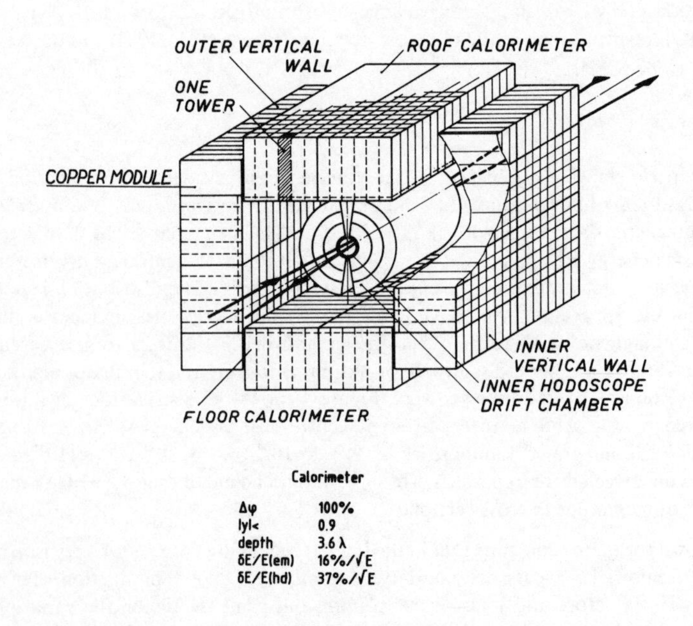

Fig. 3 The AFS experimental set-up

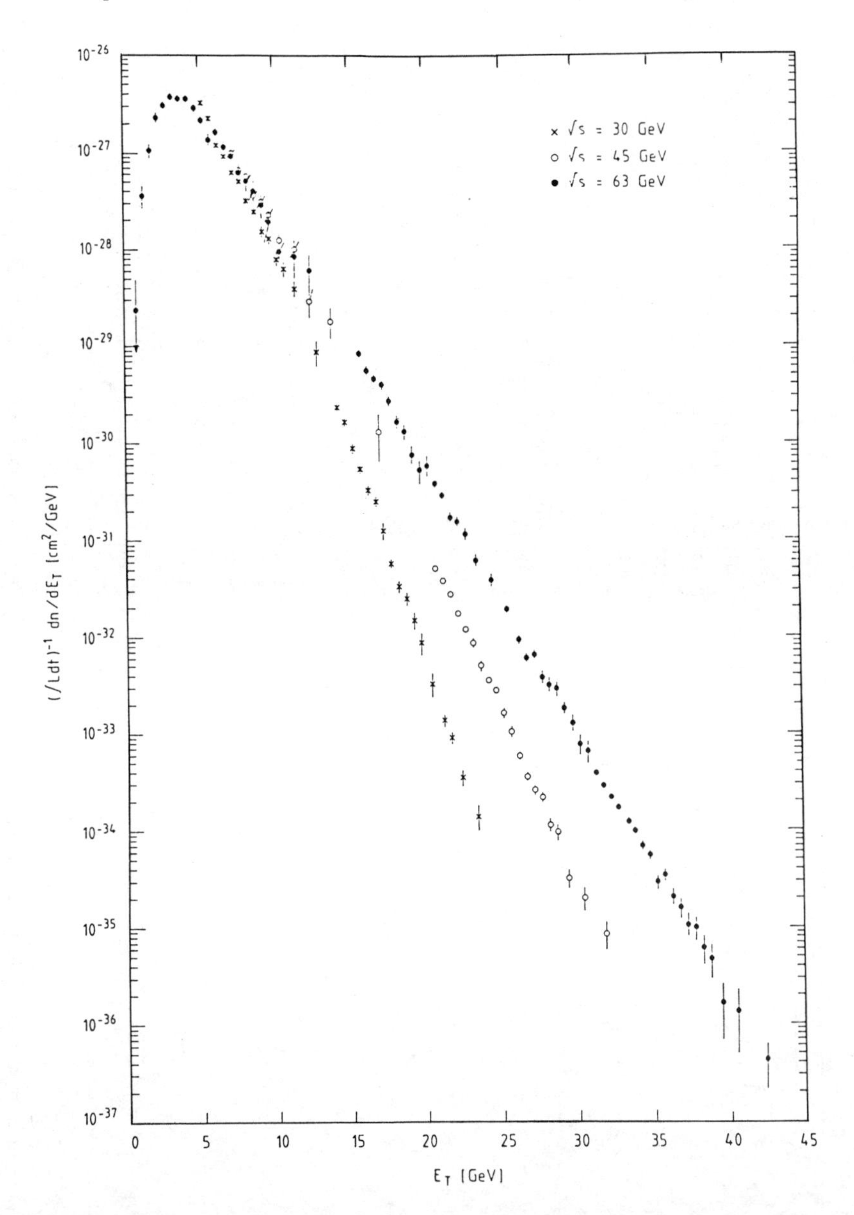

Fig. 4 Uncorrected spectra of transverse energy measured by the AFS Collaboration at $\sqrt{s}$ = 30, 45, and 63 GeV.

shape over the region from 0 to 26 GeV in E_T, we see that it does not change. It remains isotropic despite the large variation in E_T. If we now go beyond 26 GeV, we get a quick change and at 35 GeV the events are completely dominated by jets. We do not observe one continuously shifting distribution between 26 GeV and 35 GeV; instead we observe two competing distributions. This means that, for example, at E_T = 30 GeV one either gets a jet event *or* an isotropic event. The same thing is observed by the CMOR experiment (Fig. 6). Finally, I show some Lego plots (Fig. 7) of the events in the high-E_T region, $E_T > 35$ GeV, measured by the AFS experiment. Let us now examine

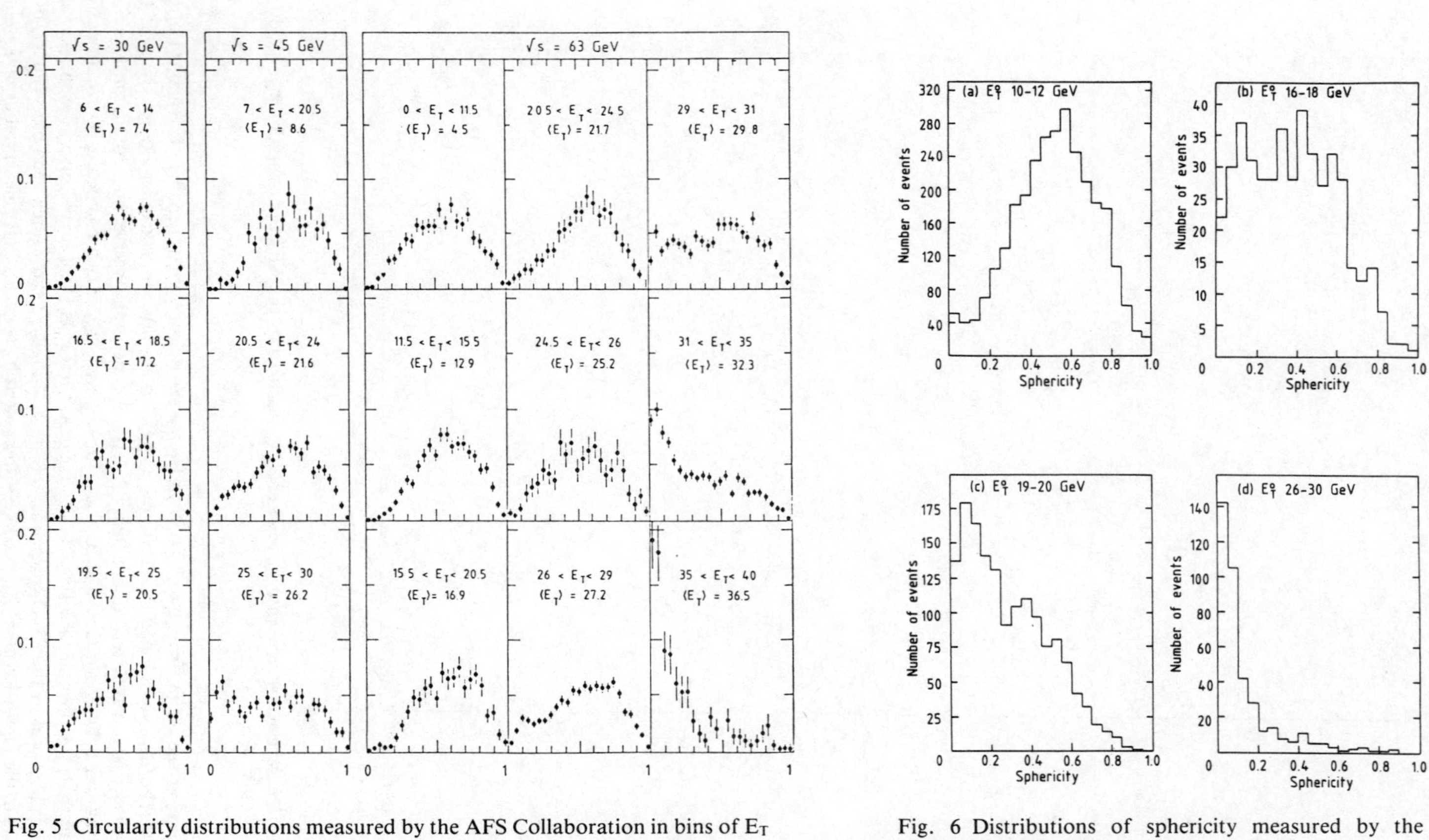

Fig. 5 Circularity distributions measured by the AFS Collaboration in bins of E_T

Fig. 6 Distributions of sphericity measured by the CMOR Collaboration

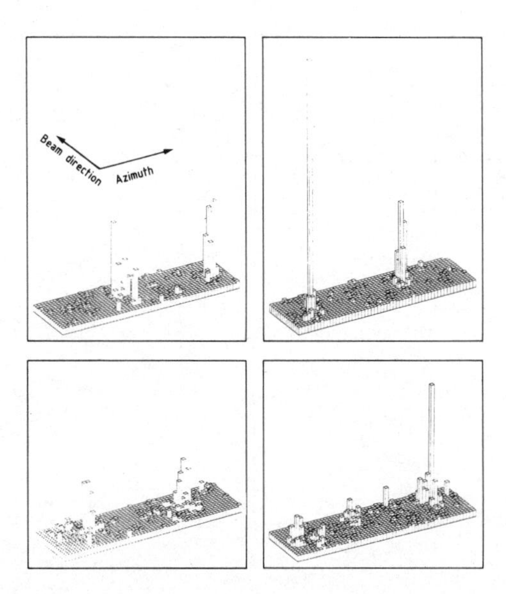

Fig. 7 Examples of events with a transverse energy in excess of 35 GeV. The transverse energy is plotted in bins of $\Delta\phi = 7.5°$, $\Delta\eta \sim 0.12$.

the structure of the jets. To select them, the AFS experiment requires $E_T > 32$ GeV and the CMOR experiment requires more than 70% of the neutral E_T in the two highest energy clusters. Most of the results shown will be from the AFS Collaboration. The CMOR measurements show the same features but are more biased since they trigger on neutral energy. To be able to measure the jet fragmentation we have to find the energy scale of the event and the jet axes.

3. Constituent Centre-of-Mass Energy and Jet Axis

The hard-scattering centre-of-mass energy $\sqrt{\hat{s}}$ is estimated by using the information from the calorimeter. Since not all the measured clusters belong to the high-p_T jets, the invariant mass of *all* clusters seen in the calorimeter will in general be an overestimate of $\sqrt{\hat{s}}$. We therefore use the ISAJET Monte Carlo [Paige and Protopopescu, 1982] with full detector simulation to estimate $\sqrt{\hat{s}}$. For the Monte Carlo events we calculate the invariant mass of all clusters with $p_T > 1$ GeV/c, which are predominantly jet particles, obtaining a conversion factor between this invariant mass and $\sqrt{\hat{s}}$. This is a correction of ~ 20%. In the data we now calculate the invariant mass of all clusters with $p_T > 1$ GeV/c, and translate this to our estimate of $\sqrt{\hat{s}}$, W, using the relation found by the Monte Carlo. The estimated error in W, including both the uncertainty in the calorimeter response and uncertainties due to model assumptions, is 6%. For the present data sample, the average value of W is found to be 28 GeV. We repeated the same procedure, taking all clusters with $p_T > 0.8$ GeV/c or $p_T > 1.5$ GeV/c, and found that the procedure is stable, i.e. the $\langle W \rangle$ found is the same. The jet axes are found by a jet-finding algorithm described in Sjöstrand (1983).

4. Fragmentation Functions

We use the charged particles in $|y| < 1$ measured in the drift chamber to study the jet fragmentation in terms of the variables x_p and z; x_p is frequently used by the experiments at e^+e^- machines and is defined there as p/p_{beam}. In our case we define x_p as $2p/W$, where p is the momentum of any

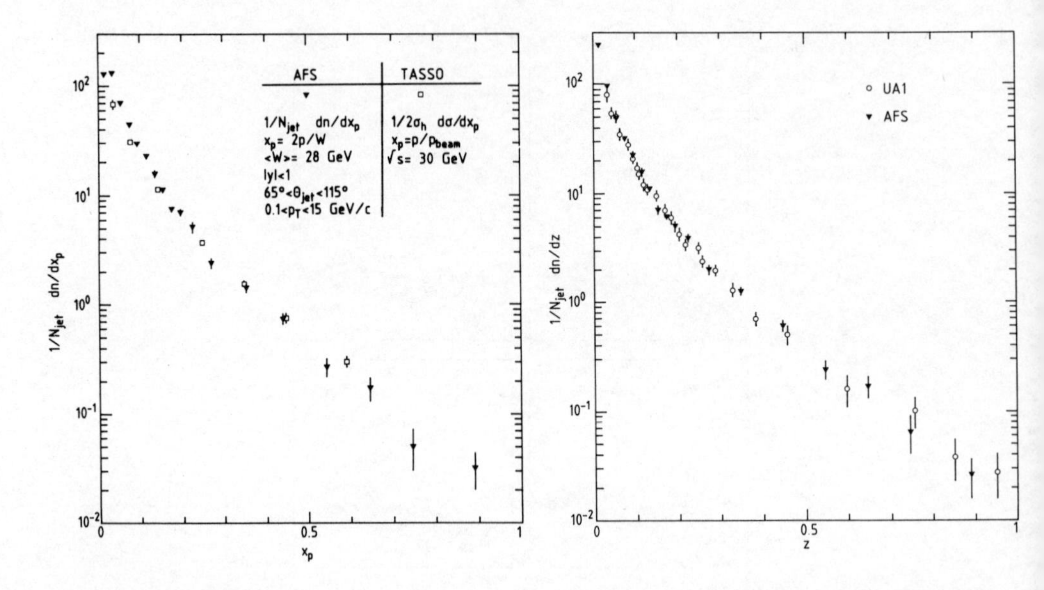

Fig. 8 x_p distribution of the charged particles in $|y| < 1$ compared with measurements in e^+e^- collisions at similar jet energy.

Fig. 9 z distribution of the charged particles in $|y| < 1$. The open points indicate a measurement at the $p\bar{p}$ Collider for jets with $E_T > 30$ GeV.

particle measured in the central drift chamber. Figure 8 shows our measured x_p distribution together with a measurement in e^+e^- by the TASSO Collaboration [Brandelik et al., 1980]. Both measurements are normalized to the number of jets, assuming two jets per event. For $x_p > 0.15$ there is a good agreement between the two data sets, while the present data show an excess of particles at low x_p, corresponding to particle momenta $\leqslant 1$ GeV/c. The agreement at high x_p indicates that the jets we measure are very similar to those produced at e^+e^- machines. The excess in the region of low x_p may be attributed to additional particles from other sources, e.g. beam jets.

Figure 9 shows the z-distribution from our data together with the measurement by the UA1 Collaboration at the $p\bar{p}$ Collider [Arnison et al., 1983]. We define z as $(2\vec{p} \cdot \hat{n}_{jet}/W)$, where $\hat{n}_{jet}$ is the jet direction found in the calorimeter. The z distributions are also normalized to the number of jets.

The two distributions are in close agreement. To the extent that the collider data is dominated by gluon jets and our data by quark jets [Horgan and Jacob, 1981], we can infer that the longitudinal fragmentation of quark and gluon jets is very similar at large values of z.

5. E_T Flow

The charged E_T flow measured in the drift chamber is shown in Fig. 10a (solid line) as a function of azimuthal angle relative to one of the jet axes (chosen randomly) measured in the calorimeter. It can be observed that perpendicular to the jet axis the E_T flow is about seven times higher than for minimum-bias events. By minimum bias we mean a reference level, i.e. the inclusive density of charged particles in pp interactions at $\sqrt{s} = (63 - \langle W \rangle)$ GeV ($= 35$ GeV). When we select the subsample of 96 events where the two jets themselves carry 32 GeV in E_T, then the E_T flow at 90° moves towards the minimum-bias reference level. The jet-E_T here is calculated as the sum of E_T in a

$\pm 40°$ region in azimuth around the found jet axis. If the region in which the jet-E_T is calculated is changed from $\pm 40°$ to $\pm 60°$ then the level at 90° increases by ~ 20%. The effect of an enhanced soft background is thus partly due to a a trigger bias when we demand $E_T > 32$ GeV in the whole calorimeter, independent of particle momentum and particle direction.

The charged E_T flow $dE_T dy$, in the region $70° < \Delta\phi < 110°$, is 3.62 ± 0.10 GeV for all events, 1.61 ± 0.26 GeV for the event sample with $(E_T^{jet\,1} + E_T^{jet\,2}) > 32$ GeV, while the minimum-bias level is 0.52 GeV.

In Fig. 10b we show again the 96 two-jet triggered events and the minimum-bias level. Here we have superimposed as a solid line the charged E_T flow distribution from an e^+e^- Monte Carlo [Sjöstrand, 1983] known to reproduce well the e^+e^- data. The e^+e^- events are generated at the same $\sqrt{s}$ as the $\langle W \rangle$ for the 96 two-jet triggered events, and we have imposed the same trigger conditions on the e^+e^- jets as we have on the jet fragmentation and the minimum-bias level.

The charged E_T flow versus the relative distance in rapidity from the jet axis (same side in azimuth, $\Delta\phi < \pi/2$) is shown in Fig. 11 for the complete data sample. Also here the minimum-bias level is indicated.

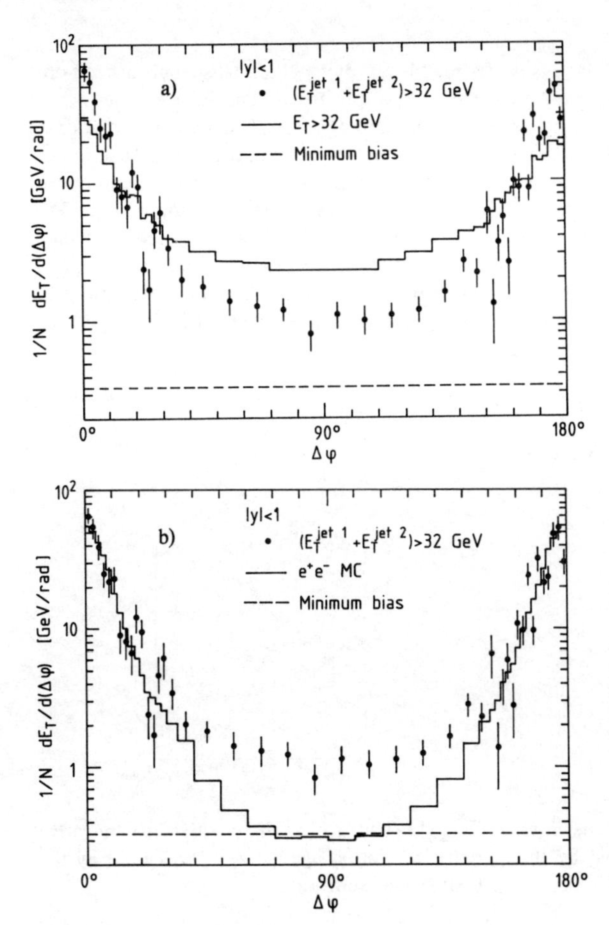

Fig. 10 Charged E_T flow plotted against the azimuthal distance from a jet axis.

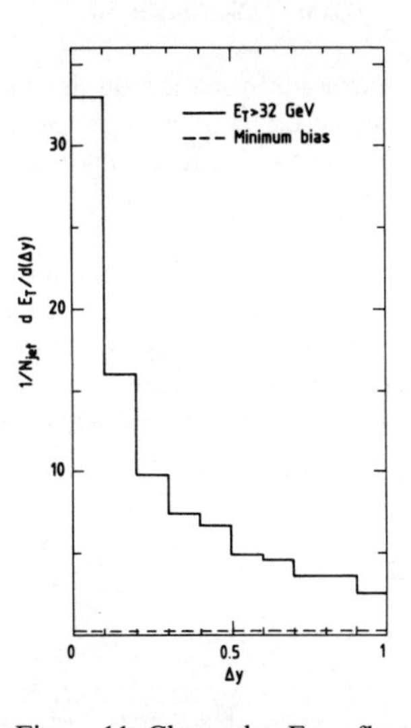

Fig. 11 Charged E_T flow plotted against the rapidity distance from a jet axis. The dashed line indicates the minimum-bias level.

6. Jet Imbalance in the Transverse Plane

The azimuthal angle between the two jets deviates, on the average, from 180° by 13.4° ± 0.5°, given by both the resolution of the apparatus and a genuine kink between the two jet axes. The latter effect is sensitive to the transverse momenta k_T of the colliding constituents. This k_T is usually considered to have an intrinsic component due to the constituents being bound inside the proton, and an additional 'hard' component due to gluon bremsstrahlung. The observed k_T will depend on the measured process, and hence on the trigger, e.g. a single jet or high-p_T hadron trigger will give a bias towards large k_T. After unfolding the apparatus resolution we measure a deviation $\langle\Delta\phi\rangle$ of 8.2° ± 0.8°. The average transverse momentum of the jets is 12.8 ± 0.8 GeV/c. Combining these numbers implies*) a mean jet–jet imbalance of 2.8 ± 0.7 GeV/c corresponding to an effective k_T of 2.3 ± 0.5 GeV/c.

It must be noted that this value of effective k_T includes the effects of the underlying event which is now well reproduced by the Monte Carlo model. These effects are, however, estimated to be small compared to the found value of k_T, mainly because of the $p_T > 1$ GeV/c requirement for energy clusters used in the jet algorithm.

7. Charge Distributions

Further information on the nature of the jets is provided by studying charge distributions. Jets of neutral origin (gluons) are expected to have the same amount of positive and negative charges, while quark jets are expected to show a positive excess since the colliding protons have two up quarks and one down quark. Figure 12 shows the ratio between positive and negative particles as a function of z. From the rising ratio we see a clear effect from the valence quarks. We have superimposed on the figure the prediction from the Lund Monte Carlo [Bengtsson, 1982] for different subprocesses. The filled circles are from data triggered on total transverse energy and the crosses are preliminary data taken with a two-jet trigger requiring large energy deposition in two localized regions. The range of $\sqrt{\hat{s}}$ is similar for the two data samples. We hope to be able to produce this plot in different regions of $\sqrt{\hat{s}}$. The CMOR experiment observes the same rising charge ratio as can be seen in Fig. 13.

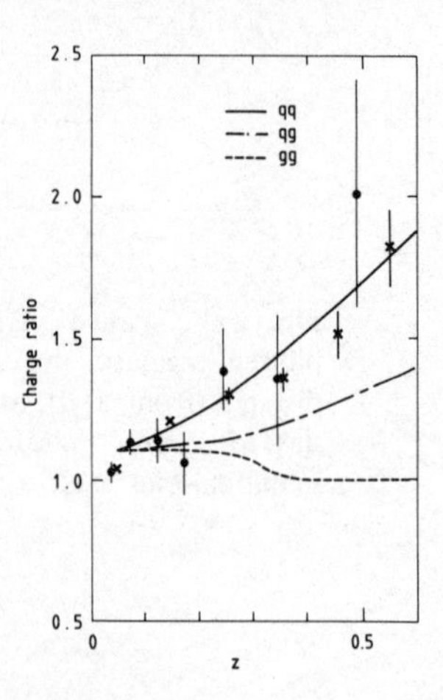

Fig. 12 The ratio of positive to negative particles plotted against z as measured by the AFS Collaboration.

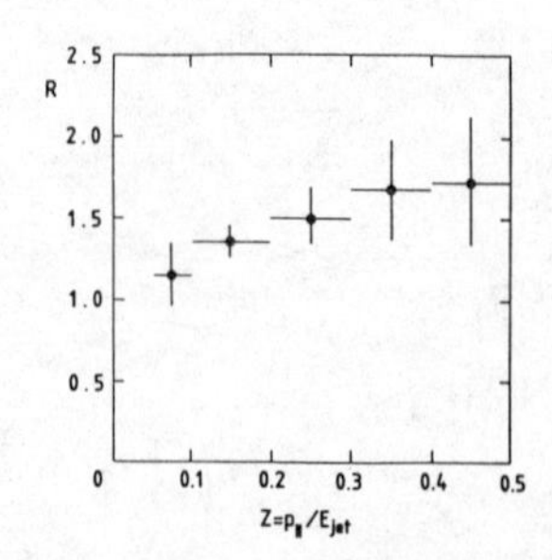

Fig. 13 The ratio of positive to negative particles plotted against z, as measured by the CMOR Collaboration.

*) $\langle\Delta\phi\rangle = (8/\pi^2)(k_T/p_{jet})$ and imbalance $= 4k_T/\pi$.

8. Fixed-Target Experiments

One does not, as we all know, observe any jets in events triggered on a calorimeter with full azimuth and large-rapidity coverage if the centre-of-mass energy is too low, and it is too low for the SPS and FNAL fixed-target experiments. Therefore, one has to introduce more specialized triggers. I will present results from two experiments that have done this, and start with those from the E609 experiment at FNAL [Arenton et al., 1984].

8.1 The E609 experiment

This experiment is shown in Fig. 14. There is a 400 GeV/c proton beam incident on a 45 cm liquid-hydrogen target, followed by a magnetic spectrometer with a magnet giving a 0.1 GeV/c transverse-momentum kick. After this spectrometer is the calorimeter wall with 132 towers pointing to the target, and after this wall there is a beam calorimeter taking the remainder of the beam energy. The calorimeters in the wall have a hadronic resolution of $70\%/\sqrt{E}$, an electromagnetic resolution of $35\%/\sqrt{E}$, and the pseudorapidity coverage is $-0.88 < \eta < 1.32$. The trigger requires at least two energy depositions in the wall with more than 1.5 GeV in E_T. This may seem to be a mild trigger, but one should bear in mind that less than one event in a thousand has a single particle with p_T greater than 1.5 GeV/c. The E609 Collaboration define the jet-E_T as the E_T in a cone of 55° half opening angle. The position of the cone is found by an iterative procedure and the 55° is chosen (from a Monte Carlo) as the angle where the cone loses the same amount of E_T from the tails of the jet fragmentation as it gains by picking up particles from the beam fragmentation.

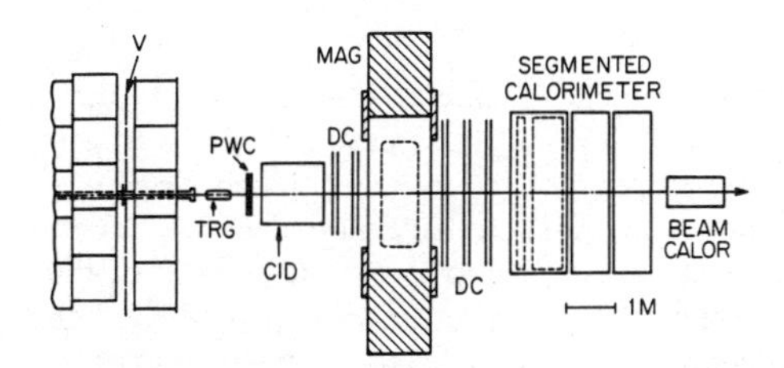

Fig. 14 The E609 experimental set-up

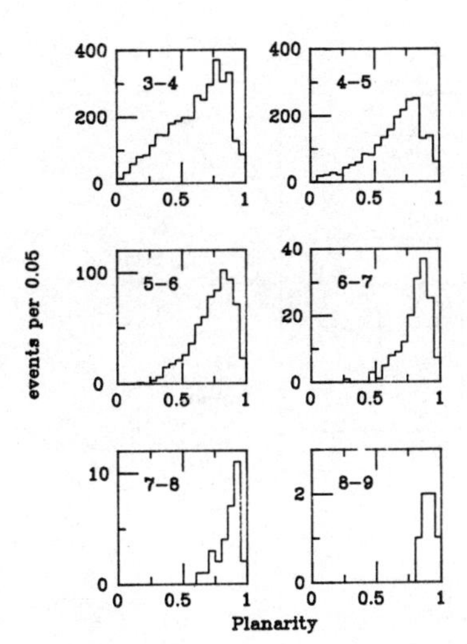

Fig. 15 Planarity distributions of the data for six bins of jet-p_T, indicated by the numbers on the plots.

Figure 15 shows the planarity distributions that experiment E609 measured in the wall as a function of the average transverse energy of the two highest E_T cones, and the distributions all peak at large values; hence the events seem to be very jet-like. This is not in contradiction with earlier fixed-target experiments, e.g. de Marzo et al. (1982), firstly because of the trigger and secondly because the distributions are plotted in bins of E_T in the previously described 55° cones and not in the whole calorimeter. Now with the jet-p_T given by these cones they can derive the di-jet cross-section shown in Fig. 16.

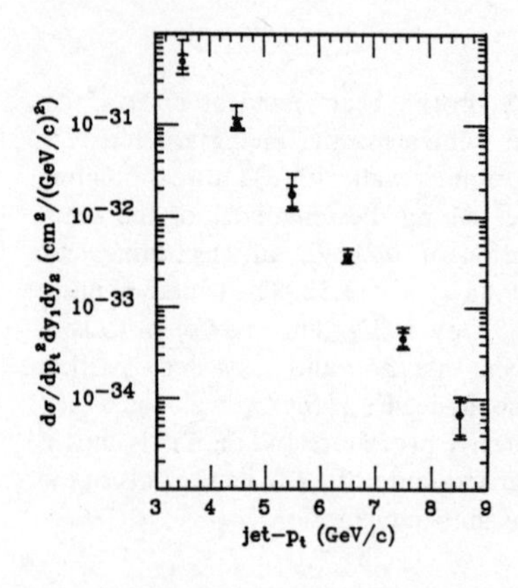

Fig. 16 The invariant cross-section for di-jets plotted against p_T. The inner error bars are statistical and the outer error bars include systematics.

8.2 The NA5 experiment (de Marzo et al., 1983)

This SPS experiment, shown in Fig. 17, analysed streamer-chamber information in their calorimeter-triggered data. The data were taken with 300 GeV/c π^- and protons on a hydrogen target inside a streamer chamber. The calorimeter downstream of the streamer chamber covers a pseudorapidity region of $-0.88 < \eta < 0.67$. The data are required to have more than 8.5 GeV in E_T in two opposite azimuthal regions of 120°. For these events they plot the charge ratio of the highest p_T particle in the event measured in the streamer chamber as a function of p_T. Figure 18 shows this ratio for both π^- p and pp data. The influence of the valence quarks is clearly observed.

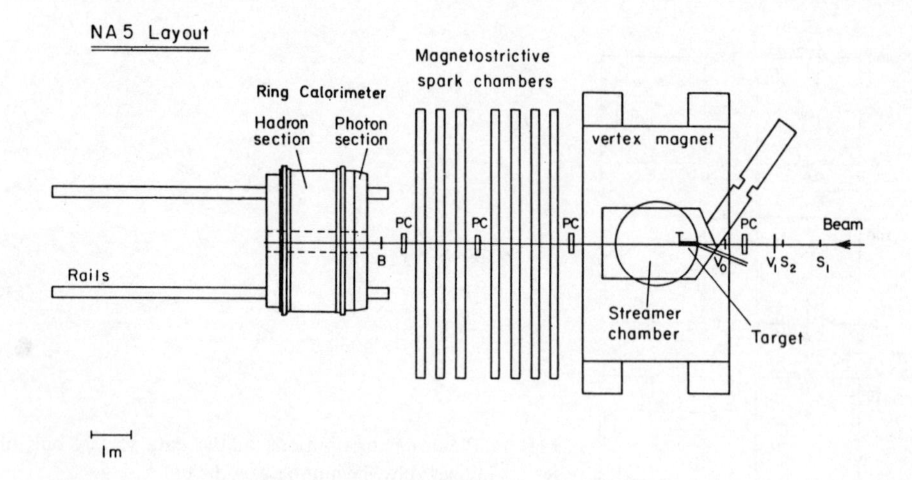

Fig. 17 The NA5 experimental set-up

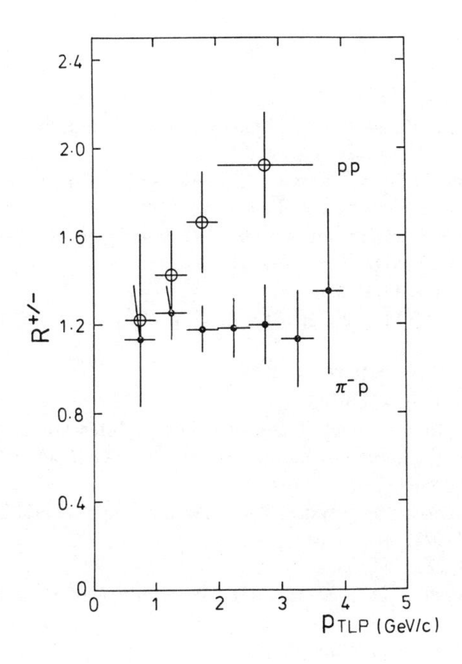

Fig. 18 Ratio of the number of events for which the highest p_T charged particle in the calorimeter acceptance is positive to that for which it is negative for π^-p (filled circles) and pp (open circles) interactions at 300 GeV.

9. Conclusions

In the high-E_T region there is a clear jet dominance observed at the ISR. The change from isotropic to jet events occurs over a very limited region in E_T. In this region of E_T it looks like a competition between two processes. The jets in the high-E_T region look like the jets at the Collider and at e^+e^- machines. The distribution of charges shows a clear dominance of valence-quark scattering. At lower $\sqrt{s}$ (fixed-target experiments) one is forced to use specialized triggers and this makes it possible to do jet physics even here. The E609 Collaboration have produced an important result by measuring a di-jet cross-section at fixed-target centre-of-mass energies.

Acknowledgements

I wish to thank the CMOR, NA5, and E609 Collaborations for the information that they have helpfully provided, and my collaborators in the AFS Collaboration for many helpful discussions. Finally, I would like to thank the organizers, and especially M. Pennington, for a very successful and friendly conference in the beautiful English University town of Durham.

References

Åkesson T et al. 1983 Phys. Lett. **128B** 354

Åkesson T et al. 1984 Properties of Jets in High-E_T Events Produced in pp Collisions at $\sqrt{s}$ = 63 GeV CERN–EP/84–56 to be published in Z. Phys. C

Angelis A L S et al. 1984 Jets in High Transverse Energy Events at the CERN Intersecting Storage Rings CERN–EP/84–46 to be published in Nucl. Phys. B

Arenton M W et al. 1984 Measurement of the Di-jet Cross-section in 400 GeV/c pp interactions ANL–HEP–PR–84–14 submitted to Phys. Rev. D

Arnison G et al. 1983 Phys. Lett. **132B** 223

Bengtsson H U 1982 The Lund Monte Carlo for High-p_T Physics, Lund Univ. preprint LU–TP 82–15

Brandelik R et al. 1983 Phys. Lett. **94B** 437

de Marzo G et al. 1982 Phys. Lett. **112B** 173

de Marzo G et al. 1983 A Study of High Transverse Energy Interactions of 300 GeV Pions and Protons on Hydrogen using a Streamer Chamber HEN–229 15 October 1983

Horgan R and Jacob M 1981 Nucl. Phys. **B179** 441

(Proc. 21st) Int. Conf. on High-Energy Physics, Paris, 1982, J. Phys. **43** (1982):
UA2 Collaboration, p C3–571.
AFS Collaboration, p C3–122.

Paige F E and Protopopescu S D 1982 ISAJET: A Monte Carlo Generator for pp and $p\bar{p}$ Interactions BNL 31987

Sjöstrand T 1983 Comput. Phys. Commun. **48** 229

Inst. Phys. Conf. Ser. No. 73: Section 2
Paper presented at VII Eur. Symp. Antiproton Interactions, Durham 1984

Latest results on high p_T jets in UA2

The UA2 Collaboration

presented by L. Fayard
LAL, Orsay, France

ABSTRACT

Jet production and fragmentation properties have been measured in the UA2 detector at the CERN SPS $\bar{p}p$ Collider. Results on the total transverse momentum of the two-jet system, on the parton density in the nucleon (structure function) and on the two-jet angular distributions are reported and compared with QCD predictions. Results on the charged particle multiplicity in jets and the energy flow are compared with extrapolations of lower energy e^+e^- data and QCD inspired models. The preliminary results of a search for $W/Z \to 2$ jets are presented.

1. INTRODUCTION

The recent unambiguous identification of jets in hadronic collisions [1-4] at the CERN $\bar{p}p$ Collider and at the ISR has enabled a detailed study of the jet production and fragmentation properties. The UA2 detector has been described in detail elsewhere[5]. The data taking and reduction have also been described in detail elsewhere[6-7]. The data reported here have been collected during the 1983 run at the CERN $\bar{p}p$ Collider and correspond to an integrated luminosity of $\int \mathcal{L}dt = 112\ nb^{-1}$. The inclusive single and two-jet production cross-sections from these data have been previously published[1] and found to be in agreement with QCD predictions. In section 2 results are presented on the two-jet production properties and compared to lowest order QCD predictions. In section 3 results on fragmentation properties of jets are presented and compared with extrapolations of lower energy e^+e^- data and QCD based models. Finally in section 4 preliminary results are presented on a search for W/Z → 2 jets and on a possible enhancement around 150 GeV.

2. TWO-JET PRODUCTION PROPERTIES

2.1 P_T of the Two-Jet System

The intrinsic transverse momentum of the partons in the nucleon and the initial state gluon bremsstrahlung are expected to contribute to the transverse momentum p_T^{jj} of the two-jet system.

Experimentally, p_T^{jj} is the vector sum of two large and opposite momenta and is therefore sensitive to instrumental effects. The two components p_η^{jj} (along the bisector of the two jets in the transverse plane) and p_ξ^{jj} (orthogonal to it and roughly parallel to the two-jet axis) are studied separately, because the effect of the detector resolution is much less on $|p_\eta^{jj}|$ than on $|p_\xi^{jj}|$.

Figs. 1a and 1b show the normalized distributions of $|p_\xi^{jj}|$ and $|p_\eta^{jj}|$ for the two-jet system. The systematic errors, due to the uncertainty on the energy scale and to the analysis procedure, are estimated to be less than ±10%. The rms of the two distributions are respectively 9.1 GeV for p_ξ^{jj} and 7.5 GeV for p_η^{jj}, with a negligible statistical error. Note however that the analysis criteria reject events with a very high value of p_T^{jj} and that instrumental effects contribute substantially to the measured widths of the p_ξ^{jj} distributions. In a perfect detector the widths of the p_ξ^{jj} and p_η^{jj} distributions are expected to be approximately equal.

The dashed lines in Fig. 1 show a QCD prediction[8], which was computed by assuming a fixed parton-parton invariant mass of 60 GeV and a uniform azimuthal distribution of p_T^{jj} in the ξ-η plane. The uncertainty results mostly from the error on the gluon structure function (gluon radiate more than quarks on the average) and is of the order of ±20%. To compare the prediction with the data, a complete Monte Carlo simulation, including all the detector details, was performed. Two-jet events were generated with the above p_T^{jj} distribution. A fragmentation

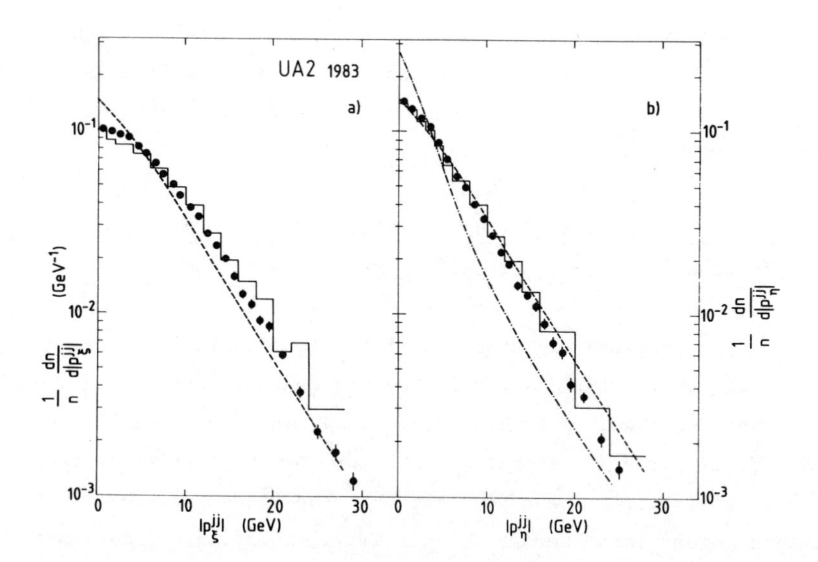

Fig. 1. a) Normalized distribution of $|p_\xi{}^{jj}|$, the component of the two-jet transverse momentum along the two-jet axis. The dashed line is a QCD prediction[8], while the histogram represents the prediction after detector smearing.
b) The same as a), but for $|p_\eta{}^{jj}|$, the component of the two-jet transverse momentum along the bisector of the two jets. The dotted-dashed line is the prediction of Ref. [8] assuming the same colour charge for gluons as for quarks.

function close to the one observed in the data[1] was used. The Monte Carlo event sample was submitted to the same trigger and analysis criteria as the real data. The corresponding distributions, shown as histogrammes in Fig. 1, are in good agreement with the experimental points. The result is sensitive to the assumption that gluons radiate more than quarks because of the larger colour charge of the gluon. As an example, the dashed-dotted line of Fig. 1b shows the distribution[8] expected in the case that gluons radiate like quarks.

2.2 Factorisation

As was pointed out in Ref. [9] since all the QCD subprocesses involving a t-channel exchange of a vector boson give rise to a similar cms angular distribution, in lowest order QCD the jet cross section is expected to approximately factorise into an overall structure function F(x) (sum of all the quark and gluon densities) and an effective parton-parton cross section $d\sigma/d\cos\theta^*$, obtained from a weighted average over all the elementary subprocesses :

$$\frac{d^3\sigma}{dx_1 dx_2 d\cos\theta^*} = \frac{S(x_1,x_2))}{x_1 x_2} \frac{d\sigma}{d\cos\theta^*} = \frac{F(x_1)}{x_1} \frac{F(x_2)}{x_2} \frac{d\sigma}{d\cos\theta^*} \tag{1}$$

where x_1 (x_2) represents the fraction of the longitudinal momentum carried by the interacting parton in the proton (antiproton). This analysis has been done with $\alpha_s = 12\pi/23\ell n(Q^2/\Lambda^2)$, where $Q^2 = (p_T)^2$ and $\Lambda = 0.2$ GeV. Equation (1) is valid to the extent that different subprocesses have a similar θ^* behaviour and that the Q^2 dependence in both the structure functions and in $\alpha_s(Q^2)$ is small. Equation (1) does not include higher order QCD corrections, which are usually described by a multiplicative factor $K < 2$[10]) and give rise to a nonzero transverse momentum for the two-jet system. In the following analysis these higher order effects will be neglected (K = 1). The approximation of a small transverse momentum for the two-jet system, $p_T^{jj} \ll m_{jj}$, which is necessary to define θ^* without kinematical ambiguities [11]), has been shown to be reasonable in the previous section. This requirement is applied by retaining events only if they have $p_T^{jj} < 20$ GeV and $\Delta\phi^{jj} > 140^o$.

As a first test of the factorisation hypothesis the ratio of the distributions of the cms scattering angle $\cos\theta^*$ for two different intervals of jet-jet invariant mass is shown in Fig. 2(a). As can be seen from Fig. 2(a) this ratio is consistent with unity as expected if factorisation holds. As a further test, the factorisation in x_1 and x_2 is tested by plotting the ratio of $S(x_1,x_2)/S(x_1,x_2 + \Delta x)$ as a function of x_1 for different intervals of x_2 (see Fig. 2(b)). For each interval of x_2 the ratio is independent of x_1 in agreement with the factorisation hypothesis.

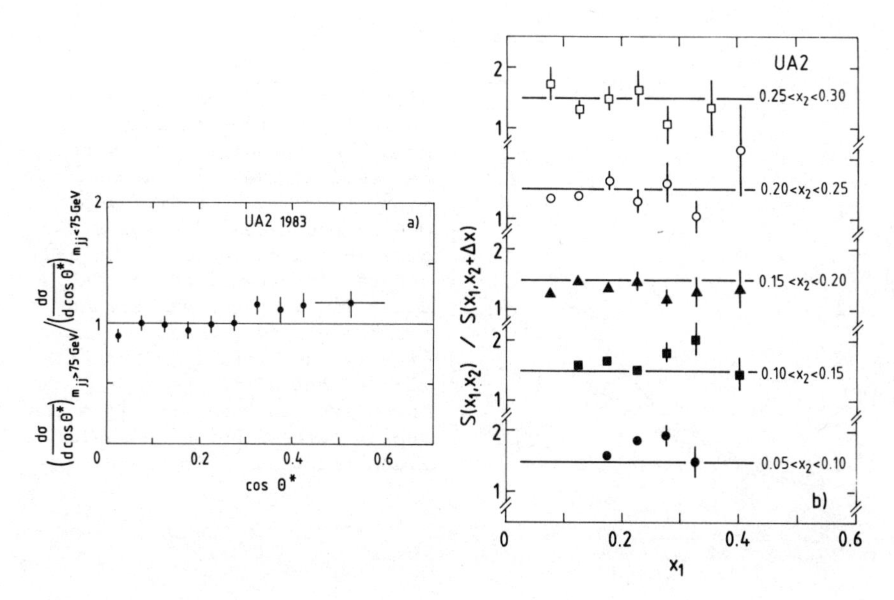

Fig. 2. a) Ratio between the angular distribution of the sample of events with m_{jj} > 75 GeV and the events with m_{jj} < 75 GeV. b) Ratio $S(x_1,x_2)/S(x_1,x_2 + \Delta x)$ (see eq. (1) in the text) for different x_2 slices and $\Delta x = 0.05$. The horizontal lines correspond to the slope of F(x).

2.3 CMS Angular Distributions

The $d\sigma/d\cos\theta^*$ distribution is computed by considering the event distribution in the $\cos\theta^*-\beta^{jj}$ plane, where β^{jj} is the velocity of the two-jet system in the laboratory. The calorimeter acceptance is uniform in a well defined domain in this plane, that depends, for each event, on the two-jet mass and on the interaction position along the beam line. A fiducial domain in the region $\cos\theta^* < 0.6$ can be defined where the acceptance is complete. Different slices of β^{jj} are considered for different regions of $\cos\theta^*$. The overall angular distribution is evaluated by normalizing the different $\cos\theta^*$ distributions to the same area in the interval where they overlap. The overall normalization is not defined by this method and is chosen by arbitrarily setting the data point at $\cos\theta^* = 0$ equal to 1. The results are shown in Fig. 3. The systematic uncertainties which depend on the relative normalization of different

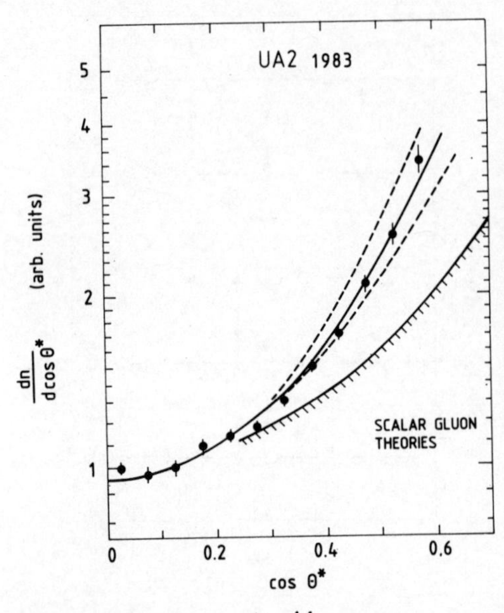

Fig. 3
Distribution of the scattering angle in the two-jet centre of mass. The normalization is defined setting the value at $\cos\theta^* = 0$ equal to 1. All the different QCD subprocesses[13] (except the pure s-channel one $q_1\bar{q}_1 \to q_2\bar{q}_2$), separately normalized to the data, lie in the area between the two dashed curves. The full line is the overall QCD prediction discussed in the text, normalized to the data. Also the region corresponding to scalar gluon theories[13] is shown.

regions in the $\cos\theta^*-\beta^{jj}$ plane are included in the errors. The systematic uncertainties due to the $\cos\theta^*$ resolution (< 0.04, independent of $\cos\theta^*$) and to the definitions of the kinematical variables adopted in this analysis are estimated to be < 5%. In the region of overlap, the data agree well with the results of the UA1 Collaboration[12], to within the statistical and systematic errors.

The observed angular distribution can be compared with the predictions of different theories. For example scalar gluon theories are clearly excluded[13]. The observed distribution is compared in Fig. 3 with the QCD prediction for the dominant parton-parton subprocesses[14] which lie in the area between the two dashed lines. The statistical and systematic errors are such that the relative importance of the different subprocesses cannot be measured. An overall QCD prediction, obtained from a sum of all the subprocesses, each weighted by a factor depending on the structure fuctions (derived from Ref. 15), gives good agreement. This curve is shown as a full line normalized to the data.

2.4 Structure Functions

The function $S(x_1,x_2)$, defined in eq. (1) is computed by weighting each event with an acceptance factor w, that accounts for the probability to observe the event :

$$w^{-1} = \int_0^{Cmax} (d\sigma/\cos\theta^*)_{QCD}\, d\cos\theta^* \qquad (2)$$

where Cmax (function of the observed values of x_1 and x_2) is the maximum detectable value of $\cos\theta^*$ in the fiducial region and $(d\sigma/d\cos\theta^*)_{QCD}$ is the QCD parametrization of the angular distribution. The effective structure function F(x) is evaluated according to eq. (1) for 10 bins of x between 0.05 and 0.60. The effects of the resolution in x and of the variation of the acceptance as a function of the vertex position have been taken into account. One should note that, if the usual definition of the higher order correction factor K were adopted, the result of the measurement would be the product $\sqrt{K}\,F(x)$. The result is shown in Fig. 4. An empirical fit (shown as a full line in Fig. 4) of the form $F(x) = A \exp(-\alpha x)$ gives $A = 6.2 \pm 0.1$ and $\alpha = 8.3 \pm 0.1$. The systematic errors, due to the resolution, to the definition of the kinematics and to the uncertainty on the luminosity, are ±30%.

The structure function agrees with the one reported in Ref. 12 in the region $x > 0.10$, to within the statistical and systematic uncertainties. Since that analysis assumed K = 2, those points (shown in Fig. 4) have been multiplied by $\sqrt{2}$.

The parton density of Ref. 15, extrapolated to Collider energies ($Q^2 = 2000\ GeV^2$), is shown in Fig. 4 as a dashed line. This structure function has been calculated assuming $F(x) = g(x) + 4/9\, q(x)$, where g(x) is the gluon structure function and q(x) is the sum over all the quark and antiquark densities. Part of the overall scale factor ($\sim$1.4) may be attributed to higher order corrections, not taken into account in the computation of F(x). The two distributions agree well in shape both in the region of small x (where F(x) is dominated by the gluon density) and of high x, where mainly valence quarks are present.

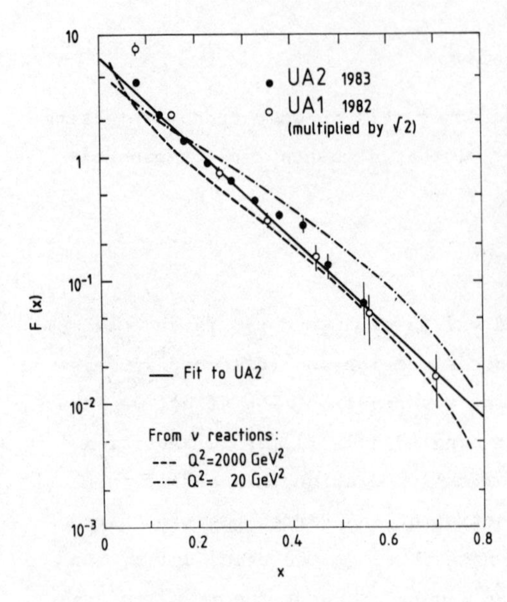

Fig. 4

Effective structure function. The full points are the results of this experiment and the open points are derived from Ref. 15. The full line represents an exponential fit to the data, while the dashed lines are computed from neutrino deep inelatic scattering[17], taking into account scaling violation effects.

3. FRAGMENTATION PROPERTIES

From the results on structure functions presented above, the jets in this energy range should be mixtures of quarks and gluons. It is therefore interesting to compare these jets with quark jets from e^+e^- data. According to QCD inspired models[16 17], gluon jets are expected to fragment differently because of the larger colour charge of the gluon. Results are presented on the charged particle multiplicity in jets and on the transverse energy flow in two-jet events.

3.1 Charged Particle Multiplicity in Jets

The charged particle multiplicity in jets was measured using the vertex detector. The analysis is restricted to the transverse plane where the track reconstruction efficiency is highest and where particles not associated with a jet are expected to contribute a uniform azimuthal distribution. Distributions of the azimuthal separation $\Delta\phi$ between all transverse tracks observed in the vertex detector and the energy centroid of the cluster having the highest transverse energy are shown in

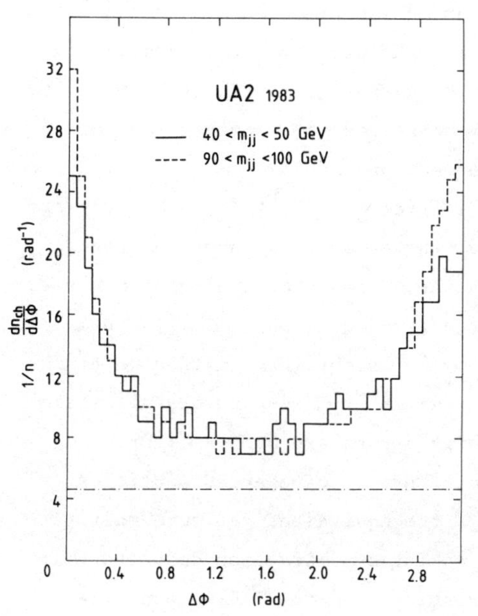

Fig. 5.

Azimuthal separation $\Delta\phi$ between the energy cluster centroid of the leading jet and all charged transverse tracks, normalised to n, the number of events. The dashed-dotted line represents the level measured in minimum bias events.

Fig. 5 for 2 intervals of m_{jj}. There are two peaks at $\Delta\phi \simeq 0$ and $\Delta\phi \simeq \pi$ as expected for two-jet events. The peak at $\Delta\phi \simeq 0$ is narrower than the peak at $\Delta\phi \simeq \pi$ because of the finite two-jet transverse momentum.

There is no unique way to define the charged particle multiplicity of jets produced in $\bar{p}p$ collisions because the relative fractions of the particles coming from the jet and coming from the "underlying event" (the part of the event which is due to the spectator parton fragments) are unknown. As a lower limit on the true jet multiplicity one can define the "jet core" multiplicity as the number of charged particles above a flat level corresponding to the value at $\Delta\phi = \pi/2$ (see Fig. 5) :

$$< n_{ch}^{jet} > = 1/2 \{\int_0^\pi [dn/d(\Delta\phi)]\ d(\Delta\phi) - \pi[dn/d(\Delta\phi)]_{\Delta\phi=\pi/2}\} \quad (3)$$

Since the angular coverage of the vertex detector is much larger than that of the central calorimeter, the losses from geometrical acceptance are negligibly small (< 1%).

The data are corrected for detector inefficiencies. The transverse track finding efficiency (measured using the forward-backward spectrometer[5]) was 94.8 ± 0.8%, the fraction of spurious tracks and tracks not coming from the primary vertex was 21 ± 1%. A correction was also made for the loss of tracks from the finite two track resolution ($\sim$ 5 mm). The effect of this correction varies from ≃12% at $\Delta\phi = \pi/2$ to ≃ 25% at $\Delta\phi = 0$. The data were also corrected for γ conversions and π^0 Dalitz decays (the effect of this correction was 2%).

The corrected jet core multiplicities $< n_{ch}{}^{jet} >$ [see Eq.(3)] are shown as a function of m_{jj} in Fig. 6. The error bars shown include a common 5% statistical uncertainty in the estimate of the two track resolution loss. The systematic error is estimated to be $\sim$ 15%. The dominant source of systematic error comes from the uncertainty in the two track resolution correction. The systematic error is mainly an overall scale error and the point to point systematic errors are much smaller.

Using the measured angular distributions of charged particles around the jet axis from e^+e^- data[18] the equivalent jet core multiplicity was evaluated for e^+e^- data by making a subtraction corresponding to the flat level at $\Delta\phi = \pi/2$. The results from the TASSO experiment are also shown in Fig. 6. Neither the data from this experiment, nor the TASSO data, were corrected for strange particle decays. The result suggests that jets from $\bar{p}p$ collisions in the mass range $40 < m_{jj} < 80$ GeV have higher mean multiplicities than one would expect from extrapolations of e^+e^- data. In order to understand this effect in more detail, the QCD parton shower model of Ref. 17 was used to predict the growth of multiplicity with m_{jj} for quark-antiquark ($q\bar{q}$) and gluon-gluon (gg) jets. The model for quark jets has been tuned to agree with e^+e^- data and therefore provides a good model for the extrapolation of e^+e^- data. The model also predicts the relative multiplicities of gluon jets compared to quark jets. The full line represents the prediction of the model for quark jets while the shaded area represents the range of predictions for gluon jets. Gluon jets are expected to have higher multiplicities than quark jets because of the larger colour charge of the gluon. Using the structure functions of Ref. 15 which are consistent with the ones measured in this experiment (see section 2), one expects that the fraction of gluon jets varies from $\sim$ 75% to $\sim$ 30% as m_{jj} varies from 40 GeV to 140 GeV. Allowing for this variation in the gluon fraction, the data are in agreement with the expectations of the model.

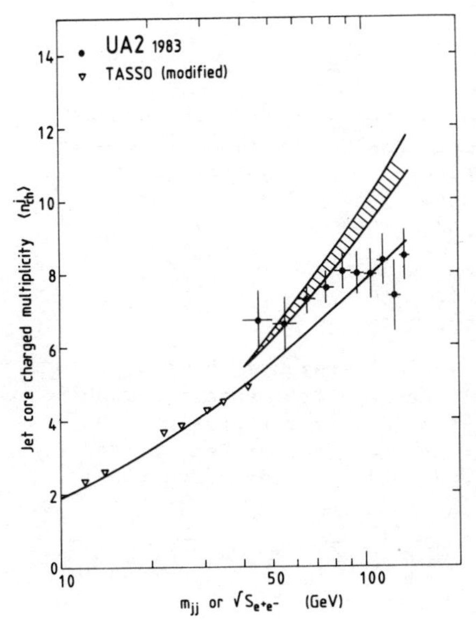

Fig. 6.

Mean charged "jet core" multiplicities n_{ch}^{jet} (see text for definition) as a function of $\sqrt{s_{e^+e^-}}$ for e^+e^- data and of the invariant two-jet mass m_{jj} for $\bar{p}p$. The solid curve is the prediction of the model of Ref. 17 for quark jets and the shaded area represents the predictions for gluon jets. The uncertainty in the predictions arises from the range of parameters for which the quark jets fit the e^+e^- data.

3.2 Energy flow

The transverse energy flow around the jet axis was measured for a clean sample of two-jet events. To reduce the energy leaking outside the calorimeter acceptance, only clusters having a pseudorapidity, $|\eta|$, less than 0.3 (central clusters) are considered in this analysis. The distribution of the azimuthal transverse energy density $dE_T/d\Delta\phi$, integrated over a rapidity interval of ±0.7 units around this central cluster is shown in Fig. 7 for three different intervals of the cluster transverse energy (E_T^j). The peaking in the energy flow is much stronger than in the charged particle flow (note the logarithmic scale in Fig. 7). The data are not corrected for the effects of calorimeter granularity and resolution. The main effect arises from the finite cell size ($\Delta\phi \times \Delta\theta = 15^o \times 10^o$) which tends to smear out the peaks. To compare the data with different fragmentation models, several Monte Carlo calculations were made. All details of the detector response were simulated for samples of events generated according to these models. The Monte Carlo event sample was submitted to the same trigger and analysis procedure as the real data. The corresponding distributions are shown as curves

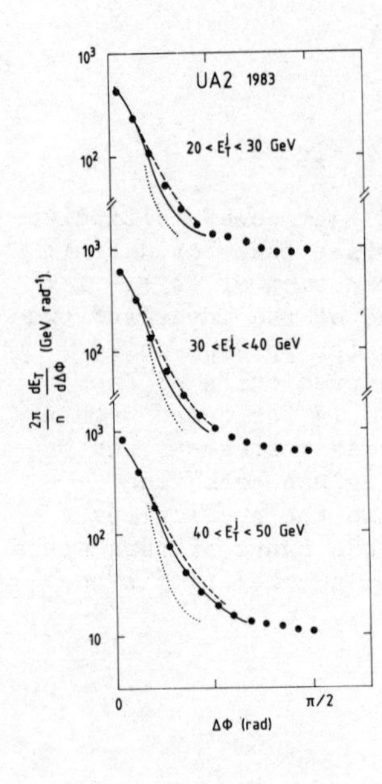

Fig. 7.

Distribution of the azimuthal transverse energy density $dE_T/d\Delta\phi$ where $\Delta\phi$ is measured with respect to the centroid of the central cluster ($|\eta| < 0.3$) for three ranges of E_T^j.

(a) $20 < E_T^j < 30$ GeV.
(b) $30 < E_T^j < 40$ GeV.
(c) $40 < E_T^j < 50$ GeV.

The dotted curve represents the predictions of the Field-Feynman fragmentation model[19]. The dashed (solid) curve represents the predictions of the model of Ref. 17 for gluon (quark) jets (see text for details).

in Fig. 7. The Field-Feynman fragmentation model[19] with a rms transverse momentum of particles with respect to the jet axis, q_T, set equal to 350 MeV produces jets which are much narrower than the data. The data can be better reproduced by models which include explicity the effects of gluon bremsstrahlung, such as that used in section 3.1. The data are bracketed by the predictions for $q\bar{q}$ and gg as expected for a mixture of quark and gluon jets.

4. TWO-JET MASS DISTRIBUTION

The dominant decay modes of the W and Z bosons are expected to result in two final state jets. Therefore it is interesting to look for structure in the invariant two-jet mass distribution. It is important to optimise the mass resolution because of the large continuum QCD background. A sample of clean events with two jets in the central calorimeter and not too much energy in the forward-backward calorimeter was selected. For each event the energy of the second of the two jets is increased to balance the transverse energy of the first jet ($P_T^{jj} = 0$). The estimated mass resolution at a mass around 80 GeV is $\sim$ 10 GeV.

The normal threshold data were only used for invariant two-jet masses, m_{jj}, above 64 GeV where there is no trigger bias (see Fig. 8). For the mass range $m_{jj} > 52$ GeV the low threshold data was used and was allowed to have a free normalisation in the fits. A first fit was made to the background region excluding the W and Z mass region ($66 < m_{jj} < 98$ GeV) of the form

$$d\sigma/dm_{jj} = A \exp(-\alpha m_{jj}) + B \exp(-\beta m_{jj}) \quad (4)$$

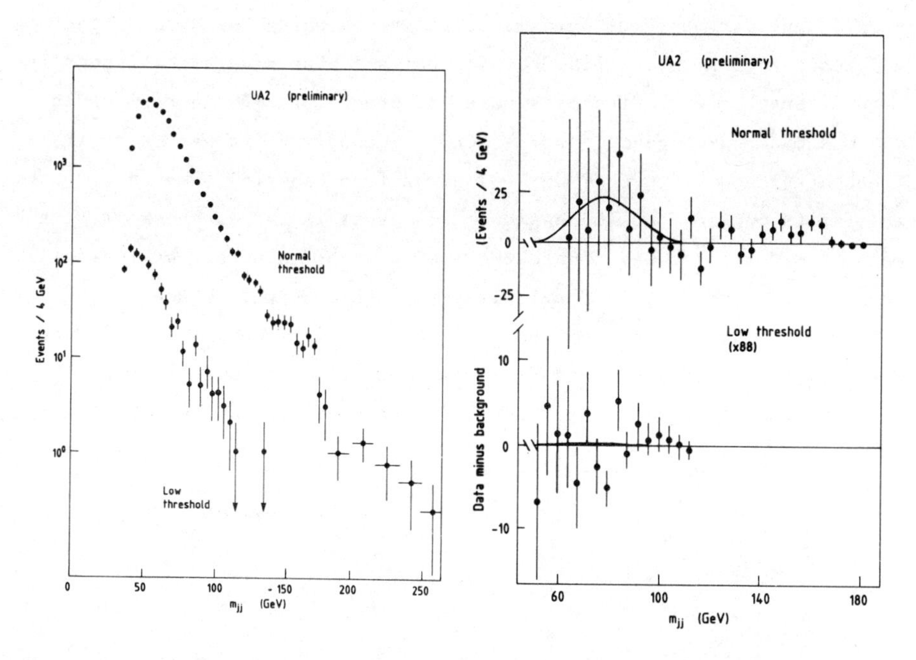

Fig. 8.
Two-jet invariant mass distribution for normal and low threshold data.

Fig. 9.
Results of the W and Z fit. See text for details.

Although there was no significant excess of events above the background a further fit was made over the full range, now including two Gaussian curves to allow for any contribution from W and Z decays into two jets. The background curve was fixed to the previous background fit. The ratio of Z^o to W was fixed at 0.41 ($N_{tot} = N_W + N_Z$; $N_Z = 0.41\ N_W$) and the ratio of the masses of the Z and the W were fixed at 1.15 as expected according to the standard model. The resolution was fixed at 10 GeV.

The result of the fit (see Fig. 9) was $N_W = 108^{+81}_{-70}$ and $M_W = 75 \pm 10$ GeV. Using the measured cross-section times branching ratio for $W \to e\nu$ decays[20] and the standard model prediction for the branching ratio of $W \to e\nu$, the expected value of $N_W \simeq 150$. Therefore the values of N_W and M_W are consistent with expectations. However since the value of N_W is not significantly different from zero it is clear that more data are needed. From the mass distribution (Fig. 8) there is some evidence for structure at $m_{jj} \sim 150$ GeV. To understand if this is statistically significant further fits were made, after rebinning the data in the high mass region ($m_{jj} > 180$ GeV) so that all bins have a statistically significant content. Firstly a pure background fit of the form of Eq. (4) was made which gave $\chi^2/\mathrm{ndf} = 40/27$. Secondly a fit was made with a free background of form of Eq. (4) and a free Gaussian. The result of the fit (shown in Fig. 10) has 69 ± 17 events in the enhancement at a mass of $M = 147 \pm 3$ GeV (where the quoted error is purely statistical). The χ^2/ndf of the fit is 28/25. The statistical significance of the enhancement was evaluated by using a maximum likelihood fit. This was done for the pure background and background plus Gaussian curves. The results from these fits (see Fig. 10) were consistent with the results

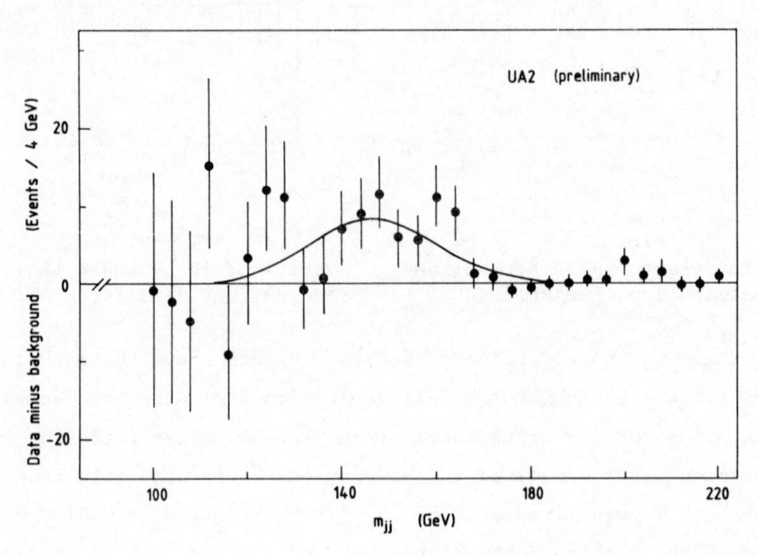

Fig. 10
Results of the high mass region fit ($m_{jj} > 64$ GeV).

of the χ^2 fits. From the ratio of the likelihoods of the fits with and without the Gaussian the probability that a statistical fluctuation could generate such a enhancement can be estimated[21] to be 0.5%. Clearly more data are needed to confirm this result.

5. CONCLUSIONS

The two-jet p_T was measured and found to be in good agreement with QCD predictions from initial state bremsstrahlung. The parton-parton cross section was shown to have the expected factorisation properties. The cms angular distribution was found to be in good agreement with lowest order QCD. The effective structure function was in agreement with the extrapolations of lower energy neutrino data. The charged particle multiplicity in jets was measured and found to exceed extrapolations from lower energy e^+e^- data. The energy flow data can be fitted by models which include gluon bremsstrahlung and in particular the data can be fitted by a mixture of quark and gluon models according to Ref. 17. A preliminary search for structure in the two-jet mass distribution showed no statistically significant evidence for $W/Z \rightarrow 2$ jets. There is some evidence for structure around $m_{jj} \sim 150$ GeV although more data are needed to confirm this observation.

REFERENCES AND FOOTNOTES

1. UA2 Collaboration, M. Banner et al., Phys. Lett. 118B (1982) 203.
UA2 Collaboration, P. Bagnaia et al., Z. Phys. C20 (1983) 117.
UA2 Collaboration, P. Bagnaia et al., Phys. Lett. 138B (1984) 430.

2. Axial Field Spectrometer Collaboration, T. Åkesson et al., Phys. Lett. 118B (1982) 185 and 193.

3. UA1 Collaboration, G. Arnison et al., Phys. Lett. 123B (1983) 115.

4. COR Collaboration, A.L.S. Angelis et al., Phys. Lett. 126B (1983) 132.

5. B. Mansoulié, The UA2 apparatus at the CERN $\bar{p}p$ Collider, proceedings 3rd Moriond workshop on $\bar{p}p$ physics, éditions Frontières, 1983, p. 609;
M. Dialinas et al., The vertex detector of the UA2 experiment, LAL-RT/83-14, ORSAY, 1983;
C. Conta et al., The system of forward-backward drift chambers in the UA2 detector, CERN-EP/83-176, submitted to Nucl. Instrum. Methods;
K. Borer et al., Multitube proportional chambers for the localization of electromagnetic showers in the UA2 detector, CERN-EP/83-177, submitted to Nucl. Instrum. Methods;
A. Beer et al., The central calorimeter of the UA2 experiment at the CERN $\bar{p}p$ Collider, CERN-EP/83-175, submitted to Nucl. Instrum. Methods.

6. UA2 Collaboration, P. Bagnaia et al., CERN EP 84/74 submitted to Phys. Lett. B.

7. UA2 Collaboration, P. Bagnaia et al., CERN EP 84/75 submitted to Phys. Lett. B.

8. M. Greco, Back-to-back jets as a test of the three gluon coupling, LNF preprint 84/21 (1984), submitted to Z. Phys. C.

9. G. Cohen-Tannoudji et al., Phys. Rev. D28 (1983) 1628 ;
B.L. Combridge and C.J. Maxwell, RL-83-095 (1983) ;
F. Halzen and P. Hoyer, Phys. Lett. 130B (1983) 326.

10. M. Furman, Nucl. Phys. B197 (1982) 413.

11. J.C. Collins and D.E. Soper, Phys. Rev. D16 (1977) 2219.

12. UA1 Collaboration, G. Arnison et al., Phys. Rev. Lett. 136B (1984) 294.

13. D. Drijard et al., Phys. Lett. 121B (1983) 433.

14. B.L. Combridge, J. Kripfganz and J. Ranft, Phys. Lett. 70B (1977) 234 ;
see also : R. Cutler and D. Sivers, Phys. Rev. D17 (1978) 196.

15. CDHS Collaboration, H. Abramowicz et al., Z. Phys. C12 (1982) 289 ; Z. Phys. C13 (1982) 199 ; Z. Phys. C17 (1983) 283 and H. Blümer, private communication ;
see also the corresponding result from the CHARM Collaboration, F. Bergsma et al., Phys. Lett. 123B (1983) 269.

16. G. Sterman, S. Weinberg, Phys. Rev. Lett. 39 (1977) 1436.
K. Shizuya, S.-H.H. Tye, Phys. Rev. Lett. 41 (1978) 787.
M.B. Einhorn, B.G. Weeks, Nucl. Phys. B146 (1978) 445.

17. G. Marchesini and B.R. Webber, Nucl. Phys. B238 (1984) 1.
B.R. Webber, Nucl. Phys. B238 (1984) 492.

18. TASSO Collaboration, M. Althoff et al., Z. Phys. C22 (1984) 307.

19. R.D. Field and R.P. Feynman, Nucl. Phys. B136 (1978) 1.

20. UA1 Collaboration, G. Arnison et al., Phys. Lett. 126B (1983) 398.
UA2 Collaboration, P. Bagnaia et al., CERN EP 84/39 submitted to Z. Phys. C.

21. W. Eadie et al., Statistical methods in experimental physics, 231, North-Holland 1971.

Paper presented at VII Eur. Symp. Antiproton Interactions, Durham 1984

Jet phenomenology

B.R. Webber

Cavendish Laboratory, Cambridge, U.K.

Abstract. Applications of perturbative QCD to jet production and fragmentation are reviewed. Models for the nonperturbative process of hadron formation are discussed. Results of combining QCD with such a model are compared with e^+e^- and $\bar{p}p$ jet data.

1. Introduction

We have heard quite a lot at this meeting about the 'exotica' of jet physics, i.e., the small number of very interesting events that do not seem to have any explanation within the Standard Model. It is my job to talk about the 'non-exotica' - to review our quantitative understanding of the large body of jet data that is not in manifest disagreement with conventional theoretical ideas.

The general picture of QCD jet production in hard hadron-hadron collisions like those at the CERN Sp$\bar{p}$S collider is illustrated in Fig. 1. It is convenient to distinguish three phases of the process:

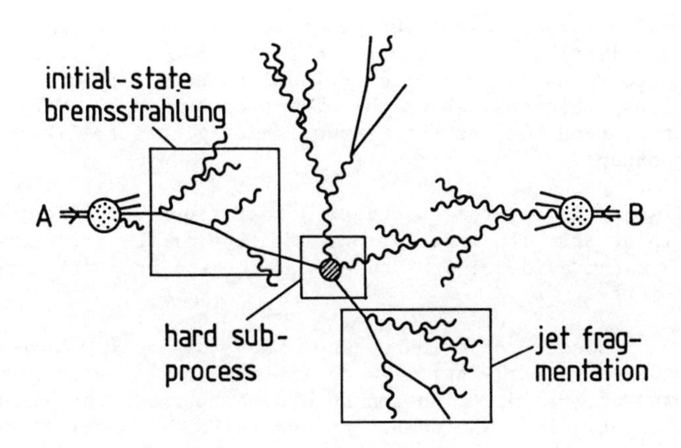

Fig.1
Hard parton scattering in a collision between hadrons A and B. In this case the subprocess is quark-gluon elastic scattering leading to the emission of a quark jet and a gluon jet.

(i) Initial-state bremsstrahlung. The partons (quarks, q, or gluons, g) participating in the hard scattering are coloured objects and they therefore radiate gluons, mainly close to the colliding beam axis, before colliding.
(ii) Hard subprocess. In the parton model (QCD in the limit of negligible strong coupling constant, α_s) this would be the whole story. In addition to elastic parton-parton scattering, inelastic strong processes such as $gg \to q\bar{q}$ and $qq \to qqg$ are possible, as well as electroweak scattering like $q\bar{q} \to$ lepton pair via $W^{\pm}$, Z^{o} or virtual photon formation.
(iii) Jet fragmentation. If the hard subprocess gives rise to outgoing partons they will radiate gluons and initiate parton cascades, giving rise to jets of hadrons at large angles to the beam axis, along the directions of the outgoing partons. The invariant mass M_{jj} of the jet system thus measures the centre-of-mass energy of the hard subprocess.

Notice that Fig.1 is incomplete in one important respect: in the real world, the outgoing partons are converted non-perturbatively into hadrons. Therefore, for some applications we need in addition to perturbative QCD a model for this 'hadronization' process. I shall try to make clear which predictions should or should not be particularly sensitive to the hadronization model adopted.

In the remainder of this talk I shall concentrate on each of the above three aspects of jet physics in turn, noting some features of the QCD predictions and the extent to which they are confirmed by experiment. I shall then discuss hadronization models and give some results of one such model,ending with a brief summary of conclusions.

2. Initial-State Bremsstrahlung

The gluons radiated by partons before they collide will carry off some transverse momentum, leading to a net transverse momentum for the hard subprocess and hence for the outgoing jets or lepton pair. The distribution [1] of this transverse momentum is shown in Fig. 2 for jet production, together with QCD predictions based on the calculations of Greco [2].

Since objects with larger colour charges emit more gluon bremsstrahlung,the width of the k_T distribution is sensitive to the identity of the colliding partons, the colour charge being proportional to $C_F = 4/3$ for quarks and to $C_A = 3$ for gluons. Figure 3 shows the effect of setting $C_A = C_F = 4/3$: the predicted distribution becomes too narrow, showing that collisions involving gluons are important.

As we shall discuss later, the observation that gluon collisions are important is in accord with expectations from the proton structure function and with independent evidence from the fragmentation properties of the SppS collider jets.

In the case of $W^{\pm}$ boson production we know that (neglecting higher-order QCD corrections) only quark-antiquark collisions should be involved, so the transverse momentum (Q_t) distribution should be narrower than that of strongly-produced dijets. The number of events [6,7] is still rather too low to check this point (see Fig.4), but the agreement with QCD looks satisfactory.

Outside the small-Q_t region, the transverse momentum distributions of weak bosons and dijets should be insensitive to the hadronization model, which will only involve momentum transfers of the order of those in soft hadronic

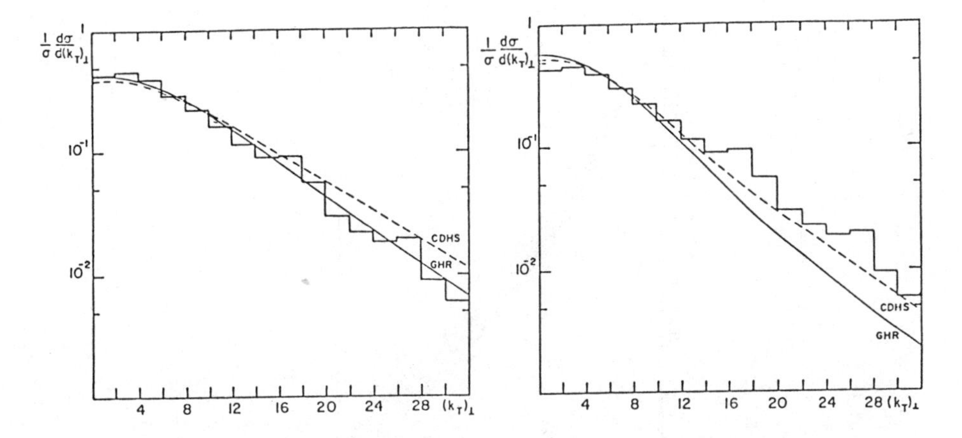

Fig.2
Preliminary UA1 data[1] on the distribution of $(k_T)_\perp$, the component of the two-jet transverse momentum perpendicular to the axis of the principal jet. The dashed and solid curves are QCD predictions[2] based on the proton structure function of Refs. [3] and [4] respectively.

Fig.3
Data as in Fig.2, predictions changed by setting the gluon colour charge equal to that of the quarks.

interactions. Ref. 5 contains a detailed study of non-perturbative effects, including those associated with the structure of the incoming hadrons, deduced from the data on lepton pair production at lower energies. The conclusion is that these effects are not very important above $Q_t \sim 4$ GeV/c. We see from Fig. 4 that this is also the point at which predictions become insensitive to acceptable variations in the hadron structure functions.

3. Hard Subprocess

As illustrated in Fig.5, QCD calculations based on two-parton hard scattering give a good semi-quantitative account of the data on jet production cross sections at the Sp$\bar{p}$S collider. The uncertainties in the predictions, amounting to a factor of 2 or more, come from our lack of knowledge of the QCD scale parameter Λ ($\simeq$ 0.1 - 0.5 GeV), of the proton structure function, and of the best

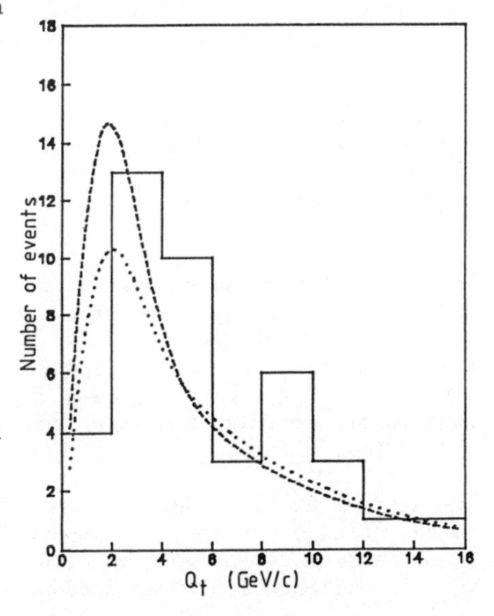

Fig.4

Transverse momentum distribution of $W^\pm$ bosons. The UA1 data [6] are compared with the QCD predictions of Davies et al. [5] for the two sets of structure functions in Ref. [8].

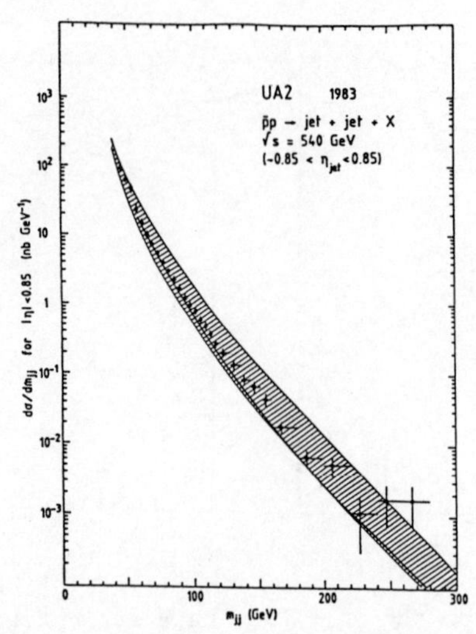

Fig.5

Two jet production cross section as a function of dijet mass. The UA2 data [9] are compared with QCD predictions of Humpert [10].

scale Q^2 at which to evaluate the running coupling constant $\alpha_s(Q^2)$ in hard scattering amplitudes. One might, for example, choose $Q^2 = M_{jj}^2$, the dijet mass squared, or $4p_T^2$, the squared sum of the jet transverse momenta.

Such ambiguities can only be resolved by computing higher-order QCD corrections to all the relevant amplitudes, a very large task which has so far been attempted only for the simplest subprocess, namely the scattering of a pair of quarks of different flavours [11,12]. In this case, the results of Ref. 12 suggest that for the single-jet inclusive cross section the best choice of scale is $Q^2 = M_{jj}^2$.

The dominant hard subprocesses are the parton-parton elastic collisions, $q\bar{q} \to q\bar{q}$, $qg \to qg$, $gg \to gg$, etc. These subprocesses have differential cross-sections $d\hat{\sigma}$ which are of similar shape, since they are dominated by single gluon exchange, which gives the Rutherford formula $d\hat{\sigma}/d\cos\theta^* \propto (\sin\theta^*/2)^{-4}$ where θ^* is the c.m. scattering angle. The Sp$\bar{p}$S data [13,14] confirm this functional form rather well at $\cos\theta^* \gtrsim 0.3$ (Fig.6), giving $(\sin\theta^*/2)^n$ where $n = -3.94 \pm 0.18$.

The approximately universal form of the parton-parton cross sections leads to the 'single effective subprocess' (SES) approximation [15], in which the cross section for production of a pair of jets with dijet invariant mass M_{jj}, in collisions between hadrons A and B at c.m. energy $\sqrt{s}$, is given by

$$\frac{d\sigma}{dM_{jj}^2 \, \cos\theta^*} = \int dx_1 dx_2 \; F_A(x_1) F_B(x_2) \delta(M_{jj}^2 - x_1 x_2 s) \; \frac{d\hat{\sigma}}{d\cos\theta^*} \quad , \qquad (1)$$

where x_1, x_2 are the momentum fractions of the colliding partons in the incoming hadrons. The functions $F_{A,B}(x)$ are effective structure functions specifying the parton densities in the colliding hadrons, weighted by their scattering probabilities. Owing to the different colour charges for quarks and gluons we find, in an obvious notation,

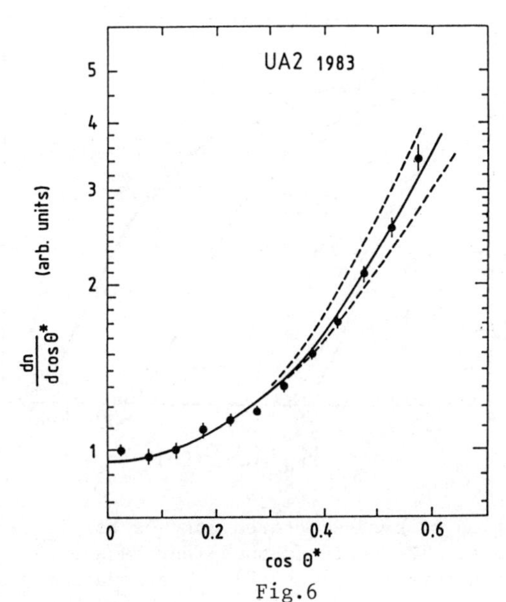

Fig.6
UA2 data [14] on hard subprocess angular distribution. The solid curve shows the expected average over subprocesses; all the important subprocesses lie between the dashed curves.

$$F_i(x) = g_i(x) + \frac{C_F}{C_A}\left[q_i(x) + \bar{q}_i(x)\right] \qquad (i = A \text{ or } B) \tag{2}$$

where $C_F/C_A = 4/9$.

The parton densities in Eq.(2) actually depend also on the momentum scale of the hard subprocess. This scaling violation follows from the existence of initial-state bremsstrahlung. The harder the subprocess, the more gluon radiation will be emitted, shifting the x distribution of the colliding partons to lower values. Taking this into account, the effective proton structure functions obtained [13,14] from Eq.(1) agree with those deduced via Eq.(2) from deep inelastic lepton-nucleon scattering at lower momentum transfers [3].

Since all hard parton scattering is elastic in the SES approximation, Eq.(2) tells us that the proportion of gluons among the scattered partons should be

$$f_g = \frac{1}{2}\left[\frac{g_A(x_1)}{F_A(x_1)} + \frac{g_B(x_2)}{F_B(x_2)}\right]. \tag{3}$$

This is shown in Fig.7 at $x_1 \simeq x_2$ for $\bar{p}p$ collisions at various values of c.m. energy $\sqrt{s}$ and dijet mass M_{jj}. We see that gluons predominate up to $M_{jj} \sim 100$ GeV at the Sp$\bar{p}$S collider, whereas at the ISR few gluons are scattered at $M_{jj} \gtrsim 30$ GeV, the lowest value at which jet production can be readily identified [16].

The gluon jet fractions in Fig.7 may be subject to errors of the order of ±0.1 arising from the SES approximation. Taking into account the uncertainties about higher-order QCD corrections mentioned at the beginning of this Section, and those in the gluonic structure function of the proton,

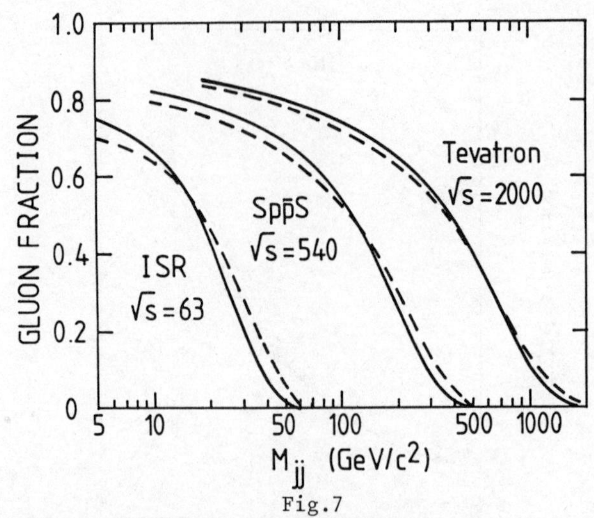

Fig.7

Gluon fraction of hard-scattered partons for $x_1 \simeq x_2$ in pp or $\bar{p}p$ collisions. The solid (dashed) curves are for Duke and Owens' set 1(2) structure functions [8], at a momentum scale $Q^2 = M_{jj}^2$.

a more careful treatment going beyond SES is unlikely to be warranted for some time.

4. Jet Fragmentation

4.1 Parton cascade

We saw in Fig.1 that following hard parton scattering we expect further gluonic bremsstrahlung from the outgoing struck partons, with radiated gluons themselves emitting further gluons or splitting into quark-antiquark pairs. This leads to a "parton cascade" picture [17,18] of the earlier stages of jet development, as illustrated again in Fig.8, this time for the hard scattering process $e^+e^- \rightarrow$ hadrons, which has the advantage of no contamination from initial-state QCD bremsstrahlung.

The parton cascade may be viewed in the following way. The outgoing partons from a hard scattering process may initially be off mass shell by amounts up to the hard process scale, Q^2. They reduce their virtuality (move closer to the mass shell) by successive parton branching. The branching can be followed perturbatively down to some scale Q_o^2 such that $\alpha_s(Q_o^2)/\pi$ is still small enough for perturbation theory to make sense, i.e., such that $Q_o^2 \gg \Lambda^2$, where Λ is the intrinsic QCD scale. If $Q^2 \gg Q_o^2$ we still have a substantial region in which perturbative predictions can be made.

The jet structure of the final state is already manifest at this stage, owing to the collinear singularities of QCD. We shall see that many jet properties are asymptotically determined in this region. Other properties depend on later developments, at scales below Q_o^2, where perturbation theory is no longer applicable. I shall discuss alternative models for this "hadronization" region, and give some results of a Monte Carlo jet simulation program which combines a careful treatment of the parton cascade [19] with a simple hadronization model [20].

An important property of the QCD cascade is the colour coherence of parton jets [21-23]. This suppresses the emission of relatively soft, wider-angle gluons, which are unable to resolve the colour sources in the jet. As a result, the successive opening angles in the parton cascade have to be uniformly decreasing (Fig.9). The effect of this on the distribution of gluon energy fractions is illustrated in Fig.10. An incoherent mechanism would give a spectrum peaking near the lower kinematic limit $x \sim Q_o/Q$. Coherence produces negative interference terms which grow rapidly with decreasing x, suppressing low-x emission. The resulting spectrum is roughly symmetrical in $\ell n x$, peaking at $x \sim \sqrt{Q_o/Q}$.

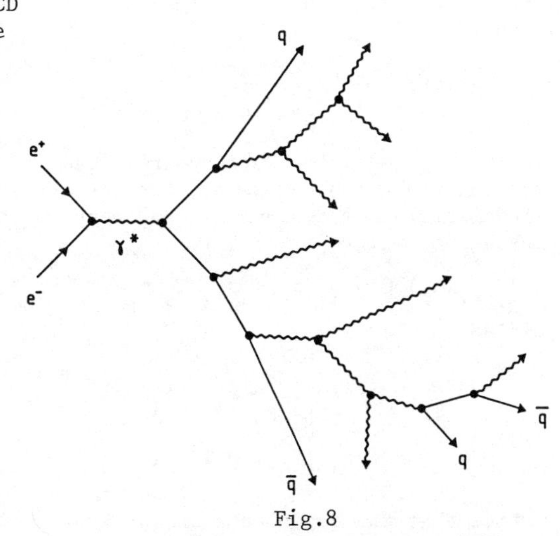

Fig.8

QCD parton cascade following e^+e^- annihilation.

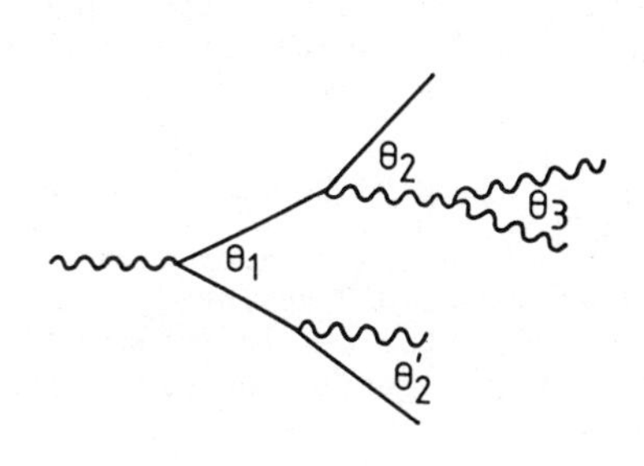

Fig.9

Angular ordering in the cascade: $\theta_1 > \theta_2 > \theta_3 \ldots$, $\theta_1 > \theta_2' \ldots$, etc.

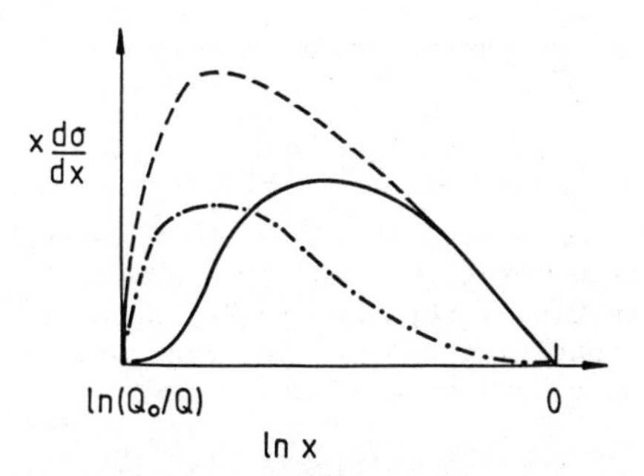

Fig.10

Gluon energy fraction distribution. Dashed = incoherent approximation; dot-dashed = minus interference terms; solid = sum = coherent distribution.

4.2 Average multiplicities in quark jets

If $D_{i \to h}(x,Q^2)$ is the fragmentation function for a jet of type i (i = q or g) fragmenting into a hadron of type h, then the moments of this function have the general asymptotic behaviour [24]

$$\int_0^1 dx\, x^{n-1}\, D_{i \to h}(x,Q^2) \sim \exp \int_{Q_o^2}^{Q^2} \frac{dq^2}{q^2}\, \gamma_n[\alpha_s(q^2)] \tag{4}$$

where γ_n is the appropriate anomalous dimension. For n>1 (more precisely, for $n-1 >> \sqrt{\alpha_s}$), γ_n has a Taylor expansion in α_s. Substituting this in (4) with

$$\alpha_s(q^2) = \frac{4\pi}{\beta_o \, \ell n(q^2/\Lambda^2)} \tag{5}$$

($\beta_o = 11 - 2N_f/3$ for N_f flavours) leads to the familiar formulae for scaling violation. However, for the average multiplicity (n = 1) infrared singularities of the anomalous dimension lead to an expansion in $\sqrt{\alpha_s}$ instead of α_s:

$$\gamma_1(\alpha_s) = a_{11}\sqrt{\alpha_s} + a_{12}\,\alpha_s + O(\alpha_s^{3/2}) \ , \tag{6}$$

giving

$$\langle n(Q^2)\rangle \propto \exp\left\{\frac{8\pi}{\beta_o} a_{11}\,[\alpha_s(Q^2)]^{-1/2}\right\}.[\alpha_s(Q^2)]^{-4\pi a_{12}/\beta_o}$$

$$.\ \{ 1 + O\,(\sqrt{\alpha_s}) \} \tag{7}$$

Notice that the dependence on the type of jet i, the type of hadron h and the cutoff scale Q_o have all factorized out in Eq.(7), leaving a universal asymptotic Q^2-dependence determined by the coefficients [21,25,26]

$$a_{11} = \sqrt{\frac{3}{2\pi}}\ ,\ a_{12} = -\frac{11}{16\pi}\,(1 + \frac{2}{27}N_f). \tag{8}$$

As shown in Fig.11 [27], Eqs.(7,8) are in excellent agreement with recent data on charged multiplicities in quark jets from e^+e^- annihilations at c.m. energies of 12 GeV and above [28,29]. The fitted value of Λ is

$$\Lambda_{MULT} = 80^{+40}_{-30}\ \text{MeV}\ , \tag{9}$$

the subscript 'MULT' being a reminder that this is a process-dependent parameter which cannot be compared with the universal QCD scale $\Lambda_{\overline{MS}}$ until the $O(\sqrt{\alpha_s})$ term in Eq. (7) has been computed. It turns out [27] that the recently-calculated coefficient a_{12} is important for good agreement between theory and experiment.

4.3 Average multiplicities in gluon jets

Equation (7) is equally valid for quark and gluon jets, but the normalization constant and $O(\sqrt{\alpha_s})$ correction will be different in the two cases. It turns out that one can compute these quantities for the ratio of average multiplicities without needing to know them for quark and gluon jets separately [26,30]:

$$\frac{\langle n\rangle_g}{\langle n\rangle_q} \sim \frac{9}{4}\left[1 - (1 + \frac{N_f}{27})\sqrt{\frac{\alpha_s}{6\pi}}\ \right]\ . \tag{10}$$

After fitting Eq. (7) to quark jets we can thus predict <n> for gluon jets, as shown in Fig.12.

To compare with data [31] on the CERN $\bar{p}p$ collider jets, we need to know the relative abundance of quark and gluon jets as a function of the dijet invariant mass, M_{jj}. This can be estimated from Fig.7. The slight discrepancy with the UA2 data is probably not significant(see Sect.3 and 6.2).

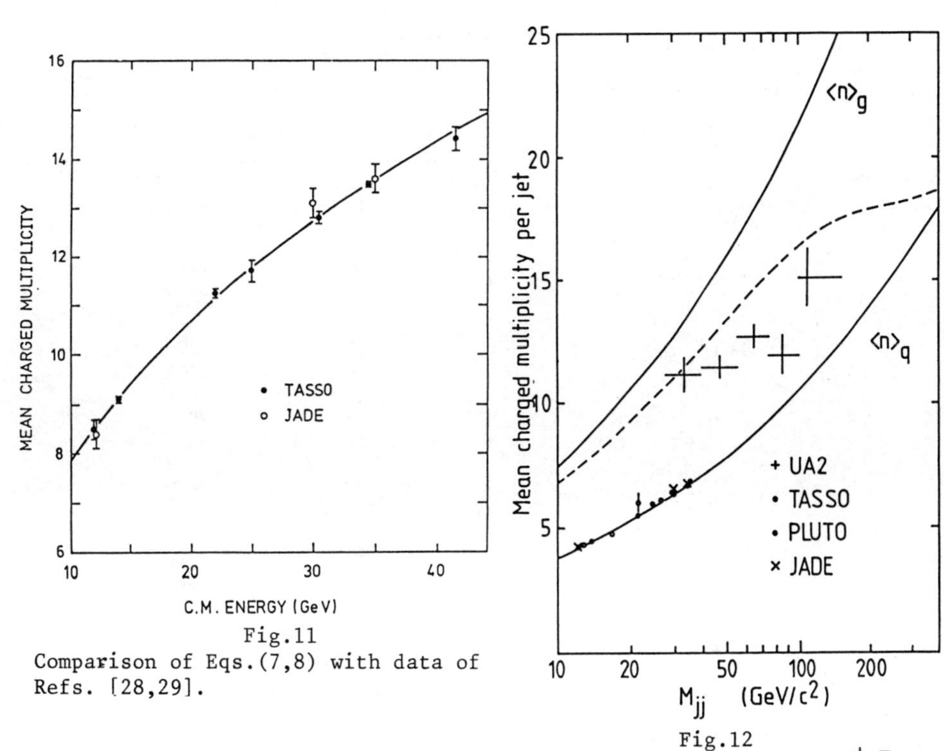

Fig.11
Comparison of Eqs.(7,8) with data of Refs. [28,29].

Fig.12
$\langle n\rangle_q$ = fit of Eq.(7,8) to e^+e^- data; $\langle n\rangle_g$ = resulting prediction of Eq. (10). Dashed curve = expected average multiplicity for $\bar{p}p$ collider jets, to be compared with UA2 data (Ref.[31].

4.4 Higher multiplicity moments

Expressions of the form (10) can also be derived for higher moments of the multiplicity distribution [30]. Results for some quantities of interest are summarized in the Table. These quantities should be asymptotically independent of the type of hadron under consideration and therefore the results for e^+e^- annihilation may be compared directly with data on the charged multiplicity distribution in this process. Since the predictions depend only on α_s [$\equiv \alpha_s(Q^2)$], one may solve for the value of α_s required for agreement between theory (to $O(\sqrt{\alpha_s})$) and experiment. The results are summarized in Fig.13: the values of α_s obtained are reasonable (between 0.1 and 0.2), corresponding to a leading-order value of Λ = 0.2-0.4 GeV. However, the corresponding $O(\sqrt{\alpha_s})$ corrections are so large (of the order of 60%) that further corrections, of order α_s and higher, are unlikely to be negligible. At least we may say, though, that the next-to-leading QCD corrections are of the desired sign and order of magnitude.

TABLE : Multiplicity Moments

$$R = A\left[1 + (B + CN_f)\sqrt{\frac{\alpha_s}{6\pi}}\right]$$

R	A	B	C
$\langle n\rangle_g/\langle n\rangle_q$	$\frac{9}{4}$	-1	$-\frac{1}{27}$
$\langle n(n-1)\rangle_q/\langle n\rangle_q^2$	$\frac{7}{4}$	$-\frac{55}{14}$	$\frac{20}{567}$
$\langle n(n-1)\rangle_g/\langle n\rangle_g^2$	$\frac{4}{3}$	$-\frac{55}{24}$	$\frac{35}{972}$
$\langle n(n-1)\rangle_{e^+e^-}/\langle n\rangle_{e^+e^-}^2$	$\frac{11}{8}$	$-\frac{5}{2}$	$\frac{20}{891}$
$\langle n(n-1)(n-2)\rangle_q/\langle n\rangle_q^3$	$\frac{289}{64}$	$-\frac{45771}{4624}$	$\frac{14327}{187272}$
$\langle n(n-1)(n-2)\rangle_g/\langle n\rangle_g^3$	$\frac{9}{4}$	$-\frac{99}{16}$	$\frac{4703}{52488}$
$\langle n(n-1)(n-2)\rangle_{e^+e^-}/\langle n\rangle_{e^+e^-}^3$	$\frac{625}{256}$	$-\frac{66891}{10000}$	$\frac{22007}{405000}$

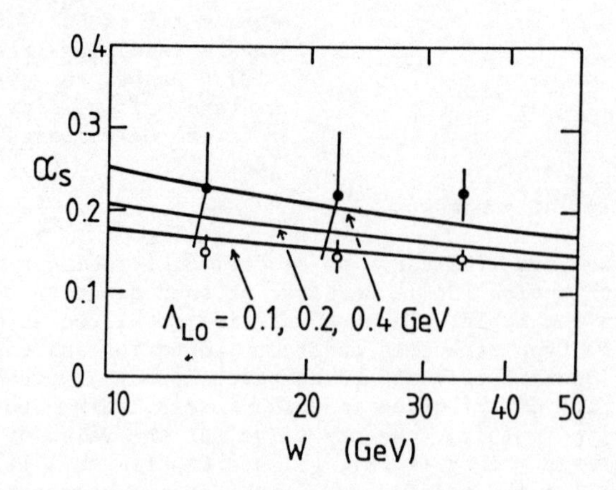

Fig.13
Values of α_s deduced from e^+e^- charged multiplicity moments [30]. Data from Ref.[29]. Solid circles are values obtained from $\langle n(n-1)\rangle/\langle n\rangle^2$; open circles are those from $\langle n(n-1)(n-2)\rangle/\langle n\rangle^3$. Curves are the leading-order expression for $\alpha_s(Q^2 = W^2)$ at the indicated values of the leading-order Λ parameter.

5. Hadronization Models

None of the predictions presented so far depends (asymptotically) on the model adopted for converting partons at scale Q_o^2 into hadrons, as long as this process only leads to momentum transfers of order Q_o or less. To make more detailed predictions, and to estimate sub-asymptotic effects, we need to supplement perturbative QCD with a hadronization model.

A type of model that is becoming increasingly popular is based on the 'pre-confinement' property of the QCD parton cascade: the outgoing partons from the cascade tend to be naturally arranged in colour singlet clusters of limited mass and size [32,33]. This may be illustrated using the Monte Carlo program described in Refs. [19,20], which generates parton cascades with proper treatment of the coherence effects discussed in Sect.4.1. The clusters were actually formed as illustrated in Fig.14, by splitting apart the colour and anticolour indices of outgoing gluons and associating each with the corresponding anticolour/colour index [34,35]. The splitting was assumed to occur via a non-perturbative $g \to q\bar{q}$ transition, isotropic in the virtual gluon rest-frame. The cutoff Q_o was 700 MeV with Λ = 300 MeV (the values used later for fitting data), so that $\alpha_s(Q_o^2)/\pi \simeq 0.3$.

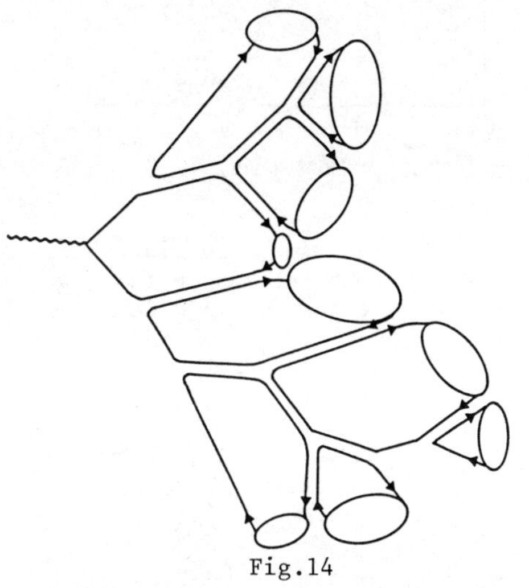

Fig.14

Colour structure of QCD cascade in Fig.8. The lines represent colour indices and the blobs colour singlet combinations.

The results, illustrated in Fig.15, show clearly that the preconfined colour singlets are universal objects, independent of the type of jets and the value of Q^2. On the average they are smaller, both in mass and size, than random combinations of final-state quarks and antiquarks. (The 'size' was defined as the minimum separation of the quark and antiquark in their c.m. frame, after associating a space-time interval q^μ/q^2 with each internal line of momentum q^μ in the cascade.)

Two plausible hadronization models that make use of the preconfinement property are illustrated in Fig.16. In the first [20,35], the universal colour singlet clusters discussed above provide the basis of hadronization. They are assumed to decay like heavy, broad meson resonances,with isotropic, quasi-two-body decay modes chosen according to density of states from meson-meson and baryon-antibaryon combinations with the correct quantum numbers. In Ref.[20] the allowed decay products are $J^P = 0^-, 1^\pm, 2^+$ mesons and 1/2+, 3/2+ baryons, which decay in turn to stable hadrons as indicated in Fig.16a.

In Fig.16b, it is assumed that semiclassical colour strings [36] are formed instead of clusters at scale Q_o. The colour structure of Fig.14 is still used, to determine how the strings connect the partons, but the non-perturbative splitting of gluons is no longer necessary. Instead, the

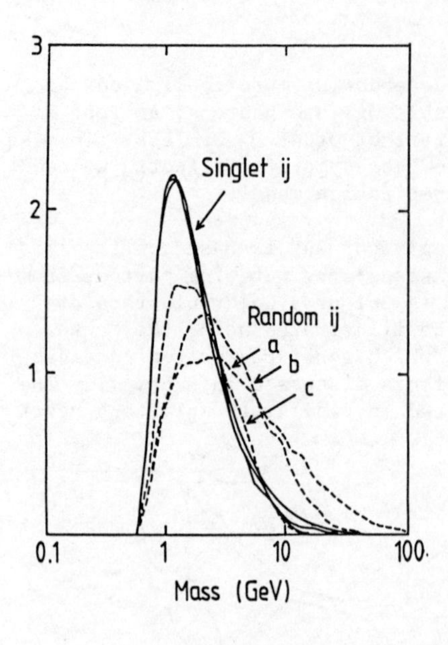

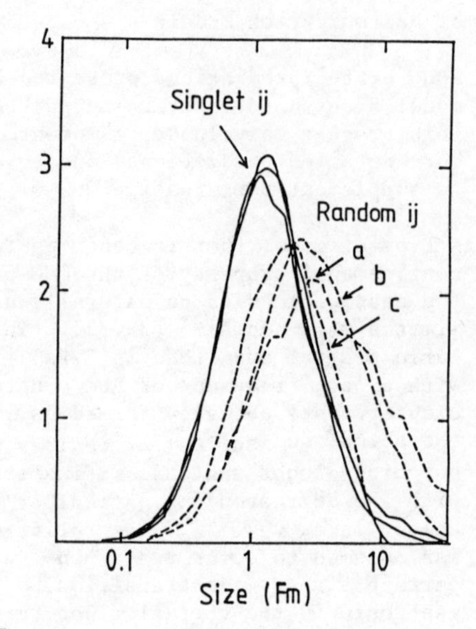

Fig.15

Mass and size distributions (arbitrary units) for $q_i\bar{q}_j$ combinations in (a) quark jets at Q = 100 GeV; (b) gluon jets at Q = 100 GeV; (c) gluon jets at Q = 20 GeV.

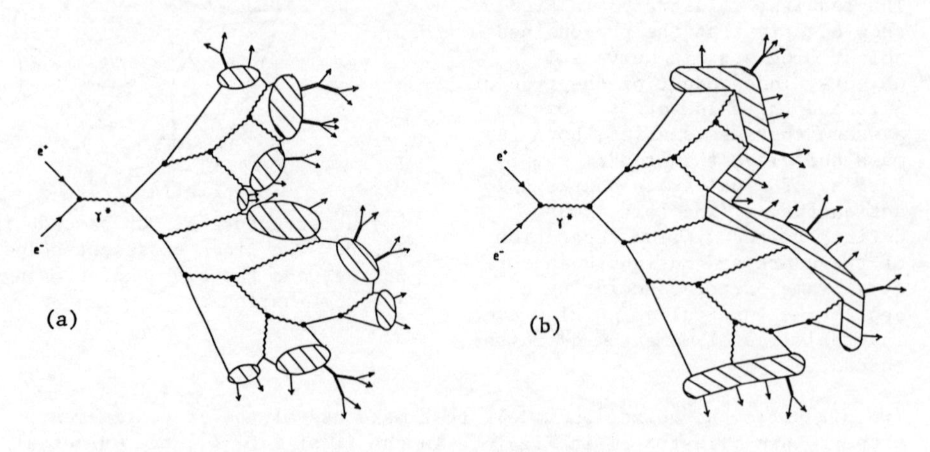

Fig.16

Models combining the QCD cascade with a mechanism for hadron formation (shaded): (a) cascade plus clusters; (b) cascade plus Lund strings.

strings split directly into hadrons ($J^P = 0^-$, 1^- mesons and 1/2+, 3/2+ baryons in the Lund string model [36]).

Given our present ignorance about non-perturbative QCD, both of these hadronization models look equally sensible. Unlike schemes based on the independent fragmentation of partons [37], they conserve energy-momentum and respect the colour structure of the QCD cascade. The 'cascade + strings' model has some additional parameters relating to the properties and splitting of the strings, but in practice the 'cascade + clusters' model also has to invoke a stringlike decay scheme for the small fraction of clusters in the high-mass tail of the distribution of Fig.15, for which the approximation of isotropic decay is unreasonable.

In the next section some results of the 'cascade + clusters' model of Ref. [20] will be discussed. Unfortunately, shortage of space prevents the presentation of results from a variety of models. Some comparative discussion may be found in Ref.[38]. The 'cascade + Lund strings' model is at present under active investigation [39]; the results so far look rather similar to those presented below. The Lund string model as conventionally formulated [36], without the QCD cascade, incorporates exact QCD perturbation theory for configurations of up to four partons, which is equivalent to including the cascade as long as Q^2 is not too large. Cascades giving more than four partons become important for $Q \gtrsim 40$ GeV, and the model with cascading then gives jets with larger angular widths and multiplicities.

6. Results from a 'Cascade + Clusters' Model

6.1 e^+e^- annihilation

The results shown in Figs. 17-21 are from a slightly updated version of the model described in Ref. [20]. (The new version is obtainable from me as a Monte Carlo program in Fortran 77, suitable for any computer installation that has the CERN program library). The parameter values used here are

$$Q_o = 700 \text{ MeV}, \quad \Lambda = 300 \text{ MeV}, \; M_f = 3.5 \text{ GeV}, \tag{11}$$

where M_f is the cluster mass "fission threshold" above which the program adopts an anisotropic, stringlike model of cluster decay. (As may be seen from Fig.15 about 1/6 of the clusters have masses above this threshold). The values (11) give a slightly higher charged multiplicity than those used in Ref. [20], in better agreement with the new data (Fig.17). On the other hand, they give hadron distributions that are slightly too soft at large x and p_T (Figs 20,21). However, given the shortage of adjustable parameters, the overall agreement with the data is quite striking.

6.2 $\bar{p}p$ collider jets

With a Monte Carlo approach we can easily impose on the model predictions the rather complicated cuts imposed to isolate the high-Q^2 jets in hadron collisions. Some results of this type are shown in Figs. 22-24. The 'jet core multiplicity' defined by the UA2 collaboration [42] eliminates about half the true total multiplicity of the jet, mostly at large angles to the jet axis, affecting gluon jets more severely than quark jets (c.f. Figs 12, 22). The data show the expected transition from gluon jet dominance at low dijet mass to quark jet dominance at higher masses, although the transition is at rather lower masses than expected on the basis of the Duke and Owens [8] structure functions (dashed curve).

For the comparisons with the UA1 data [43] on jet fragmentation at transverse energies above 30 GeV, a simple mixture of 40% quark jets and 60% gluon jets has been assumed (solid curves in Figs 23,24), which is the kind of ratio expected at dijet masses of 60-100 GeV/c^2 (see Fig.7). The charged-particle z distribution in Fig.23 (z = momentum component along jet axis divided by jet energy) clearly shows the dominance of gluon jet

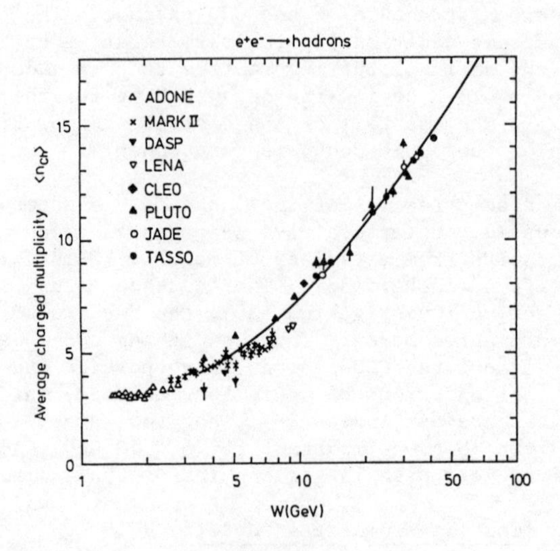

Fig.17
Model prediction of mean charged multiplicity in e^+e^- annihilation as a function of c.m. energy W, compared with data compiled in Ref. [29].

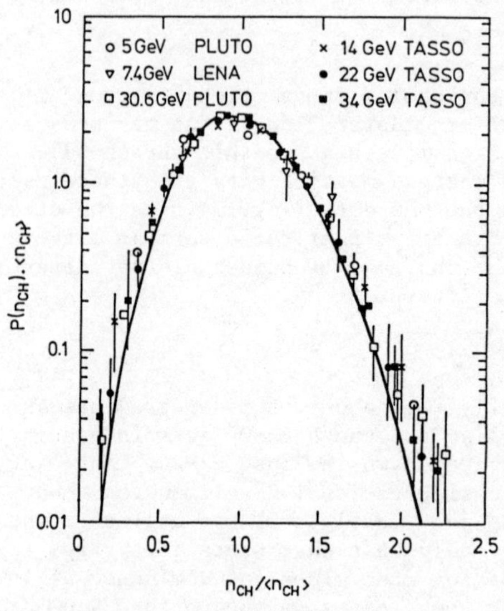

Fig.18
Multiplicity distribution in e^+e^- annihilation, plotted in 'KNO scaling' form. Data compiled in Ref.[29]; curve = model prediction at 14 GeV.

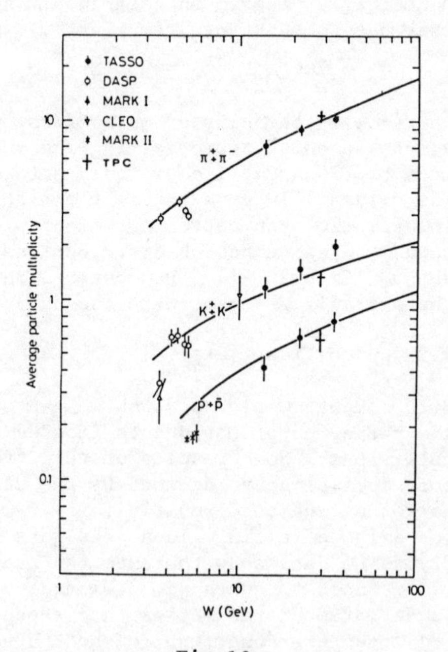

Fig.19
Predicted yields of different types of charged particles. Data from Refs. [40, 41].

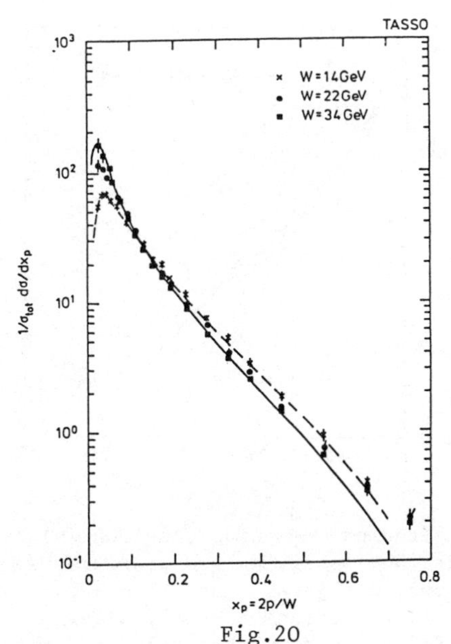

Fig.20
Predicted charged particle momentum fraction (x_p) distributions at 14GeV (dashed curve) and 34 GeV (solid curve); data from Ref. [29].

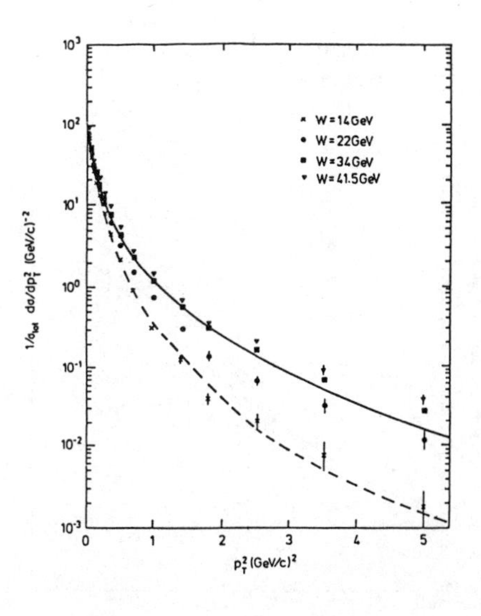

Fig.21
Predicted distributions of transverse momentum squared (relative to sphericity axis) at 14 GeV (dashed) and 34 GeV (solid); data from Ref. [29].

fragmentation at low z and of the harder quark jet fragmentation at higher z. In the case of the momentum distribution transverse to the jet axis (Fig.24), there is very little difference between the predictions for quark and gluon jets after the UA1 cuts, but again the canonical mixture agrees best with the data.

7. Conclusions

We have seen that the QCD appears to give a satisfactory semi-quantitative account of the main features of jet production in hadron collisions, provided of course we set aside the totally new phenomena that may be appearing in 'exotic' jet events. By a 'semi-quantitative' account I mean one within the margin of uncertainty, often as great as a factor of two, associated with poorly-known higher-order QCD corrections and hadron structure functions.

Unfortunately there is little immediate prospect of dramatic progress in the handling of higher-order corrections. Even if the next-to-leading corrections for all the various hard parton scattering processes were calculated, then those which turn out to be large will not be of great quantitative help, because they will simply tell us that the convergence of the perturbative expansion is poor; its sum will remain uncertain.

We can expect better progress on structure functions, especially in our understanding of the gluon distributions within hadrons. Up to now, these have been poorly known because our main tool for probing hadron structure

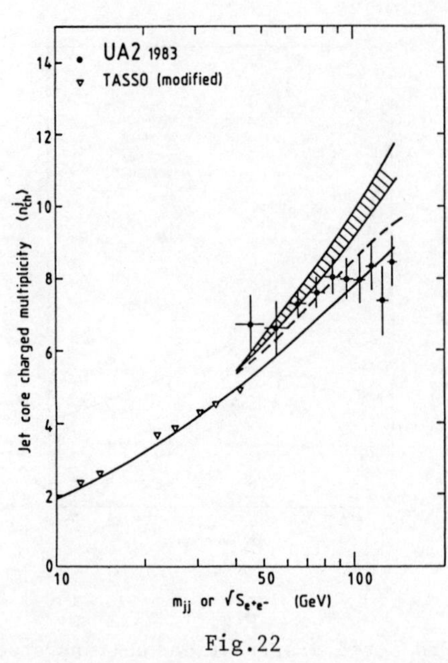

Fig.22

Jet core multiplicities, as defined by the UA2 Collaboration (Ref.[42]). Solid curve = prediction [20] for quark jets; shaded area = gluon jet predictions for a reasonable range of model parameters; dashed curve = prediction for expected mixture of quark and gluon jets.

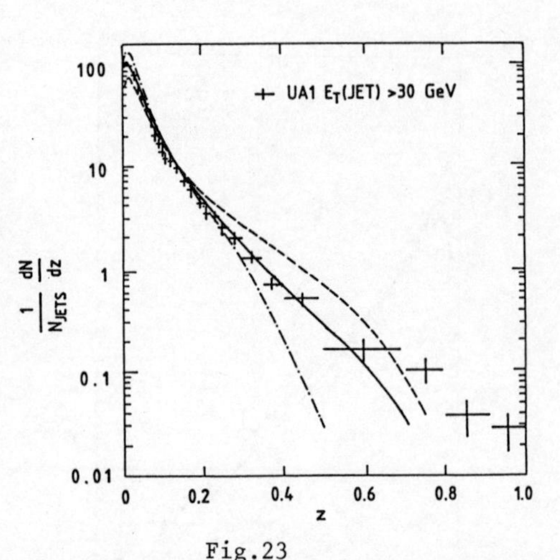

Fig.23

Longitudinal momentum fraction (z) distribution of charged particles in $\bar{p}p$ collider jets. Data from Ref.[43]. Model predictions are as follows: dashed curve = quark jets; dot-dashed = gluon jets; solid = 40% quarks + 60% gluons.

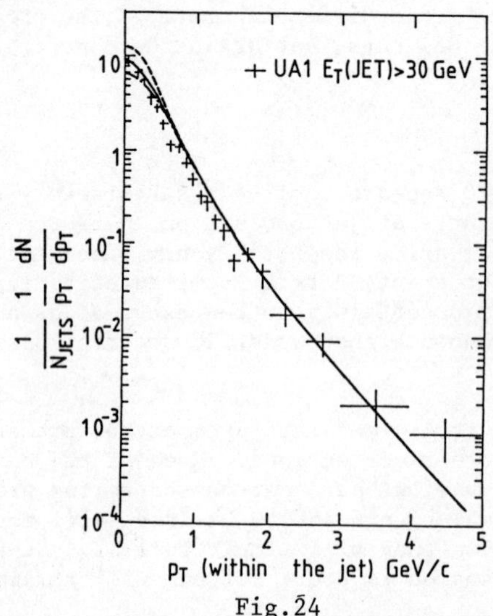

Fig.24

Transverse momentum distribution of charged particles in $\bar{p}p$ collider jets. Data and predictions as in Fig. 23 .

has been deep inelastic lepton scattering, which is only indirectly sensitive to gluons. With the steady accumulation of data on jet production (and also on heavy flavour production via gluon fusion), which measures the gluon distribution more directly, there should be real clarification. This is an important point because we have seen (Fig.7) that gluon scattering becomes more and more dominant as we go to higher energies. In the future generations of hadron colliders, gluons will constitute our main probes of the short-distance structure of matter.

Turning to the question of jet fragmentation, we saw that QCD is consistent with the hadron multiplicity data, though again there are uncertainties about higher-order corrections. Supplementing the theory with a model for hadronization, one can account for the fragmentation properties of quark and gluon jets in some detail. This means that one can now extrapolate with some confidence to predict the broad features of jets at much higher energies [39]. The question of which hadronization model is more correct (e.g. 'clusters' or 'strings') will probably depend on rather fine details such as multiparticle correlations.

One interesting subject that I have not discussed owing to limitations of space is the global structure of hard scattering events. How do the fragments of jets and initial-state bremsstrahlung merge together to give the observed event topologies? There have been a number of recent advances in understanding of such questions [44], but further theoretical work remains to be done, e.g. on QCD coherence effects when both timelike and spacelike virtual partons are present.

Acknowledgements

It is a great pleasure to acknowledge the hospitality of the CERN Theory Division during a large fraction of the past two years, and to thank many colleagues there and elsewhere (especially G. Ingelman, G. Marchesini, A. Petersen and A. Weidberg) for valuable discussions on the topic of this talk.

References

[1] M. Della Negra and W. Scott, private communication to M. Greco, cited in Ref. [2].
[2] M. Greco, in Proc. 4th Topical Workshop on $\bar{p}p$ Collider Physics, Berne, 1984.
[3] CDHS Collaboration, H. Abramowicz et al., Z. Phys. C12 (1982) 289; C13 (1982) 199; C17 (1983) 283.
[4] M. Glück, E. Hoffman and E. Reya, Z. Phys. C13 (1982) 119.
[5] C.T.H. Davies, W.J. Stirling and B.R. Webber, CERN preprint TH.3987 (1984); see also G. Altarelli et al., CERN preprint TH.3851 (1984).
[6] UA1 Collaboration, G. Arnison et al., Phys. Lett, 122B (1983) 103; 126B (1983) 398.
[7] UA2 Collaboration, G. Banner et al., Phys. Lett. 122B (1983) 476; 129B (1983) 130.
[8] D.W. Duke and J.F. Owens, Phys. Rev. D30 (1984) 49.
[9] UA2 Collaboration, P. Bagnaia et al., Phys. Lett. 138B (1984) 430.
[10] B. Humpert, private communication to UA2 Collaboration, cited in Ref. 9 .
[11] R. K. Ellis et al., Nucl. Phys. B173 (1980) 397.
[12] M. Furman, Nucl. Phys. B197 (1982) 413.
[13] UA1 Collaboration, G. Arnison et al., Phys. Lett. 136B (1984) 294.

[14] UA2 Collaboration, P. Bagnaia et al., Phys. Lett. 144B (1984) 283.
[15] F. Halzen and P. Hoyer, Phys. Lett. 130B (1983) 326;
B.L. Combridge and C.J. Maxwell, Nucl. Phys. B239 (1984) 429.
[16] CMOR Collaboration, A.L.S. Angelis et al., CERN preprint EP/84-46 (1984); AFS Collaboration, T. Åkesson et al., CERN preprint EP/84-56 (1984).
[17] K. Konishi, A. Ukawa and G. Veneziano, Nucl. Phys. B157 (1979) 45.
[18] A. Bassetto, M. Ciafaloni and G. Marchesini, Nucl. Phys. B163 (1980) 477.
[19] G. Marchesini and B.R. Webber, Nucl. Phys. B238 (1984) 1.
[20] B.R. Webber, Nucl. Phys. B238 (1984) 492.
[21] A.H. Mueller, Phys. Lett. 104B (1981) 161; B.I. Ermolaev and V.S. Fadin, JETP Lett. 33 (1981) 269; Yu. L. Dokshitzer, V.S. Fadin and V.A. Khoze, Phys. Lett. 115B (1982) 242; Z. Phys. C15 (1982) 325; Z. Phys. C18 (1983) 37; A. Bassetto, M. Ciafaloni, G. Marchesini and A.H. Mueller, Nucl. Phys. B207 (1982) 189.
[22] A. Bassetto, M. Ciafaloni and G. Marchesini, Phys. Rep. 100 (1983)201.
[23] L.V. Gribov, E.M. Levin and M.G. Ryskin, Phys. Rep. 100 (1983) 1.
[24] G. Altarelli, Phys. Rep. 81 (1982) 1.
[25] A.H. Mueller, Nucl. Phys. B213 (1983) 85; B228 (1983) 351.
[26] A.H. Mueller, Nucl. Phys. B241 (1984) 141.
[27] B.R. Webber, Phys. Lett. 143B (1984) 501.
[28] JADE Collaboration, W. Bartel et al., Z. Phys. C20 (1983) 187.
[29] TASSO Collaboration, M. Althoff et al., DESY 83-130 (Dec. 1983).
[30] E.D. Malaza and B.R. Webber, Cambridge preprint HEP 84/3 (July 1983).
[31] UA2 Collaboration, P. Bagnaia et al., Z. Phys. C20 (1983) 117.
[32] D. Amati and G. Veneziano, Phys. Lett. 83B (1979) 87.
[33] G. Marchesini, L. Trentadue and G. Veneziano, Nucl. Phys. B181 (1981) 335.
[34] K. Kajantie and E. Pietarinen, Phys. Lett. 93B (1980) 269.
[35] R.D. Field and S. Wolfram, Nucl. Phys. B213 (1983) 65;
T.D. Gottschalk, Nucl. Phys. B214 (1983) 201.
[36] B. Andersson, G. Gustafson, G. Ingelman and T. Sjöstrand, Phys.Reports 97 (1983) 31.
[37] R.D. Field and R.P. Feynman, Nucl. Phys. B136 (1978) 1; O. Mazzanti and R. Odorico, Z. Phys. C7 (1980) 61;
F.E.Paige and S.D. Protopopescu, Brookhaven report BNL 31987(1982);
S. Ritter, Z. Phys. C16 (1982) 27.
[38] T.D. Gottschalk, Caltech preprint CALT-68-1075 (= CERN preprint TH.3810 = lectures at 19th Int. School of Elem. Particle Physics, Lupari-Dubrovnik, Yugoslavia, 1983).
[39] G. Ingelman, CERN preprint TH.3969 (1984) and private communication.
[40] TASSO Collaboration, M. Althoff et al., Z. Phys. C17 (1983) 5; and references therein.
[41] TPC Collaboration, H. Aihara et al., Phys. Rev. Lett. 52 (1983) 577.
[42] UA2 Collaboration, P. Bagnaia et al., Phys. Lett. 144B (1984) 291.
[43] UA1 Collaboration, G. Arnison et al., Phys. Lett. 132B (1983) 233.
[44] G.C. Fox and R.L. Kelly, in AIP Conf. Proc. No.85 (1981) p.435;
R.D. Field, G.C. Fox and R.L. Kelly, Phys. Lett. 119B (1982) 439;
R. Odorico, Nucl. Phys. B228 (1983) 381;
R.D. Field, in Proc. 4th Topical Workshop on $p\bar{p}$ Collider Physics, Berne, 1984.

Discussion on Section 2

T. Akesson (CERN): Have you taken data without beam in the collider to estimate the cosmic ray background? Have you compared the charged E_T measured in the central detector with the total E_T measured in the calorimeter?

Speaker, J. Tuominiemi: (i) Yes, we made dedicated cosmic ray runs without the beam. (ii) We have checked carefully that the information from the central detectors is consistent with that from the calorimeters.

P. Pavlopoulos (CERN): What is the actual limit on the number of generations with the full statistics?

Speaker, J. Tuominiemi: I do not have a new number here as we are still working on it, but the number we have already given is $\leqslant 20$.

R.L. Jaffe (MIT): Have you looked for structure in three and more jet invariant mass plots?

Speaker, L. Fayard: Of course, we have looked at that, but nothing has been found.

T. Kalogeropoulos (Syracuse): What does UA1 have to say about your 150 GeV jet-jet peak?

Speaker, L. Fayard: The corresponding result from UA1 was shown this morning by Dr. Tuominiemi.

B. Andersson (Lund): You have two parameters in your jet-cascade: Q_o and Λ. Would you have to change Λ if you change Q_o, or vice-versa?

Speaker, B.R. Webber: Yes, to maintain a good fit to the data you would have to vary both Q_o and Λ together. However, the parameter dependence of the predictions is not extremely strong. For example, doubling Q_o at fixed Λ decreases the charged multiplicity by only about 10%. The thing most sensitive to Q_o is the baryon multiplicity, because increasing the mean cluster mass makes more phase space available for the decay of clusters into baryons. If we had more information about the spectrum of high-mass mesons, so that they could be included as possible cluster decay products, then they would compete with baryons and the Q_o-dependence would be reduced further.

G.L. Kane (Michigan): Could you comment on the recent claim from the TPC detector at PEP that three jet multiplicity data support the string fragmentation rather than cluster fragmentation.

Speaker, B.R. Webber: This is a very interesting point. In fact, the conclusion of the TPC group (which confirms that of the JADE group at PETRA) is that cluster fragmentation explains the stringlike effects in 3-jet events just as well as the Lund string model, provided coherence effects are included in the QCD perturbation theory. The trouble with previous cluster fragmentation schemes was that they did not do the perturbation theory sufficiently accurately. These results were presented in detail at the Lund multiparticle conference by Lina Galtieri (TPC) and Alfred Peterson (SLAC).

U. Gastaldi (CERN): Hadronization also occurs in $p\bar{p}$ annihilation at rest, where it is experimentally feasible to know everything about the final state and to select S and P initial states. Is there a hope that some quantities related to this can be computed in QCD?

Speaker, B.R. Webber: Soft hadronic physics cannot (yet) be computed in QCD from first principles, because perturbation theory is inapplicable. However, $p\bar{p}$ annihilation at rest may provide constraints on models of the hadronization process. Gottschalk [Nucl. Phys. B239 (1984) 325] has taken this idea quite far. For myself, I do not see why the soft process by which hadrons form following a hard QCD cascade should be connected with the one which occurs when you put three quarks and three antiquarks together at rest.

P.L. Braccini (Perugia): Please, would you comment on the general comparison of QCD with experiment? In your transparencies, you wrote "uncertainties in structure functions and higher order corrections imply a more careful treatment is not justified anyway". Would I be entitled to change this to imply "more careful experiments are not justified anyway"?

Speaker, B.R. Webber: As I said in my talk, I am quite pessimistic about high precision tests of perturbative QCD in hadron collisions, because we are unlikely ever to have sufficient control over higher order corrections. More precise data may nevertheless be helpful in developing models for things that we cannot calculate from first principles, such as hadronization. After all, that is what is done in most fields of physics.

Inst. Phys. Conf. Ser. No. 73: Section 3
Paper presented at VII Eur. Symp. Antiproton Interactions, Durham 1984

LEAR—Machine and experimental areas: experience and future plans after one year of operation

M. Chanel, P. Lefèvre, D. Möhl, D.J. Simon

CERN, CH-1211 Geneva 23, Switzerland

Abstract. CERN has a new machine, LEAR, devoted to low energy antiproton physics, which has been in operation for two years. This paper gives a review of the experience gained during the first year of operation and summarizes the ideas for future developments.

1. Introduction

Since the starting-up of LEAR [1, 2, 3] in July 1982, extracted antiproton beams have been delivered to the users at four momenta (1.51, 1.48, 0,61, and 0.31 GeV/c). This paper reviews the characteristics of the beam and compares them to the predictions made at the Erice Conference [2]. In addition, we discuss the tests foreseen in the coming months to prepare next year's operation.

2. Performances

2.1 Operation

LEAR [1, 2, 3] (Fig. 1) received its first protons for machine tests in July 1982 and the first antiprotons in September 1982. The ultra-slow extraction was made to work for the first time in April 1983. Since July 1983, the LEAR users have received beam during 5 major periods. All the 16 groups who are presently ready to take data have received beam during one or several of these running periods.

In 1983 we had to work with a mode of operation called "alternate mode": antiprotons were produced and accumulated in the AA for 8 to 10 hours during the night. LEAR did not receive beam as long as the $\bar{p}$ production went on. During the day $\bar{p}$ production and stacking was stopped and LEAR took beam from the AA and delivered it to the users.

Since the beginning of 1984, we have been working in a "continuous mode" where $\bar{p}$ accumulation and extraction from LEAR work in parallel: 3 to 4 PS cycles out of 6 have been used for antiproton production. Production and stacking in the AA are only interrupted for a few minutes about once every hour during the transfer of a new antiproton pulse to LEAR.

2.2 LEAR beam

LEAR was specified [1] to have a slow extraction of 15 min. This is already an enormous extrapolation from all other existing synchrotrons that have an extraction of at best a few seconds. After careful tests and improvements of the LEAR power supplies and the extraction hardware, it

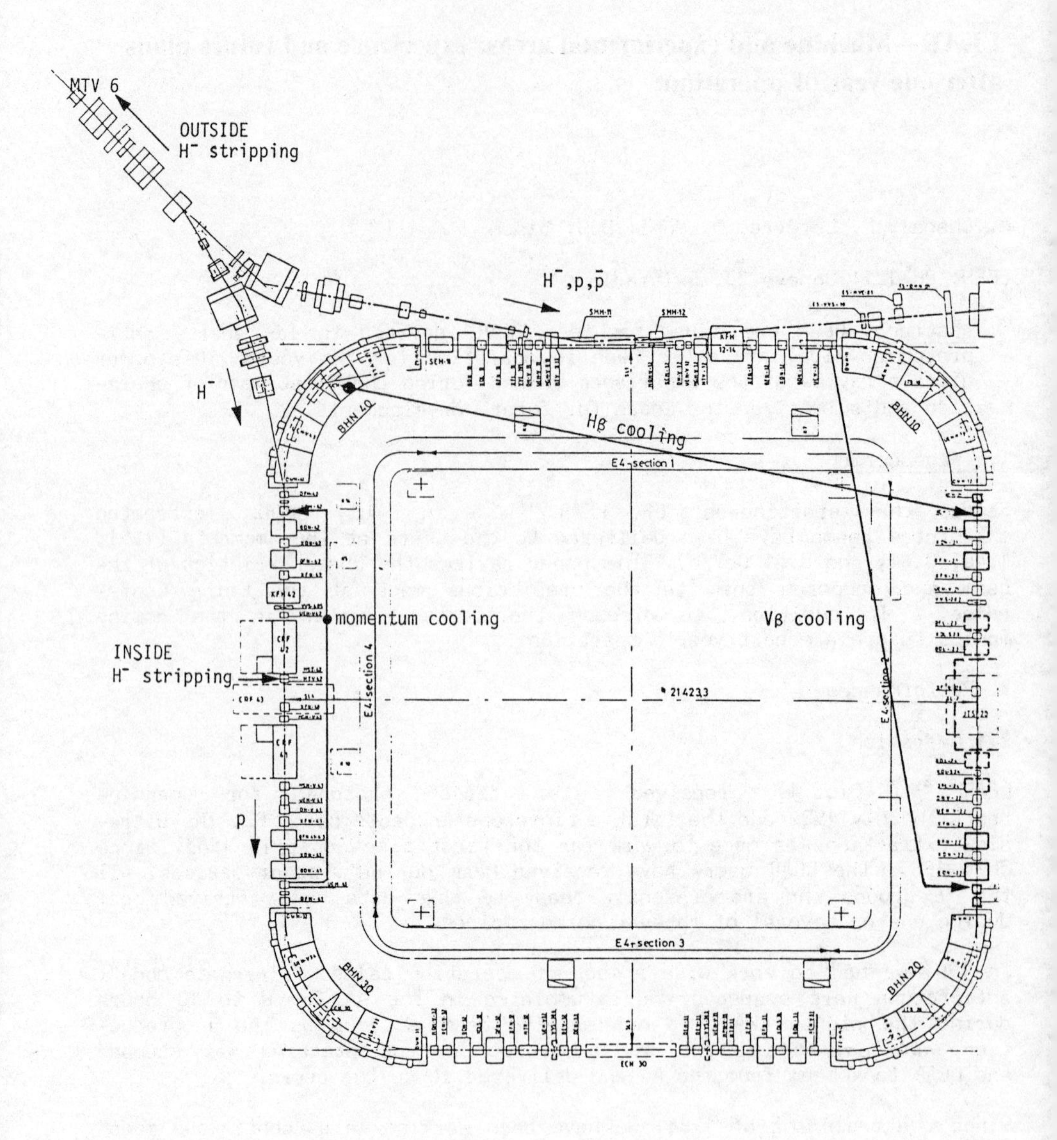

Fig. 1: General layout of LEAR

was possible to obtain extraction times as long as one hour, with an acceptable duty factor. All the particle physics runs in 1984 have been done with this extraction time as it facilitates the operation of the PS complex, which has to stop stacking only once per hour. In addition, the longer cycle has permitted improved LEAR beam quality because a longer time for cooling is acceptable. In fact, to have a good extracted beam and to permit clean deceleration, stochastic cooling was applied during 5 to 10 minutes, depending on the final momentum.

The number of antiprotons injected is typically 2 to 3×10^9. It is determined by the AA stacking rate ($3\text{-}4\times10^9$/h) and by the transfer efficiency (70%-90%). During an one hour spill 5 to 8×10^5 antiprotons per second are delivered to the experiments with an overall efficiency of 50% to 80% for ejection, beam splitting and beam transport.

2.3 Injected beam

Table 1 gives the beam characteristics just prior to extraction in the AA and after injection into LEAR. The longitudinal emittance in LEAR is given after bunch to bucket capture and adiabatic debunching. This technique reduces the $\Delta p/p$ of the beam by a factor 2.5 compared to injection without bunch to bucket matching as used in early runs. All emittances are as defined in ref. 2. Table 2 and the Schottky scans of Fig. 2 illustrate the characteristics of the beam on the 609 MeV/c injection flat-top ($\Delta p/p$ without bunch to bucket matching) prior to and after stochastic cooling.

	Operational	Best done	
AA 3.5 GeV/c			
ε_H^*	< 8	4	π mm.mrad
ε_V^*	< 8	4	π mm.mrd
A^*	0.25	0.1	eVs
LEAR 0.609 GeV/c			
ε_H^*	20	13	π mm.mrad
ε_V^*	10	7	π mm.mrad
A^*	0.5	0.2	eVs

Table 1: Normalized beam emittances at ejection from the AA and after injection into LEAR.

609 MeV/c	ε_H	ε_V	$\Delta p/p$ ($\pm10^{-3}$)
before cooling	30	15	3
after 5' cooling of $2.5\ 10^9$ particles	8	4	1

Table 2: Beam characteristics (unnormalized emittances) before and after stochastic cooling at injection energy.

Fig. 2: 5 min. cooling of 2.5 10^9 particles at 609 MeV/c

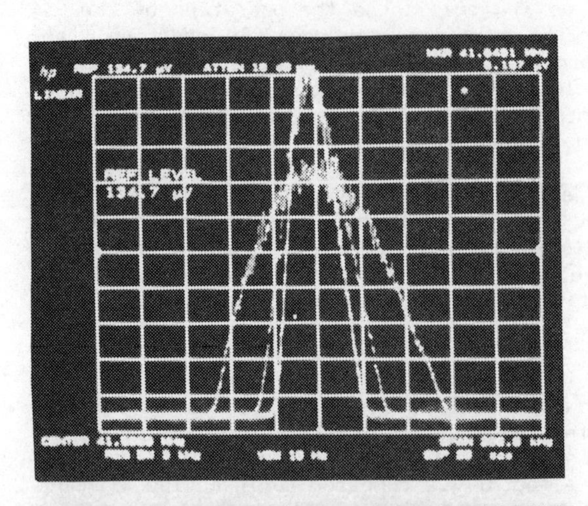

a) Longitudinal Schottky scan at the 20th harmonic of the revolution frequency. The curves represent the square root of the particle density vs. momentum, $\Delta p/p = 0.1\%$ per division. The momentum width decreases from 0.55% to 0.2% as a result of cooling.

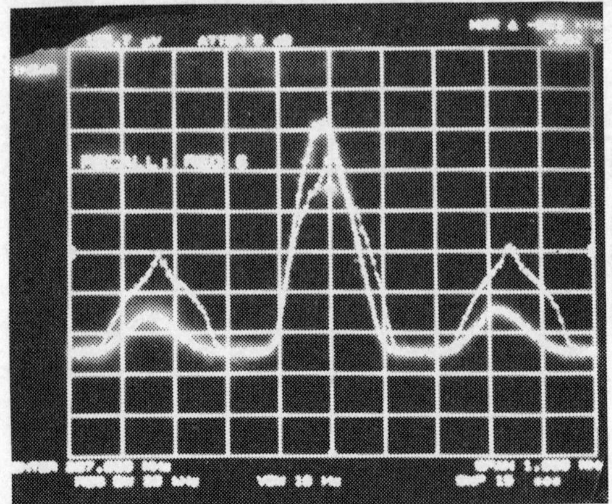

b) Horizontal Schottky scan; signal at the 100th harmonic of the revolution frequency and adjacent betatron sidebands: 100-q and 100+q where q is the non-integral part of the horizontal betatron tune. The height of the sidebands is a measure of the emittance. As a result of cooling, the emittance decreases by a factor of ~3.5.

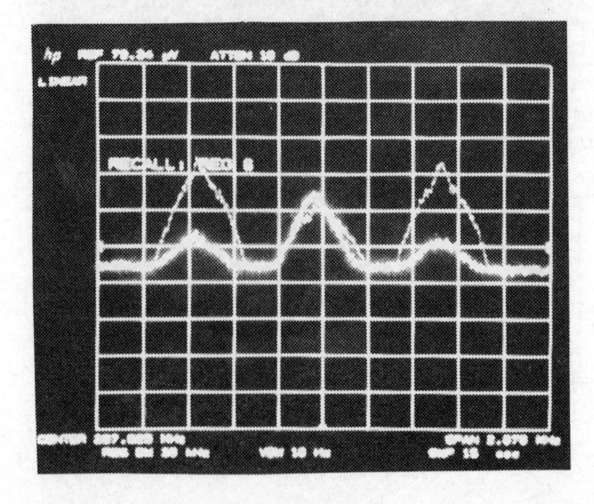

c) Vertical Schottky scan, 100th harmonic of the revolution frequency and adjacent betatron sidebands: 100-q and 100+q. As a result of cooling the vertical emittance decreases by a factor of ~5.

2.4 Extracted beams

Independent of the extraction momentum, the beam is always cooled at 609 MeV/c for about 5 min. For extraction momenta larger than 609 MeV/c no further cooling has been used as the beam damps adiabatically during acceleration. For lower momenta, additional cooling was applied at 309 MeV/c for 3 to 5 min.

The emittance characteristics of the extracted beam are shown in Table 3. They compare favourably with the values anticipated at Erice[2].

p (MeV/c)	ε_H	ε_V	$\Delta p/p$
609	~ 2	5	± 0.5
309	~ 4	10	± 1.0
1500	2	5	± 0.5

Table 3: Characteristics (unnormalized emittances in π mm.mrad and $\Delta p/p$ in % of the extracted beam

The long-term time structure of a spill is shown in Fig. 3. With careful adjustment of the RF noise used to drive the beam into the extraction resonance, this structure is fairly constant. In the earlier runs, the spill structure was strongly modulated at 50 Hz and some higher harmonics of the mains power supplies. This 50 Hz ripple has now been reduced to an acceptable level by the addition of strong noise (the "chimney" in Fig. 4a) at the position of the resonance[4] and by improving the stability of some of the LEAR power supplies. These measures have improved the duty factor from at best 30% to better than 70%.

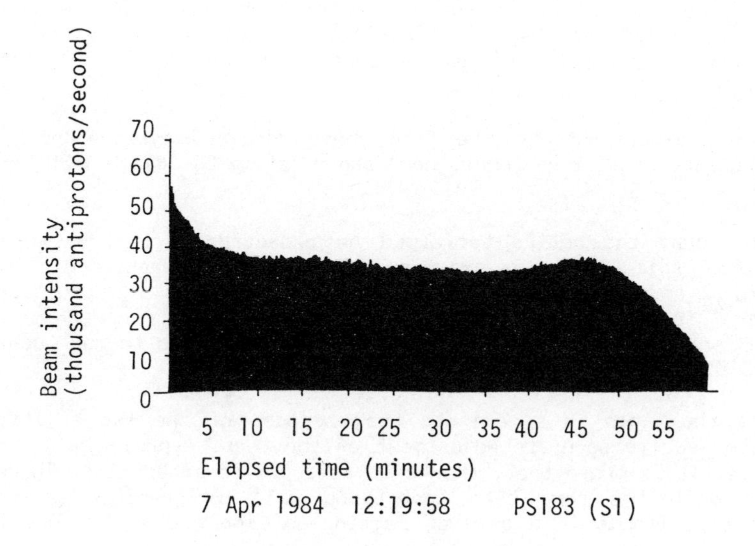

Fig. 3: Time structure of a spill in an experiment

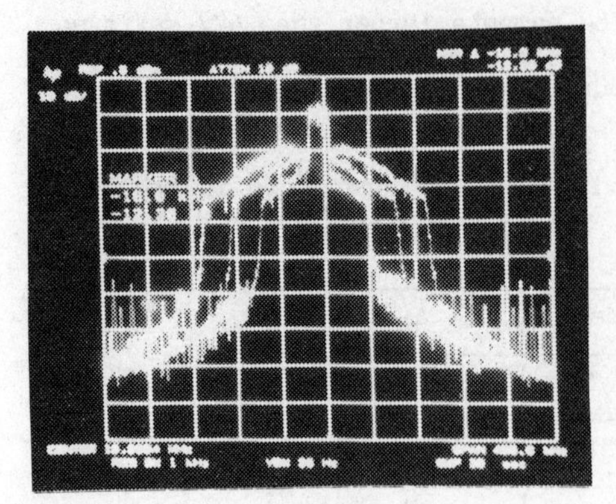

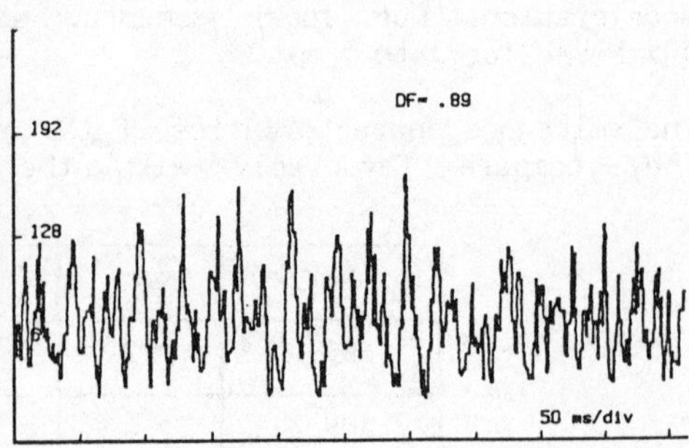

Fig. 4a: Power spectrum of the noise used to drive the beam into the extraction resonance. The "chimney" represents the additional noise at the resonance.

Fig. 4b: 50 Hz and higher frequency structure of the spill with the additional noise at the resonance. The duty factor is better than 80%.

2.5 Experimental areas

The layout (Fig. 5) was fully described at the Erice Workshop[5]). Let us recall that the 16 approved experiments are installed in 6 independent areas. The slowly extracted beam is split into 3 branches (North, Centre, South) which operate simultaneously. DC switching magnets are used to feed alternatively the two areas of each branch. Most installations were finished in Spring 1983 so that the first tests of the optics and of the experiments could be performed as scheduled in May-July 1983 with 309 MeV/c protons.

Since then, all experiments on the floor have received antiprotons and taken data during the runs mentioned above at 310, 612, 1480 and 1512 MeV/c.

In all these runs, the optics performed as expected. Fig. 6 shows an example of beam splitting. Two examples of spot size measurements are given in Fig. 7 and 8. Fig. 7 shows the first antiproton beam focussed within ~1 mm^2 (user: experiment PS177 - $\bar{p}$ at 310 MeV/c - North 1 area). Fig. 8 shows the smallest beam spot size obtained up to now (user: experiment PS185 - p at 1512 MeV/c - North 1 area).

Setting-up of the lines is a delicate task because of the two splitter magnets, which require accurate adjustment of the beam, especially at the lower energies. It is hoped that, thanks to appropriate scheduling, longer physics runs with the same beam conditions will be possible in the future. This will result in a gain of setting-up time and an economy of antiprotons.

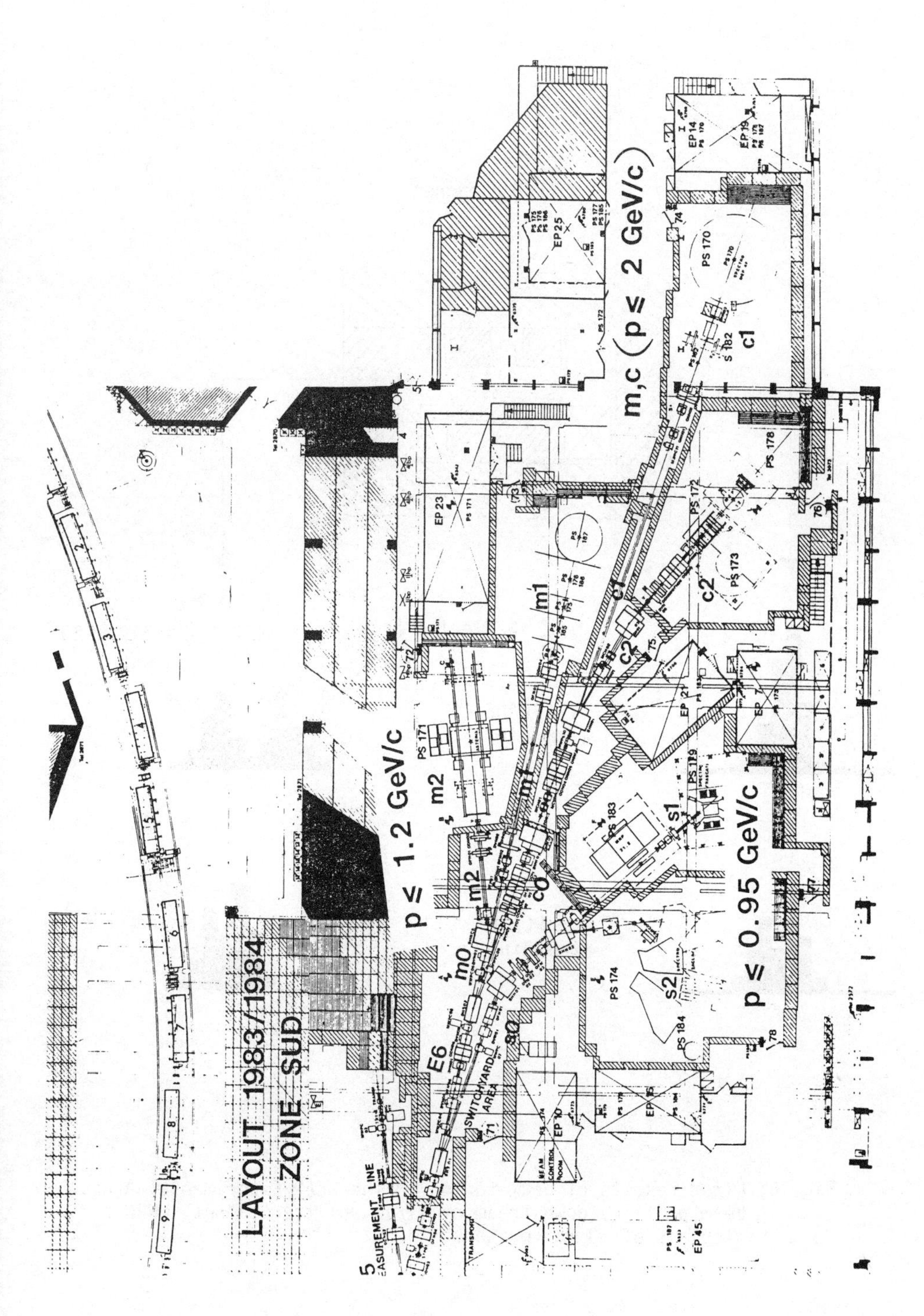

Fig. 5: LEAR Experimental Area (June 1983)

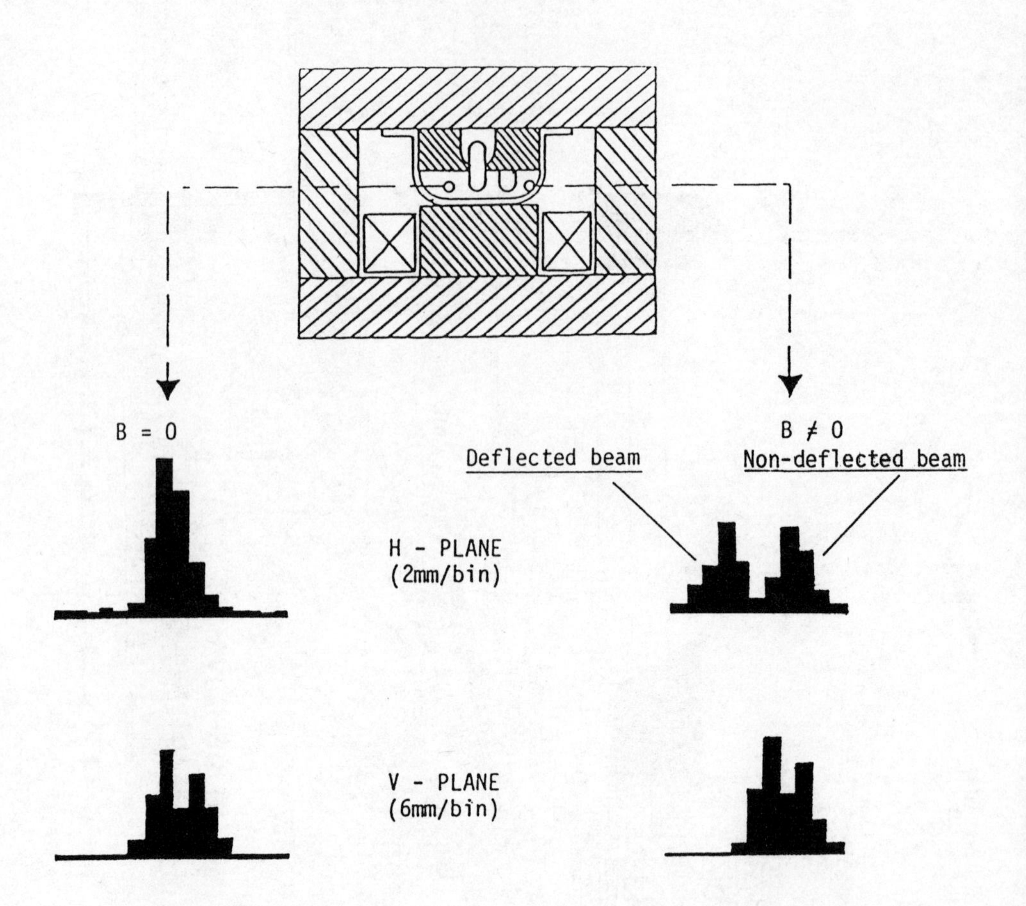

Fig. 6: First results of beam tuning in the LEAR Experimental Area. Beam profiles downstream of first splitter magnet: MWPC's displays of 50 MeV protons.

Resolution : 1mm/bin

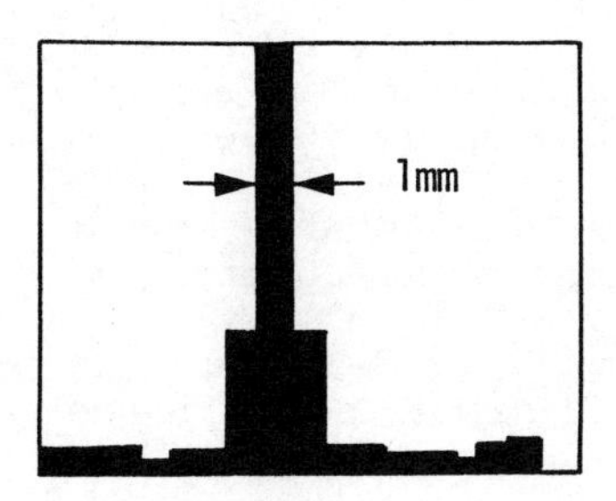

Horizontal plane

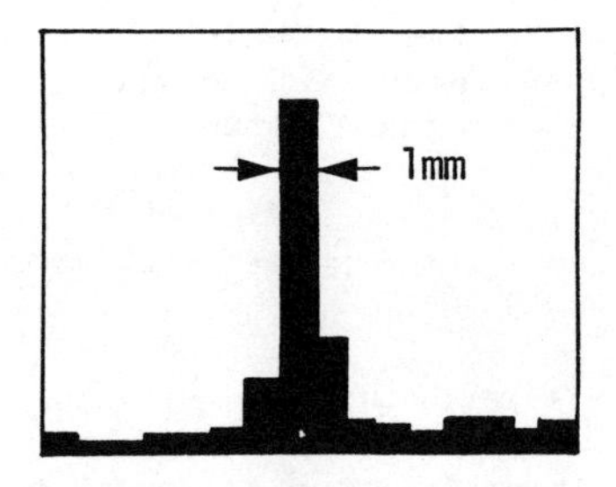

Vertical plane

Fig. 7:

Lear Experimental area
Beam profile measured at PS 177 focus with a multiwire proportional chamber

Intensity = 10^5 antiprotons/second p = 0.310 GeV/c

Resolution : 0.25mm/bin

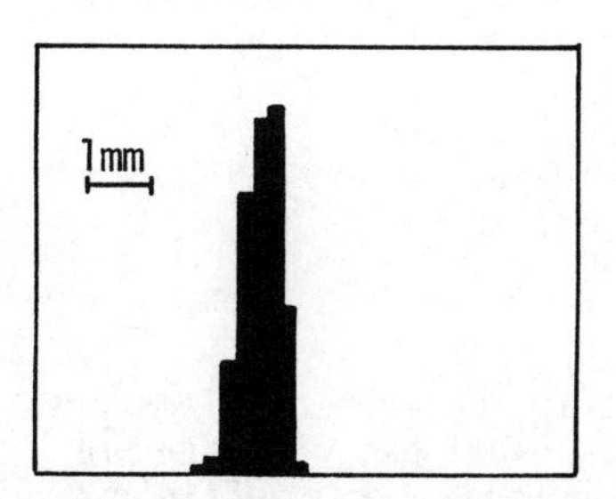

Horizontal plane

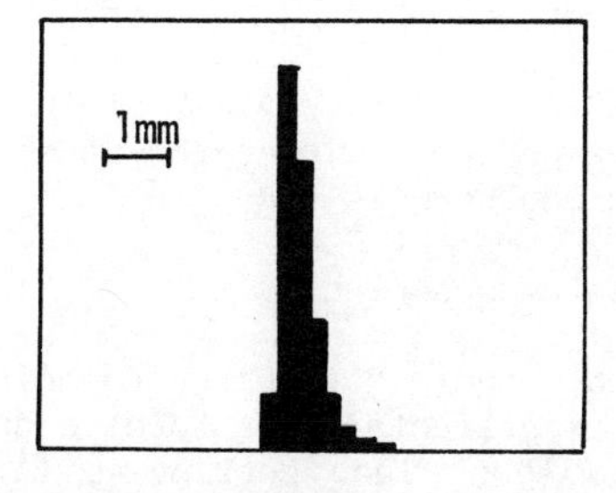

Vertical plane

Fig.8:

Lear Experimental area
Beam profile measured at PS 185 focus with the "Erlangen Wobbler"

Intensity = 10^5 antiprotons/second p = 1.5 GeV/c

3. Future plans and improvements

3.1 H^- injection

During a shutdown in October this year, the hardware for the injection of negative hydrogen accelerated in Linac 1, will be installed in LEAR (Fig. 1). In a first stage, three types of operation could be used:

a) H^- particles injected in the normal direction to test the machine in the antiproton polarity at momenta below the electromagnetic stripping limit (0.70 GeV/c).

b) H^- stripped in the injection line. This would permit the testing of the machine - e.g. the slow extraction - with protons, without having to dismantle the H^- source from the Linac 1.

c) H^- injected via a new piece of transfer line (the E3 line - Fig. 1) and stripping in LEAR. This charge exchange injection of protons turning anticlockwise provides for tests with the antiproton polarity. In addition to LEAR, the antiproton injection line could be tested by ejecting protons from LEAR to the PS. This can considerably reduce the time and the antiprotons needed to set up LEAR and its lines.

3.2 Low momenta

During this Autumn, we plan to test LEAR with protons decelerated down to 0.1 GeV/c. A series of runs is needed to study the beam behaviour and the stability of the power supplies as well as the extraction system. In addition, the transport elements of the experimental area, some of which are large 26 GeV devices, could limit the minimum momentum which can be delivered to the experiments under stable conditions. Beam tests are needed to specify the necessary modifications.

Other studies are on the way to decelerate antiprotons to momenta lower than 100 MeV/c, either in LEAR itself, or by means of a small "postdecelerator synchrotron" or a "radiofrequency quadrupole linac structure".

The Electron Cooling device, useful in the jet target option, is now in an intensive period of laboratory tests. It could be installed in LEAR during the long shutdown in 1986.

3.3 Momentum scanning

"Momentum scanning" requires changing the currents in the machine elements and the experimental area to have an extracted beam available at different momenta without long setting-up times. For this purpose, both efficient software and long machine development runs are needed. The necessary techniques will be developed and tested during the next months.

We hope that scanning in a limited number of steps and around a few strategic momenta will be available at the end of 1985. It could then be used during the SPS collider period, if parallel running of LEAR with the collider is accepted. The scanning operation will be much easier if only one experiment is served at a time because this avoids the additional difficulty of the beam splitting.

3.4 Experimental areas

Lower momenta

LEAR runs with antiprotons at 200 MeV/c will take place in the next months (August-September). Although no big problems have been encountered up to now at 310 MeV/c, we are approaching the limits of some of the transfer components used in this layout (stability of power supplies, non linearity of magnetic fields, parasitic fields, etc.). Further decreasing of the momentum in the next year will certainly require some new equipment.

Multiple Coulomb scattering produced by vacuum windows in front of the experiments becomes very severe when the antiproton momentum is decreased. When necessary, we shall consider the use of beryllium windows (0.1 mm thick aluminium windows are generally used at present).

Possible installation of a third splitter magnet [6, 7].

The South branch is limited at a maximum momentum of about 900 MeV/c by the bending power of the 30° bending magnets. For higher momenta, only 2 experiments can run simultaneously (North 1 - Centre 1 or North 1 - Centre 2).

The addition of a third splitter in the central branch (close to the "FOC1" position) together with a serious modification of the C0, C1 and C2 lines would permit simultaneous operation of the experiments installed at present in the C and N areas up to 2 GeV/c. The drawback of this operation would be the disappearance of the special layout of the C2 line, i.e. the use of a degrader (possibility to change the momentum independently of the LEAR momentum, and to perform scattering experiments).

A spare splitter is being built at the moment, but additional equipment is needed to install this splitter in the central branch (new vacuum chambers, new multiwire proportional chambers). A shutdown of ~6 weeks is necessary for this installation, which could be envisaged around mid 1985.

It should be noticed, however, that not more than 2 splitters can be used simultaneously during operation.

Control improvements

The "Beam Control Room" (EP10 barrack) offered to the users to adjust their beam lines (downstream of the splitter magnets) is often overloaded. Therefore, it is envisaged for 1985 to install another control point in the South Hall Extension. In addition, the computer-control of these lines will be made easier.

An automatic (or semi-automatic) system may be envisaged to match the beam ejected from LEAR with the switchyard line; although difficult to realize because our detectors tend to destroy the beam, such a control would make the full experimental layout insensitive to small variations of the beam ejected from LEAR.

4. Conclusion

To conclude one can say that the CERN Low Energy Antiproton Facility has worked so far successfully and that no major difficulties have been encountered during the first year of operation. Nevertheless, LEAR is a very demanding machine. One of the most delicate problems is the stability of the LEAR bending magnet and quadrupole power supplies as the ultraslow extraction requires formidable tolerances. As an example, $5\ 10^{-5}$ fluctuations (at 100 MeV/c, this corresponds to $2.5\ 10^{-6}$ of the maximum power supply current) reduce the duty factor to less than 50%. Much further work is, therefore, needed to assure that the full range of design performances can be covered and the delicate modes of operation required by our users can be realized.

5. References

1) Lefèvre P., Design study of a facility for experiments with low energy antiprotons (LEAR), CERN/PS/DL 80-7.
2) Lefèvre P., Construction of the LEAR facility: status report, published in "Physics at LEAR with Low-Energy Cooled Antiprotons", Proceedings of a Workshop on Physics at LEAR, held in Erice, Sicily, May 9-16, 1982 (p. 15-26).
3) Lefèvre P., "LEAR", lecture given to the "CERN Accelerator School", PS/LEA/Note 84-7.
4) Hardt W., Moulding the noise spectrum for much better ultraslow extraction, PS(DL)LEAR Note 84-2.
5) Simon D.J., The LEAR experimental area, published in "Physics at LEAR with Low-Energy Cooled Antiprotons", Proceedings of the Erice Workshop, Erice, Sicily (p. 53-68).
6) Bradamante F., University of Trieste, memorandum to PSCC, CERN/PSCC 84-23, March 1984.
7) Hoffmann L. and Simon D.J., Splitter magnet for LEAR experimental area, PS/MU/EP/Note 84-1, March 1984.

Inst. Phys. Conf. Ser. No. 73: Section 3
Paper presented at VII Eur. Symp. Antiproton Interactions, Durham 1984

First observation of *K* x-rays of antiprotonic hydrogen atoms

ASTERIX Collaboration

(Presented by Rolf Landua)

S. Ahmad, C. Amsler, R. Armenteros, E. Auld, D. Axen, D. Bailey, S. Barlag, G. Beer, J.C. Bizot, M. Caria, M. Comyn, W. Dahme, B. Delcourt, M. Doser, K.D. Duch, K. Erdman, F. Feld, U. Gastaldi, M. Heel, B. Howard, R. Howard, J. Jeanjean, H. Kalinowsky, F. Kayser, E. Klempt, R. Landua, G. Marshall, H. Nguyen, N. Prevot, L. Robertson, C. Sabev, U. Schaefer, R. Schneider, O. Schreiber, U. Straumann, P. Truöl, B. White, W. Wodrich and M. Ziegler

Abstract. X-rays from the K and L series of $\bar{p}p$ atoms formed in hydrogen gas at NTP have been observed in association with annihilations into neutral particles.

Valuable information about the strong $\bar{p}p$ interaction close to threshold can be obtained from the study of L(nd→2p) and K(np→1s) x-rays emitted during the atomic cascade of $\bar{p}p$ atoms. The strong interaction is expected both to shift and broaden the energy levels calculated from the purely electromagnetic interaction. Theoretical models predict, for instance, a negative shift and broadening of the 1s level of about 1 KeV (1). Also a splitting between the singlet and triplet s states may be expected because the $\bar{p}p$ strong interaction is known to be spin-dependent. Although several experiments have been performed in the past (2) to look for L and K lines in $\bar{p}p$ atoms formed in hydrogen, only one (3) has observed L x-rays. Three experiments (4a,b,c) are currently searching for x-rays at the CERN LEAR (5) facility. In this contribution we report the results of one of them (4a). L and K x-rays have been found in the study of $\bar{p}p$ annihilations into neutral particles. These annihilations have been studied first to minimize the background from inner bremsstrahlung (6).

The experimental details are as follows:
An incident $\bar{p}$ beam of 308 MeV/c momentum and $\Delta p/p = 2\cdot10^{-3}$ and of average intensity $2\cdot10^{4}$ $\bar{p}$/sec was moderated and entered the detector, which is described in more detail in ref. 7. An

antiproton stop in the gaseous hydrogen target is defined by the coincidence of two beam counter scintillators T_1 and T_2 and the anticoincidence of two scintillators T_3 and T_4. The cylindrical scintillator T_3 surrounds T_1 and T_2 and vetoes annihilations occurring in the vicinity of the target entrance region. T_4 is placed at the end of the target and vetoes on antiprotons which did not stop. At 308 MeV/c momentum about 5% of the beam $\bar{p}$'s stopped in the target.

X-rays emitted in the cascade of $\bar{p}p$ atoms are detected in the central chamber acting as an X-ray Drift Chamber (XDC, Fig.1), described in more detail in ref. 8. The cylindrical target region, containing H_2 gas at NTP, is separated from the counter region by a 6μ thickness aluminized mylar foil, which serves as central cathode for the drift field. X-rays traversing the mylar foil can convert in the counter gas and produce electrons which drift to one of the 90 sense wires. The charge collected by one of the sense wires is digitized in 32 nsec time bins, its position along the symmetry axis of the XDC measured by charge division. The data were taken in a six hour period beam time triggering on a $\bar{p}$ STOP signal, a single hit wire and no hit on adjacent wires in the XDC, and no hit in the two innermost chambers of the seven MWPCs surrounding the XDC.

In the offline analysis of the 600 000 raw events obtained the following cuts were set:

- Exactly one hit wire in the XDC
- A pulse shape corresponding to an x-ray pulse whose shape is well known from source events
- conversion of the x-ray in the region z=-20 ... +17 cm (the center of the XDC is at z=0) and in a distance of at least 1.5 cm from the mylar foil.

The first two cuts were applied to suppress events with annihilation into charged particles, the other cuts are made to avoid the contamination arising from annihilations in the material surrounding the H_2 gas target (scintillators and mylar foil).

Possible background arising mainly from inner bremsstrahlung, radiation from undetected charged particles, target gas contamination and random coincidence with calibration sources have been carefully studied and the conclusion reached that their contributions are very small.

Fig. 2a shows the x-ray spectrum after all cuts have been applied; part of the spectrum has been enlarged by a factor 20 to visualize better the structure of the spectrum above 4 keV. Fig. 2b shows the detection efficiency for x-ray as a function of energy after all cuts have been applied.

The dominating peak between 0.9 and 3.5 keV originates from the L series of antiprotonic hydrogen and confirms the results of Auld et al. (3). The total L yield was determined from the x-ray spectrum in coincidence with annihilation into charged prongs and is 13 ± 2% per antiproton stop. The region above 4.0 keV shows a smooth background with a broad peak in a region between 7.0 and 10.5 keV. We attribute this peak to the K series of x-rays populating the 1s state.

A complete analysis of this K series energy region taking into account the effect of the strong interaction on the two hyperfine states $^1S_0$ and $^3S_1$ (their splitting, shifting and/or broadening) would have to consider both the isospin and spin dependence of the potential. The present statistics do not allow such an analysis. The following simple fit to the data in the K line region has been made: two gaussian distributions have been assumed with a line width corresponding to the experimental resolution (20% FWHM at 5.5 keV) and separated by 1.74 keV, which is the calculated energy difference between the K_α and K_β transitions. The best fit yields an energy shift of the 1s level of E = -0.5 ± 0.3 keV and a relative yield $Y(K_\alpha)/(Y(K_\alpha) + Y(K_\beta)) = 0.4 \pm 0.1$.

Assuming that the branching ratios into neutral particles to be equal from s and p states, the relative yield of K_α x-ray transitions to L x-ray is $Y(K_\alpha)/Y(L_{tot}) = (2 \pm 1) \cdot 10^{-2}$.

These results are fully compatible with the predictions of Richard and Sainio (1), but they only have to be considered as a first step towards a more complete picture of the complex structure of the $\bar{p}p$ atom ground state.

References

(1) B.O. Kerbikov, Proc. 5th European Symposium on Nucleon-Antinucleon Interactions (Bressanone, 1980) (CLEUP, Paduva, 1980) p. 423
(Reference to earlier work can be found herein)

W.B. Kaufmann, Phys. Rev. 19C (1979) 440
J.M. Richard and M.E. Sainio, Phys.Lett. 110B (1982) 349
A.M. Green and S. Wycech, Nucl.Phys. A377 (1982) 441
M.A. Alberg et al., Phys.Rev. D27 (1983) 536

(2) R.E. Welsh, in Proceedings of the 7th International Conference on High Energy Physics and Nuclear Structure, Zürich 1977 (ed. M.P. Locher) (Birkhäuser Verlag, Basel und Stuttgart, 1977) p. 95
M. Izycki et al., Z.Phys. 297 (1980) 1

(3) E. Auld et al., Phys.Lett. 77B (1978) 454

(4) S. Ahmad et al., Protonium spectroscopy in: Physics at LEAR with Low Energy Cooled Antiprotons, edited by U. Gastaldi and R. Klapisch, p. 109
J. Davies et al., Precision survey of x-rays from $\bar{p}p$ ($\bar{p}d$) atoms, op.cit. p. 143
D. Gotta, X-ray spectroscopy of $\bar{p}$ hydrogen in the cyclotron trap, op. cit. p. 165

(5) R. Rueckl and C. Zupancic, subm. to Phys.Lett. B

(6) P.Lefevre et al., The CERN Low Energy Antiproton Ring (LEAR) project. 11th Int.Conf. on High Energy Acceleration, Geneva, Switzerland, 7-11 July 1980 (Basel, Switzerland: Birkhäuser Verlag 1980) p. 819-23

(7) S. Ahmad et al., The ASTERIX detector, to be submitted to Nucl. Instr. and Methods

(8) U. Gastaldi et al., Construction and operation of the spiral projection chamber, to be submitted to Nucl. Instr. and Methods

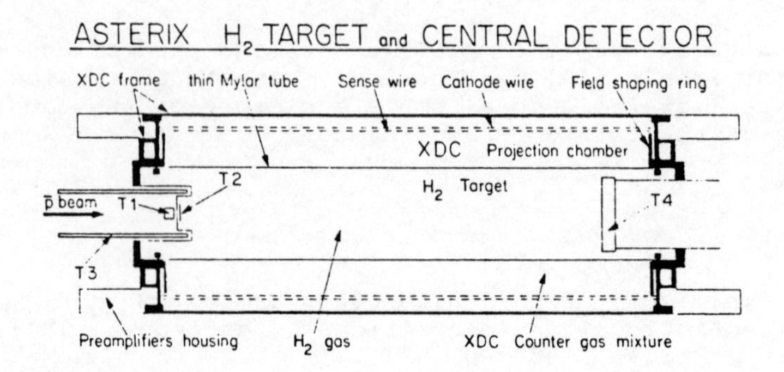

Fig. 1 The X-Ray Drift Chamber (XDC)

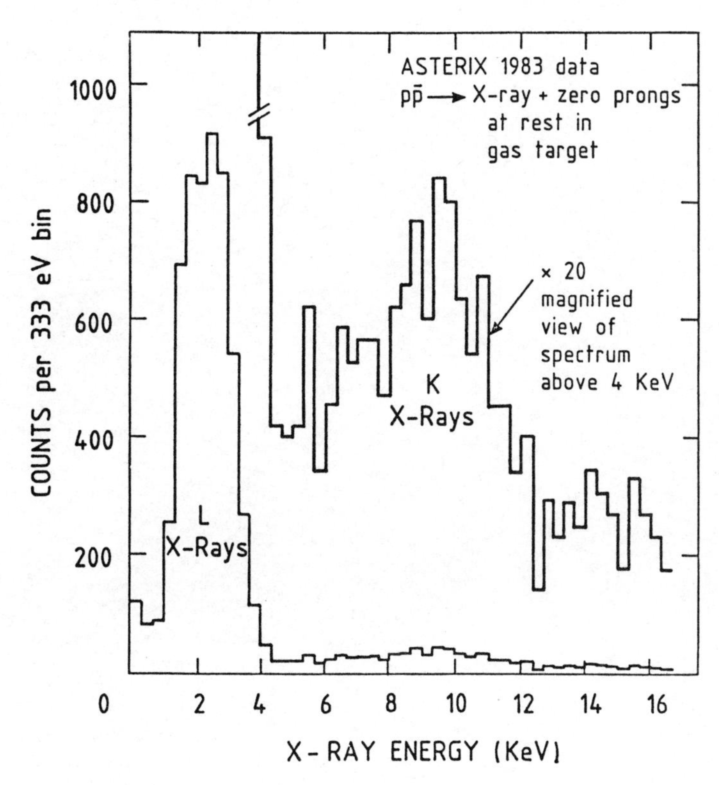

Fig. 2 a X-Ray Energy Spectrum (Zero Prong Data)

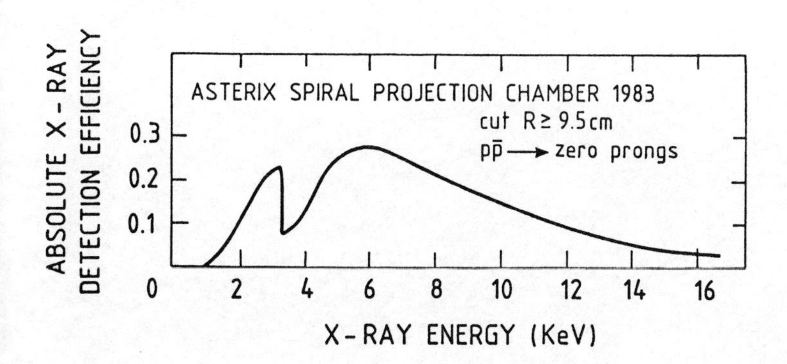

Fig. 2 b X-Ray Detection Efficiency (Zero Prong Data)

Inst. Phys. Conf. Ser. No. 73: Section 3
Paper presented at VII Eur. Symp. Antiproton Interactions, Durham 1984

X-rays from antiprotonic hydrogen, deuterium and helium

J D Davies, T P Gorringe, J Lowe, J M Nelson, S M Playfer, G J Pyle, G T A Squier
Physics Department, University of Birmingham
C A Baker, C J Batty, S A Clark, A I Kilvington, J Moir, S Sakamoto
Rutherford Appleton Laboratory
R E Welsh, R G Winter
College of William & Mary, USA
E W A Lingeman
NIKHEF, Amsterdam

Abstract. Antiprotons from the LEAR facility at CERN were stopped in targets of gaseous H_2, D_2 and He. We report observations of K and L series X-rays from $\bar{p}p$ and L series X-rays from $\bar{p}d$ using a SiLi detector. For $\bar{p}$ He X-rays shifts, widths and yields of inner transitions are reported.

1. Introduction

We are concerned with X-rays from a $\bar{p}$ cascading in an atom in which it has replaced an electron. For $\bar{p}p$ (protonium) QED gives L X-ray energies between 1.7 and 2.9 KeV and 9.4 to 12.3 KeV for K lines. Conventional potential models predict that the strong interaction will shift (corresponding to smaller X-ray energies) and broaden the 1S level of protonium by ½ to 1 KeV with a 0.2 KeV hyperfine splitting (RIC82). The main purpose of this experiment is to detect $\bar{p}p$ (and $\bar{p}d$) K X-rays and then to measure their shift and width.

The major problem is the very low yield of K X-rays. $\bar{p}$ are captured into high n, high ℓ orbits but stark mixing during inter-atomic collisions transfers them to high n, low ℓ orbits, from whence they are annihilated before the cascade reaches the inner transitions. Also the 2p annihilation width is expected to be 20-40 meV (RIC82) compared with a radiation width of 0.4 meV. At high densities, $\sim \rho$(liquid), one expects the total yield to be of order 10^{-5} and inversly proportional to ρ. For gas densities $\leqslant 10\rho_{STP}$ the total yield (BOR82) varies slowly, somewhere about 10^{-3}.

2. Apparatus

The experimental arrangement is shown in figure 1. A large Aℓ flask contains gas at atmospheric pressure and can be maintained at any temperature from 30° K to 300° K. The S2 beam at LEAR delivered (1-6) 10^4p/sec. at 308 MeV/c with $\Delta p/p = 2.10^{-3}$ and negligeable contamination. A target temperature of 30°K was chosen so as to stop 82% of the beam in the gas. The X-ray window consists of 6μ mylar supported on Aℓ strips. The 300mm^2 x 5mm SiLi X-ray detector has pulsed optical

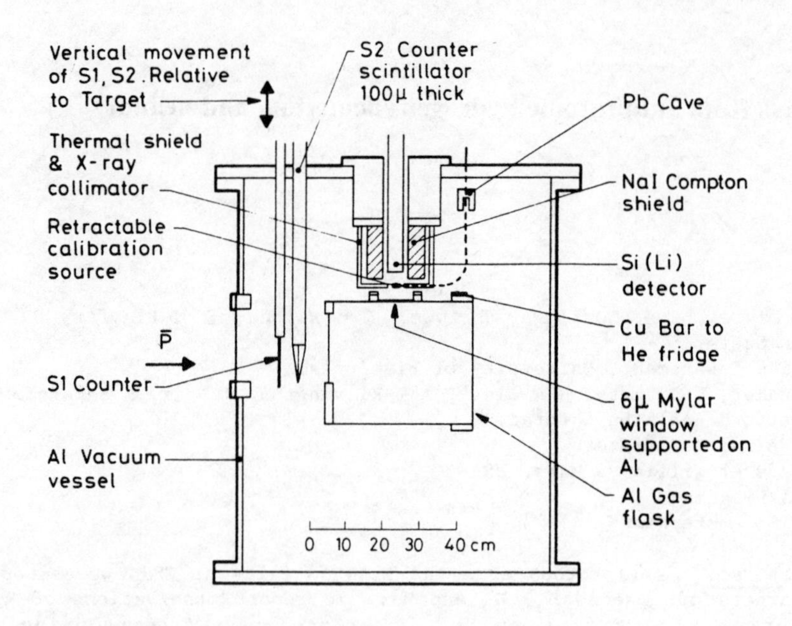

Fig.1. Schematic diagram of Target and Counters

feedback with an in-beam resolution of 320 eV FWHM at 6.4 KeV. The large continuum X-ray background was consistent with Compton scattering of 50-500 KeV γ-rays in the SiLi crystal and was partially suppressed by the NaI(Tl), 'Polyscint' annulus around the crystal. High purity Aℓ, 1mm thick, was used for flask, radiation shield and detector end-cap and these collimated the SiLi to view $\bar{p}$ only when stopping in the gas.

On average a $\bar{p}$ annihilation in H_2 or He gives $3\pi^{\pm}$ and $2\pi^0$. The former can each leave 4 MeV in the SiLi. This does not contribute to the X-ray spectrum directly but gives a high optical reset rate which required extensive modification of the SiLi electronics (DAV82). 3 scintillator counter telescopes orthogonal to the beam viewed $\bar{p}$ annihilations inside the flask. An Aℓ degrader of variable thickness was used to maximize the counts in the central telescope and hence optimize the $\bar{p}$ stopping distribution.

3 He Data

$\bar{p}$-He spectra were measured frequently, eg. figure 2 which improves the pre-LEAR world data dramatically. The various lines from N_α at 1.8 KeV through the 11-19 KeV L series provide much useful information, viz the energy calibration and resolution of the SiLi under beam conditions; estimates of the suppression of the continuum background using the NaI annulus (x5 at 10 KeV to x3 at 3 KeV); the 25 ± 5% self-veto of good events by annihilation products in the NaI crystal; the dependence of incident $\bar{p}$ - SiLi timing on X-ray energy. The data permits the first

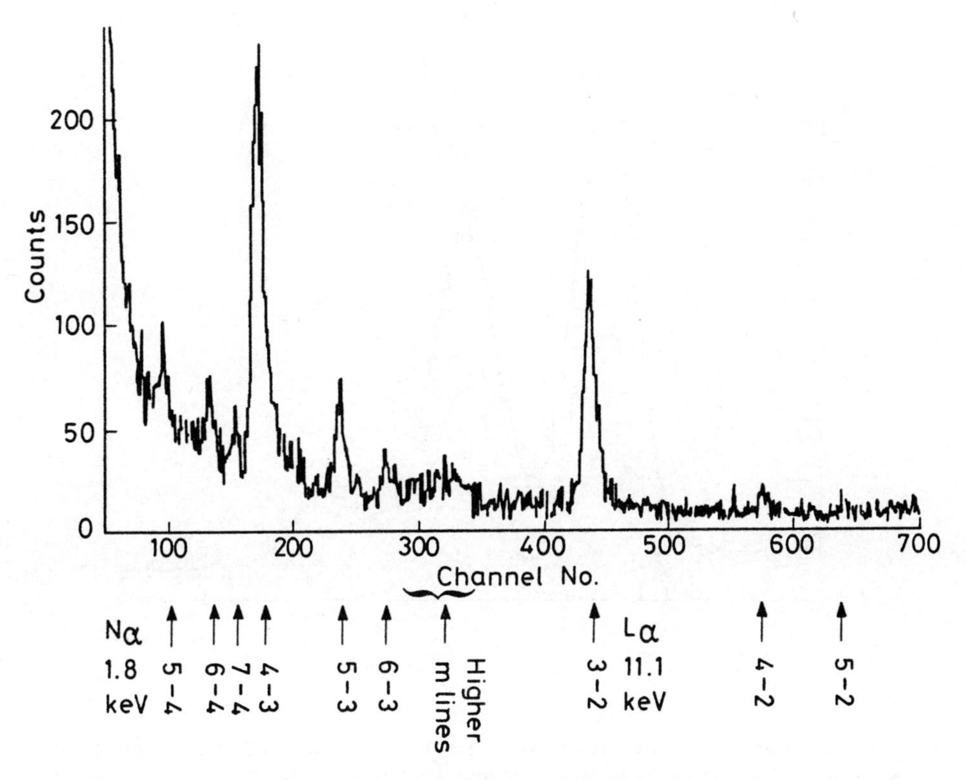

Fig. 2 Spectrum of X-rays from $\bar{p}$ He atoms in a gas at 1 atm. and 30°K from 6.10^7 incident $\bar{p}$.

comparison of measured line yields with cascade calculations as a function of density, eg for L_α at 1, 3 and 10 x ρ_{STP} yields of 9 ± 1%, 5.2 ± 1.2% and 4.6 ± 0.5% were measured and 9.1%, 7.7% and 5.5% are predicted. (LAN84). Both sets of numbers are preliminary; the theoretical calculations have not used our latest values for ε_{2p}, Γ_{2p} and Γ_{3D}.

There is conflict between the previous direct measurements (POT78 and BAI83) of the strong interaction shift and broadening of the 2p level of $\bar{p}^4$He. The earlier result (POT78) allows strongly bound states of $\bar{p}^4$He. (DUM84). None are predicted for the later results (BAI83) which are, however, in agreement with optical model predictions. 8 hours of beam provided the L_α data of figure 3 - with simultaneous calibration by the K_α, K_β lines of a ^{75}Se source - and gave ε_{2p} = -7.4 ± 4.9 eV with Γ_{2p} = 35 ± 15 eV; the relative yields of the M and L_α transitions gave Γ_{3D} = 2.4 ± 0.5 meV. These new values of much improved accuracy are in good agreement with optical model predictions and suggest that any strongly bound $\bar{p}$ He state would have a width > 100MeV.

4. H_2 and D_2 Data

Figures 4a and 4b show Compton-suppressed X-ray spectra from 1.12 x 10^9 $\bar{p}$ stopped in H_2 gas and 1.17 x 10^9 $\bar{p}$ stopped in D_2. Both were obtained with gas at 30°K and 1 atm. pressure and have one contaminant line, $\bar{p}$Aℓ(8 → 7)

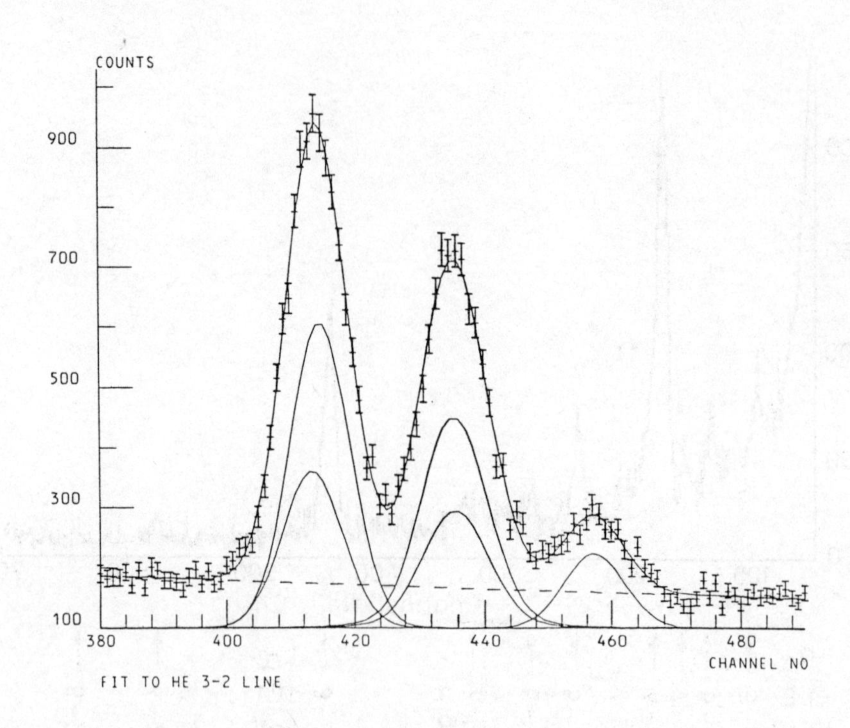

Fig. 3 Measured X-ray spectrum in the region of the $\bar{p}$ He 3 → 2 transition. The left hand peak is due to the K_{α_1} and K_{α_2} lines from the ^{75}Se calibration source and the right hand peak due to the K_β line. The central peak corresponds to the $\bar{p}$-He 3 → 2 doublet lines. The energy dispersion was 27.7 eV/channel.

at 19.5 KeV; an upper limit of 10% of the 8 → 7 intensity can be placed on $\bar{p}$ Aℓ lines of lower energy (POT84). Now X-rays above 5 KeV arrive within a 200ns time band started by the incident $\bar{p}$ in a thin (100μ) scintillator. This band has been used as a software cut on the data of figure 4* to completely eliminate late pulses which come from energetic particles creating charge trapped in regions of the SiLi not accessible to X-rays. A 1.6μs gate had been used to allow for rise-time slew and noise broadening of X-rays down to 1.5 KeV.

The data were fitted with a least-squares, multi-parameter program. A single exponential gave the best fit to the background above the L-line region to which was then added multi Voigt profiles; the widths of the Gaussian components were determined by the detector resolution while the centroids and widths of the Lorentzian components were free parameters. In the K-line region of the $\bar{p}$p data there is only one peak, at 11.7 ± 0.1 KeV with width 1.1 ± 0.3 KeV from 300 ± 100 counts corresponding to 0.05 to 0.1% yield; this is not inconsistent with the above predictions for shift, width and yield of the unresolvable higher K transitions.

*The time-cut was applied to the data after the Durham Conference.

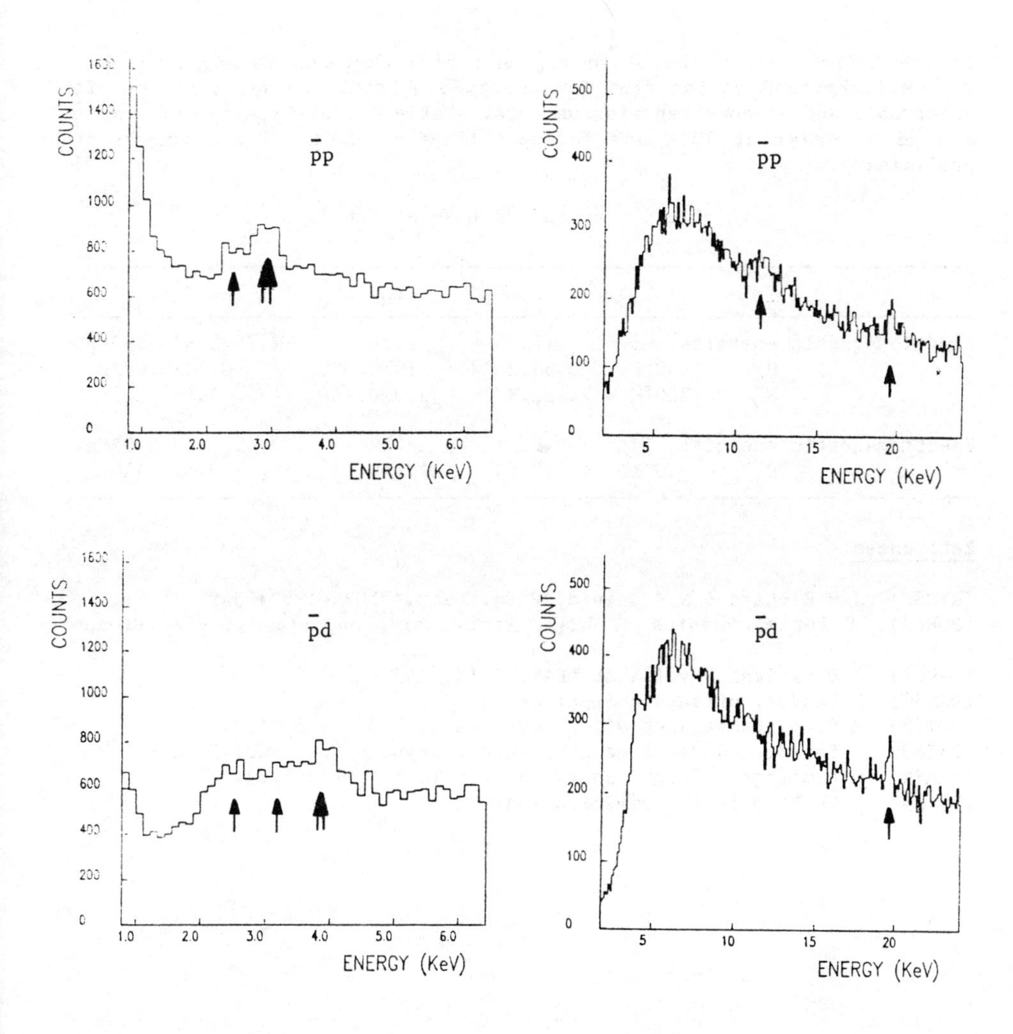

Fig. 4a and 4b. $\bar{p}p$ and $\bar{p}d$ X-ray spectra from gases at 1 atm. and 30°K. Low energy data (on left) have a 1.6μs time-window; high energy X-rays are within a 200ns window. Arrowed are $\bar{p}$ Aℓ(8→7) at 19.5 KeV in both spectra, a possible $\bar{p}p$ K X-ray peak at 11.7 KeV and various L-lines.

For the $\bar{p}d$ system QED energies are up by $\frac{4}{3}$ but there are no predictions of shifts, widths or yields other than the greater size and mass of the deuteron giving much lower yields than for protonium. There are no visible peaks in the K-line region of the $\bar{p}d$ spectrum. That the single-exponential fits to the $\bar{p}p$ and $\bar{p}d$ backgrounds have the same exponent strengthens the argument that the above peak at 11.7 KeV comes from hydrogen and is the first observation of $\bar{p}p$ K X-rays.

In the L-line region the above exponential background is augmented by a random background coming from low energy electronic noise, together with electronic and window-transmission cuts. Table I lists yields for some $\bar{p}p$ and $\bar{p}d$ L X-rays at 30°K and for $\bar{p}p$ L lines at 300°K. These numbers are preliminary.

Table I $\bar{p}p$ and $\bar{p}d$ L X-ray Yields

			L_α	L_β	$\Sigma L_{\geq\gamma}$
Electromagnetic energies, $p\bar{p}$			1.74	2.35	2.78,2.87,2.93KeV
	H_2	30°K	<0.66(2σ)%	0.19±0.08%	0.34±0.07%
	H_2	300°K	7.4±2.9%	2.0±0.9%	3.6±1.2%
Electromagnetic energies, $\bar{p}d$			2.31	3.12	3.49,3.69,3.82KeV
	D_2	30°K	4.5±0.5%	1.1±0.2%	3.5±0.4%

References

(RIC82) J M Richard & M E Sainio, Phys. Lett. 110B (1982) 349.
(BOR82) E Borie, Physics at LEAR, Proc. Workshop, Erice (1982) Plenum, 185.
(DAV82) J D Davies, Physics at LEAR, Ibid, 143.
(LAN84) R Landua, private communication.
(POT78) H Poth, R Abela et al., Phys. Lett. 76B (1978) 523.
(BAI83) S Baird, C J Batty et al., Nucl. Phys. A392 (1983) 297.
(DUM84) O Dumbrajs, Czech. Jour. Phys. (To be published).
(POT84) H Poth, private communication.

Study of $\bar{p}p \rightarrow e^+e^-$ reaction at rest

G. Bardin[4], G. Burgun[5], R. Calabrese[1], G. Capon[2], R. Carlin[3], P. Dalpiaz[1], P.F. Dalpiaz[1], J.P. de Brion[5], J. Derre[5], U. Dosselli[3], J. Duclos[4], J.L. Faure[4], F. Gasparini[3], M. Huet[4], C. Kochowski[5], D. Lafarge[5], S. Limentani[3], G. Marel[5], A. Meneguzzo[3], E. Pauli[5], F. Petrucci[1], M. Posocco[3], M. Savriè[1], A. Schuhl[4], G. Simone[6], L. Tecchio[6], C. Voci[3]

1 - Istituto de Fisica dell'Università, Ferrara and Istituto Nazionale di Fisica Nucleare, Bologna.
2 - CERN, Geneva.
3 - Dipartimento di Fisica and Istituto Nazionale di Fisica Nucleare, Padova.
4 - Departement de Physique Nucleaire des Hautes Energies, CEN, Saclay.
5 - Departement de Physique des Particules Elémentaires, CEN, Saclay.
6 - Istituto di Fisica Superiore and Istituto Nazionale di Fisica Nucleare, Torino.

Abstract. Preliminary results on e^+e^- pair production in $\bar{p}p$ annihilation of PS170 experiment at LEAR at CERN are presented.

In what follows we report the preliminary analysis of the PS170 experiment (P. Dalpiaz et al 1979; J.P. de Brion et al 1980) at LEAR, obtained with a 300 MeV/c incident $\bar{p}$ momentum. We look at the reaction

$$\bar{p}p \rightarrow e^+e^- \tag{1}$$

and we normalize the results with

$$\bar{p}p \rightarrow \pi^+\pi^-(K^+K^-)$$

The aim of the experiment is the measurement of the proton electromagnetic formfactors in the time-like region - (P.F. Dalpiaz 1984) and a detailed vector meson spectroscopy (J. Duclos 1984).

Our apparatus (G. Bardin et al 1984a) is located in the C1 line of the LEAR beam. It is constituted by:

- a C magnet (1.3 Tesla with a 40 cm. gap and 1 m. diameter);
- a liquid hydrogen target, 30 cm. long, surrounded by multiwire proportional chambers with cathode analogic read-out, inside the poles of the magnet;
- a 28-cell gas Cerenkov counter ($\hat{C}$);
- 100 drift tubes
- two hodoscopes (H1, H2) with 90 scintillators and 120 phototubes;

- an electromagnetic calorimeter (9 planes of limited streamer tubes with lead absorbers).

The rejection power against hadrons is 10^3 for the Cerenkov counter and 10^3 for the calorimeter. After the analysis of data the rejection power for a two body event is better than 10^{10}, which is largely sufficient to discriminate the reaction (1) from $\bar{p}p \to$ any, the branching ratio being 3×10^{-7} (at rest). The momentum resolution is about ± 2%.

A beam trigger (G. Bardin et al 1984b) enables a triple coincidence (H1. H2 . $\hat{C}$) for electron identification and a simple coincidence (H1 . H2) for hadron identification. A correlation between two particles is required to trigger the electronics.

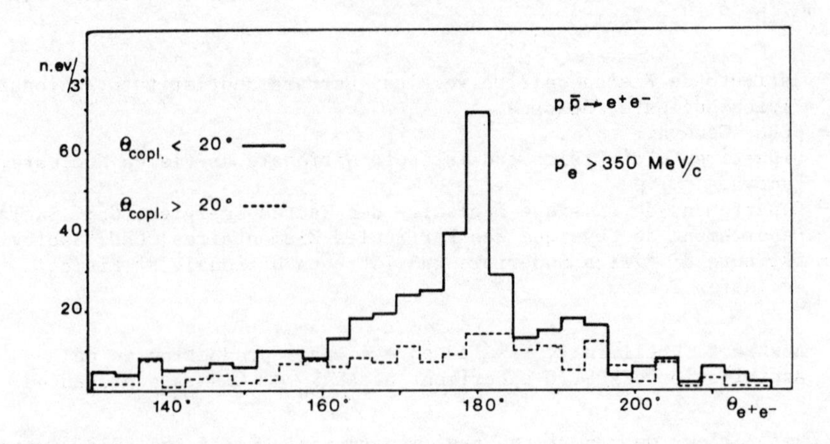

Fig. 1 Opening angle of electron-positron pairs.

Fig. 1 shows the opening angle distribution of electron-positron pairs with a cut on the electron's momentum (p_e > 350 MeV/c): the full line shows events with coplanarity angle < 20^0 and the dashed line shows events with coplanarity angle > 20^0. In the plot we have 598 events.

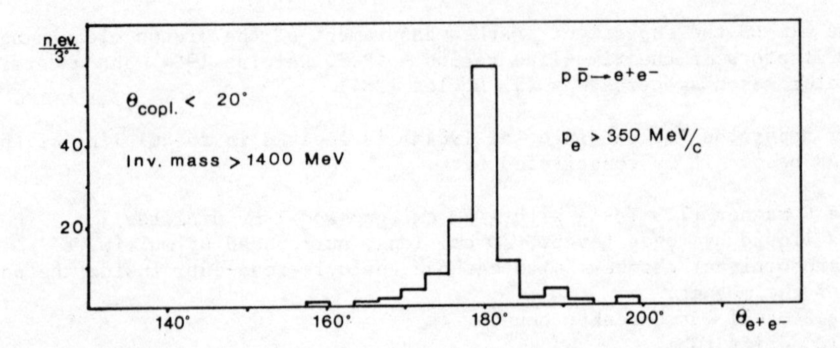

Fig. 2 Opening angle of e^+e^- pairs of events selected on total energy.

Fig. 2 shows the distribution of the opening angle of 118 electron-positron pairs, with coplanarity angle < 20^0, p_e > 350 MeV/c, and the invariant mass of the electrons pairs > 1400 MeV. The distribution shows a collinear peak of about 97 events at 180^0, and we can also recognize about 21 events in the range 155^0-178^0, attributed to the antiproton interaction in flight.

Similar spectra can be obtained with hadron trigger, and the normalization gives us the correct branching ratio $BR = \Gamma(\bar{p}p \rightarrow e^+e^-)/\Gamma(\bar{p}p \rightarrow h^+h^-) \sim 10^{-4}$, as expected (Bassompierre et al 1983).

Presently the analysis and the data taking of the experiment are in progress.

References

Bardin et al 1984a Physics at LEAR with Low-Energy Cooled Antiprotons, ed. by U. Gastaldi and R. Klapisch, p.347, Plenum Press.
Bardin et al 1984b Fundamental Interactions in Low Energy Systems, ed. by P. Dalpiaz and G. Fiorentini, Plenum Press.
Bassompierre et al 1983 Il Nuovo Cimento 73A, 347.
J.P. De Brion et al 1980 Proposal CERN/PSCC/80-95/PSSC/P25.
P. Dalpiaz et al 1979 CERN/PSCC/79-56/PSSC/17.
P.F. Dalpiaz 1984 Physics at LEAR with Low-Energy Cooled Antiprotons, ed. by U. Gastaldi and R. Klapisch, p.329, Plenum Press.
J. Duclos 1984 Physics at LEAR with Low-Energy Cooled Antiprotons, ed. by U. Gastaldi and R. Klapisch, p.339, Plenum Press.

Inst. Phys. Conf. Ser. No. 73: Section 3
Paper presented at VII Eur. Symp. Antiproton Interactions, Durham 1984

Results from PS 172: $\bar{p}p$ total cross-sections and spin effects in $\bar{p}p \rightarrow K^+K^-$, $\pi^+\pi^-$, $\bar{p}p$ above 200 MeV/c

C.I. Beard[4], R. Birsa[2], K. Bos[3], F. Bradamante[2], D.V. Bugg[4], A.S. Clough[5], S. Dalla Torre-Colautti[2], S. Degli-Agosti[1], G.A. Edgington[4], M. Giorgi[2*], J.R. Hall[4], E. Heer[1], R. Hess[1], J.C. Kluyver[3], R.A. Kunne[3], C. Lechanoine-Leluc[1], L. Linssen[3], A. Martin[2], Y. Onel[1], A. Penzo[2], D. Rapin[1], P. Schiavon[2] and A. Villari[2]

1. D.P.N.C., University of Geneva, Geneva, Switzerland.
2. I.N.F.N., Trieste and University of Trieste, Trieste, Italy.
3. NIKHEF-H, Amsterdam, The Netherlands.
4. Queen Mary College, London, Great Britain.
5. University of Surrey, Guildford, Surrey, Great Britain.

* At present Research Associate, CERN, EP division, Geneva.

Presented by S. Dalla Torre-Colautti and L. Linssen

Abstract

We report on the results of a $\bar{p}p$ high statistics total cross-section measurement between 388 and 599 MeV/c in small momentum steps showing no evidence for the S-meson and on the preliminary data obtained measuring the elastic $\bar{p}p$ differential cross-section at small scattering angles at 204, 233, 272 and 545 MeV/c.

1. Introduction

The experiment PS172 at LEAR (CERN, SING collaboration) has a variety of experimental aims [1]:

i) measurement of the $\bar{p}p$ total cross-section;
ii) measurement of the elastic $\bar{p}p$ scattering in the Coulomb-nuclear interference region;
iii) measurement of the carbon analysing power in small-angle $\bar{p}$ scattering as a preliminary to further experiments (construction of a polarized $\bar{p}$ beam);
iv) study of the two-body channels $\bar{p}p \rightarrow \bar{p}p$, $\pi^+\pi^-$, K^+K^- measuring $d\sigma/dt$ and the polarization parameters using a polarized target.

In this paper we report on the definitive results of the total cross-section (paragraph 2) and on the preliminary results of the elastic $\bar{p}p$ scattering at small angles (paragraph 3).

2. Results of the $\bar{p}p$ total cross-section measurement from 388 to 599 MeV/c in small momentum steps

For some years there has been a controversy about the existence of a resonance called the S(1936) meson, a baryonium candidate, at a $\bar{p}$ momentum of about 500 MeV/c. The situation was reviewed recently by Kamae [2]. In the hope of resolving some of this controversy, we have remeasured the $\bar{p}p$ total cross-section, σ_{TOT}, from 388 to 599 MeV/c at LEAR (the low energy antiproton storage ring at CERN) with good absolute and statistical accuracy and in small steps of momentum. This measurement has the advantage over previous experiments of a high intensity $\bar{p}$ beam free of contaminating particles and of small diameter and good momentum resolution.

The experimental set-up is shown schematically in Fig. 1.

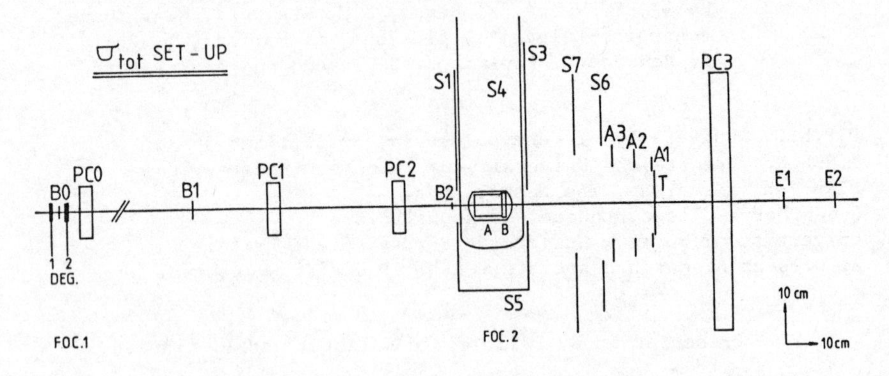

Fig. 1 Schematic layout of the experimental set-up.

It has been designed to overcome the disadvantages of scintillator measurements of total cross-sections as enumerated, for example, in Refs. [10] and [11].

The beam signal B was obtained from three 0.5 mm thin NE102A scintillators (B0, B1, B2) in coincidence, the last defining a beam spot of 1 cm maximum diameter. The last two defined a cone well clear of the walls of the hydrogen target (A) and small enough that multiple scattering made no measurable contribution to losses from the smallest transmission counter T. The beam was extracted from LEAR at 610.8 MeV/c, and the lower momenta were obtained using a carbon degrader (DEG) of variable thickness at an intermediate focus. Downstream a symmetric, achromatic beam line refocusses the beam to a spot of 0.5 cm diameter at B2. The beam intensity varied through a 60 min. spill typically from $1.2 * 10^5$ to $1.5 * 10^4$ $\bar{p}$/sec. Time of flight measurements over 20 m (B0-B1) agreed with calculated momenta following the degrader to within 2 MeV/c. The width of these timing spectra corresponded to a momentum spread of $\sigma = 4$ MeV/c. The coincidence between B0 and B1 excluded the tiny ($\leqslant 0.1\%$) contamination of pions and muons from the degrader. The direction and position of the incoming beam was monitored continuously by two multiwire proportional counters, PC1 and PC2, each with horizontal and vertical sense wires.

The measurements were done with a liquid hydrogen target, 8.33 ± 0.04 cm long and 4 cm diameter, containing a cylindrical insert of thin mylar, of diameter 3.5 cm, to reduce boiling. The target could be emptied for background measurements. The temperature of the liquid (and gas in empty target runs) was monitored by accurately calibrated germanium resistors. Due to the uncertainty in the effective length and the density of the target the measured total cross-sections have an absolute normalisation error of 0.7%. The momentum loss in the target varied from 10 MeV/c at 599 MeV/c (the maximum momentum at the centre of the target) to 14 MeV/c at 500 MeV/c and 25 MeV/c at 388 MeV/c. Data were taken in steps of momentum of about 10 MeV/c, but 5 MeV/c near 500 MeV/c. At this momentum the attentuation in the final beam counter, air and the empty target was 0.017, and attenuation in the target liquid was 0.056.

The transmitted beam was measured in thin (1.5 mm) NE102A scintillator counters (see Fig. 1) in air light guides, a circular one (T) of 6 cm radius and three overlapping concentric annular ones (A_{1-3}) of external radii 8.5, 9.9 and 10.8 cm; this arrangement was chosen to minimise annihilations as antiprotons passed through the array. The signals from these detectors were added electronically via OR-gates to form T_i (e.g. $T_3 = T + A_1 + A_2$) and a set of four transmissions $B.T_i/B$. These spanned uniformly a range of momentum-transfer squared $|t|$-values typically 15×10^{-4} to 75×10^{-4} $(GeV/c)^2$. The efficiency of the in-beam detector T was monitored continuously using beam particles defined by the telepscope $B.E_1.E_2$.

A box of counters S_{1-5} surrounded the target except for (i) a hole at the top, where the target was suspended from a reservoir, and (ii) 6 cm diameter holes at the beam entrance and exit. The angular range between the downstream hole and the transmission array was covered by two movable annular counters S_{6-7}. The sum of the counter outputs, S, recorded predominantly annihilations into charged pions. It was used as a veto to derive a second set of transmissions, $B.T_i.\bar{S}/B$, with a slope dictated by elastic scattering only.

After correcting for detection efficiency and for accidental coincidences and vetoing, partial cross-sections were obtained in separate analyses of the two sets of transmissions. These were corrected for single Coulomb scattering and Coulomb-nuclear interference, using formulae quoted most recently by Cresti et al. [3]. For ρ, the ratio of the real and imaginary parts of the forward elastic scattering amplitude, we used a linear fit to their experimental results: $\rho = 1.333p - .662$, where p is the laboratory momentum in GeV/c. Corrections were also made for small full-empty target momentum differences and the hydrogen gas present in the empty runs. Total cross-sections were determined by linear extrapolation to $|t| = 0$. Excellent agreement (to $\leqslant 0.3$ mb generally) was obtained between the two separate analyses. At every momentum subsets of data have been checked and agree within statistical fluctuations. The absolute magnitude of the results, when corrected for the value ρ used, agrees within 0.3% with that of Sumiyoshi et al. [10], is $\sim$ 2% higher than that of Hamilton et al. [8], within normalizationm uncertainty, and is $\sim$ 3% higher than that of Nakamura et al.[11]. The trend of the data is very smooth (Fig. 2). Statistical errors are smaller than 0.5%. Results are fitted to the function a + b/p (solid curve Fig. 2). Values of a = 65.8 mb and b = 53759 mb MeV/c are found with a total χ^2 of 40.9 for 22 degrees of freedom.

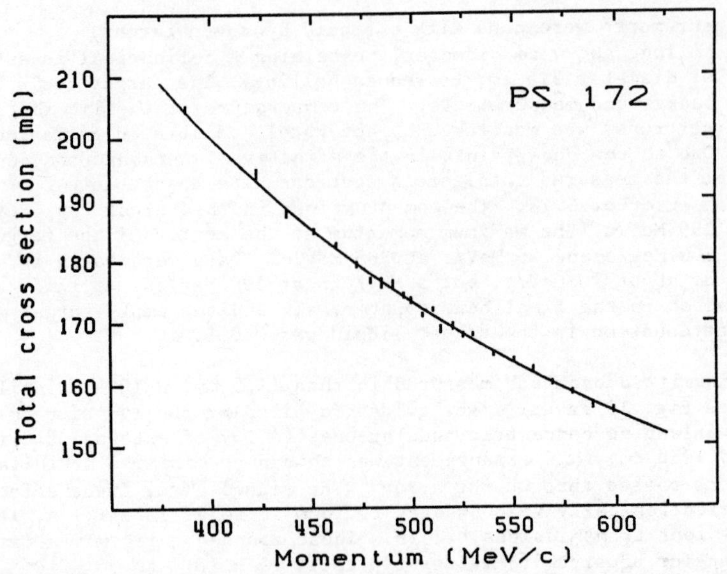

Fig. 2 $\bar{p}p$ total cross-section as a function of laboratory momentum. The solid curve shows the result of a fit to the function a + b/p, where a = 65.78 ± 1.71 mb, b = 53759 ± 845 mb MeV/c, χ^2 = 40.9.

Adding a Breit-Wigner of large width with the fixed parameters of Hamilton et al. and refitting, χ^2 deteriorates to 111. Adding in quadrature an estimated 0.25% point-to-point systematic uncertainty, due to variation in the effective target length and beam momentum uncertainty, χ^2 improves to 21.1 for 22 degrees of freedom. Adding to this function a Breit-Wigner of narrow width, folded with the mass resolution due to beam momentum spread and energy loss in the target ($\sigma \sim 1.4$ MeV/c²), χ^2 does not improve.

The limit at 90% confidence level for the strength, width * height (c * Γ), of a possible resonance with Γ = 4 MeV/c² as a function of momentum is shown in Fig. 3. As a result of this calculation we quote an upper limit of 2 mb MeV/c² for the strength of a resonance narrower than our experimental resolution (Γ = 3.5 MeV/c²) near 500 MeV/c. This limit can be compared with previous results in Table 1. All experiments mentioned in Table 1 refer to total cross-section measurements, except ref. [6], which refers to the sum of the charged annihilations cross-section and the elastic cross-section. Refs. [4] to [8] actually found a resonance in their data, with given parameters. Refs. [9] to [12] calculate an upper limit for a possible resonance at 90% confidence level around 500 MeV/c. These limits are given in column 7.

We intend to continue measurements of the total cross-section over a broader momentum range if and when beam time is available at LEAR.

Table 1

Summary on the status of the S-meson in $\bar{p}p$ total cross-section measurements

Year	Experiment	Ref.	p (MeV/c)	Mass (Mev/c²)	Width Γ (MeV/c²)	Strength,c*Γ (mbMeV/c²)
1974	Caroll et al.	[4]	475	1932±2	9^{+4}_{-3}	162±25
1976	Chaloupka et al.	[5]	491	1936±1	$8.8^{+4.3}_{-3.2}$	93±22
1977	Brückner et al.	[6]	505	1939±3	≤ 4	46±12
1979	Sakamoto et al.	[7]	489	1936±1	2.8±1.4	41±23
1980	Hamilton et al.	[8]	505	1939±2	22±6	66±24
1980	Kamae et al.	[9]			≤3	<39
1982	Sumiyoshi et al.	[10]			≥10	<10
1984	Nakamura et al.	[11]			≤4	<24
1984	This experiment	[12]			≤3.5	<2

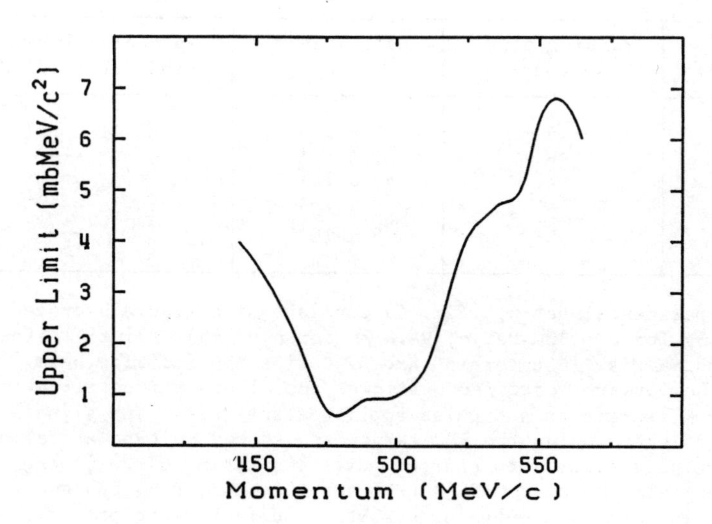

Fig. 3 Upper limit at 90% confidence level for the strength (height * width) of a resonance with width Γ = 4 MeV/c² as a function of momentum

3. Measurement of the phase of the elastic $\bar{p}p$ scattering amplitude at 204, 233, 272 and 545 MeV/c

The measurement of $d\sigma/dt$ of $\bar{p}p$ elastic scattering at small angles can provide both the real and imaginary part of the dominant non-flip amplitude using the interference between the nuclear and Coulomb amplitudes. This information is interesting at low energy because of the existence of a long unphysical region (annihilation cut for $s < 4\ m^2$) and possibly of quasi-bound states (mass $\leqslant 2$ m) and resonances (mass $\geqslant 2$ m).

Since Kaseno et al. published their results for the measurement of $\rho = \mathrm{Re}\ f_N\ (t = 0)\ /\ \mathrm{Im}\ f_N\ (t = 0)$ at $p = 697$ MeV/c [13], it has been stated [14] that the standard dispersion relation calculations were not able to reproduce the ρ behaviour versus momentum at low energy and that it was necessary to introduce in the calculation an extra high mass pole term. New sets of data refs. [15], [11], [3] and [16] measuring ρ down to about 400 MeV/c seem to confirm the inadequacy of the standard dispersion relation calculations, even if there is not complete agreement among the results of the different measurements. New data were required, in particular investigating the momentum region below 300 MeV/c, where the ρ behaviour is more sensitive to the scattering amplitude structure near threshold and in the unphysical region.

We have measured the small angle elastic scattering at three momenta between 204 and 272 MeV/c and at $p = 545$ MeV/c. Table 2 gives the summary of the data. The measurements at 204 and 233 MeV/c were achieved by degrading the primary LEAR beam; the data at 545 MeV/c are obtained as a by-product of the C analysing power measurement.

Table 2
Data Summary

p (MeV/c)	running time (h)	useful beam	$\bar{p}$ p elastic events ($10^{-3} < \lvert t \rvert < 10^{-2}$ GeV^2)
204.	5	7×10^6	1×10^4
233.	26	62×10^6	---
272.	8	37×10^6	4×10^4
272.	5	32×10^6	---
545.	66	5×10^6	---

The experimental set-up (Fig. 4) consists of a liquid hydrogen target (11.5 mm long for $p < 300$ MeV/c, 94.8 mm for $p = 545$), scintillation counters and MWPC's. Counters B0 and B2 define the incoming beam. Counter R detects the forward scattered particles and allows $\bar{p}$ identification through TOF measurements and pulse amplitude analysis. The signals from the veto counters surrounding the LH2 target are stored on tape and allow to reject $\bar{p}$ annihiliations into charged particles taking place in the target region. Two telescopes of MWPC's (PC1 and PC2; PC3, PC4, PC5 and PC6) measure the incoming and outgoing tracks. The fast micro-processor ESOP rejects on-line the tracks in the forward direction according to the hits in the last chamber of the second telescope.

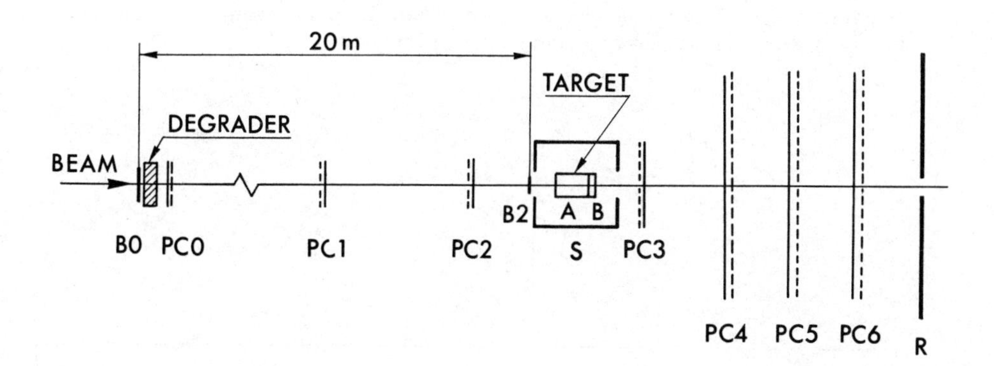

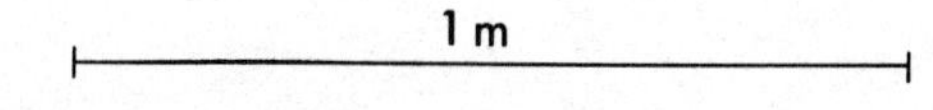

Fig. 4 Experimental set-up

We have collected samples of events for the full target and for the empty target (about the same statistics).

In the off-line analysis we have selected events requiring:

- pulse amplitude in the R counter larger than a threshold value adjusted to select $\bar{p}$,

- reconstructed vertex lying in the target fiducial volume.

Before performing the subtraction of the spectra obtained with full and empty target, we have smeared the empty target data adding randomly to each event the effect of multiple scattering in the liquid hydrogen.

The geometric acceptance of the apparatus has been evaluated using a Monte Carlo calculation: for example, at 272 MeV/c the acceptance is constant and equal to 100 % between $|t| = 10^{-3}$ and $|t| = 10^{-2}$ $(GeV/c)^2$ and larger than 50% between $|t| = 0.3 * 10^{-3}$ and $|t| = 13 \times 10^{-3}$ $(GeV/c)^2$. A second correction factor is applied to the data at $|t|$ smaller than 10^{-3} $(GeV/c)^2$ to account for the on-line cut of ESOP (50% efficiency at $t = 0.3 \times 10^{-3}$ $(GeV/c)^2$).

The preliminary spectra obtained at 272 and 204 MeV/c (after applying the acceptance corrections) are presented in Fig. 5. ρ will be obtained by fitting to these spectra the theoretical $d\sigma/dt$ computed using the standard formulas (see, for example, ref. [11]) and folded with the experimental resolution (due essentially to the multiple scattering in the target material: $\sim 1.4°$ at 272 MeV/c). We have to recall that, in this process, two parameters are needed: the slope of the nuclear scattering amplitude and the total cross-section. This experiment is not very sensitive to the former (due to the angular acceptance) so we will use an extrapolation of the existing data [11], while for σ_{TOT} we will use the fitted parameters of our measurement described in paragraph 2 and anyhow further measurements at lower momenta are planned.

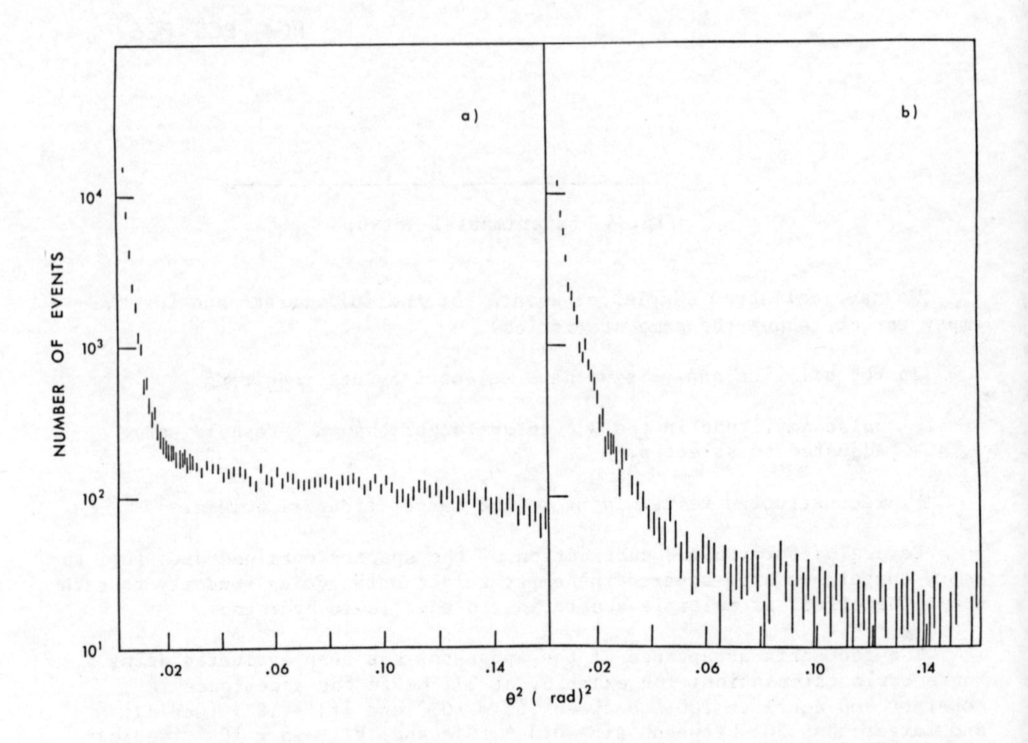

Fig 5 Preliminary $d\sigma/d\theta^2$ distributions at 272 MeV/c (a) and 204 MeV/c (b)

REFERENCES

[1] J. Bailey et al., CERN/PSSC/80-76, 21 July (1980).

[2] T. Kamae, Nucl. Phys. A374 (1982) 250.

[3] M. Cresti et al., Phys. Lett. 132B (1983) 209.

[4] A.S. Caroll et al., Phys. Rev. Lett. 32 (1974) 247.

[5] V. Chaloupka et al., Phys. Lett. 61B (1976) 487.

[6] W. Brückner et al., Phys. Lett. 67B (1977) 222.

[7] S. Sakamoto et al., Nucl. Phys. B158 (1979) 410.

[8] R.P. Hamilton et al., Phys. Rev. Lett. 44 (1980) 1182.

[9] T. Kamae et al., Phys. Rev. Lett. 44 (1980) 1439.

[10] T. Sumiyoshi et al., Phys. Rev. Lett. 49 (1982) 628.

[11] K. Nakamura et al., Phys. Rev. D29 (1984) 349.

[12] A.S. Clough et al., preprint Surrey University, to be published in Phys. Lett. B (1984).

[13] H. Kaseno et al., Phys. Lett. 61B (1976) 203.

[14] H. Kaseno, Nuovo Cimento 43A (1978) 119.
W. Grein, Nucl. Phys. B131 (1977) 255.

[15] H. Iwasaki et al., Phys. Lett. 103B (1981) 247.

[16] T. Takeda et al., A measurement of the real-to-imaginary ratio of the pp and pN forward scattering amplitudes, UT-HE-84/08, July (1984), presented at this conference.

Inst. Phys. Conf. Ser. No. 73: Section 3
Paper presented at VII Eur. Symp. Antiproton Interactions, Durham 1984

Measurement of p̄p cross-sections at low p̄ momenta

W. Brückner, H. Döbbeling, D. von Harrach, H. Kneis, S. Majewski, M. Nomachi, S. Paul, B. Povh, R. D. Ransome, T.-A. Shibata, M. Treichel and Th. Walcher

Max-Planck-Institut für Kernphysik and Physikalisches Institut der Universität Heidelberg, Heidelberg, Germany

Abstract

In a recent experiment at the Low Energy Anti-Proton Ring (LEAR) at CERN, the p̄p differential elastic and charge exchange (CEX) cross sections as well as the annihilation cross section to charged and neutral pions have been measured between 150 and 600 MeV/c. A description of the experiment and some preliminary results are presented.

I. Introduction

The proton-antiproton reaction has long been of particular interest. Until recently, however, experiments had been limited to either higher energies, with 400 Mev/c being the effective lower limit, or stopped anti-protons. Almost nothing was known of the reaction between zero and 400 MeV/c (a kinetic energy of 100 MeV), the energy range in which much of what we know of the nucleon-nucleon interaction was learned. This region became accessible with the completion of LEAR (Gastaldi 1984) at CERN. LEAR provides antiproton beams of high intensity (up to 10^6 p̄/s) and high quality ($\Delta p/p$ less than 0.5% and no pion contamination) at momenta between 100 MeV/c and 2000 MeV/c. The first physics runs at LEAR occurred during November and December of 1983, with two fixed beam momenta of 600 and 300 MeV/c. A second run at 300 MeV/c was made during April, 1984.

In this paper we report on experiment PS 173, which aims to measure the differential elastic and charge exchange cross sections over the full angular range, and the annihilation to charged and neutral pions in the momentum range of 150 MeV/c to 1000 MeV/c (a kinetic energy range of 12 to 430 MeV).

II. Physics Motivation

The N̄N interaction is a promising tool for studying the strong interaction. It provides a complement to the NN interaction, because the long range part of the N̄N force can be related to the NN force via the G-parity transformation. Furthermore, it provides another channel, the annihilation, not available in the NN interaction. Another peculiar aspect of the N̄N system is that the total baryon number is zero.

Through the annihilation channel we expect to learn about the quark confinement radius. Although QCD is widely accepted as the theory of strong interactions, the nature of confinement is poorly understood. One would like to distinguish between interactions involving meson exchanges and those involving quark-gluon interactions. This distinction cannot be made in the nucleon-nucleon case, especially at low momenta, because regardless of the interaction there remain two nucleons in the final state. In contrast, the nucleon-antinucleon system has two distinct final states: the nucleon-antinucleon final state and the multiple meson final state. The first case (elastic or CEX in the $\bar{p}p$ system) should be dominated by meson exchanges, while the second case (annihilation) must involve quark-quark interactions. Thus by studying these final states and their momentum dependence, it should be possible to extract some information regarding the ranges of the $\bar{N}N$ interaction and of the quark confinement.

The second feature of the $\bar{N}N$ system, baryon number zero, allows access to possible states inaccessible from the NN system. A possibility which has been extensively discussed is that of exotic mesons, for example the so-called baryonium or glueballs. Baryonium e.g. would be a $qq\bar{q}\bar{q}$ state which should couple strongly to the $\bar{N}N$ system, while glueballs consist solely of gluons. Interest in baryonium increased greatly when experiments reported structure in the total and annihilation cross sections. Later experiments, however, gave contradictory results and reported little or no structure. This caused interest in baryonium to fade. However, another series of experiments with better resolution again reported structure, although narrower and weaker (Walcher 1984). It is expected that LEAR will finally give sufficient intensity and high enough quality beam to settle this question.

Elastic and CEX provide information about the $\bar{N}N$ system in a similar fashion to the NN system. The $\bar{N}N$ system differs from the NN system in that all combinations of spin and isospin are allowed, and inelasticity exists even at the lowest momenta. This makes the system much more difficult to unravel and a phase shift analysis of the quality that exists in NN scattering is impossible with today's means.

Two measurements which can be done now are the elastic and CEX differential cross sections. The elastic cross section allows a determination of ρ, the ratio of the real to imaginary part of the scattering amplitude at zero degrees, via the interference of the Coulomb and strong amplitudes. The ρ parameter is sensitive to resonance type behavior. The CEX cross section can be related to np CEX through crossing relations. The np CEX is characterized by a sharp peak with a width characteristic of one pion exchange interfering with a constant background. Measurements of pp CEX scattering show a similar shape (Bogdanski 1981, Tsuboyama 1983, and Nakamura 1984), although they also indicate some possible structure not seen in the np cross sections. More precise measurements of CEX scattering at low momenta should give some insight into the nature of the background and the long range part of the NN force. The elastic and CEX can be combined to give some information on the isospin dependence of the interaction, since the elastic scattering is proportional to the square of the sum of the isospin zero and isospin one amplitudes while the CEX scattering is proportional to the square of the difference.

III. Experimental Setup

Fig. 1 shows a horizontal cross section of the apparatus in the scattering plane and in Fig. 2 a view from the side. The main components are a beam defining detector (SD), a cylindrical multi-wire proprotional chamber, a forward hodoscope array (FHD), an array of anti-neutron detectors (ANC), upper, lower and back hodoscopes (UHD, LHD and BHD), and lead-glass detectors (Pb G). Two scintillation detectors are in the beam line (BA1 and BA2) for vetoing the beam and for making small angle

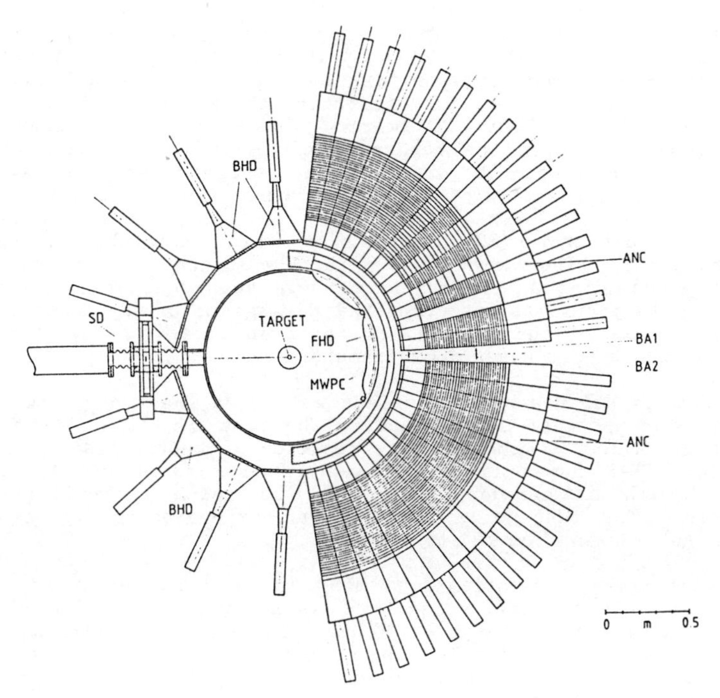

Fig. 1 Section of the apparatus viewed from above, in the scattering plane.

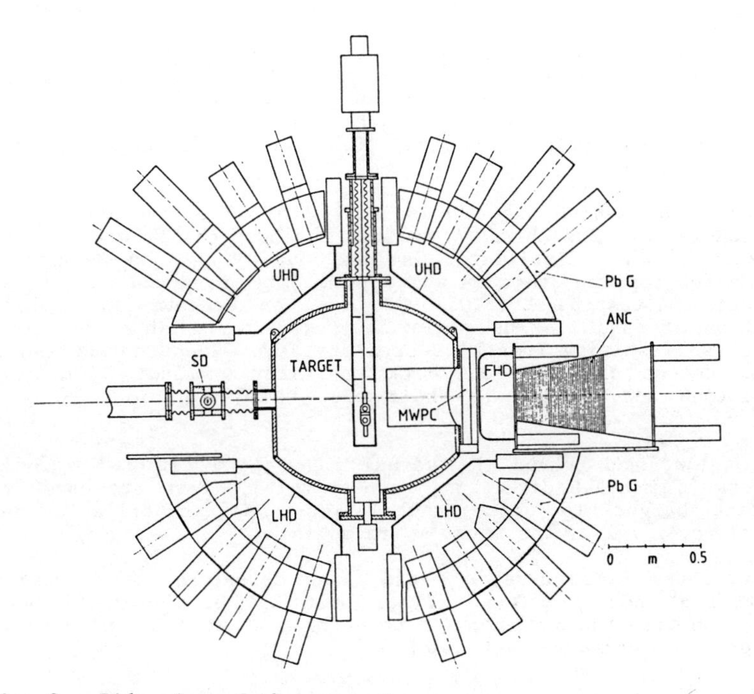

Fig. 2 Side view of the apparatus.

scattering measurements. The target consists of a 20 mm cell of liquid hydrogen, a 6 mm cell of liquid hydrogen, identical empty cells, and a 5 mm thick 3 mm diameter carbon target. In addition, not shown here, there is a carbon degrader, of variable thickness, and a scintillation detector, both about 20 meters upstream. The degrader allows measurements between the rather widely separated primary momenta which have been available to date. The upstream detector serves for a measurement of the time of flight for a precise momentum determination.

The beam detector (SD) consists of two phototubes viewing a thin scintillator. The scintillator thickness can be between 25 and 200 micron, depending on the momentum range being measured. SD is about 85 cm from the focus (target) thus the divergence of the beam can be limited by the choice of the diameter of the scintillator to ± 10 mr. The beam spot at the target has a 5 mm diameter.

The beam is defined by a coincidence between the two photomultipliers of SD and the upstream scintillator. The time of flight between these two detectors allows a momentum determination of better than 0.5%. As stated earlier, LEAR has only run at two fixed momenta so far, 600 and 300 MeV/c, with a degrader being used to reach intermediate and lower momenta. Thus, depending on the amount of degrading, the beam intensity on target varies from 10^5 to 10^3 $\bar{p}$/s.

The multi-wire proprotional chamber is used to measure the scattering angle for elastic scattering. It is a cylindrical chamber covering ± 75^0, with an active height of 30 cm. It contains two wire planes inclined at ± 12^0. The scattering angle can be determined with an angular resolution of 1^0, which includes the finite target spot and beam divergence.

The FHD consists of 32 scintillators, each 5^0 wide at a distance of 66 cm from the target. They allow for particle identification by time of flight (TOF) and energy loss. They can thus be used to distinguish between an elastically scattered proton or anti-proton, pions resulting from annihilation in the target, and pions resulting from an anti-proton annihilation in the detectors or vacuum tank wall. Our energies are below the single pion production threshold, so all pions result from an anti-proton annihilation somewhere. The time resolution of the detectors is about 0.5 nsec.

Starting 14 cm behind the FHD is a ring of 32 anti-neutron detectors. These consist of 50 slabs of 4 mm iron and 50 slabs of 6 mm scintillator interleaved. The scintillator is read out by two wave-length shifter bars, one on top the other on the bottom. By using the scintillator/WLS combination Altustipe UV15105/390B, we can achieve an overall time resolution of about 1.7 ns, allowing a separation of anti-neutrons and gamma rays (resulting from π^0 decays) by TOF. These detectors are also used to detect charged pions from annihilations and, of course, also anti-protons and protons if their energy is sufficient to enter the module.

Above, below, and behind the vacuum tank are additional scintillation detectors (8 in each direction, respectively). These are used primarily to detect charged pions from annihilations. All scintillation detectors in total cover 74% of the total solid angle.

Finally, there is an array of lead glass detectors for detecting gamma rays from π^0 decays. These cover about 25% of the solid angle. In conjunction with the anti-neutron detectors, about 35% of the solid angle is covered for gamma ray detection.

The signature for an elastic scatter is then a high pulse height at the right TOF in one of the forward hodoscopes. Protons from large angle scattering can be distinguished from anti-protons by the presence of pions in the other detectors. If the anti-proton is scattered forward pions will be seen with a late TOF. If the proton is scattered forward pions with the timing of annihilation in the target will be seen from the backward scattered anti-proton (which is so slow that it annihilates in the target). For angles near 90^0 center of mass both the proton and anti-proton are seen in kinematical coincidence.

CEX is identified by a hit in an anti-neutron detector with the right TOF and no associated forward hodoscope or wire chamber hit. We also require that no other hodoscope fire with a timing associated with an anti-proton annihilation in the tank wall (an important source of background at some angles). A TOF window of 6 ns also eliminates essentially all of the cosmic ray backgound.

The absorption efficiency of the anti-neutron detectors can be estimated from the known anti-neutron absorption cross sections. This gives about 90% absorption. Though the cross sections are not well known, their error is propagated little because of the thickness of the detectors. In order to better determine the efficiency, we have replaced the iron in some modules with styro-foam, as indicated in Fig. 2. Thus one module has no iron, one has 1/4, then 1/2, and finally 3/4 of the normal amount. By comparing the left and right sides, we will be able to determine the efficiency. At this time, we do not have sufficient data to make this determination.

Annihilations are identified by pions at the correct TOF in any of the scintillation detectors or the lead glass. As stated earlier, these detectors cover about 74% of the total solid angle. If we take multiplicity distributions from bubble chamber measurements, we can estimate the approximate efficiency of the apparatus for detecting annihilations to charged pions. This turns out to be about 93%, due to the high multiplicity of annihilations (about 90% have more than 4 pions). Since the multiplicity slowly increases with energy in the range of our measurements, there should basically no energy dependence for annihilation detection for our measurement. Once again, we can easily separate annihilations which occur in the target from those which occur in the SD detector, or in the beam detectors, by TOF.

IV. Results

The results presented here have to be considered as preliminary because in the short runs performed not all systematic errors could be studied. Also, we will take more data during 1984 and 1985, enabling us to make checks of the systematics of the apparatus and improve the statistics of the measurements.

First, shown in Fig. 3 is the differential cross section for elastic scattering at 290 MeV/c. A comparison of this with Fig. 4, the differential cross section for pp scattering illustrates some important features. The $\bar{p}p$ cross section drops off smoothly and is about a factor of 10 lower at 90^0 cm than at 20^0. In contrast, the pp cross section has a pronounced dip near 20^0 then rises slightly and is nearly flat to 90^0.

The region around 20^0 is where the Coulomb and nuclear scattering amplitudes are nearly equal in magnitude. The interference of the two amplitudes in this region affects the shape of the cross section and is strongly dependent on ρ. Fig. 3 shows a fit to the cross section. We find that a ρ value close to zero ($-.055 \pm .016$) gives the best fit. The value of ρ for pp scattering is 1.5, giving the dip, as shown in Fig. 4.

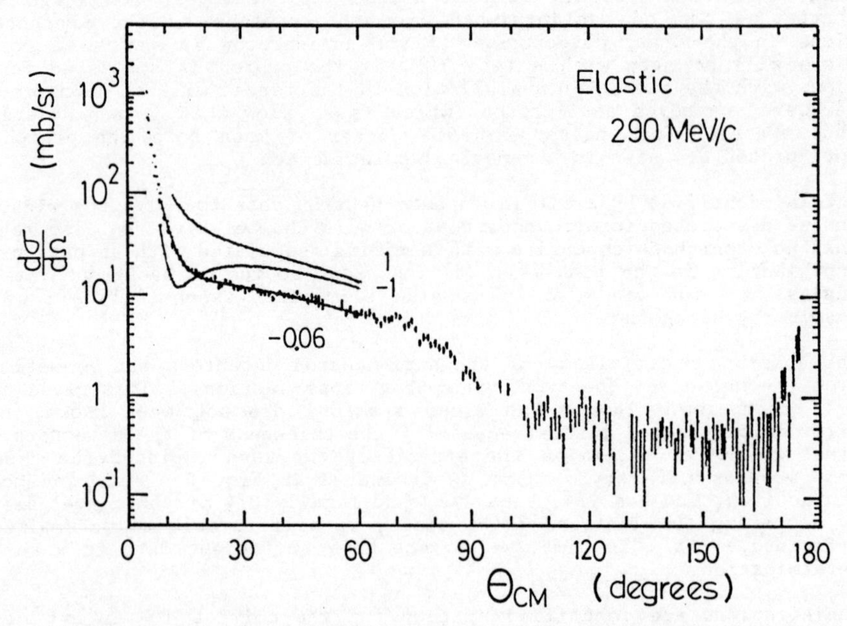

Fig. 3 The differential elastic cross section at 290 MeV/c vs. the cm scattering angle. The best fit curve is calculated using the standard parameterisation for σ_{tot}=237±3, b= 44.9±2 $(GeV/c)^{-2}$ and ρ = -0.055±0.016. Two other curves using the same σ_{tot} and b values and ρ = 1 and -1 are shown to illustrate the sensitivity to the ρ parameter.

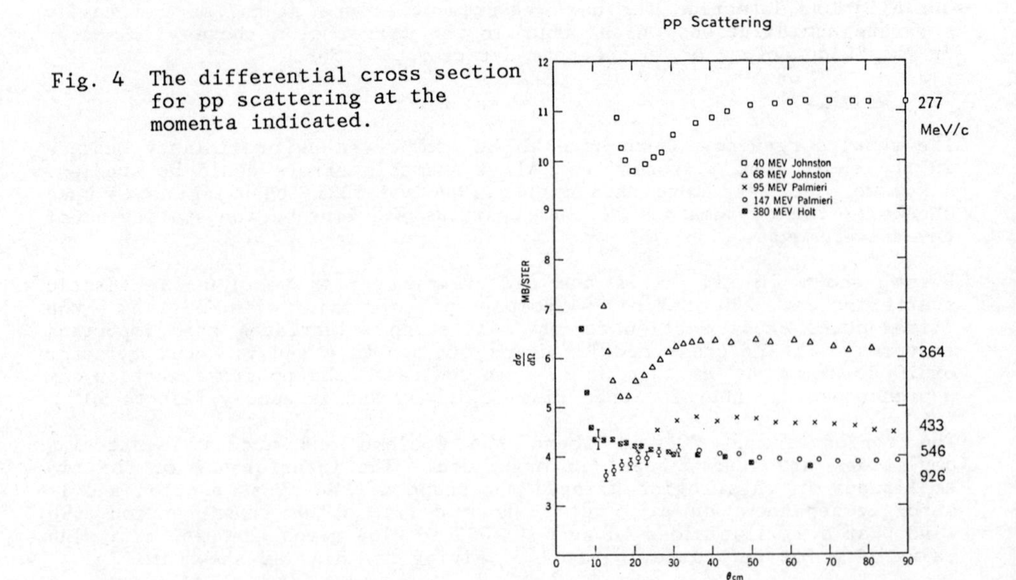

Fig. 4 The differential cross section for pp scattering at the momenta indicated.

The trend of earlier measurements at higher momenta had been toward much more negative values of ρ (Cresti 1983). Dispersion relation calculations which fit the previous data had indicated a value of ρ between -.5 and -1. at this momentum (Iwasaki 1981).

The drop off of the cross section in $\bar{p}p$ scattering indicates that s-wave scattering does not dominate the cross section. We have done a fit to the cross section using an expansion of Legendre polynomials. Fig. 5 shows the energy dependence of the first 4 coefficients. The importance of $\ell \geq 2$ partial waves decreases sharply with decreasing momentum, while p-waves remain important (about 50% of the cross section) to as low as 225 MeV/c. This is in sharp contrast to the pp case, where s-wave scattering gives 73% of the cross section at 290 MeV/c and 90% and 225 MeV/c.

Next, Fig. 6 shows the differential cross section for CEX at 290 MeV/c. The most striking feature of the cross section is that it is strongly forward peaked. This provides further support that at 290 MeV/c p-wave scattering is still comparable to s-wave scattering. In the np system at this momentum s-wave scattering constitutes about 90% of the cross section. A comparison of the CEX and elastic cross sections shows that at forward angles the elastic cross section is about 3-4 times larger than the CEX. This indicates that the two isospin amplitudes have different amplitudes and different relative contributions of the partial waves.

Fig. 5 The Legendre polynomial coefficients fit to the elastic data as a function of momentum.

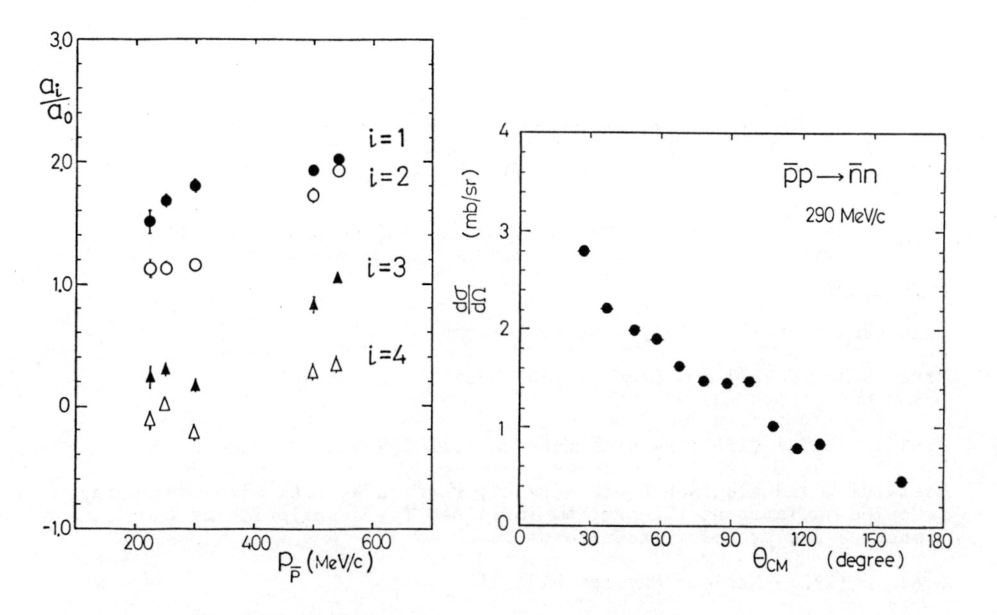

Fig. 6 The differential charge exchange cross section at 290 MeV/c.

Finally, we have the annihilation cross section to charged pions, shown in Fig. 7. It was in this cross section that two previous measurements reported a narrow enhancement, often called the "S-meson" (Walcher 1984). The current measurement is not in disagreement with these measurements, but until more checks of the systematics of our apparatus can be made no firm statement can be given on this enhancement.

Otherwise, the measurement shows a 1/p behaviour in the region above 400 MeV/c. However, at the lower momenta, between 150 and 290 MeV/c, there seems to be a rise in cross section slightly steeper than the simple empirical 1/p extrapolation would indicate. The cross sections at lower momenta are also in disagreement with the Bryan and Phillips (1968) predictions, which gave a larger elastic cross section and a less steep rise of the annihilation cross section at the lower momenta.

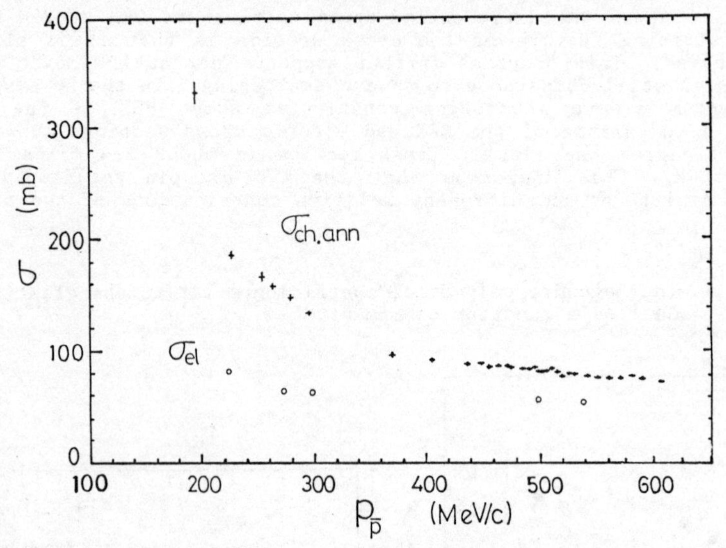

Fig. 7 The annihilation cross section to charged pions and total elastic cross section as a function of momentum.

References

Bogdanski M et al. (1981) Physics Letters 62B 117

Bryan R A and Phillips R J N (1968) Nucl. Phys. B7 481 and Nucl. Phys. B5 201

Cresti M et al (1983) Physics Letters 132B 209

Gastaldi U and Klapisch R eds. (1984), Physics at LEAR with Low-Energy Cooled Antiprotons (Plenum:New York) see for description of the machine and proposed experiments.

Grein W (1977) Nuclear Physics B131 255

Iwasaki H et al. (1981) Physics Letters 103B 247

Nakamura K et al. (1984) Physical Review Letters 53 885

Tsuboyama T et al. (1983) Physical Review D28 2135

Walcher Th (1984) Physics at LEAR with Low-Energy Cooled Antiprotons, Gastaldi U and Klapisch R eds. (Plenum:New York) p 375

Annihilation mechanisms

I S Shapiro

Lebedev Physical Institute, Moscow, USSR

Abstract. It is shown that the $\bar{N}N$ annihilation from continuum is significantly enhanced by strong $\bar{N}N$ nuclear attraction and therefore large annihilation cross-section can coexist with narrow $\bar{N}N$ bound and resonant states.

1. Introduction

I consider here only the annihilation of low energy $\bar{N}N$ pair ($\bar{N}N$ c.m. momenta $\kappa \ll m$, m is the nucleon mass) *) . For this momentum range all annihilation phenomena are strongly affected by nuclear $\bar{N}N$ interaction, whereas the initial annihilation transition (IAT) itself may be presented by few quantities which are practically independent on kinematical variables (energy, momentum transfer and so on). Even the magnitudes of these constants have nothing common with "visible" values of annihilation cross-sections and widths of $\bar{N}N$ nuclear or $\bar{p}$-atomic states.

Therefore the problem of understanding the low energy $\bar{N}N$ physics reduces mostly to finding the right way of introducing some properly choosed annihilation parameters into the $\bar{N}N$ interaction amplitudes.

The key for uncovering the annihilation mechanism is the fact that the annihilations distances are always of the order of 1/m. The IAT amplitudes depending on k/m (A) are expected to be smooth functions at low energies. In contrast the nuclear $\bar{N}N$ Green function (G) varies essentialy within the momentum interval $\Delta\kappa \sim 1/R$ ($R \gg$ 1/m is the radius of nuclear forces). Therefore the details of A's κ - dependence are not very significant. It is only important that A must be a smooth function compared to G . An approximate factorization relation follows this [1]. As usually the observable quantities are obtained by integrating the products like GA and because of the feature mentioned above we have:

---

*) Let me emphasis that the epithet "low energy" belongs only to the $\bar{N}N$ c.m. energy. The low energy $\bar{N}N$ physics in this sense can appear also in production processes at high energies.

$$\text{N},\ \bar{\text{N}}:\quad (G)\,[A] = \int G(\kappa,\kappa')A(\kappa,\kappa')\,d\kappa' = A(\kappa,\bar{\kappa})\int G(\kappa,\kappa')\,d\kappa' \tag{1}$$

with
$$0 \leq \bar{\kappa} \lesssim 1/R$$

Strictly speaking the G in eq.(1) is also a functional of A. If the relation (1) does not hold it means that the very details of $N\bar{N}$ interaction at small distances affect the $N\bar{N}$ scattering even at low energies *). If so, the potential approach becomes meaningless. I don't belive that it is true for real $N\bar{N}$ physics, and I make this remark only because several calculation schemes were published which seems to be inconsistent from the aspect mentioned above. The validity of the factorization relation is a crutial point for each theory in which the potential approach is used [2-3].

On my opinion a reliable calculation of A from the "first principles" is hardly possible on the present stage of our knowledge neither in the quark nor in the hadron fields models. In the last case, for instance, it is necessary to calculate a sum of diagrams like one shown below

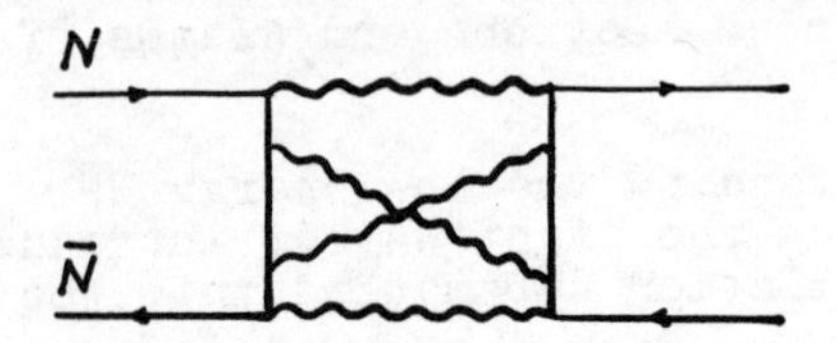

Fig. 1
Annihilation diagram

taking into account graphs with infinite number of internal boson lines in the S-channel. It is easy to see that such graphs are important for the real part of the annihilation interaction "potential" V_a : they contain the t-exchange with heavy "baryonium" mesons (with masses close to 2m) strongly coupled to $N\bar{N}$ (it is clear that we meet here the well known boot-strap problem). Instead of this only graphs which contribute to $Im\,V_a$ are usually treated (and mostly with two pion S-channel exchange). Going to the current quark $N\bar{N}$ annihilation models I would like to remark that if such models are combined with potential approach for the nuclear $N\bar{N}$ interaction than it seems to be the same as to introduce fiting parameters into IAT amplitude A (as I mentioned above at low $N\bar{N}$ energies the observable annihilations cross-sections are not sensetive to the A's κ - dependence).

*) This may be caused by nearthreshold poles appeared as a result of very strong and shortranged (~ 1/m) attraction [12-13].

2. Nuclear forces and annihilation

The range of nuclear forces R

$$R \simeq 1\,fm \simeq 10\,r_a$$

Annihilation from bound nuclear $N\bar{N}$ - states must be reduced - the orbit is out of the annihilation region:

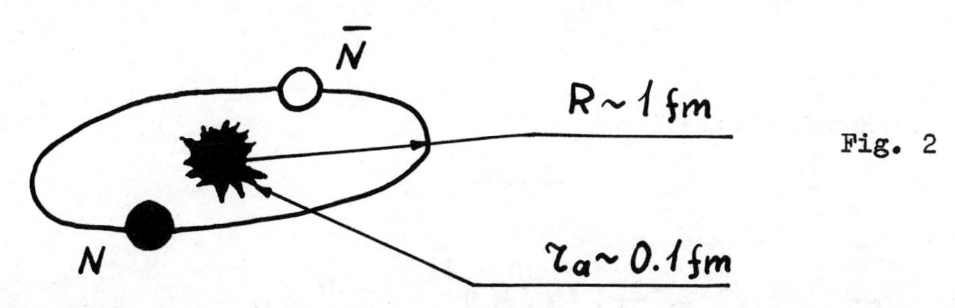

Fig. 2

Annihilation from continuum states may be enhanced by nuclear attraction:

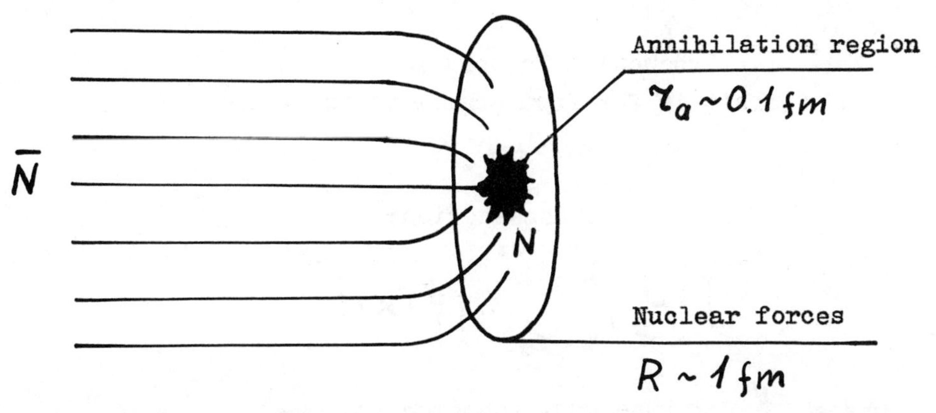

Fig. 3

3. Factorization relations

From $r_a/R \ll 1$ it follows the "separation" of nuclear forces and annihilation dynamic quantities [2-3].

Bound $N\bar{N}$ state annihilation width Γ_a

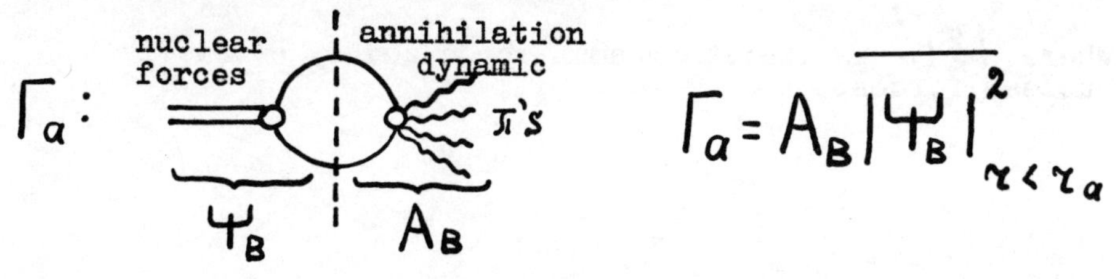

$$\Gamma_a = A_B \overline{|\Psi_B|^2}_{r < r_a}$$

Annihilation cross-section σ_a

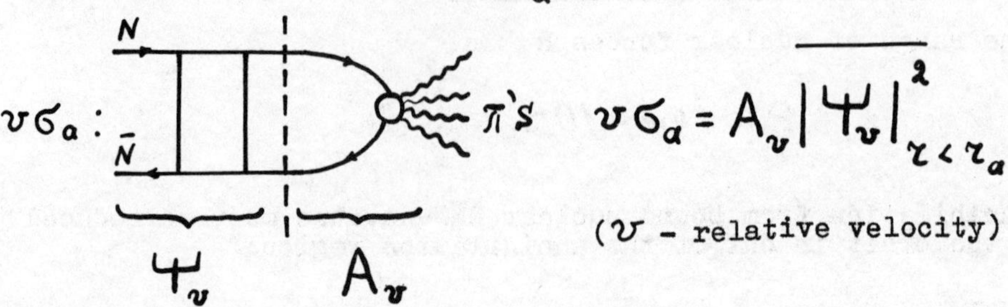

$$v\sigma_a = A_v \overline{|\Psi_v|^2_{r<r_a}}$$

(v - relative velocity)

3.1 Rough estimations

Width:

$$\Gamma_a \simeq (A_B R^{-2}) \cdot \frac{(r_a/R)^{2L}}{[(2L+1)!!]^2} \cdot R^{-1} < R^{-1} (\sim 200\ \mathrm{MeV})$$

reduction factors < 1
(orbit is far from annihilation region)

Cross-section:

$$|\Psi_v|^2_{r<r_a} \simeq |f(k)|^{-2}$$

where f is the Jost function

With the near-threshold levels and $v \ll 1$

$$v\sigma_a = A_v |f|^{-2} \gg A_v$$

where $|f|^{-2}$ is the enhancement produced by attractive nuclear forces.

3. 2 A well-known example: two photon e^+e^- annihilation and positronium width Γ_a

$$\Gamma_a = \underset{\substack{\uparrow \\ A}}{\frac{4\pi\alpha^2}{m^2}} \cdot \underset{\substack{\uparrow \\ |\Psi_B|^2}}{\frac{m^3\alpha^3}{8\pi}} = \frac{\alpha^5}{2} m$$

The reduction factor $\alpha^3 \sim 10^{-6}$! (- orbit far from the annihilation region).

$$v\sigma_a = \underset{\substack{\uparrow \\ A}}{\frac{4\pi\alpha^2}{m^2}} \cdot \underset{\substack{\uparrow \\ \text{Coulomb attraction enhancement}}}{\frac{2\pi\alpha}{v}} \cdot \underset{\substack{\uparrow \\ \text{spin factor}}}{\frac{1}{4}} \quad , \quad v \ll 1$$

3.3 Interrelations $\Gamma_a \rightleftarrows v\sigma_a$

$$v \ll 1 \ , \quad B \ll m \Longrightarrow$$

$$A_B \simeq A_v \Longrightarrow$$

$$\Gamma_a = v\sigma_a \overline{|\Psi_B|^2_{r<r_a}} \cdot |f|^{-2}$$

The main feature: annihilation of low-energy $N\bar{N}$ pairs is strongly affected by nuclear forces. This is calculable because of small ratio $r_a/R \sim 0.1$ [1].

4. <u>A quantitative proof by coupled channel calculations with realistic $N\bar{N}$ nuclear forces [4]</u>

The annihilation channel - two spinless noninteracting bosons with ρ - meson masses = 760 MeV.

The $N\bar{N}$ OBEP with L- dependent cut-off 0.55÷0.68 fm [5].

The L- dependent transition "potential":

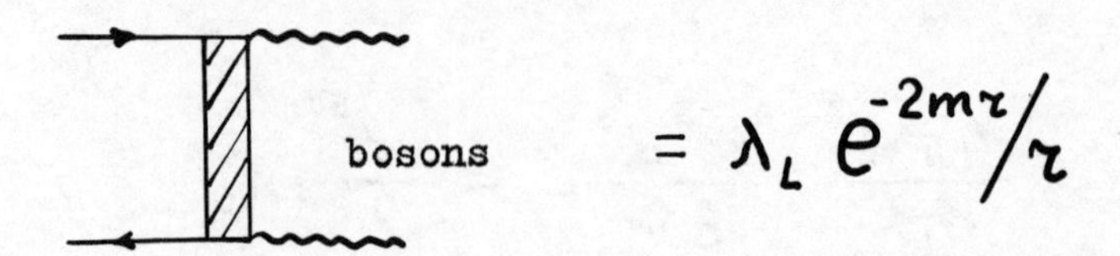

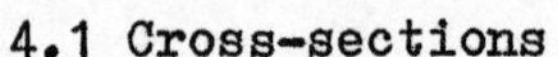

(for orbital momenta L = 0, 1, 2; λ = 86, 86, 100).

4.1 Cross-sections

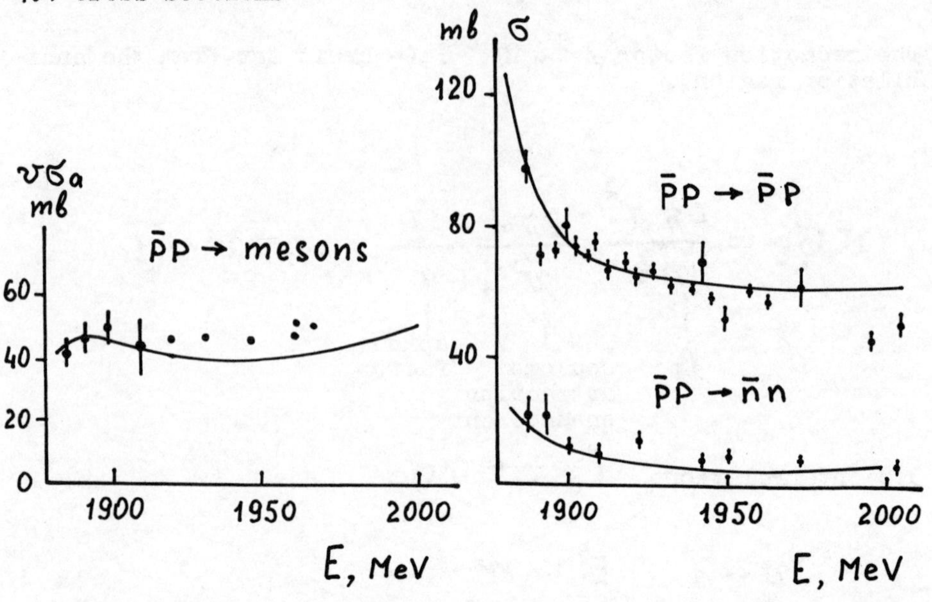

Fig. 4 Annihilation, elastic and charge-exchange cross-sections . Solid lines - coupled channel calculations [4]

4.2 Influence of nuclear forces

$\bar{\sigma}_a$ - cross-section with nuclear interaction "switched off"

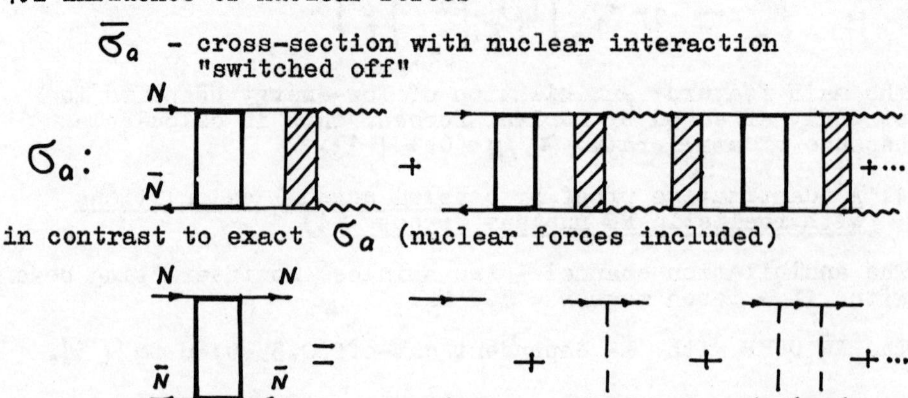

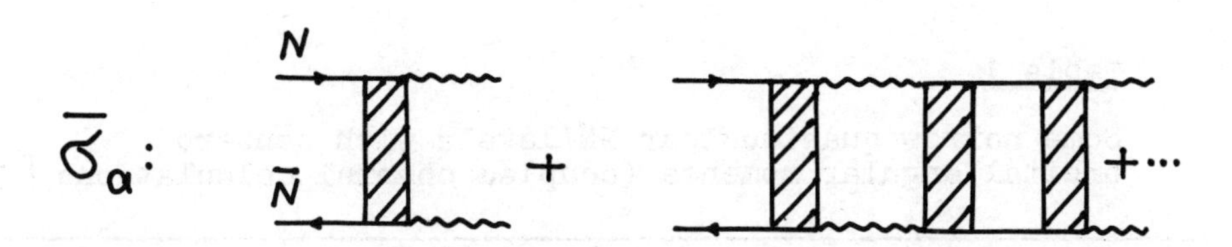

The annihilation cross-section is enhanced by nuclear attraction.

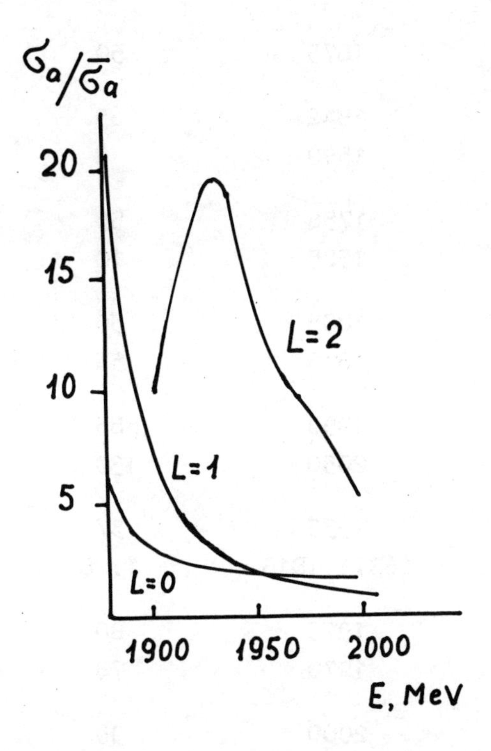

Fig. 5 Enhancement factors (σ_a - nuclear forces included, $\bar{\sigma}_a$ - nuclear forces "switched off")

Some narrow resonances could be well pronounced in partial-wave cross-sections and Argand plots, but not in the total excitation curve.

Table 1

Some narrow quasinuclear $N\bar{N}$ levels with nonzero orbital angular momenta (coupled channel calculations [4])

$^{2S+1}L_J$	$I^G J^P$	M MeV	Γ MeV
$^1P_1$	$1^+ \ 1^+$	1875	50
^3P_o	$1^- \ 0^+$	1842	65
	$0^+ \ 0^+$	1590	8
$^3P_1$	$1^- \ 1^+$	1756	55
	$0^+ \ 1^+$	1585	10
$^3P_2$	$1^- \ 2^+$	1885	50
	$0^+ \ 2^+$	1875	55
$^1D_2$	$1^- \ 2^-$	1990	85
	$0^+ \ 2^-$	2050	130
$^3D_1$	$1^+ \ 1^-$	1935	37
	$0^- \ 1^-$	1637; 1815	1; 8
$^3D_2$	$1^+ \ 2^-$	1975	80
	$0^- \ 2^-$	1970	76
$^3D_3$	$1^+ \ 3^-$	2000	95
	$0^- \ 3^-$	2000	96
$^3F_2$	$0^+ \ 2^+$	1937	3.4

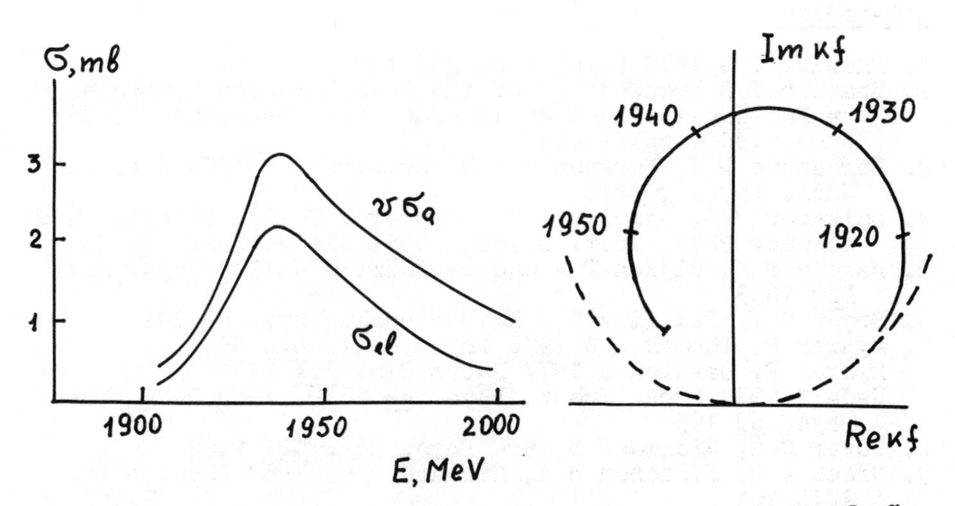

Fig. 6 Partial cross-sections and Argan plot for $J^P I^G = 1^- 1^+$ ($^3D_1$ (1935) state, Γ = 37 MeV)

4. Concluding remarks

The phenomenological model demonstrated above takes into account only the main physical features of $N\bar{N}$ annihilation (the strong nuclear attraction together with the short range "proper" annihilation interaction). The results are in striking contradiction with optical model and some other (N/D Born approximation) calculations.

I would like to emphasize that the results presented here do not pretend to describe the real baryonium spectrum quantitatively. The main result demonstrated on the figures is the coexistence of large cross-sections for annihilation from continuum states and small annihilation widths of $N\bar{N}$ bound and resonant states*). Let me repeat once more that this is possible because strong nuclear attraction is combined with small (compared to nuclear forces range) annihilation distances. For the enhancement of the cross-sections both properties are important. For the bound states widths the smallness of annihilation distances is significant. A number of optical model [6-11] and some other (Born N/D method) calculations [12-13] describe the cross-section data equally well. But the baryonium levels width given by such calculations are greater than our ones by order of magnitude. In both cases the attraction is reduced by the calculation scheme (therefore the imaginary part of the optical potential or the IAT amplitude must be increased to fit the cross-sections data, and this leads to large widths of the bound states). The coupled channel model regarding to calculation scheme is the same as field theory.

*) For resonant states in the table 1 the total (annihilation + elastic) widths are given.

References

1. Shapiro I S 1978 Phys. Rep. 35C 129
2. Shapiro I S Proceedings of the 5-th European Symposium on Nucleon-Antinucleon Interactions. Bressanon 23-28 June, 1980, pp589-602
3. Bogdanova L N, Markushin V E, Shapiro I S 1979 Sov. J. Nucl. Phys. 30 248
4. Dalkarov O D, Shapiro I S, Tyapaev R T 1984 Preprint N°21 Lebedev Phys. Inst. Moscow; 1984 Pis'ma ZhETF 39 38
5. Nagels M M, Rijken T A and de Swart J J 1979 Phys. Rev. D20 1633
6. Bryan R A, Phillips R J N 1968 Nucl. Phys. B5 201
7. Myhrer F, Thomas A W 1976 Phys. Lett. 64B 59
   Myhrer F, Gersten A 1977 Nuovo Cim. 37A 21
   Ueda T 1979 Prog. Theor. Phys. 62 1670; 1980 Prog. Theor. Phys. 63 195
8. Dover C B, Richard J M 1980 Phys. Rev. C21 1466
9. Green A M, Niskanen J A, Richard J M. 1983 Phys. Lett. 121B 101
10. Maruyama M, Ueda T 1981 Nucl. Phys. A364 297
11. Green A M, Sainio M E, Wycech S 1980 J. Phys.G: Nucl. Phys. 6 L17
    Green A M, Sainio M E Ibid. 1375
    Green A M 1982 J. Phys.G: Nucl. Phys. 8 485
    Cote J, Lacombe M, Loiseau B, Moussallam B, Vinh Mau R 1982 Phys. Rev. Lett. 48 1319
    Vinh Mau R 1982 Nucl. Phys. A374 3
    Moussalam B 1983 Nucl. Phys. A407 413
12. Badalyan A M, Polykarpov M I, Simonov Yu A 1978 Phys. Lett. 76B 277
    Van Doremalen J C H, Van der Velde M, Simonov Yu A 1979 Phys. Lett. 87B 315
    Van Doremalen J C H, Simonov Yu A, Van der Velde M 1980 Nucl. Phys. A340 317
13. Simonov Yu A, Tjon J A 1979 Nucl. Phys. A319 429

Inst. Phys. Conf. Ser. No. 73: Section 3
Paper presented at VII Eur. Symp. Antiproton Interactions, Durham 1984

Antiprotonic atom spectroscopy at LEAR (PS 176)

H Koch[a], G Büche[a], A D Hancock[a], J Hauth[a], Th.Köhler[a], A Kreissl[a], H Poth[a], U Raich[a], D Rohmann[a], Ch Findeisen[b], J Repond[b], L Tauscher[b], A Nilsson[c], S Carius[c], M Suffert[d], S Charalambus[e], M Chardalas[e], S Dedoussis[e], T von Egidy[f], F J Hartmann[f], W Kanert[f], J J Reidy[g], M Nicholas[g]

a KfK und Universität Karlsruhe, Karlsruhe, Germany
b Universität Basel, Basle, Switzerland
c Research Institute for Physics, Stockholm, Sweden
d CRN et Université Louis Pasteur, Strasbourg, France
e University of Thessaloniki, Thessaloniki, Greece
f Technische Universität München, D8046 Garching, Germany
g University of Missisippi, University, Missisippi 38677, USA

Abstract. The X-ray spectroscopy of antiprotonic atoms has started at Lear with measurements on the targets (mainly isotopically pure) ^{16}O, ^{17}O, ^{18}O, ^{19}F, ^{23}Na, and ^{138}Ba. The strong interaction effects, energy shifts, Lorentzian broadening, intensity reductions, could be determined with unprecedented accuracy. In ^{138}Ba a first hint to a spin dependence of the effects was observed.

1. Introduction

The motivations to do spectoscopy on antiprotonic atoms are fourfold:
(i) Strong interaction effects
When antiprotonic X-ray transitions occur between states of high quantum numbers n, l, j, the intensities and energies of the X-ray-lines are determined by purely electromagnetic effects. The widths of the lines are Gaussian due to the finite energy resolution of the detectors. When the antiproton approaches the nucleus, the situation changes drastically: Strong interaction effects (scattering and annihilation) decrease the intensity of the last observable transition, cause an energy shift (ε) of the line as compared to its electromagnetic value and a Lorentz-broadening (Γ_{low}) due to the strong annihilation in the lower level of the transition. From the intensity reduction the annihilation width (Γ_{up}) of the upper level of the transition can be derived. These strong interaction effects (ε_{low}, Γ_{low}, Γ_{up}) give information on the details of the $\bar{p}$-nucleus interaction when compared with appropriate calculations. Of particular importance in this respect are comparisons of pure isotopic targets because here the effect of one or more additional nucleons is explicitly seen and can be used to get information on the low energy elementary $\bar{p}N$-interaction. Another very interesting aspect is the determination of the spin-orbit part of the $\bar{p}$-nucleus interaction, which seems feasible with a high statistics measurement of the members of a fine-structure multiplett.

(ii) Mass and magnetic moment of the antiproton.
The energies of higher transitions in the X-ray cascade are mainly determined by the mass of the antiproton, the splitting of the lines by its magnetic moment. Both quantities can be determined from high statistics measurements with an accuracy considerable better than in previous experiments. Taken that the antiproton mass is equal to the proton mass, the precise energy measurements can be interpreted in terms of an upper limit for an eventual long range component (predicted by some QCD-calculations) of the strong interaction.

(iii) Residuals after the $\bar{p}$-nucleus annihilation
The distribution of the residual nuclei after the annihilation of the antiproton by its capture into the nucleus gives a global picture of the low energy $\bar{p}$-nucleus interaction. These distributions can be found either by the observation of prompt (or delayed) γ-rays or neutrons emitted from the usually excited residual nuclei. Further information comes from the observation of electronic X-rays from the residual nuclei due to electron rearrangements after the $\bar{p}$-induced spallation process.

(iv) $\bar{p}$-nucleus bound states
Depending on the details of the $\bar{p}$-nuclear interaction the chance exists that the antiproton and the nucleus form a bound system of considerable long lifetime. Such states would have binding energies greater than several MeV and could be detected by the registration of high energy Gammas emitted during the transition of the system between an atomic and a deeply bond nuclear state. The chance to find such states which are sufficiently narrow will be greatest in light nuclei.

2) Experimental set up

The experimental set up is shown in Fig. 1.
It is designed such that X-rays from the antiprotonic cascade ($\rightarrow$500 keV), nuclear γ-rays ($\rightarrow$6 MeV), TOF-neutron spectra and high energy X-rays ($\rightarrow$1 GeV) can be detected simultaneously. The X-rays were measured by two Si(Li)- and two pure Ge-detectors, respectively, covering different energy ranges with optimal energy resolution. The γ-rays were measured with a large volume Ge-diode for lower energies and with a 12"x 10" NaI-crystal for energies up 1 GeV. The targets were mounted on a ladder which allowed a quick target exchange even during a spill. For some of the measurements the target region was enclosed in a Helium bag in order to minimize $\bar{p}$-stops in air, giving rise to Nitrogen- and Oxygen background lines.
The experiment was set up in the North Branch of the Lear experimental area. A beam of 300 MeV/c momentum was used. Under best conditions 1-2 x 10^8 $\bar{p}$'s could

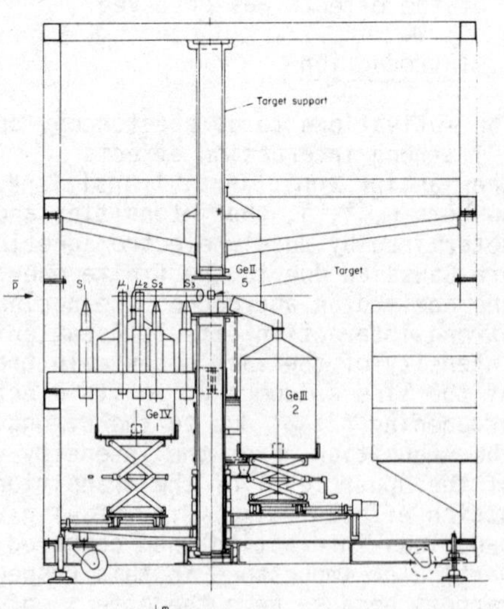

Fig. 1 Experimental set up

be stopped during one spill ($\approx$ 1 hour). The width of the range curve (taken with a target of 100 mg/cm^2 thickness) was about 120 mg/cm^2, the width of the beam spot less than 2 x 2 cm^2.

The data taking worked very efficiently in spite of the high counting rates delivered by the detectors. The energy- and time spectra of all detectors were stored in a large 3 M-byte memory controlled by a 68000 Motorola microprocessor. Coincident events were registered in an event-to-event mode. After the spill the data were transferred to a PDP-11-computer, where a quick data analysis and the storage of the data was performed.

3. Results

The targets (all isotopically pure, except for ^{17}O) measured so far are: ^{16}O(2); ^{17}O(4); ^{18}O(2); ^{19}F + ^{23}Na(5); ^{138}Ba(11). Survey measurements were done on ^{44}Ca(1.5); ^{208}Pb(1). The numbers in brackets mean the number of spills for these measurements. All spectra show an excellent peak/background ratio and nearly no disturbing lines from materials arount the target. Many transitions are clearly seen, which fixes all eventually necessary cascade corrections. As an example

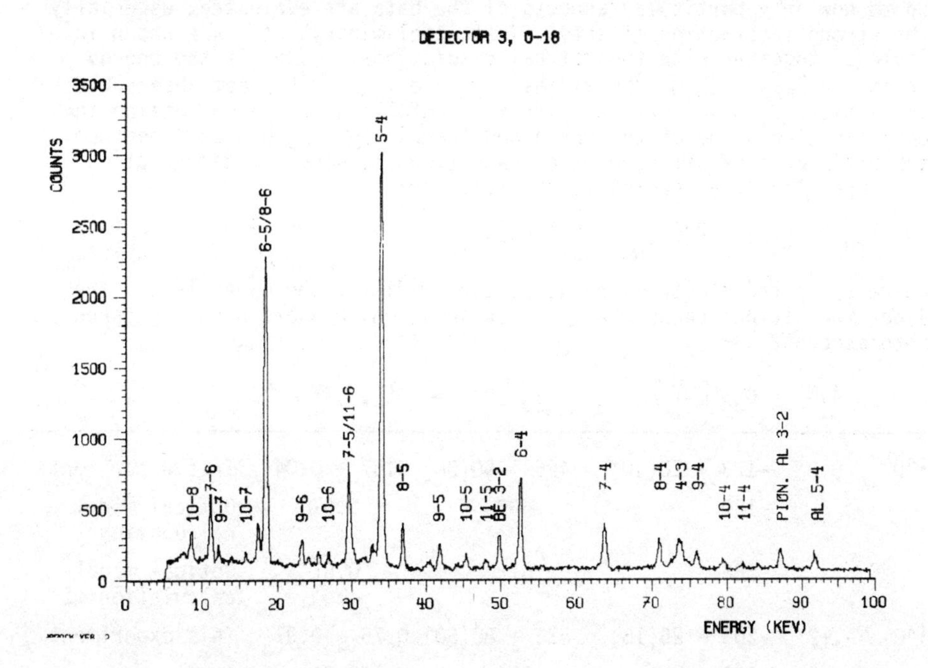

Fig. 2 $\bar{p}$-^{18}O-spectrum

the spectrum of $H_2{}^{18}O$ is shown in Fig. 2, measured within two hours. Fig. 3 shows the last observable transition (4 $\rightarrow$ 3) in ^{18}O, which was used for the extraction of the strong interaction effects.

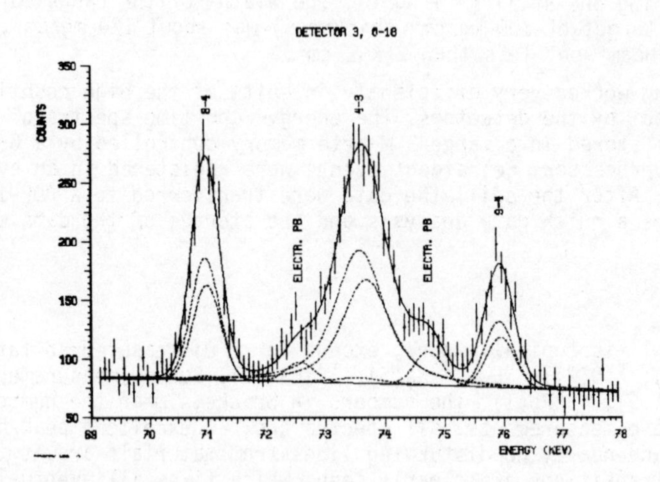

Fig. 3 $\bar{p}$-^{18}O-spectrum, 4-3 line

Up to now only particular aspects of the data are evaluated, especially the strong interaction effects. First, preliminary data are shown in Table 1, together with theoretical predictions. Listed is the energy shift $\varepsilon = E_{exp} - E_{QED}$, the widths Γ_{low} and Γ_{up} of the last observable transition. Γ_{low} was extracted from an unfolding of the Lorentzian and Gaussian line shape of the broadened transition, Γ_{up} was derived from the ratio of intensities of the transitions feeding and depopulating the upper level, according to the relation[1])

$$\Gamma_{up} = \Gamma_x \left[{}^{I}pop/_{I_{Depop.}} - 1 \right]$$

where Γ_x is the electromagnetic partial width of the upper level. The Auger transitions cause minor corrections, which where properly taken into account.

	Z,N	ε_{3d} [eV]	Γ_{3d} [eV]	Γ_{4f} [eV]	
^{16}O	8,8	-123 ± 20(10)	496 ± 50(30)	0.57 ± 0.04	This experiment
		-120	463	0.57	Optical model calculation I
		-186	512	0.67	Optical model calculation II
^{18}O	-,+2	-205 ± 25(15)	636 ± 80(60)	0.75 ± 0.07	This experiment
		-147	524	0.71	I
		-131	541	1.36	II
^{19}F	+1,+2	-450 ± 50	1360 ± 200	1.84 ± 0.10	This experiment
		-336	1714	3.85	I
		+45	2235	6.10	II

Table 1

The errors given in brackets are the statistical ones, the errors in front of the brackets can be regarded as upper limits. It is expected that with ongoing analysis and new data which are already taken the final errors will become considerably smaller than the statistical ones given in the Table. Irrespective of this, very clean isotopic effects have been determined. Their significance is already now a factor of 3-4 greater than in previous measurements[2]. A detailed analysis of the data in terms of elementary $\bar{N}N$-interactions, differences in the $\bar{p}p$ and $\bar{p}n$-effective interaction in nuclei and nucleon distribution on the nuclear surface is left to theorists who have already made first predictions [3]. A simple analysis, however, was made in terms of two different optical potentials I and II. Both were of the form $V_{op.} \propto \bar{a} \cdot \rho(r)$, with ρ being the proton-distribution in the nuclei. The neutron distribution was put equal to the proton distribution. The effective scattering length $\bar{a}$ was put to $\bar{a}$ = (2.0+i1.9) fm (I) and $\bar{a}$ = (4.0+i0.6)fm (II). The first case corresponds to a potential with strong absorption, the second case to shallow absorbtive interaction. The potenials were introduced into the Klein-Gordon-equation (spin was neglected) and solved for the complex energy values. The results given in Table 1 show that potential I fits the data considerably better than potential II, which rules out the ambiguities in the $\bar{p}$-nucleus interactions recently brought up[4].

A good candidate to look for spin-effects of the $\bar{p}$-nucleus strong interaction is the 8-7 transition in $\bar{p}$-^{138}Ba. Part of the Ba-spectrum is shown in Fig. 4.

The elektromagnetic fine structure splitting between the two main components is calculated to be 2.1 keV. A preliminaring fit to the 8-7 structure, however, shows, that the energy difference is about 1.9 keV, which means that a spin-dependent line shift has been seen for the first time in exotic atoms. In a further analysis and with more data it will be tried, to get unambiguous results for the spin-dependent shifts and widths of the two main components of the n = 7, l = 6, j = 13/2, 11/2 evels.

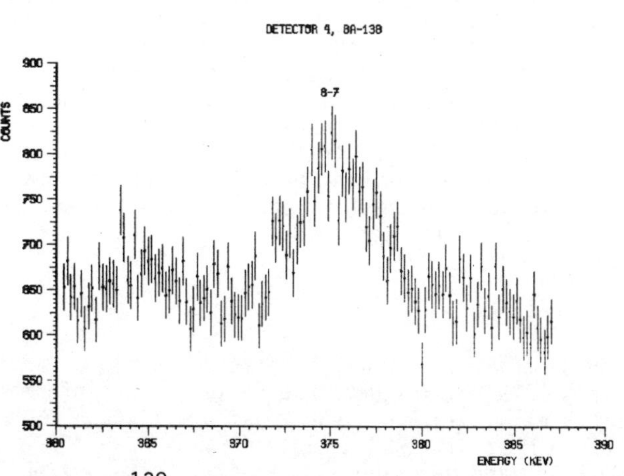

Fig. 4 $\bar{p}$-^{138}Ba-spectrum. 8-7 transition.

Fig. 5 shows the 10 → 9 transition in $\bar{p}$-^{208}Pb, measured during one hour. This target belongs to the best candidates to look for a high accuracy determination of the $\bar{p}$-magnetic moment. Its influence is clearly seen in the splitting of the line. In higher statistics measurements in the future, a simultaneous analysis of this (contains tiny strong interaction effects) and higher transitions will allow a drastic improvement in the accuracy of the $\bar{p}$-magnetic moment.

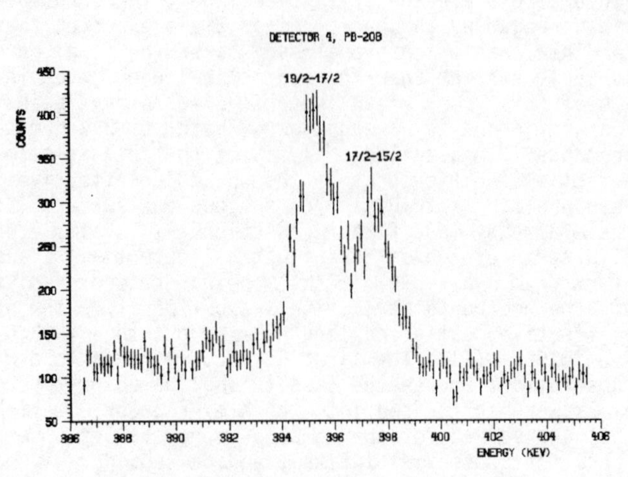

Fig. 5 $\bar{p}$-^{208}Pb-spectrum. 10-9 transition

4. Conclusions

Already the first hours of $\bar{p}$-atom spectroscopy at Lear have shown, that these data are not only useful in getting information about the global features of the $\bar{p}$-nucleus interaction but also to investigate more detailed effects, like the spin-dependence. Measurements on selected isotopes now have reached such a precision that the disentangling of $\bar{p}$-annihilations on protons and neutrons seems feasible. The high accuracy measurements on higher transitions in the cascade can be interpreted in terms of improved values for the mass and the magnetic moment of the antiproton and in terms of upper limits for a hypothetical long range strong interaction component.

References

1)Koch H et al. 1969 Physics Letters 29B 140
2)Poth H 1979 Physics Data Report 14-1, Fachinformationszentrum Karlsruhe
Poth H et al. 1978 Nuclear Physics A294 435
3)Green A M et al. 1983 Nuclear Physics A399 307
Szuzuki T and Narumi H 1983 Physics Letters 125B 251
4)Wong C Y et al. 1984 Physical Review C29 574

Inst. Phys. Conf. Ser. No. 73: Section 3
Paper presented at VII Eur. Symp. Antiproton Interactions, Durham 1984

Antiprotonic molybdenum: resonant coupling of atomic and nuclear states and the impact of annihilation on the nucleus (PS 186)

T von Egidy[a], W Kanert[a], F J Hartmann[a], H Daniel[a], E Moser[a], G Schmidt[a], J J Reidy[b], M Nicholas[b], M Leon[c], H Poth[d], G Büche[d], A D Hancock[d], H Koch[d], Th Köhler[d], A Kreissl[d], U Raich[d], D Rohmann[d], M Chardalas[e], S Dedoussis[e], M Suffert[f], A Nilsson[g]

a Technische Universität München, D8046 Garching, Germany
b University of Mississippi, University, Mississippi 38677, USA
c Los Alamos National Laboratory, Los Alamos, New Mexico 87545, USA
d KFK und Universität Karlsruhe, D7500 Karlsruhe, Germany
e University of Thessaloniki, Thessaloniki, Greece
f CRN et ULP, Strasbourg, France
g Research Institute for Physics, Stockholm, Sweden

Abstract. The coupling by the E2 nuclear resonance effect of atomic and nuclear states in antiprotonic atoms was observed in ^{94,98}Mo by an attenuation of the (n=7 → n=6) transition. In ^{100} Mo the effect manifests itself by an almost complete attenuation of the (n=8 → n=7) transition and a broadening of the (n=9 → n=8) line. Strong interaction widths were determined for n=7 and n=6 levels. The distribution of residual nuclei after the $\bar{p}$ annihilation in Mo isotopes was obtained by activation analysis showing that more than thirty nucleons may be emitted.

1. Introduction

Antiprotonic atoms and the $\bar{p}$ annihilation in nuclei yield important information on the interaction of antimatter with matter. The group of experiment PS 186 at LEAR (CERN, Geneva) is concentrating on the following topics:

1. the E2 nuclear resonance effect in antiprotonic atoms predicted by Leon (1976),
2. the impact of $\bar{p}$ annihilation on nuclei, as determined from the distribution of residual nuclei,
3. detection of charged particles emitted after $\bar{p}$ annihilation in nuclei (not yet measured),
4. strong interaction widths and shifts of the last observable antiprotonic levels, and
5. the antiprotonic cascade which is discussed in a separate contribution to these proceedings (Hartmann et al. 1984).

2. Experiment

The first measurements were performed with ^{92,94,98,100}Mo targets because resonant coupling of atomic and nuclear states was predicted for ^{94}Mo and ^{100}Mo and because isotope and shell effects might be observed in the antiproton interaction with a series of nuclei. The experimental details are described in another contribution to these proceedings (H Koch et al. 1984). The slow antiproton beam of LEAR (300 MeV/c, flux between 10^4 and 10^5 $\bar{p}$/s) was stopped in the target (area 25x30 mm^2, thickness 100 mg/cm^2). Prompt x and γ ray spectra were detected with various Ge and Si detectors. Each target was exposed to the $\bar{p}$ beam for two to four hours and then within a few minutes brought to other Ge and Si detectors in a low background area in order to measure the induced radioactivity. This radioactivity yielded the distribution of residual nuclei.

3. The E2 nuclear resonance effect

Configuration mixing of $\bar{p}$-atomic and nuclear states via quadrupole interaction takes place if a nuclear excitation energy, in particular the energy of the first $I^\pi=2^+$ state in even-even nuclei, closely matches a $\Delta\ell = 2$ atomic transition. As the strong absorption width strongly increases with decreasing ℓ, this E2 nuclear resonance effect can be observed by an attenuation of one or two atomic transitions (Leon 1976).

For the usual case of weak coupling the mixed wave function can be written

$$\Psi = \sqrt{1-\alpha^2}\ \Phi(n,\ell,0^+) + \alpha\Phi(n-2,\ell-2,2^+)$$

with a small admixture coefficient

$$\alpha = \pm\langle n-2,\ell-2,2^+|H_Q|n,\ell,0^+\rangle\ /[E(n-2,\ell-2,2^+)-E(n,\ell,0^+)].$$

The electric quadrupole interaction is

$$\langle H_Q\rangle = \pm\frac{1}{2}e^2Q_o\langle\frac{1}{r^3}\rangle \times \text{angular momentum factors}$$

with the nuclear quadrupole transition strength Q_o.
For close coupling in the case of a very good energy match the Schrödinger equation has to be solved for the Hamiltonian

$$H = H_O + H_Q = \begin{bmatrix} E_1 & \delta \\ \delta^* & E_2 \end{bmatrix},$$

where E_1 and E_2 are complex energy eigenvalues of H_O and δ is the matrix element of H_Q.
The situation in antiprotonic ^{94}Mo and ^{100}Mo is shown in Fig. 1. Weak

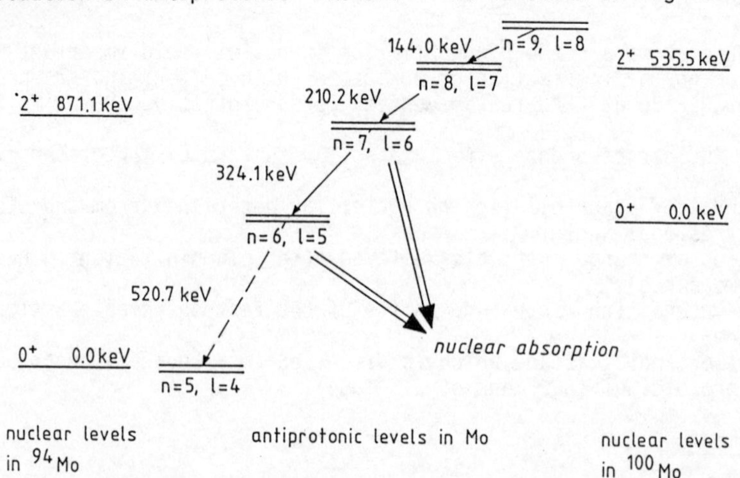

Fig. 1 The lowest antiprotonic levels in Mo and the first nuclear levels in ^{94}Mo and ^{100}Mo which are relevant for the E2 resonance effect.

coupling is expected in ^{94}Mo because $E(2^+) = 871$ keV while $E(7\rightarrow5) = 845$ keV, and also in ^{98}Mo with $E(2^+) = 787$ keV. Close coupling is predicted for ^{100}Mo because $E(2^+) = 535.5$ keV matches very closely with $E(8\text{-}6) = 534.3$ keV. In these cases the (7-6) transition and the (8-7) transition, respectively, will be attenuated. Since the n=5 level is involved in the resonant coupling in ^{94}Mo and ^{98}Mo, information on the width of this level which is normally not populated can be obtained. The close coupling in ^{100}Mo also causes a broadening of the (9-8) line.

4. Strong interaction widths and resonance attenuations

The relevant parts of the spectra given in Fig. 2 demonstrate that the

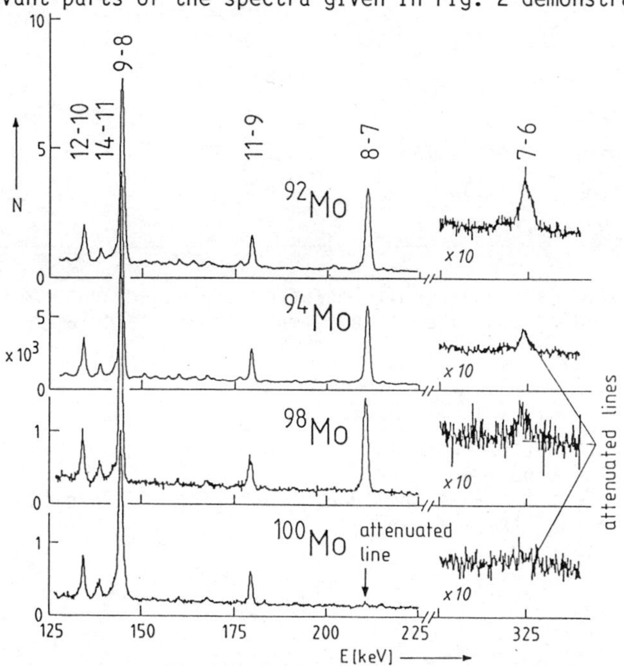

Fig. 2 Part of the spectra measured with ^{92,94,98,100}Mo targets. A high purity Ge detector (5 cm² area x 1.3 cm thickness) was used. The lines attenuated by the E2 resonance effect are indicated.

$(7\rightarrow6)$ transition is attenuated in all Mo isotopes except for ^{92}Mo, where the energy of the first 2^+ state is 1.51 MeV. The $(8\rightarrow7)$ transition in ^{100}Mo is almost completely attenuated; the remaining intensity is due to the presence of other Mo isotopes in the target. The most important results are listed in Table 1 and compared with theoretical predictions. The widths of the n=6 levels were obtained from the line widths of the (7-6) transitions. The strong interaction width Γ_{SI} of the n=7 level in ^{92}Mo was determined according to Koch (1969) from the ratio of the population I_{pop} and depopulation $I(7\rightarrow6)$ of this level assuming a calculated radiative width $\Gamma_{rad}(7\rightarrow6) = 5.35$ eV:

$$\Gamma_{SI}(n=7,\ell=6) = \Gamma_{rad}(7\rightarrow6)[I_{pop}/I(7\rightarrow6)-1].$$

Auger transitions may be neglected in this part of the $\bar{p}$ cascade.

Table 1. Widths and attenuations of x ray transitions in $\bar{p}$-Mo isotopes

Target	^{92}Mo	^{94}Mo	^{98}Mo	^{100}Mo
Isotopic purity of target	98.33 %	93.85 %	97.24 %	95.94 %
Strong interaction width of n=6(keV)	2.0±0.2	1.6±0.3	2.2±0.7	-
Calculated by Suzuki (1983)(keV)	1.13	1.17	1.27	1.31
Strong interaction width of n=7(eV)	13.6±0.7			
Calculated by Suzuki (1983)(eV)	14.0	14.8	16.5	17.3
Measured intensity of (7-6)*)	100.0±2.4	38.6±1.6	52.7±5.9	25.9±3.5
Expected intensity without atten.*)	100.0	91.8	84.9	79.5
*) arbitrary units				
Attenuation of (7-6)		0.62±0.02	0.39±0.08	0.70±0.05
Calculated attenuation (Leon 1976)		0.67		0.83
Measured attenuation of (8-7)				0.998±0.006
Calculated attenuation (Leon 1976)				0.98

Table 1 shows that measured strong interaction widths are not completely in agreement with calculations. The attenuations "a" are calculated with the formula

$$a = (I_o - I_{meas})/(c \cdot I_o)$$

where I_o is the expected intensity without attenuation and c is the enrichment. The attenuations calculated by Leon (1976) are three standard deviations above the experimental values.
Leon (1976) also predicted that the (9-8) transition in ^{100}Mo is broadened. Fig. 3 illustrates that this line really is broader in ^{100}Mo than in ^{98}Mo.

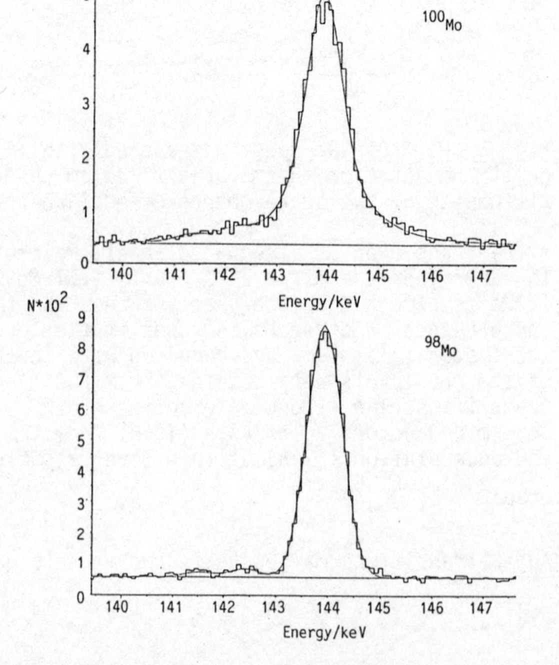

Fig. 3 The antiprotonic (9-8) line in ^{98}Mo and ^{100}Mo showing the broadening in ^{100} Mo by the resonant mixing of atomic and nuclear states. In ^{98}Mo the fine structure splitting and two additional lines were taken into account, in ^{100}Mo the structure given by Leon instead of the fine structure.

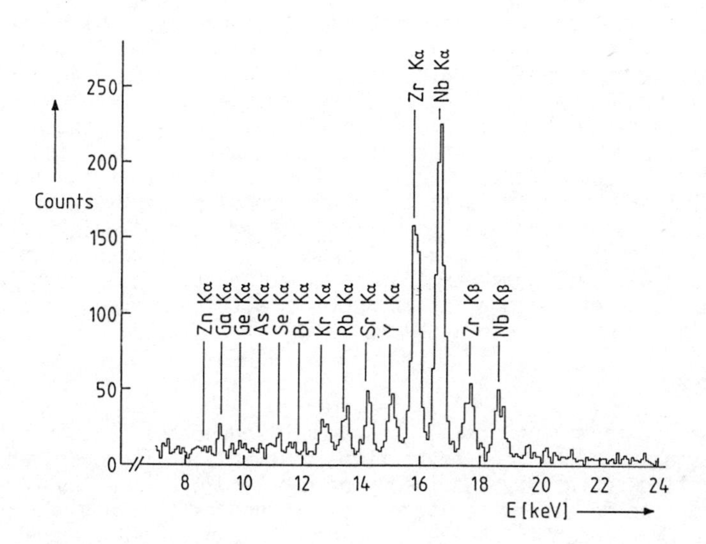

Fig. 4 Electronic x-rays of a ^{98}Mo target emitted 13 days after the irridiation with antiprotons. The spectrum was measured with a Si(Li) detector.

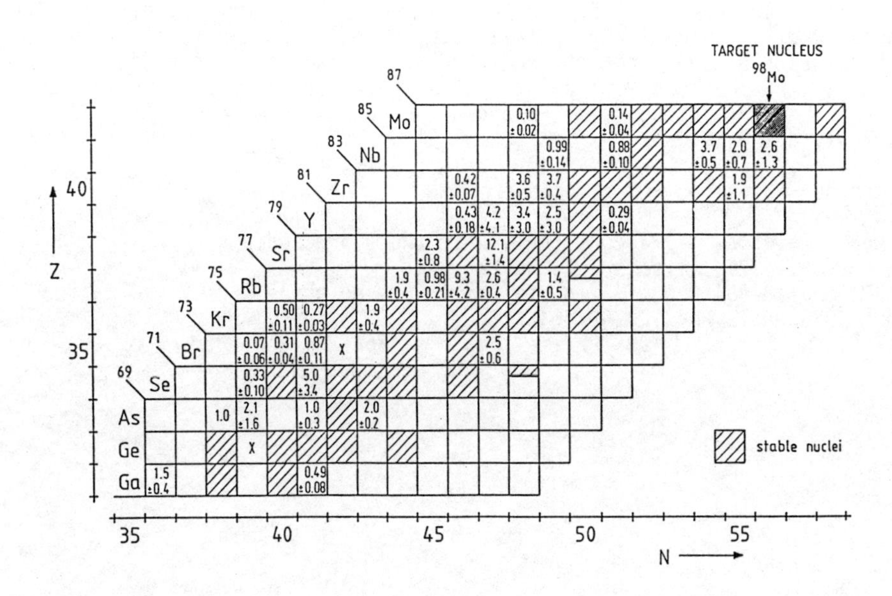

Fig. 5 Section of the nuclide chart with the distribution of residual nuclei after $\bar{p}$ annihilation in ^{98}Mo. Relative abundances of residual nuclei are given as determined by activation analysis. x indicates that electronic x-rays after K-capture of this isotope were observed.

5. Distribution of residual nuclei after annihilation

As an example of the spectra measured after the irradiation with antiprotons, Fig. 4 shows electronic x-rays and thus the range of elements and isotopes decaying by K-capture which are produced following $\bar{p}$-annihilation in ^{98}Mo. Relative abundances of radioactive isotopes are given in the section of the nuclide chart displayed in Fig. 5. Not all of these isotopes are necessarily produced during the annihilation but some may also be produced later as decay products. There are indications that the isotopes near ^{96}Nb originate at least partially from secondary reactions of fast neutrons, protons and pions with the target rather than $\bar{p}$ annihilation. The maximum of the distribution is near ^{85}Sr and ^{83}Rb. The lightest isotope identified is ^{67}Ga. This demonstrates that the impact of the annihilation causes the emission of up to about 30 nucleons with an average of about 15 nucleons.

6. Conclusions

The effect of configuration mixing of $\bar{p}$-atomic and nuclear states by E2 resonance was observed for the first time in several Mo isotopes, and is reasonably well described by the theory of Leon (1976). It allows one to obtain information on otherwise hidden levels such as n=5 in $\bar{p}$-Mo. This effect may occur frequently enough to need to be considered whenever strong interaction widths are deduced from antiprotonic x-ray intensities.
The $\bar{p}$ annihilation creates nuclei with more than 30 nucleons less than the target nucleus.

Acknowledgement

We wish to thank Profs. G. Fricke and L. Schellenberg for providing the Mo targets, Dr. B. Jonson for essential help during the activation measurements and H. Angerer, H. Hagn and P. Stoeckel for technical assistance. Furthermore we thank the LEAR staff for providing the intense and beautiful antiproton beam. Financial support of the German BMFT is gratefully acknowledged.

References

Hartmann F J et al. 1984 Contribution to these proceedings
Koch H et al. 1969 Physics Letters 29B 140
Koch H et al. 1984 Contribution to these proceedings
Leon M 1976 Nucl. Phys. A260 461
Suzuki T and Narumi H 1983 Physics Letters 125B 251 and priv. comm.

PS 184: a study of p̄–nucleus interaction with a high resolution magnetic spectrometer

D. Garreta, P. Birien, G. Bruge, A. Chaumeaux, D.M. Drake*), S. Janouin, D. Legrand, M.C. Lemaire, B. Mayer, J. Pain and J.C. Peng*)

DPHN/ME, CEN, Saclay, France

M. Berrada, J.P. Bocquet, E. Monnand, J. Mougey and P. Perrin

DRF, CEN, Grenoble, France

E. Aslanides and O. Bing

CRN, Strasbourg, France

A. Erell, J. Lichtenstadt and A.I. Yavin

Tel Aviv University**), Tel Aviv, Israel

Abstract. The aim of the experiment PS184 at LEAR is to study a few simple and well-defined channels of the p̄–nucleus interaction using SPES II, a high-resolution magnetic spectrometer with large solid angle and momentum acceptances [Thirion and Birien]. Results of elastic scattering from ^{12}C, ^{40}Ca and ^{208}Pb, inelastic scattering from ^{12}C, and (p̄,p) knock-out reaction from ^{12}C, ^{63}Cu and ^{209}Bi, are presented. An optical-model analysis of the elastic-scattering data as well as microscopic KMT-type calculations have been performed.

1. INTRODUCTION

Before the advent of LEAR in 1983 very little was known about the p̄–nucleus interaction at low energy. From the experimental point of view the data were scarce and of rather poor quality, consisting mainly of bubble-chamber data [Agnew et al., 1957], a few reaction cross-sections [Aihara et al., 1981] and level widths and shifts from X-ray studies of antiprotonic atoms [Batty, 1981]. The analysis of these data yielded non-unique p̄–nucleus optical potentials [Wong et al., 1984] leading to elastic scattering angular distributions [MacKellar et al., 1984] with very different behaviours at sufficiently large angles (see Fig. 1). From the theoretical point of view very large ambiguities also existed which were leading to p̄–nucleus optical potentials with a real part ranging from strongly attractive to repulsive values [Bouyssy and Marcos, 1982; Auerbach et al., 1981; Niskanen and Green, 1983]. Therefore, the main purpose of the elastic-scattering measurements (which, in contrast with other recent measurements [Nakamura et al., 1984; Sakitt, 1984], cover a wide angular range and have elastically scattered antiprotons well resolved, with no pion contamination) was to set constraints on the p̄–nucleus potential. This would in particular supply some information on the possibility of nn̄ oscillations [Dover et al., 1983]. Using microscopic calculations they may also provide a test of the elementary N̄N amplitudes. Inelastic scattering from collective states also sets constraints on the p̄–nucleus potential when analysed in terms of coupled channel calculations. When unnatural parity states are concerned it provides a sensitive test of the spin and isospin components of the N̄N elementary amplitudes [Dover et al., 1984]. The purpose of the (p̄,p) knock-out reaction was to observe possible bound or resonant states of an antiproton and a nucleus which would be formed in a way similar to that of the (K$^-$, π) "recoilless" hypernuclei production. The width of such states is predicted to be very large [Green and Wyceh, 1982; Wong et al., 1984] but their observation, speculative as it might be, would provide very useful information on the inner part of the p̄–nucleus potential.

*) Permanent address: Los Alamos National Laboratory, New Mexico, USA. Supported in part by the US Department of Energy.

**) Supported by the Israel Fund for Basic Research.

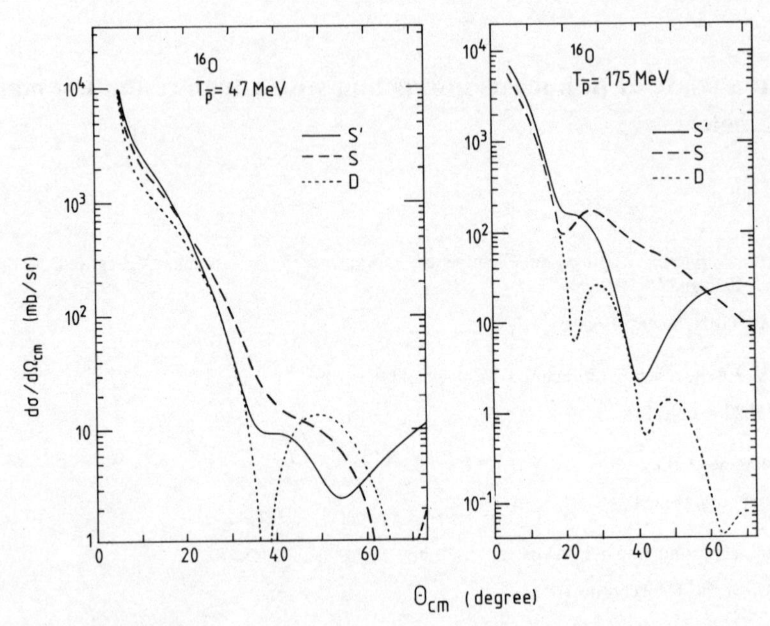

Fig. 1 $\bar{p} + {}^{16}O$ elastic scattering angular distributions calculated with S-type (S and S′) and D-type (D) optical potentials of MacKellar et al. [1984].

2. EXPERIMENTAL SET-UP (See Fig. 2)

The incident antiprotons (with an intensity ranging from 2×10^4 to 10^5/s) were counted by a 0.36 mm thick scintillator (S_1) located 25 cm in front of the target. Scattered antiprotons, or outgoing protons, were momentum analysed in the magnetic spectrometer SPES II, which has a momentum resolution of 5×10^{-4}, a solid angle of 30 msr and a momentum acceptance of ±18%. They were detected in three multiwire proportional chambers (MWPCs) [Chaminade et al., 1974] and a scintillator hodoscope located near the focal plane. Pions produced by annihilation in the target were discarded by time-of-flight measurement. Information from the MWPCs was used to compute the scatttering angle and the excitation energy of the residual nucleus. For elastic and inelastic scattering the full angular acceptance was divided into 1.67° bins. The energy resolution was about 1 MeV (FWHM) and the overall angular resolution, including multiple scattering in the target, varied from about 2° for the C and Ca targets to about 3° for the Pb target. The uncertainty on the absolute scattering angle was 0.2°. The uncertainty on the absolute normalization is 10%.

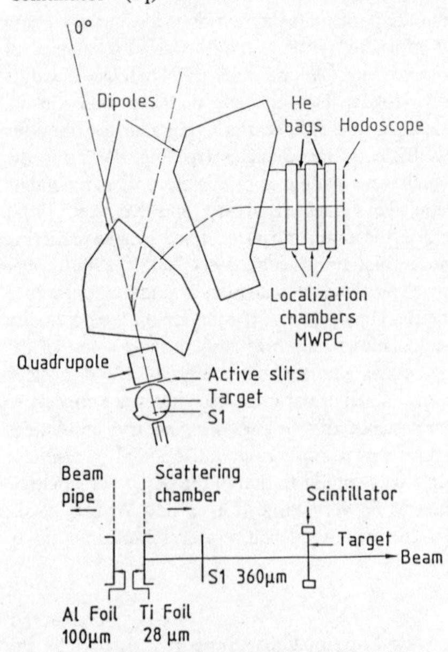

Fig. 2 PS184 experimental set-up.SPES II is represented at 0° scattering angle.

3. ELASTIC SCATTERING

Angular distributions of antiproton elastic scattering are shown in Fig. 3 for ^{12}C at 46.8 MeV [Garreta et al., 1984a] and in Fig. 4 for ^{12}C, ^{40}Ca, and ^{208}Pb at 180 MeV [Bruge, 1984; Garreta et al., 1984b] (solid dots). One can immediately see that they exhibit oscillatory behaviour typical of a diffraction pattern, similar to that calculated with a D-type potential shown in Fig. 1 and very different, at sufficiently large angle, from that corresponding to an S-type potential or that of proton elastic scattering (open circles) also shown for comparison. This agreement between our data and strongly absorptive potential predictions is confirmed by an optical-model analysis performed using the ECIS code of Raynal [1981] with an optical potential parametrized by a Woods-Saxon geometry, with volume absorption and with no spin orbit, typical examples of which are shown as solid curves in Figs. 3 and 4. Although equally good fits were achieved with optical potentials having quite different geometries, they all had similar values V(R) and W(R) (shown in Table 1) at a distance R, radius of strong absorption [Barrett and Jackson, 1977], and are all strongly absorbing, that is $|W(R)| \gtrsim 2|V(R)|$. This indicates that a necessary condition for orbiting does not exist [Auerbach et al., 1981; Kahana and Sainio, 1984]. When the geometrical parameters of the real and imaginary parts are assumed to follow those of the point charge distribution, corrected for the interaction range, taken as a Yukawa function with $\mu^{-1} = 0.6$ fm, V_0 and W_0 become well determined. These values are also shown in Table 1, where we can see that $V_0 < 70$ MeV

Table 1
Real and imaginary potentials at the radius of the strong absorption and calculated reaction cross-sections. Also shown are the strengths of the real and imaginary potentials obtained by using a geometry derived from that of the charge distribution (see text).

Target	$E_{\bar{p}}$ (MeV)	R (fm)	V(R) (MeV)	W(R) (MeV)	σ_R (mb)	V_0 (MeV)	W_0 (MeV)
^{12}C	46.8	3.7	−3.5 ± 1.5	−8.5 ± 1	600 ± 30	35 ± 4	77 ± 4
^{12}C	179.7	3.3	−7.8 ± 1.5	−19.6 ± 2	500 ± 25	44 ± 4	96 ± 2
^{40}Ca	179.8	4.94	−6.2 ± 1.5	−13.3 ± 2	990 ± 50	43 ± 4	119 ± 3
^{208}Pb	180.3	8.15	−5.4 ± 1.5	−10.2 ± 2	2670 ± 140	60 ± 6	152 ± 2

and $W_0 \gtrsim 2\ V_0$. These results remove the ambiguity [Wong, 1984] in the $\bar{p}$–nucleus interaction, by rejecting the shallow (S-type) imaginary potentials. Moreover, the depth V_0, which for ^{12}C does not show the strong energy dependence predicted by some models [Bouyssy and Marcos, 1982; Niskanen and Green, 1983], is shallower than that calculated in the relativistic mean-field approach [Bouyssy and Marcos, 1982], and provides a lower limit for the $n\bar{n}$ oscillation time [Dover et al., 1983], $\tau_{\bar{n}n}$, of about 3×10^7 s. Despite the optical-model ambiguities, the reaction cross-sections are well determined in the present analysis, and their values are given in Table 1. The decrease with incident energy of both R and σ_R, as seen in the table, is consistent with the energy dependence of the $\bar{p}$N cross-section. At 180 MeV the reaction cross-section can be represented by the expression $\sigma_R = \pi(a + r_0 A^{1/3})^2$, with $a \simeq 0.65$ fm and $r_0 \simeq 1.44$ fm. Our determination of σ_R for ^{12}C agrees with that of Nakamura et al. [1984], whereas the value quoted by Aihara et al. [1981] is 20% lower.

Results of microscopic calculations are also shown in Figs. 3 and 4. The dashed and dotted curves represent KMT-type calculations done with the $\bar{p}$N amplitudes of Dover and Richard [1982] or of the Paris potential [Coté et al., 1982], respectively. The proton density was taken from electron-scattering analysis and the neutron density from scattering of high-energy protons and kaons. Both predictions, which have no free parameters, agree with the data reasonably well. This agreement is somewhat surprising in view of the necessary conditions for KMT calculations to be valid. We note that a recent Glauber-type calculation [Dalkarov and Karmanov, 1984], also agrees with the data at 46.8 MeV. A possible explanation for these agreements is that the elementary $\bar{p}$N scattering is forward peaked, a condition favourable to multiple-scattering calculations. Our results also agree with the predictions of von Geramb et al. [1984] (not shown), whose method was originally developed for nucleon–nucleus scattering with the nuclear matter approach [von Geramb, 1979]. The agreement of the prediction by Niskanen and Green [1983] with the 46.8 MeV data all but disappears at 180 MeV [Niskanen, 1984]. We note that the predicted ratio V/W is very high. The disagreement could be attributed to the too early truncation of the elementary amplitude (only s and p waves) and to the use of the local t matrix at an energy which is too high [Niskanen, 1984].

Elastic scattering from ^{7}Li, ^{40}Ca and ^{208}Pb at 47 MeV, and ^{16}O and ^{18}O at 180 MeV have also been measured and are being analysed.

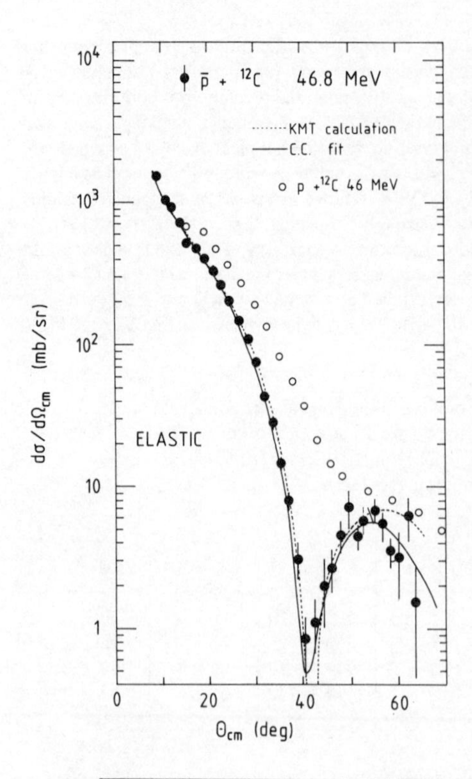

Fig. 3 Differential cross-sections for p̄–elastic scattering from ^{12}C (solid circles). The cross-sections for proton elastic scattering are also shown for comparison (open circles). The dotted curve is a KMT calculation (see text) using N̄N amplitudes of Coté et al. [1982]. The solid curve results from a coupled-channel fit to the data with the following parameters: (V_0, W_0, r_{0v}, a_v, r_{0w}, a_w) = (25 MeV, 61 MeV, 1.17 fm, 0.61 fm, 1.2 fm, 0.51 fm).

Fig. 4 Differential cross-sections for p̄–elastic scattering from ^{12}C, ^{40}Ca, and ^{208}Pb (solid circles). The cross-sections for proton elastic scattering are also shown for comparison (open circles). The dashed and dotted curves are KMT calculations (see text) using N̄N amplitudes of Dover and Richard [1982] and Coté et al. [1982], respectively. The solid curves result from an optical-model fit to the data (see text) with the following parameters: V_0 = 30 MeV, r_{0v} = 1.225 fm, and r_{0w} = 1.1 fm for all three targets, and W_0 = 118, 124, 172 MeV, a_v = 0.514, 0.572, 0.672 fm, and a_w= 0.500, 0.590, 0.649 fm for C, Ca, and Pb, respectively.

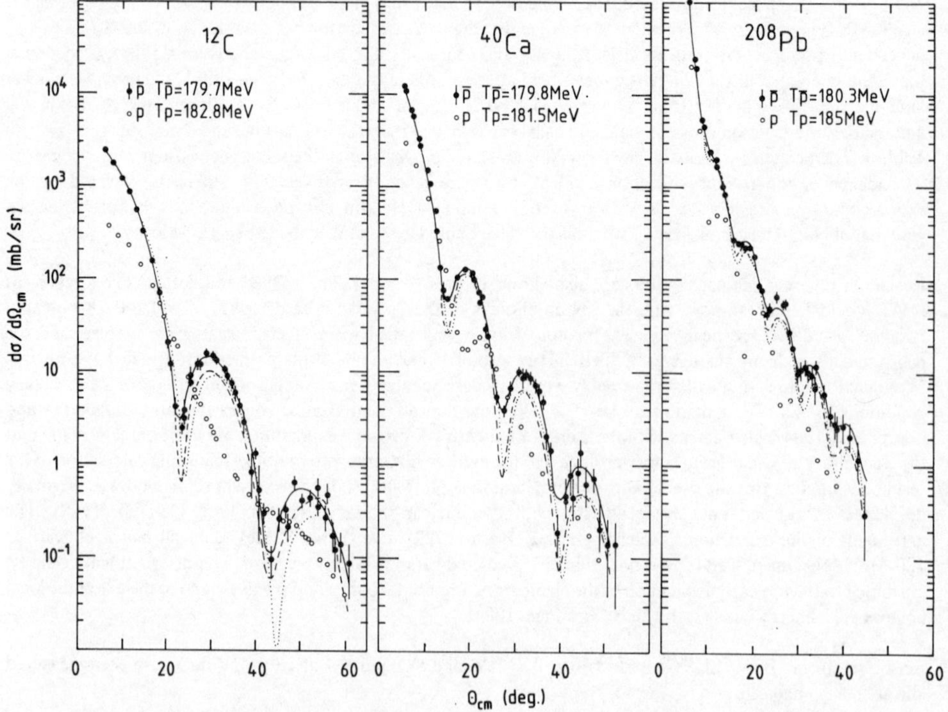

4. INELASTIC SCATTERING

Inelastic scattering angular distributions measured from ^{12}C at 46.8 MeV and 180 MeV are displayed in Fig. 5. For the 4.44 MeV, 2^+ state, it is clear that they are typical of a diffractional pattern, the oscillations being out of phase with those of the elastic scattering. Coupled channel calculations performed with the ECIS code reproduce well the data (solid curves) with deformation lengths β_{2N} R_{2N} with values extracted from proton inelastic scattering [Satchler, 1967]. At 46.8 MeV it was shown [Garreta et al., 1984a] that inelastic scattering sets further constraints on the determination of V_0 compared to what it is when using elastic scattering alone.

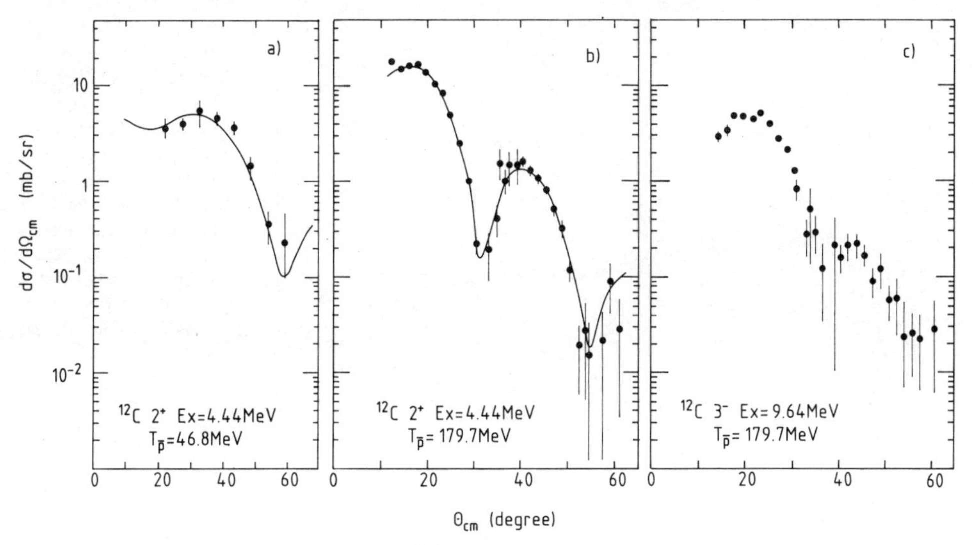

Fig. 5 Differential cross-section for $\bar{p}$ + ^{12}C inelastic scattering. In (a) each point corresponds to a measurement integrated over an angular range of 7.2°. In (b) and (c) the angular resolution is the same as for elastic scattering (see text). The solid curves result from a coupled channel fit to the data with the following parameters: (a) the optical-potential parameter are those of Fig. 3 and the deformation length β_{2N} R_{2N} = − 1.6 fm; (b) (V_0, W_0, r_{0v}, a_v, r_{0w}, a_W) = 40.7 MeV, 176 MeV, 1.116 fm, 0.522 fm, 1 fm, 0.487 fm and β_{2N} R_{2N} = − 1.71 fm.

Inelastic scattering from the two 1^+ unnatural parity states of ^{12}C at 12.7 MeV (T = 0) and 15.1 MeV (T = 1) excitation energy have also been measured at 180 MeV at forward angle and are being analysed. These measurements are very important because it has been shown [Dover et al., 1984], that the corresponding cross-sections are very sensitive to the spin and isospin components of the elementary amplitudes. In particular, at 180 MeV and forward angle, the ratio between the cross-sections to the 12.7 MeV and 15.1 MeV states is predicted [Dover, 1984a] to be almost an order of magnitude larger with the Paris amplitudes than with the Dover and Richard amplitudes.

5. THE ($\bar{p}$,p) REACTION

Energy spectra of protons coming from scintillator, ^{12}C, ^{63}Cu and ^{209}Bi targets bombarded with the 180 MeV antiproton beam of LEAR are shown in Figs. 6 and 7 [Garreta et al., 1984c]. They are well reproduced (solid curves) by a Maxwellian distribution $d^2\sigma/d\Omega dE = C\sqrt{E}\exp(-E/T)$, where T is associated with an "effective temperature". The values of C and T which fit the data are given in Table 2. These energy spectra of protons emitted after antiproton annihilation in a nucleus have been calculated by several groups [Clover et al., 1982; Cahay et al., 1982 and 1983; Iljinov et al., 1982], using an intranuclear cascade (INC) model. In particular, Clover et al. have made the calculation for $\bar{p}$+ ^{12}C annihilation at 600 MeV/c. Their result, plotted as the dashed curve in Fig. 6b, is in good agreement with the data in the overall magnitude. However, the predicted slope is somewhat steeper than that of the data. This corresponds to the difference between the value of 62 MeV deduced from the calculation [Clover, 1982], and the measured value of 86 MeV. The nearly isotropic angular dependence in the $^{12}C(\bar{p}$,p) reaction is consistent with the cascade calculation. From the measured ($\bar{p}$,p)

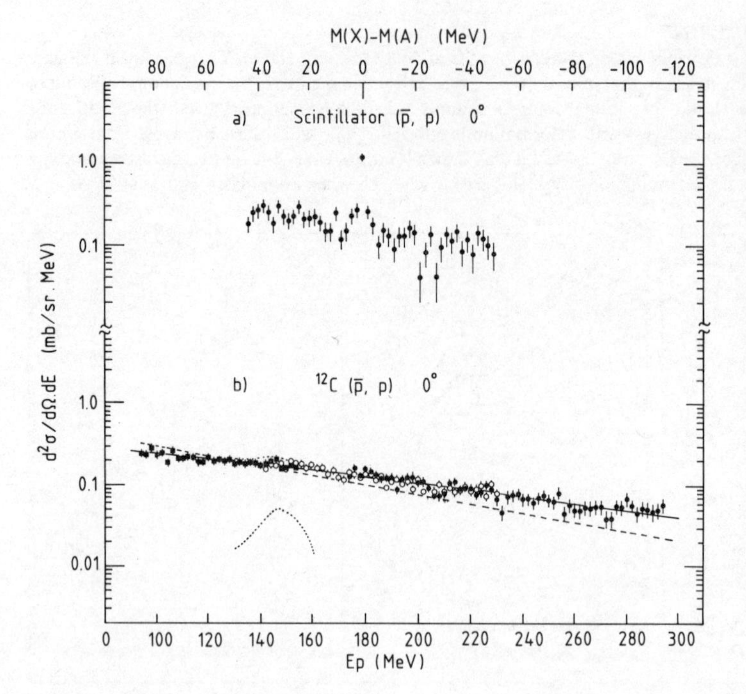

Fig. 6 Proton spectra for the A($\bar{p}$,p)X reaction at $T_{\bar{p}}$ = 180 MeV and 0° (a) for scintillator and (b) for carbon targets. The double differential cross-section is plotted versus the proton kinetic energy and the mass difference [M(X) − M(A)]. The sharp peak near 180 MeV in (a) corresponds to elastic scattering from hydrogen. Also shown are an INC calculation (dashed line) and a Maxwellian distribution best fit (solid line). A calculation for the quasi-free cross-section is indicated by the dotted curves.

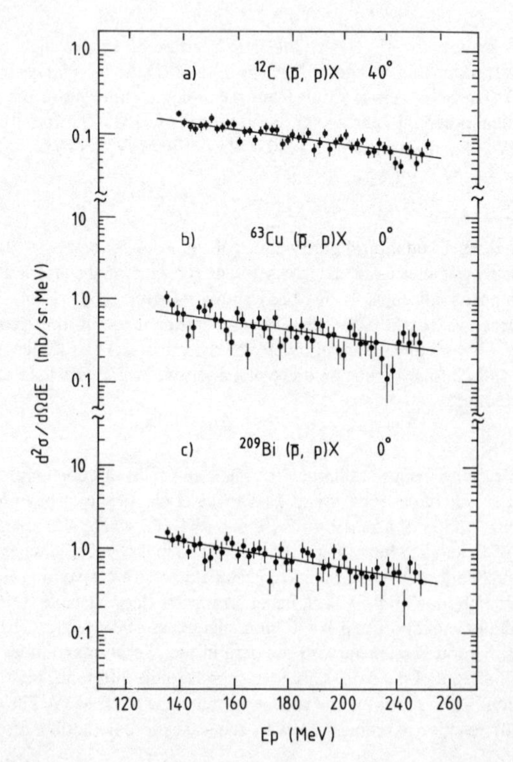

Fig. 7 The ($\bar{p}$,p) spectra from ^{12}C at 40°, and from ^{63}Cu and ^{209}Bi at 0°. The solid curves are best fits assuming a Maxwellian distribution.

Table 2
Parameters resulting from the best fits to the proton spectra with the expression
$d^2\sigma/d\Omega dE = C\sqrt{E}\exp(-E/T)$

Target	θ_{lab} (degrees)	T (MeV)	C (μb/sr · MeV$^{3/2}$)
^{12}C	0	86 ± 1.5	80
^{12}C	40	77 ± 6	75
^{63}Cu	0	69 ± 10	405
^{209}Bi	0	69 ± 7	770

cross-section for ^{12}C, ^{63}Cu and ^{209}Bi, at 180 MeV proton energy, we deduce a mass dependence $A^{0.63}$, in good agreement with the $A^{0.67}$ dependence we deduce from Clover et al. [1982]. The narrow peak observed in Fig. 1a near 180 MeV comes from the 180° $\bar{p}p$ elastic scattering. The sharpness of this peak reflects the good energy resolution (~ 1 MeV) in the present experiment. The c.m. cross-section of this reaction is measured to be 0.67 ± 0.10 mb/sr, in good agreement with a previous experiment [Alston-Garrjost et al., 1979]. We calculated semiclassically the cross-section of the quasi-free $p(\bar{p},p)\bar{p}$ reaction with protons of the target nucleus for the $p_{3/2}$ proton shell of ^{12}C assuming that ^{11}B recoils with momentum opposite to the Fermi momentum of the proton before collision. The shape of the momentum distribution of the proton was taken to be uniform sphere of k_F = 220 MeV/c [Moniz et al., 1971], and the effective number of $p_{3/2}$ shell protons contributing to the quasi-free process was estimated by Bouyssy [1984] to be 0.5. The calculated cross-section is shown by the dotted curve in Fig. 6b. The spectra of Figs. 6 and 7 offer no clear evidence of any peak which could be attributed to a $\bar{p}$–nucleus state. Although a direct comparison cannot be made with the calculation of Heiselberg et al. [1983], their prediction of a peak cross-section of 0.3 mb/sr · MeV for $^{16}O\ (\bar{p},p)_{\bar{p}}{}^{15}N$ at $E\bar{p} \approx 100$ MeV appears to be too large and is not supported by the present measurement. But the statistics of the present measurements are rather poor. Measurements from ^{6}Li and scintillator targets have been performed with higher statistics and are being analysed.

6. CONCLUSIONS

We have found that the real and imaginary potentials are well determined by elastic scattering only at the nuclear surface, where $|W(R)| \gtrsim 2|V(R)|$. If we assume a Woods-Saxon geometry the real potential is found to be attractive but shallow. If the geometry is derived from that of the charge distribution, V_0 and W_0 are well determined, with $V_0 < 70$ MeV and $W_0 > 2\ V_0$. These results indicate that the surface of the nucleus is not transparent to antiprotons up to 180 MeV, even though 180 MeV antiprotons seem to probe the nucleus more deeply than 46.8 MeV ones. Our results do not support the orbiting idea [Auerbach et al., 1981; Kahana and Sainio, 1984], the relativistic mean field approach [Bouyssy and Marcos, 1982], and one microscopic calculation [Niskanen and Green, 1983; Niskanen, 1984]; they agree with the conclusions of Batty et al. [1984], that shallow imaginary potentials should be ruled out, and are in fair agreement with several microscopic calculations. Comparison of elastic scattering from ^{16}O and ^{18}O should be a sensitive test of the $\bar{p}n$ elementary amplitude. Inelastic scattering from collective states sets further constraints on the determination of the optical potential. Inelastic scattering from unnatural parity states will be a powerful test of the elementary $\bar{N}N$ amplitudes.

We have also reported the results of the first $(\bar{p},p)$ experiment intended to search for $\bar{p}$–nucleus states. No evidence of such states has been observed in the present experiment. Further measurements with improved statistics will provide more sensitive limits on the cross-sections of such states. The gross features in the $(\bar{p},p)$ spectra can be explained by INC calculations. Protons emitted after antiproton annihilation are a major source of "background", which complicates the task of finding $\bar{p}$–nucleus states. These background protons will be less abundant with lighter target nuclei such as ^{3}He and ^{6}Li, because of the A dependence of the emission cross-section. Other reactions such as $A(\bar{p},\Lambda)_{\bar{\Lambda}}(A-1)$ could also be contemplated [Peng, 1983; Dover, 1984b] since the cascade background would be absent. Unfortunately the cross-section of the $p(\bar{p},\Lambda)\bar{\Lambda}$ reaction is small. Finally, one could also consider detecting $\bar{p}$-nucleus resonant states [Auerbach, 1981] by measuring excitation functions of the $(\bar{p},\bar{p})$ or $(\bar{p},\bar{p}')$ reactions. Further experimental efforts are definitely required in order to search for $\bar{p}$-nucleus states.

7. REFERENCES

Agnew L E et al. 1957 Phys. Rev. **108** 1545
Aihara H et al. 1981 Nucl. Phys. **A360** 291 and references therein
Alston-Garrjost et al. 1979 Phys. Rev. Lett. **43** 1901
Auerbach E H, Dover C B and Kahana S 1981 Phys. Rev. Lett. **46** 702
Barrett R C and Jackson D F 1977 Nuclear sizes and structure (Oxford: Clarendon Press) p. 263
Batty C J 1981 Nucl. Phys. **A372** 433 and references therein
Batty C J, Friedman E and Lichtenstadt J 1984 Optical potentials for low-energy antiproton-nucleus interactions, Rutherford Appleton Laboratory, Didcot
Bouyssy A and Marcos S 1982 Phys. Rev. Lett. **114B** 397
Bouyssy A 1984 Private communication
Bruge G 1984 report DPh-N Saclay No. 2136
Cahay M, Cugnon J, Jasselette P and Vandermeulen J 1982 Phys. Lett **115B** 7
Cahay M, Cugnon J and Vandermeulen J 1983 Nucl. Phys. **A393** 237
Chaminade R, Durand J M, Faivre J C and Pain J 1974 Nucl. Instrum. Methods **118**, 477
Clover M R, De Vries R M, DiGiacomo N J and Yariv Y 1982 Phys. Rev. **C 26** 2138
Coté J et al. 1982 Phys. Rev. Lett. **48** 1319
Dalkarov O D and Karmanov V A 1984 Lebedev Inst. preprint No. **77**
Dover C B 1984a Private communication
Dover C B 1984b Proc. 10th Int. Conf. on Few Body Problems in Physics Karlsruhe 1983 (Amsterdam: North Holland) vol. 1
Dover C B, Gal A and Richard J M 1983 Phys. Rev. **D 27** 1090
Dover C B and Richard J M 1982 Phys. Rev. **C 25** 1252
Dover C B et al. 1984 Antinucleon-nucleus inelastic scattering and the spin dependence of the $\bar{N}N$ annihilation potential, Orsay preprint IPN
Garreta D et al. 1984a Phys. Lett. **135B** 266, and **139B** 464
Garreta D et al. 1984b Preprint CERN–EP/84–93
Garreta D et al. 1984c Preprint CERN–EP/84–92
Green A M and Wyceh S 1982 Nucl. Phys. **A377** 441
Heiselberg H, Jensen A S, Miranda A and Oades G C 1983 Phys. Lett **132B** 279
Iljinov A S, Nazaruk V I and Chigrinov S E 1982 Nucl. Phys. **A382** 378
Kahana S H and Sainio M E 1984 Phys. Lett. **139B** 231
MacKellar A D, Satchler G R and Wong C Y 1984 Z. Phys. **A316** 35
Moniz E J et al. 1971 Phys. Rev. Lett. **26** 445
Nakamura K et al. 1984 Phys. Rev. Lett. **52** 731
Niskanen J A 1984 Private communication
Niskanen J A and Green A M 1983 Nucl. Phys. **A404** 495
Peng J C 1983 Proc. 3rd LAMPF II Workshop, Los Alamos, July 1983, Report LA–9933–C, p. 531
Raynal J 1981 Phys. Rev. **C 23** 2571
Sakitt M 1984, paper presented at this conference
Satchler G R 1967 Nucl. Phys. **A100** 497
Thirion J and Birien P Le spectromètre SPES II, Rapport DPHN/ME, Saclay
Von Geramb H V 1979 Microscopic Optical Potentials, Hamburg, 1978, Lecture Notes in Physics (Berlin: Springer Verlag) vol. 89
Von Geramb H V, Nakano K and Rikus L 1984 Microscopic analysis of antiproton scattering from carbon, Universität Hamburg
Wong C Y, Kerman A K, Satchler G R and MacKellar A D 1984 Phys. Rev **C 29** 574

Inst. Phys. Conf. Ser. No. 73: Section 3
Paper presented at VII Eur. Symp. Antiproton Interactions, Durham 1984

PS 187—a good statistics study of antiproton interactions with nuclei: preliminary results

J. W. Sunier, K. D. Bol, M. R. Clover, R. M. DeVries, N. J. DiGiacomo, J. S. Kapustinsky, P. L. McGaughey, G. R. Smith‡ and W. E. Sondheim

Physics Division, Los Alamos National Laboratory,
Los Alamos, NM 87545, USA

and

M. Buenerd, J. Chauvin, D. Lebrun and P. Martin

Institut des Sciences Nucléaires, 38026 Grenoble, Cedex, France

and

J. C. Dousse, Université de Fribourg, 1700 Fribourg, Switzerland

Abstract: The goals of LEAR experiment PS187 are reviewed. The current status of the experiment is discussed, and final data on a search for $\bar{p}$-nucleus bound states formed by the reactions $^{28}Si(\bar{p},p)$ and $^{28}Si(\bar{p},d)$ are presented.

1. Introduction

The LEAR experiment PS187 has four main goals: (a) measure with good statistics, over a broad range of momenta and angles, the production of light charged particles ($\pi^{\pm}$, $K^{\pm}$, p, d, t, ...) following the annihilation of antiprotons on a variety of nuclei spanning the whole range of available nuclear masses; (b) study the Bose-Einstein correlations ($\pi^+\pi^+$, $\pi^-\pi^-$) to determine the space-time structure of the pion source when an antiproton annihilates in a nucleus; (c) search for evidence of antiproton-nucleus bound or resonant states using the $A(\bar{p},p)_{\bar{p}}(A-1)$ reaction; (d) develop techniques to identify central annihilations. The study of the inclusive spectra aims at establishing a much needed data base to determine whether or not anything unusual (above and beyond $\bar{p}$-N, N-N, π-N and Δ-N interactions) occurs in the $\bar{p}$-nucleus annihilations. The range of momenta ($0.1 < p < 1.5$ GeV/c) and angles ($0 < \theta_{LAB} < \pm 180^{\circ}$) simultaneously covered by the experiment will give rather complete double differential cross sections ($d^2\sigma/d\Omega dE$) with few normalization problems. These data are to be compared to the predictions of Intra Nuclear Cascade (INC) calculations [Clover et al., Cahay et al.]. The search for possible antiproton-nucleus bound states aims at obtaining information on the $\bar{p}$-nucleus potential complementary to those obtained in scattering experiments [Garetta et al.]. The possibility of tagging central annihilations is of particular interest. One expects that significant differences between the nucleon-nucleon and nucleon-nucleus annihilation processes could show up in such events. In particular, the study of the K^+ spectra as a function of the annihilation radius has a special significance. Because of the low K^+-N interaction cross section, a K^+ travels relatively undisturbed through the nuclear medium and may carry cleaner information on the annihilation process inside nuclei than is the case with pions.

2. Experimental Arrangements

A broad range spectrometer, depicted in Fig. 1, was developed for the experiment, and tested at LAMPF and TRIUMF [Sunier et al., DiGiacomo et al.]. It consists of a circular dipole magnet, 81 cm in diameter. The bulk of the data was acquired with a pole gap of 12.7 cm and a magnetic field of 1.2 Tesla. Targets were placed in a vacuum chamber, either at the center of the magnet or, as in the ^{28}Si ($\bar{p}$,p) experiment, at its edge (effective field boundary). Six detector modules consisting of a pair of x,y position sensitive gas counters, 10 cm apart from each other, together with plastic scintillators (ΔE) and a plastic threshold Cherenkov detector [Sondheim et al.] surround the pole gap. These detector modules provide position, time of flight, energy loss and Cherenkov information. Together with the momentum and scattering angle obtained from the reconstruction of the particle trajectories, they allow for the identification of π, p, K, d, t. The beam is counted by a thin scintillator (S_1), and event triggers are defined by the combination $S_1 \cdot \bar{S}_2 \cdot \Delta E \cdot \bar{S}_3$. Inter-detector coincidences are identified by tag bits in the event stream. Up to two intra-detector coincidences can be detected and identified in two of the detector modules (1 and 6). The electronics and data acquisition system

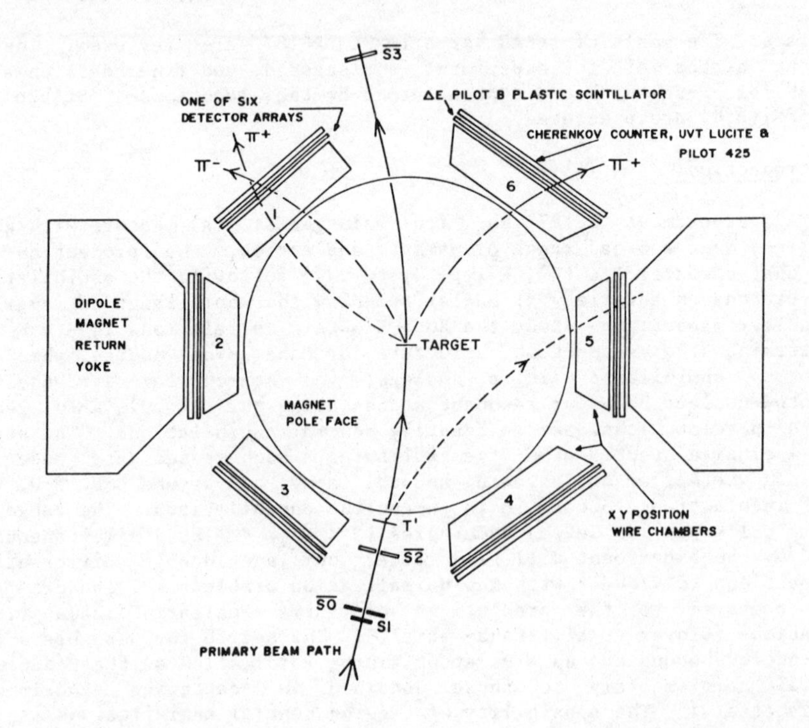

Fig. 1. Schematic diagram of the Calliope spectrometer.

are capable of acquiring data at a sustained rate of 1000 events per second, with minimal dead time.

3. Experiment Status and Run Statistics

The experiment received beam of 608 MeV/c during two periods, in November 1983 and June 1984. A total flux of 5×10^{10} $\bar{p}$ was obtained to complete the experiment. Inclusive and inter-detector correlation data were taken with ^{28}Si, ^{89}Y and ^{238}U targets (3×10^9 $\bar{p}$ each) as well as with ^{12}C and ^{208}Pb (2×10^9 $\bar{p}$ each). The ^{28}Si ($\bar{p}$,p) reaction used 4×10^9 $\bar{p}$. The bulk of the correlation data was taken on ^{28}Si with 3.3×10^{10} $\bar{p}$. The inclusive data, in the preliminary stage of analysis, contain some 10^5 K^+. The correlation data have about 15,000 $\pi^+\pi^+$ pairs and 25,000 $\pi^-\pi^-$. The analysis of the ^{28}Si($\bar{p}$,p) reaction is completed.

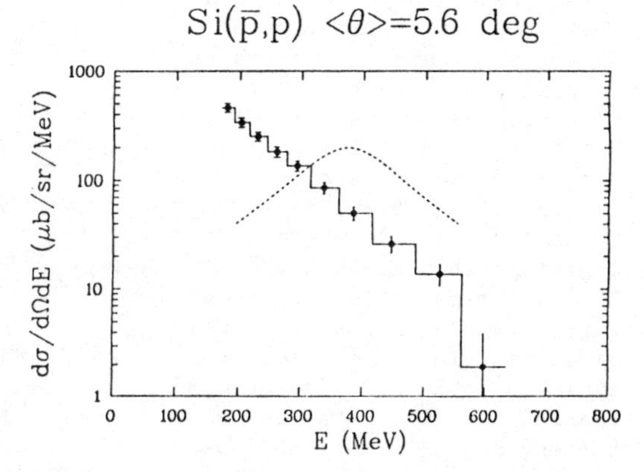

Fig. 2. Differential cross section of protons emitted at forward angle (0 to 9°) following the interaction of 180 MeV antiprotons on ^{28}Si. The dashed curve is an extrapolation of the calculation of Heiselberg et al. (^{16}O) to the incident $\bar{p}$ energy of 180 MeV.

4. Search for $\bar{p}$-Nucleus Bound States via the ^{28}Si($\bar{p}$,p) and ^{28}Si($\bar{p}$,d) Reaction

It has been theoretically suggested [Heiselberg et al., Wong et al.] that antiproton-nucleus bound states could exist, given certain assumptions about the strengths and ranges of the real and imaginary $\bar{p}$-nucleus optical potential. We report here the results of an experiment designed to test these conjectures using the ^{28}Si($\bar{p}$,p) and ^{28}Si($\bar{p}$,d) knock-out reactions. After knocking out a proton (deuteron) with little momentum transfer to the nucleus, the antiproton can fall into a quasibound state prior to annihilating in the nucleus. The angular and momentum distributions of the outgoing protons (deuterons) should reflect the energy and width of such states. Our experiment was designed to look for deeply bound states (20 to 200 MeV). An active target in the form of a 25 mm diameter, 1 mm thick intrinsic Silicon detector, was placed at the effective field boundary of the Calliope magnet (position T′ in Fig. 1). With this geometry, the momentum resolution varies from 5 to 10% (FWHM) in the momentum range 600 to 1200 MeV/c. The proton and deuteron spectra were recorded in the two forward detectors, at averaged lab angles of 5.6 and 43.7°, respectively. These detectors covered the angular range (variable with momentum) of -6 to 15° and 36 to 51°. The active Si target allowed for the rejection of background events (scattering or annihilations in the walls of the vacuum chamber) by requiring an energy deposition of ΔE(Si) > 300 keV in the target. The loss of efficiency due to this cut-off is less than 1%, and the spectra obtained are very clean. The data were binned in $\Delta\theta$ = 3 degree and Δp = 10 MeV/c bins. No evidence for broad structures was found anywhere. The spectra shown in Figs. 2 and 3 represent a subset of the data, rebinned according to the momentum resolution of the spectrometer. These spectra exhibit the characteristic statistical shape of proton and deuteron spectra resulting from $\bar{p}$ annihilation in the nucleus. The fit of the data, in the momentum range 600 < p <1200 MeV/c, to a cross section of the form $d^2\sigma/d\Omega dE$ = Const • exp(-E/T), where T is a characteristic temperature and E is the particle kinetic energy, results in the parameters given in Table I. The upper limit (3 standard deviations) placed by our data on a $\bar{p}$-nucleus bound state following knock-out of a proton is

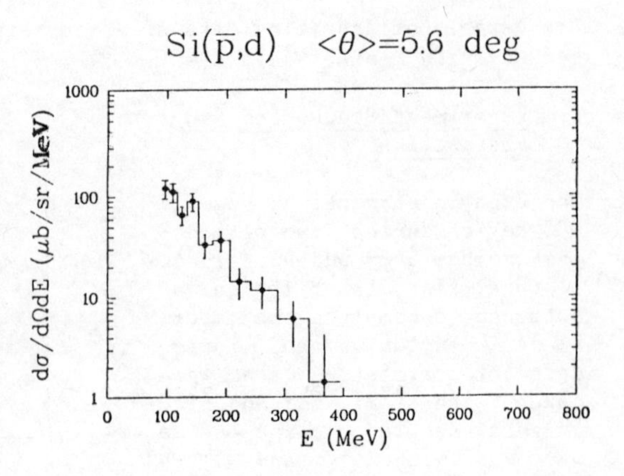

Fig. 3. Same as Fig. 2 for deuterons.

Table I. Fit of the ^{28}Si($\bar{p}$,p) and ^{28}Si($\bar{p}$,d) cross sections to the function $d^2\sigma/d\Omega dE$ = C • exp(-E/T).

Reaction	θ_{Lab} (degree)	T (MeV)	C (μb/sr/MeV)
^{28}Si($\bar{p}$,p)	5.6	96 ± 5	2830
	43.7	96 ± 5	2090
^{28}Si($\bar{p}$,d)	5.6	67 ± 8	530
	43.7	65 ± 8	480

60 μb/sr MeV at E_p = 250 MeV and 15 μb/sr MeV at E_p = 450 MeV. This is an order of magnitude lower than the prediction of Heiselberg et al. The observed proton spectra are in good agreement with INC predictions [Clover et al.] both in magnitude and shape, indicating that the annihilation process dominates the proton yields.

5. Use of an Active Si Target to Tag Central Annihilations

The use of an active target, like a Si detector, besides offering a clean way of scaling the beam intensity on target and efficiently rejecting most of the background triggers, promises to also allow tagging of central annihilations. Using the results of INC calculations, one can calculate the amount of charged particle energy deposited in the Silicon detector, following the annihilation of $\bar{p}$ in ^{28}Si. The result of such a calculation is shown in Fig. 4, where the various contributions of $\bar{p}$, pions as well as direct (cascade) and evaporation protons to the Si pulse-height are detailed. It is clear from these data that, at large pulse-heights, the spectrum is dominated by cascade and evaporation protons, thereby reflecting the amount of excitation deposited in the nucleus by the annihilation process. INC calculations [Clover et al.] clearly correlate this excitation energy to the annihilation radius. The effect of a software Si pulse-height cut on the angle integrated π^+ momentum spectrum from ^{28}Si is shown in Fig. 5. One clearly sees the enhancement of the "secondary" low energy pion bump in the spectrum when a large Si pulse height is required. This bump corresponds to pions which were created deep in the nucleus and "thermalized" by subsequent Δ-N and π-N interactions. Further model calculations to explicitly show the Si pulse height dependence upon annihilation radius are in progress.

The authors acknowledge the assistance of K. Kilian, P. LeFevre and the LEAR staff in the timely completion of the experiment. This work was supported by the U.S. Department of Energy.

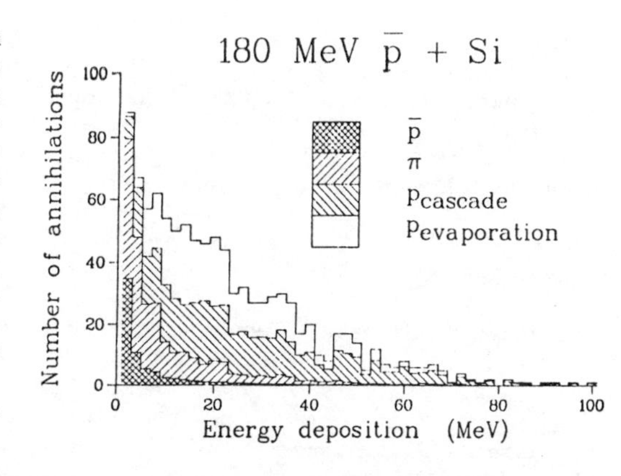

Fig. 4. Contribution of $\bar{p}$, π and p to the energy deposited in a 1 mm thick, 25 mm diameter Si detector, following the calculated (INC) interaction of 180 MeV $\bar{p}$ with Silicon.

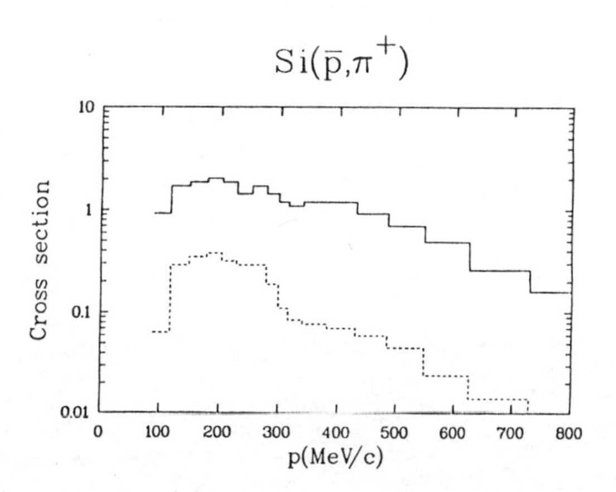

Fig. 5. Angle integrated momentum distribution of π^+ following the $\bar{p}$ annihilation in ^{28}Si. The solid curve is for all events. The dashed curve is for events leaving more than 7 MeV in the target.

References

‡Permanent address: TRIUMF, Vancouver V6T 2A3, British Columbia, Canada.

Cahay M., Cugnon J. and Vandermeuten J. 1983, Nucl. Phys. A393, 237.

Clover M. R., DeVries R. M., DiGiacomo N. J. and Yariv Y. 1982, Phys. Rev. C26, 2138.

DiGiacomo N. J. et al. 1984, submitted to Phys. Rev. C.

Garetta D. et al. 1984, Phys. Lett. 135B, 266.

Heiselberg H., Jensen A. S., Miranda A. and Oades G. C. 1983, Phys. Lett. 132B, 279.

Sondheim W. E. et al. 1984, Nucl. Instr. and Methods 219, 97.

Sunier, J. W. et al. 1982, Bull. Am. Phys. Soc. 27, 525 and 1984, submitted to Nucl. Instr. and Methods.

Wong C. Y., Kerman A. K., Satchler G. R. and Mackellar A. D., Phys. Rev. C29, 441.

Inst. Phys. Conf. Ser. No. 73: Section 3
Paper presented at VII Eur. Symp. Antiproton Interactions, Durham 1984

Investigation of the $\bar{p}p \to \bar{\Lambda}\Lambda$ reaction near threshold at LEAR

P.D. Barnes[1], R. Besold[3], P. Birien[6], B.E. Bonner[5], W. Breunlich[8], W. Dutty[4]
R.A. Eisenstein[1], G. Ericsson[7], W. Eyrich[3], R. Frankenberg[3], G. Franklin[1],
J. Franz[4], N. Hamman[4], D. Hertzog[1], A. Hofmann[3], T. Johansson[7], K. Kilian[2],
C. Maher[1], R. Müller, G. Nicklas[4], H. Ortner[3], P. Pawlek[8], S.M. Polikanov[2],*
E. Rössle[4], H. Schledermann[4], H. Schmitt[4], W. Wharton[1], and P. Woldt[3]

Carnegie-Mellon Univ.[1] - CERN[2] - Erlangen-Nürnberg Univ.[3] - Freiburg Univ.[4]-
Los Alamos National Lab.[5] - Saclay CEN DPhN[6] - Vienna Inst. Radiumforschung
und Kernphysik[8] - *: now at GSI Darmstadt

Abstract. The motivations for CERN experiment PS185 are reviewed and prospects for future developments are summarized.

1. Introduction

Studies of the $\bar{p}p \to \bar{\Lambda}\Lambda$ reaction near threshold (P_{lab} = 1435 MeV/c) offer an excellent opportunity to examine the dynamics of $s\bar{s}$ quark pair creaction (Barnes 1981, 1982). Since the ud ($\bar{u}\bar{d}$) pair couple to spin and isospin zero, the s($\bar{s}$) quarks carry the spin of the $\Lambda(\bar{\Lambda})$. Therefore by studying the polarization of the $\Lambda\bar{\Lambda}$ pair, one can examine directly the quantum numbers of the $s\bar{s}$ pair at the point of their creaction. In particular previous measurements are consistent with the $\Lambda\bar{\Lambda}$ and hence $s\bar{s}$ pair being created in a spin triplet state almost to the exclusion of spin singlet (see references in Barnes 1981). In experiment PS185 we will measure cross sections, polarizations and spin correlations in the final state for the $\bar{p}p \to \bar{\Lambda}\Lambda$ reaction. This will provide stringent tests of quark and/ or meson exchange model calculations of this process (Genz 1983).

2. Experimental setup and results

During May, 1984 we received 100 hours of antiproton beam at two momenta: 1.48 and 1.51 GeV/c. A total of 4.5 . 10^{10} $\bar{p}$'s were focussed to a spot size of 1.2 x 0.4 mm^2 FWHM on a 2.5 mm thick CH_2 target. The experimental arrangement is shown in Fig. 1. Hyperons produced in the target enter a MWPC and drift chamber stack where the charged decay $\bar{\Lambda} \to \bar{p}\pi^+$ and $\Lambda \to p\pi^-$ are detected and tracked. The baryons are kinematically constrained to a narrow forward cone which is intercepted by the scintillator hodoscope. The deflection of the charged tracks in the solenoid field is measured by drift chambers in order to distinguish the Λ and $\bar{\Lambda}$ vertices. A veto scintillator box (S2, S3) surrounding the target and the hodoscope are used to form the "charged-neutral-charged" on-line trigger (S1.$\overline{S2}$.$\overline{S3}$. Hodoscope).

A first inspection of tracks reconstructed from the chamber coordinates demonstrates that we can clearly identify the $\bar{\Lambda}\Lambda$ events (Fig. 2). By restricting our attention to that class of events having all four charged

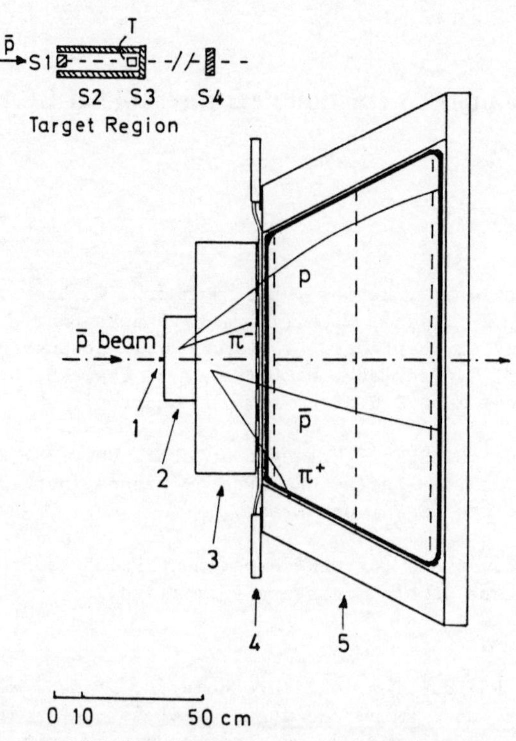

Fig 1 Experimental setup for PS185. 1: target, 2: Proportional chambers 3: drift chambers, 4: hodoscope, 5: baryon number identifier. An example of a "perfect event" is indicated. The target region is given in a magnified view with T = target, S1-4: scintillation counters.

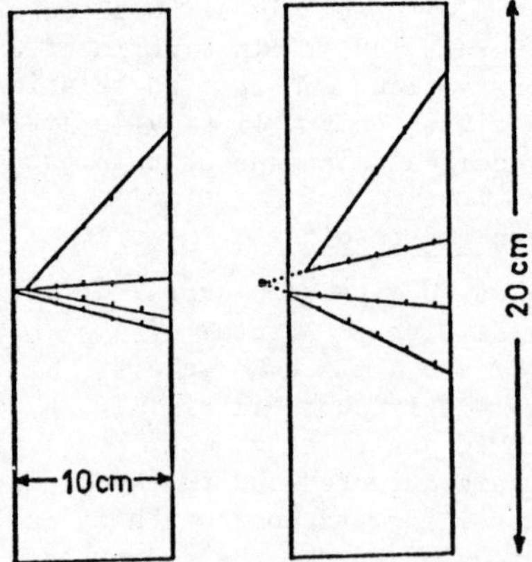

Fig 2 On-line display of double vertex event in two orthogonal projections as observed in the MWPC stack (item 2 in Fig. 1). The beam momentum was 1.51 GeV/c. The event is consistent with $\bar{\Lambda} \to \bar{p}\pi^+$ and $\Lambda \to p\pi^-$ decays of the neutral hyperons.

tracks going forward into the drift chamber stack, we observe that the events identified as $\bar{\Lambda}\Lambda$ vertices follow the expected kinematic characteristics of this reaction. The analysis is proceeding; we should soon have cross-sections and polarizations to report. We expect that about 1000 $\bar{\Lambda}\Lambda$ events have been accumulated in our May run.

3. Outlook

In addition to the $\bar{\Lambda}\Lambda$ channel, other $\bar{Y}Y$ thresholds are available for study at LEAR: $\bar{\Lambda}\Sigma^{o} + \Lambda\,\bar{\Sigma}^{o}$ and $\bar{\Sigma}\Sigma$ at momenta around 1700 and 1900 MeV/c respectively. In order to extract spin correlation information, we need an order of magnitude increase in our $\bar{\Lambda}\Lambda$ sample. In addition it appears feasible to do Λ and $\bar{\Lambda}$ scattering off free protons using a slight modification of our present apparatus (Schmitt 1982). Our apparatus is also capable (Kilian 1984) of investigating the narrow state observed by Mark III at Spear (Einsweiler 1983) in radiative ψ decays:

$\psi \rightarrow \gamma + \xi \rightarrow K\bar{K}$ with $M(\xi) = M(K\bar{K}) = 2220 \pm 15$ MeV and $\Gamma = 30 \pm 15$ MeV. The speculations about this state range from glueballs and Higgs particles to a prosaic $^{3}F_{2}\,s\bar{s}$ meson (Godfrey, 1984).

There are many fascinating questions that are amenable to investigation by our threshold detector. The only questionable aspect at this time is whether or not LEAR will be provided sufficient $\bar{p}$'s to allow the successful completion of the many experiments in progress there.

References

Barnes P et al. (1981) CERN/PSCC/81-69/P49.

Barnes P et al. (1982) CERN/PSCC/82-57/M122.

Einsweiler K F et al. (1983) SLAC-P4B 3202.

Genz H and Tatur S (1983) Preprint TKP 83-21 Karlsruhe Univ.

Godfrey S et al. (1984) Physics Letters 141B 439.

Kilian K et al. (1984) CERN/PSCC/84-26/M192.

Schmitt H et al. (1982) Proc. of Physics at LEAR with low-energy cooled Antiprotons (Plenum: New York) Gastaldi and Klapisch eds. p. 489.

Low energy antiproton interaction studies at Brookhaven

M. Sakitt, V. Ashford, M. E. Sainio and J.F. Skelly
Brookhaven National Laboratory, Upton, New York, USA, 11973

R. Debbe, W. Fickinger, R. Marino and D.K. Robinson
Case Western Reserve University, Cleveland, Ohio, USA, 44106

Abstract

Results on antiproton quasielastic scattering on Al, Cu and Pb for two incident momenta 514 and 633 MeV/c are presented. Differential cross sections for these reactions were combined with other published data and analyzed with a nonrelativistic optical model. Absorption cross section results from our experiment are in good agreement with predictions from our optical model fits. Preliminary results are presented from $\bar{p}$ elastic scattering on hydrogen for the forward angular region in the momentum range from 360 to 652 MeV/c. We obtain small values of the ratio of the real to imaginary part of the forward scattering amplitude in this momentum interval.

1. Introduction

We are presenting results from an antiproton experiment at the Brookhaven National Laboratory AGS in the low energy separated beam line. The main purpose of the experiment was to study $\bar{p}p$ interaction in the region of the putative S meson. During a period of several months, various data runs with a hydrogen target were carried out with incident momentum of 360 to 652 MeV/c. The momentum settings were chosen to give overlapping bins from the energy loss in the hydrogen target.

The purpose of the experiment was to study the momentum dependences of the following topics:

a) total cross section
b) elastic differential cross section
c) annhilation cross section
d) the ratio of the real to imaginary part of the forward scattering amplitude.

In addition, we made a short excursion into nuclear physics and ran 20 hours with antiprotons on Al, Cu and Pb targets at 514 and 633 MeV/c. Data were taken for

a) quasielastic differential cross section
b) absorption cross section.

Our results on the nuclear quasielastic scattering, preliminary results on σ_{abs} on nuclei and preliminary results on the ratio of the real to imaginary parts of the forward scattering amplitude are presented.

2. Experimental Layout

The experimental setup is shown in Fig. 1. The partially separated beam was tagged electronically with S1, S2 and S3 used as both time of flight and dE/dx counters. In addition, a lucite Cerenkov counter Č provided a a pion veto. The resulting antiproton beam identification was 99% pure and was used as the experimental trigger. The incoming beam was measured in four triplet drift chambers (DC1-12) each consisting of an x, u and v plane. The outgoing scattered particle was measured in 18 doublet drift chambers (DC13-30), six each for x, u and v.

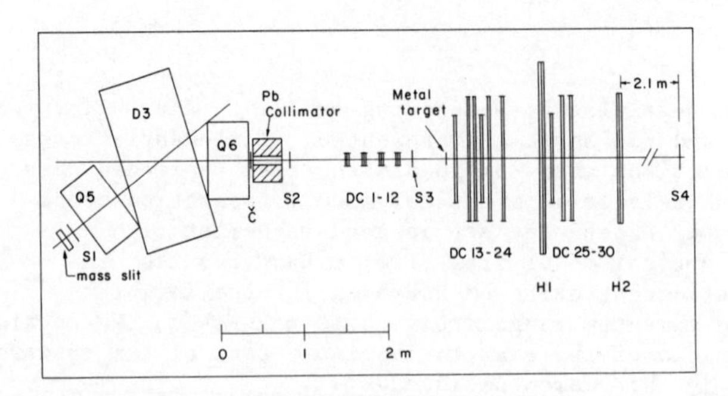

Fig. 1. Experimental layout.

The scattered particle also passed through one of two hodoscopes, H1 and H2 in Fig. 1. Each of these hodoscopes consisted of two orthogonal sets of elements, with a square hole in the center of H1, to allow more precise measurement of the forward going particles. The photomultipliers on all of the elements had their time of flight and pulse heights recorded. A minimum of three photomultipliers was required to distinguish between outgoing pions from annihilations and elastically scattered antiprotons.

Both the time of flight and the pulse height were used to identify the antiprotons with a resulting tagging efficiency of about 95% per plane. Time of flight distributions are shown in Figure 2. Further details of both the time of flight and pulse height criteria are given in Ashford et al. (1984).

3. Quasielastic $\bar{p}$ Nucleus Scattering

A summary of the various nuclear physics runs is shown in Table I.

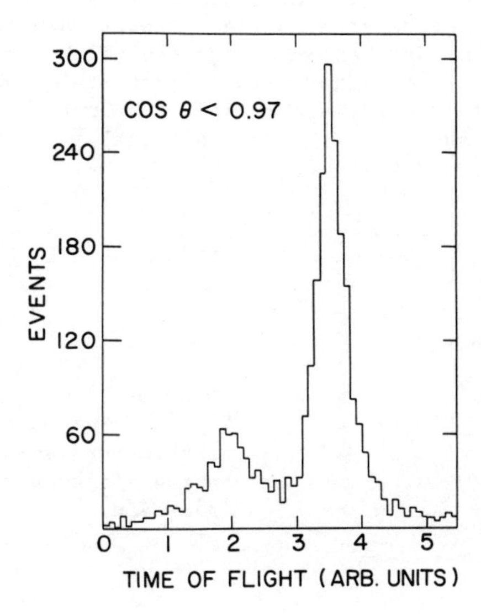

Fig. 2. Time of flight distribution for forward secondary tracks in the downstream hodoscope.

Table I

TRIGGER TOTALS AND TARGET THICKNESSES

Target	Thickness (cm)	Total Triggers 514 MeV/c	633 MeV/c
Aluminum	0.299	144,680	145,932
	0.686	72,361	73,531
Copper	0.099	145,445	146,511
	0.284	72,737	72,932
Lead	0.081	145,656	146,557
	0.244	71,385	73,265

To help understand systematics we made measurements with two target thicknesses for each nucleus. The differential cross sections determined from the two thicknesses were in agreement at angles greater than three times the mean multiple scattering angle expected for the thick target. We therefore merged the thick and thin target data in that region and used the thin target data alone in the forward region. Typical differential cross section results are shown in Fig. 3; additional distributions are shown in Ashford et al. (1984).

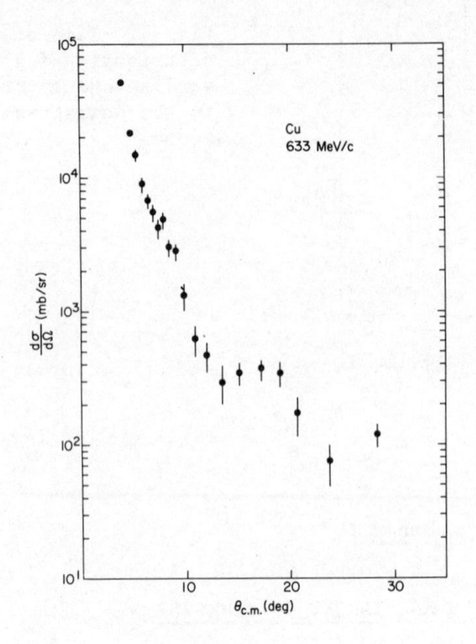

Fig. 3. The quasi-elastic angular distribution for copper at 633 MeV/c.

We analyze the elastic cross sections in terms of a nonrelativistic optical potential model. We take the potential to be of the form

$$V_{opt}(r) = - V_o f_R(r) - iW_o f_I(r)$$

where f_R and f_I have the Woods-Saxon shape

$$f_{R,I}(r) = \left[1 + \exp\left(\frac{r - R_{R,I}}{a_{R,I}}\right)\right]^{-1}$$

We therefore have six parameters to fit, the two strengths V_o and W_o, the two radii R_R and R_I, and the two diffuseness parameters a_R and a_I. We have included in our analysis recent data for ^{12}C, ^{27}Al and ^{63}Cu at energies about 110, 150 and 190 MeV from Nakamura et al. (1984) and for ^{12}C at 46.8 MeV from Garreta et al. (1984). In principle the sizes of the nuclei can be determined from the diffraction minima in the pure elastic differential cross section. It is important to note, however, that the data of Ashford et al. (1984) and Nakamura et al.

(1984) are quasielastic rather than pure elastic scattering. Since the outgoing tracks in these experiments are not momentum analyzed the data include inelastic scattering to low lying nuclear states.

It is known, in the case of ^{12}C, that the inelastic processes can fill in the diffraction minimum (Garreta et al., 1984). Therefore, in order to study the A dependence of the strengths of the potential we fix the well shape parameters (Auerbach, et al., 1981).

$R_R = 1.3\ A^{1/3}$ fm; $R_I = 1.1\ A^{1/3}$ fm;

and

$a_R = a_I = 0.52$ fm.

We have used the program A-THREE (Auerbach, 1978) to fit the data to the above potentials. Since two of the experiments include quasielastic data, we restrict the fitted region to the forward region where the pure elastic is expected to be dominant. That limit varied between 10 to 14 degrees depending on the nucleus. In our preliminary fitting we saw no energy dependence of the strength parameters for any of the nuclei in the energy region measured. Therefore we have merged the different energies and fit each nucleus as one set of data. The resulting strength parameters, shown below, are all strongly absorptive with an attractive real well.

Table II

Optical Model Well Depth Parameters for C, Al, Cu and Pb.

	V_o (MeV)	W_o (MeV)
^{12}C	25	91
^{27}Al	24	96
^{63}Cu	27	121
^{208}Pb	25	173

We have made an estimate of the absorption cross section from our experimental data via the transmission technique and compared the results with the predictions based on these optical parameters. Using the downstream drift chambers to define a series of transmission counters subtending decreasing solid angles, we can extrapolate to zero degrees, correcting for the elastic scattering, obtain the absorption cross section. In Fig. 4 we show one of the extrapolation plots. The errors in our absorption cross sections are dominated by uncertainties in the various efficiencies and not by the extrapolation procedure. Overall, our experimental results agree with the predictions of the optical model fits to the elastic scattering data and are shown in Table III.

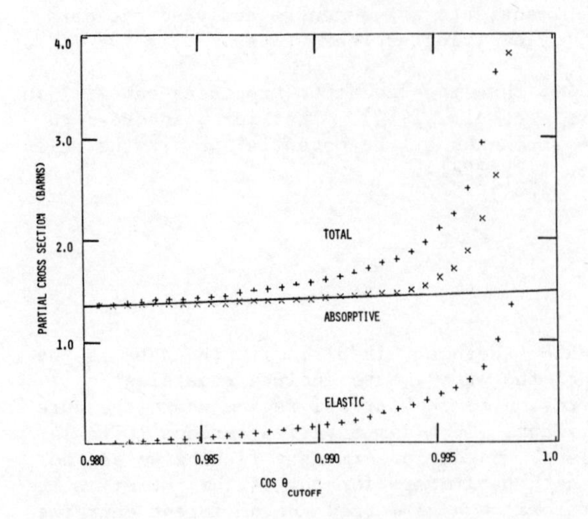

Fig. 4. Extrapolation plot for absorption cross section for Cu at 514 MeV/c. Total and elastic cross sections are represented by pluses and the inelastic cross section, which is the difference, is represented by crosses.

Table III

Element	P(MeV/c)	Global Fit	Experimental Results
			Absorption Cross Section (Barns)
Al	514	0.73	0.77 ± .06
Al	633	0.68	0.62 ± .05
Cu	514	1.21	1.42 ± .10
Cu	633	1.14	1.10 ± .10
Pb	514	2.51	2.1 ± .70
Pb	633	2.36	1.4 ± .25

4. Ratio of Real to Imaginary Part of the $\bar{p}p$ Forward Scattering Amplitude

We present some preliminary results on measurements of the forward elastic antiproton scattering in hydrogen. The separation of outgoing antiprotons from annihilation pions was carried out in the same manner as was described for the nuclear data. We estimated less than 1% contamination in the forward region. Since we are studying the shape of the differential cross section, it is important for the target empty correction to take into account the different multiple scattering for the two different samples.

We have differential cross section data at 11 momenta between 360 and 652 MeV/c; a typical angular distribution is shown in Fig. 5.

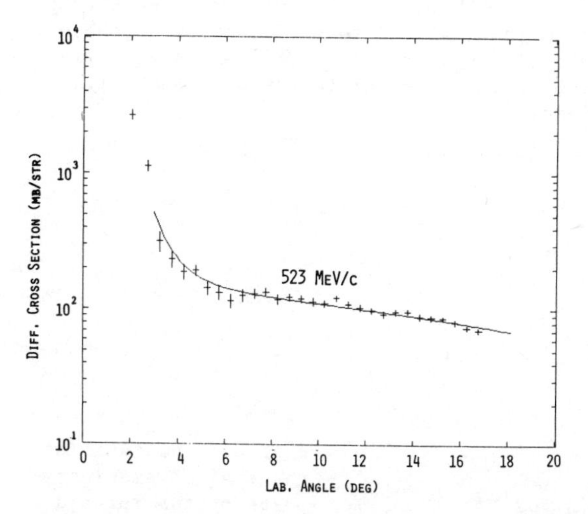

Fig. 5. Forward elastic differential cross section for $\bar{p}p$ scattering. The curves represent the fits described in the text.

To extract the ratio of the real to imaginary part of the forward elastic scattering amplitude we use the standard formalism. We divide the effective scattering into three pieces, a Coulomb term, a nuclear term and an interference term. Explicitly, we have

$$\frac{d\sigma^C}{dt} = 4\pi \left(\frac{\alpha\, hc}{\beta t}\right)^2 F(t)^2$$

$$\frac{d\sigma^I}{dt} = \frac{\alpha\sigma_{TOT}}{\beta t} \cdot F(t) \exp\left(-\frac{1}{2} bt\right) (\rho\cos\delta - \sin\delta)$$

and

$$\frac{d\sigma^N}{dt} = \frac{1}{\pi} \left(\frac{\sigma_{TOT}}{4\ hc}\right)^2 (1+\rho^2) \exp(-bt)$$

where $-t$ is the four-momentum transfer, α is the fine structure constant, β is the velocity of the incident $\bar{p}$ in the laboratory, b is the slope parameter of the $\bar{p}p$ diffraction peak, F(t) is the Coulomb form factor of the proton, $F(t) = (1 + t/0.71)^{-4}$, δ is the phase of the Coulomb amplitude and ρ is the ratio of the real to imaginary parts of the forward scattering amplitude. We have neglected, in this preliminary analysis, as is usually done, any spin dependent effects in the forward direction.

Rather than attempt to fit the three parameters σ_{TOT}, b and ρ, we use the parameterization for σ_{TOT}: $\sigma_{TOT} = A + B/p$
with A = 61.2 and B = 53.4 (Nakamura, et al., 1984) and only fit for ρ and b. Our fits include the effects of the angular resolution and multiple scattering. It should be noted that a change in σ_T of ± 2.0% produces a uniform change in the fitted values of ρ and b of roughly ∓ 0.04 and ± 1.5 $(GeV/c)^{-2}$, respectively. Scaling our data by ± 2.0% changes ρ and b by roughly ± .03 and ∓ 1.0 $(GeV/c)^{-2}$.

In Fig. 6 we show our preliminary results from the fits for ρ throughout the region from 360 MeV/c to 652 MeV/c. In Fig. 7 we show the corresponding b values from these fits. Our results are consistent with small values of ρ, between 0 to 0.2 and, with no significant structure throughout this momentum region. These results tend to be flatter than those reported by earlier experiments (Iwaski, et. al., 1981) which gave rise to speculation of high-mass pole terms in order to get agreement with dispersion relation calculations.

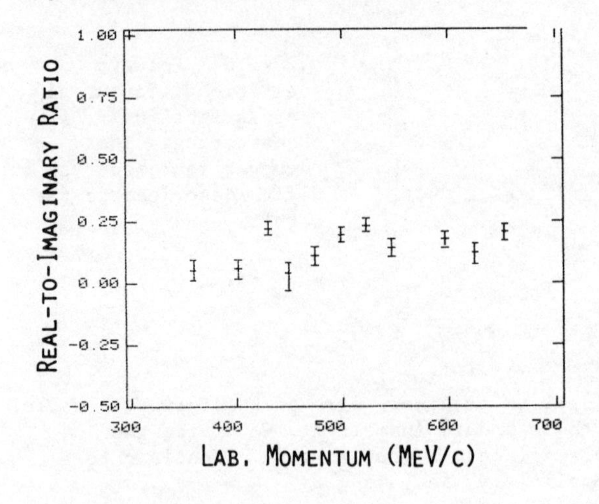

Fig. 6. The ratio of the real to imaginary parts of the forward scattering amplitude for $\bar{p}p$ elastic scattering.

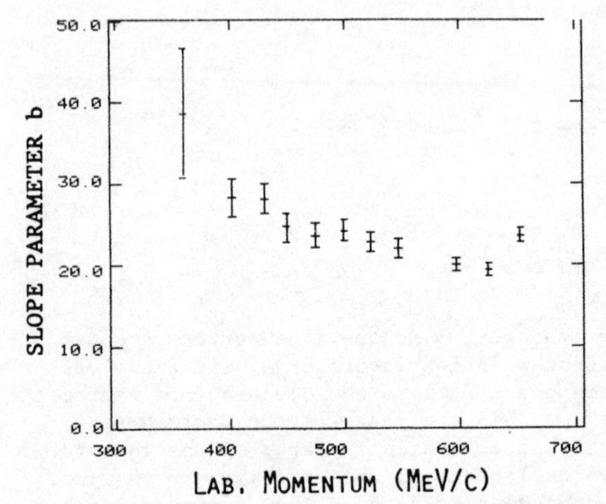

Fig. 7. The slope of the forward $\bar{p}p$ elastic nuclear cross section.

This work was supported by the U.S. Department of Energy under contract DE-AC02-76-CH00016 and by the National Science Foundation under contract PHY80-20418.

6. References

Ashford, V. et al., Sept. 1984, Phys. Rev. C, in press.
Auerbach, E.H., 1978, Comp. Phys. Commun. 15, 165.
Auerbach, E.H. et al., 1981, Phys. Rev. Lett 46, 702.
Garreta, D. et al., 1984, Phys. Lett. 135B, 266.
Iwaski, H. et al., 1981, Phys. Lett 103B, 247.
Nakamura, K. et al., 1984, Phys. Rev. Lett. 52, 731.
Nakamura, K. et al., 1984, Phys. Rev. D 29D, 349.

What antiproton physics can tell us about nuclear physics

A.M. Green

Research Institute for Theoretical Physics, University of Helsinki,
SF-00170 Helsinki 17, Finland

Abstract. The talk is in two parts. The first part discusses the potential use of nuclear targets in the current LEAR experiments. The second part reviews current models for $N\bar{N}$ annihilation.

1. Introduction

After accepting the invitation to talk on the subject of "What Antiproton Physics can tell us about Nuclear Physics" I realized that such a topic quickly becomes very speculative with the implication that a primary use of LEAR is for the study of nuclear structure. In fact, it appears that the use of nuclei as targets will probably at first give more information about the basic $N\bar{N}$ interaction than about nuclear structure. In view of this I would like to include in my talk a second part which could, for example, be entitled "How can Nuclear Physics help us with Antiprotons?" This will emphasize problems in the basic $N\bar{N}$ interaction that, in my opinion, are of more immediate concern for the understanding of low energy antiproton physics and also illustrates how techniques, well developed in nuclear physics, can be of help with antiprotons.

Under the first title the topics to be discussed are:-

a) Selectivity in the excitation of nuclear levels.
b) Antiproton-nucleus optical potentials.
c) Nuclear density distributions.

The second part of the talk will concentrate on the more fundamental problems concerning models for nucleon-antinucleon annihilation at low energies, in particular:-

a) Baryon exchange versus quark models.
b) Quark rearrangement with $N\bar{N}$, $N\bar{\Delta}$ and $\Delta\bar{\Delta}$ configurations.
c) $N\bar{N}$ bound states.

2. Selectivity in the excitation of nuclear levels

The advent of a new nuclear probe always leads to the question "What can be done now that could not be done before?" and usually the first answer is that, possibly, new nuclear states can be excited or that there may be a selectivity that emphasizes certain known nuclear levels.

The first category of state - the possible new ones - is an illustration of the speculation mentioned earlier. With LEAR the appropriate example is the search for the production of a quark-gluon plasma inside the nucleus (De Vries 1984). Such a question is confronted with two main problems:-

a) Theoretically the indications are that the ≈ 2 GeV of energy released by the $\bar{p}$ annihilation leads to an energy density that is too small and lasts too short a time for any Nuclear→Quark-gluon plasma phase transition to occur. Of course, it has to be admitted that, if such a transition does indeed occur under these unfavourable circumstances, then the subject is of more interest than the observation of the transition in conditions where it might be expected i.e. during the collision of relativistic heavy ions. However, it should be added that a recent $\bar{p}p$ collider experiment at CERN failed to see any clear evidence for the onset of "new physics" emerging from the 540 GeV of energy released (Carlson 1984). But here the argument could be used that, inspite of the large increase of energy density compared with LEAR, the $\bar{p}p$ system does not have enough degrees of freedom to apply the ideas of a plasma, which is basically a many-body concept.

b) Experimentally, if the existence of the plasma does not lead to a very characteristic signal, then the whole subject gets complicated by the need to unravel the effects of "normal" nuclear states. An example of such a problem taken from recent history is the story of pion condensates, which were essentially pion plasmas inside the nucleus. This can be briefly summarized as follows. Theory suggested that they may exist even at normal nuclear density ρ_0. However, they were not observed. Therefore, theory now suggested that they might "almost occur" - the idea of a precursor to the condensates - with the actual condensate only arising at densities ≈ 2-3 ρ_0. Experiment then saw an enhancement in the cross-section $^{12}C(e,e')^{12}C(1^+,T=1,15.1$ MeV$)$ - something that would be characteristic of the precursor phenomena being looked for. However, this observed enhancement was only with respect to a crude theoretical description. As the latter was improved (Suzuki et al. 1981) by the inclusion of tensor correlations - a "normal" nuclear physics procedure for introducing the effects of the virtual pions generated from the NN potential - the relative enhancement disappeared. The moral of this story is that it is not sufficient to demonstrate that an exotic phenomenon can explain some experimental observation. First it must be shown that "normal" ideas cannot be equally successful - a notoriously difficult task. In addition to this above example of pion condensates from low energy nuclear physics, there is also the famous problem of dibaryon resonances and their confusion with NΔ configurations. Likewise, over the next few years we shall, presumably, be witnessing a similar struggle between the "exoticists" and the "normalists" in the interpretation of the EMC effect (Aubert et al. 1983) and also any quark-gluon plasma "signals" that are observed.

The second category of state - those selectively excited by $\bar{p}$'s - will probably be most useful for determining the form of the $\bar{p}$-Nucleus interaction and indirectly the basic $N\bar{N}$ interaction. Already there are definite suggestions in this direction. For example, Dover et al. (1984) show that, using impulse approximation, two currently popular phenomenological models for the $N\bar{N}$ interaction lead to $^{12}C(\bar{p},\bar{p}')^{12}C(1^+,T=0,12.7$ MeV$)$ cross sections that differ by an order of magnitude. The larger cross section occurs for the Paris potential (Côte et al. 1982) which has an annihilation component containing a much stronger spin dependence than that of Dover and Richard (1980) with which it is compared. Preliminary results given by Garreta at this conference suggest that the observed excitation of the 12.7 MeV state is, in fact, more in line with that expected from the Paris potential. As will be seen later in section 6, this favouring of the Paris potential goes contrary to some current theories for the $N\bar{N}$ annihilation mechanism. However, as discussed in the next sections the steps between the excitation of a nuclear state and the extraction of the underlying $N\bar{N}$ interaction in free space is by no means trivial, so that the analysis of Dover et al. (1984) may be misleading.

3. Antiproton-nucleus optical potentials

The discussion in the previous section implies that, to some extent, the features of the $N\bar{N}$ interaction may be preserved in the $\bar{p}$-Nucleus interaction. This is probably most reliable for $E(\bar{p},lab.) \gtrsim 100$ MeV where impulse approximation is expected to be valid. However, at lower energies or for antiprotonic atoms more detailed theories for the optical potential are necessary. Already there have been several attempts to carry out such a programme (Niskanen and Green 1983 and Von Geramb et al. 1984). These are based on folding with the nuclear density $\rho(r')$

$$V(r,E) \propto \int d^3r' \; \rho(r') \; t(\underline{r}-\underline{r}',E) \tag{3.1}$$

where t is the $N\bar{N}$ t-matrix derived from some "favourite" $N\bar{N}$ potential. As E decreases the complications of t being off-the-energy-shell and the presence of many-body effects make the evaluation of V(r,E) more and more difficult. Even so, for scattering these initial attempts have been surprisingly successful as seen in fig. 1. Of course, some might say that the agreement is not unexpected, since the $\bar{p}$-Nucleus interaction is basically that of a black-sphere and the various microscopic theories are simply reflections of this. However, eventhough the $\bar{p}$-Nucleus interaction is relatively more absorptive than, for example, the K-Nucleus interaction, there are indications (Batty 1983) that, on an absolute scale, the $\bar{p}$-Nucleus interaction is still far from the black-sphere limit. It should be added that the method of Niskanen and Green (1983) for evaluating V(r,E) is appropriate only for low energies and is found to be inadequate for $E(\bar{p},lab.) \gtrsim 100$ MeV. Furthermore, in the antiprotonic atom situation two alternative approaches (Green and Wycech 1982 and Green et al. 1983) leading to completely different forms for $V(r,E<0)$, predict atomic level shifts and widths that differ considerably from each other - for more details see the talk and discussion of Koch (PS176) at this conference. One of the important differences between these two approaches is the use of a separable versus local form for the basic $N\bar{N}$ interaction. This indicates that perhaps the $N\bar{N}$ interaction should be better understood before reliable theories can be made about more complicated situations involving many nucleons.

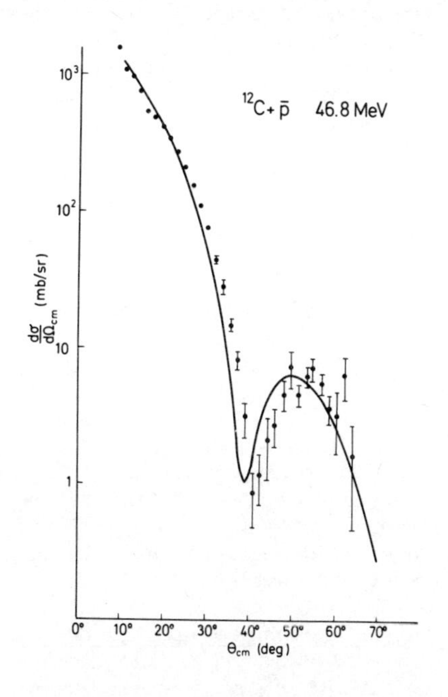

Fig. 1 $\bar{p} + {}^{12}C$ scattering at 46.8 MeV Niskanen and Green (1983) versus the experiments of Garreta et al. (1984).

4. Nuclear density distributions

One of the hopes for antiprotonic atoms is that the level shifts (ΔE) and widths(Γ) will lead to an understanding of the form of the nuclear density $\rho(r)$ at the surface of nuclei. However, this involves the sequence of relationships

$$\Delta E, \Gamma \leftrightarrow V(r,E) \leftrightarrow \rho(r)$$

where the second step involves the unfolding of eq. (3.1). Unfortunately, the latter requires a knowledge of $t(r,E)$ with all its attendant complications of being off-the-energy-shell and the inclusion of many body effects (Green and Wycech 1982 and Green et al. 1983). Furthermore, t is based on some $N\bar{N}$ potential and at present this is not well determined. To make matters worse, those nuclei that have S=0 or T=0 - the situation for the most popular targets - do not have a first order contribution to $V(r,E)$ from the OPEP component of the $N\bar{N}$ interaction. Therefore, the result depends strongly on the TPEP and all its inherent uncertainties. Another problem is that the TPEP has an important component that involves $N\bar{\Delta}$ intermediate states. Since this is mainly a tensor-like process e.g. $N\bar{N}(^1S_0) \rightarrow N\bar{\Delta}(^5D_0)$, it is - compared with the corresponding NN case - enhanced due to the coherent effect of π-plus ρ-meson exchange. The presence of these intermediate $N\bar{\Delta}$ components in nuclear states different from that of the original nucleus could well complicate the extraction of the nuclear density in a model independent manner. It is likely that ΔE and Γ could be used more effectively to give information about t, and so indirectly information about the $N\bar{N}$ interaction, than about ρ.

This ends the first part of my talk. I hope I have not appeared too pessimistic. However, my main aim has been to advocate caution and to not expect from LEAR a rapid series of reliable model independent results concerning nuclear properties.

In my opinion a subject where LEAR may be more successful in the nearest future is in the understanding of the basic $N\bar{N}$ interaction at low energies. This then leads me naturally into the second part of my talk - "How can Nuclear Physics help us with Antiprotons?" The result will be a mixture of conventional nuclear physics and quark physics. Since here our main interest is in low energy phenomenon, the discussion of quark physics involves soft-gluon situations - the hardest regime for applying QCD. Of course, many of the problems raised will be old ones left behind by earlier generations of particle physicists. Hopefully, these physicists can now guide us in directions that, in the absence of more complete theories, are the most reasonable.

5. Baryon exchange versus quark models in the description of $N\bar{N}$ annihilation

At present there are basically two approaches for describing low energy $N\bar{N}$ annihilation. These are best summarized by the traditional baryon exchange diagrams of fig. 2 and the quark diagrams of fig. 3.

The most complete calculation of the baryon exchange mechanism is by Moussallam (1983). He has calculated - in a distorted wave approach - fig. 2a for the cross sections $N\bar{N}\to M_1M_2$, where the M_i are π,η,ρ,ω and ε, and fig. 2b for the case where the M_i are only π's. Since he finds that fig. 2b is comparable to fig. 2a, to make his work complete he should also calculate the processes presented by fig. 2c. A method for doing this will be discussed in section 6. The problem with this approach is the massive dependence on the $N\bar{N}M_i$ form factors - a point that, so far, has only been treated in a phenomenological manner. This dependence is not unexpected, since the model is based on structureless particles, and so leads naturally to a radial dependence of essentially the form $\exp(-Mr)/r$, where M is either the nucleon or Δ-mass. If this is now moderated by two form factors corresponding to the size of the nucleon, i.e. $\simeq 0.6$ fm compared with $1/M \simeq 0.2$ fm, then the radial dependence generated by these form factors dominates. In view of this, it is hard to see how the approach can ever be justified without resorting to the more microscopic descriptions involving quarks.

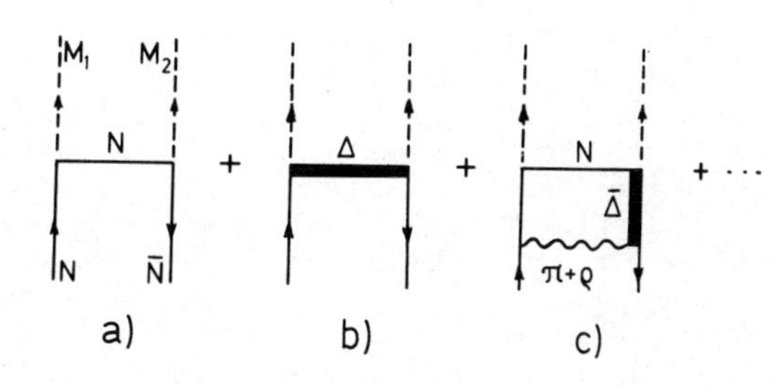

Fig. 2 Baryon exchange models for $N\bar{N}$ annihilation
a) N-exchange
b) Δ-exchange
c) $N\bar{N}\to N\bar{\Delta}\to M_1M_2$

Those quark models now being discussed most actively are illustrated in fig. 3. Two immediate questions arise:-

a) How do these relate to the baryon exchange model?
b) Which (if any) of the processes depicted is dominant - and does the series converge?

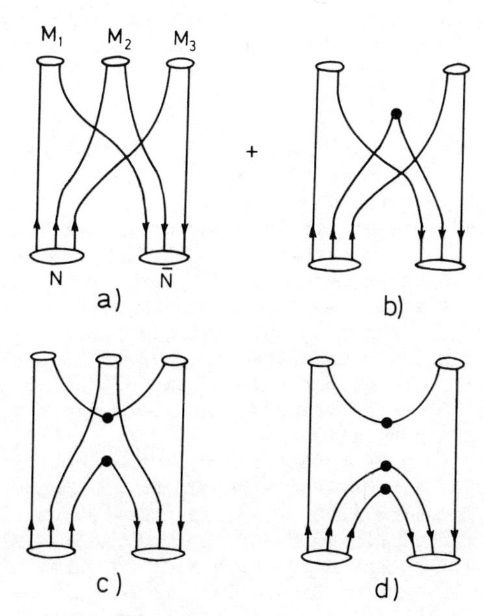

Fig. 3. Quark diagram models for $N\bar{N}$ annihilation

The answer to the first question - namely, the connection between figs. 2a, b and figs. 3b,d - must have been discussed by earlier generations of particle physics. However, recent attempts to resolve this seem to be indicating that the connection is less straightforward than first expected (Dover and Fishbane 1984 and Furui 1984). The second question involves a comparison of fig. 3a with c and fig. 3b with d. So far this has only been carried out by taking ratios of experimental branching probabilities (Logan et al. 1977, Genz 1983, Dover and Fishbane 1984 and Furui and Faessler 1984) and the results have been inconclusive. However, there are many criticisms that can be raised with this type of approach.

(i) All of the analyses are incomplete, since the authors do not allow for processes involving the $\Delta(1236)$ as, for example, in fig. 2c. In the next section it will be shown that these $N\bar{\Delta}$ and $\Delta\bar{\Delta}$ configurations can play a very important role.

(ii) Some of the earlier works (Logan et al. 1977 and Genz 1983) tried to draw conclusions by assuming (incorrectly) that the wavefunction of the N (and $\bar{N}$) could be factorized in a single term as spin x flavour.

(iii) Another complication that can arise is in the treatment of orbital angular momentum. So far, discussions have only concerned processes involving s-wave mesons such as $\pi,\eta,\rho,\omega,\eta'$. However, the inclusion of p-wave mesons (i.e. the f,A_1,A_2,B,H etc. - already seen in $N\bar{N}$ annihilation) in conjunction with the $^{13}P_0$ model - depicted by black dots in fig. 3 - will probably necessitate a more careful treatment for the coupling of the spin to the orbital angular momentum.

(iv) These approaches often depend on branching ratios that are extremely small and poorly known e.g. $p\bar{p} \rightarrow \pi^0\pi^0$ is taken to be 0.06±0.05 %. Only when these methods confront branching ratios that are significantly larger than 1 % can they claim to be discussing the most important annihilation mechanisms.

(v) Some of the observed branching ratios for "stopped" $\bar{p}$'s do not necessarily occur for $N\bar{N}$ in relative <u>S-waves</u>. The importance of P-waves, in spite of Stark mixing in $p\bar{p}$ atoms, has been pointed out by Borie (1984). Also, as seen in the next section, if the "stopped" $\bar{p}$'s are not annihilated at rest then P-waves already give 10 % effects at $E(\bar{p},\text{lab.}) \simeq 1$ MeV and are comparable to S-waves at 10 MeV.

(vi) The assumption that, in particular, pions are created only by rearrangement mechanisms is not true. The N and $\bar{N}$ quark cores are surrounded by sizeable meson clouds e.g. the 0.8 fm RMS charge radius of the proton is often broken up into two contributions - 0.6 fm from the quark core and the remainder from the meson cloud. Also these meson clouds give rise to the long range part ($\gtrsim$ 0.8 fm) of the $N\bar{N}$ interaction, which is known to have important effects, and so should not be neglected as a source of pions in $N\bar{N}$ annihilation.

(vii) In connection with the comparison of the various mechanisms in fig. 3, attempts have been made to ignore some contributions (e.g. Furui and Faessler (1984) fig. 3a compared with fig. 3c) by invoking large N_c arguments. These state that coplanar diagrams (e.g. fig. 3c) are of $O(N_c)$ - N_c being the number of colours, whereas non-coplanar diagrams (e.g. fig. 3a) are of O(1). Therefore, fig. 3a should be negligable compared with fig. 3c. However, these same arguments (Witten 1979) predict that the annihilation $N\bar{N} \rightarrow$ Mesons is "forbidden" being of order $\exp(-aN_c)$ compared with the scattering $N\bar{N} \rightarrow N\bar{N}$, which is of $O(N_c)$. This shows that such arguments are not applicable in the case of low energy $\bar{p}$ physics, since experimentally $\sigma(N\bar{N} \rightarrow N\bar{N})$ is <u>less than</u> σ(annihilation).

In view of the above list of criticisms it is clear that more theoretical input is needed before reliable statements can be made about the relative importance of the annihilation processes of fig. 3.

6. Quark rearrangement with $N\bar{N}$, $N\bar{\Delta}$ and $\Delta\bar{\Delta}$ configurations

In the previous section the models were designed for an understanding of specific $N\bar{N}$ annihilation channels. For these the initial state interactions of the $N\bar{N}$ were either ignored (Furui and Faessler 1984), treated as free parameters (Maruyama and Ueda 1981) or included with the aid of a phenomenological $N\bar{N}$ potential (Moussallam 1983). Therefore, in an attempt to develop a more consistent model, the Helsinki group assumes that $N\bar{N}$ annihilation is dominated by the rearrangement process of fig. 3a, with more details of the following being found in Green and Niskanen 1984 a,b,c. This automatically generates an effective $N\bar{N}$ potential which can then be added to a standard meson exchange potential. The resulting model for calculating particular $N\bar{N}$ annihilation channels now contains an $N\bar{N}$ initial state interaction that is consistent with the basic annihilation mechanism. To be more specific, the effective $N\bar{N}$ potentials have the energy dependent separable form for ℓ=0,1 and 2

$$V_S(r,r') = -\lambda\, I_S(TS,E)\exp[-\tfrac{3}{4}\beta(r^2 + r'^2)] \qquad (6.1)$$

$$V_P(r,r') = -\lambda\, I_P(TS,E)\,\beta\, r\, r' \exp[-\tfrac{3}{4}\beta(r^2 + r'^2)] \qquad (6.2)$$

$$V_D(r,r') = -\tfrac{2}{5}\lambda\, I_D(TS,E)\beta^2 r^2 r'^2 \exp[-\tfrac{3}{4}\beta(r^2 + r'^2)] \qquad (6.3)$$

where β is the size parameter of the meson quark wavefunctions - taken to be of Gaussian shape - and λ is the overall strength - essentially a free parameter. Ignoring intermesonic interactions I(TS,E) can be expressed in the form

$$I(TS,E) \propto \sum_{M_1M_2M_3} \int \frac{d^3k_1 d^3k_2\; F(M_1M_2M_3,k_1,k_2)}{T(k_1)+T(k_2) + \sum_i M_i - 2M - E} \qquad (6.4)$$

where k_1,k_2 are the two momenta needed to describe the three meson system $M_1M_2M_3$, E is the $N\bar{N}$ c.m. energy and the M_i are the meson masses - including their half widths. The I's are different for V_S, V_P and V_D, since the possibilities for the M_i are different in the three cases. For V_S the M_i are the s-wave mesons π,η,ρ,ω or η'. On the other hand, for V_P only two of the M_i are these s-wave mesons with the third being one of the p-wave mesons H,B,ε,D,f,δ,A_1 or A_2 and for V_D two of the M_i are now p-wave mesons. Furthermore,the I's are different for the $N\bar{N}$, $N\bar{\Delta}$ and $\Delta\bar{\Delta}$ configurations. For V_P and V_D it is crucial that, in the denominator of I(TS,E), the M_i contain their half widths.

The TS and E dependence of the I's are themselves of interest. It turns out that both the real and imaginary parts of I(TS,E) are only weakly dependent on T and S - see fig. 4 for the case of Im I_S(TS,E). At E($\bar{p}$,cm)=0 this can be expressed as

$$2.87 - 0.12\underline{\sigma}_1\cdot\underline{\sigma}_2 + 0.08\,\underline{\tau}_1\cdot\underline{\tau}_2 - 0.03\,\underline{\sigma}_1\cdot\underline{\sigma}_2\underline{\tau}_1\cdot\underline{\tau}_2$$

which is qualitatively different to the phenomenological model of Paris (Côte et al. 1982) and indicates that hopes for the preferential excitation of nuclear states discussed in section 2 will not be fulfilled. In fact this approximate state independence is much more like the earlier potential of Dover and Richard (1980). Furthermore, if the separable potentials V_S and V_P are converted into equivalent local forms, the resulting real components are quantitatively close to that of Dover and Richard (1980) - see Green and Niskanen (1984c) and Green et al. (1984) for more details. Fig. 4 also shows that the energy dependence of Im I in the range -400 MeV$\lesssim E \lesssim 400$ MeV is approximately of the form exp[AE] where $A \simeq 3$ GeV^{-1} for $\ell=0$ and $\simeq 2$ GeV^{-1} for $\ell=1$. It would be interesting for someone to produce a phenomenological potential assuming such an energy dependence.

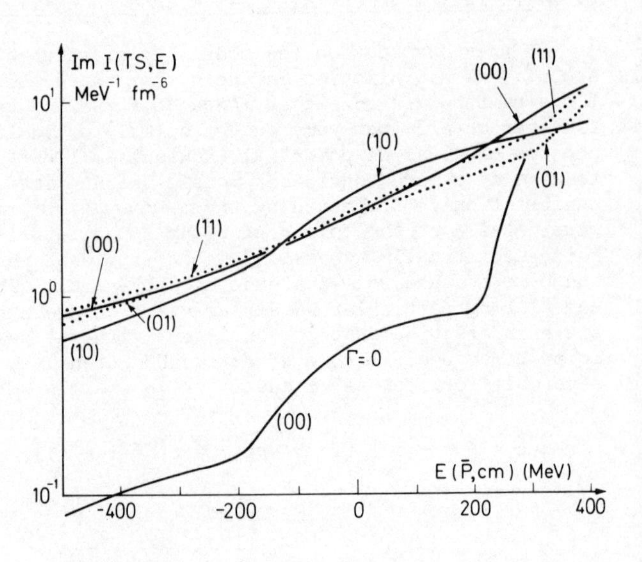

Fig. 4 Im I_S(TS,E) for the different (TS). Upper lines use $\Gamma_\rho = 154$ MeV and $\Gamma_\omega = 10$ MeV. The lowest line uses $\Gamma_\rho = \Gamma_\omega = 0$

The Helsinki approach not only considers annihilation through $N\bar{N}$ configurations but also induced $N\bar{\Delta}$ and $\Delta\bar{\Delta}$ configurations. This is depicted in fig. 5 when the $N\bar{N}$ are in an initial S- or D-wave.

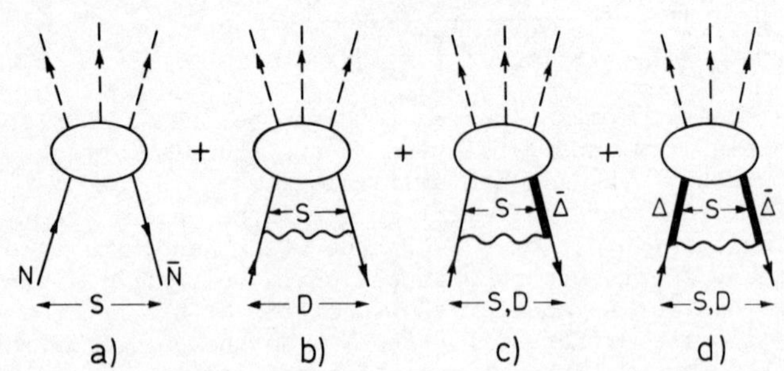

Fig. 5. The "oval" represents quark-antiquark rearrangement as in fig. 3a
a) Alternative form of fig. 3a
b) Effect of $N\bar{N}$ tensor potential
c) Effect of $\pi+\rho$ transition potential $N\bar{N}\rightarrow N\bar{\Delta}$
d) Effect of $\pi+\rho$ transition potential $N\bar{N}\rightarrow \Delta\bar{\Delta}$

Similar processes are also included for initial P- or F-waves. The method for incorporating these effects is by way of coupled Schroedinger equations for $N\bar{N}$ scattering. For example, if the $N\bar{N}$ are in an initial $^{33}S_1$ state then the corresponding coupled equations are schematically

$$[-\frac{d^2}{dr^2} + V_C + V_S(N\bar{N})-E]\psi(S,N\bar{N}) = -V_T\psi(D,N\bar{N}) - V_{Tr}\psi(N\bar{\Delta}) - V_{Tr}\ \psi(\Delta\bar{\Delta})$$

$$[-\frac{d^2}{dr^2} + \frac{6}{r^2} + V_C + V_D(N\bar{N}) - E]\psi(D,N\bar{N}) = -V_T\psi(S,N\bar{N}) - V_{Tr}\psi(N\bar{\Delta}) - V_{Tr}\psi(\Delta\bar{\Delta})$$

$$[-\frac{d^2}{dr^2} + (\Delta-M) + V_S(N\bar{\Delta}) - E]\ \ \psi(N\bar{\Delta}) = -\ V_{Tr}\psi(N\bar{N})$$

$$[-\frac{d^2}{dr^2} + 2(\Delta-M) + V_S(\Delta\bar{\Delta}) - E]\psi(\Delta\bar{\Delta}) = -\ V_{Tr}\psi(N\bar{N}) \quad (6.5)$$

Here $V_{C,T}$ are essentially the central (C) and tensor (T) meson exchange potentials of Dover and Richard (1980) but with the attraction due to the $N\bar{N} \rightarrow (N\bar{\Delta} + \Delta\bar{\Delta}) \rightarrow N\bar{N}$ mechanisms (now treated explicitly) removed in V_C to avoid double counting. The $V_{S,D}(N\bar{N}$ or $N\bar{\Delta}$ or $\Delta\bar{\Delta})$ are analogous to eqs. (6.1) and (6.3) and are of separable form in addition to being complex and energy dependent. The $V_{Tr}(N\bar{\Delta}$ or $\Delta\bar{\Delta})$ generate by means of $\pi+\rho$ exchange $N\bar{\Delta}$ and $\Delta\Delta$ configurations from the incident $N\bar{N}$ channel.

To the non-nuclear physicist this whole procedure may seem very complicated and full of uncertainties. However, it should be pointed out that the corresponding model has been very successful in the understanding of the imaginary part of NN phase shifts (Green et al. 1978, Green and Sainio 1979, 1982), and for the reaction $p+p \rightarrow d+\pi^+$ with $E(p,lab.) \lesssim 1$ GeV in describing detailed polarisation cross sections which involve delicate cancellations between various scattering amplitudes - see for example Niskanen (1978). This gives us confidence that, for example, the transition potentials $V_{Tr}(r,\pi+\rho)$ for $N\bar{N} \rightarrow N\bar{\Delta}$ are reliable. Of course, in the $N\bar{N}$ case, due to the G-parity transformation, the π and ρ meson effects are additive in the tensor component. The above feedback from nuclear physics is what prompted me to introduce my second title "How can Nuclear Physics help us with Antiprotons?" The result of this calculation is shown in fig. 6. Several points should be made about this figure:-

a) At $E(\bar{p},lab.) \simeq 100$ MeV the model yields about 80 % of the observed annihilation cross section - decreasing to $\simeq 60$ % at 400 MeV

b) P-wave annihilation dominates over S-wave for all $E(\bar{p},lab.) \gtrsim 10$ MeV. When this observation is combined with the Stark mixing calculations of Borie (1984), which indicate that in protonium upto 40 % of the cascades could take place at the 2p-level, it suggests that the observed branching ratios for "stopped" $\bar{p}$'s in hydrogen could well have a major P-wave component. In fact, experimentally there are indications that P-wave annihilation is much larger than earlier thought (Tauscher et al. 1983).

c) Much of the P-wave annihilation is due to $\bar{\Delta}$ configurations. For example at $E(\bar{p},lab.) = 100$ MeV the $N\bar{N}$, $N\bar{N}+N\bar{\Delta}$ and $N\bar{N}+N\bar{\Delta}+\Delta\bar{\Delta}$ cross sections are 15, 35 and 50 mb respectively. Of even more importance is the effect of the meson widths. If these widths are set to zero, then the $N\bar{N}+N\bar{\Delta}+\Delta\bar{\Delta}$ P-wave cross section drops to less than 10 mb.

The model leading to fig. 6 could include many more contributions in addition to those contained in eqs. (6.1) to (6.5). For example, fig. 3a will also give contributions when:-

a) $N\bar{N}$(D-wave) $\rightarrow M_1^d M_2^s M_3^s$ where M_1^d is a d-wave meson e.g. $\omega(1670)$ with J=3, and $M^s_{2,3}$ are either π,η,ρ,ω or η'. This could well have an effect for $E(\bar{p},lab.) \lesssim 400$ MeV since the widths of the M^d are large.

b) The p-wave meson configurations $M_1^p M_2^s M_3^s$ and $M_1^p M_2^p M_3^s$ can also be coupled with the $^{13}P_0$ quark-pair creation (or annihilation) model since the p-wave meson ε has the vacuum quantum numbers. This $^{13}P_0$ model (see

Le Yaouanc et al. 1975) has been successful in describing meson decays $M_1 \rightarrow M_2M_3$ and also the baryon-meson vertices $B\bar{B}M$ in terms of a single parameter γ. If the process is now viewed as in fig. 3b, then there is first a rearrangement proportional to the $\sqrt{\lambda}$ of eqs. (6.1) to (6.3), followed by a propagation incorporated as the energy denominator $1/2m_q$ (where m_q is the mass of a constituent quark $\simeq$ 300 MeV) and, finally, the $^{13}P_0\, q\bar{q}$ annihilation takes place with factor γ. The overall contribution of fig. 3b to the effective $N\bar{N}$ potential is then of the form

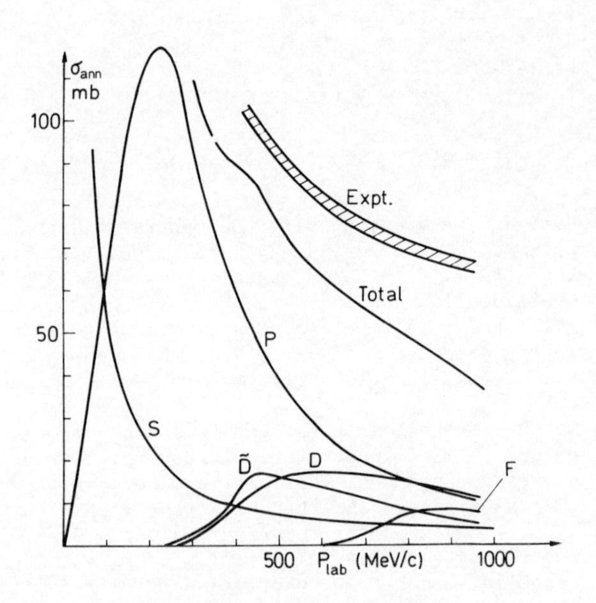

Fig. 6 The annihilation cross section as a function of laboratory momentum. S,P,D, D̃ and F refer to the incident $N\bar{N}$ orbital angular momentum. D contains only intermediate s-wave mesons, whereas D̃ contains two p-wave mesons.

$$V(^{13}P_0) \propto \lambda(\frac{\gamma}{2m_q})^2 \exp[-\frac{3}{2}\beta(r^2 + r'^2)] \tag{6.6}$$

i.e. a separable potential with a shorter range than the $V_{S,P,D}$ in eqs. (6.1) to (6.3). This is now being incorporated in the above coupled channels procedure.

The outstanding problem in the above model is to calculate the value of λ, the overall rearrangement factor, appearing in eqs. (6.1) to (6.3) and (6.6). Arguments are given in Green and Niskanen (1984 a,b,c) for considering rearrangement as a non-perturbative effect related to the change in quark confining potentials when going from $N\bar{N}$ to $M_1M_2M_3$. In addition, it is shown that the value of λ ($\simeq$ 10-20 MeV fm) which maximizes the annihilation cross section - the situation given in fig. 6 - can also be understood qualitatively. Furthermore, this value of λ also seems to be preferred for the few known branching ratios $N\bar{N}$(S-waves) $\rightarrow M_1^sM_2^sM_3^s$ (Niskanen et al. 1984). Much work needs to be done for a better determination of λ. However, this will not be easy since the problem is essentially one of soft-gluon QCD.

7. $N\bar{N}$ bound states

Since the $N\bar{N}$ potential is more attractive than the NN potential, it is reasonable to expect $N\bar{N}$ bound states analogous to the deuteron. Experimentally there is evidence (Richter et al. 1983, Adiels et al. 1984 and Angelopoulos et al. 1984) for interesting states below the $N\bar{N}$ threshold. But the interpretation of these as either baryon bound states $(N\bar{N})_B$ or exotic $q^2\bar{q}^2$ states is not clear. The problem with the first alternative is that standard phenomenological potentials, such as that of Dover and Richard (1980), always lead to $N\bar{N}$ bound states with widths that are orders of magnitude larger than those observed ($\lesssim$ 30 MeV). If part of the Dover-Richard annihilation potential is replaced by the explicit treatment of the inelastic process

$$N\bar{N} \to (N\bar{N})_B\pi \to N\bar{N}$$

it is possible to reduce these widths to $\approx$ 150 MeV still much larger than experiment (Green and Sainio 1980, Green 1982). However, with the annihilation potentials developed in the last section it is easy to get $N\bar{N}$ states bound by 100 or 200 MeV with widths $\lesssim$ 30 MeV. (Niskanen and Green 1984) This is especially so if a reasonable energy dependence is included in the widths for the mesons of the form

$$\Gamma(s) = \frac{m}{\sqrt{s}} \left(\frac{q}{q_0}\right)^3 \exp[-\frac{1}{3\beta}(q^2-q_0^2)]\ \Gamma(m^2) \qquad (7.1)$$

where $s = 2(q^2+\mu^2)$ and q_0 corresponds to $s=m^2$ - m being the meson mass and $\Gamma(m^2)$ the experimental width. The reason for these much smaller widths compared with the earlier works follows directly from the strong energy dependence of I(TS,E) depicted in fig. 4 and this is simply a reflection of the opening of more and more annihilation channels as E increases. The results are shown in fig. 7 where it is seen that even S-wave bound states can be quite narrow e.g. for $\lambda \approx$ 20 MeV fm a state $N\bar{N}(^{11}S_0)$ is at -150 - 3 i MeV.

The experimental confirmation and study of these states is of great importance for the understanding of the $N\bar{N}$ interaction. So far the only experiments have been using stopped $\bar{p}$'s in hydrogen or ^{4}He with the observation of the subsequent monochromatic γ-rays (Richter et al. 1983 and Adiels et al. 1984) or $\pi^{\pm}$ (Angelopoulos et al. 1984). The next stage is to extract the quantum numbers of these states. In addition to refining the above experiments this identification could possibly be achieved with standard nuclear physics mechanisms such as the pick up reaction $A(\bar{p},b)(A-1)$ where b is $(N\bar{N})_B$. Preliminary calculations by Moalem et al. (1984) indicate that with $\bar{p}^4$He the resultant cross sections should be measurable.

Of course, $N\bar{N}$ states bound by $\approx$ 200 MeV have a peak in their relative wavefunction at about r $\approx$ 0.8 fm - a distance at which quark effects begin to enter. However, it could well be that the three $q(\bar{q})$ cluster making up the $N(\bar{N})$ does not get significantly distorted until smaller distances. In this way the $N\bar{N}$ wavefunction would be of use down to much smaller values of r.

8. Conclusion

As discussed in sections 2-4 there seems, in my opinion, little hope that in the near future $\bar{p}$'s will be of help in nuclear physics for determining nuclear structure properties. It is much more likely that nuclear physics will first throw light on the basic $N\bar{N}$ interaction. This interaction was then the theme for sections 5-7 where it was shown that nuclear physics techniques - in particular coupled channel Schroedinger equations - seem to have a future. This showed that $N\Delta$ and $\Delta\Delta$ configurations must be treated explicitly and that $N\bar{N}$ P-wave effects are important down to surprisingly low energies - dominating for $E(\bar{p},\text{lab.}) \lesssim 10$ MeV.

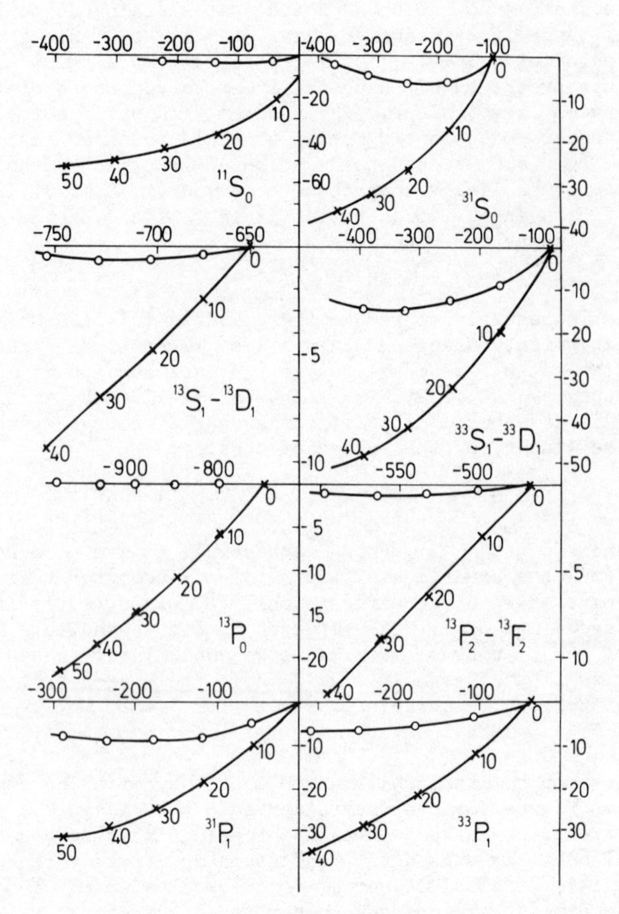

Fig. 7 Bound state energies as a function of λ, with $\lambda \approx 20$ preferred. Crosses with $\Gamma(m^2)$ and circles with $\Gamma(s)$ of eq. (7.1)

At present there are several competing theories for $N\bar{N}$ annihilation. Probably the $N\bar{N}$ scattering data that will become available in the near future will not be sufficient to distinguish between these alternatives. This may only be realized by detailed comparison with specific $N\bar{N}\to M_1M_2$.. cross sections and branching ratios for the in-flight and "stopped" situations. Unfortunately, very little is documented in a reliable way for these reactions.

References

Adiels L et al. 1984 Phys. Lett. 138B 235

Angelopoulos A et al. 1984 CERN preprint EP 184-47

Aubert J J et al. 1983 Phys. Lett. 123B 275

Batty C J 1983 Nucl. Phys. A411 399

Borie E 1984 Physics at LEAR-Ettore Majorana Int. Science Series Vol. 17 eds. U. Gastaldi and R. Klapisch (New York: Plenum) pp 185-189

Carlson P 1984 Proc. of the IVth International Workshop on pp Collider Physics Bern March 1984

Côte J, Lacombe M, Loiseau B, Moussallam B and Vinh Mau R 1982 Phys. Rev. Lett. 48 1319

DeVries R M and DiGiacomo N J 1984 Physics at LEAR-Ettore Majorana Int. Science Series Vol.17 eds. U. Gastaldi and R. Klapisch (New York:Plenum) pp 543-560

Dover C B and Richard J M 1980 Phys. Rev. C21 1466
Dover C B and Fishbane B 1984 Orsay preprint IPNO/TH 84-8
Dover C B, Franey M A, Love W G, Sainio M E and Walker G E 1984 Indiana preprint Phys. Lett. in press
Furui S and Faessler A 1984 to be published in Nucl. Phys.
Furui S 1984 Proc. of PANIC conference Heidelberg July 1984 contribution C36
Garreta D et al. 1984 Phys. Lett 135B 266
Genz H 1983 Phys. Rev. 28D 1094
Green A M 1982 J. Phys. G: Nucl. Phys. 8 485
Green A M and Sainio M E 1979 J. Phys. G: Nucl. Phys. 5 503
Green A M and Sainio M E 1980 J. Phys. G: Nucl. Phys. 6 1375
Green A M and Sainio M E 1982 J. Phys. G: Nucl. Phys. 8 1337
Green A M and Wycech S 1982 Nucl. Phys. A377 441
Green A M and Niskanen J A 1984a Nucl. Phys. A412 448
Green A M and Niskanen J A 1984b Helsinki preprint HU-TFT-83-54 to be published in International Review of Nuclear Physics Vol. I ed. T.T.S. Kuo (Singapore: World Scientific)
Green A M and Niskanen J A 1984c Helsinki preprint HU-TFT-84-19 to be published in Nucl. Phys. A
Green A M, Niskanen J A and Sainio M E 1978 J. Phys. G: Nucl. Phys. 4 1055
Green A M, Stepien-Rudzka W and Wycech S 1983 Nucl. Phys A399 307
Green A M, Niskanen J A and Wycech S 1984 Phys. Lett. 139B 15
Le Yaouanc A, Oliver L, Pène O and Raynal J C 1975 Phys. Rev. D11 1272
Logan R K, Kogitz S and Tanaka S 1977 Can. J. Phys. 55 2059
Maruyama M and Ueda T 1981 Nucl. Phys. A364 297
Moalem A, Niskanen J A and Green A M 1984 Helsinki preprint HU-TFT-84-31
Moussallam B 1983 Nucl. Phys. A407 413
Moussallam B 1984 to be published in Nucl. Phys.
Niskanen J A 1978 Nucl. Phys. A298 417
Niskanen J A and Green A M 1983 Nucl. Phys. A404 495
Niskanen J A and Green A M 1984 Helsinki preprint HU-TFT-84-24 to be published in Nucl. Phys. A
Niskanen J A, Kuikka V and Green A M 1984 Helsinki preprint HU-TFT-84-33
Richter B et al. 1983 Phys. Lett. 126B 284
Suzuki T, Hyuga H, Arima A and Yazaki K 1981 Phys. Lett. 106B 19
Tauscher L et al. 1983 Proc. VIth European Symposium on $N\bar{N}$ interactions, Santiago de Compostela (Spain 1982) Annales de Fisica 79 24
Von Geramb H V, Nakano K and Rikus L 1984 University of Hamburg preprints
Witten E 1979 Nucl. Phys. B160 57

Discussion on Section 3

T. Kalogeropoulos (Syracuse): How many antiprotons have you delivered to experiments so far?

Speaker, M. Chanel: For physics purpose there have been between 10^{12} and 3.10^{12} antiprotons.

H. Koch (KFK and Univ. Karlsruhe): How many antiprotons can be extracted from LEAR running in parallel with the SPS-Collider?

Speaker, M. Chanel: Assume: (i) that the collider has 20% antiproton losses, (ii) the accelerator chain faults are shared equally between the collider and LEAR, (iii) that one PS proton cycle is killed during LEAR transfer of 10 minutes and together with other losses, then one can expect LEAR to receive 10% of the antiprotons created every 90 minutes. If the AA reaches the accumulation rate of 6.10^9 $\bar{p}$/hour, then LEAR can have 10^9 $\bar{p}$ per 1.5 hour. It will be more efficient for all these antiprotons to be given to one experiment, which can receive 70% of that number for a one hour spill and have 12 spills a day, but wait for future decisions and improvements!

L. Pinsky (Houston): Do you attribute the large number of vertices from outside the target to multiple scattering?

Speaker, R. Landua: Yes, we have now developed a trigger which vetoes on more than 90% of these stops in Argon.

G. Karl (Guelph/Oxford): How does the shift you measure compare with theoretical expectations?

Speaker, R. Landua: Theoretical predictions vary from 0.3 to 1.0 keV for the shift of the 1S level. With the present accuracy none of these predictions can be excluded.

D. Geesaman (Argonne): What is the ultimate precision you hope to achieve in the K_α energy measurement?

Speaker, R. Landua: Around 100 eV.

M. Mandelkern (UC Irvine): How does your result for the K_α energy compare with that reported at Barr in 1978?

Speaker, R. Landua: The data presented at Barr showed only very weak evidence for K-lines in $\bar{p}p$. The energies given would be in contradiction with those presented here.

G.A. Smith (Penn State): How many X-rays do you detect per second in the beam?

Speaker, R. Landua: If we trigger on X-rays and have a primary intensity of 3.10^4 $\bar{p}$/sec, we could detect

Intensity × stops in H_2 × Total X-ray yield × Average detector efficiency = $3.10^4 \times 0.1 \times 0.2 \times 0.1 = 60$ X-rays/sec.

T. Kalogeropoulos (Syracuse): How do the intensities, particularly those of the K- to L-lines, compare with cascade calculations?

Speaker, R. Landua: Present models of the $\bar{p}p$ cascade contain phenomenological parameters for the Stark mixing and approximate the initial de-excitation steps only crudely. Predictions are therefore always given within a wide range which accommodates our yields well. The most pessimistic calculation for the strong interaction width of the 2P level gives a branching ratio of 10^{-2} for radiative K_α transitions from the 2P level. Our data give a branching ratio a little above this value.

J. Vandermeulen (Liege): Can the pressure in the target be varied?

Speaker, R. Landua: No. The H_2 gas target has a pressure of 1.001 atmospheres and is at room temperature.

A.M. Green (Helsinki): What can you say about $\Gamma(1S)$?

Speaker, R. Landua: The K peak in our spectrum is broader than the experimental width. However, the K_β contribution to the spectrum must be known before we can make any statement about the width. We hope to get this information from the coincidence spectra with higher statistics.

U. Gastaldi (CERN): Would you comment on the width of the K-line candidates you have in the $p\bar{p}$ X-ray spectrum?

Speaker, J.D. Davies: If we identify the lower energy peak as K_α, then it has a width considerably greater than any "potential" model prediction given at Erice. Additionally, Γ_{2P} would be much less than the 20-40 MeV also predicted by some of these models.

A. Lundby (CERN): To get your 1000 events per momentum, you seem to need ∿ 20 days of stopped antiprotons. What about 1 GeV/c?

Speaker, P.F. Dalpiaz: In the next runs, we will concentrate our efforts on antiproton momenta below 1 GeV and then turn to higher momenta the second time.

P.L. Braccini (Perugia): Do you also plan to study very low mass e^+e^- pairs? Your experiment should be very good for studying this interesting subject.

Speaker, P.F. Dalpiaz: There are many interesting physics topics which could be studied with PS170. However, the pressure on running time at LEAR means it is presently unthinkable to consider subjects other than those in the proposal. Low energy electron pairs is a difficulty in $\bar{p}p$ annihilation, because of the large background of γ's from π^o's and from Dalitz pairs.

D. Garreta (Saclay): Where does the beam spread of 2% come from?

Speaker, L. Linssen: The degrader.

D. Garreta (Saclay): A comment. It would be interesting to convolute your energy distribution and a signal, which has the form of a bump followed by a dip (as reported by the Heidelberg-Saclay collaboration in Erice in 1982 in the excitation function of the annihilation cross-section) to put an upper limit on the strength of such a signal which would be compatible with your experiment.

P. Pavlopoulos (CERN): Can you stress what are the main improvements of your set-up over previous experiments?

Speaker, L. Linssen: A transmission experiment was considered unsuitable for low momentum total cross-section measurements as there are too few possibilities to check on the quality of the recorded data. We feel we have overcome these doubts with our careful design.

First of all, the incoming beam is very well defined by the set of three 0.5 mm thin beam counters, which have reflection light guides instead of the conventional plexiglass light guides, which can cause false triggers. The last two beam scintillators define a cone well clear of the walls of the hydrogen target.

A very important improvement is that two multiwire proportional chambers, PC1 and 2, monitor the location and direction of the $\bar{p}$ beam in front of the target continuously during a measurement. For the transmitted beam, a third multiwire proportional chamber (PC3) monitors the beam distribution after the transmission counters. Its mean multiple scattering angle can then be checked, so that we can tell whether outgoing beam particles with large multiple scattering angles (up to 4σ) are still detected in T (see R.P. Hamilton et al, Phys. Rev. Lett. 44 (1980) 1182).

A final important improvement to the set-up is the transmission counters themselves. The configuration with one disc (T) and three annular counters reduces the amount of material in the outgoing beam considerably. The transmission counters are equally equipped with reflection light guides. The most suitable scintillation material (NE102A) and photomultipliers (XP2020) were chosen. On top of these instrumental improvements there is, of course, the beam of LEAR with its clean, high intensity compared to the previously used secondary $\bar{p}$ beams from a production target.

K. Nakamura (KEK): I would like to comment that at KEK we have obtained new data on the real-to-imaginary ratio for the forward $\bar{p}p$ amplitude between 367 and 764 MeV/c. The ratio is close to zero at low momenta and increases slowly with momentum. This new result will be presented in the poster session and in the proceedings.

T-A. Shibata (Heidelberg) question to K. Nakamura: The new results on the ρ parameter differ from your published data (Iwasaki et al, Nakamura et al). Why is this?

K. Nakamura (KEK): It appears that our new data show less momentum dependence than the previous result (Iwasaki et al). However, one should consider the systematic error of ± 0.15 associated with the new data, which mainly originate from the uncertainties in the $\bar{p}p$ total cross-section. Then these two results are consistent within the quoted errors.

M. Sakitt (Brookhaven): What is your mass resolution for the annihilation cross-section measurements?

T. Walcher (Speaker): Our mass resolution is determined by the energy loss straggling in the degraders. It is determined by time of flight of the antiproton over a distance of 20 metres. It is 1.0 MeV/c^2 (FWHM).

D. Geesaman (Argonne): The elastic angular distribution shows some pronounced high frequency oscillations at large angles which seem to be statistically very significant. Do you believe these oscillations?

T. Walcher (Speaker): These oscillations are due to the cuts which separate the elastically scattered antiprotons from annihilation pions not being correctly adjusted. They will disappear in the final analysis.

G. Karl (Guelph/Oxford): Doesn't the rise at low momenta of the $\bar{p}p$ annihilation come from a (1/v) effect?

T. Walcher (Speaker): At very low momenta ($<$ 100 MeV/c as it appears from our data) where one expects s-wave scattering to dominate, the well-known 1/v law will occur. However, at higher energies where more partial waves contribute, the 1/p dependence is purely empirical.

A.M. Green (Helsinki): What does your parameter dependence say about the range of annihilation?

T. Walcher (Speaker): We have used two models to derive the range of the annihilation, r_W. With an optical model potential V + iW, where V and W are square wells, one gets $r_V \simeq 1.8$ fm and $r_W \simeq 1.0$ fm. With the boundary condition model of Dalkarov and Myher, one gets a critical radius of $r_c \simeq 1.1$ fm.

G. Karl (Guelph/Oxford): What accuracy level can you reach for the magnetic moment of the antiproton?

Speaker, H. Koch: We hope to improve by an order of magnitude with respect to the present value.

T. Kalogeropoulos (Syracuse): What kind of sensitivity do you expect in the search for long range QCD forces?

Speaker, H. Koch: A long range potential of the form $V = \lambda/r_o \, (r/r_o)^N$ with $\lambda = 0.1$, $r_o = 1$ fm, $N = 5$ would give an observable effect of about 20 eV at nuclei of $Z = 30$ which is easily detectable with present techniques. Further measurements (critical absorber techniques, crystal spectrometer) could lead to a considerable improvement. References: G. Fiorentini et al, Pisa Preprint IFUPTh10/82; K. Heitlinger et al, KFK Karlsruhe, Internal Report.

D. Hertzog (Carnegie-Mellon): Can you explain the large difference in amplitudes of the components in the Pb fine structure doublet?

Speaker, H. Koch: There is no explanation yet. The statistical weight would be 170 : 152, different from the ratio seen in the spectrum.

A.M. Green (Helsinki): For the record I would like to point out that calculations have been made by Green, Stepien-Rudzka and Wycech in Nucl. Phys. A399 (1983) 307. The results in eV are as follows for $\Delta E(\varepsilon_R, \Gamma/2)$:-

		Local approx.	Separable approx.	Expt.
^{16}O	3d	-89, 277	-115, 141	-123, 218
	4f	0.18, 0.47	0, 0.20	—, 0.29
^{18}O	3d	-177, 357	-176, 216	-205, 318
	4f	-0.08, 0.82	-0.2, 0.49	—, 0.37

J.D. Davies (Birmingham): Were your isotopes in the same chemical state?

Speaker, T. von Egidy: Yes, all targets were metal.

G.C. Phillips (Rice Univ.): What fraction of $\bar{p}$ captures in nuclei produce multi-nucleon emission?

Speaker, T. von Egidy: Isotopes of Ga, for instance, might be produced after $\bar{p}$ annihilation at ^{98}Mo with abundances between 0.1 and 1% according to preliminary evaluations of residual nuclei distributions.

D. Geesaman (Argonne): Do you have an example of the ^{12}C inelastic spectra where you see the 12.7 MeV state?

Speaker, D. Garreta: I have only the very first spectrum taken for ^{12}C at 300 MeV/c at $\theta = 25^o$ where the 2^+ (4.44 MeV), 3^- (9.63 MeV) and possibly the 1^+, $T = 1$ (15.1 MeV) are seen. On-line analysis of ^{12}C spectra at 10° and 600 MeV/c indicate that we see the 1^+, $T = 0$ (12.7 MeV) and $T = 1$ (15.1 MeV) states with cross-sections of the same order of magnitude which, according to a calculation of Dover et al, indicates a preference for the Paris group elementary amplitudes; the Dover-Richard amplitudes predict a cross-section for the 12.7 MeV an order of magnitude lower.

G.C. Phillips (Rice Univ.): Do you have evidence for quasi-elastic $\bar{p}$ scattering from the Fermi jittering nucleons within the nucleus?

Speaker, D. Garreta: In the case of ^{12}C, because of the large binding energy of the proton, the quasi-elastic reaction extends over a large excitation energy range (of the order of 40 MeV). This, together with the small number of effective protons (about 0.5), produces a small double differential cross-section consistent with our measurement. In the case of ^{6}Li, we only have on-line analysis information, but it seems that the quasi-elastic peak was clearly visible.

R. Cester (Torino): What is a typical, instantaneous, luminosity of your experiment and what is the sensitivity you estimate for your ξ search?

Speaker, B. Bonner: If the cross-section for ξ production is 1 μb, we expect to see on the order of 50 events in the data we have so far.

B. Andersson (Lund): A comment and a question: It is known from low energy reactions, e.g. $\pi^- p \to K\Lambda$, $K^*\Lambda$, $K^*\Sigma$, K^*Y^*, etc, that there a very large relative polarisations; all of them related to the transversity frame. The Lund group has made a set of predictions for you based upon these observations. Are you able, apart from looking at the S and P waves, to check such predictions too?

Speaker, B. Bonner: We certainly will measure the polarizations of the Λ and $\bar{\Lambda}$s as well as their spin-correlations.

T. Walcher (MPI, Heidelberg): What are the values of σ_{tot} you need and why do you believe that ρ is not sensitive to the slope parameter b if you use these fixed σ_{tot}?

Speaker, M. Sakitt: We fixed the cross-sections, at each momentum, using the formula shown in the text. That formula does a good job of representing the world's information on the cross-section. In our fits to ρ and b the results are calculated and the values of b are consistent with those of other groups.

G.C. Phillips (Rice Univ.): What are the signs of the potentials - are they attractive?

Speaker, M. Sakitt: The results for the real potential well correspond to an attractive potential. This is in agreement with the results from other groups.

D. Garreta (Saclay): Have you an idea of the accuracy you have in the determination of the real and imaginary potential depths?

Speaker, M. Sakitt: The values of the potential depths could be changed by 10 to 20 MeV. That is only a rough estimate of our accuracy.

K. Nakamura (KEK): Is it possible to calculate the momentum dependence of the annihilation cross-sections of exclusive two body channels, such as $\bar{p}p \to \pi^-\pi^+$ and $\bar{p}p \to K^-K^+$?

Speaker, A.M. Green: In the near future, we will be able to compute $\bar{p}p \to \pi^-\pi^+$ for $\bar{p}p$ in a relative P-wave since this requires a single $\bar{q}q$ annihilation. However, $\bar{p}p \to K^-K^+$ and $\pi^-\pi^+$, with the $\bar{p}p$ in a relative S-wave, require the introduction of three $\bar{q}q$ vertices. If we wish to interpret the single $\bar{q}q$ vertex process as being proportional to $\gamma\sqrt{\lambda}/2m_q$ as discussed in the talk, then the three $\bar{q}q$ vertex process will be much more difficult to calculate. However, the momentum dependences of $\bar{p}p \to \pi^-\pi^+$ and K^-K^+ should be different as each $\bar{q}q$ vertex makes the overall vertex of shorter range.

J.D. Davies (Birmingham): Will you please make predictions for $\bar{p}p$ and $\bar{p}d$ atom shifts and widths with your model? Essentially, there are no $\bar{p}d$ predictions!

Speaker, A.M. Green: The calculation of $\bar{p}p$ atomic levels is on our agenda. However, our earlier bound state programme needs modifying to deal with the case of Coulomb + local + separable + tensor forces + $N\bar{\Delta}$ and $\bar{\Delta}\Delta$ configurations. So far we only treat the local + separable + tensor force complications when making the $N\bar{N}$ bound state predictions of this talk. I agree that $\bar{p}d$ atomic levels should be calculated, but this requires the development of some 3-body theory. As far as I know, only S. Wycech of Warsaw has seriously thought about this problem.

Inst. Phys. Conf. Ser. No. 73: Section 4
Paper presented at VII Eur. Symp. Antiproton Interactions, Durham 1984

Antimatter in the universe

N.C. Rana# and A.W. Wolfendale.

Physics Department, University of Durham, Durham City, U.K.
(# On leave from Tata Institute of Fundamental Research, Bombay, India).

Abstract. A brief survey is given of the case for and against the existence of antimatter in the Universe on a large scale. This is followed by an examination of the situation in cosmic ray studies where positron results are interesting and an important excess of low energy anti-protons has been detected. A variety of explanations for the anti-protons are put forward, and an indication is given of the way in which a distinction should be possible in the future.

1. Theoretical Possibilities

1.1 Introduction

It is well known from laboratory experiments that the laws of physics are equally applicable to particles as well as their anti-particles, except for two cases known so far in which this particle-antiparticle symmetry is minimally violated. Nevertheless, one might expect that the Universe as a whole would also respect the same symmetry. Assuming that the Universe has started with an exactly equal amount of matter and antimatter and that it respects the above symmetry very precisely, the standard hot big bang model of the Universe would then predict a residual η of 1 particle-antiparticle pair (baryonic) per 10^{18} photons surviving the eventual annihilation (Steigman, 1976). However, when the total matter component (hadronic) is observationally compared with the total diffuse photon background, the ratio of the two number densities η turns out to be $\sim 10^{-9\pm1}$. It is clear then that any baryon-symmetric model of the early Universe will in the first place need a mechanism to suppress the annihilation rate, so as to generate $\eta \sim 10^{-9}$, rather than $\sim 10^{-18}$.

Furthermore, it is by now more or less certain that there cannot reside much antimatter in the solar system, in the form of antistars within 30 parsecs of the Sun, nor even in the form of tenuous gas within our Virgo cluster of galaxies. The reason is that if it did, too big a gamma-ray background would have been generated as a result of $p\bar{p}$ annihilation at the matter-antimatter interfaces. The overall fraction of antimatter to matter on the scale of the cluster of galaxies ($\ell_c \sim 10^{25}$cm), if not hiding in the form of anti-neutron stars or black-holes, turns out to be less than 1×10^{-6} (Tang and Fang, 1984). So any viable symmetric cosmology has to justify some kind of matter-antimatter separation on large enough scales of length before they could naturally undergo annihilation. Moreover, in order to avoid the spectral distortion of the 3K radiation background, and also excessive production of γ-rays, substantial remixing of these separated domains is almost prohibited at any subsequent epochs.

1.2 Early models of symmetric cosmology and their limitations

Omnes (1969, 1970, 1971) proposed mechanisms for particle-antiparticle separation into spatial domains in a symmetric cosmology, this separation having occurred during the chaotic period of hadronic phase transition at a temperature $T \gtrsim 350$ MeV, just prior to the period of annihilation. Aly et al. (1974) modelled a plausible case for no remixing of such domains, once separated, by simple diffusion. Steigman (1976) has examined all these early models in detail and pointed out various limitations. For example, depending on the degree of correlation between the separated domains, which is one of the free parameters of the model, the final ratio η could be anything between 1 and 10^{-18}. Furthermore, for some realistic choices, Steigman showed that a preferred value of η could be either around 10^{-18} or around unity, but much less likely to be $\sim 10^{-9}$.

There are however, simpler reasons to argue against such models, which we would now like to frame in the following way.

Four factors should be kept in mind; (i) Baryon-antibaryon domain separation should take place at a time no later than the hadronic phase transition (say at temperature T_c), because it is very difficult to think of such a process occurring subsequent to this era. (ii) These domains have subsequently been expanding with the general expansion of the Universe and are now no smaller than ℓ_c. (iii) The interdomain regions must show some kind of voids, otherwise they will be producing too many annihilation gamma rays at the interface. (iv) Different domains must show a widely different range of values of η, which could be anything between 10^{-18} and 1. Since this happens prior to the primordial nucleosynthesis, the amount of primordial helium (Y_p) would vary from domain to domain, ranging from 0 to 50% by mass fraction, as Y_p is very sensitive to η.

Combining (i) and (ii), the mass of each domain around $T = T_c$ should be no less than

$$M_c \sim \frac{4}{3}\left(N_f \frac{a}{c^2} T_c^{\,4}\right)\left(\ell_c^{\,3}\left(\frac{T_c}{3K}\right)^{-3}\right) \sim 10^2\, N_f\, a\, \ell_c^{\,3} T_c/c^2,$$

where ℓ_c is the present domain dimension, N_f is the effective number of particle species (counting unity for the photons) and a is Stefan's radiation constant. It means that $M_c \sim 10^{53}$g for $T_c = 3 \times 10^{12}$K and $\ell_c = 10^{25}$cm (galaxy cluster). Such a separate domain can never be formed as its (then) Schwarzschild radius has to be much much bigger than the required size of the domain.

Secondly, although one finds evidences for voids on large scales, there is no evidence for any anomalous chemical abundance of He^4 across the domain boundaries. Furthermore, distant quasars seem to show practically the same abundance of He^4 as in the Sun or in any other nearby galaxy. How is it that the different domains develop the very same value of η within possibly a factor of 2?

1.3 Later developments

In view of the above difficulties, one might ask the following question. Should the Universe be necessarily baryon symmetric? The answer comes out to be negative. Joshimura (1978) has given a demonstration as to how an asymmetric Universe might have evolved out of an initially symmetric hot and expanding Universe. He has used the grand unification theories (GUTS) that unify the electroweak and the strong interactions, in the context of a hot big bang and has assumed a minimal CP violation, which has been observed in laboratory experiments. The result is a satisfactory generation of a value of η in the range of $10^{-8\pm4}$, without producing any residue of antimatter in excess of its canonical fraction of $\sim 10^{-18}$.

The application of the GUTs to the very hot early Universe has now become an established fashion. All outstanding problems of all the possible cosmologies, it seems, can find some 'reasonable' answers under 'reasonable' assumptions. An assumption is here called 'reasonable' if it has not so far been disqualified by experiment. One of the most popular GUTs is the so-called minimal SU(5) gauge theory, which has, however, made one and only one bold prediction , namely the precise lifetime of the nucleons — and failed. It is therefore perhaps premature to write off the baryon-symmetric models.

If one assumes that the CP-violation is hard, the Universe might evolve to a globally baryon asymmetric state, the one originally advocated by Joshimura above in the context of the GUTs. But if this CP-violation is assumed to be spontaneous, and could occur in the style of supercooling and exactly matching with the time of SU(5) inflationary phase transition, the Universe may end up with islands of matter and antimatter on arbitrarily large scales (Kuzmin et al. 1981, Mohanty and Stecker 1984). From the point of view of the inflation of bubbles, each bubble would represent a definite value of CP as well as a definite excess of baryons (or antibaryons). Once again, to claim that we live in a baryon symmetric Universe, we would require a large number of CP-domains within the visible horizon of our Universe. When it is compared with the standard inflationary scenario, the latter duly resembles the inflation of a single such bubble into the whole of the observable part of our Universe.

1.4 Is the Universe symmetric?

Notwithstanding the problems with even the latest developments of the symmetric cosmology, the difficulty of justifying the same magnitude of η for all the observable CP-domains and the global uniformity of He^4 abundance, it is wisest to keep an open mind on the problem and to devise experimental tests.

The most obvious is by way of cosmic rays which may move from one island of matter (antimatter) to another. The motion of the islands as a whole is normally about a thousand times slower than the speeds of the cosmic rays. Thus, if there are a number of antimatter islands within the range of the visible horizon of our Universe, our galaxy is now old enough to receive a flux of cosmic ray antimatter from each of them. There is a caveat at the highest energies, however, say for $E > 10^{18}$eV, where the main bulk of the cosmic rays are severely attenuated by the 3K microwave background radiation (if that exists universally). Therefore, a careful study of the antimatter component of cosmic rays at lower energies can observationally tell whether the Universe is really baryon asymmetric or not. Such an analysis is presented in the following sections.

2. Cosmic Ray Evidence

2.1 General remarks

It is appropriate to comment that antimatter was first discovered in the cosmic radiation, viz. the positron, first identified by Anderson and Neddermeyer in 1933 following its prediction by Dirac in 1928. More recently, antiprotons have been detected by at least three groups. Anti nuclei of greater mass have not yet been unambiguously recorded although an anti-triton (resulting from the fragmentation of $\overline{He}$) may have been seen by Apparao et al. (1983).

In what follows we discuss e^+ and $\bar{p}$ in turn.

2.2 The situation with positrons

Probably the most interesting observation is that of the 0.51 MeV γ-ray line from e^+e^- annihilation in the general direction of the Galactic Centre. Figure 1 gives the details. A remarkable feature is the time variability of the source (the linear dimension is by no means clear, the angular resolution of the detectors being several degrees at least).

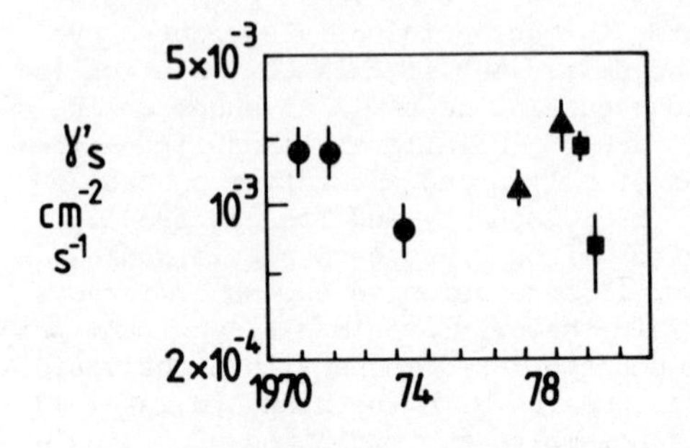

Figure 1. Flux of e^+e^- annihilation γ-rays from the general direction of the Galactic Centre. Three independent groups have measured fluxes and found clear evidence for variability (see Houston and Wolfendale, 1982, for details).

The emission needed to explain the observation is $\sim 2 \times 10^{43} e^+ s^{-1}$ or about 200 times the likely emission from a pulsar. Clearly a very energetic and unusual object is responsible. Conventional cosmic rays are not the cause in the sense that if the e^+ were secondaries in the chain $\pi^+ \rightarrow \mu^+ \rightarrow e^+$ the resulting γ-rays from the associated π^o-mesons would have been seen very strongly (a CR intensity at the G.C. of some 200 x that locally is required). Instead, and inevitably, explanations in terms of black holes tend to be favoured - e.g. Lacy et al. (1979, 80) invoke a black hole of several $10^6 M_\odot$. The actual mechanism could be that proposed by Thorne (1974) and Lovelace et al., (1979) involving rotation with associated strong electric fields, photon production and e^+e^- cascades.

The situation with the directly observed e^+ component is interesting. Observations extend to about 20 GeV and there is rough agreement between the observed energy spectrum and that expected on the basis of secondary production in the interstellar medium. Extensive calculations have been made by Giler et al. (1977) and others. The phrase 'rough agreement' is used because in fact the observed positron spectrum is somewhat lower (by up to a factor $\sim$ 2) than expected using the 'grammage' traversed in the ISM from data on the relative numbers of primary and secondary cosmic ray nuclei. It is likely that improvements to the propagation model can explain the difference.

An interesting feature is the observation that the e^+/e^- ratio is near unity for energies below 30 MeV falling to $\sim$ 10% above 1 GeV; the high value is unexpected.

2.3 Antiprotons

2.3.1 Experimental Data

The situation with anti-protons is more interesting. Figure 2 summarises the results.

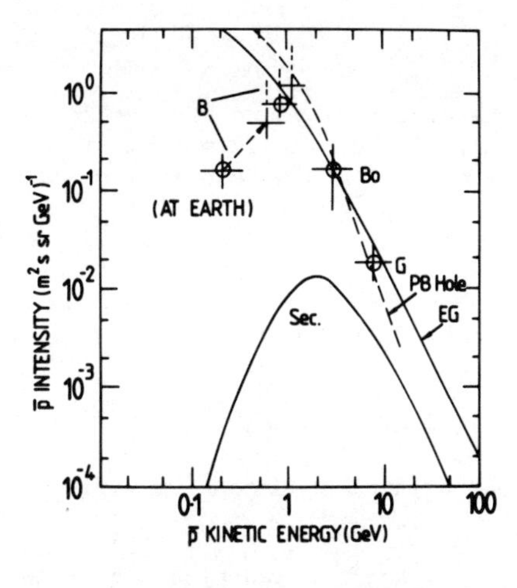

Figure 2. Summary of measurements of intensity of cosmic ray anti-protons (after Kiraly et al., 1981). B : Buffington et al. (1981); Bo : Bogomolov et al. (1979), G : Golden et al. (1979). 'Sec' denotes conventional expectation for secondaries from CR interactions in the interstellar medium. 'P B Hole' and 'EG' represent (normalised) predictions for the primordial black hole and extra-galactic origin models described in the text. The upward arrow for the 'B' intensity represents allowance for solar modulation to give the implied interstellar intensity.

Before discussing the possible explanation of the data attention should be drawn to the important lowest energy point (Buffington et al., 1981). Correction is necessary for the effect of solar modulation if we are to find the corresponding interstellar intensity. Kiraly et al. (1981) have addressed this problem with the result shown. The correction displaces the intensity upward and to the right; the central point has been derived using the best-estimate modulation parameters and the others represent likely limits.

2.3.2 Comparison with expectation for $\bar{p}$ as secondaries

The most conservative explanation is of course, in terms of secondaries from CR - ISM nucleus interactions. A number of workers have calculated appropriately and there is now a consensus (see Kiraly et al.) that the observed intensities are considerably in excess of expectation for conventional propagation models. The most serious discrepancy is seen to be at the lowest energy where the observed intensity is two orders of magnitude higher than predicted. Even with the adoption of a 'closed Galaxy' model, in which no cosmic rays escape there is still a short fall at the lowest energy and there are many problems in other areas (notably for positrons) with such a model.

Other explanations in which the $\bar{p}$ are secondaries have concentrated on making the $\bar{p}$ lose energy near their sources, so that the observed spectral shape is achieved, but there are severe difficulties, not only in absolute intensity but also in the secondary γ-rays produced at the same time. It seems to us that these models are untenable.

2.3.3 Galactic Centre explosions

An attractive idea is to have $\bar{p}$ originate in explosions in the Galactic Centre. Khazan and Ptuskin (1977) made the case for cosmic ray production in general occurring in explosions in the GC and in fact, as time goes on, such a model is becoming increasingly attractive. Said et al. (1981) invoked GC production for $\bar{p}$, pair production from exploding massive objects being the mechanism. Here, the GC protons would be augmented by low energy protons accelerated by more normal processes (e.g. supernovae) so that the observed local $\bar{p}/p$ ratio is generated.

2.3.4 Evaporating primordial Black Holes

Kiraly et al. followed up the ideas of Hawking (1974) and Carr (1976) and considered the possibility of $p\bar{p}$ (and $n\bar{n}$) pairs coming from PBH. The idea is that a PBH radiates like a black body of mass m and temperature $\Theta = hc^3/(8\pi Gmk) = 10^{26}m^{-1}K$, where m is in g, so that for $m \sim 10^{14}g$, $\Theta \sim 10^{12}K$, i.e. $10^8 eV$, and particle antiparticle pairs are expected. Now a PBH with a mass of this order has a mean life for evaporation of about the Hubble time, i.e. these primordial black holes are evaporating now. For the important mass range $10^{13.5} < m < 10^{14}g$ the fractions of particles emitted are (Carr, 1976), respectively,

$$\text{gravitons}, \gamma, \nu, e^{\pm}, p^{\pm}(n\bar{n}) \approx < 1\%, 11\%, 40\%, 25\% \text{ and } 12\%$$

and we see immediately that $\bar{p}$ will appear without an excessive yield of electrons and γ-rays.

The energy spectrum expected is roughly $(E + E_o)^{-3}$ - see Figure 2 - and this is of reasonable shape. Absolute predictions are not possible but normalisation of $\bar{p}$ allows an absolute $e^{\pm}$ spectrum to be derived. The result (Kiraly et al.) is

$$N(E_e) = 8.4E_e^{-3} m^{-2} s^{-1} sr^{-1} \text{ GeV}^{-1}$$, neglecting energy losses (which are not expected to be important until several GeV).

An important consequence of the PBH model now follows because the above magnitude of $N(E_e)$ allows the explanation of a previous difficulty concerning the flux of low energy Galactic γ-rays.

The problem is that the electron intensity in the interstellar medium in the region of 100 MeV is between 3 and 10 times what is found by extra-polating back the observed electron spectrum from the region above 100 MeV where it can be measured with some degree of certainty. Now, whereas at 1 GeV the extra PBH flux of electrons is only 10% of observation, the flux at 100 MeV is ∿ 3 x higher than the extrapolated-observed value, viz it is just in the γ-ray-required region. A further point in favour is that we expect at 100 MeV $e^+/e^- \simeq 1$, again close to observation.

The PBH model must be taken seriously.

2.3.5 Extragalactic antiprotons

The final possibility to be considered is that the excess $\bar{p}$ are of extra-galactic origin. Such a situation should be investigated despite the protestations in Section 1, the possibility of a retarding Galactic wind and slow diffusion arising from magnetic fields in extragalactic space.

Stecker and Wolfendale (1984) have taken the idea of EG origin a step further by drawing attention to a possible experimental check. Figure 3 illustrates the idea.

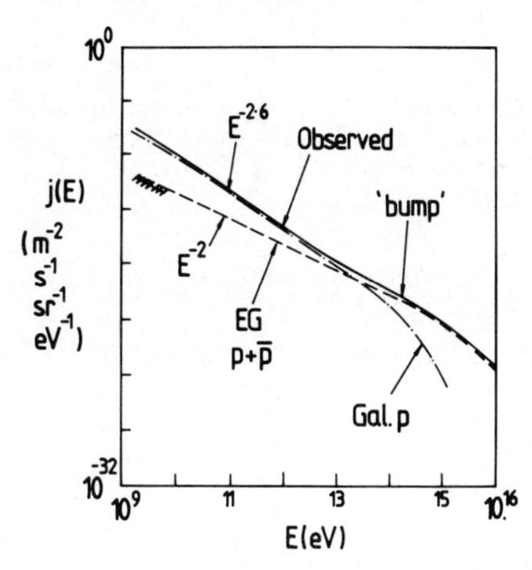

Figure 3. The extragalactic origin model for anti-protons of Stecker and Wolfendale (1984). 'Observed' represents the total cosmic ray spectrum with its 'bump' in the range 10^{14} - 10^{15}eV. Galactic protons predominate below about 3.10^{13}eV and a 50-50 mixture of extragalactic p and $\bar{p}$ at higher energies. Stecker and Wolfendale stress the need to measure the shape of the $\bar{p}$ spectrum at higher energies than hitherto as a way of supporting or destroying the extragalactic origin model.

Many observations of Galactic nuclei, both primary and secondary (by 'primary' being meant those nuclei accelerated in sources, 'secondary' refers to the products of fragmentation in the ISM) have shown that the Galactic lifetime of CR varies with energy as, roughly, $E^{-0.6}$ (e.g. Ormes and Protheroe, 1983). This means that, whereas the observed spectral shape of the dominant proton component has the form $E^{-2.6}$ (to E $\approx 10^{15}$eV), the production spectrum is probably close to $E^{-2.0}$. If, as is likely, E^{-2} is a common form of generation spectrum, and thus applies in anti-galaxies to the acceleration of antiparticles, the spectrum of arriving EG $\bar{p}$ (and p) should also have the form $E^{-2.0}$.

Normalising such an EG spectrum to that observed below 10 GeV (i.e. 2 x the measured $\bar{p}$ flux), the situation is as shown in Figure 3. An intriguing point - and perhaps a significant one - is that extrapolation of the EG p + $\bar{p}$ spectrum takes us to the bump in the measured spectrum just above 10^{14}eV (see Kempa et al., 1974, and later papers justifying the claim for a bump). Thus, it might well be that 10^{14}eV marks the energy where Galactic and EG intensities are similar. At higher energies both EG and G spectra must steepen - the reason for this is not known but a possibility is the onset of γ p reactions near the sources or, in the case of Galactic particles, increased diffusive losses.

the possibility of photon interactions is also very relevant to the question of why no anti helium nuclei have been detected ($\overline{He}$/He $\lesssim$ 2 x 10^{-5}) - such interactions are necessary. In fact, they are likely because in order to explain the absolute intensity of EG p and $\bar{p}$ more energetic galaxies (anti galaxies) are necessary than our own Galaxy and such energetic G($\bar{G}$) will have higher photon fields. To be specific, the EG flux is about 25 x what would be expected if all G($\bar{G}$) were equal in CR output.

The test of the extragalactic origin of $\bar{p}$ is now clear. It is to measure the shape of the spectrum of $\bar{p}$ to higher energies than hitherto. Measurements to 100 GeV are technically feasible and would suffice. The observation of the prescribed E^{-2} spectrum would mean that an extragalactic source for the bulk of the anti-proton flux in cosmic rays would need to be taken very seriously. Conversely, the observation of an E^{-3} spectrum would stimulate further work on the primordial black hole model.

3. Conclusions

Whether or not there is symmetry of matter and anti-matter on a Universal scale is still problematical but the detection of interestingly large fluxes of cosmic ray positrons and antiprotons makes their study of considerable contemporary value.

References

Aly J J et al. 1974 Astron. Astrophys. 35 271
Apparao K M V et al. 1983 preprint TIFR, Bombay
Bogomolov E A et al. 1979 Proc. Int. Cosmic Ray Conf. (Kyoto) 1 330
Buffington A et al. 1981 Astrophys. J. 248 1179
Carr B J 1976 Astrophys. J. 206 8
Giler M, Wdowczyk J, and Wolfendale A W 1977 J. Phys. A 10 843
Golden R L et al. 1979 Phys. Rev. Lett. 43 1196
Hawking S W 1974 Nature 248 30
Houston B P, and Wolfendale A W 1982 Vistas in Astronomy 26 107
Kempa J, Wdowczyk J, and Wolfendale A W 1974 J. Phys. A7 1213
Khazan Y M, and Ptuskin V S 1977 Proc. Int. Conf. 2 4
Kiraly P, Szabelski J, Wdowczyk J, and Wolfendale A W 1981 Nature 293 120
Kuzmin V A, Shaposmikov M E, and Tkachev I I, 1981 Phys. Lett. 105B 167
Lacy J H et al. 1979 Astrophys. J. Lett. 227 L17
Lacy J H, Townes C H et al. 1980 Astrophys. J. 241 132
Lovelace R V E, MacAuslan J, and Burns M, 1979 'Particle Acceleration Mechanisms in Astrophysics' (Ed. J Arons, C Max and C McKee) Am. Inst. Phys. 399
Mohanty A K, and Stecker F W, Phys. Lett. B. (in the press)
Omnes R 1969 Phys. Rev. Lett. 23 38; 1970 Phys. Reps. C3 1; 1971 Astron. Astrophys. 10 228
Ormes J F, and Protheroe R J 1983 Astrophys. J. 272 756
Said S S, Wolfendale A W, Giler M, and Wdowczyk J 1982 J. Phys. G8 383
Stecker F W, and Wolfendale A W 1984 Nature 309 37
Steigman G 1976 Ann. Rev. Astron. Astrophys. 14 339
Tang T B, and Fang L Z 1984 Vistas Astron. 27 1
Thorne K 1974 Astrophys. J. 191 507
Yoshimura M 1978 Phys. Rev. Lett. 41 281

Inst. Phys. Conf. Ser. No. 73: Section 4
Paper presented at VII Eur. Symp. Antiproton Interactions, Durham 1984

Particle physics and cosmology

P. J. E. Peebles

Joseph Henry Laboratories, Princeton University, Princeton, N. J.

Abstract. The new particle physics has generated some of the most stimulating ideas to have come into cosmology in many years. I have chosen for discussion here the status from the astrophysical side of two of these ideas, the inflation scenario and the possibility that galaxies are made of weakly interacting particles left over from the hot Big Bang.

1. Introduction

This is a review of what particle physics has done for cosmology lately. The infusion of new ideas has been enormously stimulating. It has not yet been translated into hard results that one would want to enter into the next edition of the handbook of Astrophysical Quantities, but that is reasonable enough because it is not easy to establish firm results in cosmology. In fact, we do not have a "standard model" in cosmology that reconciles in a natural way all the more commonly accepted clues. That is at least in part due to differences in opinion on how observational and theoretical programs in progress will turn out, though of course we must also be prepared to learn that some of our "established" clues, old or new, are wrong. The main message of this review is that cosmology does have some interesting things to say about particle physics but that by and large we are not yet sure what they are.

2. Is the Big Bang Cosmology Right?

The interaction between cosmology and particle physics is interesting only to the extent that we can believe we have a valid world model, so it is well to begin by recalling why it is thought that the hot Big Bang cosmology is pretty well founded. We start with the assumption that the universe is very nearly homogeneous and isotropic in the average over scales comparable to the horizon ($cH^{-1} \sim 10^{10}$ light years). This agrees with the accurate isotropy of counts of radio and X-ray sources, and, to poorer available accuracy, counts of galaxies. (For details see Peebles 1980, § 2). Homogeneity follows if our galaxy is not in a favoured position. A more direct test for possible radial gradients in the space density of galaxies is obtained from counts of galaxies as a function of redshift. Useful limits from published data extend only to $\sim$ 10% of the horizon cH^{-1} (Kirshner et al. 1983), but deeper surveys are in progress.

In general relativity theory a homogeneous and isotropic mass distribution has to expand or contract, the relative velocities of particles at separation r varying as

$$v = Hr \quad . \tag{1}$$

This with $H > 0$ agrees with observations of galaxy redshifts as a function of distances. Hubble's constant H (which is a function of world time) fixes the time-scale for expansion up to a factor on the order of unity (that depends on the mass density and cosmological constant). Unfortunately H is not well known because it is difficult to fix the extragalactic distance scale. Within the range of values of H under discussion the expected age of the universe is in the range 7 to 20 billion years, and it is an encouraging coincidence that the stellar evolution ages of the oldest star clusters are in this range, at $\sim 15 \pm 4$ billion years (Harris et al. 1983).

If the young universe were hot as well as dense it would relax to local thermal equilibrium, producing blackbody radiation, which would be present, cooled by adiabatic expansion, in the universe today. The signature would be isotropic flux of radiation with spectrum very close to Planckian. Measurements of the cosmic background radiation at wavelengths from 20 cm to 800 μ are consistent with a blackbody spectrum at $T = 2.75 \pm 0.1$ K (Richards 1984). Most people consider this to be tangible evidence that the universe really did expand from a density well above the present value because no one has been able to see how a blackbody spectrum could have been produced by relaxation in the universe as it is now. What is more, knowing the present temperature of the universe we can attempt to trace its thermal history. This led to a notable result: the expected abundance of helium left over from the hot Big Bang is $\sim$ 23 to 27% by mass, consistent with the helium abundance in the Sun, and, so far as the observations reveal, with the helium abundances in the gas in other galaxies and in the oldest stars in our own galaxy. The abundances of He^3 and deuterium produced in the Big Bang are sensitive to the mean density of baryons, which, as will be discussed, is not very well known. The computed abundances are consistent with the observations if the baryon density is about that seen in galaxies (Yang et al. 1984). This is particularly notable because it has proved difficult to see how deuterium could be produced in the observed abundance in known astrophysical processes in the present universe. Thus helium and deuterium seem to be further tangible evidence of a hot relativistic Big Bang cosmology.

3. Inflation

The homogeneous distribution of gas atoms in a container is an end state, the result of statistical relaxation. The homogeneity of the universe calls for a different interpretation because, if we are to believe general relativity theory, the large-scale distribution of mass is gravitationally unstable. That means the universe must be growing more clumpy, and must have been exceedingly close to homogeneous in the distant past. The most elegant explanation of why the universe started out so smooth is the inflation scenario invented by Guth and elaborated by him, Linde, Steinhardt and others. (For a review see Guth 1983.) There are, however, some side-effects of inflation that may or may not be Good Things. Homogeneity is obtained by an exponential expansion phase that is supposed to stretch all gradients present before inflation out to scales well beyond the present horizon. The same process had to suppress the curvature of surfaces of homogeneity, so we are forced to a cosmologically flat world model. Less straightforward, but still generally accepted, is the conclusion that the density fluctuations driven by quantum noise during inflation are Gaussian with power spectrum $P = |\delta_k|^2 \propto k$, so cutoffs of the power law spectrum can be put at enormously large and small wavelengths.

The restriction on space curvature means that the expansion rate H has to be related to the mean mass density ρ and the present value of the cosmological constant Λ by the equation

$$H^2 = \frac{8}{3}\pi G\rho + \frac{\Lambda}{3} \quad . \qquad (2)$$

The value of Λ allowed by observational limits on H and ρ is equivalent to an energy $\sim$ 3 meV. Most people are reluctant to believe such a small energy could be relevant, preferring to assume Λ is negligibly small. That means that ρ must be very close to the Einstein-de Sitter value ρ_c,

$$\rho_c = 3H^2/(8\pi G) \quad , \qquad (3)$$

so the density parameter is $\Omega = \rho/\rho_c = 1$. The bounds on Ω, Λ and space curvature provided by the classical cosmological tests based on apparent magnitudes m(z) of standard galaxies as a function of redshift z do not usefully test these arguments because it proves difficult to separate cosmology from the effects of evolution of the intrinsic galaxy luminosities. The situation is more hopeful now because it has become feasible to measure field galaxy counts as a function of redshift and apparent magnitude,

$$dN = f(m,z)\,dm\,dz \quad . \qquad (4)$$

As discussed by Loh and Spillar (1984), surveys planned and in progress should yield useful estimates of f(m,z) at $0.03 \lesssim z \lesssim 1$. This function of two variables affords some separation of evolution and cosmology, and may yield bounds on the cosmological parameters that will test inflation. Meanwhile, the best available measure of Ω comes from the dynamics of galaxy clustering. We can measure mass concentrations by the observed motions of galaxies, which imply accelerations v^2/r, which must be balanced on the average by the peculiar gravitational field $g \sim v^2/r$, from which we can derive the clustered part of the mass. If mass clusters the way galaxies do this gives a measure of ρ. The test has been applied under a fairly broad set of conditions and consistently yields $\Omega \sim 0.1$ to 0.2 (Peebles 1984a). This could mean that inflation is wrong, or that inflation is right and Λ dominates $G\rho$ in equation (2), or that the test is wrong because the dynamical measure missed a smoothly distributed mass component.

There is some indirect evidence against the last possibility. This is the observation that the galaxy distribution approximates a scale-invariant clustering hierarchy, one of Mandelbrot's fractals, the characteristic galaxy density scaling as $n(r) \propto r^{-\gamma}$, $\gamma = 1.77 \pm 0.04$, from the size of galaxies out three orders of magnitude or so to $r \sim 10$ Mpc where the clustering pattern starts to blend into the nearly homogeneous distribution observed on large scales (Peebles 1980). If the universe were dominated by a homogeneous mass component it would tend to disrupt this scaling. On the dense, small end of the clustering pattern crossing times are short so the pattern surely is gravitationally bound. On the large end we see diffuse objects like the Local Supercluster that under the present hypothesis would be nearly freely expanding. This would imply that the observed scaling of the clustering hierarchy is only a transient effect, which seems unreasonable. Of course, this argument must be made quantitative - how rapidly is the scaling broken? - and sufficiently detailed and reliable computation are not yet available.

One way to suppress the breaking of scaling is to accept $\Omega \cong 0.1$ and assume that equation (2) is satisfied by a non-zero cosmological constant,

$$\Lambda = 3(1 - \Omega)H^2 \quad . \tag{5}$$

This is not welcomed by particle physicists because, as we noted above, the wanted value of Λ does not seem "natural," but then the condition that the present value of Λ must be small enough to avoid manifest problems with cosmology is not well understood either so we might be justified in setting this problem to one side for the moment. A peculiar and attractive feature of the Λ-model is that the expected velocity field is small, comparable to that in an open (negative space curvature) model with $\Lambda = 0$ and the same value of Ω, which agrees with observed galaxy velocities, while the growth of density fluctuations is large, comparable to that of the Einstein-de Sitter model ($\Lambda = 0$, zero space curvature), which is scale invariant. This is seen in linear perturbation theory and in the non-linear spherical clustering model (Peebles 1984a), but again we do not yet have believable computations of the details of the observable effects of Λ on the galaxy clustering pattern and dynamics.

It is conceivable that the astronomical observations will lead us to an open universe with $|\Lambda| \ll 3H^2$, $\Omega \ll 1$. If so the inflation scenario as an explanation of homogeneity would be ruled out. Also, it would exacerbate the clustering scaling problem because in this model the growth of density fluctuations is quite strongly suppressed. The open model also shares with the Λ-model the unattractive property that we live at a special epoch, at least on a logarithmic scale, at which space curvature (or Λ) has just become dominant over the density term $8/3\ \pi G\rho$ in fixing the expansion rate H.

The standard inflation model also predicts that the primeval density fluctuations that triggered the formation of galaxies and clusters of galaxies were a random Gaussian process with spectrum $P \propto k$ (where k is the wave number). This means the mass autocorrelation function $\xi(r)$ has to go negative at large lag r (Peebles 1982a). A useful test of the behavior of ξ at large r, checking at least that it is small if not negative, should be possible in intermediate depth galaxy redshift surveys now in progress. The Gaussian character of the primeval perturbations can be tested at least in principle by studying the distribution of the mass fluctuations we have ended up with. There is some evidence that there may be problems with an initially Gaussian distribution (Peebles 1983) but again better theoretical analysis is needed!

4. Dark Matter

Coupled to the problems with the mean mass density is the dark matter puzzle. It has been known since the 1930s that the seen mass in the galaxies in a great cluster is inadequate by a factor of 30 or so to hold the cluster together by gravity against the "pressure" of the galaxy motions. It seems unlikely that the clusters are falling apart, because they would dissolve well within the Hubble time H^{-1}, so the presumption is that the mass of the cluster is dominated by dark matter between the galaxies. More recently people have found a similar effect in isolated spiral galaxies (Rubin 1983). The circular velocity of rotation of the gas in the disc of the galaxy tends to be very nearly constant, independent of distance from the center. If, as we almost always presume, Newtonian mechanics is an adequate approximation, that means the mass within distance r of the galaxy scales as

$$M(< r) \sim r \quad . \tag{6}$$

This is quite contrary to the distribution of starlight, which is more strongly concentrated around the galaxies. What is the nature of the dark mass?

In our stellar neighborhood the number dn/dm of stars per increment of mass m increases with decreasing mass all the way down to the limit of detectability at $m \sim 0.08$ Solar masses, but the variation is slower than m^{-2} at small m so the net mass integral is converging at the small (and large) mass end. But it certainly is conceivable that in the outer parts of galaxies dn/dm is a little steeper, varying faster than m^{-2}, that is, that the dark mass is in stars with masses below the threshold for nuclear burning, or "Jupiters."

A constraint on the possible contribution of Jupiters to Ω comes from nucleosynthesis in the hot Big Bang. As mentioned above, the mean mass density ρ_B in baryons can be adjusted so that element production in the hot Big Bang at least roughly agrees with the observed abundances of helium, deuterium, and Li^7, nuclei that are not readily produced in astrophysical processes in the present state of the universe. The wanted value of ρ_B is fixed by the present radiation temperature T_o which seems to be well known. (ρ_B and T_o together fix the thermal history of the universe.) The critical Einstein de-Sitter density ρ_c depends on the expansion rate H (eq. 3), which is not so well known. Within the range of values of H currently considered the contribution of the wanted value of ρ_B to the density parameter $\Omega_B = \rho_B/\rho_c$ is (Yang <u>et al.</u> 1984)

$$0.01 \lesssim \Omega \lesssim 0.1 \quad . \tag{7}$$

The upper bound (which corresponds to a small value for the expansion rate H) is within the range of values of Ω indicated by galaxy clustering dynamics. If this coincidence were confirmed by further observations it would be a pleasing concordance of nucleosynthesis and dynamics, suggesting the dark matter is nothing more romantic than stars with too little mass to shine by nuclear burning.

Getting $\Omega_B = 1$ is more problematic. The computation of nucleosynthesis in the Big Bang does assume the baryon distribution in the early universe is smooth, like the radiation. If the baryons were placed in clumps in a smooth radiation background then where the baryon density is high the local helium production would be increased (because more neutrons are captured before they can decay) and deuterium reduced (because it is more efficiently burned to helium). Observations in our galaxy and nearby ones point to low baryon density, which is not the easy direction. Rees (1984) has noted that one can get around this by supposing that the matter in the dense spots all was incorporated into star remnants or the like whose element abundances are unobservable. Still, the situation is not all that promising because it would be reasonable to expect that the observed matter also came from a broad range of primeval baryon densities, which would tend to destroy the concordance of predicted and observed helium, deuterium and Li^7 abundances. Thus the indication is that if $\Omega \sim 1$, or if $\Omega \sim 0.1$ and H is relatively large, then the universe is dominated by non-baryonic weakly interacting dark mass.

Particle physicists have supplied us with a host of candidates for the dark matter. If particle physics could show that a particular particle really does exist and does have the parameters that would make it a significant contribution to Ω in the standard hot Big Bang model it would have a revolutionary effect on the subject. Lacking that, we must proceed by weighing the apparent advantages and problems with each candidate. I will discuss here only two cases, neutrinos with a mass of a few tens of eV and weakly interacting matter with negligible primeval pressure (axions or massive "ions").

Neutrinos are particularly attractive because we know they really do exist. What is more, there is a pleasing coincidence in the wanted neutrino mass. If the neutrinos thermally decouple from the radiation in the hot Big Bang while they are relativistic then the present number density per neutrino type (with two spin states) is $n_\nu \sim 100\ \text{cm}^{-3}$. (This is fixed by the present radiation temperature.) If these neutrinos are to make an interesting contribution to the mean mass density of the universe the neutrino mass m_ν must be on the order of 30 eV. If the same neutrinos are to account for the dark mass in galaxies then we know the wanted mass density and particle velocities in a galaxy (neutrinos would move freely, like stars) so we have a lower bound on m_ν from the bound on the density of neutrinos in phase space. The result is quite close to the cosmological mass estimated above. (This was noted by Cowsik and McClelland 1973, and worked out in more detail by Tremaine and Gunn 1979).

There are also some problems with the neutrino picture. The primeval thermal relaxation that fixes the residual neutrino number density also fixes the neutrino velocity dispersion at high redshift, and that multiplied by a characteristic expansion time fixes a smoothing length for the freely moving neutrino. The predicted comoving coherence length of the mass distribution before galaxy formation is (Peebles 1982b, Bond and Szalay 1983)

$$r_\nu \sim 10(\Omega h^2)^{-1}\ \text{Mpc} \quad , \tag{8}$$

where h is Hubble's constant H in units of 100 km s^{-1} Mpc^{-1} and the density parameter Ω is supposed to be dominated by the neutrinos. Even for the largest "reasonable" values of these parameters, $\Omega \sim 1$ and $h \sim 1$, the mass enclosed by this coherence length exceeds some 100 galaxy masses. Thus we must arrange for several things. The first generation of objects has to fragment to make galaxies. The fragments have to move apart to produce the galaxies like ours in the weakly clustered field. And the fragmentation has to preserve pockets of low entropy neutrinos for the dark mass in galaxies, which, as the coincidence mentioned above shows, has to have phase space density close to the primeval value. Melott (1983) showed that in a one dimensional collapse the coarse-grain density in phase space near the plane of symmetry is little affected by the collapse. I still am betting that a realistic three dimensional collapse is so messy that the coarse-density increases everywhere, which would be a Bad Thing, but in time N-body model computations should be able to settle the point! The model also implies that galaxies are fragments, that galaxies form in associations that look rather like dense rich clusters of galaxies. I think this is a Bad Thing because most galaxies today are not in clusters, but rather are in poor and dilute systems like our Local Group in the Local Supercluster. (For details see Peebles 1984b).

A way out of both problems is to imagine that the universe commenced with a smooth sea of radiation and neutrinos in which there was the occasional lump of low pressure matter. As the universe expands the neutrinos would cool and accrete around the lumps. Lumps with mass $\sim 10^9$ Solar masses would acquire neutrino halos at least roughly similar to the dark halos seen around spirals. Because the accretion is gentle the entropy in the central parts would not be increased much, consistent with the phase space constraint. And protogalaxies could form before protoclusters, depending on how we place the lumps, which of course is ad hoc. It is not clear what the lumps might be made of. Baryons are not attractive for this purpose if one would like to think particle physics says baryons were produced in a fixed abundance relative to photons (though there are contrary opinions; Senjanović and Stecker 1980). Loops of cosmic strings (Vilenkin and Shafi 1983) would tend to move around too much, and it is not clear whether a loop would last long enough to act as a seed for accretion. All of this is unappealing, but then the ingenuity of particle theorists is legendary so the present lack of a theoretical motivation for the accretion picture perhaps is not discouraging.

Another way out is to imagine that the dark mass is axions or massive "inos" that have negligible primeval pressure. That eliminates the smoothing mentioned above and so permits a hierarchical development of structure, big objects forming after little ones, which, I suspect, agrees better with what the observations are telling us (Peebles 1984b).

One feature of this axion model is specific enough to merit special mention. A characteristic Jeans length is defined here not by the primeval pressure of the dark matter, which we are supposing is negligible, but by the pressure of the hydrogen. The coherence length defines a hydrogen mass $M_J \sim 10^6$ Solar masses. Provided the primeval spectrum of density fluctuations scales about as $P \propto k$, which is the prediction of the standard inflation model, the first generation of hydrogen gas clouds would have masses $\sim M_J$, and would form at redshifts $Z_i \sim 30$ to 100. (Z_i depends on the normalization of the power spectrum P, which can be fixed by the observed large-scale fluctuations in the galaxy distribution, as is discussed in Peebles 1984c.) At redshift Z_i the hydrogen in the gas clouds would have a temperature on the order of 100 K. The clouds would tend to radiate energy (by trace amounts of molecular hydrogen) and so contract and heat up until the temperature reaches $\sim 10^4$ K and the gas is collisionally ionized. When this happens the energy loss by thermal bremsstrahlung is rapid so the cloud collapses freely, and, I presume, fragments into stars. It will be noted finally that the cloud commenced as a mixture of hydrogen and the dark "inos." The latter are supposed to be weakly interacting and so would be left as a dilute massive dark halo around the final star cluster. Do such star clusters exist? We have a candidate, in globular star clusters (Peebles and Dicke 1968; Peebles 1984c). If we can convince ourselves that globular clusters really do have dilute massive dark halos it will be rather strong evidence for the "ino" picture. If, on the other hand, we can establish that globular star clusters really did not form this way it will be a distinct problem for the "ino" picture: where are the objects that ought to have formed at the primeval Jeans mass?

Studies of the details of how galaxies might form in this "ino" picture have commenced (Blumenthal et al. 1984), with not unpromising first results. It is too soon to say, however, whether this is because the "ino" picture has been less thoroughly examined than the neutrino case.

References

Blumenthal G R, Faber S M, Primack J R and Rees M J 1984 Nature
Bond JR and Szalay A S 1983 Ap. J. 274 443
Cowsik R and McClelland J 1973 Ap. J. 180 7
Guth A 1983 The Very Early Universe ed G W Gibbons, S W Hawking and S T C Siklos (Cambridge: Cambridge University Press)
Harris W E, Hesser J E and Atwood B 1983 Ap. J. 268 L111
Kirshner R P, Oemler A, Schechter P L and Shectman S A 1983 A. J. 88 1285
Loh E D and Spillar E J 1984 Bull. Amer. Astron. Soc. 16 489
Melott A L 1983 Ap. J. 264 59
Peebles P J E 1980 The Large-Scale Structure of the Universe (Princeton NJ: Princeton University Press)
Peebles P J E 1982a Ap. J. 263 L1
Peebles P J E 1982b Ap. J. 258 415
Peebles P J E 1983 Ap. J. 274 1
Peebles P J E 1984a Ap. J. 284
Peebles P J E 1984b Science 224 1385
Peebles P J E 1984c Ap. J. 277 470
Peebles P J E and Dicke R H 1968 Ap. J. 154 891
Rees M J Proc. Conf. on Groups and Clusters of Galaxies Trieste Sept. 1983
Richards P 1984 Proc. I/O Fermilab Conference Batavia May 1984
Rubin V C 1983 Science 220 1339
Senjanović G and Stecker F W 1980 Phys. Lett. 46B 285
Tremaine S and Gunn J E 1979 Phys. Rev. Lett. 42 407
Vilenkin A and Shafi Q 1983 Phys. Rev. Lett. 51 1716
Yang J, Turner M S, Steigman G, Schramm D N and Olive K A 1984 Ap. J. 281 493

This research was supported in part by the National Science Foundation.

Inst. Phys. Conf. Ser. No. 73: Section 4
Paper presented at VII Eur. Symp. Antiproton Interactions, Durham 1984

Interaction of $\bar{p}$ with ^{4}He and Ne nuclei at 49 and 180 MeV

F. Balestra[5], Yu.A. Batusov[1], G.Bendiscioli[4], M.P.Bussa[5], L.Busso[5], I.V.Falomkin[1], L.Ferrero[5], V.Filippini[4], G.Fumagalli[4], G.Gervino[5], A.Grasso[5], C. Guaraldo[2], E.Lodi Rizzini[4], A.Maggiora[2], C.Marciano[4], D.Panzieri[5], G.Piragino[5], G.B.Pontecorvo[1], A.Rotondi[4], M.G.Sapozhnikov[1], F.Tosello[5], M.Vascon[3], G.Zanella[3] and A.Zenoni[4].

1 - Joint Institute for Nuclear Research, Dubna, USSR

2 - Laboratori Nazionali di Frascati dell'INFN, Frascati, Italy

3 - Dipartimento di Fisica dell'Università di Padova, and INFN Sezione di Padova, Padova, Italy

4 - Dipartimento di Fisica Nucleare e Teorica dell'Università di Pavia, and INFN Sezione di Pavia, Pavia, Italy

5 - Istituto di Fisica Generale dell'Università di Torino, and INFN Sezione di Torino, Torino, Italy.

(presented by G. Bendiscioli)

Abstract. $\bar{p}$-^{4}He and $\bar{p}$-Ne reaction cross sections have been measured at 48.7 and 179.8 MeV using a self-shunted streamer chamber in a magnetic field. Charged prong multiplicity (for ^{4}He and Ne), branching ratios and ^{3}He production probability (for ^{4}He) are given. Comparison between $\bar{p}$-^{4}He and existing $\bar{p}$-^{2}H data is performed.

1. Introduction

In a previous paper (Balestra F et al 1984a) we reported some results on the inelastic interaction of 180 MeV antiprotons with ^{4}He nuclei and then we discussed some possible consequences of them in cosmological problems (Balestra F et al 1984b). Now we give similar results on ^{4}He at 49 MeV and on Ne at 49 and 180 MeV.

The research was performed by means of a streamer chamber exposed to the antiproton beams of the LEAR facility. The apparatus and its operation are described in detail in Akimov Yu K et al 1984 and here we recall only that the chamber has a sensitive volume of 70 x 90 x 18 cm^3, is filled with gas at 1 atmosphere (target transparencies: about 15 and 70 mg/cm^2) and is placed in an electromagnet (B equal to 0.8 and 0.6 T at the two energies).

The data we report were obtained simply by counting the inelastic events and the charged prongs, distinguishing, when possible, the negative particles from the positive ones.

The events considered are those with their vertices in the central region of the sensitive volume (55 cm long), where the vertex distribution is flat and the scanning efficiency was near 100%.

Considering the energy loss by ionization in the walls and in the scintillator detectors along the transport beam line, the effective beam energies were 48.7 and 179.8 MeV. The results have to be considered as preliminary, as they concern only samples of the events recorded. Here we present:

at 48.7 MeV 466 events in ^{4}He and 319 in Ne;
at 179.8 MeV 1097 events in ^{4}He and 699 in Ne.

For completeness, also the results given in Balestra F et al 1984a are included.

2. Total reaction cross section

The total reaction cross section includes all the reaction channels, but the elastic one; i.e. charge exchange, inelastic scattering (break-up, fragmentation, etc.) and annihilation. The first values obtained are given in Table 1.

Table 1: Total reaction cross section (mb) on ^{4}He and Ne at 49 and 180 MeV

	49 MeV	180 MeV
^{4}He	286 $\pm$ 13	233 $\pm$ 7
Ne	791 $\pm$ 44	612 $\pm$ 23

The quoted errors are the statistical ones, a further 2.5% uncertainty due to scanning efficiency, target transparency and beam counting must be added. Also, we note that the Ne gas is a mixture of isotopes where, however, ^{20}Ne is the most abundant one (91%).

The Ne cross sections agree well with recent data on C, Al and Cu obtained in a trasmission experiment (Nakamura K et al 1984) and with optical model calculations fitting elastic scattering of antiprotons on C and Ca nuclei (Garreta D et al 1984) (see figg. 1 and 2, where the energy is in the center of mass system).

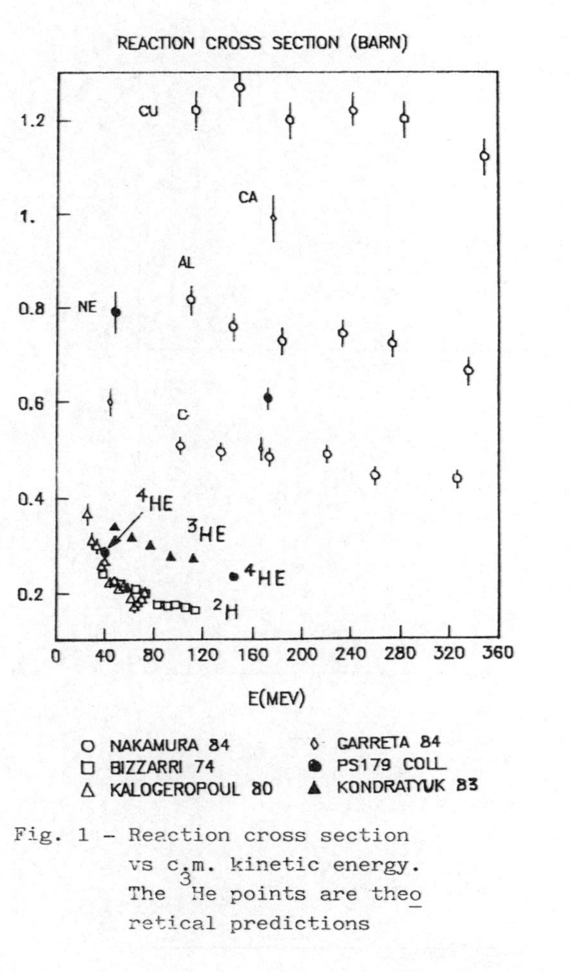

Fig. 1 – Reaction cross section vs c.m. kinetic energy. The ^{3}He points are theoretical predictions

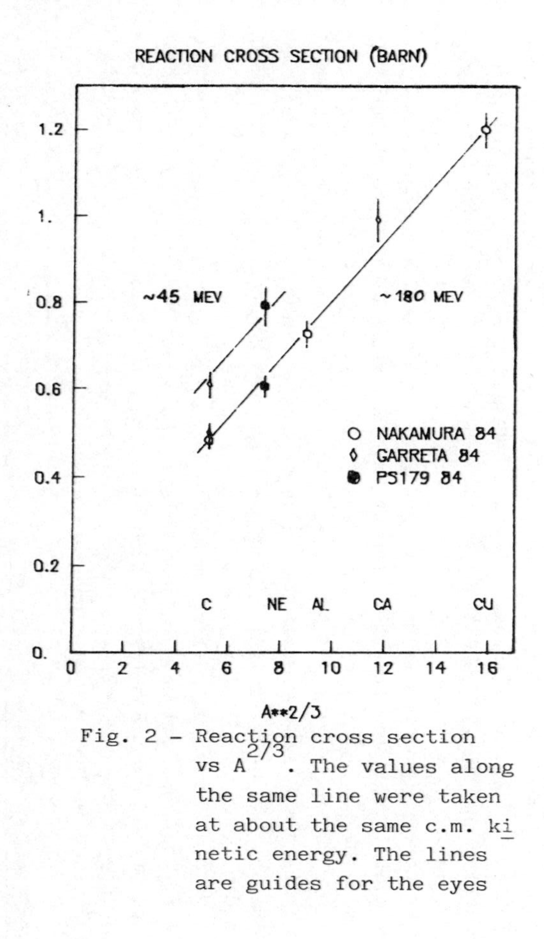

Fig. 2 – Reaction cross section vs $A^{2/3}$. The values along the same line were taken at about the same c.m. kinetic energy. The lines are guides for the eyes

The ^4He cross sections are lower than those for C, as expected, but seem too low if compared with deuterium data (Bizzarri R et al 1974; Kalogeropoulos T and Tzanakos G S 1980), particularly at 35 MeV in the c.m. (see fig. 1).

We note that in deuterium only annihilation is considered and break-up reaction is not. On the other hand we see that from 170 MeV to 40 MeV (c.m.) the ^4He cross section increases of 24%, i.e. about as much as for C (20%) and for Ne (29%), while for ^2H there is an increase of about 55%.

This seems to indicate that some unexpected effect in the $\bar{p}$ - ^2H interaction exists at low energies, as, for example, a broad resonance under 200 MeV/c incident momenta (Grach I and Shmatikov M 1982).

^4He reaction cross section data seem to be low also if compared with theoretical predictions on the annihilation of $\bar{p}$ on ^3He and ^3H nuclei (Kondratyuk L A and Shmatikov M 1983) (see fig. 1). This fact might be a consequence of the compact structure of ^4He in respect of the loose structure of ^2H, ^3H and ^3He, but an inadequacy of the theory must not be excluded.

^4He reaction channels

The cross sections for each reaction channel, defined by the number of charged prongs, are given in Tab. 2. The errors are statistical.

Tab. 2 - ^4He cross sections (mb) for various reaction channels, defined by the number of charged prongs

Number of charged prongs	Number of π^-	Energy (MeV) 48.7	Energy (MeV) 179.8
1	0	14.4 ± 2.5	16,4 ± 1.9
2	0,1	19.6 ± 3.4	9.8 ± 1.4
3	0,1	91.2 ± 7.5	72.9 ± 3.9
4	2	42.4 ± 4.0	20.0 ± 2.0
5	2	91.0 ± 7.5	90.6 ± 4.4
6	3	12.9 ± 2.8	5.5 ± 1.1
7	3	17.8 ± 3.3	17,6 ± 1.9
8	4	0,6 ± 0,6	0,4 ± 0.3
Total reaction cross section (mb)		286.6 ± 13.3	233.3 ± 7.0

The charged prongs multiplicity distributions are given in fig. 3.

Inelastic events with production of ^{3}He are the only ones with an even number of prongs and are easily recognized (Balestra F et al 1984a). Hence, the cross section for ^{3}He production is immediately obtained by adding the even prong values in Tab. 2: at 48.7 MeV $\sigma(^{3}He) = 75.6 \pm 6.8$ mb; at 179.8 MeV $\sigma(^{3}He) = 35.7 \pm 2.8$ mb. The increase of this cross section as the energy decreases has to be expected as ^{3}H and ^{3}He are the only nuclei that can be produced in a $\bar{p}-{}^{4}He$ annihilation in absence of initial or final state interactions. Hence, going towards the threshold for inelastic processes (about 25 MeV in the laboratory system) the production of these nuclei should be more and more favoured, the contribution from other channels being mainly due to the pionic final state interaction which varies slowly with energy.

The cross sections for the production of ^{3}He, ^{3}H and ^{2}H are of interest also in cosmological problems, particularly concerning the existence of antimatter in our universe. We discussed this problem in the light of our data in Balestra F et al 1984b.

From Tab. 2, we can calculate also the cross sections for the production of 0,2 and 3 negative pions, which may be compared with the corresponding quantities for $\bar{p}-{}^{2}H$ interaction.

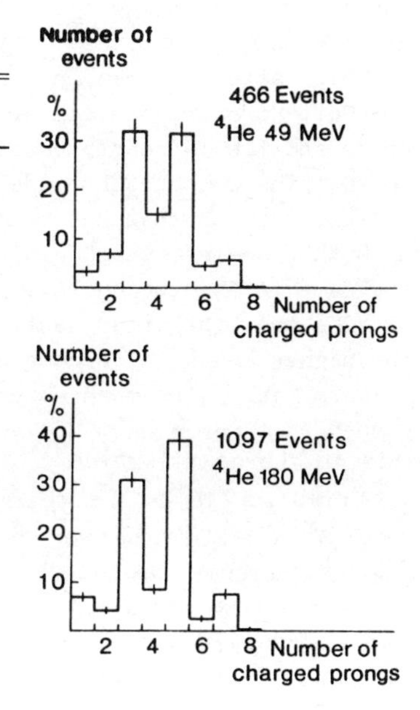

Fig. 3 - Charged prongs multiplicity of $\bar{p}-{}^{4}He$ inelastic events

Tab. 3 - Cross sections (mb) for the production of 0,2,3 negative pions, together with the relative branching ratios, for ^{4}He (present experiment) and ^{2}H (Bizzarri R et al 1974)

Energy (MeV) ($\bar{p}-{}^{4}He$ data)	Number of π^-	Cross section (mb)		Branching ratios (%)	
		^{4}He	^{2}H	^{4}He	^{2}H
48.7	0	14.4 ± 2.5	15.5 ± 1.5	5.9 ± 1.4	8.2 ± 0.9
	2	133.5 ± 9.1	137.5 ± 5.4	76.4 ± 5.1	75.8 ± 4.7
	3	30.7 ± 4.3	28.9 ± 2.4	17.6 ± 2.5	15.9 ± 1.5
179.8	0	16.4 ± 1.9	11.6 ± 0.7	10.9 ± 1.2	9.7 ± 0.6
	2	110.6 ± 4.9	86.8 ± 2.4	73.6 ± 3.2	72.0 ± 2.7
	3	23.2 ± 2.2	21.4 ± 1.2	15.4 ± 1.5	17.7 ± 1.1

In Tab. 3 our $\bar{p}-{}^4He$ data, taken at laboratory energies of 48.7 and 179.8 MeV are compared with the $\bar{p}-{}^2H$ data of Bizzarri R et al 1974 taken at laboratory energies of 57.4 and 170.5 MeV respectively.
The 48.7 and 57.4 MeV data correspond to the same c.m. energy (39 MeV), whereas the 179.8 and 170.5 MeV data correspond to c.m. energies of 144 and 114 MeV respectively.

This latter discrepancy is due to a gap in $\bar{p}-{}^2H$ data between 170.5 and 433 MeV. However, one can infer, from the existing data, that the $\bar{p}-{}^2H$ cross section is nearly constant above 170 MeV (Flaminio V et al 1979), so that the comparisons made in Tab. 3 are meaningfull for both energies.

Tab. 3 shows that the branching ratios are the same within the errors at both energies, which could indicate that the annihilation mechanism is the same in the two nuclei, indipendently of their structures. It is surprising that the cross sections, which differ on the whole about 25% at the higher energy, become equal at the lower one.

From Tab. 2 the mean number of charged prongs per event $\langle N_c \rangle$ and the mean number of π^- per event $\langle N_{\pi^-} \rangle$ are obtained. $\langle N_c \rangle$ is about 4: we will comment on this in the next section.

As far as $\langle N_{\pi^-} \rangle$ is concerned, we observe that at present we cannot distinguish π^- from $\bar{p}$, so that we cannot separate annihilation from inelastic scattering. If we assume that all the inelastic events having 2 or 3 prongs are annihilations, all they produce one π^-; if we assume that all are inelastic scattering, they do not produce π^-. Thus $\langle N_{\pi^-} \rangle$ is between 2.3 and 2.7 at both energies. The upper value is equal to that found in deuterium (Bizzari R et al 1974). Finally, the inelastic scattering cross section is at most 35 ÷ 38% of the total reaction cross section.

4. $\bar{p}$ - Ne charged prongs multiplicity

In fig. 4 the charged prongs multiplicities for the $\bar{p}$ - Ne interaction are given.

The mean number of charged prongs per event is about 6.6 at both energies. This value, together with that for ${}^4He$ (~4), is higher than those found in photografic emulsions (~7.7) (Ekspong A G et al 1961) and in carbon (~ 4) (Agnew L E et al 1960), the different atomic number being taken into account. They are higher also than the values obtained from an intranuclear cascade model calculation by Clover M R et al 1982 concerning C (~3.5) and U (~4.7) nuclei.

Since in the bubble chamber experiment of Agnew L E et al 1960 tracks shorter than 2 mm (corresponding to protons with energy smaller than 10 MeV) were neglected and in the intranuclear cascade calculation protons with energy less than 40 MeV were not considered, we think that we observed a higher relative multiplicity owing to the low density of the streamer chamber targets. We can see tracks shorter than 0.5 cm too, which, for instance, can be 0.5 protons or 2 MeV alpha particles.

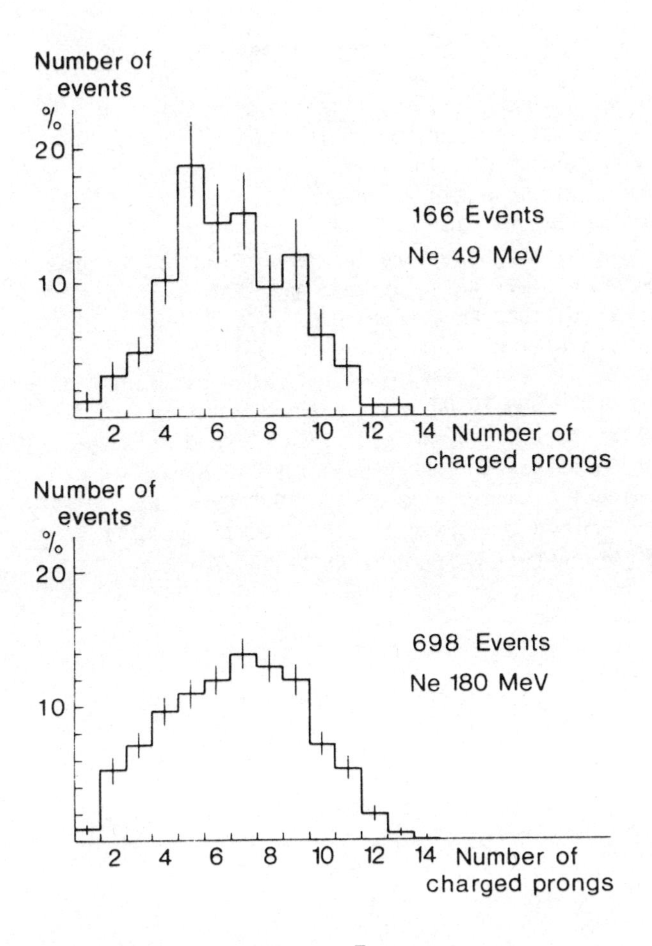

Fig. 4 - Charged prongs multiplicity of $\bar{p}$ - Ne inelastic events

5. Conclusions and perspectives

New informations on the $\bar{p}$-nucleus interaction have been obtained by an analysis based on very simple operations on inelastic reactions: counting of events and charged prongs and electric charge recognition.

The geometrical reconstruction of the events and the kinematical analysis, which are in progress, will allow us to obtain elastic and inelastic scattering cross sections, particle recognition, energy and angular distributions. Moreover, we will acquire new data on ^{4}He at 200 MeV/c, below the inelastic scattering threshold and on ^{3}H at different energies.

References

Agnew L E et al 1960 Phys. Rev. 118 1371

Akimov Yu K et al 1984 LNF 84/20 (P) and Nucl. Instr. and Meth., in press

Balestra F et al 1984a CERN-EP/84-68, submitted to Phys. Lett.

Balestra F et al 1984b Lettere del Nuovo Cimento, in press.

Bizzarri R et al 1974 Nuovo Cimento A22 225

Clover M R et al 1982 Phys. Rev. C26 2138

Ekspong A G et al 1961 Nucl. Phys. 22 353

Flaminio V et al 1979 Compilation of cross-section, III, p and $\bar{p}$ induced reactions, CERN-HERA 70-03

Garreta D et al 1984 Phys. Lett. 135B 226 and this Symposium

Grach I and Shmatikov M 1982 Report ITEP-12 Moscow

Kalogeropoulos T E and Tzanakos G S 1980 Phys. Rev. D 22 2585

Kondratyuk L A and Shmatikov M 1983 Sov. J. Nucl. Phys. 38 216

Nakamura K et al 1984 Phys. Rev. Lett. 52 731

Paper presented at VII Eur. Symp. Antiproton Interactions, Durham 1984

Measurement of the antineutron mass

M Cresti, L Peruzzo, G Sartori

Dipartimento di Fisica "Galileo Galilei" dell'Università, Padova, Italy
Sezione di Padova dell'Istituto Nazionale di Fisica Nucleare

We present here the preliminary results of a measurement of the antineutron mass, performed by analysing the kinematics of the charge exchange reaction:

$$\bar{p} + p \rightarrow \bar{n} + n \tag{1}$$

at low $\bar{p}$ momenta (270 MeV/c $\lesssim p \lesssim$ 570 MeV/c), in the CERN 2 metre Hydrogen Bubble Chamber. Momentum and direction of the incoming antiproton were measured directly and the directions of the two outgoing neutrals were determined by detecting their successive interactions in the liquid hydrogen.

To our knowledge, no measurement of the antineutron mass exists to date. In fact, the antineutron is generally "seen" in experiments either as a "missing neutral" coming from an antiproton interaction, or as a neutral particle annihilating with a nucleon. In the first case the lack of direct information about its momentum and energy does not easily allow a precise measurement of its mass. In the second, the total energy of the secondaries emitted in the annihilation usually exceeds 2 GeV and is seldom completely measurable. A measurement of the antineutron mass would then be affected, when possible at all, by very large errors.

The "expected" identity of the antineutron mass with that of the neutron cannot therefore be verified, from existing data, with an accuracy better than several MeV/c^2. On the other hand, the analysis of Reaction 1 allows a good determination of the antineutron mass, as the kinetic energy in its centre of mass is, at our $\bar{p}$ momenta, always smaller than 80 MeV in the initial state, and is even lower in the final state as the neutron is heavier than the proton (and if the antineutron is heavier than the antiproton). Therefore, even a small change in the antineutron mass would alter significantly the Q value of the reaction, producing measurable changes in the kinematics of the final state.

In the course of a measurement of low energy antiproton-proton cross sections[1] we collected, at Padova, more than 300,000 examples of antiproton interactions in liquid hydrogen, at momenta below 600 MeV/c. More than 20,000 of these were "0-prong", i.e. possible charge-exchange interactions. Selecting those that occurred in a fiducial volume within the bubble chamber, we were left with 14,000 candidates of Reaction 1, with potential paths in liquid hydrogen for the two neutral secondaries longer than $\sim$10 cm, i.e. large enough to give a significant probability to detect their successive interactions.

About three quarters of 0-prong interactions, at our momenta, are charge exchange reactions giving rise to antineutrons of a few hundred MeV/c. The probability for such antineutrons to annihilate with a proton before leaving the visible region of the chamber, is, for our sample, about one in four.

A large fraction of these annihilations are in 3 or 5 charged secondaries. About 2,000 3 or 5-prong "neutral stars" should then be visible in our sample. A careful scan of the region downstream of the point where the antiproton is seen to disappear, produced 1909 examples of such stars.

The neutron produced in Reaction 1 has roughly one probability in ten to scatter elastically off a proton leaving a visible recoil track. However, the bubble chamber is crossed during its sensitive time by a large number of neutral particles associated with the antiproton beam but not related to the identified charge exchange reaction. These particles can interact with the protons in the liquid hydrogen, leaving recoil tracks. For this reason we found for each 0 prong interaction and associated neutral star several proton tracks that could be, at first sight, associated with them.

On the other hand, since Reaction 1 is a two-body interaction, the momenta of the three particles (the antiproton, the antineutron and the neutron) should lie in a plane. Furthermore, as the four particles involved in Reaction 1 are of roughly equal masses, the laboratory angle Θ between the two outgoing neutrals should be about 90°.

To select good candidates for our purpose we therefore measured roughly on the scanning table the antiproton track, the vertex of the neutral star and the starting point of the "assumed" recoil proton track for all the events (4500) and plotted Θ versus the "coplanarity" C defined as:

$$C = \left| \frac{\vec{u}_{\bar{p}} \times \vec{u}_{\bar{n}} \cdot \vec{u}_n}{[1-(\vec{u}_{\bar{p}} \cdot \vec{u}_{\bar{n}})^2]^{1/2}} \right| \tag{2}$$

where the $\vec{u}$'s are the unit vectors in the direction of the three moving particles. C is the absolute value of the cosine of the angle between the line of flight of the neutron and the normal to the plane defined by the incoming antiproton and the outgoing antineutron. C should be 0 for coplanar events.

The two dimensional plot Θ vs C showed an accumulation of $\sim$300 events for $C \leqslant 0.3$ and for $75^\circ \leqslant \Theta \leqslant 92^\circ$, giving evidence that a two body reaction, with particles of masses in the final state roughly equal to those in the initial state, was indeed taking place. The events in this peak were then remeasured more accurately and we finally selected events with $C<0.3$ and with the neutral particles paths longer than 5 cm. The length cutoff was applied in order to increase the accuracy in the determination of angles.

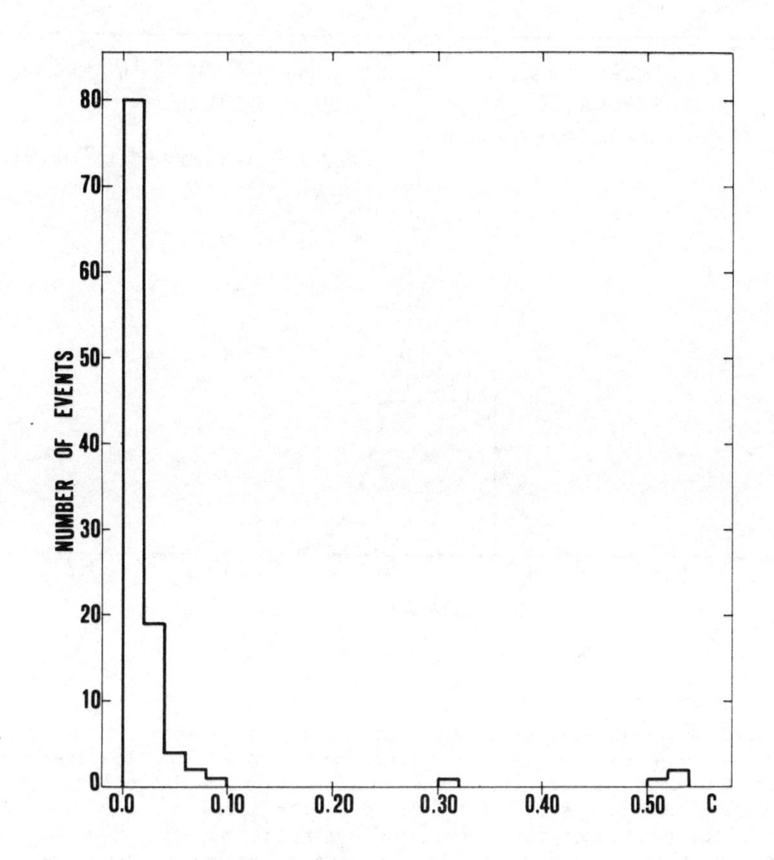

Fig. 1

The 115 events surviving these cuts were finally measured very carefully at CERN with the Erasme measuring projector. The resulting C distribution is shown in Fig. 1, where the two-body "signal" is clearly visible over a small background.

For the coplanar events we then computed the antineutron mass from the measured $\bar{p}$ momentum and direction and from the measured direction of the two outgoing neutrals. We computed also the neutron momentum and, from the measured direction of the proton track, the expected momentum of the recoil proton. To insure good accuracy in this prediction we imposed a cutoff of 0.8 cm on the length of the proton track. We then compared this predicted momentum with the momentum measured from the proton range, finding in each case compatibility within errors.

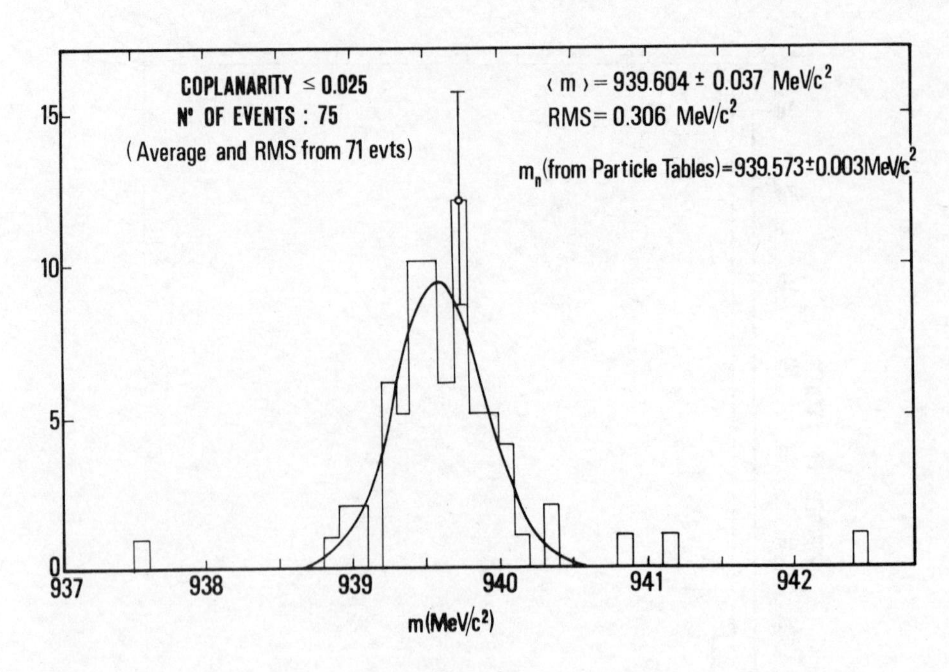

Fig. 2

Fig. 2 shows the distribution of $m_{\bar{n}}$ for the 75 events with C<0.025. Neglecting the 4 events in the tails, the r.m.s. of the distribution is ±0.306 MeV/c^2, the average value of the mass is 939.604 MeV/c^2 with a statistical uncertainty of ±0.037 MeV/c^2. The solid curve is a Gaussian, drawn with these parameters. From the figure it appears that the method used can give a determination of the antineutron mass with good precision.

To give a reliable measurement of the antineutron mass and of its uncertainty, one needs to estimate correctly the measurement errors, both to improve the selection of candidates and to give weighted averages and errors. A careful estimate of the size of the systematic errors is also necessary. We are at present working on both the above points.

We would like to thank the authors of Ref. 1 for the use of their film and data. We gratefully acknowledge the invaluable help of our scanning staff and of our programmers Mr. G. Pasquali and Dr. C. Pinori. The accuracy of the results, as shown in Fig. 2, depends strongly on the quality of the measurements performed with Erasme. For this we are in debt with the Erasme staff at CERN, and in particular with Dr. Peter Schmid, to whom go our thanks.

(1) Allen P et al 1980 Proc. 5th European Symposium on Nucleon-Antinucleon Interactions (Bressanone) p 175. See also M Cresti et al 1983 Phys. Lett. 132B 209.

Inst. Phys. Conf. Ser. No. 73: Section 4
Paper presented at VII Eur. Symp. Antiproton Interactions, Durham 1984

Do protons decay?

P J Litchfield

Rutherford Appleton Laboratory

Abstract. The experimental status of proton decay is reviewed after the Leipzig conference. A brief comparative description of the currently active experiments is given. From the overall samples of contained events it can be concluded that the experiments are working well and broadly agree with each other. The candidates for proton decay from each experiment are examined. Although several experiments report candidates at a higher rate than expected from background calculations, the validity of these calculations is still open to doubt.

1. Introduction

To stay in the spirit of this meeting I should be discussing anti-proton decay. However, despite the miracles that have been worked at CERN, the product of antiproton intensity and containment time is still some 20 orders of magnitude below that required to reach interesting levels. I am thus constrained to invoke CPT and instead consider proton decay. This written contribution follows generally the presentation given at the conference but in order to preserve topicality I have included results given at the Leipzig conference that took place a few days after the end of the meeting.

In section 2 I shall give a general, and brief, description of the currently active experiments and a comparison of their merits and demerits. Then in section 3 I shall discuss the general features of the data from all the experiments on the contained (i.e. mostly neutrino induced) events. These give a valuable indication of the quality of the data and enable a comparison to be made between the various experiments. In addition, of course, these events are the background to possible proton decay events and a good understanding of them is vital to the discrimination of a small signal hidden amongst them. In section 4 I will show and discuss the candidate events that the various groups have presented. Section 5 addresses the question of whether these are sufficient to establish nucleon instability and the crucial ingredient of the reliability or otherwise of the background calculations. Finally section 6 will summarise the present status and offer some pointers to the future.

2. Description of the Experiments

There are seven active nucleon decay experiments. Three are water cherenkov detectors, four are iron tracking calorimeters. The general properties of the experiments are by now well known so I will not attempt a detailed description of each but will just list their main strengths and weaknesses. Firstly, however, I will give a brief comparison of the two techniques.

2.1.1. Water Cherenkovs

The main impetus behind the development of big water cherenkov detectors for proton decay was the SU(5) prediction that the $e^+\pi^0$ mode would be dominant and within experimental reach. This results in a totally electromagnetic final state with large amounts of cherenkov light developed by the $\beta = 1$ electrons produced in the showers. For such modes the detector is an homogeneous, fully sensitive, total energy calorimeter with a resolution that depends essentially only on the light collection efficiency. The good directionality of the light (only between 10% and 20% being scattered out of the cherenkov cone) enables the back to back nature of this mode to be easily recognised and, in general, the presence of a production vertex to be determined. For other modes the good sensitivity to electron showers enables the low energy delayed electrons from muon decays to be detected with a high efficiency. Water is also cheap and contains free protons in case proton decays in a heavy nucleus should be either inhibited or the final state particles so strongly absorbed as to make the events unrecogniseable. On the other hand the relatively low density of water means that the ratio of fiducial mass to total mass is low for present day detectors.

For detecting charged mesons, however, water cherenkovs are less impressive. At one extreme kaons from proton decay are always below cherenkov threshold and are thus invisible. At the other, for muons, the light levels are much reduced below those from the same energy showers and there is a correction to be applied to the visible energy to allow for the presence of the cherenkov threshold. More importantly, since the vertex resolution in present detectors is no better than 50 cms, any scatter of a charged pion within this region (corresponding to almost a collision length) will go undetected. Scatters at greater distance will serve to confuse the already non-trivial pattern recognition in such a device.

Lastly, but a fact that may ultimately set the limit to the detection of nucleon decays in such a device, because the detector cannot be scaled down it is virtually impossible to expose them to an accelerator neutrino beam and thus observe known background events in the actual detector.

2.1.2. Iron Calorimeters

The main advantages of iron calorimeters are their common sensitivity to all charged particles, their much finer resolution and their ability to produce bubble-chamber-like pictures of events. Pion or kaon interactions occuring after 3 - 4 samples from the vertex (approximately 10% of a collision length) can be clearly identified both avoiding confusion and distinguishing them from muons. Their modularity makes it possible to expose them to a neutrino beam thus giving much greater confidence in their background rejection estimates. Since μ^- are captured rather than decaying as in water some measure of charge identification is available.

Their in-principle disadvantage is that they are not homogeneous and fully senitive but rely on sampling the tracks between iron sheets. This results in a total energy resolution and energy threshold for showers that is worse than in a water cherenkov, though the energy measurement of charged particles from range will be better. The resolution will also be dependent on the direction of the track relative to the plates. Nuclear effects in iron are likely to be more serious than in oxygen.

The major in-practice disadvantage is that instrumenting a large calorimeter is expensive in money and effort. This means that the presently operating calorimeters are small and/or crude devices. They either have thick iron plates (KGF, NUSEX) such that the number of samples on a proton decay track is low or by just registering hits, with no further information (NUSEX, FREJUS) the knowledge of track direction is lost. The time resolution of present devices is poor relative to the cherenkovs so that $\mu \to e$ identification is not good and so is the spacial uniformity with tracks that lie nearly parallel to the iron plates being poorly measured. However the next calorimeter to come into operation (SOUDAN 2) by having a more uniform geometry, finer sampling and measuring track ionisation will overcome most of these disadvantages and, while maintaining the intrinsic advantages of the calorimeter, will approach the capabilities of the water cherenkovs in everything other than the substantially lower threshold and total energy resolution for electromagnetic showers and the pure mass of the IMB detector.

2.2. Detectors

2.2.1. IMB

The first large detector to report results, this experiment has 3,000 tons of fiducial mass observed by 5" phototubes around the walls of the cubical detector located in the Morton salt mine (LOS84). The light collection is not as good as in the other water cherenkovs, only between 1% and 2% of the surface being covered with photocathode. The timing resolution of 1nsec. on each tube enables the time of arrival of the light at each phototube to be used in the pattern recognition. A coarser time scan up to 5μsec. after the trigger enables delayed μ decays to be identified with a 62% efficiency. Approximately 200 days of running have been analysed giving an exposure of 1800 ton years but about double that amount is now on tape.

2.2.2. KAMIOKA

This experiment (KOS84) has a 3,000 ton cylinder of water located in the Kamioka mine, Japan, yielding a fiducial mass of approximately 900 ton. Its major advantage is its photocathode coverage. The 20" phototubes cover 20% of the total surface giving an impressive threshold of 12 MeV for electromagnetic showers. The timing resolution is 150 nsec so that no pattern recognition timing is available but $\mu \to e$ decays can be recognised with a 70% efficiency. The much greater light collection compared with IMB (14:1) enables clear cherenkov rings to be observed (see figure 2 later in this report) which is not always the case with IMB. Also the structure of the rings enables electromagnetic showers to be distinguished with a high degree of probability from the cleaner and more uniform rings produced by mesons. Data from 274 days of running have been reported giving an exposure of 660 ton years.

2.2.3. HPW

This experiment (APR84) consists of a 900 ton cylinder of water surrounded by a veto shield of proportional wire tubes located in Park City, Utah. The phototubes are dispersed throughout the volume of the detector and the walls of the cylinder are reflective. This improves the light collection and total energy resolution considerably for a given size of phototube but makes pattern recognition extremely difficult as a given photon could be arriving after several reflections. However it makes a very good and efficient detector of delayed μ decays and the main thrust of the analysis has gone towards detecting events with more than one μ decay. The information on the production vertex is, at the moment, limited to a measurement of the anisotropy of the light emission. 360 days of data have been analysed giving an exposure of 500 ton years.

2.2.4. KGF

This was the first of the current round of experiments to report results and the first to make a claim to observe proton decays (NAR84). It is located in the Kolar gold field, India, and consists of a sandwich of 1.2 cm iron plates with large (10 x 10 cms^2) proportional counters. Each layer gives only one coordinate with the result that the number of hits in each projection is rather small, of the order of 4 to 5 hits on a typical proton decay track. However the tubes do measure pulse height so some attempt at the determination of track directionality and residual range can be made and the total energy measurement is better than if only hits are counted. No $\mu \to e$ determination is possible. The track definition and resolution is too coarse to give confidence in the interpretations that are made. The experiment has been running for 3 years giving an exposure of 180 ton years.

2.2.5. NUSEX

The NUSEX experiment is located underneath Mt. Blanc and is of approximately the same size as KGF with slightly thinner iron plates (1.0 cms) (FIO84). However its sensitive elements are 1 x 1 cm^2 resistive limited streamer tubes from which two coordinates are read out by means of the signals induced on orthogonal copper strips on either side of the tube layer. More than double the amount of spacial information is thus recorded for each track than for KGF but no pulse height information is obtained. The time resolution for a hit is 100 nsec and $\mu^+ \to e^+$ decays are detected with an efficiency of 35%. NUSEX is still rather small and coarse however the extra hits compared with KGF make the events rather easier to interpret. The unique feature of the NUSEX detector is that it has been exposed in a υ beam at CERN thus enabling a real comparison of the data obtained underground with the expected background. NUSEX has now run for over 2 years yielding an exposure of 220 ton years.

2.2.6. FREJUS

The FREJUS detector, in a road tunnel between France and Italy, is the first of the large iron calorimeters (JUL84). It is topologically of the same design as NUSEX but less coarse, with 0.3 cm thick plates and 0.5 x 0.5 cm^2 flash tubes as the active element. The flash tubes are triggered by Geiger tubes interleaved every eight layers. Each tube gives only one coordinate and no pulse height information. By delaying the HT pulse for the flash tubes, hits due to the electron from μ decays will be visible but they must be recognised by pattern recognition, not timing, and thus the efficiency is likely to be low. With this granularity the number of hits per track is good. The main problems with the detector are likely to be the lack of any information on the directionality of the tracks and the relatively coarse trigger which could produce problems in triggering on modes such as $p \to K^+\upsilon$. The experiment is currently being installed; 400 tons are now working and the full 900 tons should be on the air by the end of 1984. To date the integrated exposure time is 30 ton years.

2.2.7. SOUDAN 2

SOUDAN 2, which will be installed in the Soudan iron mine in Minnesota, is the most highly instrumented calorimeter so far proposed. It is a massive time projection chamber. Ionisation deposited in 1.4 cm diameter resistive plastic tubes embedded in a honeycomb matrix of 0.16 cm thick corregated steel plates is drifted up to 50 cms to crossed anode and cathode planes. The charge flowing in each anode and cathode is digitised every 150 nsec thus measuring the pulse height as well as defining the three dimensional coordinates of each tube crossing. The honeycomb geometry very much reduces the problem of the poor energy measurement that results in the other calorimeters when a track runs nearly parallel to the iron plates. The ionisation measurement will determine track direction and give some particle identification. The time structure of the readout will enable μ decays to be recognised. The triggering will be done over a much smaller volume than FREJUS enabling lower energy thresholds to be placed on all particles. The first production modules of the detector are being constructed and the first data should be available 2 years from now.

3 General features of the data

All of the experiments are observing contained interactions at a rate of around 100 events per kiloton year, due to υ produced in the hadronic cascades initiated by cosmic ray interactions in the upper atmosphere. The calculation of this rate in a given experiment is a complicated procedure. It depends on:

(a) the detailed energy spectrum and composition of the cosmic rays

(b) the high energy physics of their interactions

(c) the geomagnetic latitude of the detector since the particles are appreciably deflected by the earth's magnetic field.

(d) the detector geometry.

The best estimate to date has been made by Gaisser (GAI84). Table 1 shows the calculated and observed rates from the different experiment.

Table 1

Expt.	No of contained events at Leipzig	Contained events/ kton year(1) Observed	Predicted	Number of candidates	Exposure (ton-years)
NUSEX	22	153 ± 39	142	2	220
KGF	16	88 ± 20	110	4	180
FREJUS	3	-	-	0	30
IMB	169	120 ± 20	131	8	1800
KAMIOKA	90	187 ± 24	215	4	660
HPW(2)	5	-	-	3	500

(1) Numbers calculated by Gaisser at the time of the Park City proton decay meeting, corrected for detection efficiency and different energy thresholds.
(2) Events with ⩾ 2 muon decays.

The events can be subdivided into their topology for the NUSEX and KAMIOKA experiments. It should, of course, be remembered that the prong counts determined in the two experiments need not correspond due to the different thresholds for the various particle types.

Table 2

Track multiplicity		KAMIOKA Observed	KAMIOKA MC background	NUSEX Observed	NUSEX Expected
1		58	80.1	12	12.3
	electrons	17	30.3	3	-
	mesons	41	49.8	9	-
2		12	11.9	7	7.5
3		8	5.6	3	1.7
4		2	0.9	0	0.5
5		3	0.8	0	0

The KAMIOKA numbers now have an electron energy cut of 100 MeV. The expected numbers for NUSEX were calculated from table 3 of BAT83, describing the NUSEX υ test data, and normalised to 22 events.

It can be seen that the agreement between the predictions and experiments are reasonable though the normalisation of the KAMIOKA Monte Carlo appears slightly high and there might be a tendency to find more high multiplicity (⩾ 3 prong) events than expected. (13 found against 8.3 expected for KAMIOKA and 3 found against 2.2 expected for NUSEX). This would be particularly so for the KAMIOKA data if it was renormalised using the one prongs which are predominantly υ elastic scatters rather than attempting an absolute normalisation.

The IMB data has been used to search for ν oscillations by comparing the number of upward and downward going one prong events. Within the relatively poor statistics the numbers are in agreement with the predicted distributions assuming no ν oscillations.

To conclude, the experiments are observing atmospheric neutrinos at rather closely the expected rate and distribution. Apart from a statistically weak excess of high multiplicity events there is no obvious discrepancy in either the topology or total energy distributions. This sets a reasonable upper limit to the number of detectable proton decays of any type of between 10 and 20 per Kiloton year. Thus unless nucleons have large branching ratios into currently unobservable modes (e.g. multi neutrino modes) a lower limit of around 10^{31} years can be set for the total lifetime (ALL84). A further implication is that new experiments with masses below 1 Kiloton are not competitive.

4 Candidates for nucleon decay

In this section I will discuss the candidates for nucleon decay experiment by experiment.

4.1. HPW

HPW have so far limited themselves to searching for contained events with multiple muon decays. They have found 5 events with vertices located more than 5σ from the edge of the detector, 4 with two muon decays, 1 with three (CLI84). They judge two of the two muon events as likely to be ν events. Of the other two, one they estimate as unlikely to be a ν interaction because of its low value of anisotropy (0.18 events background at 90% confidence) and the second because it has a low visible energy (0.54 events background at 90% confidence). The three muon event has not yet had a quantitative estimate of the background made but is thought unlikely to be a ν interaction because of the low rate for $\nu N \to \mu\pi\pi X$.

4.2. IMB

Although IMB have more information on the production reaction than HPW, they are still unable to reconstruct cherenkov rings as well as Kamioka. Up to the present the selection of nucleon decay candidates has been based on the two dimensional plot of total visible energy (E_c), defined as the number of photons observed, and the anisotropy (A) defined as the vector sum of the unit vectors from the vertex to all hit phototubes. Figure 1a shows the plot for their data and 1c for a Monte Carlo sample of ν interactions (SUL84). In this plot a single track event should lie along a line with $A \approx \cos\Theta_c = 0.75$, when Θ_c is the Cherenkov angle, and back to back nucleon decay events should have low values of A. To search for candidates for any decay mode they generate Monte Carlo decay events and plot them on the E_c versus A plot (Figure 1b). They then select events in a region that maximises the number of selected Monte Carlo decays and minimises the neutrino background events. They have carried out this process for 37 different nuclear decay modes and find 8 events that are candidates for multi-track final states, most of them for more than one mode. However they find essentially the same number of events from their ν background sample and they conclude that they have no evidence for a signal over the ν background.

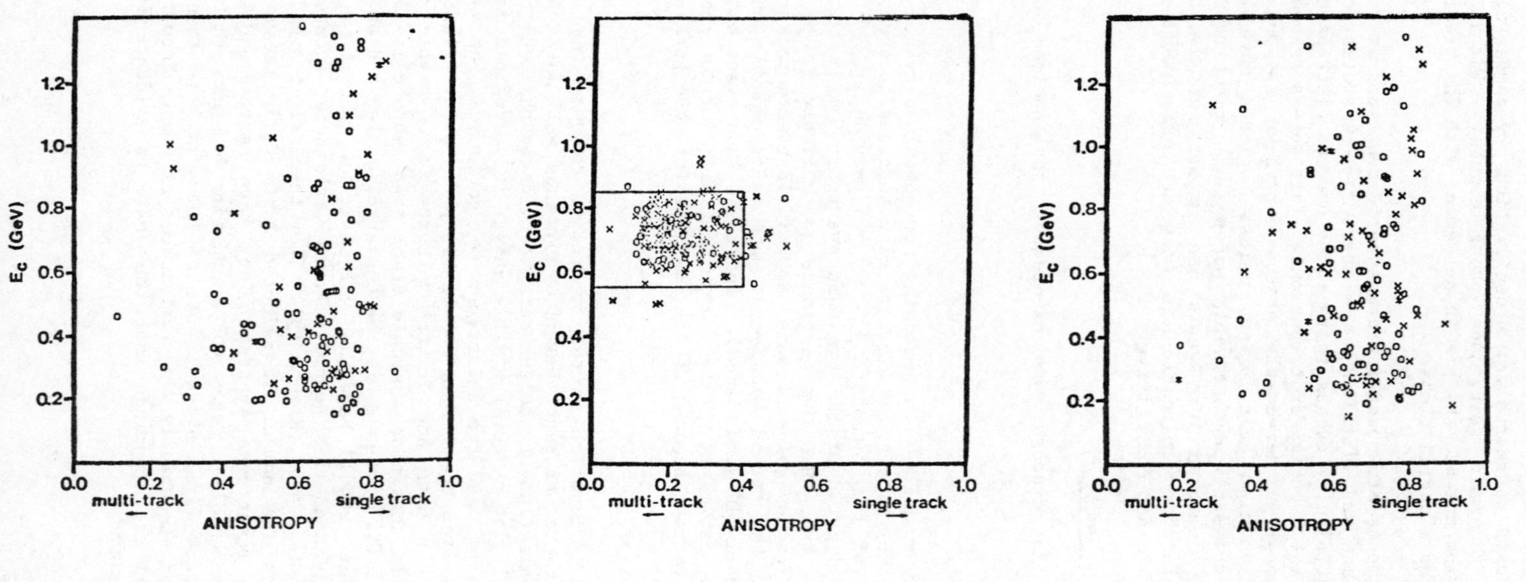

Fig. 1(a) E_C vs. A for the 169 contained events from 204 live days of data. Events with 0,1, and >2 identified $\mu\to e$ decays are indicated by 'o','x', and '*', respectively.

Fig. 1(b) E_C vs. A for a representative simulated nucleon decay mode, $p\to\mu^+\eta^o$; $\eta^o\to\gamma\gamma$. The square box cuts indicate the region in which candidates are accepted for this mode.

Fig. 1(c) E_C vs. A for a simulation of 204 days of atmospheric neutrino interactions. The number of background events for each mode was determined from a 6.5 year ν simulation.

4.3. KAMIOKA

The superior light collection of the KAMIOKA experiment enables the individual cherenkov rings to be observed and tracks reconstructed (KOS84). (Figure 2). In order to select candidates for nucleon decay they consider essentially the same set of final states as IMB. For each hypothesis they assign particle types to the various tracks, requiring consistency with showering (S) type or mesonic (M) type-rings, and calculate the total mass and residual momentum (Δp) of the event. They then require that the total mass be consistent with the nucleon mass, $\Delta p < 400$ MeV/c and submasses (e.g. the $\pi^+\pi^-$ mass in $p \to \mu K^0 \to \mu\pi^+\pi^-$) be consistent with the required values. From their 90 events they find 4 which are consistent with nucleon decay. Making the same analysis and selection on the υ background generated events they can calculate the numbers expected in the absence of nucleon decay. Figure 2 shows these events and Table 3 lists their parameters.

Table 3

Event	Mass (GeV/c^2)	Δp (GeV/c)	Final states	Number of background events Expected	90% conf
1	0.72	0.32	$\upsilon\rho^+ \to \upsilon\pi^+\pi^0$, $\mu^+\rho^- \to \mu^+(\pi^-)\pi^0$, $(\mu^+)\rho^- \to (\mu^+)\pi^-\pi^0$	0.3	< 0.76
2	0.72	0.21	$\upsilon\rho^0 \to \upsilon\pi^+\pi^-$, $(\mu^+)\rho^0 \to (\mu^+)\pi^+\pi^-$, $\mu^+K^0 \to \mu^+\pi(\pi)$	0.2	< 0.56
3	0.98	0.38	$\mu^+K^0 \to \mu^+\pi^0\pi^0$, $\mu^+\eta \to \mu^+\gamma\gamma$, $\mu^+e^+e^-$, $\mu^+\gamma\gamma$	0	< 0.33
4	0.88	0.25	$e^+\rho^-$, $e^+\omega^0$	0.3	< 0.76

Event 3 had no background events in 5 years simulated υ background. Particles in brackets are unobserved.

4.4. KGF

The KGF experiment claim 4 candidates with a background of less than 1 event. However the granularity of the detector is so poor that interpretation of the events and their separation from background is very difficult.

Figure 2: Displays of the KAMIOKA events

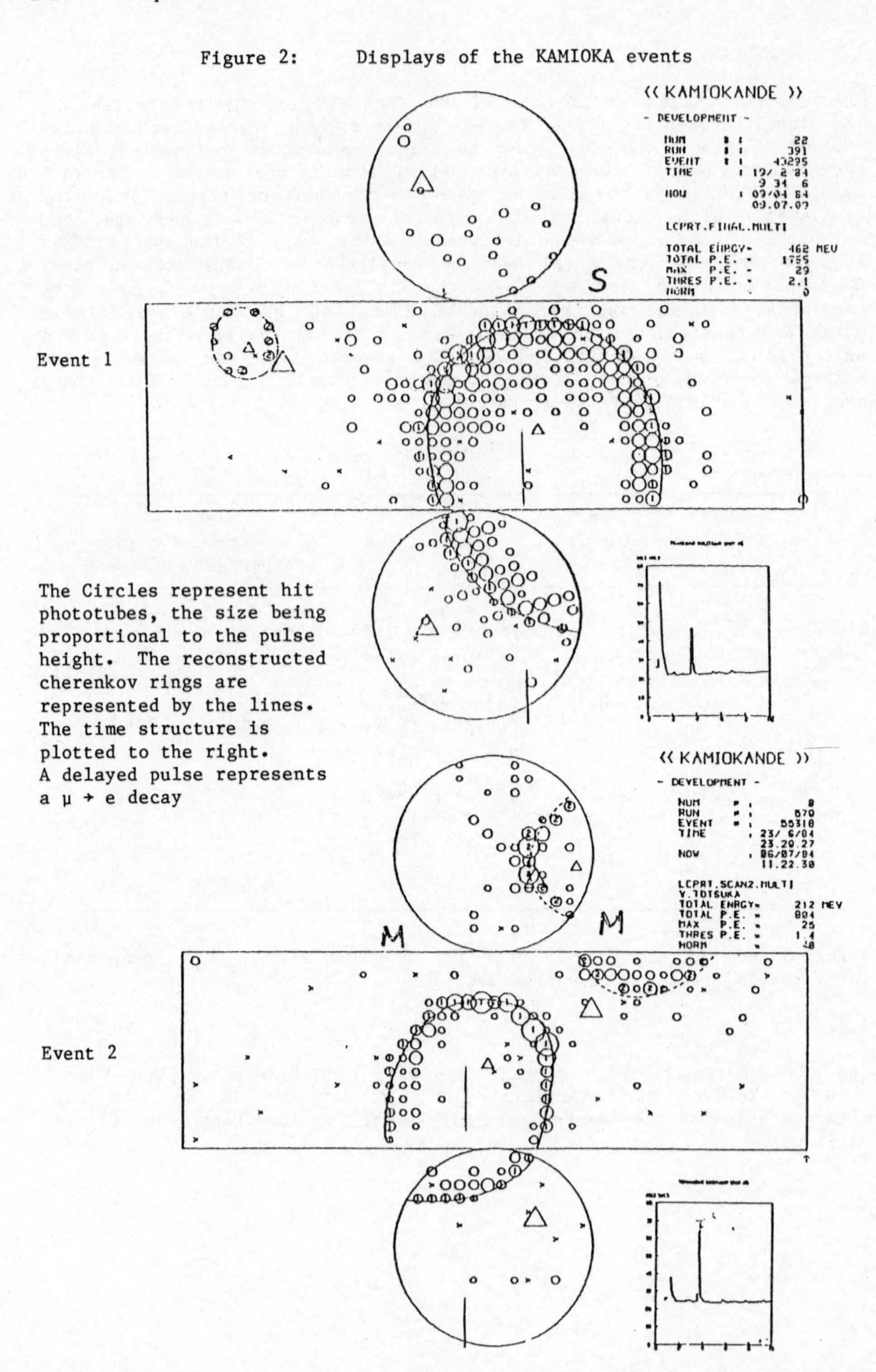

The Circles represent hit phototubes, the size being proportional to the pulse height. The reconstructed cherenkov rings are represented by the lines. The time structure is plotted to the right. A delayed pulse represents a $\mu \rightarrow e$ decay

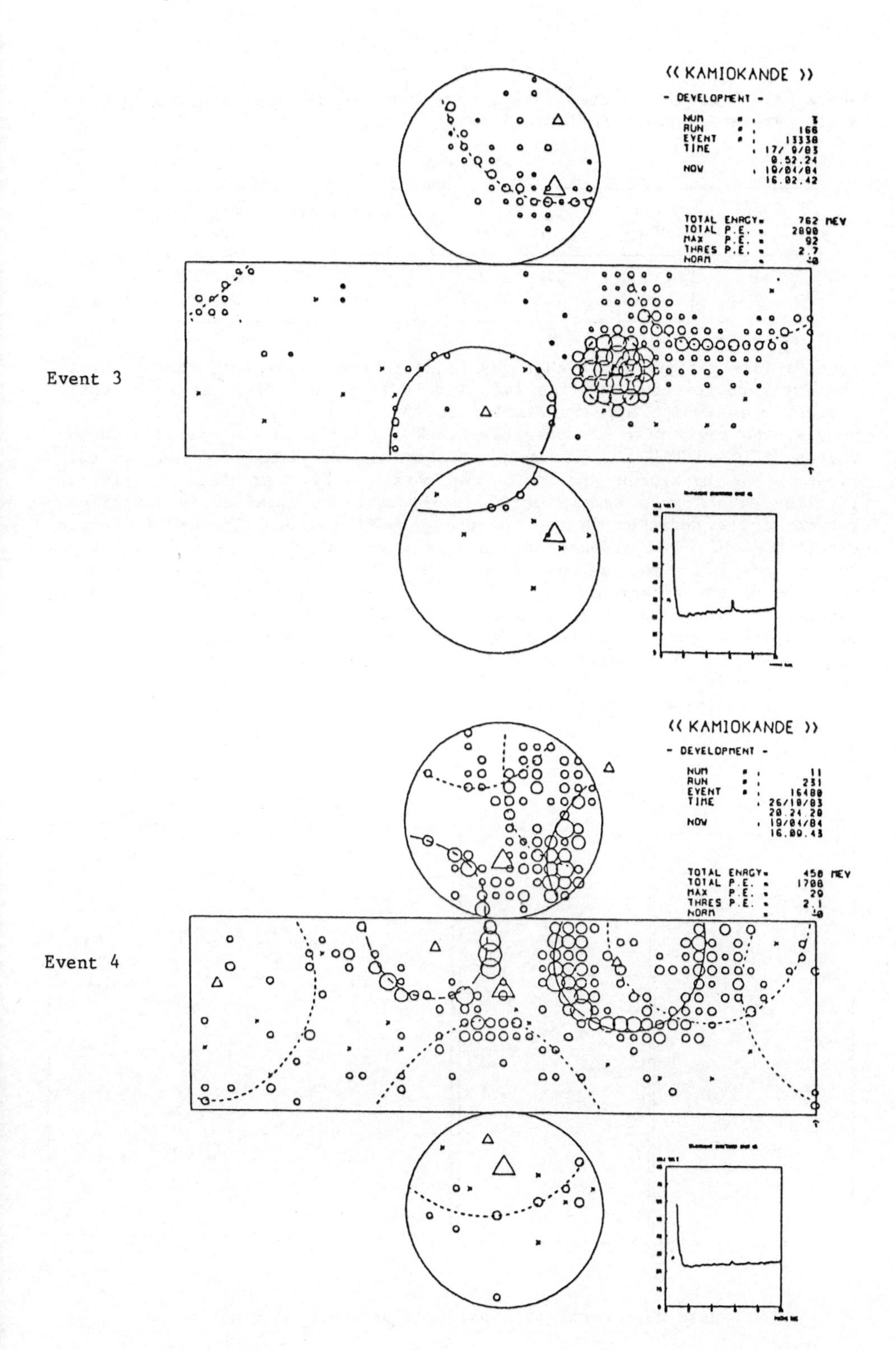
Event 3
((KAMIOKANDE))
- DEVELOPMENT -
NUM # : 3
RUN # : 166
EVENT # : 13338
TIME : 17/ 9/83
0.52.24
NOW : 10/04/84
16.02.42
TOTAL ENRGY= 762 MEV
TOTAL P.E. = 2800
MAX P.E. = 92
THRES P.E. = 2.7
NORM =
Event 4
((KAMIOKANDE))
- DEVELOPMENT -
NUM # : 11
RUN # : 231
EVENT # : 16480
TIME : 26/10/83
20.24.20
NOW : 19/04/84
16.00.43
TOTAL ENRGY= 458 MEV
TOTAL P.E. = 1708
MAX P.E. = 29
THRES P.E. = 2.1
NORM =

4.5. NUSEX

NUSEX have 2 candidate events (FIO84). They are shown in Figures 3a,b and their parameters given in Table 4.

Table 4

Event	Mass GeV/c^2	Δp GeV/c	Final state	Background (events)
1	1.0 ± 0.2	0.4 ± 0.2	$\mu^+K^0 \to \mu^+\pi^+\pi^-$	0.07 ± 0.05
2	1.1 ± 0.2	0.25 ± 0.15	$e^+\pi^0$	0.18 ± 0.06

Event 1 is the first event that was found in their detector, just prior to the Paris conference, and has been widely shown and discussed. Recently another event which appears similar to this event was found but it has a poor stereo angle with one possible track appearing to run parallel to the plates (FIO84, CUN84). New cuts to clean up the data have removed this event. The background estimate was obtained by searching for similar topology events in a sample of 400 υ interactions obtained by exposing a subset of the detector in the low energy CERN υ beam. Event 2 is purely electromagnetic. It appears not to be a single shower because of the kink in one view and the shower structure. They have made a preliminary estimate of the background by analysing 219 showers produced by 1.5 GeV/c electrons. They select showers that a) have a kink on at least one view or a three shower pattern or b) have ambiguous directionality. This yields a background estimate of 0.18 ± 0.06 events. They believe that this is a conservative estimate and that more detailed examinations of the shower structure will lower it.

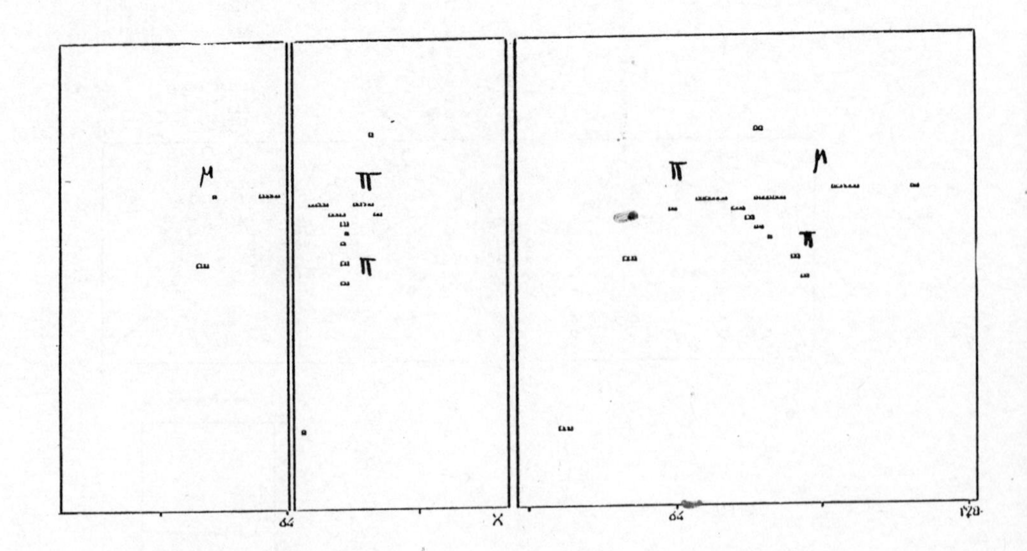

Figure 3a. Event 1, candidate for $p \to \mu^+K^0 \to \mu^+\pi^+\pi^-$

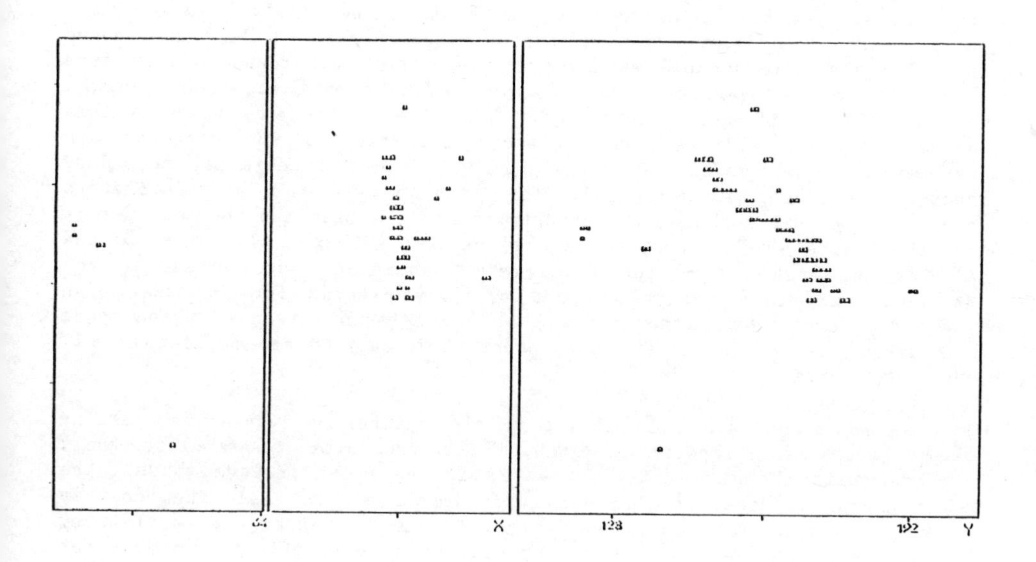

Figure 3b. Event 2, candidate for $p \rightarrow e^{+}\pi^{0}$

4.6. FREJUS

All of the three contained events so far obtained are consistent with υ interactions.

5 Do protons decay?

Four experiments claim to observe candidates for proton decay at a level higher than would be expected from the background due to υ interactions. One experiment finds candidates but ascribes them all to υ background. The limitations on the ability of the HPW and KGF experiments to reconstruct events means that one probably cannot definitively establish nucleon decay from these experiments alone. The KAMIOKA and NUSEX experiments have the greatest ability to study the events in detail and together they have 6 candidates where only 1.1 would be expected. The IMB experiment in which the background matches the signal uses a much cruder selection method. Can these apparently conflicting results be reconciled?

Table 1 lists the numbers of candidates in each experiment and also the exposure time. It can be seen that, apart possibly from KGF and HPW, the rate of production of candidates is broadly consistent between the experiments. In fact, as described in Section 3, in essentially all respects the experiments agree remarkably well amongst themselves apart possibly from their final conclusions. The crucial point is clearly the estimation of the background level. Is this being done correctly in all experiments?

There are two main routes for the background estimations. The first, and probably most reliable, has been followed so far only by NUSEX. This involves exposing the detector in a ν beam designed to have approximately the energy spectrum of atmospheric neutrinos (essentially a PS or AGS ν beam). When a candidate is found a simple search through the beam induced events gives a reasonable estimate of how often background of this type will occur. Exposures to π,μ and e beam enables the typical characteristics of such tracks to be determined. The advantages of this method is that all secondary effects specific to the detector are automatically accounted for (e.g. nuclear absorption effects, secondary scatters, detector geometry etc). The disadvantages are that calorimeters are not isotropic and thus the orientation of the beam and the detector is important (the NUSEX detector was only exposed with the plates at 90^0 and 45^0 to the beam), that the accelerator beam does not contain ν_e and therefore showering events are not properly catered for and that high statistics are required to ensure that all relevant areas of phase space are probed since the interesting events are only a few percent of all neutrino events.

The second method involves taking bubble chamber ν events and passing them through the experiment Monte Carlo generator to simulate their appearance in the real detector. This is the route followed by all the water cherenkov detectors because it is impossible to scale them down in size so as to be able to expose a test module in a beam and, so far, by KGF and FREJUS because they have not had the time or effort available for the exposure. The difficulties of this route are legion. Firstly there is the purely practical difficulty of obtaining complete records of the bubble chamber data and knowing exactly what the events represent. Then the effects of the bubble chamber liquids have to be unfolded and the effects of the detector material reproduced. Finally the problems of statistics and lack of ν_e events also apply here.

A further problem with the methods so far adopted by IMB and KAMIOKA is that the background estimate is hypothesis dependent. It is quite possible for an event to have no background under one hypothesis and appreciable background under another. Since there is no way of knowing the correct hypothesis the proper background estimate cannot be found. What is required is a means of estimating a background level just from the topology of the event, including energy measurements and any particle identifications available.

Despite all these difficulties there seems to be agreement, from both methods and different groups (including a contribution from an independent source DER84) that if individual track information is used together with overall total energy and residual momentum, then the neutrino background in many modes is still small at the current levels of sensitivity. If this is correct and the background estimates of NUSEX, KAMIOKA and HPW are reliable then it is difficult to avoid the conclusion that we may be on the verge of establishing nucleon instability. On the other hand, if this is correct then nearly all of the IMB candidates must be nucleon decays. What has happened to the ν events that they believe should occupy the appropriate regions of the energy versus an isotropy plot?

One final argument has been put forward against the interpretation of the NUSEX and KAMIOKA events as nucleon decays. This is that if they are real then the much more massive IMB experiment should have many similar events. That is only true if there is no limit on the total number of events expected. If the total proton lifetime was such as to give an average of one event per 200 ton years of exposure, consistent with the numbers in Table 1, and the proton had ten 10% branching fractions then in principle IMB would have one event in each final state and KAMIOKA or NUSEX one event in a few of the final states. Clearly the hypothesis that NUSEX had observed one $e^+\pi^0$ event would not necessarily imply that IMB should have observed nine.

6 Summary and the Future

It is difficult to appreciate that it was only two years ago that the first events from the current round of nucleon decay experiments were reported at the Paris conference, so great have been the advances since then. The overwhelming response must be to congratulate the experimental teams who have installed and made to work these very sophisticated detectors in inhospitable environments. Not only do the experiments work but they show an impressive measure of agreement, one with the other.

The one strong negative conclusion that has arisen so far is that the model which generated all the interest in proton decay, naive SU(5), is in deep trouble. It predicted a lifetime divided by branching ratio for the $e^+\pi^0$ mode of less than 10^{31} years to be compared with the best current lower limit (LOS84) of 1.5×10^{32} years.

There are no irrefutable nucleon decay events. The best probabilities so far are the $\mu\eta$ candidate of KAMIOKA and the μK^0 candidate of NUSEX. 21 events have been put forward worldwide as possibilities. The crucial question is are they background ν events or are they real? Almost the entire world sample of bubble chamber ν events in the relevant energy region is now in use for background estimation and it is not clear how this process can be significantly improved. It may well be that the limitation on the water cherenkov detectors will be their reliance on this data. If a good fraction of the present candidates are nucleon decays then they point towards many different decay modes with a tendency towards high multiplicity final states and a high ratio of decays producing strange quarks.

What of the future? More events are required from the present detectors. A single event in any channel is never going to be conclusive. The experiments are being upgraded. IMB is adding wavelength shifter plates in front of the present tubes and afterwards will double their number of tubes, improving their light collection by a factor of 4 and bringing them closer to KAMIOKA's sensitivity. KAMIOKA is adding fast timing and a veto shield to their present detector. They are also discussing a 22 kiloton water cherenkov. This could push the limits for easily detectable modes such as $e^+\pi^0$ to well over 10^{33} years but since the present problem in most modes is probably not lack of events but uncertainty of interpretation it is doubtful that such a detector will answer the most important questions. By the end of this year FREJUS will be fully installed and calorimeter events will be being produced at a rate comparable with the water cherenkov events. Apart from considerably finer sampling however they have no qualitative improvement in background rejection to NUSEX and they have not, at present, the advantage of a ν exposure. We shall have to wait another two years before the only experiment that offers significantly better background rejection, SOUDAN 2, comes on the air.

It now appears unlikely that the discovery of nucleon decay will be a short spectacular event like the discovery of the W and Z bosons. Rather it will take painstaking work over a number of years by the most sophisticated detectors we can devise and some measure of confimation using widely different techniques to obtain a concensus as to whether they do it or whether they don't!

References

ALL84 W W M Allison. Proceedings of the 1984 proton decay conference, Park City, Utah (to be published).

APR84 E Aprile-Giboni et al, Paper submitted to the Leipzig International conference, July 1984.

BAT84 G Battistoni et al, Nuclear Instruments and Methods 219, 300 (1984).

CLI84 D Cline, (HPW collaboration). Presentations to the Leipzig International Conference, July 1984.

CUN84 D Cundy, private communication.

DER84 M Derrick et al, Argonne preprint ANL-HEP-PR-84-23, May 1984.

FIO84 E Fiorini, (NUSEX collaboration) presentation to the Leipzig International Conference, July 1984 and G Battistoni et al, PL118B, 461 (1982).

GAI84 TK Gaisser. Proceedings of the 1984 proton decay conference, Park City, Utah (to be published) and T K Gaisser et al, PRL51, 223 (1983).

JUL84 S Jullian, (FREJUS collaboration) presentation to the Leipzig International Conference, July 1984.

KOS84 M Koshiba, (KAMIOKANDE collaboration) presentation to the Leipzig International Conference, July 1984.

LOS84 J M LoSecco (IMB collaboration), presentation to the Leipzig International Conference, July 1984.

NAR84 V S Narasimham (KGF collaboration), presentation to the Leipzig International Conference, July 1984.

Discussion on Section 4

R.L. Jaffe (MIT): In an asymmetric universe what is the expected flux of antiprotons from primordial sources?

Speaker, A.W. Wolfendale: Presumably, it is of the order of $10^{-18}/10^{-8}$, viz 10^{-10}, of the proton flux. These are simply the antiprotons which have escaped from the primordial annihilation.

W. Mitaroff (Vienna): How can one distinguish He from anti-He by remote observation?

Speaker, A.W. Wolfendale: To my knowledge, one cannot. Observation of cosmic ray nuclei (and antinuclei) is required. But it is not understood why, in a baryon symmetric cosmology, a distant domain of antimatter should also develop the very same magnitude of baryon excess and therefore the same abundance of primordial antihelium; thus, observation of large areas having unusually high, or low, He/H ratios could be remote observation of anti-He.

U. Gastaldi (CERN): Is there any experimental evidence that matter and antimatter attract gravitationally?

Speaker, A.W. Wolfendale: I'm afraid that I don't know of any.

S. Loucatos (Saclay): Why is the axion pressure lower than that of neutrinos?

Speaker, P.J.E. Peebles: I am told that that is because the primeval de Broglie wavelength of the axions is set by the horizon at axion formation, which is red-shifted to quite a large value.

R. Fong (Durham): I think you have been a little unfair about the lack of observational evidence for homogeneity. The number - magnitude counts for galaxies seem to scale to quite deep depths and the amplitude of the covariance function for galaxies scale to even greater depths, before we even begin to see any need to introduce luminosity evolution for galaxies. It would also be somewhat unnatural if the isotropy of microwave background were not a consequence of homogeneity, in that inhomogeneity should generally give rise to observable distortions in background.

Speaker, P.J.E. Peebles: I think we're in agreement: the evidence for homogeneity is quite strong. But still a skeptic could point out that the variation of galaxy counts with optical flux density is strikingly close to a power law but with the wrong index. In terms of apparent magnitude the observations give

$$\frac{dN}{dm} \sim 10^{0.43m} ,$$

while a homogeneous distribution in Euclidian space predicts $10^{0.6m}$. We can argue that this is "really all right" because the counts are boosted at the bright end by the galaxy concentration in the Local Supercluster, depressed at the faint end by the cosmological redshift which shifts the galaxy light into the infrared, but still this does weaken dN/dm as a test for homogeneity. I did not mention the isotropy of the microwave background because this radiation is not produced by seen objects so its use as a test for the distribution of these objects is more indirect than that of the X-ray background, say.

T. Bressani (Torino): What effect does the error on the incident $\bar{p}$ momentum have on the final value of the $\bar{n}$ mass?

Speaker, M. Cresti: The $\bar{p}$ momentum, in our experiment, is known with an uncertainty of ± 5 to ± 15 MeV/c, depending on the momentum. This leads to an uncertainty on the mass of ± 0.03 to 0.12 MeV/c^2 (depending on the kinematic situation) on a single measurement.

A.W. Wolfendale (Durham): Has the contribution to the background from neutrons (secondary to neutrinos) been considered?

Speaker, P.J. Litchfield: The backgrounds for IMB and Kamioka are calculated using the ν events in heavy liquid bubble chambers. Therefore, in general, secondary neutron interactions should be seen and such events should be present in the background samples. However, it must be admitted that neutral current events with no charged particles at the production vertex and with a subsequent neutron interactions would be difficult to recognise and there would be some deficiency in the background estimates in this case. Of course, the events observed in the ν exposure of the NUSEX calorimeter will contain these events and the background should be correctly estimated.

J.D. Davies (Birmingham): Can you say anything about $\bar{n}n$ oscillations?

Speaker, P.J. Litchfield: All the proton decay experiments have looked for $\bar{n}n$ oscillation candidates where 2 GeV of energy is released and as far as I am aware no good candidates exist. Water Cherenkovs are not very good at looking for high multiplicity events containing lots of charged pions, so definitive results on $\bar{n}n$ oscillations will probably have to await the large iron calorimeters.

T. Kalogeropoulos (Syracuse): Don't all proton decay experiments need a few antiprotons to calibrate them?

Speaker, P.J. Litchfield: Iron calorimeters can be calibrated in test beams of e, μ, π, p and ν. Water Cherenkovs cannot be calibrated in the same way because they cannot be scaled down. Any source of known particles underground would be a god-send to them!

Inst. Phys. Conf. Ser. No. 73: Section 5
Paper presented at VII Eur. Symp. Antiproton Interactions, Durham 1984

Current issues in meson spectroscopy

G. Karl

Dept. of Physics, Univ. of Guelph, Guelph, Ontario, N1G 2W1, Canada
Dept. of Theoretical Physics, Oxford University, Oxford, England

1. Introduction

I am indebted to Mike Pennington and the organizing committee of this Conference for the opportunity to review with you (my prejudices in) meson spectroscopy. This is an interesting time for our subject and I shall first review very briefly what previous issues in meson spectroscopy have been solved so that we may get a good vantage point. Then we shall discuss a number of current issues and single out what I think the main issue is now: namely gluonic degrees of freedom in mesons. I shall end up discussing the issue of energy dependence of OZI forbidden processes, which is connected with some gluonium candidates.

2. History

Meson spectroscopy starts with Yukawa's proposal that nuclear forces are mediated by a field - the pion. For the first ten years after his proposal the problem was to find the pion, and for a while there was some confusion between the muon and the pion. What do we learn? That good ideas are confirmed by experiment, but also that on occasion we may confuse one observation with the wrong theory for that observation. It can also happen today, and it may be relevant I believe to all the confusion with gluonia. Later on, when accelerators became available many mesons were found and it became clear that mesons have large families with the pion being the lightest one. The way to understand such large families is to see that they are all composites of some 'elementary' constituents. And so Gell-Mann and Zweig[1)] invented quarks: the constituents of hadrons. The quark model only became widely accepted after the discovery of the first family of heavy mesons J/ψ and ψ' when it became very clear that the spectrum of this system is very similar to that of positronium. So instead of electrons and positrons we have quarks and antiquarks. And instead of the photon we have gluons. So the theory underlying all of hadron spectroscopy was guessed by Fritzsch, Gell-Mann and Leutwyler[2)]: QCD. The generalization of QED involved appears to be slight - instead of electric charge we have SU_3^c charges. But the change which results is substantial - the gluons carry color and this makes the theory unusual: vacuum polarization has the wrong sign, asymptotic freedom and confinement come about, and many other things which we would like to know but don't yet know. So this is our present status: meson spectroscopy has an underlying theory but it's not yet fully connected to it. Perhaps an analogy is to say that in principle superconductivity is the consequence of QED but nevertheless it took a long time after the discovery of quantum mechanics to understand the logic of superconductivity. Eventually B, C and S succeeded in understanding the connection, but it took a while. In the meantime people had to use approximate models to just have a very rough idea of what was going on in superconductors.

Actually, many people[3)] believe that there is a detailed connection between BCS and meson spectroscopy, that instead of superconductivity where there is perfect diamagnetism, the confinement of quarks is a dual phenomenon with perfect (color) dielectrics being responsible for the electric field being expelled from the vacuum. Meson models, which explain how massless quarks give rise to massless pseudoscalars but massive vector mesons are just being invented by people at Orsay[4)] and Princeton.[5)] Although I don't understand most details, I believe that this work is important since it is constructing a bridge between the fundamental Lagrangian of the QCD theory and the data of meson spectroscopy.

To summarize: We probably know the Lagrangian underlying hadron spectroscopy but we can't use it since we don't understand the structure and properties of the vacuum. We probably still need some clues from experiment, and to understand experiment we need to use as crutches very simple mechanical models - strings, bags, potentials.

3. Conventional Mesons

These are the mesons composed of a quark and an antiquark. The best cases are heavy quarkonia J/ψ and the $c\bar{c}$ family, Υ and the $b\bar{b}$ family and also the $t\bar{t}$ family. Here nonrelativistic models work at their best[6)] and we have a chance to disentangle the spin dependence of the forces, and the spatial dependence of the spin independent part. The best surprise is no surprise: the spin independent confining forces appear universal, ie flavor independent just as one would guess if the degree of freedom responsible for these forces (color) is independent of flavor, the same for all quarks. This is a very good thing for QCD, even though the force itself is hard to compute from QCD. The details of the spin dependent forces are being analysed from the hyperfine structure of heavy quarkonia.[7)]

The nice thing is that all of this can be applied even to light mesons if one is sufficiently daring[8)], and it works reasonably well. Although many experts express surprise and disapproval, it should not be surprising that the quark model works for the hadrons for which it was invented, namely light mesons and baryons. Of course other models, like bags, strings, flux tubes also work reasonably well. There are many examples one could discuss of detailed calculations of mesonic spectra. At the moment the spectrum of mesons in the mass range of 1-2.5 GeV is an important subject of debate.

I shall only discuss one example, the state ξ (2.22 GeV) observed at SLAC in the radiative decay $J/\psi \to \gamma K\bar{K}$, which is very narrow, the observed width being less than about 50 MeV. A conventional explanation for this state was proposed by Godfrey, Kokoski and Isgur[8)] who suggest that it is an $s\bar{s}$ $^3F_2$ state. The mass fits very well a detailed dynamical model for mesons by Godfrey and Isgur while the observed width is compatible with simple estimates. These authors predict that the ξ resonance should also be observed in the $K^*\bar{K}$ and $\eta\eta'$ channels. More experiments are needed to confirm or refute this proposal but the general point that one expects a great many radial and orbital excitations of lower lying mesons is quite clear and one should devote work to identify these states. This work is very important since we cannot identify any exotic states unless we know first where the ordinary radial and orbital excitations are to be found.

4. Gluonia

While the photon does not carry electric charge, its QCD analogues the gluons do carry color charge. So it has been proposed right away that gluons form bound states by themselves - gluonic mesons[9)], which are now called gluoniums. There are two kinds of models in which these states have been described, perturbative ones and nonperturbative ones. In the perturbative models one talks about 2 gluon bound states, 3 gluon bound states, etc., while in nonperturbative models the number of gluons is so large that one is talking instead of strings without quarks at their ends - so called string loops. It is interesting that the spectrum of states is rather similar in both kinds of models, although, in my opinion the exact mass scale of these states is very uncertain. The uncertainty in masses is much greater than one would infer from published results which tend to be biased by attempts at identification with observed resonances. The only general statement that must be true is that these states should have masses of the same order as ordinary hadrons. In making attempts at identification we have to keep in mind that gluonia will mix with ordinary mesons of the same J^{PC} and other quantum numbers - ie isoscalars and SU_3 singlets. Depending on the extent of mixing the identification may prove to be difficult, and in fact there is little consensus achieved at present. An additional difficulty comes from the lack of reliable information on the way gluonia are supposed to decay. This is rather model dependent, with perturbative and nonperturbative models giving different answers. I would expect that in perturbative models the less mixing there is the narrower the states will be since mixing and decay have a common origin. The situation is not clear in nonperturbative models, which may nevertheless be a better representation of these states. The only reliable diagnostic tool is the coupling of these states to the electromagnetic field which by definition can only proceed through mixing with quark-antiquark pairs. However data on the radiative decay of gluonium candidates is very hard to obtain since the relevant branching ratios are very small. So at the moment the situation is still rather unclear.

There are two important classes of candidates for gluonium. In both classes one is considering OZI forbidden decays in which the mass of the final state can vary over a certain range, and one is looking for resonances which must be formed through an intermediate gluonic state. One class of states is observed in radiative J/ψ decays, where mesons are produced through a two gluon intermediate state. The observed "new" states are $i(0^-)$ at 1470 MeV in $K\bar{K}\pi$, $\theta(2^+)$ at 1700 MeV in $\eta\eta$ and $K\bar{K}$ and ξ (probably 2^+) at 2220 MeV in $K\bar{K}$. As discussed above the ξ may have a mundane interpretation[8)] as a quark antiquark state $s\bar{s}$ $^3F_2$. There is some recent data on the radiative decay $i \to \gamma\rho$ which appears to have a large rate, of the order of 1 MeV. If this is true then the i must have a large $(u\bar{u} + d\bar{d})$ component. There is no information available as yet on the radiative decay of the $\theta(1700)$, but a particular difficulty for its interpretation as a gluonium state is the small branching ratio for its decay into the $\pi\pi$ channel. Naive considerations would lead to this branching ratio being of the same order as $K\bar{K}$ for a gluonium candidate, on the basis of flavor independence. A preferred coupling to $K\bar{K}$ and $\eta\eta$

would suggest for a 2^{++} state an important $s\bar{s}$ component. Thus the radiative decay $\theta \to \gamma\phi$ would be an important diagnostic to ascertain the composition of this state. Even from this brief discussion it should be clear that we urgently need good data on the radially and orbitally excited quarkonia in the mass range 1-2.5 GeV before we can make any conclusive interpretation of the gluonium candidates. Also we need more information on the mechanism of decay of ordinary quarkonia before we can interpret data on the new gluonium candidates; for example, my conclusions on the θ(1700) mentioned above are strongly dependent on the assumption that the decay takes place entirely through the $q\bar{q}$ component. This assumption need not be true.[10)]

The other gluonium candidates come from the OZI forbidden reaction $\pi^- P \to \phi\phi N$, observed at BNL.[11)]. Here there are two broad enhancements in the s-wave $\phi\phi(J^P = 2^+)$ and one in the d-wave $\phi\phi$ $(J^P = 2^+)$. Since the $\phi\phi$ final state is composed entirely of strange quarks, and there are no strange (valence) quarks in the initial state, the process must proceed through a multigluon intermediate state. Therefore it is natural to interpret resonances in the $\phi\phi$ as reflecting either resonances of the gluonic medium[11)] or resonances of an $s\bar{s}$ intermediate state.[12)] Here we will describe a possible interpretation[13)] in which the observed mass bumps are not even resonances but a reflection of quantum oscillations in the rate. This phenomenon[14)] can be a consequence of sequential quark pair creation. The main physical assumption is that quark-antiquark pairs are created in sequence when the final state is formed:

$$u\bar{u} \to gg \to s_1\bar{s}_1 \to s_1(\bar{s}_2 s_2)\bar{s}_1 \to \phi\phi$$

The rate of production of the final state $\phi\phi$ is a strongly oscillating function of the mass system. The rate of conversion of the intermediate $s_1\bar{s}_1$ pair into the final state involves the overlap integral between the initial and final state

$$\Gamma(s_1\bar{s}_1 \to \phi\phi) \simeq k_f \left| \int_0^\infty \psi_{in}(E)\, r^3 dr \left[\int_0^1 d\sigma\, \phi_1(\sigma r)\phi_2((1-\sigma)r) \right] j_f(k_f r/2) \right|^2$$

where ϕ_1, ϕ_2 are the radial wavefunctions of the two ϕ mesons, j_f is a spherical Bessel function appropriate to the f partial wave of the $\phi\phi$ system. The incident wavefunction for the initial $s_1\bar{s}_1$ pair is taken to be the semiclassical form

$$\psi_{in}(E) = \frac{1}{r} e^{i(k_o r - Kr^2/4)}$$

for an s-wave pair and similar forms for other partial waves. Here k_o is the initial energy of an s quark, K - the string tension and k_f the momentum of the ϕ in the final state. The energy dependence of Γ comes through K_o and k_f and leads to very broad oscillations. The minima correspond to zeros of $j_f(k_f r_t/2)$ where the r_t is the value of r at the classical turning point for the first $s\bar{s}$ pair. In the stationary phase approximation the second pair is always created when the first pair is at its classical turning point in the string potential between the quark s_1 and antiquark $\bar{s}_1$. The broad mass bumps so generated show a good resemblance to the experimental data. Even if the mechanism described above is physical, there is no prohibition for real resonances to occur at the masses favored by this mechanism. However it is unlikely a priori for several resonances to correspond to several mass bumps generated by the overlap integral. The crucial test for a genuine resonance is that it couples to several decay channels. For example if the $\phi\phi$ mass bumps are real resonances they should also be seen in channels like $K\bar{K}$, $K^*\bar{K}$ etc. Of course, those channels are not OZI forbidden with π^-p as the entrance channel, however they could show up in radiative J/ψ decay: $J/\psi \to \gamma K\bar{K}$ where they are OZI forbidden. There is, as yet no sign of peaks in this process which correspond to the $\phi\phi$ peaks; only the $\theta(1700)$ and $\xi(2220)$ are observed in $J/\psi \to \gamma K\bar{K}$. The process $J/\psi \to \gamma\phi\phi$ has not been observed either (except in the region of η_c), but it is rather probable that when it will be observed the $\phi\phi$ pair will be in $J^P = 0^-$, just like the analogous $J/\psi \to \gamma\rho\rho$, $\gamma\omega\omega$. This corresponds to p-wave in the $\rho\rho$, $\omega\omega$, and $\phi\phi$ channels and if the mechanism of sequential quark pair creation is responsible for the production of bumps these will be at higher masses than in the s-wave states, since the zeros of j_1 are higher than those of j_o.

An interesting question concerns the predominance of a particular J^P channel in a given reaction. This comes about through an interplay between the mechanism of production of the initial and final state. For example in the process $u\bar{u} \to gg \to s\bar{s} \to \phi\phi$ near the $\phi\phi$ threshold it is natural that s waves dominate in the $\phi\phi$ channel. This constrains the final J^P to 0^+_- or 2^+. On the other hand at these energies ($E_{CM} \sim 2$ GeV) the process $u\bar{u} \to gg$ is predominantly through the $M = \pm 1$ component of the initial state, which in the gg channel has $J \geq 2$. So the state which is produced is predominantly 2^+. In the radiative J/ψ decay the initial quarks are charmed which are much heavier, and the predominant channel in $c\bar{c} \to gg$ is M=0 which in fact prefers the 0^- final state, if possible, as in $J/\psi \to \gamma\rho\rho$. It is because of this that we think that $J/\psi \to \gamma\phi\phi$ will also be dominated by the 0^- channel.

5. Conclusions

The issue of gluonic degrees of freedom in meson spectroscopy is the dominant issue of the subject. As we have tried to outline it ties together several aspects of meson physics - production, decay, ordinary mesons, fake mesons. We have not stressed that theoretical progress is limited also by our lack of understanding of this important degree of freedom. Meson spectroscopy had a glorious past and is assured of an excellent future. There is no doubt also that $p\bar{p}$ machines have an important role to play in the development of this subject.

Acknowledgements

I have enjoyed very much the actual conference and learned a great deal. I am indebted to the organizers for their hospitality. I have also enjoyed useful conversations on the topic of sequential pair creation with N. Isgur, R. Jaffe, J. Paton, and my collaborators W. Roberts and N. Zagury. The hospitality of the Department of Theoretical Physics, Oxford was also much appreciated.

References

1. Gell-Mann, M 1964 Phys. Letters 8 214, Zweig, G 1964 CERN preprint TH412
2. Fritzsch, H, Gell-Mann, M and Leutwyler, H 1973 Phys. Letters 47B 365
3. ´t Hooft, G 1982 Physica Scripta 25 133

4. Le Yaouanc, A, Oliver, L, Pène, O and Raynal, J C 1984 Phys. Rev. D29 1233
5. Adler, S L and Davis, A C Chiral Symmetry Breaking in Coulomb Gauge QCD, preprint
6. For a review, see Quigg, C and Rosner, J L 1979 Phys. Reports 56C 167
7. For a review, see Rosner, J L 1983 Experimental Meson Spectroscopy p. 461
8. Godfrey, S and Isgur, N 1984 Toronto preprint
Frank, M and O'Donnell, P J (1984) Phys. Rev. 29D 921
Frank, M and O'Donnel, P J (1983) Phys. Lett. 133B 253
Godfrey, S, Kokoski, R and Isgur, N 1984 Phys. Lett. 144B 439
9. Fritzsch, H and Gell-Mann, M 1972 Proceedings of the XVIth International Conference on High Energy Physics, Batavia, vol 2 p 135; Freund, P and Nambu, Y 1975 Phys. Rev. Lett. 34 1645
10. See e.g. Schnitzer, H J 1982 Nucl. Phys. B207 131 for a different point of view
11. Lindenbaum, S J 1981 Nuovo Cimento 65A 222; Etkin et al 1982 Phys. Rev. Lett 49 1620
12. Ono, S, Pène, O and Schöberl, F May 1984 Tokyo preprint UT-431, interpret one of the states as a $^3F_2$ resonance of the $s\bar{s}$ system
13. Karl, G, Roberts, W and Zagury, N 1984 Oxford preprint submitted to Physics Letters B
14. Karl, G and Roberts, W 1984 Oxford preprint to appear in Physics Letters B

$\bar{p}p$ annihilation at rest from atomic p-states

ASTERIX Collaboration

CERN[1], Mainz[2], München[3], Orsay (LAL)[4], TRIUMF-Vancouver-Victoria[5], Zürich[6]

S. Ahmad[4], C. Amsler[6], R. Armenteros[1], E.G. Auld[5], D. Axen[5], D. Bailey[1], S. Barlag[1], G. Beer[5], J.C. Bizot[4], M. Caria[6], M. Comyn[5], W. Dahme[3], B. Delcourt[4], M. Doser[6], K.D. Duch[2], K. Erdmann[5], F. Feld[3], U. Gastaldi[1], M. Heel[2], B. Howard[5], R. Howard[5], J. Jeanjean[4], H. Kalinowsky[2], F. Kayser[2], E. Klempt[2], R. Landua[1], G. Marshall[5], H. Nguyen[4], N. Prevot[4], L. Robertson[5], C. Sabev, U. Schaefer[3], R. Schneider[2], O. Schreiber[2], U. Straumann[2], P. Truöl[6], B.L. White[5], W.R. Wodrich[3], M. Ziegler[2].

(Presented by C. Amsler)

Abstract. First results in $\bar{p}p$ annihilation in hydrogen gas from atomic p states are presented. Annihilation into $\rho\pi$ occurs mainly from the isospin I = 0 state of the $\bar{p}p$ system. The rate for $\bar{p}p \to K^+K^-$ relative to $\bar{p}p \to \pi^+\pi^-$ is 0.15 ± 0.03. No narrow state associated with monoenergetic pion emission is seen. The 5σ upper limit is 2 x 10^{-3}.

1. Introduction

Antiproton-proton annihilation at rest has been previously studied in liquid hydrogen. Antiprotons stopping in hydrogen are captured in high n orbitals of the $\bar{p}p$ atom. In liquid hydrogen collisions with neighbouring H_2 atoms induce Stark mixing of levels with different angular momenta. The admixture of L = 0 states leads to annihilation from a high atomic s level,for which the centrifugal barrier vanishes. In hydrogen gas, on the other hand, collisional effects are less important compared to radiative transitions. A large contribution from p (L = 1) orbitals is therefore expected (Borie and Leon, 1980). Also, the lower density enhances the yield of radiative transitions to the 2p level (L X-rays) which was measured to be 13 ± 3 % in hydrogen gas at NTP (parallel paper on X-rays presented at this conference by R. Landua). From the 2p state the $\bar{p}p$ system annihilates with a probability of more than 96 % thus suppressing transitions to the 1s state. Annihilation from p orbitals is expected to enhance the production of meson resonances with higher spins. In particular $\bar{p}p$ bound states with higher angular momenta are more likely to be narrow and are therefore more likely to be observed. This is due to the centrifugal barrier which reduces the $\bar{p}p$ overlap and inhibits annihilation. Thus far no data exist for annihilation at rest in hydrogen gas.

In this paper we concentrate on the final states and show results for the exclusive channels $\bar{p}p \rightarrow \pi^+\pi^-\pi^0$, $\pi^+\pi^-$, K^+K^-, and for multiprong final states $\bar{p}p \rightarrow$ 4 prongs + missing mass (MM). A by-product of the experiment is a measurement of the inclusive pion spectrum in $\bar{p}p \rightarrow \pi^{\pm}X^{\mp}$ and a search for narrow states $X^{\mp}$ produced by monochromatic π emission.

2. Apparatus

The apparatus is shown in fig. 1. Antiprotons from LEAR enter a solenoidal magnetic spectrometer (0.8 T) with a momentum of 308 MeV/c. After passing through a copper degrader, they stop in a 76 cm long 16 cm diameter gaseous hydrogen target (NTP). Stopping antiprotons are defined by a scintillation counter telescope followed by a veto counter. Typically 10^4 $\bar{p}$/s enter the telescope. An argon/ethane projection chamber (XDC) surrounds the target. The separation is provided by a 6 μm aluminized mylar membrane at high voltage transparent to X-rays with energies larger than 1 keV. X-rays are converted in the XDC and their energies measured by pulse height analysis on 90 signal wires running parallel to the beam axis. The coordinate along the wire is determined by charge division. Charged annihilation products usually fire several wires and can be distinguished from X-rays by their characteristic long pulse shape, while X-ray conversions are well localized in space.

The momenta of the outgoing charged particles are measured by seven cylindrical MWPC (C1, C2, Q1, Q2, P1, Q3, P2). The anode signal wires run parallel to the beam. The inner and outer cathode surfaces are made of helicoidal wires. The minimum transverse momentum required to reach Q3 is 100 MeV/c. The solid angle subtended by this chamber is 2π.

The momentum resolution for tracks reaching Q3 is $\Delta p / p$ [FWHM] = $0.13\ p + 0.005 / p$ (p in GeV/c). This resolution improves substantially for exclusive annihilation channels for which constraining kinematics fits can be performed.

Photons are detected by conversion in a 5 mm thick lead converter inserted before Q3 and between planar endcap MWPC's.

The annihilation vertex is determined from the outgoing tracks with a precision of σ = 2.5 mm.

A fast trigger requires a charged multiplicity of at least two in C1 and C2. The fraction of events associated with a detected L X-ray is about 5 %. For the data presented here we used however a hardware trigger that enhances annihilation from the 2p level: at least

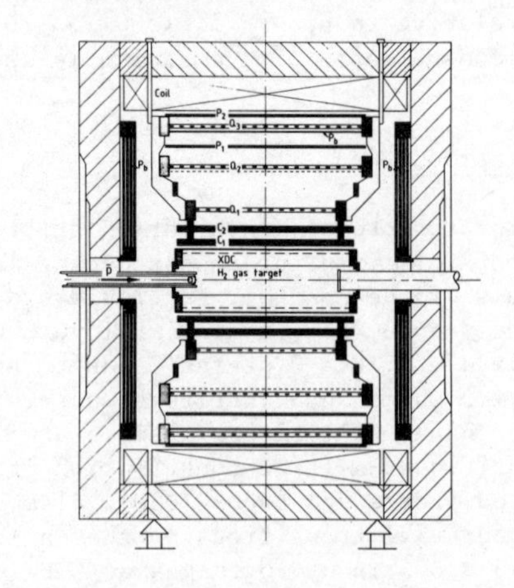

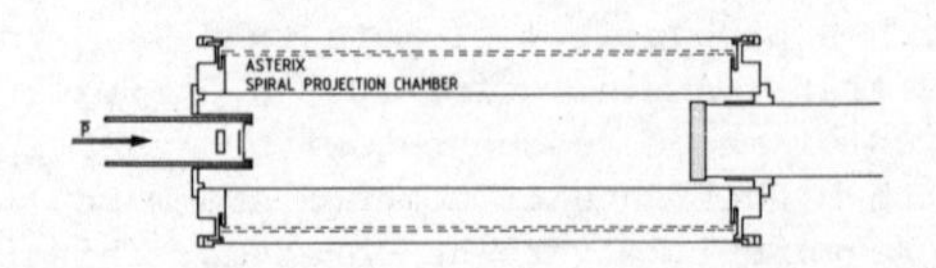

Fig. 1: Side view of the spectrometer (top); target and projection chamber (bottom).

one isolated cluster firing a single wire with no signal on neighbouring wires is required. This trigger enriches the sample of events with associated L X-rays to 25 %.

3. $\bar{p}p \to \pi^+\pi^-\pi^o$

Fig. 2 shows the missing mass recoiling against two charged particles with opposite charges from a sample of 800,000 annihilations in H_2. Signals from $\pi^+\pi^-\pi^o$, $\pi^+\pi^-\eta$ and $\pi^+\pi^-\rho$ (or ω) are clearly visible above the background stemming from $\bar{p}p \to \pi^+\pi^-MM$ with more than one neutral pion. A 1C kinematics fit is then applied to select $\pi^+\pi^-\pi^o$ events. The dashed line in fig. 2 gives the contribution of the events satisfying the fit.

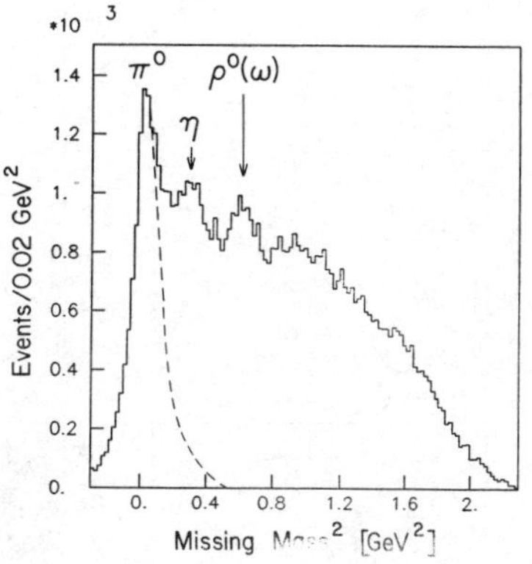

Fig. 2: Missing mass distribution recoiling against $\pi^+\pi^-$.

The invariant masses m^2 $(\pi^+\pi^-)$, m^2 $(\pi^+\pi^o)$, m^2 $(\pi^-\pi^o)$ and the corresponding Dalitz plot are shown in fig. 3a - d. Apart from the ρ production one observes a strong signal from $\bar{p}p \to f\pi^o$, $f \to \pi^+\pi^-$. Fig. 3e shows for comparison the m^2 $(\pi^+\pi^-)$ invariant mass as observed in a bubble chamber experiment (Foster et al., 1968). The ratio of f/ρ^o production was 12 ± 4 %, while our data give a ratio of about 50 %. Annihilation from s orbitals was found to be dominant in liquid hydrogen, the $\rho\pi$ channel coming mainly from the 3S1 $\bar{p}p$ state. A satisfactory fit to the Dalitz plot was found with no p wave contribution. Since the f is a spin 2 meson one expects the production of $f\pi^o$ to be favoured from L $(\bar{p}p)$ = 1. Our data suggest therefore a strong p state annihilation in H_2 at NTP.

For the production of ρ^o only the 3S1 and 1P1 $(\bar{p}p)$ waves contribute due to conservation of C parity. These are isospin I = 0 states of the $\bar{p}p$ system. Annihilation into $\rho^o\pi^o$ is selected by accepting only events in the ρ^o band $m_\rho - 2\Gamma_\rho < m\ (\pi^+\pi^-) < m_\rho + 2\Gamma_\rho$ where m_ρ = 769 MeV and Γ_ρ = 154 MeV. The angular distribution of $\pi^\pm$ in the ρ^o rest frame is shown in fig. 4a. The distribution is symmetric around 90°. The broad bump at $\cos\theta$ = 0.65 is due to interference with the $\rho^\pm$ bands. A Monte Carlo simulation of the channel $\bar{p}p \to \rho^o\pi^o, \rho^o \to \pi^+\pi^-$, has been performed. Fig. 4b shows the expected angular distribution from the 3S1 state, varying as $\sin^2\theta$ apart from the constructive interference bump. For the 1P1 state the angular distribution is isotropic (fig. 4c). Comparison with data indicates clearly a dominant p state contribution.

In addition to the 3S1 and 1P1 states the 1S0, 3P1 and 3P2 I = 1 states may contribute to $\bar{p}p$ annihilation into $\rho^\pm\pi^\mp$. Fig. 5a shows the angular distribution of the $\pi^\pm$ in the $\rho^\pm$ rest frame. The slight slope is due to detector acceptance. The three enhancements are produced by interference with the ρ^o, the f and the destructive interference with the $\rho^\mp$ bands. The simulation without f production (fig. 5b - 5f) indicates again that the dominant contribution to $\rho\pi$ is 1P1.

We have performed a full Dalitz plot analysis. We conclude that 40 % of

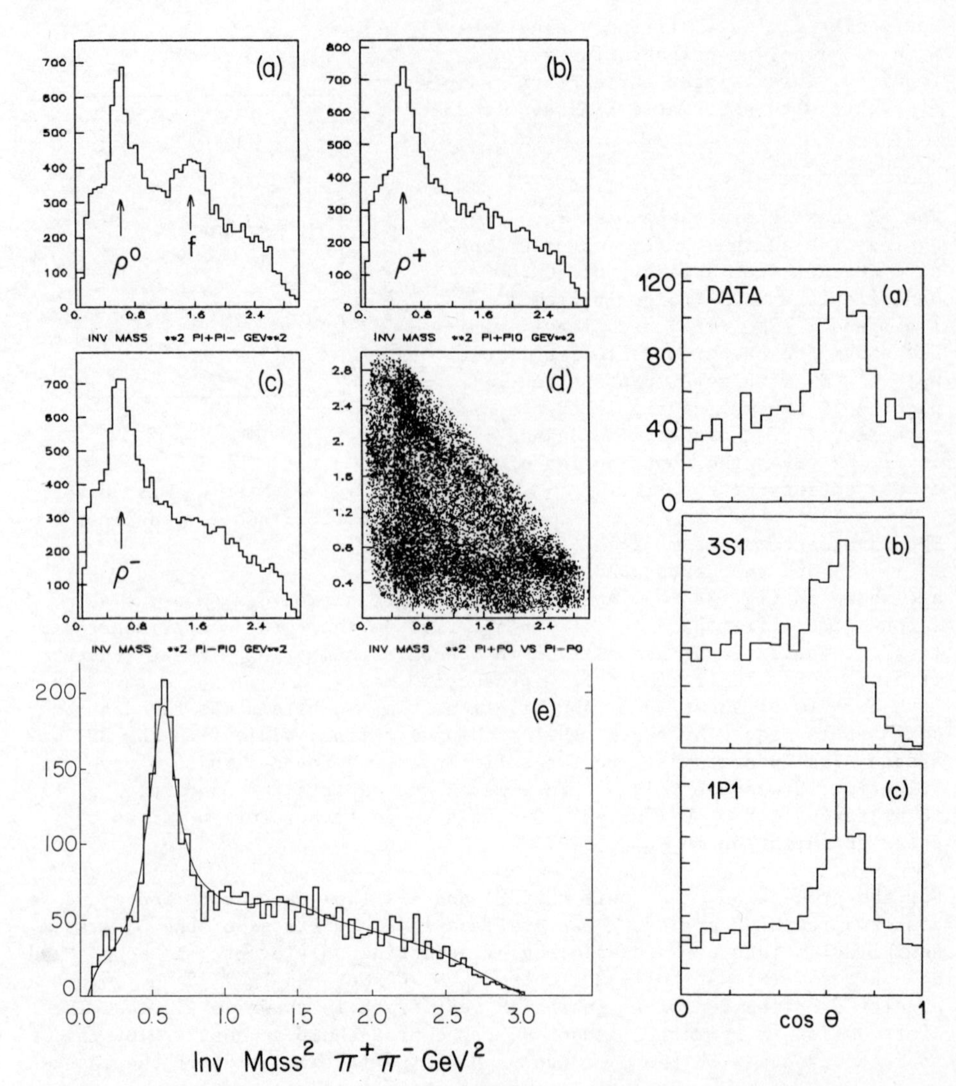

Fig. 3: Invariant 2π mass distributions (a - d) for $\pi^+\pi^-\pi^o$ in gaseous H_2; result in liquid H_2 (e) (from Gavillet (1968)).

Fig. 4: $\cos\theta$ distribution in the ρ^o rest frame (data) (a) and Monte Carlo simulation (b - c) for 3S1 and 1P1 $\bar{p}p$ states.

the channel $\bar{p}p \rightarrow \pi^+\pi^-\pi^o$ is nonresonant. The resonant $\rho\pi$ contribution proceeds dominantly via 1P1 (> 90 %) with a small admixture of 3P1 and 3P2. We do not find any S wave contribution with an upper limit of 15 %.

As mentionned earlier 3S1 dominates annihilation into $\rho\pi$ in liquid hydrogen (Foster et al., 1968), while our data shows mainly 1P1 in gaseous hydrogen. We conclude therefore that $\bar{p}p \rightarrow \rho\pi$ is dominantly produced from the I = 0 $\bar{p}p$ system.

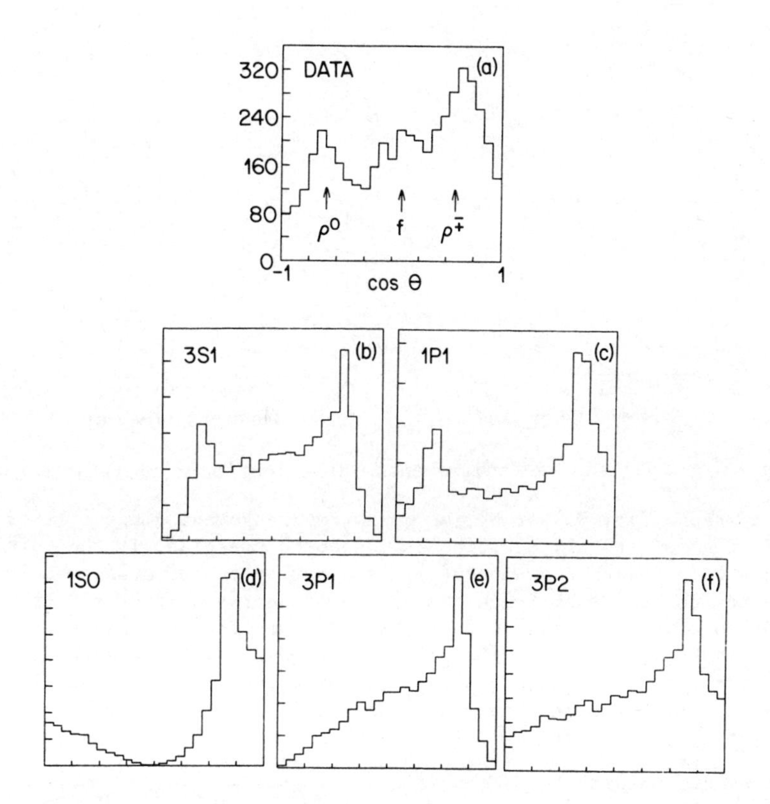

Fig. 5: cosθ distribution in the $\rho^{\pm}$ rest frame (a) (data), and Monte Carlo simulation (b - f) for all contributing $\bar{p}p$ states (without f production).

4. $\bar{p}p \rightarrow \pi^+\pi^-, K^+K^-$

These channels, although rather rare (the branching ratio is a few 10^{-3}), are easy to reconstruct: the events appear as single tracks traversing the whole detector. Events with two collinear tracks are therefore submitted to a global fit which improves the resolution and allows a good momentum separation between K's (798 MeV/c) and π's (928 MeV/c). To have a clean sample of p state annihilation we select only collinear events associated with a L transition to the 2p state. The X-ray spectrum (Fig. 6a) shows the 3d → 2p L_α transitions and the unresolved nd → 2p (n > 1) L_β - L_∞ transitions. Events passing the global fit and associated with a detected X-ray in the energy range 1.1 to 3.0 keV are shown in fig. 6b. After correction for K decay we arrive at a ratio of branching ratios $R = \Gamma(\bar{p}p \rightarrow K^+K^-)/\Gamma(\bar{p}p \rightarrow \pi^+\pi^-) = 0.15 \pm 0.03$. The corresponding ratio for liquid hydrogen, where s wave annihilation dominates, is 0.28 ± 0.03 (compilation by Armenteros et al., 1980).

This result could be explained by an annihilation model proposed by Dover and Fishbane (1984). Annihilation into two mesons is described by the anni-

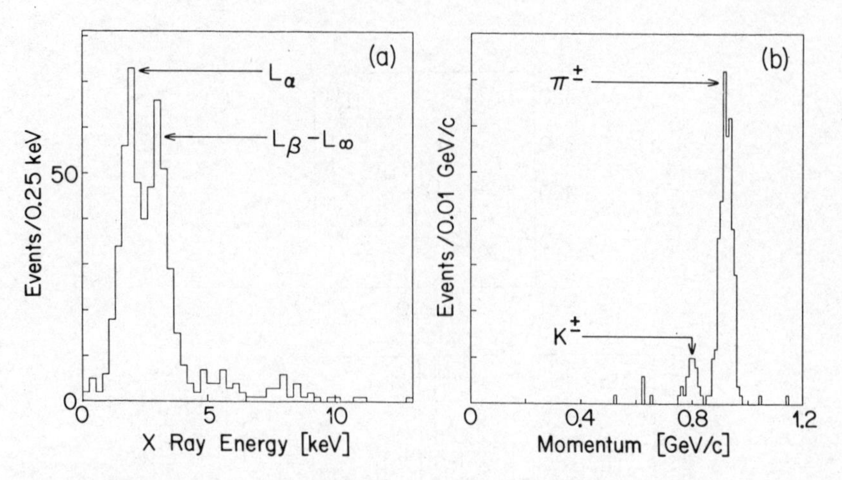

Fig. 6: X-ray spectrum (a) and momentum distribution of collinear events(b)

hilation graph (fig. 7a) or by the quark rearrangement diagram (fig. 7b). For K^+K^- the former only contributes due to $s\bar{s}$ creation. If the initial $\bar{p}p$ state has relative angular momentum L = 0 parity conservation requires the final state to be L = 1. Since the spectator quarks carry L = 0 graph b does not contribute to $\pi^+\pi^-$ either. Simple quark counting and correcting for phase space yields R = 0.38 in fair agreement with the experimental value. If the initial state has L = 1, parity requires L = 0 (or 2) in the final state. Both graphs then contribute to $\pi^+\pi^-$ and R becomes smaller as demonstrated by our data.

For all events passing the global fit, with or without L X-ray coincidence, we find R = 0.14 ± 0.02. This confirms that annihilation in gas at NTP is dominated by p state annihilation. Given the limited statistics in this data sample we find that s wave annihilation does not contribute more than 25 % with 95 % probability.

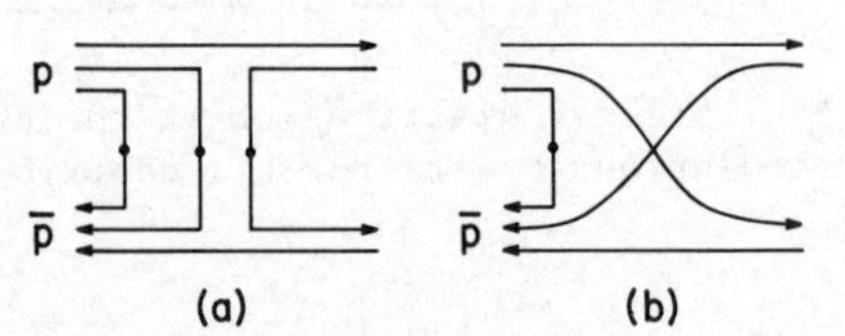

Fig. 7: Quark diagrams for annihilation into two mesons.

5. $\bar{p}p \to$ 4 prongs + MM

Fig. 8a shows the missing energy distribution for four prong events. All particles are assumed to be pions. The peak at zero missing mass contains the events $\bar{p}p \to 2\pi^+2\pi^-$. The pion momentum distribution of the events satisfying a 4C kinematics fit to $\bar{p}p \to 2\pi^+ 2\pi^-$ is shown in fig. 8b. Only pions reaching Q3 were considered for maximum resolution. The dashed curve is a Monte Carlo simulation assuming phase space. The $A_2^\pm$ meson recoiling against $\pi^\mp$ ($A_2 \to \rho\pi$, $\rho \to \pi\pi$) is clearly visible.

Making no assumption on the identity of the particles, we plot the total momentum (fig. 9a). Again, the peak above zero stems from annihilation into exactly four prongs. Fig. 9b shows the total energy of the events with a total momentum of less than 25 MeV/c, assuming pions. The peak at two nucleon masses corresponds to exactly four pions, while the structure between 1.2 and 1.6 GeV/c corresponds to $\bar{p}p \to K^+K^-\pi^+\pi^-$.

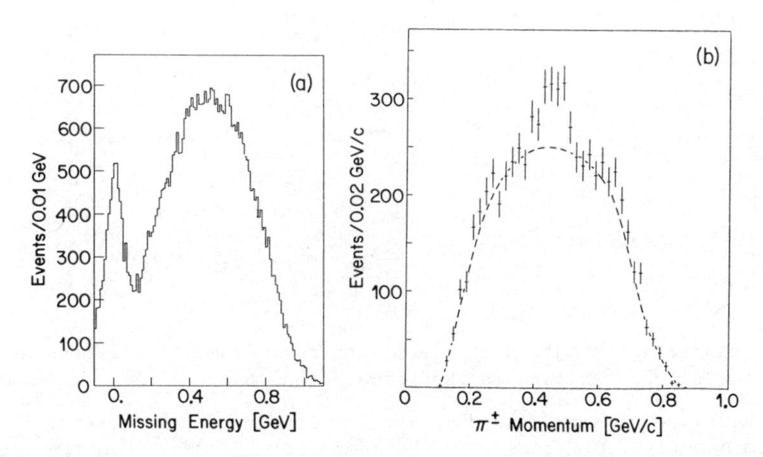

Fig. 8: Missing energy distribution for $2\pi^+ \; 2\pi^-$ MM events (a) and $\pi^{\pm}$ momentum spectrum for $\bar{p}p \to 2\pi^+ \; 2\pi^-$ (b).

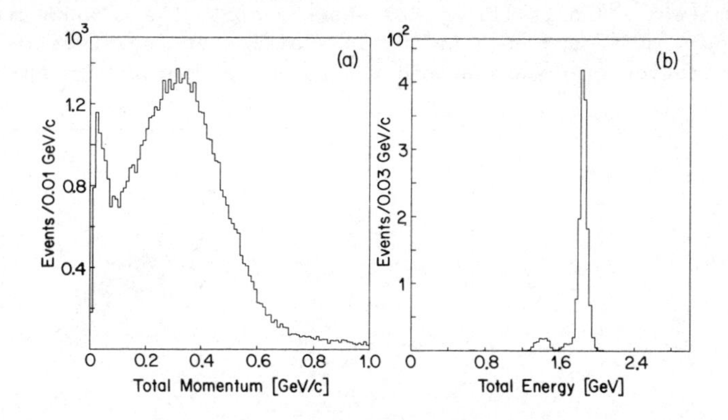

Fig. 9: Total momentum P distribution for four prong events (a) and total energy distribution for P < 25 MeV/c, assuming pions (b).

A search for the channel $2\pi^+ \; 2\pi^-\gamma$ illustrates the γ detection capability of the apparatus. We select events with four prongs and a missing mass of less than 300 MeV. In addition the detection of at least one photon is required. The angular distribution of the γ in the rest frame of the missing momentum (assuming a π^o) is shown in fig. 10a. The flat distribution is due to γ's from π^o decay. The backward peak is produced by noise (wrongly identified γ's). The signal from $2\pi^+ \; 2\pi^-\gamma$ is expected in the forward direction as shown by a Monte Carlo simulation (fig. 10b) of a sample of $2\pi^+ \; 2\pi^-\pi^o$ and $2\pi^+ \; 2\pi^-\gamma$ events. In the last five bins of fig. 10a we observe an excess of 23 ± 12 events. We do not consider this excess as significant and quote an upper limit of 4×10^{-3} for $\bar{p}p \to 2\pi^+ \; 2\pi^-\gamma$, using as a guideline the branching ratio of 19 % for $\bar{p}p \to 2\pi^+ \; 2\pi^-\pi^o$ in liquid hydrogen (Armenteros et al., 1980).

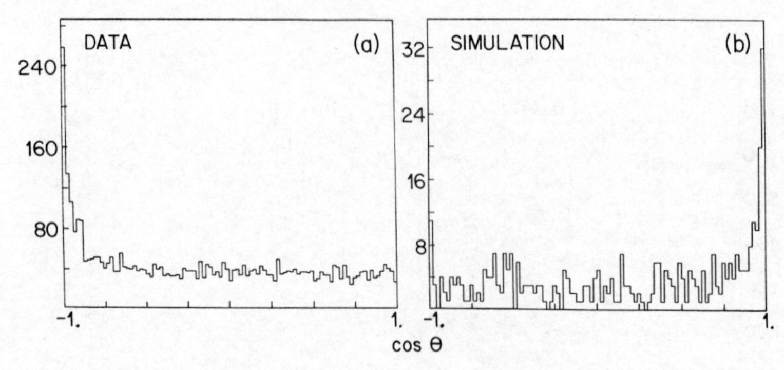

Fig. 10: Angular distribution of γ's in the rest frame of the missing momentum for $2\pi^+ 2\pi^-$ MM data (a) and simulation (b).

Fig. 11 shows the missing mass distribution for $2\pi^+ 2\pi^-$ MM events. The peak comes from exactly four pions, the shoulder from $2\pi^+ 2\pi^-\pi^0$ and the tail from events with more than one π^0. After performing a 1C kinematics fit to $\bar{p}p \to 2\pi^+ 2\pi^-\pi^0$ we plot the $\pi^+\pi^-\pi^0$ invariant mass versus the invariant mass of the remaining $\pi^+\pi^-$ (four combinations per event, fig. 12a). The projection on the m $(\pi^+\pi^-\pi^0)$ axis (fig. 12b) shows clearly the ω and η mesons produced in $\bar{p}p \to \omega\pi^+\pi^-$ and $\bar{p}p \to \eta\pi^+\pi^-$. In addition one observes in the ω band a concentration of events around the ρ mass stemming from $\bar{p}p \to \rho\omega$.

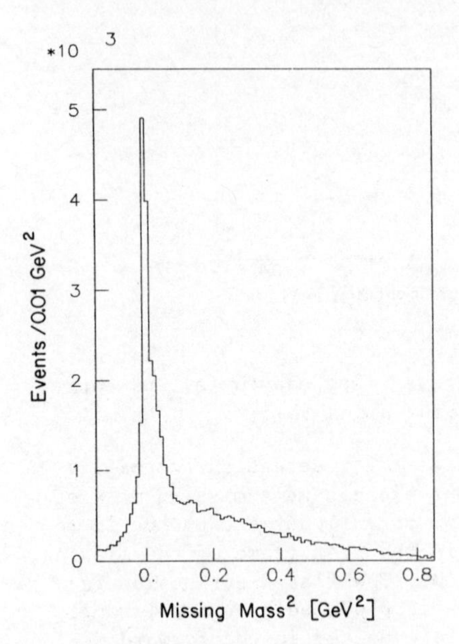

Fig. 11: Missing mass distribution in $\bar{p}p \to 2\pi^+ 2\pi^-$ MM.

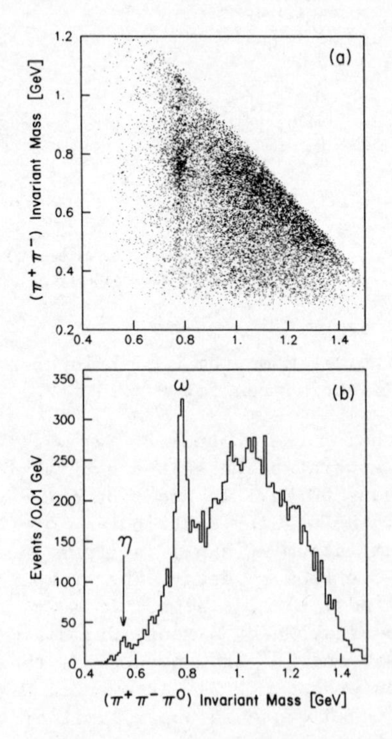

Fig. 12: 3π vs 2π invariant mass distribution.

6. Search for narrow states in $\bar{p}p \rightarrow \pi^{\mp} X^{\pm}$

We have analyzed 800,000 annihilations in hydrogen requiring a unique vertex in a fiducial volume of 7 cm radius and 72 cm length. The momentum distribution of the charged pions is shown in fig. 13a. Below 300 MeV/c events are grouped in 2 MeV/c bins and above 300 MeV/c in 10 MeV/c bins. Only pions reaching Q3 are considered for optimum resolution. The resolution is indicated by horizontal bars. The lower half of the spectrum is fitted by a 6th order polynomial (χ^2 = 77 for 88 degrees of freedom) and the higher half with a 5th order polynomial and a Breit-Wigner (χ^2 = 58 for 41 degrees of freedom). The peak observed at 460 MeV/c is the $A_2^{\pm}$ meson (M = 1320 ± 2, Γ = 99 ± 4 MeV) seen previously in the channel $\bar{p}p \rightarrow 2\pi^+ 2\pi^-$. Fig. 13b shows the residuals obtained from the data by subtracting the fitted curves. No structure is observed.

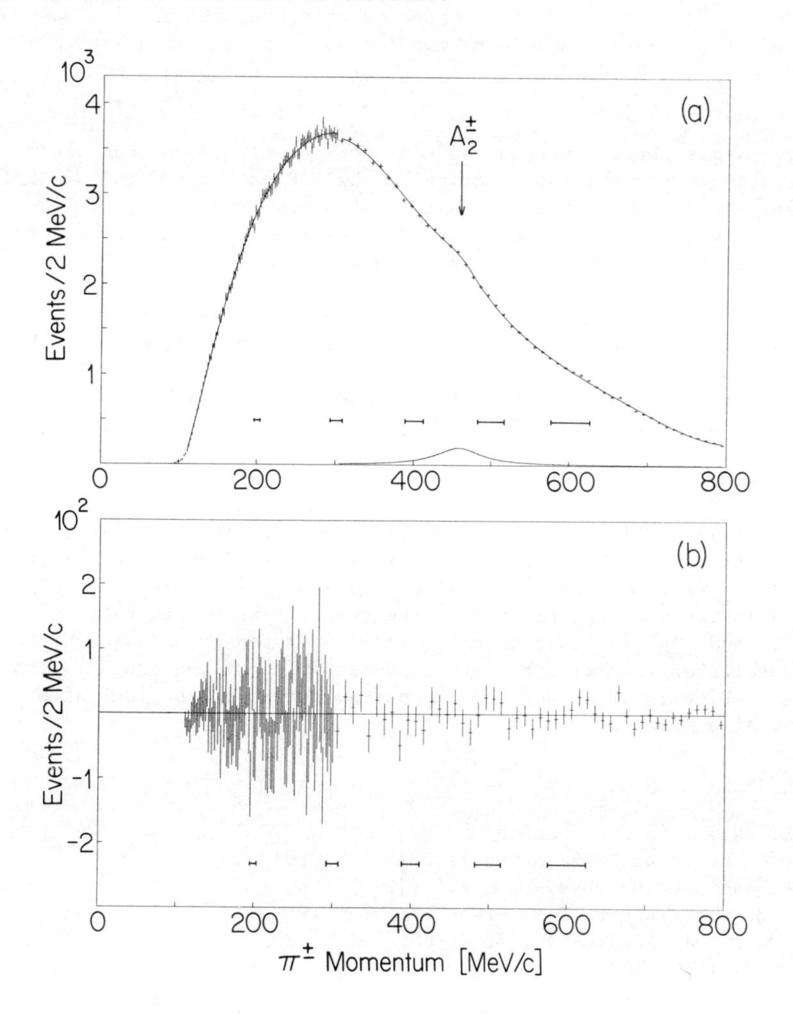

Fig. 13: Inclusive $\pi^{\pm}$ momentum distribution (a) and residuals (b) obtained by subtracting the fitted curves in (a).

The branching ratio for the production of narrow states $X^{\pm}$ associated with $\pi^{\mp}$ emission from p orbitals, narrower than our experimental resolution and leading to a 5σ signal in our data is given in Table 1 as a function of mass. Since we do not observe any 5σ peak the quoted numbers are upper limits (Ahmad et al., 1984).

M [MeV]	Γ[MeV]	Upper limit
1665	7	2.7×10^{-3}
1584	13	3.0×10^{-3}
1446	33	3.4×10^{-3}
1109	85	3.1×10^{-3}

Table 1: 5σ upper limit for the existence of narrow states with mass M and width less than Γ associated with monochromatic $\pi^{\pm}$ emission from p atomic states.

The upper limit can be improved by considering all tracks reaching Q3 without requiring a vertex. Annihilation in the XDC gas is reduced to 15 % by rejecting events with large pulses from stopping $\bar{p}$ in argon. No structure is observed either. This procedure increases the statistics by a factor of 2 and reduces the quoted upper limits in Table 1 by $\sqrt{2}$.

The observation of a narrow state at 1620 MeV in liquid hydrogen was reported at this conference by D. Schultz et al. With their quoted branching ratio of 3×10^{-3} we would observe a 8σ peak in our data. We could not however comment on the existence of this state if it would be produced dominantly from s orbitals.

7. Conclusions

The first LEAR running period has enabled us to collect the first data on $\bar{p}p$ annihilation at rest in gaseous hydrogen. We find that annihilation from p atomic states dominates at NTP. The production of $\rho\pi$ occurs mainly from the isospin I = 0 of the $\bar{p}p$ system. The relative branching ratio for $\bar{p}p \to K^+K^-$ and $\pi^+\pi^-$ is lower than in liquid hydrogen by a factor of two. These results may contribute to the understanding of the annihilation mechanism. Finally we do not observe narrow states in the inclusive $\pi^{\pm}$ spectrum at a level of 2×10^{-3}.

References

Ahmad S. et al. in preparation
Armenteros R. et al. CERN proposal PS 171 (1980)
Borie E. and Leon M. Phys. Rev. A21 (1980) 1460
Dover C.B. and Fishbane P.M. Orsay Preprint IPNO / TH 84 - 8 (1984)
Foster M. et al. Nucl. Phys. B6 (1968) 107
Gavillet P. These 3eme cycle, Grenoble (1968)

Paper presented at VII Eur. Symp. Antiproton Interactions, Durham 1984

Investigations on baryonium and other rare p̄p annihilation modes using high resolution π^0 spectrometers (PS 182)

The Basle[1)]-CERN[2)]-Stockholm[3)]-Thessaloniki[4)] Collaboration

L. Adiels[3)], G. Backenstoss[4)], I. Bergström[3)], S. Carius[3)], S. Charalambous[4)], M. Cooper[4)], C. Findeisen[1)], K. Fransson[3)], D. Hadjifotiadou[4)], A. Kerek[3)], K. Papastefanou[4)], P. Pavlopoulos[2)], J. Repond[1)], L. Tauscher[1)], D. Tröster[1)], C. Williams[1)] and K. Zioutas[4)]

(Presented by P. Pavlopoulos)

1. Introduction

In the search for heavy, narrow exotic states such as baryonium or glueballs, carried out so far, various methods of high- and low-energy production, low-energy formation, and e^+e^- reactions were used. In particular, a search for baryonium states below threshold ($m_X \lesssim m_{p\bar{p}}$) has shown some evidence [Richter et al., 1983; Brando et al., 1984] for a few states. However, the results of these measurements suffer from poor statistics. Because of the relatively low level of confidence in these results it would be highly desirable to compare them with new measurements of better statistics, now available at the CERN Low-Energy Antiproton Ring, LEAR, and with independent measurements based on techniques hitherto not used, which the quality of LEAR now makes feasible.

Among the possible methods which can be applied in the baryonium bound state research at LEAR, we consider the study of the reaction

$$p\bar{p} \rightarrow X + \pi^0 \tag{1}$$

as the most promising. The reason is that in a stop experiment the initial state is already a well-defined bound $p\bar{p}$ state. The transition from this state to a baryonium state by emission of a neutral particle populates both isospin 0 and isospin 1 baryonia. Moreover, the definite C-parity of the neutral pion would indicate, or rather exclude, some quantum numbers of these exotic states. The combined information of our results with the results of inclusive charged pion measurements can really clarify the existence of such exotic states and enlighten our understanding of low-energy QCD.

We have performed an experiment to study the reaction (1) by means of an experimental set-up especially tailored for measuring monoenergetic π^0's, allowing for both angular distribution measurements and high-resolution spectroscopy. Here we report on the preliminary results from our actual measurements of the inclusive π^0 and η spectra following the antiproton annihilation at rest.

1) Institute of Physics, University of Basle, Switzerland.
2) CERN, Geneva, Switzerland.
3) Research Institute for Physics, Stockholm, Sweden.
4) Nuclear Physics Department, University of Thessaloniki, Greece.

2. The Experiment

The experiment took place at the C1 beam of LEAR. At a momentum of 300 MeV/c and with a momentum spread of better than 10^{-3}, we stopped in our 0.8 g/cm^2 thick liquid-hydrogen target a total number of 3×10^9 antiprotons. A thin scintillator of 1 mm with a radius of 1 cm identified the arrival of the antiprotons.

In Fig. 1 is sketched the apparatus which fulfils the requirements of our experiment. It consists mainly of two identical bismuth germanate (BGO) detector systems and an array of 4×40 pieces of lead-glass detectors. Bismuth germanate is a scintillation crystal ($Bi_4Ge_3O_{12}$). Such a big BGO detector (Fig. 2) has never before been used in high-energy spectroscopy. Each BGO detector is composed of seven modules of hexagonal cross-section. Each module has a diameter of 6.4 cm and a length of 20 cm, and is equipped with its own photomultiplier. An improved position resolution of $\pm 1°$ for the central module is achieved by a hodoscope placed in front of it, consisting of seven small BGO crystals. Each of these is $1 \times 5 \times 1.2$ cm^3 and is equipped with one photomultiplier. The compactness of the detector and the chemical and mechanical durability of BGO [Pavlopoulos et al., 1982] make the entire system very versatile and easy to handle. The two BGO detector systems can be positioned at any angle and any distance relative to the target. The performances of the BGO detectors were tested at CERN (150–700 MeV/c) and at SIN (36–240 MeV/c) in an electron beam. The homogeneity and the efficiency of light collection turned out to be equivalent to those of NaI. The resolution [Adiels et al., 1984] was found to vary with energy as

$$\text{FWHM} = (2.8 \pm 0.2)\% / \sqrt[4]{E(\text{GeV})} \qquad (2)$$

for electrons entering the central module.

The lead-glass array consists of 160 truncated rectangular pyramids so as to provide a cylindrical geometry, where the gammas of the practically point-like target always enter the calorimeter radially. The solid angle of each array module is of the order of 10^{-3}, and its energy resolution was $\Delta E/E = 10\%/\sqrt{E(\text{GeV})}$ above an energy threshold of 20 MeV. We chose an arrangement that

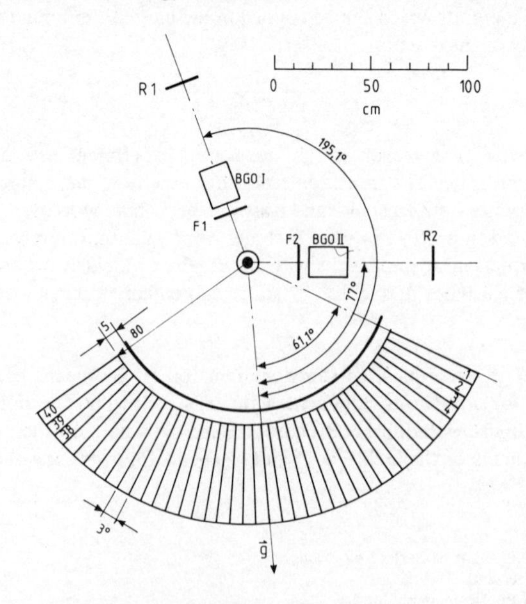

Fig. 1 The experimental set-up

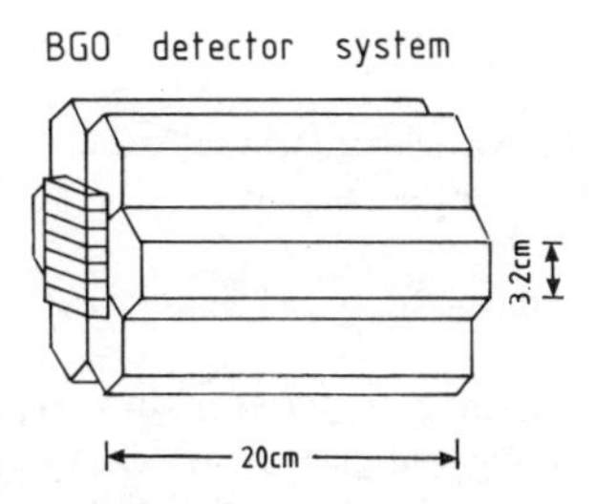

Fig. 2 The BGO detector assembly

would keep a constant angular resolution of 3° and would measure simultaneous relative angles up to 120°. The range of the angles relative to a BGO system can span from 17° to 180°. This leads to an energy-dependent energy resolution and to an energy-dependent acceptance for the π^0 spectroscopy. Each group of four counters was vetoed for charged particles individually by a system of double scintillation counters.

The intercalibration of the BGO systems was performed by the use of thorium radioactive sources, one being placed on each hexagonal crystal. The 2.4 MeV γ-line of this source was continuously monitored on-line. Moreover, the peak of minimum ionizing charged pions, triggered by two scintillators at the front and the back of the BGO crystals, provided a continuous monitoring of the calibration stability of the detectors. Thus possible small changes in the calibration gains could be corrected off-line. Furthermore, the monoenergetic line at 129 MeV, owing to the reaction

$$\pi^- p \rightarrow n\gamma \tag{3}$$

of annihilation pions stopping in the target, gave a very nice check of the absolute calibration.

The intercalibration of the lead-glass module could be done by the minimum ionizing peak of the cosmic radiation at the beginning and at the end of each run period. Radioactive ^{241}Am sources glued on small scintillators provide a peak corresponding to 100 MeV equivalent owing to the scintillation light produced from the 5 MeV α particles. Each module of the lead-glass array was equipped with such a source, which allowed us to monitor continuously the gain stability. A further on-line and off-line check of their intercalibration was the matching of the shapes of the γ-ray spectra measured by each individual module. The absolute calibration of the lead-glass array was adjusted and checked from the position of the π^0 invariant mass of events having both γ's enter the lead-glass calorimeter.

The high $\bar{p}$ flux of LEAR could result in an unrealistically high data flux through the on-line computer to the magnetic tape. In order to reduce this to a manageable rate, we have developed components that allow us to cut the count rate at different levels. A 'logic unit' was constructed which is capable of defining whether an event in the lead-glass array is topologically acceptable or not. The answer can be given after a delay of only 35 ns and could thus be implemented in the master trigger. In addition, an intelligent crate controller (ICC) has been developed [Tröster, 1984], which allows for a selective readout of 160 lead-glass analog-to-digital converters (ACDs) together with a fast pattern decoding. Moreover, this allows for data transfer, interrupt handling, and bus arbitration, whilst being completely CAMAC compatible. Using a 64 kword long FIFO buffer we arrived at less than six transfers to the VAX computer every second. The rate problems could thus be solved, particularly in view of the unfortunate fact that we never profited from a full LEAR intensity.

3. π^0 Spectroscopy

The two γ-rays from the decay of a particle of mass m (for example, the π^0 and η) and total energy E are strictly correlated by their kinematical relation

$$m^2 = 2E_{\gamma_1}E_{\gamma_2}(1 - \cos\theta_{\gamma_1\gamma_2}) \,. \tag{4}$$

In Fig. 3a we display a typical invariant mass spectrum of two photons, i.e. one arriving in one of the BGO detectors and the other in the lead-glass array. The two peaks corresponding to the π^0 and η neutral decays are clearly visible. The mass resolution obtained for this combination was ~ 20 MeV. The background below the peaks, so-called 'combinatorial background', arises from γ's belonging to different π^0's. This type of background can be obtained from the data too and accordingly subtracted by combining γ-rays of different events. The study of this background allows the uncovering of systematic effects related to the imperfections of the detection system. The mass spectrum of this combinatorial background is illustrated in Fig. 3b.

Sufficiently accurate measurements of all three quantities in Eq. (4) define the particle uniquely ($E = E_{\gamma_1} + E_{\gamma_2}$; $\vec{p} = \vec{p}_{\gamma_1} + \vec{p}_{\gamma_2}$) and allow the determination of its momentum by a 1C-fit. The correct position of the π^0 invariant mass (Fig. 3a) makes us confident about the precision of our calibration. By selecting events lying inside a window of ± 20 MeV around the π^0 mass, we have performed a kinematical fit with the constraint of requiring an invariant mass equal to the neutral pion mass. From relativistic kinematics it can be seen that there exists a very pronounced angular correlation between the two γ-rays; most of these are emitted into a narrow angular interval just above a minimum angle, $\theta^{min}_{\gamma_1\gamma_2}$, thus forming the so-called Jacobian peak, where both γ-rays have nearly equal energies (symmetric decay). It can easily be shown that such decays provide the best

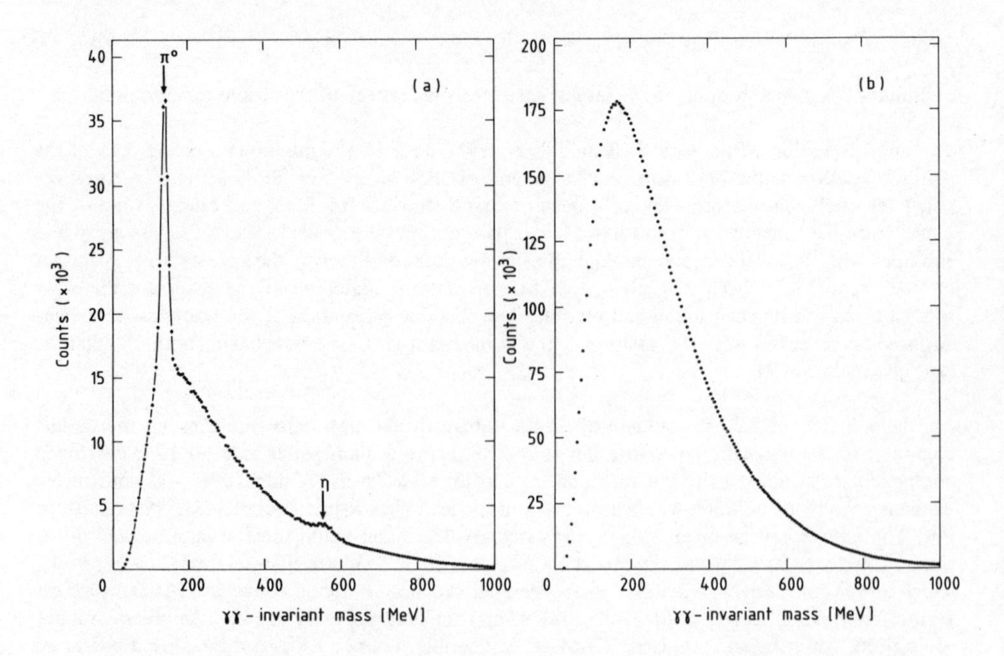

Fig. 3 The invariant mass spectrum for two photons coming (a) from the same event and (b) from different events

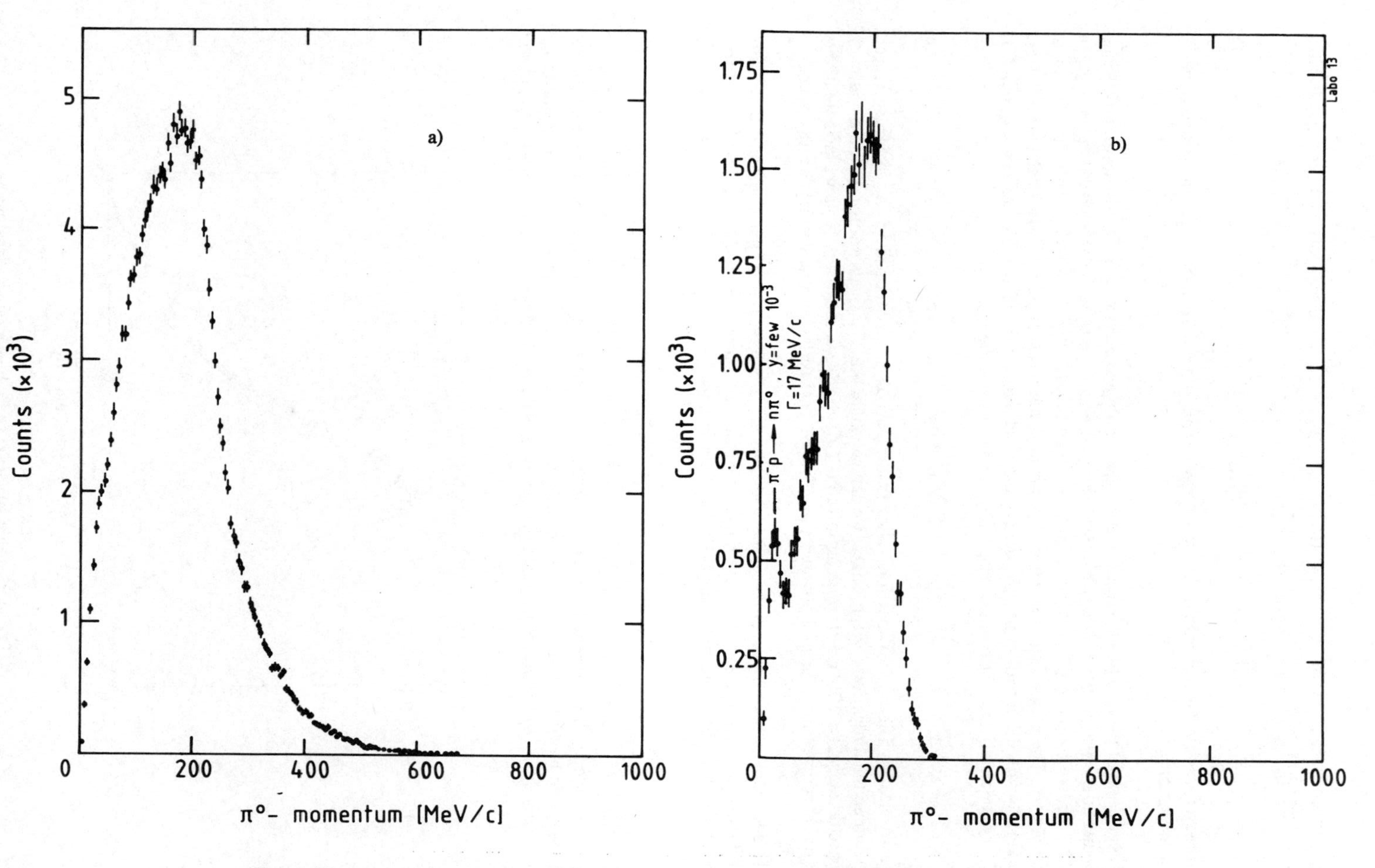

Fig. 4 The momentum spectrum of low-energy neutral pions (a) before and (b) after the cuts

momentum resolution. We thus expect to have an improved signal by applying a cut for symmetric decays given by

$$(E\pi^0/2)(1 - \beta/4) \leq E_{\gamma_1}(E_{\gamma_2}) \leq (E\pi^0/2)(1 + \beta/4)\,. \tag{5}$$

Moreover, such a cut being a physical one provides an additional criterion for the structure under consideration.

The position of the two BGO detector systems have been arranged such as to provide three different momenta spectra of different sensitivities. The first BGO spectrometer was placed at an angle of 122° relative to the lead-glass array such as to provide a high sensitivity for momenta up to 180 MeV/c. The π^0 spectrum for this detector combination is shown in Fig. 4a for the raw data and in Fig. 4b with the cut for symmetric decays and a subtraction of the combinatorial background. It is very interesting that the monoenergetic neutral pions from the reaction

$$\pi^- p \rightarrow n\pi^0 \tag{6}$$

are so clearly visible. This reaction arises from charged pions stopped in the hydrogen target. Assuming the dimensions of our target, we can calculate that the yield of this reaction amounts to approximately 2×10^{-3} per stopped antiproton. This clear peak at 28 MeV/c shows the potentiality of our experimental set-up and provides a measure of its sensitivity for at least low-energy pions.

The spectrum of both γ-rays entering the lead-glass array is sensitive to high π^0 momenta owing to the possibility of having small relative angles. The momentum spectrum of such events is illustrated in Fig. 5, where the monoenergetic peak from the reactions

$$\begin{aligned} p\bar{p} &\rightarrow \pi^0\varrho^0 \\ p\bar{p} &\rightarrow \pi^0\omega^0 \end{aligned} \tag{7}$$

is pronounced. The two reactions (7) cannot be separated since they have the same recoiling π^0 momentum. However, a shape analysis of this structure can separate the two reactions owing to the extremely different widths of the ϱ and the ω meson. Such an analysis is already under way and will

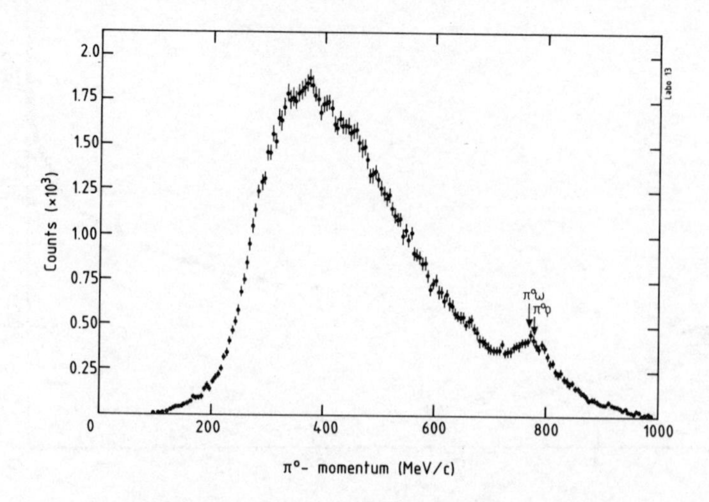

Fig. 5 The momentum spectrum of high energetic neutral pions

provide the yield of the $\pi^0\omega$ annihilation channel, which so far has only been observed indirectly [Backenstoss et al., 1983]. This spectrum (Fig. 5) gives a measure of the sensitivity and the quality of our apparatus at high π^0 momenta.

Finally, the second BGO spectrometer has an angle of 77° relative to the lead-glass array, being dedicated to the spectroscopy of neutral pions with a momentum around 200 to 300 MeV/c. This spectrum is of particular interest in the search for baryonium candidates (Fig. 6) and shows some structures, in particular around 225 MeV/c and 280 MeV/c. The meson masses corresponding to these possible structures coincide very accurately with the $\varrho'(1600)$ and f′ mesons. Although the width of the structure at 280 MeV/c can easily be accommodated with the width of the f′ meson, the structure at 225 MeV/c seems to be narrower than the width of the ϱ' meson. However, it is too early to come to a definite interpretation of this structure owing to the need for further statistics and systematic analysis. We should just like to mention that the reflection of the width of the ϱ' in the momentum spectrum of π^0's can be completely distorted by the limited phase space available in the $p\bar{p}$ annihilation at rest.

Summarizing, we conclude that the measurements of the inclusive π^0 spectrum show two expected structures, one at 28 MeV/c from the Panofsky reaction and one at 800 MeV/c from the annihilation reactions with a ϱ and an ω meson recoiling. Furthermore, the structures at 225 MeV/c and 285 MeV/c need more statistics and analysis in order to be properly interpreted. The sensitivity reached in the search for exotic states can put a limit of a yield of 4×10^{-3} for the 5σ effect in the region around 200 MeV/c.

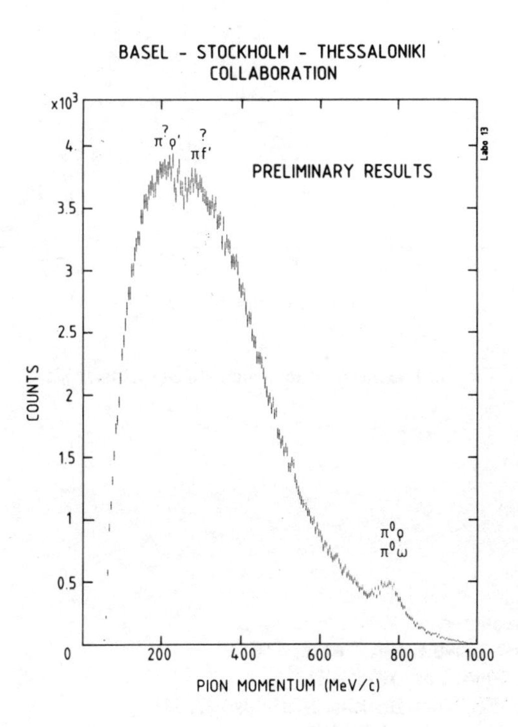

Fig. 6 The momentum spectrum of neutral pions dedicated to the momentum region of 200–300 MeV/c

4. η Spectroscopy

By applying the same arguments as for the detection of neutral pions, we can perform simultaneous measurements of the inclusive η spectrum using the same experimental set-up. A preliminary momentum spectrum of the η mesons is illustrated in Fig. 7, where the peak corresponds to monoenergetic η's from the reaction

$$\begin{aligned} p\bar{p} &\rightarrow \eta\varrho^0 \\ p\bar{p} &\rightarrow \eta\omega\,. \end{aligned} \tag{8}$$

The η spectroscopy has hardly been investigated, since most of the existing $p\bar{p}$ annihilation data come from hydrogen bubble-chamber experiments. The inclusive η spectrum although it may not be so exciting for the search for exotic states, can be highly interesting for theoretical models of the annihilation mechanism by studying channels such as $\pi^0\eta$, $\eta\eta$, $\eta\eta'$, etc.

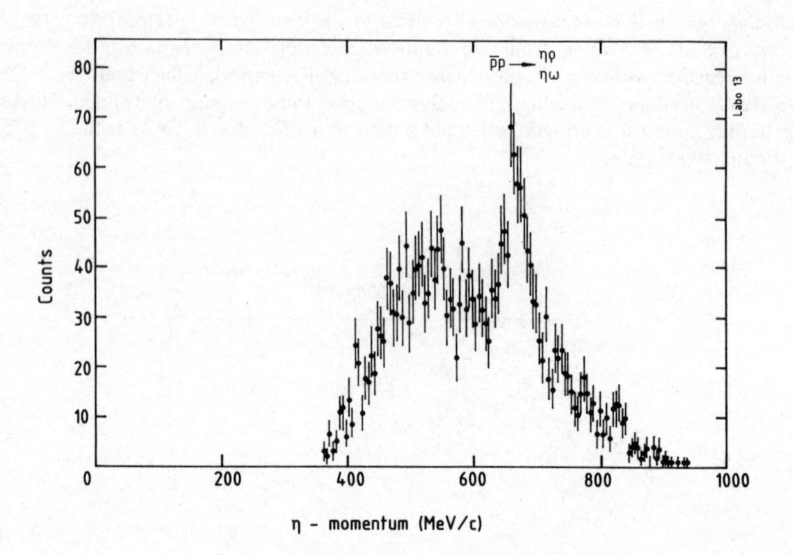

Fig. 7 The momentum spectrum of inclusive η's

References

Adiels L. et al., to be published.
Backenstoss G. et al. (1983), Nucl. Phys. **B228,** 424.
T. Brando et al. (1984), Phys. Lett. **139B,** 133.
Pavlopoulos P. et al. (1982), Nucl. Instrum. Methods **197,** 331.
Richter B. et al. (1983), Phys. Lett. **126B,** 284.
Tröster D.A. (1984), BST-PINK PANTHER, An Intelligent CAMAC Crate Controller, report CERN 84-03.

Inst. Phys. Conf. Ser. No. 73: Section 5
Paper presented at VII Eur. Symp. Antiproton Interactions, Durham 1984

A search for structure in the charged meson spectra from proton–antiproton annihilations at rest

A. Angelopoulos[1], A. Apostolakis[1], T. Armstrong[4], B. Bassalleck[3], C. Benulis[5], J. Biard[4], P. Denes[3], N. Graf[3], R. Hill[3], N. Komninos[3], R.A. Lewis[4], M. Mandelkern[2], W.K. McFarlane[5], B.Y. Oh[4], P. Papaelias[1], S.M. Playfer[4], J.L. Press[5], R. Ray[2], H. Rozaki[1], L. Sakelliou[1], D. Schultz[2], J. Schultz[2], G.A. Smith[4], M. Soulliere[4], M. Spyropoulou-Stassinaki[1], R.D. Von Lintig[5], J. Whitmore[4] and D.M. Wolfe[3]

(Presented by D. Schultz)

1 Nuclear Physics Laboratory, University of Athens, Athens, Greece
2 Dept. of Physics, University of California, Irvine, CA, USA
3 Dept. of Physics, University of New Mexico, Albuquerque, NM, USA
4 Dept. of Physics, Pennsylvania State University, University Park, PA, USA
5 Dept. of Physics, Temple University, Philadelphia, PA, USA

Abstract. LEAR experiment PS183 is described and preliminary results are presented. The experiment studies the inclusive spectra of charged mesons and gammas produced in the annihilations of antiprotons stopped in liquid hydrogen. There is evidence of a narrow structure in the momentum spectra of both positive and negative mesons. The line occurs at a momentum of 200 MeV/c with an approximate yield of 3×10^{-3} per annihilation. As there is no known process which could produce such a line, we interpret it as evidence for a new state; the C(1620).

1. Introduction

Experiment PS183, in the south branch of the LEAR experimental area, uses a single arm magnetic spectrometer to measure the inclusive spectra of charged mesons and gammas coming from the annihilation of antiprotons brought to rest in liquid hydrogen. The antiproton, upon stopping, is captured into a quasiatomic state of high principle quantum number and high angular momentum. From bubble chamber data it is known that Stark mixing of this initial state with states of lower angular momentum is quite strong. Although we are not able to identify the angular momentum of the state in which annihilation finally occurs, we believe that most of the annihilations are S-wave.

In the following section the apparatus we use to measure the momentum of the annihilation products will be described. Following that will be preliminary results coming from the charged meson data that were taken in December 1983. As will be described, those data are contained on twenty magnetic tapes (3×10^6 events). Since that time we have collected two more sets of data, one in April and one in August 1984, contained on 350 and 425 data tapes respectively (10^8 events total). The data on gammas from the annihilations were taken during the April and August runs and are currently under analysis.

2. The apparatus

A plan view of the apparatus as used in December 1983 is shown in fig. 1. The antiproton beam (309 MeV/c) enters from the lower left. The beam is well contained within a 1 cm diameter, with an energy resolution of 10^{-3} and a typical rate of 3×10^4. The beam is steered onto the liquid hydrogen target through a double plane proportional chamber with 1 mm wire spacings. The antiprotons pass through two scintillation counters (S1, S2) which define the beam and are used for timing. They then enter the 3 cm diameter by 20 cm long target and come to rest approximately at the target's centre. Surrounding the target are 20 scintillation counters (M array) which, with an "end cap" counter, subtend .91 of 4π from the target centre. Particles entering the spectrometer from the target must pass through a scintillation counter (T counter) which is used to determine whether the particle comes promptly from the annihilation or whether it leaves the target at a later time.

On either side of the target are a pair of identical drift chamber triplets (R1, R2, N1, N2). Each of the X (vertical), U (+30°), and V (-30°) planes has an active area of 50 cm by 150 cm. Together the NDC and RDC pairs can be used to find the annihilation vertex and show that it is well within the liquid hydrogen volume. The RDC pair alone is used to identify those particles coming from the stopping region in the X dimension, and with the identical P1 and P2 drift chamber triplets to reconstruct the trajectories of charged mesons which pass completely through the magnet.

Inside the magnet aperture, in which the field is perpendicular to the plane of the diagram, are seven proportional wire chambers (B, C1, C2, D1, D2, E1, E2). All seven chambers are similar, with a 27 cm by 126 cm active area and 3 mm wire spacings. The C1, C2, D2, and E2 chambers have three anode planes, X (vertical), Y (horizontal), and U (45°), and are 73 mm thick. The B, D1, and E1 chambers have X anode wires only and are 41 mm thick. The B, C1, and C2 chambers track particles which turn 180° in the magnetic field (40-100 MeV/c at 7.5 kG), while the D1 and D2 or E1 and E2 chambers track particles of higher momentum (120-400 MeV/c). In December 1983 only the X wires of the proportional chambers were instrumented.

On either side of the B wire chamber there is a scintillation counter (V, Q counters) which is double ended, 14 cm in height, 80 cm long, and 6 mm thick. Between the V counter and the B chamber there is a 5%

radiation length lead converter which is also 14 cm by 80 cm. Between the R drift chambers and the magnet aperture are six double ended scintillation counters (A hodoscope), each 21.5 cm wide, which are used to tag trajectories which turn 180° degrees in the magnet and exit through the front. This requires sensitive timing (0.8 ns) between the Q counter signal and the signal from the A counter to discriminate returning tracks (a "late A") from tracks entering the spectrometer. Two single ended scintillation counters (D and E) cover the active areas of the D2 and E2 chambers and five double ended scintillation counters (P hodoscope) cover the active area of the P drift chambers.

The event trigger is formed in two stages. The first stage requires a coincidence between the signals from the S counters and the Q counter, and any combination of P, D, E, or late A signals. This first stage gates the CAMAC ADC's and TDC's and triggers the second stage of the trigger. The second stage of the trigger logic occurs in CAMAC programmable logic units. It requires a specific pattern of V, D, E, P, and late A signals to generate a specific type of trigger. A "side-side gamma" trigger, for example, requires hits in the D and E counters, no hit in the V counter, and no P or late A signals. The second stage triggers the data collection system to read the data and prepare an event packet to be written to tape. If any stage's requirements are not met, the CAMAC systems are cleared and the trigger is reset. This paper deals only with "trapped" events which require one D or E signal, a V signal, and no P or late A signals. All other event types (returning mesons, mesons which go through the P hodoscope, double mesons, and gammas) are being analyzed.

Upon receiving an event trigger the proportional chambers and the drift chambers use dedicated CAMAC to collect and read out data. Using separate CAMAC crates with specialized crate controllers, these systems sort and compress the data from the chambers. The data are then sent via separate busses to CAMAC buffers housed in a crate with a LeCroy CAB intelligent crate controller. The TDC, ADC and event topology information from the scintillators is also sent to the CAB controlled crate. The CAB collates the event's data and forms an event packet. The event packets are stacked into buffers in the CAB memory according to event topology. When a buffer becomes full it is sent to a PDP 11/60 computer to be written to tape. The CAB can process 2000 events per second in this way, and write 300 events per second to tape.

In addition to controlling the tape drives, the PDP computer unpacks event buffers and makes histograms of parts of the data. This allows chamber and counter performance to be monitored on-line. The entire data collection system is continuously monitored by a Hewlett Packard desktop computer. The HP computer also serves to program the CAMAC logic units and initialize all CAMAC systems before each run.

3. The data

In December 1983 we completed our first data collection period. We received a total of 44 hour-long spills, 20 of which were used for tuning and debugging the apparatus, and 24 used for writing data onto tape. The off-line analysis which followed uncovered problems in the data sample requiring corrections. Data taken with the spectrometer magnet turned off contain particle trajectories going straight through the drift and proportional chambers. This set of data was used to check the surveyed positions of the chambers. Discrepancies of a few millimetres were found that were due to errors in installation, and to inadvertent movement after surveying. After these discrepancies were checked by resurveying the chambers, modifications to the analysis routines were made to correct them.

The most severe cut to the data is due to faulty proportional chamber electronics. During many of the spills a set of contiguous wires in one of the chambers stopped working. This failure required the replacement of a chamber amplifier card or a CAMAC module. The data were scanned before the analysis and classified by severity of wire chamber failure. For final analysis only the most conservative set, those data which contained no dead or hot wires, was used. This choice was made to avoid spurious effects which might be associated with using an imperfect data sample. It reduced the data sample to the equivalent of five hours of beam time. The effects described below are also observed in the full data sample.

Two other cuts were placed on the data sample to improve its quality, a vertex cut and a cut on the fiducial volume in the magnet. The particles' trajectories were found from the locations of the wires hit in the B, C2, and D2 or E2 chambers. This trajectory was extrapolated to the beam line to obtain the vertex position to a few centimetres. The cut on this vertex position was mild and was intended only to eliminate those tracks which originate up or downstream of the target. The fiducial cut was made to eliminate trajectories which do not traverse much of the magnet volume and so have low momentum resolution. Particles seen in the E chambers on the right-hand side of the magnet were required to have passed through the left two-thirds of the Q counter, and those in the D chamber, the right two-thirds.

4. The results

Fig. 2 shows the inclusive momentum spectra for positively charged mesons (a), and negatively charged mesons (b) between 150 and 300 MeV/c. The solid lines are fits to the data which will be described below. Two significant enhancements are seen in the positive spectrum, and one is seen in the negative spectrum.

Table 1 contains a list of expected lines in the momentum spectra coming from known processes, notably from charged kaons stopping in and around the hydrogen target. The expected strengths of these background signals is somewhat uncertain due to the lack of published data on the kaon momentum spectrum below 100 MeV/c from proton-antiproton interactions at rest. None of these lines can account for the effect at 191.5 MeV/c in the negative spectrum, fig. 2(b).

The enhancement at 229.2 MeV/c in the positive spectrum (fig. 2(a)) matches in both position and strength what is expected from the decay at rest, $K^+ \rightarrow \mu^+\nu$ (τ = 12.4 ns). Fig. 3 shows the momentum spectrum for positive particles which pass through the the T counter 2 ns or more after the proton-antiproton annihilation occurred. The data in fig. 3 are fit to a third order polynomial and two Gaussians centred at 227.9 ± 0.3 and 195.3 ± 0.6 MeV/c. The positions of these peaks match the momenta (after energy loss in the target and counters) of muons and pions from K^+ decay. The ratio of their strengths, approximately 3.8 : 1, is as expected from the K^+ decay branching ratio and the pion's probability of decay or interaction in the material in the spectrometer. A similar graph of the momentum spectrum for late negative mesons shows no significant structure. Negative kaons which come to rest are captured, forming hypernuclear states or reacting with hydrogen (table 1).

TABLE 1

Possible lines due to background processes

Charge	Momentum[(a)] (MeV/c)	Process	Predicted area (events)
Positive	163	$K^-p \rightarrow \Sigma^-\pi^+$	131 ± 30
	171	$\bar{K}^op \rightarrow \Sigma^o\pi^+$	< 1
	175	$\Sigma^+ \rightarrow n\pi^+$	14 ± 4
	196	$K^+ \rightarrow \pi^+\pi^o$	160 ± 50
	197	$K^o_s \rightarrow \pi^+\pi^-$	< 1
	229	$K^+ \rightarrow \mu^+n$	600 ± 150
Negative	171	$K^-p \rightarrow \Sigma^+\pi^-$	59 ± 15
	183	$\Sigma^- \rightarrow n\pi^-$	< 1
	197	$K^o_s \rightarrow \pi^+\pi^-$	< 1

(a) The momenta quoted include energy loss in the apparatus

The spectra in fig. 2 were combined, less the regions between 186-204 and 224-236 MeV/c in the positive spectrum and between 186-198 MeV/c in the negative spectrum, and fit to a seventh order polynomial. The results of this fit were used as a background for the two spectra separately. The omitted regions in the positive spectrum were fit to the background plus three gaussian peaks centred near 192, 197 and 229 MeV/c, the latter two constrained to be separated by 32.3 MeV/c and to have a ratio of areas of 3.8 : 1. The three gaussians were also constrained to have widths of 2.4, 2.4 and 2.8 MeV/c respectively, consistent with the observed width of the 229 MeV/c calibration line shown in fig. 3. The omitted region in the negative spectrum was fitted to a single gaussian of fixed width. The ratio of the areas of the gaussians near 192 MeV/c in the positive and negative spectra is 1.65 ± 0.61, consistent with unity. The fits are shown as solid lines in fig. 2 and the results of the fits are given in table 2. The χ^2 per degree-of-freedom of these fits are 79/72 for the positive spectrum and 103/73 for the negative.

TABLE 2

Results of fits to Gaussian line shapes plus background for the data of figs 2(a,b); see text for details

Charge	Momentum (MeV/c)	Sigma (MeV/c)	Events	#SD
Positive	192.2 ± 0.9	2.4	416	3.2
	196.9 ± 1.4	2.4	213	1.5
	229.2 ± 0.7	2.8	810	5.5
Negative	191.5 ± 0.6	2.4	688	5.2

5. Conclusions

We are unable to associate the lines at 192 MeV/c with any known process. We interpret these lines, corresponding to a pion momentum of 200 MeV/c (after correction for 8 MeV/c loss before reaching the C1 chamber), as evidence for a bound state in the proton-antiproton system: $\bar{p}p \rightarrow C^{\pm}\pi^{\mp}$. The mass of the charged C recoiling off the pion is 1620 ± 1 MeV/c². Its width is less than 2.3 MeV/c², limited by our resolution. The yield per annihilation of the state is determined to be 3×10^{-3}, based on a Monte-Carlo study of the spectrometer's acceptance in this topology at this momentum.

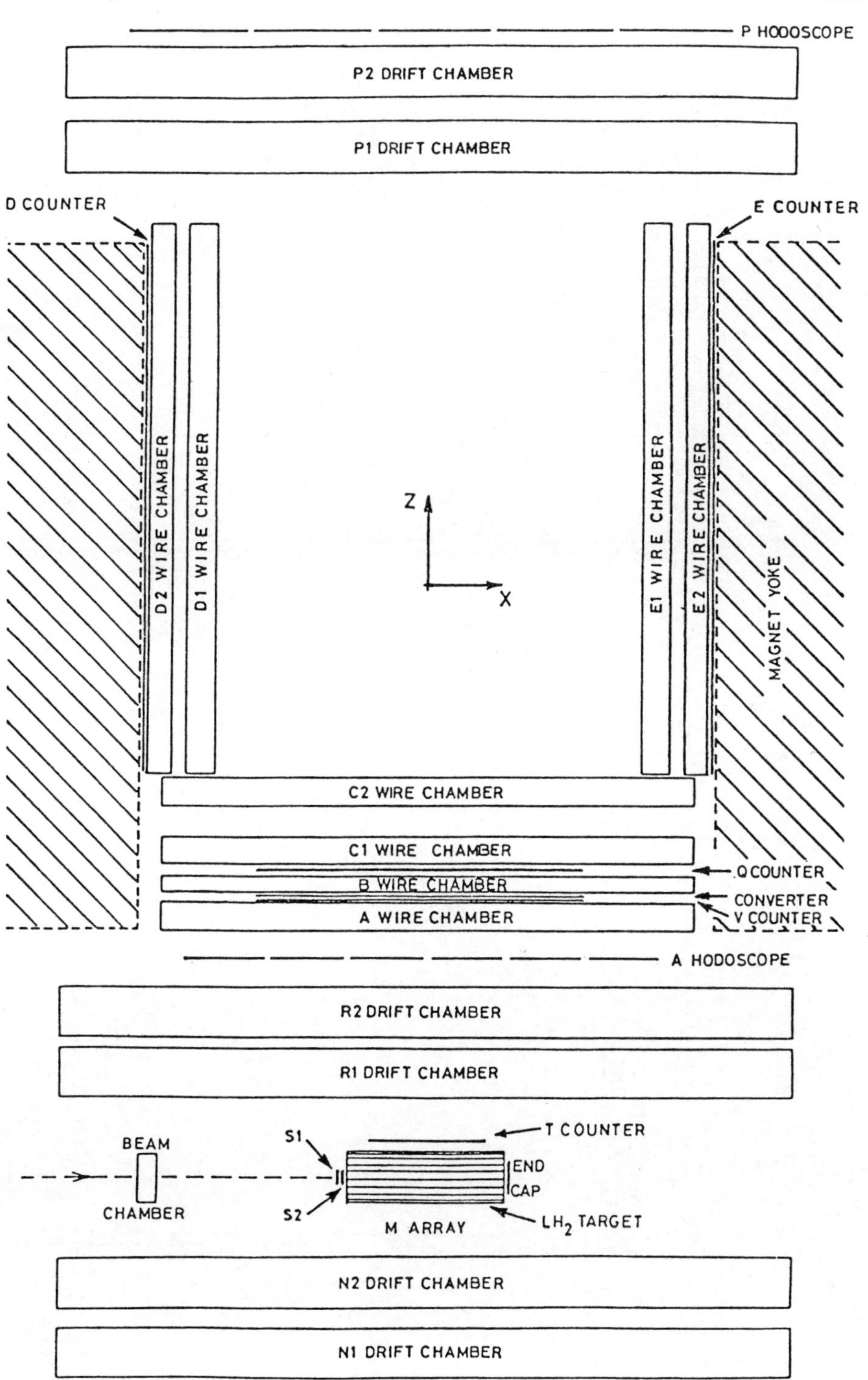
P HODOSCOPE
P2 DRIFT CHAMBER
P1 DRIFT CHAMBER
D COUNTER
E COUNTER
D2 WIRE CHAMBER
D1 WIRE CHAMBER
Z
X
E1 WIRE CHAMBER
E2 WIRE CHAMBER
MAGNET YOKE
C2 WIRE CHAMBER
C1 WIRE CHAMBER
Q COUNTER
B WIRE CHAMBER
CONVERTER
A WIRE CHAMBER
V COUNTER
A HODOSCOPE
R2 DRIFT CHAMBER
R1 DRIFT CHAMBER
BEAM
CHAMBER
S1
S2
T COUNTER
END
CAP
M ARRAY
LH_2 TARGET
N2 DRIFT CHAMBER
N1 DRIFT CHAMBER

Fig. 1

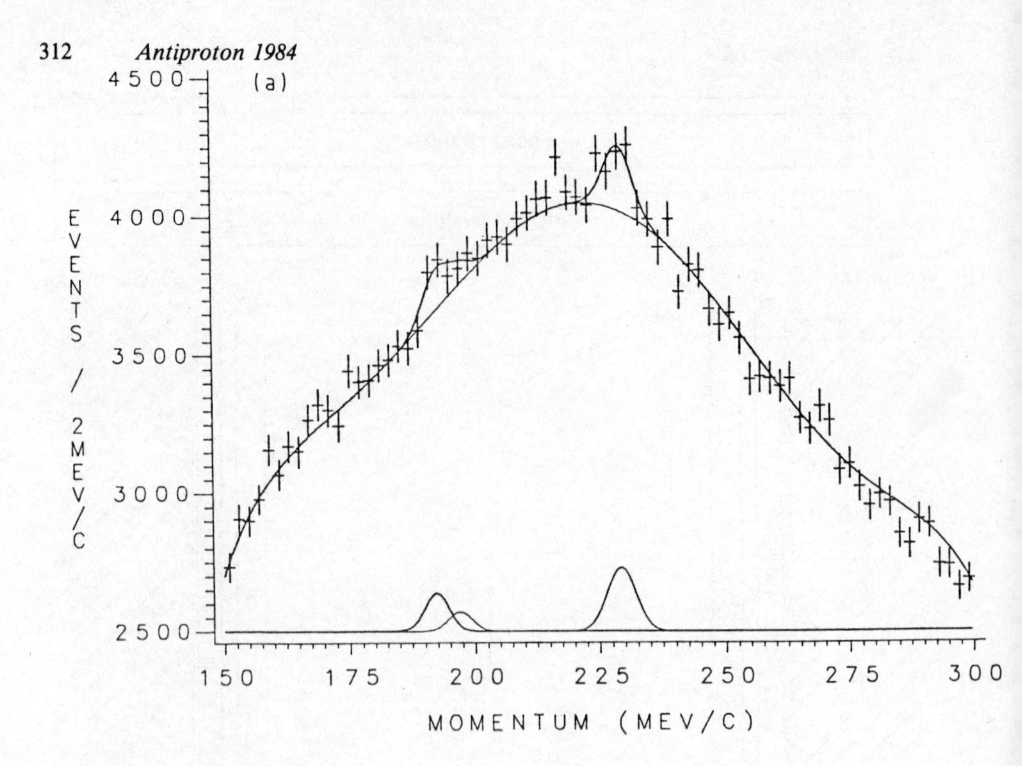
(a)
EVENTS / 2MEV/C
4500
4000
3500
3000
2500
150
175
200
225
250
275
300
MOMENTUM (MEV/C)

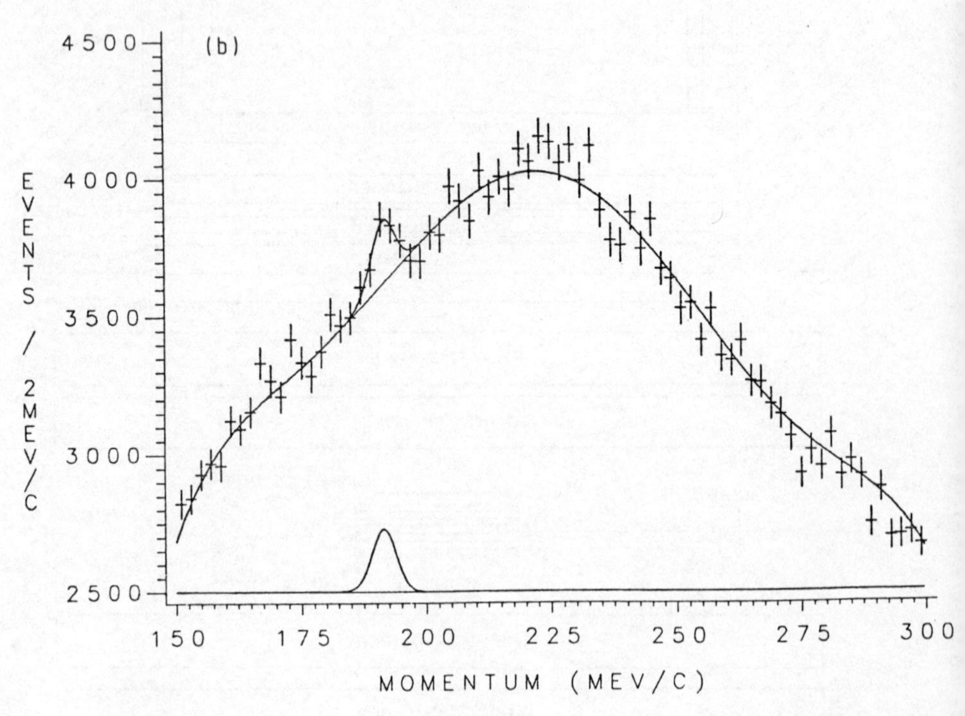
(b)
EVENTS / 2MEV/C
4500
4000
3500
3000
2500
150
175
200
225
250
275
300
MOMENTUM (MEV/C)

Fig. 2

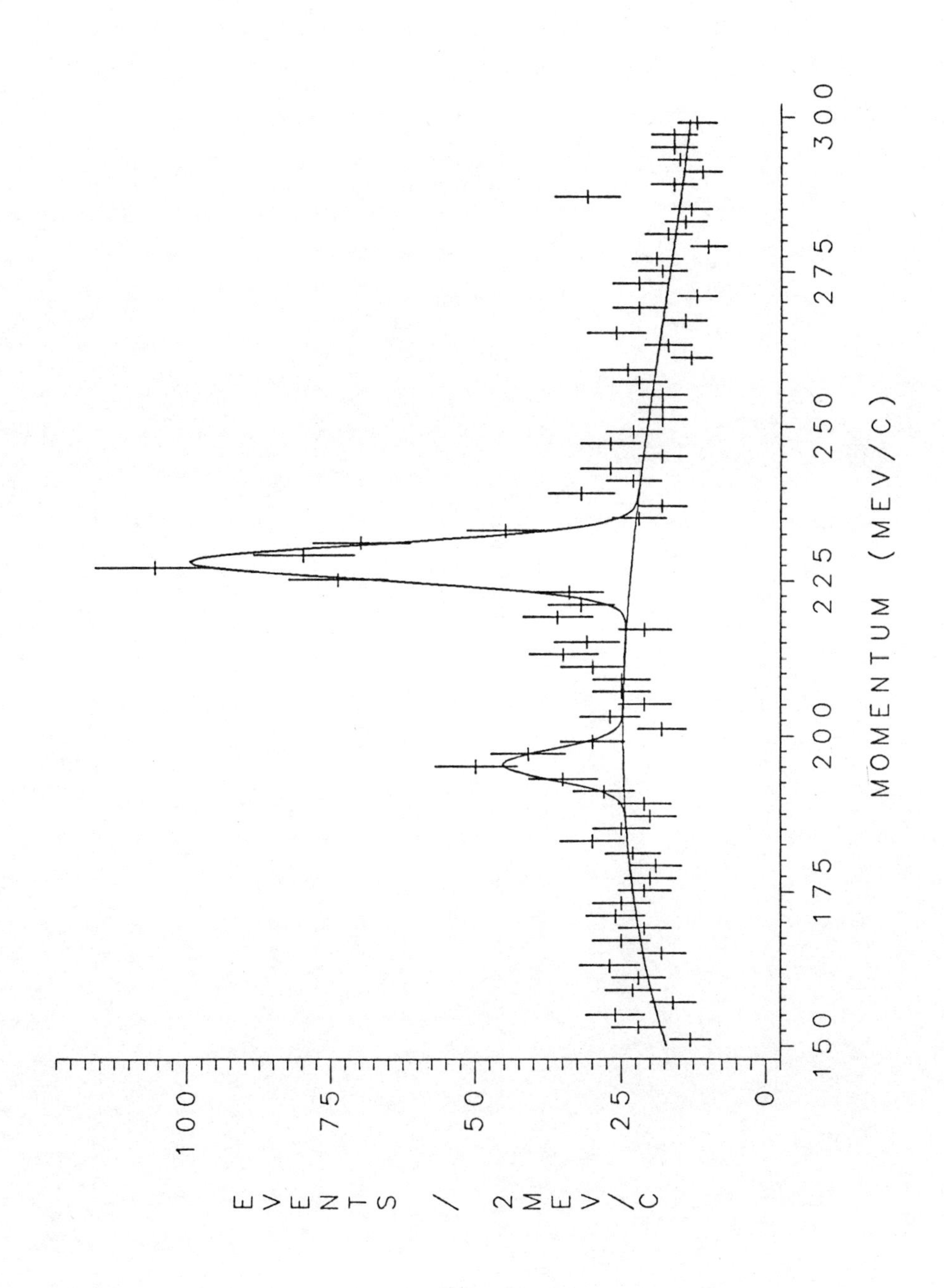
EVENTS / 2MEV/C
100
75
50
25
0
150
175
200
225
250
275
300
MOMENTUM (MEV/C)

<u>Fig. 3</u>

In search of p̄N bound states

D.Bridges[5], H. Brown[2], I. Daftari[7], R. Debbe[3], W. Fickinger[3], L. Gray[1], A. deGuzman[7], T. Kalogeropoulos[7], R. Marino[3], D. Peaslee[6], Ch. Petridou[7], D.K. Robinson[3], G. Tzanakos[4], R. Venugopal[7].

Bloomsburg[1]-Brookhaven[2]- Case Western Reserve[3]- Columbia[4]- Lemoyne[5]
Maryland[6]- Syracuse[7]

Abstract. The inclusive $\pi^{+,-}$ spectra from $\bar{p}d$ annihilations at rest have been measured in a high statistics experiment. The π^- shows a pronounced peak at $\sim$ 350 MeV/c which fits to at least one resonance recoiling against the π^-:mass X(1485±11) MeV and width $82\,^{+27}_{-19}$ MeV. A similar but less pronounced peak is seen in the π^+ implying I=1. Another peak is also observed in both $\pi^{+,-}$ spectra at $\sim$ 180 MeV/c, a region of rapidly changing acceptance. Other features of these structures are presented.

1. Introduction

We report first observations from AGS experiment E-772. The objective of this experiment is a search for $\bar{N}N$ bound states formed in $\bar{p}d$ interactions at rest. The search is based on studies of high statistics charged pion and proton inclusive spectra. Existence of such states would show as peaks in these spectra. Today we present the pion spectra.

2. The Experiment

2.1 Spectrometer-Beam

The measurements were made with the single arm magnetic spectrometer described elsewhere [Lowenstein et al.]. Fig. 1 shows those features of the spectrometer relevant to the present work. The spectrometer is a large (92" horizontal, 28" vertical) apperture magnet covered with the drift chambers RDC-UDC and PDC. The central field for this work is 7 KG and has been mapped to an accuracy of 1:10^3. The spectrometer is installed in a partially separated antiproton beam.

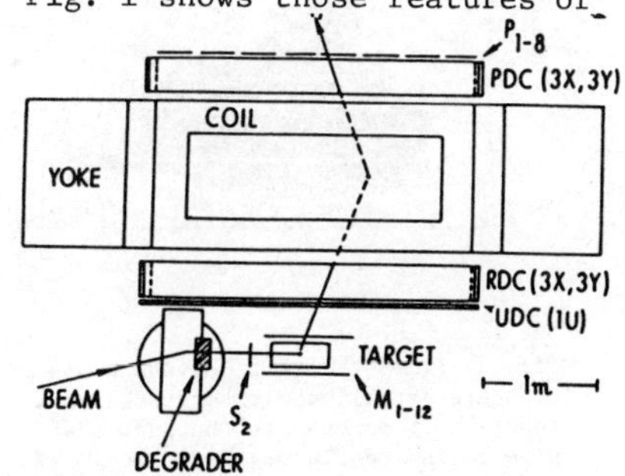

Fig. 1. Plan view of the spectrometer. The UDC-RDC and drift chambers determine the entrance and exit from the magnetic field region. The vertical gap of the magnet is 28 inches. The $M_{3,4}$ and P_i counters are used for trigger and time-of-flight. Data were collected with both magnet polarities.

Antiprotons of 600 MeV/c, after passing thru a degrader, are brought to rest in a 50 cm long 20 cm diameter liquid deuterium target. About 2000 $\bar{p}$ stop per 2×10^{12} protons. About 15% of the antiprotons interact in flight. Antiprotons are identified clearly by TOF between S_2 and an upstream hodoscope and the S_2 pulse height.

2.2 Trigger-Data

The trigger required coincidence of an incoming antiproton with M-counters which shadow the target to the magnet apperture and any of the eight P counters. The $\bar{p}\cdot M$ coincidence was tight (5 ns) while the P was 80 ns. The experiment ran with the magnet current direction reversed regularly. About 4.5 and 3.7 million triggers were recorded corresponding to the two polarities.

2.3 Data Reduction

Tracks were made by "matching" straight segments in RDC-UDC and PDC chambers [Petridou]. 30% of the triggers yielded tracks going through the spectrometer and pointing to the target (Fig. 2). Their momenta were determined by the empirical relation $p=p_o + (PQ)p_1$ where the polarity (P) and charge (Q) take the values ±1. The terms p_o and p_1, functions of the track parameters in RDC and PDC, were determined by Monte Carlo simulations using the field map.

The identification of the particles is based on their time-of-flight (TOF) between the M and P counters. Fig. 3 shows the π^+ and proton bands while Fig. 4 shows selection of π^- and π^+ by TOF.

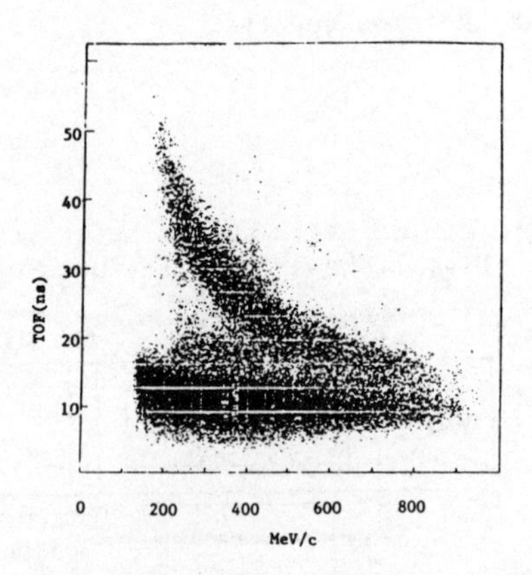

Fig. 3. Time-of-flight between M and P counters for all positive tracks as a function of momentum. The proton and pion bands are clearly separated. The corresponding figure for π^- shows only the π^- band.

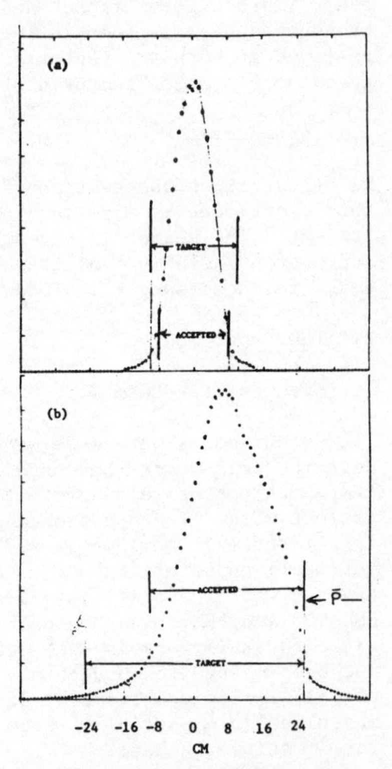

Fig.2. Intersection of tracks recon- in RDC-UDC with a vertical plane at the middle of the target and parallel to the beam. The non-Gaussian shape (b) is a consequence of a double image of the production target re- sulting to a two component monen- tum distribution.

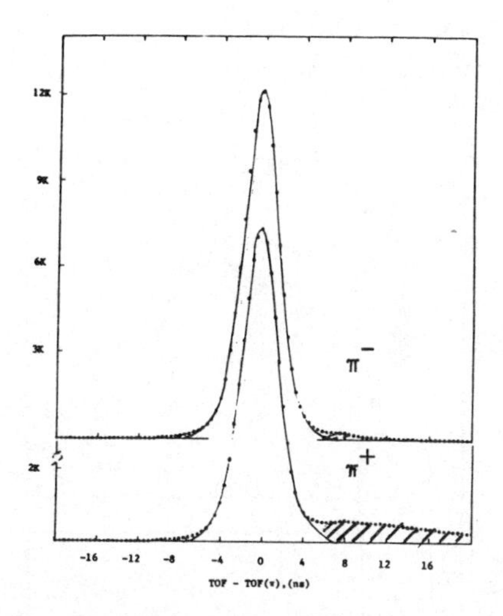

Fig. 4. Selection of $\pi^{+,-}$ by TOF. The tracks under the curve have been selected as pions.

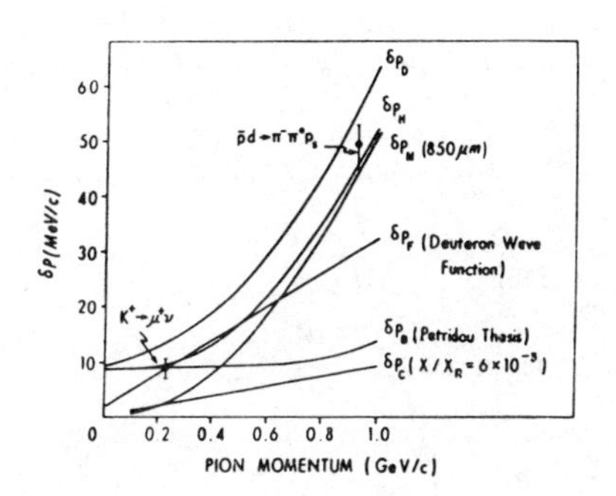

Fig. 5. Momentum resolutions expected in this experiment. δP_C, δP_B, δP_M, δP_F are the contribution of the coulomb scattering, momentum parametrization, measuring errors, and Fermi motion fudging. The observed widths of the $K^+ \to \mu^+ \nu$ (see Fig. 6) and $\pi^- \pi^o p_s$ are in agreement with expectations.

2.4 Resolution

The actual momentum resolution expected in this spectrometer is shown in Fig. 5 as δP_H which is equal to $[(\delta p_C)^2 + (\delta p_B)^2 + (\delta p_M)^2]^{\frac{1}{2}}$. The Coulomb scattering in the chamber contributes δp_C while the measuring precision of 850 μm in the drift chambers contributes δp_M. Further, the formula which calculates the momentum contributes δp_B. Our estimated resolution is in agreement with the width of the peak observed in secondary K^+(stops) $\to \mu^+ \nu$ decay (Fig. 6). In hydrogen a unique pion momentum corresponds to a unique missing mass. On the other hand, in deuterium a unique missing mass of the annihilation products will result in a recoiling Doppler shifted pion momentum due to the 'spectator' nucleon. This smearing is represented by δp_F and should be included in the evaluation of the missing mass error. Thus we expect an overall error in deuterium δp_D which is in agreement with the observation of the $\bar{p}d \to \pi^-(\pi^o p_s)$ in this experiment.

2.5 Acceptance

The detection efficiency as a function of momentum has a sharp cut off which depends on the magnetic field and the efficiency of reconstruction in PDC at large incident angles. Our main observations and conclusions in the paper are independent of the details of this efficiency because they depend on π^- and π^+ comparison. Apart from the interactions of pions from the vertex up to the P counters which may depend on charge, the detection efficiency for π^+, π^- will be the same for the spectra obtained by combining the two polarities, normalized to the same number of stopping antiprotons. This is confirmed by the identity of the π^+ and π^- angular distributions (Fig. 7).

Limited Monte Carlo studies of the acceptance show that it can be parametrized by

$$\varepsilon \propto 1 - \exp[-(p-p_o)/p_1 - ((p-p_o)/p_2)^2] \quad (1)$$

where p_o, p_1, p_2 are parameters. Detail studies of the acceptance in the region of the rapid increase are continuing.

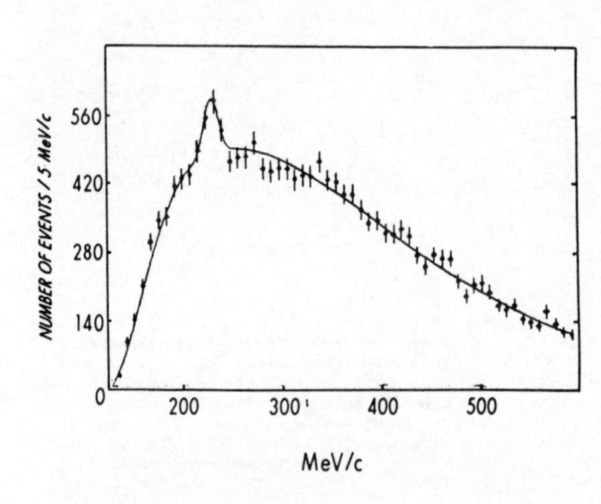

Fig. 6. π^+ spectrum of events with > 5 ns between S_2 and M counters for a fraction of our data. The peak is at 237±1.6 MeV/c and its width is 9.7±1.7 MeV/c. This peak corresponds to secondary K^+ stops in the target and decaying to μ^+ (236 MeV/c) + ν. The peak is present in both current polarities.

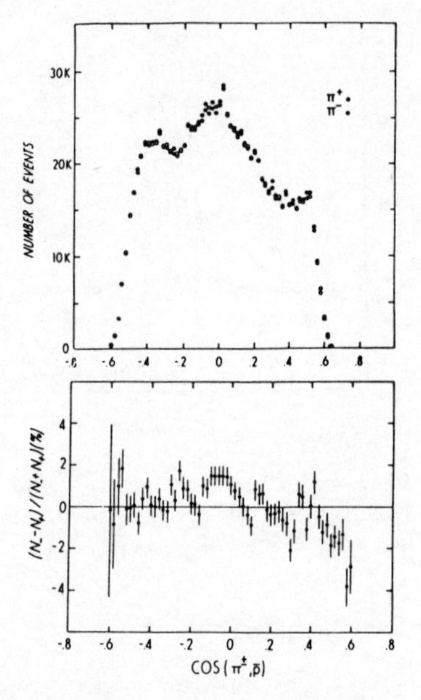

Fig. 7. Angular distributions of $\pi^{-,+}$ with respect to the beam. The π^+ events of polarity ±(-) have been normalized to the π^- of polarity -(+). The similarity of the distributions shows that the detection efficiencies are identical. The forward-backward asymmetry is due to the off center stopping of the beam.

3. Results

Fig. 8 shows the π^- and π^+ inclusive spectra obtained in this experiment where corrections have been made for losses due to interactions between the vertex and the P counters. Except for a small effect due to hydrogen the correction is the same for π^+ and π^-. The pions go through 10 cm of deuterium on the average, 5/32" of Al, 1/4" of scintillator, 35" of chamber gas (70/30 argon to ethane) and 50" of He which fills the magnet gap. Cross-sections equal to $1/2[\sigma(\pi^+p) + \sigma(\pi^-n)]A^{2/3}$ were used for the complex nuclei. The maximum correction, 9%, occurs at 270 MeV/c; half of which is due to deuterium.

The ratio of the π^- to the π^+ events is 1.38 and is in good agreement with that observed, 1.33±0.04, in the bubble chamber [Kalogeropoulos]. Since the acceptance and the interaction corrections are charge independent, differences between the π^- and π^+ spectra cannot be instrumental. The spectra are markedly different showing a pronounced structure in the π^- spectrum in the region 300 - 400 MeV/c. This difference can be better appreciated in the difference of the spectra from the same "background" distribution (Fig. 9) discussed below.

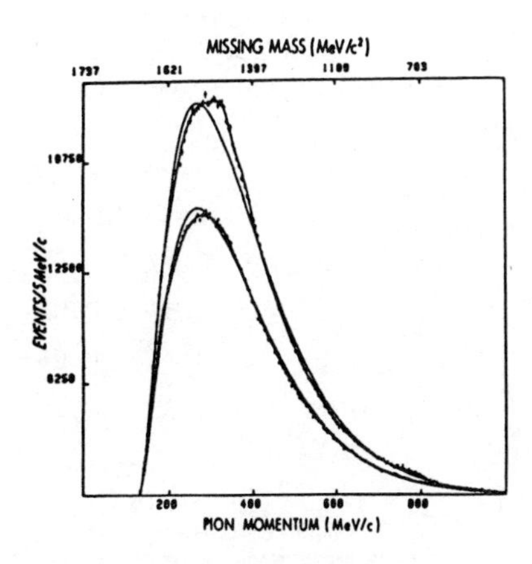

Fig. 8. The π^-(upper), π^+(lower) inclusive spectra after correction for losses due to interactions. The curves represent fits with and without resonances as described in the text.

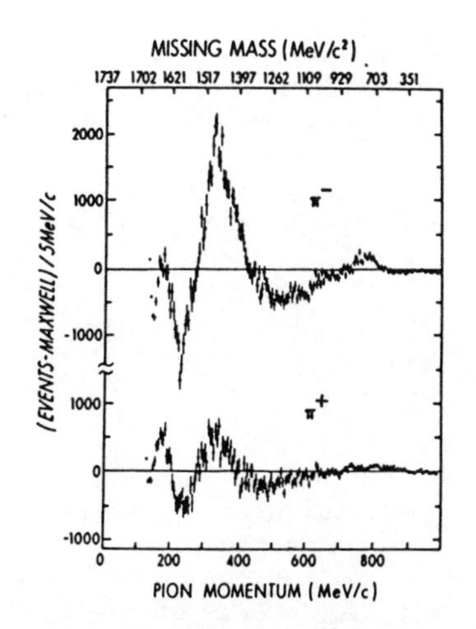

Fig. 9. Differences between the fitted background without resonances and the measured π^-, π^+ spectra of Fig. 8.

4. Interpretation

To fit the spectra a "phase space" distribution and the detection efficiency are needed. Inclusive π^+, π^-, K^o, Λ spectra from $\bar{p}d$ at rest annihilations have been studied using bubble chamber data [Roy]. These spectra are described well by Maxwell-Boltzman distributions by adjusting the temperature (T). A suppression of pions at low momenta is observed which needs a current factor (p/E)2 [Orfanidis]. We thus consider as "phase space" the distribution

$$dN/dp \propto (p/E)^2 \; p^2 \; \exp(-E/T) \tag{2}$$

where $E^2 = m^2 + p^2$. Using this form for "background" and the detection efficiency given by (1) a fit was attempted using the computer optimization program MINUIT [James]. The temperature (T) and the efficiency parameters p_o, p_1, p_2 were optimized. In Fig. 8 the results of these fits are shown for the optimized parameters T, [p_o, p_1, p_2] equal to 91 MeV [131, 116, 90] MeV/c respectively. Clearly our high statistics data do not fit to this structureless interpretation. The structures needed are clear in Fig. 9: one at $\sim$ 180 MeV/c, one or two at $\sim$ 350 MeV/c and one at $\sim$ 800 MeV/c corresponding to a missing ρ. We thus introduce resonances in the missing mass and attempt to fit the spectra by adding resonances to (2). The resonances are of the form: $BW = 1/[M - M_o)^2 + (\Gamma/2)^2]$ where M is the missing mass calculated from the pion momentum assuming at rest annihilation and zero spectator momentum. The width $\Gamma = (\Gamma_D{}^2 + \Gamma_R{}^2)^{\frac{1}{2}}$ where Γ_D depends on δP_D (see Fig. 5) and Γ_R is the natural width of the resonance. The results of the fits are shown in Fig. 8, Fig. 10 (by subtracting the "background" contribution given by eq. 2 from the data), and Table I.

The errors quoted correspond to variations of unity in χ^2/DF. The width Γ/D varies from 25 to 52 MeV/c^2 for masses from 1650 to 1400 MeV/c^2.

Table I. Results of the best fit to the pion momentum spectra. π^-, π^+ multiplicities of 1.74,1.31 have been used [Kalogeropoulos]. Errors are statistical and do account for possible systematic ones due to the specific parametrization.

$M(MeV/c^2)$	$\Gamma(MeV/c^2)$	BR in Percent $BR(\pi^- + X^{+,o})$	$BR(\pi^+ + X^-)$	$\frac{BR(\pi^+ + X^-)}{BR(\pi^- + X^o)}$
$1647 \pm ^{5}_{4}$	$44 \pm ^{10}_{7}$	8.8 ± .1	8.5 ± 0.1	0.97 ± 0.02
1485 ± 11	$82 \pm ^{27}_{19}$	5.4 ± 0.1	2.0 ± 0.1	0.37 ± 0.02
$1416 \pm ^{27}_{26}$	$125 \pm ^{77}_{40}$	3.6 ± 0.4	0.4 ± 0.3	0.11 ± 0.08
773 ± 4	188 ± 7	0.8 ± 0.03	0.3 ± 0.03	0.38 ± 0.04

The natural width Γ_R may thus be much smaller than the Γ of the table. The branching ratios are substantial and one to two orders of magnitude larger than the "typical" $\pi\rho$, $\pi\pi$ branching ratios.

Structure in the π^+ spectrum implies $I_X = 1$ for the recoiling state X, if $I_X = 2$ states are excluded. In this case the ratio $BR(\pi^+ + X)/BR(\pi^- + X^{o,+})$ can vary from 1/3 to 1 depending on the contributions of the $I_{N\bar{N}} = 0$ and 1 initial states. The ratio 1 corresponds to pure $I_{N\bar{N}} = 0$ contributions and 1/3 to pure $I_{N\bar{N}} = 1$. These bounds are satisfied by our results and indicate that the X(1647) comes from $I_{N\bar{N}} = 0$ while X(1485) from $I_{N\bar{N}} = 1$. A peak seen in π^- but not in π^+ implies $I_X = 0$. The X(1416) is not required in the π^+ spectrum and may therefore be an I-scalar in which case it may be identified as the E.

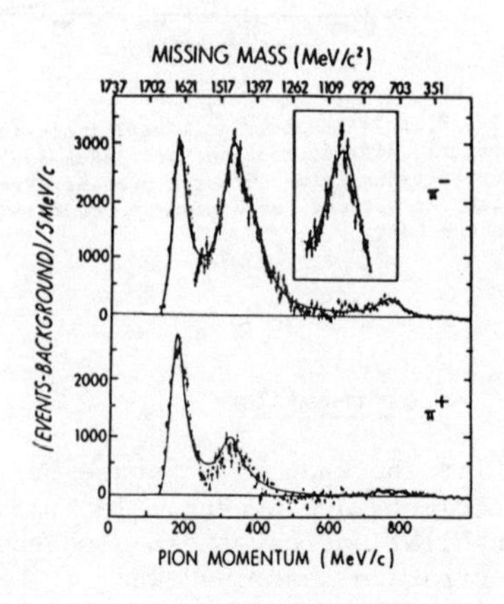

Fig. 10. Results of fits with four resonances after background subtraction. Insert shows the fit of the 350 MeV/c region to a single Breit-Wigner.

5. Discussion

Similar structure in the 300 - 400 MeV/c region of the inclusive π^- pion spectrum from $\bar{p}d$ annihilations at rest has been observed in a bubble chamber experiment [Roy]/ This effect was particularly associated with $\bar{p}n$ annihilations and in agreement with our $I_{N\bar{N}} = 1$ assignment (Fig. 11). Furthermore, the five pion annihilation channel $\bar{p}n \rightarrow 3\pi^- 2\pi^+$ [Roy, Bettini] at rest shows evidence for a $\rho^o\rho^o$ enhancement at M=1410 and $\Gamma = 90$ MeV/c^2. Also the $\bar{p}p \rightarrow 2\pi^- 2\pi^+ \pi^o$ [Defoix] indicates the possibility of a double peaked $\rho^o\rho^o$ structure at M = 1425, 1500 and Γ = 50, 40 MeV/c^2. The $\rho^o\rho^o$ interpretation in the five pion channels is difficult to prove and, if true, implies $I_X = 0$ which is incompatible with our $I_X = 1$ assignment for the dominant component of the observed structure.

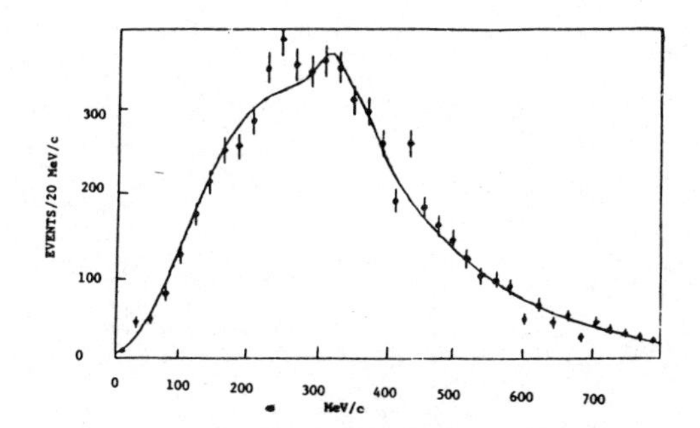

Fig. 11. Fit of the $\bar{p}n \rightarrow \pi^- + X$ bubble chamber data of Ref. 4 with the temperature and the (1416) and (1875) resonance parameters obtained in the fits of Figs 8, 10.

6. Conclusions

We observe a strong enhancement at 300-400 MeV/c in the π^- inclusive spectrum from $\bar{p}d$ annihilations at rest. This has been interpreted in terms of at least one resonance recoiling against the π^- with a mass and width 1485 ± 85 and 82^{+27}_{-19} MeV/c^2. Analysis of the π^+ spectrum reveals similar, but less pronounced, structure. This indicates an isovector ($I_X = 1$). The relative branching ratios imply production from $I_{N\bar{N}} = 1$ initial states. An additional structure observed in a region of rapidly changing acceptance requires further investigation. A shoulder in the π^- spectrum at M = 1416 MeV/c^2 needs verification from improved statistics.

References

*This work has been supported by the U.S. National Science Foundation and the Department of Energy.

Bettini, A. et al., Nuovo Cimento 42, 695 (1966).
Defoix, C. and Espigat, P., Proc. of the Symposium on Antinucleon-Nucleon Interactions (Liblice-Prague). CERN 74-18 Report (L. Montanet Editor).
James, F. and Roos, M., Computer Physics Communications 10, 343 (1975).
Kalogeropoulos, T. and Tzanakos, G.S., Phys. Rev. 22D, 2585 (1980).
Lowenstein, D.I. et al., in Proceedings of the Fourth European Antiproton Symposium, Volume 2, Barr, 1978, edited by A. Fridman (Editions du CNRS, Strasbourg, France), p. 669.
Orfanidis, S.J. and Rittenberg, V., Nucl. Phys. B56, 561 (1973).
Petridou, Ch., Ph.D. Dissertation, Syracuse University, 1983 (unpublished).
Roy, J. Proceedings of the Fourth $\bar{N}N$ International Symposium, Syracuse, 1975, edited by T. E. Kalogeropoulos and K. C. Wali, Vol. I, p. III-1.

Inst. Phys. Conf. Ser. No. 73: Section 5
Paper presented at VII Eur. Symp. Antiproton Interactions, Durham 1984

Preliminary results from antineutron–proton annihilation cross-section measurements from 100–500 MeV/*c* at the AGS

D. Lowenstein
Brookhaven National Laboratory, Upton, New York

C. Chu, E. Hungerford, T. Kishimoto, B. Mayes, L. Pinsky, L. Tang, Y. Xue
University of Houston, Houston, Texas

T. Armstrong, C. Elinon, C. Hartman, A. Hicks, R. Lewis, W. Lochstet, G. Smith
Penn State University, University Park, Pennsylvania

J. Clement, J. Kruk, B. Moss, G. Mutchler
Rice University, Houston, Texas

W. von Witsch
University of Bonn, Bonn, West Germany

M. Furic
University of Zagreb, Zagreb, Yugoslavia

Abstract. Preliminary results from an antineutron-proton annihilation measurement in the range of laboratory momenta between 100-500 MeV/c is reported. These preliminary data show some structure around an invariant mass of 1898 MeV. However, considerable data analysis remains before this structure can be associated with interactions in the $\bar{n}p$ final state system.

1. Introduction

This paper contains the preliminary results of an antineutron annihilation cross-section measurement between 100-500 MeV/c lab momentum undertaken at the AGS during the first half of 1984. Also included is a very preliminary total cross-section measured by the transmission technique. Prior to a discussion of the experimental technique and the presentation of the data, some comments about the motivation for doing antineutron physics in this momentum region are presented.

One may point out that since the $\bar{p}p$ system is a mixture of two isospin states one will ultimately need several isospin independent results to unfold any details seen in $\bar{N}N$ interactions. These can come from either $\bar{n}p$ or $\bar{p}n$ studies. However, the lack of pure neutron (or $\bar{p}$) targets means that the $\bar{p}n$ data must come from $\bar{p}$ beams incident on deuterium. Coulomb effects make this measurement difficult as one moves to lower momenta due to the proximity of stopping particles. Further, the presence of the proton, requiring a model dependent evaluation of the data, makes this approach undesirable. However, one may produce an $\bar{n}$ beam from an incident $\bar{p}$ beam by charge exchange, so that the $\bar{n}p$ interaction maybe investigated without the complexity of "spectator" particles.

2. Apparatus

As an initial attempt to investigate the $\bar{n}p$ reaction in the low momentum region below 400 MeV/c, the apparatus shown in Fig. 1 was constructed. It consists of three principle parts. First, the AGS Low Energy Separated Beam II (LESB II) is focused on a stack of twenty-1/4" scintillators, referred to as the source, which is designed to range the $\bar{p}$'s in the beam, and produce $\bar{n}$'s via in-flight charge exchange. This scintillator stack is surrounded by a veto box containing three layers of a lead-scintillator sandwich for the purpose of vetoing any $\bar{p}$ annihilations in the source. The trigger requires a beam $\bar{p}$ and no more than one scintillator coincidence in the veto box. One veto box scintillator is allowed in the trigger to preclude vetoing events where the neutron, resulting from the charge exchange, triggers one of the scintillators. Figure 2 shows the beam $\bar{p}$ momentum spectrum and Figure 3 shows the $\bar{p}$ "disappearance"(either by annihilation, charge exchange, or scatters out of the source) as a function of scintillator number. The spike in the last scintillator is due to the high momentum tail in the beam spectrum, which allows some $\bar{p}$'s to penetrate the entire stack, as well as earlier annihilations with sufficient forward forward charged particles produced to m'mic a forward moving $\bar{p}$. The veto box information is ignored for these data. Figure 4 shows the histogram of $\bar{n}$ triggers as a function of source scintillator number for "pure" events (i.e. <u>no</u> veto box hits). Our total $\bar{n}$-trigger rate is slightly more than 1% of the incident $\bar{p}$ flux. Given the known differential cross-sections for $\bar{p}p$ charge exchange, this number is compatable with a significant supression of the charge exchange from the carbon in the source scintillator. However, due to the greater angular spread in the charge exchange differential cross-section from carbon, as well as the details of the source/veto geometry and trigger conditions, one cannot yet quantify this suppression.

One and a half meters downstream of the source there is a 60 ℓ cylindrical liquid hydrogen target 50 cm long and 40 cm in diameter. This target is surrounded by 12 "barrel slat" scintillators, each "slat" viewed by photomultiplier tubes on either end. Outside these scintillators are a sequence of flat drift chambers, designed to determine the vertex of pions from any $\bar{n}$ annihilation in the liquid hydrogen. We estimate from Monte Carlo calculations that our net efficiency for finding a vertex from an $\bar{n}$ annihilation in the liquid hydrogen is ~50%. This includes the $\bar{n}p$ annihilation charge multiplicity, the detector acceptance, and particle decay. For the preliminary data reported here no angle corrections have been applied to the drift times for non-normal incidence tracks. The vertex resolution attainable under these circumstances can be estimated from the image of a thin scintillator paddle located in place of the Hydrogen target. Figure 5 shows a projection from the vertex reconstruction of the thickness dimension of a 0.64 cm thick scintillator. This result suggests an rms value of ~1 cm. We expect to improve this resolution by a factor of 2 but the present value is acceptable for these preliminary results.

A calorimeter is located downstream of the target and consists of 12 planes of crossed scintillator hodoscopes and crossed 1.27 cm diameter drift tubes coupled by two's for read-out. A 2.54 cm Al absorber is included in all but the first three modules, which in total represents typically 2-1/2 interaction lengths to the $\bar{n}$'s. All of the scintillators in the experiment have both ADC and TDC outputs recorded on magnetic tape.

3. Annihilation Cross-Section

From the frequency of $\bar{n}p$ annihilation as a function of $\bar{n}$ momentum, where the momentum is determined by time-of-flight between the last source scintillator and the reconstructed vertex; and from the $\bar{n}$ flux as a function of momentum passing through the target as estimated by the calorimeter, one can calculate the annihilation cross-section. These two raw spectra are shown in Figure 6 with different relative normalizations. The calculated cross-section is shown in Figure 7. The curve shown is an arbitary renormalization by (2/3) of the well known empirical fit: 35 + 35/p (GeV/c) to the $\bar{p}p$ annihilation cross-section. Also shown are the previous data of Gunderson et. al.[1] Figure 8 presents the same data plotted as $\beta\sigma$ vs. Kinetic Energy. Also shown in Figure 8 is an arbitrarily renormalized calculation by Dover.[2]

The appearance of the structure at about 1898 MeV invariant mass must be regarded as extremely preliminary at this time and the structure is subject to change. It may be noted that Kalogeropolous and Tzanakos[3] have previously reported a structure in $\bar{p}d$ annihilation at the same invariant mass. However, their effect is evident only when the recoil momenta of the "spectator" proton is above 170 MeV/c. It is not clear, therefore, just what correlation their measurement has with any effect that may be present here.

4. Total Cross-Section

By taking calorimeter data with both target empty and target full one can calculate a total cross-section by the transmission method. At this time considerably less calorimeter data has been analyzed than drift chamber annihilation data. Further, evidence exists that a significant amount of background remains in the calorimeter data from beam related effects. These data therefore contain large absolute systematic errors and poor relative statistics. Figure 9 shows the target full and target empty yields as a function of n momentum. The total cross-section is proportional to the log of the ratio of these two curves. Figure 10 shows the calculated total cross-section. No corrections have been made for forward elastic scattering in these data. This crude estimate has a dip in the cross-section at the same place where the annihilation cross-section shows an enhancement. Present indications are that any discrepancy is more likely due to the low statistics and the background events present in the total cross-section.

5. Conclusions

The apparatus has performed well and at a minimum it has been demonstrated that accelerator antineutron beams and counter experiments are quite feasible With these results current plans exist to extend these data down to a laboratory Kinetic Energy below 1 MeV. We would like to recognize support from the U. S. Department of Energy and the National Science Foundation and to acknowledge the enthusiastic, competent and essential support of the entire AGS staff at Brookhaven National Laboratory.

+Work supported in part by: U. S. DOE AS05-76ERO, U.S. DOE-A-S05-81ER40032, and U.S. NSF PHY 82-19518.

References

1. B. Gunderson, et. al., Phys. Rev. D 23, 587 (1981).
2. C. Dover, private communication.
3. T. E. Kalogeropolous and G. S. Tzanakos, Phys. Rev. Lett. 34, 1047 (1975)

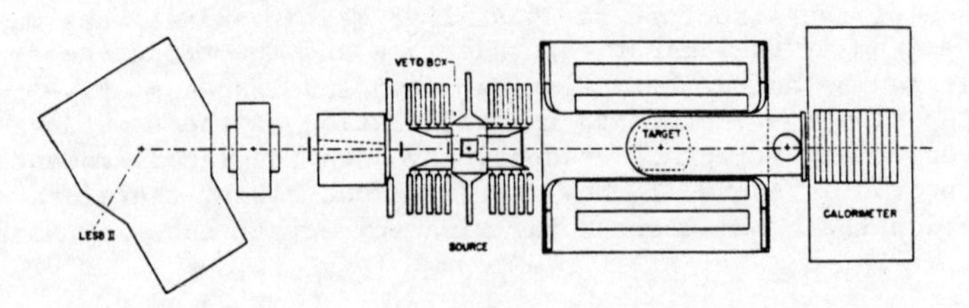

Figure 1

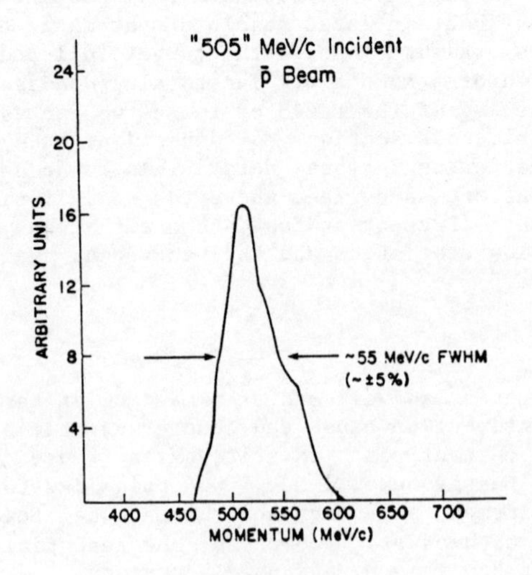

Figure 2

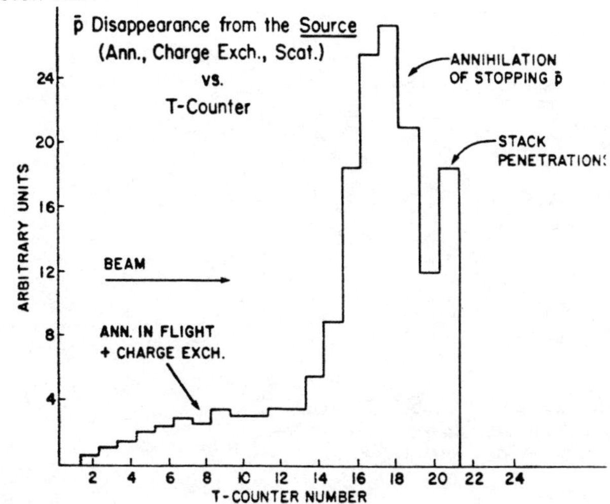

Figure 3

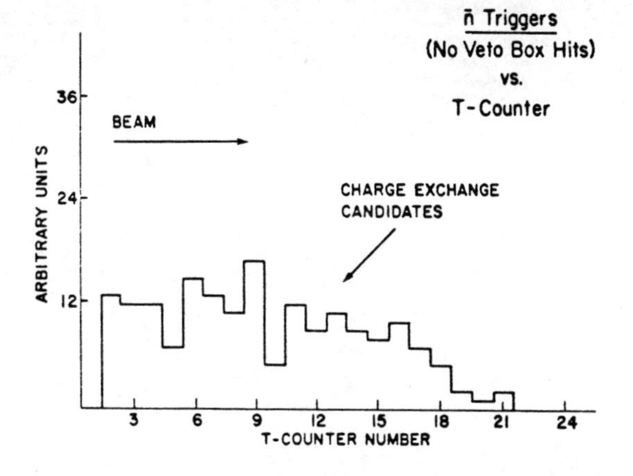

Figure 4

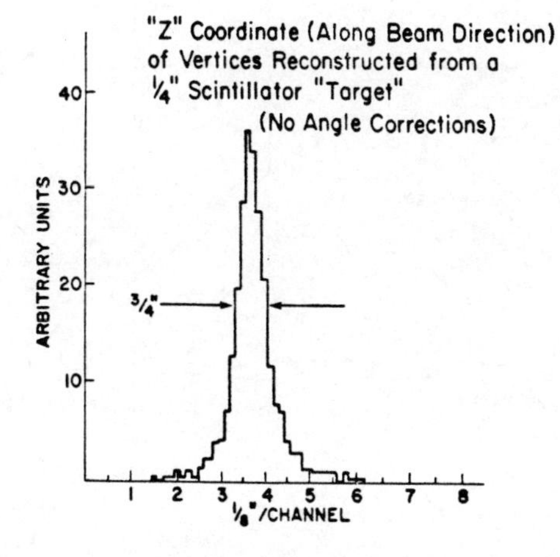

Figure 5

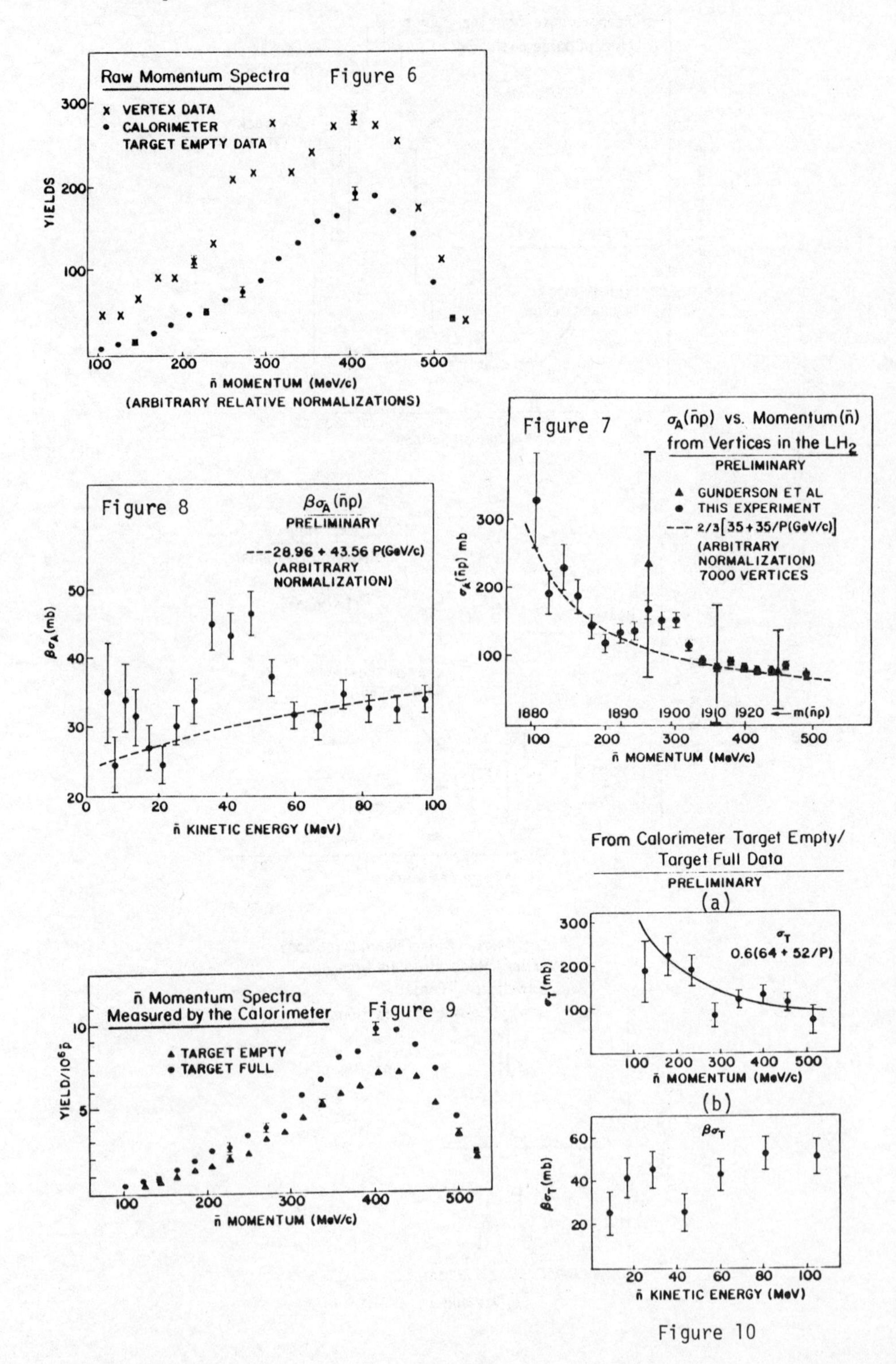

Raw Momentum Spectra
Figure 6
VERTEX DATA
CALORIMETER
TARGET EMPTY DATA
YIELDS
n̄ MOMENTUM (MeV/c)
(ARBITRARY RELATIVE NORMALIZATIONS)
Figure 7
σA(n̄p) vs. Momentum(n̄)
from Vertices in the LH2
PRELIMINARY
GUNDERSON ET AL
THIS EXPERIMENT
2/3[35+35/P(GeV/c)]
(ARBITRARY NORMALIZATION)
7000 VERTICES
σA(n̄p) mb
m(n̄p)
n̄ MOMENTUM (MeV/c)
Figure 8
βσA(n̄p)
PRELIMINARY
28.96 + 43.56 P(GeV/c)
(ARBITRARY NORMALIZATION)
βσA(mb)
n̄ KINETIC ENERGY (MeV)
From Calorimeter Target Empty/ Target Full Data
PRELIMINARY
(a)
σT
0.6(64 + 52/P)
σT(mb)
n̄ MOMENTUM (MeV/c)
(b)
βσT
βσT(mb)
n̄ KINETIC ENERGY (MeV)
n̄ Momentum Spectra Measured by the Calorimeter
Figure 9
TARGET EMPTY
TARGET FULL
YIELD/10⁶p̄
n̄ MOMENTUM (MeV/c)

Figure 10

Inst. Phys. Conf. Ser. No. 73: Section 5
Paper presented at VII Eur. Symp. Antiproton Interactions, Durham 1984

Antiproton results from KEK

K Nakamura† and T Tanimori§

†National Laboratory for High Energy Physics (KEK), Oho, Ibaraki 305, Japan
§Department of Physics, University of Tokyo, Tokyo 113, Japan

Abstract. Recent results for the $\bar{p}p$ reactions into two-body final states, $\bar{p}p \to \bar{p}p$, $\bar{n}n$, $\pi^+\pi^-$ and K^+K^-, are reported. The differential cross sections of these reactions were measured at 390, 490, 590, 690 and 780 MeV/c. Evidence has been found for a resonant state in the K^+K^- channel. It has a mass of 1935±15 MeV, a width $\lesssim$ 40 MeV, and possible quantum numbers of $J^{PC}=2^{++}$, $I^G=0^+$ or 1^-.

1. Introduction

We report some selected topics from our recent $\bar{p}p$ experiment at KEK. To begin with, however, we would like to introduce the antiproton experiments so far performed at KEK, because perhaps what is going on at KEK is not very well known in Europe. Table 1 summarizes these experiments. Of these, experiments E33, E74/I and E74/II were performed by a University of Tokyo group and collaborators. The results presented in this talk are from E74/II. In addition, measurement of the real-to-imaginary ratios of the $\bar{p}p$ and $\bar{p}n$ forward amplitudes between 367 and 764 MeV/c (from E74/I) and measurement of the reaction $\bar{p}C \to \bar{n}X$ at 590 MeV/c (from E74/II) have been reported in a poster session at this Symposium (Takeda et al 1984, Nakamura et al 1984a).

Measurements of the $\bar{p}p$ elastic scattering and charge-exchange reaction have been motivated by the investigation of the antinucleon-nucleon ($\bar{N}N$) interactions at low momenta. In view of the recent development of the $\bar{N}N$ potential models, it is important to supply good data to determine, or to improve, the phenomenological part of the $\bar{N}N$ potential. In particular, the charge-exchange reaction is among the least studied $\bar{p}p$ reaction channels despite its importance in determining the detailed structure of the $\bar{N}N$ scattering amplitudes. Previously, $\bar{p}p$ elastic differential cross sections were measured with high statistics only above 690 MeV/c (Eisenhandler et al 1976). For the charge-exchange reaction below 1 GeV/c, the measurements of the differential cross sections have poor statistics except the one at 700-760 MeV/c by Bogdanski et al (1976).

The reactions $\bar{p}p \to \pi^+\pi^-$ and K^+K^- have been measured primarily in search of high-mass meson resonances. Annihilation into two pseudoscalar mesons has many advantages for this purpose in spite of small cross sections: (i) Signals are simple and clean; (ii) There is a strong restriction on the quantum numbers of the mesons to be formed; and (iii) Partial-wave analyses can be done if polarization parameters are measured in addition to the differential cross sections. Indeed, the partial-wave analyses of the $\pi\pi$

Table 1. Antiproton experiments performed at KEK.

Exp.	Collab.	Measured quantities	Momentum (GeV/c)	Publications
E33 (CTR)	Tokyo-Hiroshima	$\sigma_T(\bar{p}p)$ (Search for a narrow resonance)	0.4-0.74	PRL 44('80)1439 PR D 29('84)349
		Forward $d\sigma/d\Omega(\bar{p}p)$	0.41-0.72	PL 103B('81)247 NP A, to be pub.
		$\sigma_{abs}(\bar{p}$-C, Al, Cu)	0.49, 0.6	NP A 360('81) 291
E62 (H_2 BC)	KEK-Kobe-Nara Women's Univ.-Niigata-Osaka City Univ.	$\bar{p}p \to K^\circ_S X$, $K^\circ_S K^\circ_S X$, $K^{*\pm}X$	3-4.5	ZP C 23('84)369
		$\bar{p}p \to \Lambda X$, $\Lambda\bar{\Lambda}X$, $\Sigma^\pm(1385)X$	3-4.5	ZP C, to be pub.
		$\bar{p}Ta \to \Lambda X$, $\bar{\Lambda}X$, $K^\circ_S X$	4	submitted for pub.
E68 (CTR)	Fukui-INS-KEK-Kyoto Sangyo-Meisei-Osaka-Tokyo Metropolitan Univ.	$\bar{p}p$ at rest $\to \gamma X$ $\pi^\circ X$ nX γnX		Analysis stage
E74/I (CTR)	Tokyo-KEK-Tsukuba	$\sigma_T(\bar{p}p)$ (Search for a	0.33-0.96	PRL 49('82)628
		$\sigma_T(\bar{p}d)$ broad resonance)	0.33-0.96	(not published yet)
		Forward $d\sigma/d\Omega(\bar{p}p, \bar{p}d)$	0.37-0.76	these Proceedings (preliminary)
		σ_{abs}, $d\sigma/d\Omega(\bar{p}$-C, Al, Cu)	0.47-0.88	PRL 52('84)731
E74/II (CTR)	Tokyo-Tsukuba	$d\sigma/d\Omega(\bar{p}p \to \bar{p}p)$	0.39-0.78	these Proceedings (preliminary)
		$d\sigma/d\Omega(\bar{p}p \to \bar{n}n)$	0.39-0.78	PRL 53, to be pub.
		$d\sigma/d\Omega(\bar{p}p \to \pi^+\pi^-)$	0.39-0.78	these Proceedings (preliminary)
		$d\sigma/d\Omega(\bar{p}p \to K^+K^-)$	0.39-0.78	these Proceedings (preliminary)
		$d\sigma/d\Omega(\bar{p}C \to \bar{n}X)$	0.59	these Proceedings

annihilation channels between 1 and 2 GeV/c (Martin and Pennington 1980, Martin and Morgan 1980) led to a rich spectrum of possible resonant waves. Previous high-statistics measurements of the $\bar{p}p \to \pi^+\pi^-$ and K^+K^- differential cross sections were performed by Eisenhandler et al (1975) between 0.79 and 2.43 GeV/c. No statistically significant data exist below 790 MeV/c.

2. Experiment

The measurement was performed with a low-momentum separated beam (K3) from the 12 GeV proton synchrotron at KEK. The differential cross sections ($d\sigma/d\Omega$) were measured at five beam momenta, 390, 490, 590, 690 and 780 MeV/c.

The momentum spread of the beam was ±2.5%. The experimental arrangement is shown in Fig. 1. A 17.5-cm long liquid-hydrogen target was surrounded by a five-layer cylindrical drift chamber (CDC). The target and CDC were placed in a magnet with the applied magnetic field of 2.5 kG at the center. The $\bar{p}$ beam was defined by the counters C1 (not shown in Fig. 1), C2 and C3, and its trajectory was determined by two MWPC's and CDC. Outside of the magnetic field were located four-layer planar drift chambers DC1-DC4. On the surface of the magnet pole pieces were attached scintillation counters (called the pole-face counters, not shown in Fig. 1) in order to detect charged annihilation products. In the forward and backward directions, two walls of TOF counters TF and TB were located. The $\bar{n}$'s were detected by iron-scintillator sandwich counters NB. An NB counter module consisted of 7 layers of 1-cm thick scintillator interleaved between iron plates with a total thickness of 24.8 cm.

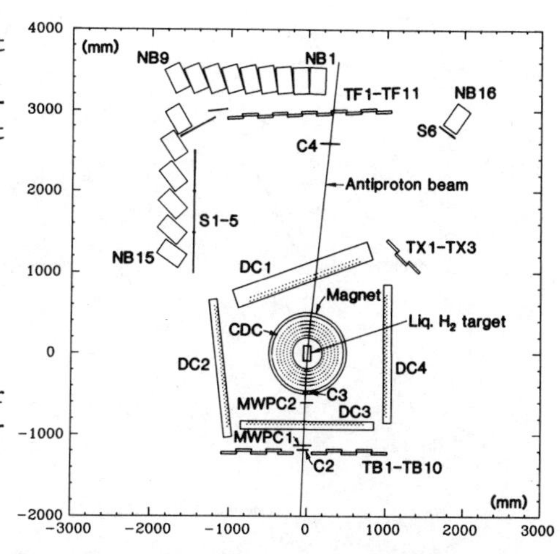

Fig. 1. Experimental arrangement. C2-C4, trigger counters; MWPC1 and MWPC2, multiwire proportional chambers with bi-dimensional readout; CDC, five-layer cylindrical drift chamber; DC1-DC4, four-layer drift chambers; TF1-TF11, TB1-TB10 and TX1-TX3, TOF counters; NB1-NB16, iron-scintillator sandwich counters; S1-S6, scintillation counters.

The events were triggered by $(C1 \cdot C2 \cdot C3)_{\bar{p}} \cdot \overline{C4}$, where $(C1 \cdot C2 \cdot C3)_{\bar{p}}$ represents the incident $\bar{p}$ defined by the counters C1, C2 and C3, and $\overline{C4}$ eliminated non-interacting $\bar{p}$ events. This loose trigger requirement was adopted because the intensity of the $\bar{p}$ beam was low (typically 200 $\bar{p}$'s per 10^{12} primary protons at 590 MeV/c) and several reaction channels were concurrently measured.

3. Elastic Scattering

For the measurement of $\bar{p}p$ elastic scattering, we mainly used the information from the drift chambers CDC and DC1 and the TOF counters TF1-TF11. For the forward and backward elastic scatterings, one of the final-state particles does not have sufficient kinetic energy to emerge from the target. For such cases, it was required that either a proton or an antiproton be identified by one of the TOF counters TF1-TF11. In the intermediate angular region, both proton and antiproton have sufficient kinetic energies (except at 390 MeV/c) to be observed in the CDC, and the elastic events were unambiguously identified by the angular correlation of the two charged tracks. No TOF information was required in this latter case.

The preliminary results obtained for the elastic-scattering $d\sigma/d\Omega$ are shown in Fig. 2. The error bars represent the statistical errors only. The systematic errors amount to ±3% for the forward and backward data points shown by the open diamonds, and ±5% for the data points in the intermediate angular region, shown by the open squares. At 390 MeV/c, the data are not shown in the intermediate angular region, because both proton and antipro-

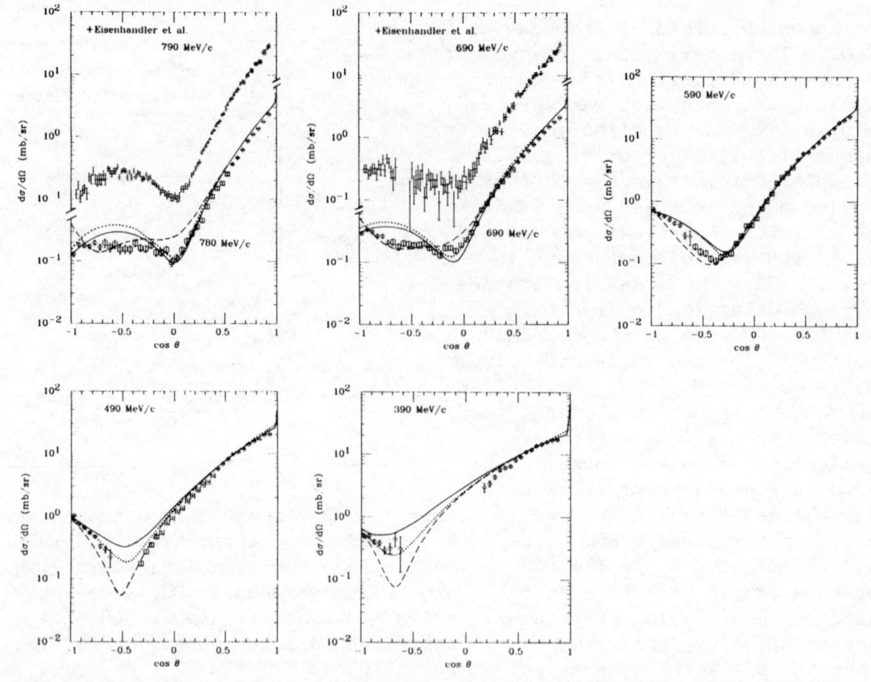

Fig. 2. The differential cross sections for $\bar{p}p$ elastic scattering. The data at 780 and 690 MeV/c are compared with those of Eisenhandler et al (1976). The curves show the predictions of the Nijmegen model (solid curves), the boundary-condition model (dashed curves) and the Paris model (dotted curves).

ton have too low kinetic energies to be observed reliably. Our results are compared with those of Eisenhandler et al (1976) where their data are available. At 690 MeV/c, both data agree well. At 780 MeV/c, both data agree in most of the angular regions, but there is some discrepancy around $\cos\theta \sim -0.5$.

The curves in Fig. 2 show predictions of some $\bar{N}N$ potential models: Nijmegen model (Timmers et al 1984), Paris model (Côté et al 1982) and boundary-condition model (Dalkarov and Myhrer 1977, Mizutani 1984) with a boundary radius of 0.5 fm. Clearly, none of these models completely accounts for the data, and this fact indicates the necessity for improving the models.

Figure 3 shows $d\sigma/d\Omega$ for the backward elastic scattering. Our results agree well with the high-statistics data of Alston-Garnjost et al (1979). The curves in this figure show the results of the potential-model calculations. Apparently, both Paris model and Nijmegen model exhibit good agreement with the data, but this only reflects the fact that these data were used in the fits for determining the parameters involved in these models.

4. Charge-Exchange Reaction

The identification of $\bar{n}$'s from the charge-exchange reaction $\bar{p}p \to \bar{n}n$ was performed on the basis of the pulse-height and TOF information from the NB counters. Additionally, we required the presence of no charged-particle trajectory in the CDC other than the incident $\bar{p}$, and no pole-face counter hits. The $\bar{n}$ detection efficiency of the NB counters was estimated to be about 73% with slight momentum dependence, from the measured detection efficiency for $\bar{p}$'s and a Monte Carlo simulation of the energies deposited by $\bar{p}$'s and $\bar{n}$'s annihilating in the NB counters. For further details of the event selection and applied corrections, see Nakamura et al (1984b).

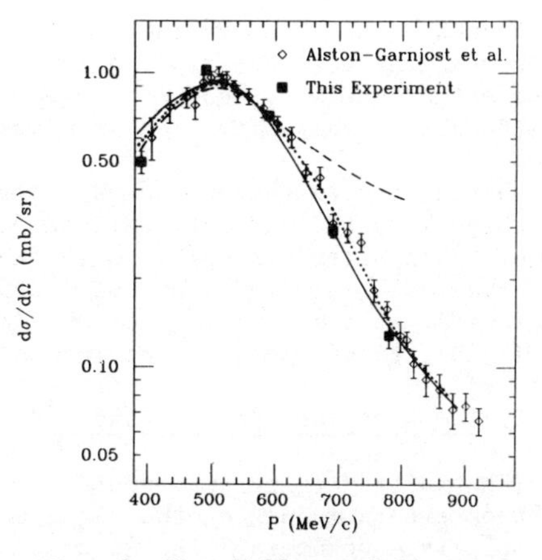

Fig. 3. The differential cross section for $\bar{p}p$ backward elastic scattering as a function of incident momentum. The curves are coded as in Fig. 2.

The angular distributions obtained are shown in Figs. 4a-4e. The error bars represent the statistical errors. The total systematic error amounts to ±8.5%. The angular distributions at 390, 490, 690 and 780 MeV/c clearly show the existence of a forward dip. At 590 MeV/c, there is a shoulder instead of a dip due perhaps to statistical fluctuation. Theoretically, the existence of the forward dip is explained by the interference of the pion-exchange amplitude with a coherent background amplitude (Leader 1976). The observation of such a structure below 1 GeV/c was first reported by Bogdanski et al (1976).

The predictions of the potential models are shown by the curves in Figs. 4a-4e. All three models predict the existence of the forward dip. The position and depth of the dip seem to be reasonably reproduced by the

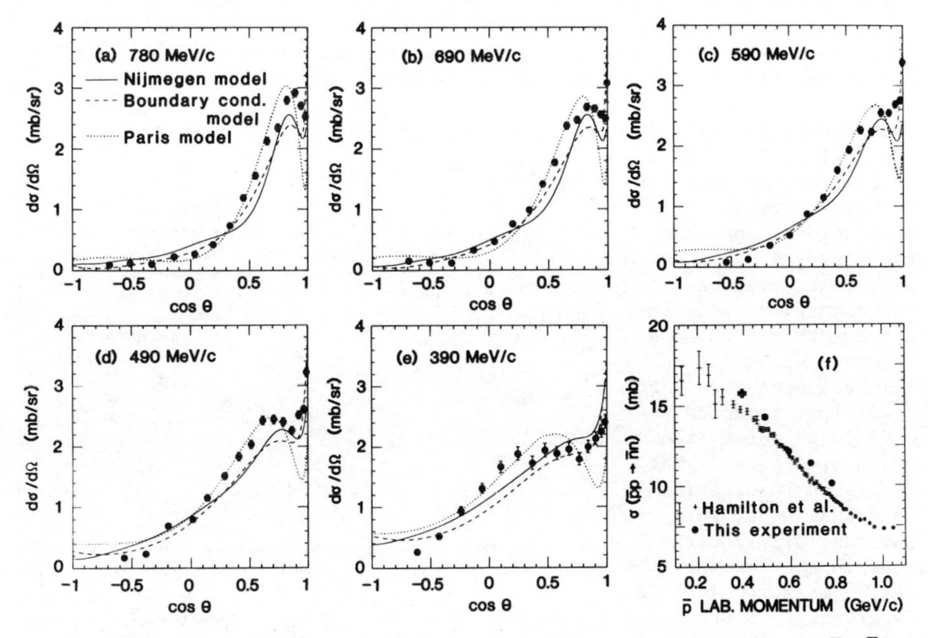

Fig. 4. (a)-(e) The differential cross sections for the reaction $\bar{p}p \to \bar{n}n$. The curves are coded as in Fig. 2. (f) The integrated charge-exchange cross section data are compared with the data of Hamilton et al (1980).

Nijmegen model and by the boundary-condition model. However, the Paris model exhibits better agreement with the data except in the dip region. As a matter of fact, none of these models completely explains the behavior of the data. This situation is similar to the case for elastic scattering.

The integrated charge-exchange cross section is shown in Fig. 4f and compared with the data of Hamilton et al (1980). The error bars indicate the statistical errors only. Our data are somewhat higher than the data of Hamilton et al (1980) except at 590 MeV/c. However, considering the systematic errors of ±8.5% for our data and ±5% for the data of Hamilton et al (1980), these data are not inconsistent.

5. The Reactions $\bar{p}p \to \pi^+\pi^-$ and K^+K^-

The two-body ππ and KK events were discriminated from multi-body annihilation events mainly by the requirements of coplanarity and no hits in the pole-face counters. The identification of the ππ and KK events was done by the TOF information at forward and backward angles, and by the difference in the opening angles in the intermediate angular region. The consistency of the two methods was checked in the angular regions where both methods can be used.

Below, we first present the $\bar{p}p \to \pi^+\pi^-$ results, and then the $\bar{p}p \to K^+K^-$ results. It should be noted that these results are quite preliminary. A detailed estimation of the systematic errors has yet to be done, but probably the data shown below have ±5 ∿ 10% of systematic errors.

The integrated cross section $\sigma(\pi^+\pi^-)$ is shown in Fig. 5 together with other data (Bizzarri et al 1969, Mandelkern et al 1971, Nicholson et al 1973, Eisenhandler et al 1975, Sai et al 1983). Our results are consistent with the other data except at 490 and 590 MeV/c where our data are somewhat higher than the other data. The momentum dependence of $\sigma(\pi^+\pi^-)$ looks smooth and there is no indication of resonance peaks.

The angular distributions for the reaction $\bar{p}p \to \pi^+\pi^-$ are shown in Fig. 6. The angle θ is defined by that of the negative outgoing particle. As the incident momentum decreases, a strong forward peak develops. These angular distributions were fitted to a Legendre expansion up to the 6th order. The resulting expansion coefficients normalized by A_0 are shown in Fig. 7. These results indicate that a J=1 wave substantially contributes in the low-momentum region measured in this experiment. Since the Legendre expansion coefficients A_5 and A_6 do not show significant

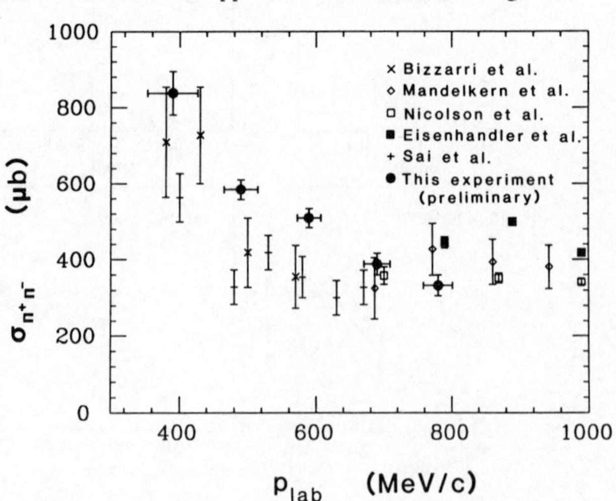

Fig. 5. Cross section for the reaction $\bar{p}p \to \pi^+\pi^-$.

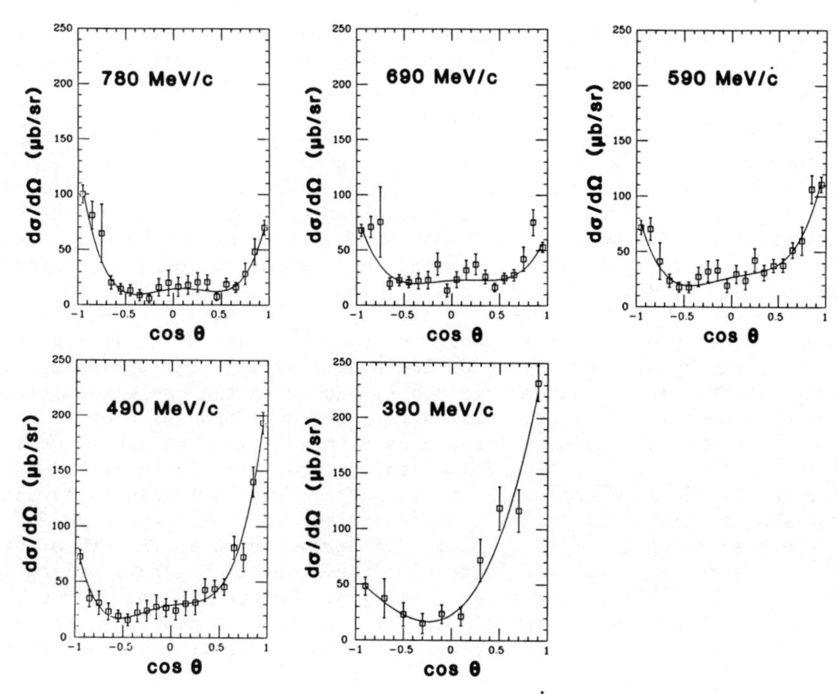

Fig. 6. Preliminary results for the $\bar{p}p \to \pi^+\pi^-$ differential cross sections.

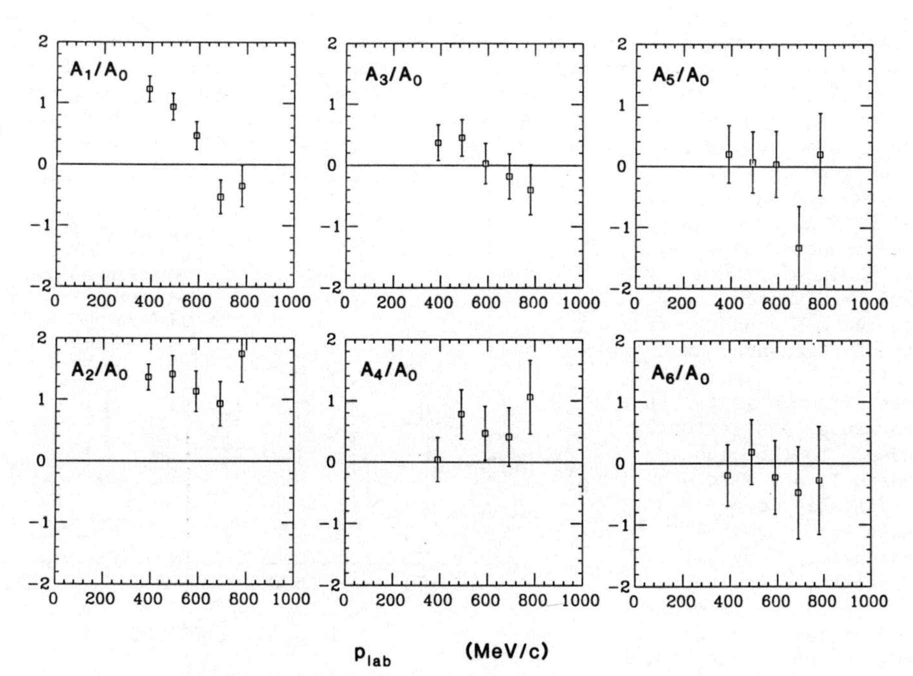

Fig. 7. Legendre expansion coefficients normalized by A_0 for the reaction $\bar{p}p \to \pi^+\pi^-$.

contribution in this momentum region, the angular distributions were fitted to a Legendre expansion up to the 4th order, and the results are shown by the curves in Fig. 6.

Figure 8 shows the integrated cross section $\sigma(K^+K^-)$ for the reaction $\bar{p}p \to K^+K^-$ together with the results obtained by other groups (Bizzarri et al 1969, Mandelkern et al 1971, Nicholson et al 1973, Eisenhandler et al 1975, Sai et al 1983). There is a clear enhancement in our data at 490 MeV/c. The statistical significance of this enhancement is more than 5σ. The data of Bizzarri et al (1969) show a similar enhancement at somewhat different momentum, $\sim$430 MeV/c.

The angular distributions for the reaction $\bar{p}p \to K^+K^-$ are shown in Fig. 9. At 490 MeV/c, the angular distribution looks quite symmetric, while at other momenta, it exhibits an asymmetric shape, peaked in the backward direction. The angular distributions were fitted to a Legendre expansion up to the 6th order, and the resulting normalized expansion coefficients are shown in Fig. 10. It is seen that A_2 and A_4 significantly contribute around 490 MeV/c, and A_1 and A_3 become zero at 490 MeV/c. The coefficients A_5 and A_6 do not show statistically significant contribution. The curves in Fig. 9 show the results of the fit to a Legendre expansion up to the 4th order. We also studied the effect of increasing the order of Legendre polynomials included in the fit. However, the quality of the fit did not improve, indicating that Legendre polynomials higher than the 4th order are not needed to fit the K^+K^- angular distributions. All these results indicate that a J=2 wave dominantly contribute at 490 MeV/c.

If the structure observed in the reaction $\bar{p}p \to K^+K^-$ at 490 MeV/c is interpreted as a meson resonance with J=2, its quantum numbers are determined to be $J^{PC}=2^{++}$ and $I^G=0^+$ or 1^-. Then, it can also decay into the $K^0_S K^0_S$ and $K^0_L K^0_L$ channels. Although the mass and width are not very well determined because the enhancement was observed at only one momentum, we quote a mass of 1935 $\pm$15 MeV and a width of $\Gamma \lesssim 40$ MeV for this state. The mass of this state coincides with that of the S resonance, which was a candidate for a baryonium state but has not been confirmed in recent $\bar{p}p$ total cross section measurements (for references, see Nakamura et al 1984c). Because of the reason stated below, we shall distinguish the 1935 MeV state observed in the K^+K^- channel from the S resonance, and call it λ. A baryonium state formed in the $\bar{p}p$ s-channel must strongly couple to the elastic channel, and also substantially decay into multi-body annihilation channels. However, because the cross section for $\bar{p}p \to \lambda \to K^+K^-$ is about 200 μb (see Fig. 8) and the 90%-confidence-level upper limit for a possible resonance at $\sim$500

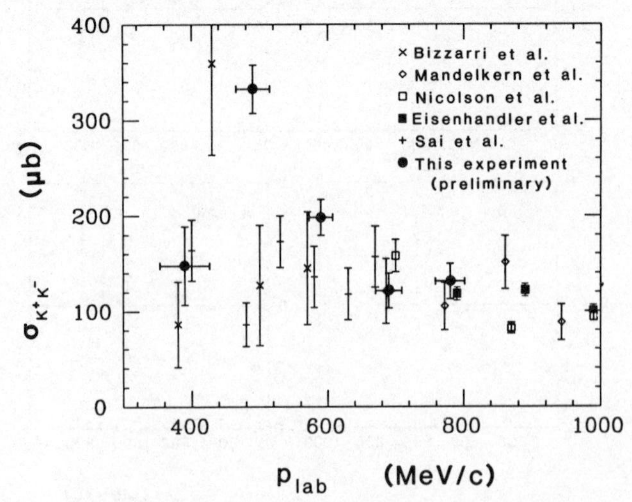

Fig. 8. Cross section for the reaction $\bar{p}p \to K^+K^-$.

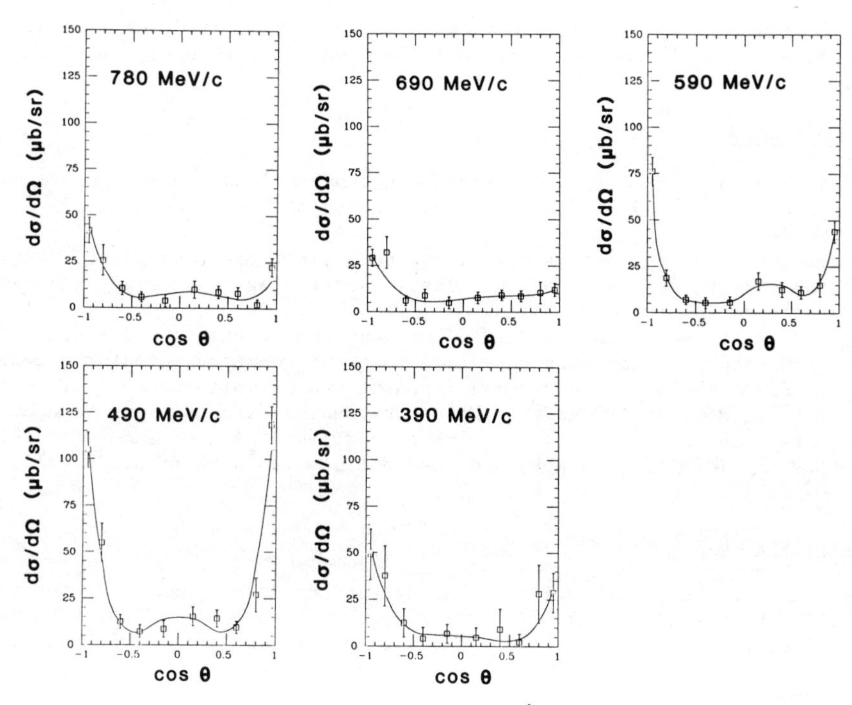

Fig. 9. Preliminary results for the $\bar{p}p \to K^+K^-$ differential cross sections.

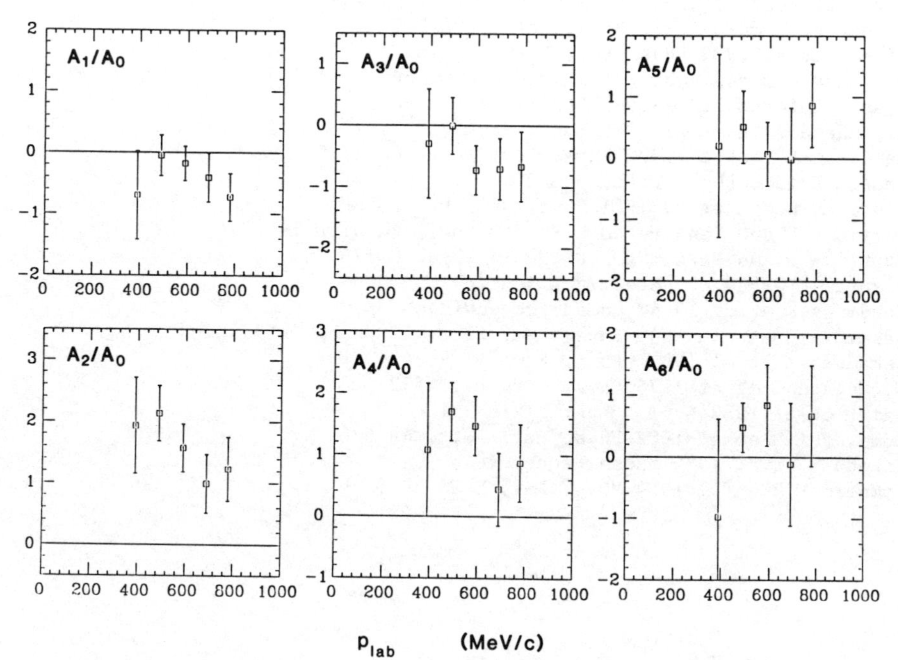

Fig. 10. Legendre expansion coefficients normalized by A_0 for the reaction $\bar{p}p \to K^+K^-$.

MeV/c with Γ=10 ∿ 50 MeV is reported to be about 1 mb (Sumiyoshi et al 1982), probably λ does not substantially decay into channels other than $K\bar{K}$.

6. Conclusions

(1) We have presented the differential cross sections for the reactions $\bar{p}p \to \bar{p}p$, $\bar{n}n$, $\pi^+\pi^-$ and K^+K^- at incident momenta of 390, 490, 590, 690 and 780 MeV/c.
(2) None of the presently available $\bar{N}N$ potential models completely explains the observed behavior of the elastic-scattering and charge-exchange differential cross sections.
(3) The existence of the forward dip in the charge-exchange differential cross sections has been confirmed over the measured momentum range.
(4) Evidence has been found for a resonant state, which we call λ, in the K^+K^- channel at 490 MeV/c. No corresponding structure has been observed in the $\pi^+\pi^-$ channel. The properties of λ are: a mass of 1935±15 MeV, a width of $\lesssim$ 40 MeV, and possible quantum numbers of $J^{PC}=2^{++}$, $I^G=0^+$ or 1^-.

Acknowledgments

It is a pleasure to acknowledge the work of our colleagues for Experiment E74/II. We wish to thank Dr. T Mizutani, Dr. M Lacombe and Dr. P H Timmers for sending us their results of calculations with $\bar{N}N$ potential models.

References

Alston-Garnjost M et al 1979 Phys. Rev. Lett. 43 1901
Bizzarri R et al 1969 Nuovo Cimento Lett. 1 749
Bogdanski M et al 1976 Phys. Lett. 62B 117
Côté J et al 1982 Phys. Rev. Lett. 48 1319
Dalkarov O D and Myhrer F 1977 Nuovo Cimento 40A 152
Eisenhandler E et al 1975 Nucl. Phys. B 96 109
Eisenhandler E et al 1976 Nucl. Phys. B 113 1
Hamilton R P et al 1980 Phys. Rev. Lett. 44 1179
Leader E 1976 Phys. Lett. 60B 290
Mandelkern M A et al 1971 Phys. Rev. D 4 2658
Martin A D and Pennington M R 1980 Nucl. Phys. B 169 216
Martin B D and Morgan D 1980 Nucl. Phys. B 176 355
Mizutani T 1984 private communication
Nakamura K et al 1984a these Proceedings
Nakamura K et al 1984b Phys. Rev. Lett. 53 to be published
Nakamura K et al 1984c Phys. Rev. D 29 349
Nicholson H et al 1973 Phys. Rev. D 7 2572
Sai F et al 1983 Nucl. Phys. B 213 371
Sumiyoshi T et al 1982 Phys. Rev. Lett. 49 628
Takeda T et al 1984 these Proceedings
Timmers P H et al 1984 Phys. Rev. D 29 1928

Formation of charmonium states in antiproton–proton annihilation

C.Baglin[1], S. Baird[2], G. Bassompierre[1], G. Borreani[8], J.C. Brient[1]
C. Broll[1], J.M. Brom[7], L. Bugge[5], T. Buran[5], J.P. Burq[4],
A. Bussière[1],A. Buzzo[3], R. Cester[2], M. Chemarin[4], M. Chevallier[4],
B. Escoubes[7], J. Fay[4], S. Ferroni[3], C. Girard[1], V. Gracco[3],
J.P. Guillaud[2], K. Kirsebom[5], B. Ille[4], M. Lambert[4], L. Leistam[2],
A. Lundby[2], M. Macri[2], F. Marchetto[8], L. Mattera[3], E. Menichetti[8],
B. Mouellic[2], N. Pastrone[8], L. Petrillo[6], M.G. Pia[3], M. Poulet[1],
A. Pozzo[3], G. Rinaudo[8], A. Santroni[2], M. Severi[6], G. Skjevling[5],
B. Stugu[5], F. Tomasini[3], U. Valbusa[3].

Annecy (LAPP)[1]- CERN[2]- Genova[3]- Lyon (IPN)[4]- Oslo[5]-Roma[6]-
Strasbourg[7]- Torino[8] Collaboration

Abstract: We have performed an experiment at the CERN-ISR to study the direct formation of charmonium states in antiproton-proton annihilation. Preliminary results on η_c, χ_1, χ_2 states are presented.

1. Introduction

The study of the charmonium spectrum started in 1974 after the discovery of J/PSI and its interpretation in terms of cc bound states. The spectrum of states below the charm open threshold is shown in fig. 1.

The existing information on mass, width and decay branching ratios of charmonium states, comes from experiments at e^+e^- colliders, where only $J^{PC} = 1^{--}$ states can be directly formed. In a formation experiment the precision on the measurement of the resonance parameters depends uniquely on the knowledge of the energy of the initial state and is correspondingly good.

At e^+e^- machines the (cc) states not carrying the quantum numbers of the photon can only be reached through decay from higher 1^{--} states. In this case the production rate is reduced by the decay branching ratio, and the precision on the measurement of the resonance parameters rests on the capability

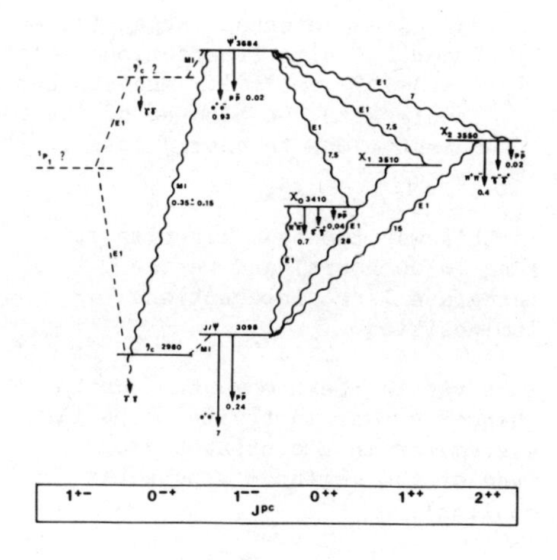

Fig. 1 Charmonium spectrum

of the detector to identify and measure the decay products. As a consequence, while the parameters of the ψ states have been precisely measured [1], the knowledge of ^3P (χ),^1S (η_c) states is still too limited to provide a stringent test of the theoretical predictions, and other states (^1P,$^1D_2$,$^3D_2$), expected to be narrow, are as yet undetected. The existence of a coupling of charmonium states to the pp system implies that they can be formed in the pp annihilation.

The pp channel allows to study any cc state and it is specially interesting to reach, in a direct formation experiment, states not directly accessible in the e^+e^- annihilation channel. Until now this very interesting property was offset by the difficulty of designing a source comparable in luminosity and energy definition to the one provided by an e^+e^- collider. The intense p beam now available at CERN with the A.A. facility and their precise momentum definition obtained with the stochastic cooling system have made these experiments feasible.

In this contribution we report <u>preliminary</u> results of an experiment performed at CERN-ISR where the direct formation of charmonium states has been studied in pp annihilation.

The source of interactions

The requirements to study charmonium states formed in pp annihilation were met by:

1. the use of the p complex of CERN (Fig. 2) which is able to provide intense p beams well defined in momentum,

2. use of ring 2 of ISR where the p beam is injected. Here a stochastic momentum cooling system [2] further reduce the $\Delta p/p$ to ± 5 10^{-4}. The setting of the beam momentum is known and reproducible with ± 1 MeV/c in the range of momenta used to form the charmonium states;

3. use of an internal target [3] built using the technics of molecular beams. A high interaction luminosity is then obtained $L = 3 \cdot 10^{30}$ $cm^{-2}s^{-1}$. The stochastic cooling system acting on the beam counteracts the blow-up of the beam dimensions and the degradation of momentum due to energy loss in the multiple traversal of the gas target.

In this way the beam lifetime is long (>100 hours) and we can integrate large interactions luminosities.

Moreover the beam momentum can be changed either in finite steps (RF system) or in a continuous mode (use of the momentum stochastic cooling).

This gives the possibility to generate the excitation curve of the resonance under study.

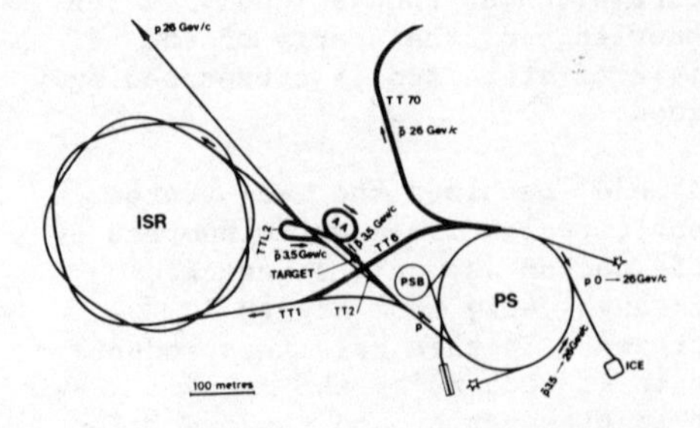

<u>Fig. 2</u> Antiproton complex of CERN

Experimental method

We inject the p beam in ring 2 of ISR, we tune the beam momentum in such a way that the c.o.m. energy is in the region of the mass of the resonance we want to study. We construct the excitation curve changing in appropriate steps the beam momentum. At each step we determine the number of events for the selected decay channel.

The resonance parameters are determined using the knowledge of the beam momentum which as already mentioned is known with a good accuracy.

To optimize the signal-to-background ratio we have designed the experimental apparatus to detect electromagnetic final states. Specifically we are tagging

ψ	state	through its	e^+e^-	decay
χ_1, χ_2	states	through their	$\psi\gamma$	decay
η_c	state	through its	$\gamma\gamma$	decay

The detector

The detector is a two-arm non-magnetic spectrometer complemented by a large-acceptance veto system and a silicon detector telescope to monitor the source luminosity.

Fig. 3 shows in a schematic way the apparatus. The transverse jet target with its sophisticated vacuum system and the ISR beam pipe impose severe space limitations. The detection system consists of two symetrical arms allowing a e^+e^- or $\gamma\gamma$ geometrical acceptance of the order of 0.10 (of 4π).

Each of the two identical arms consists of an upstream section for tracking charged particles followed by an electromagnetic calorimeter, segmented both transversally and longitudinally, to optimize γ/π° and $e^\pm/\pi^\pm$ separation.
Each arm is a truncated pyramid with trapezoidal bases. Its acceptance ranges in polar angle from 17° to 66° and covers 45° in azimuthal angle. Details on the structure and performance of the charged particle telescope and of the electromagnetic calorimeter are given in Table I.

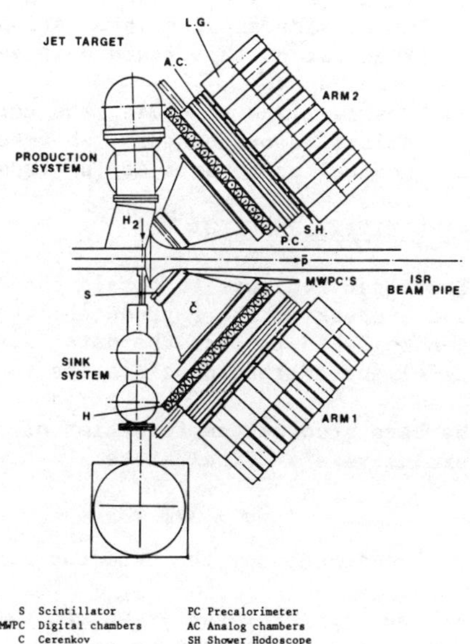

Fig. 3 Schematic view of experimental apparatus

The veto system surrounds the two arms ranging from 1.8° to 77° in polar angle with full azimuthal coverage, It consists of two parts.

a) A system of 4.6 X_0 lead-scintillator sandwiches (neutral veto) to detect charged particles and

TABLE 1
a) Charged particle detector

	Label	d*) (cm)	Transverse structure	Longitudinal structure	Performance
Scintillator	S_i	29.0	1 counter		
MWPC		30.0	2 mm pitch	3 planes (x,u,y)	$\epsilon_{pl} \sim 0.95$ $\Delta x = \Delta y = 6.0$ mm
Freon Cherenkov	C_i	37.0	2 cells, $\Delta\phi$=22.5°	45.0 cm radiator	$<N>_{pe} \sim 5.5$
MWPC		82.5	2 mm pitch	3 planes (x,u,y)	$\epsilon_{pl} \sim 0.95$ $\Delta x = \Delta y = 6.0$ mm
Scintillator	H_θ	91.5	10 count. $\Delta\theta$=4.9°		
Scintillator	H_ϕ	93.0	6 count. $\Delta\phi$=7.5		
			b) Calorimeter		
					$\Delta x = \Delta y$ for shower centre of gravity
Precal. θ		101.0	29; $\Delta\theta$ = 1.7°	4.6 X°	
Precal. ϕ			14; $\Delta\phi$ = 3.2°	2.5 mm Pb/ 5.0 mm scint.	~ 8 mm
Analog readout chambers		111.0	strips, 10 mm pitch	4 x (x,y)	~ 2 mm
Lead glass		133.5	66 counters 15 x 15 cm^2	10 X°	

*) d = distance from target centre

γ-rays. The threshold for γ ray detection is $E_\gamma \simeq 30$ MeV. This structure is subdivided into 50 elements read out individually to provide angular information on the detected particle. As will be seen later, this feature is very useful in the detection of ψγ final states.

b) Upstream and shielding the neutral veto system is a structure of scintillator counters (charged veto) to separate out the charged from neutral components coming from the target region.

PRELIMINARY RESULTS

The first goal was to obtain the absolute energy calibration of the ISR, and a check of the reproducibility of the energy setting. This was accomplished through the detection of J/ψ formation, which provides an excellent calibrator since its mass is known to ± 100 keV[1].

We have studied the formation of J/ψ through the detection of the exclusive e^+e^- final state:

$$\bar{p}p \to J/\psi \to e^+e^- \qquad (1)$$

The acceptance of the detector for channel (1) is ~ 10%.

At the trigger level we select events by requiring an ARM1 * ARM2 coincidence where ARMi is a coincidence of four scintillator hodoscopes (s_i, H_θ, H_φ, SH) and an athmospheric pressure FREON 13 Cherenkov counter to tag electrons. Fig. 4 shows an invariant mass plot for all events in which two tracks reconstruct to a good vertex in the target area. The

events populating the low energy part of the spectrum are due predominantly to contaminations to the electrons sample coming from π°'s (DALITZ decays or γ converted in the vacuum chamber walls) and from δ rays or accidentals associated to $\pi^\pm$ (which are all below threshold.)
Requiring the kinematic of the events to be compatible with (1) and no extra energy deposition in the detector, we obtain a background free J/ψ data sample (Fig. 5).

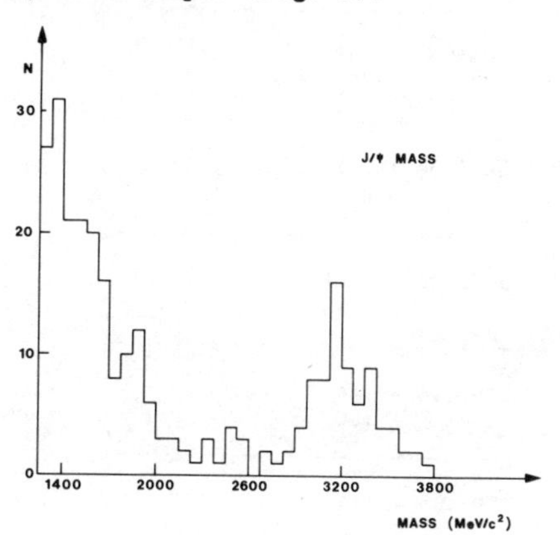

Fig. 4 Invariant J/Psi mass (before cuts).

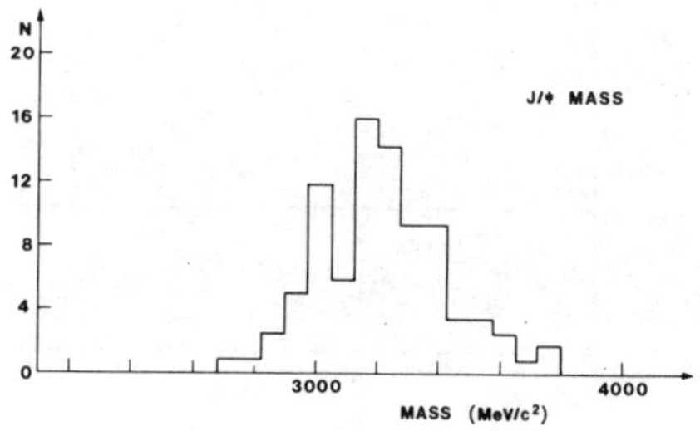

Fig. 5 Invariant J/Psi mass (see text)

Data at the J/ψ formation energy where accumulated in various runs in the course of our experiment to check the stability of the results. The summary (for an integrated luminosity of ~ 70 nb^{-1}) is given in TAB II.

The data are analyzed with a maximum likelihood calculation which finds the best mass value (Column 3 in tab. II) compatible with the distribution of events as a function of the beam momentum profile characteristics.

Due to the limited statistics for each run we cannot exclude small fluctuations from one time to the other but the value integrated over the total data sample agrees well with the particle data table value confirming the measurement of the absolute momentum done with accelerator techniques, and the estimate of the precision in the measurement.

χ states

We have performed energy scans to study χ_1 and χ_2 states through the formation/decay chain:

$$\bar{p}p \rightarrow \chi_i \rightarrow J/\psi + \gamma \rightarrow e^+e^- + \gamma \qquad (2)$$

Our goal was to determine the parameters (mass, partial and total width) of these resonances through the measurement of the excitation curve.

The trigger requirements and off-line analysis chain are the same as for the J/ψ study. Fig. 6 shows the events distribution as a function of the invariant mass, M_{12} reconstructed from the two charged particles (assumed to be e^+e^-) in the detector arms.

Table II (Different J/Psi runs)

Run #	J/ψ Events	J/ψ Mass
1	15	$3096.96 {}^{+.29}_{-.29}$
2	11	$3097.37 {}^{+.25}_{-.23}$
3	12	$3095.50 {}^{+1.10}_{-0.50}$
4	19	$3096.70 {}^{+0.44}_{-0.50}$
5	35	$3097.14 {}^{+.44}_{-.50}$
All	92	$3096.95 {}^{+.14}_{-.15}$
Table value		$3096.93 {}^{+.089}_{-.084}$

This distribution shows an enhancement in the J/ψ mass region. The background level is higher than in the J/ψ formation data due to the fact that χ formation rate is on the average 1/20 th that of the J/ψ.

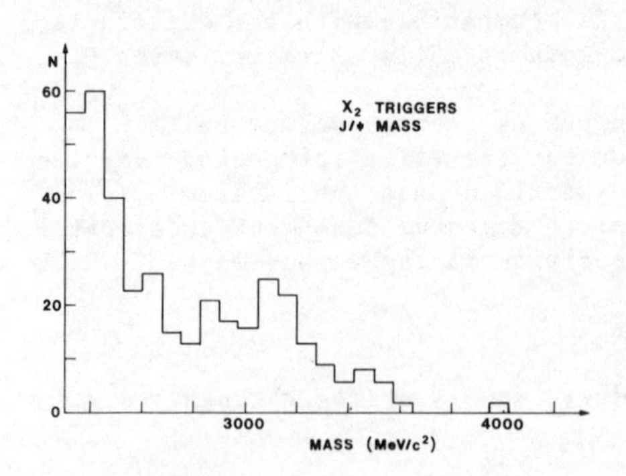

Fig. 6 Invariant J/Psi mass for χ_2 events (before cuts)

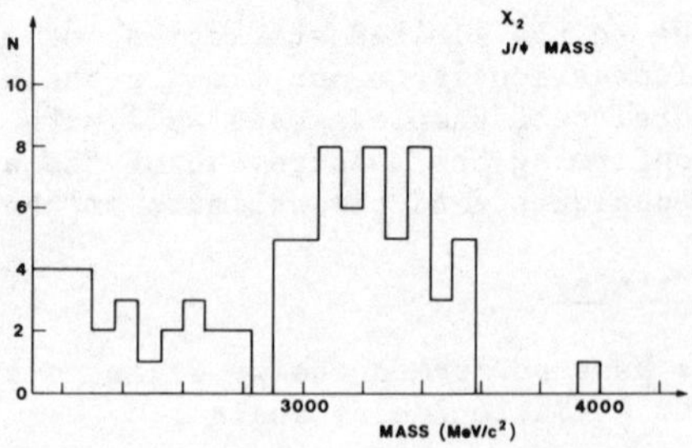

Fig. 7 Invariant J/Psi mass for χ_2 events (see text)

We applied to the data a further analysis which uses the analogic information from the trigger counters and the longitudinal development of the showers to suppress the above mentioned backgrounds from π°'s and charged hadrons. The invariant mass distribution after cuts, is given in Fig. 7. We accept in the final J/ψ inclusive sample only events with an invariant mass ≥ 2.75 GeV/c². The total efficiency of the cuts is estimated to be ~ 90%. From the measured direction of the two charged tracks (e^+e^-) and with the mass assignment of reaction (2), we solve the momentum/energy conservation equation system to obtain the direction of the γ and the momentum of γ, e^+, e^-. It is assumed that the beam momentum vector is known. With this assumption the system of equations can be solved leading to two possible solutions. For each of the two solutions we determine from the predicted γ direction which detector should have been hit by the γ. For an event to be accepted we require agreement between the calculation prediction and the counter information. A summary of the data taken at the χ formation energy are given in Tab. III.

Table III (χ Data)

State	Limit of scan (MeV)	$\int L\,dt$ (nb^{-1})	# Events
χ_1	3509.6 - 3514.3	505	28
χ_2	3554.0 - 3561.5	775	46

The excitation curves for χ_1 and χ_2 are given respectively in Fig. 8 and Fig. 9.

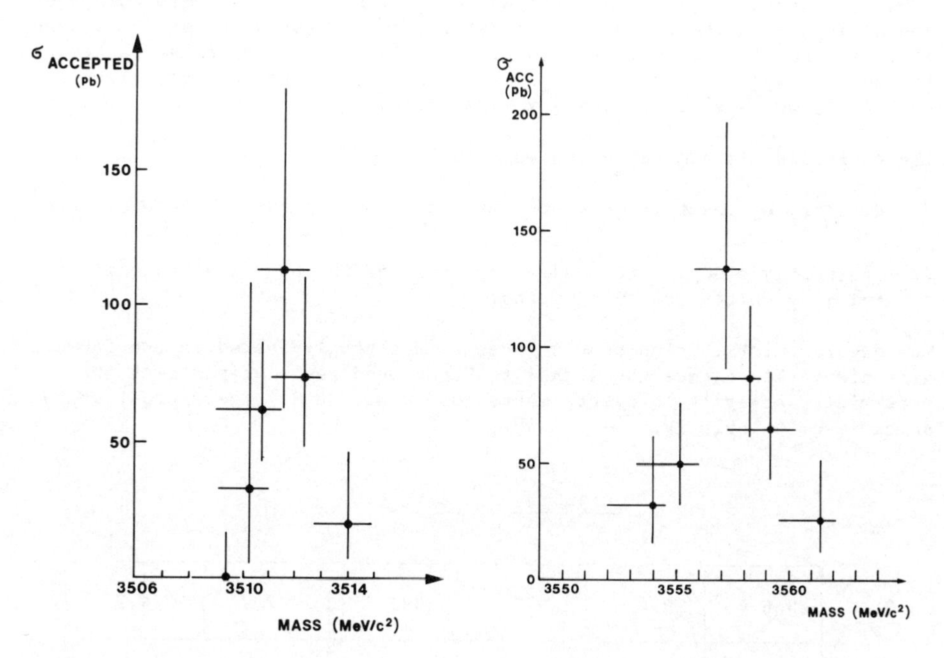

Fig. 8 χ_1 Excitation curve

Fig. 9 χ_2 Excitation curve

The vertical error bars are statistical, while the horizontal bars represent the beam profile approximated by the sum of two half gaussian joining at the peak value.

The residual background level for fully reconstructed events has been directly measured by applying the same analysis to a sample of data corresponding to an integrated luminosity of 1017 nb^{-1} taken, in the c. of m. energy range $3.519 < \sqrt{s} < 3.529$ GeV/c^2. We have found that only 2 events are compatible with the formation/decay chain (2) corresponding to very low background level.

We are at present in the process of doing a carefull statistical analysis of the data to extract all information on the resonance parameters.

The preliminary results for the masses and partial widths to pp are:

$$M_{\chi_2} = (3556.8 \pm .4 \pm .4)\ MeV/c^2 \quad \Gamma_{\bar{p}p} = 208^{+75}_{-35}\ eV$$

$$M_{\chi_1} = (3511.4 \pm .3 \pm .4)\ Mev/c^2 \quad \Gamma_{\bar{p}p} = 62 \pm 15\ eV$$

η_c state

We have performed an energy scan in the η_c region to investigate the $p\bar{p} \to \eta_c \to \gamma\gamma$ channel.

Candidate events were selected from the neutral trigger requiring a single shower in each arm, with the correct two-body kinematics and no energy deposition in the veto system. The efficiency of the different cuts were inferred from the J/ψ data analysis. In this case an important background is due to assymetric π° decays, coming mainly from the two-body reaction $pp \to \pi^\circ\pi^\circ$, which simulate the γ-γ kinematics.

The background is evaluated in two ways:

1. directly by measuring the $p\bar{p} \to \gamma\gamma$ cross section outside the resonance.

2. indirectly studying the characteristics of the $p\bar{p} \to \pi^\circ\pi^\circ$ channel for fully reconstructed 4γ's events.

The angular distribution of $\pi^\circ\pi^\circ$ events is strongly peaked in the forward direction. To enhance the signal to background ratio we restrict our preliminary analysis to events where $\cos\theta$ c.m. < 0.3. The present status is summarized in Table IV.

Table IV (η_c data)

E (MeV)	2966.6	2974	2978	2983.6	2987.2	3024.2
$\int L\,dt$	102	31	104	195	118	102
Events	2	0	6	9	3	2

Table IV presents together with the luminosity the number of events passing all cuts at the different energies. A maximum likelihood analysis was performed to determine the most probable value of the product BR ($\eta_c \rightarrow$ pp). BR ($\eta_c \rightarrow \gamma\gamma$). A background of the form a + b ($E_{\eta c}$ - E) was assumed where b, the energy dependence factor was inferred from $\pi^\circ\pi^\circ$ cross section.

The total width and the energy position of the resonance were taken from the Cristal Ball analysis.

$$\Gamma\eta_c = 11.5 \text{ MeV}, \quad M\eta_c = 2984 \text{ MeV/c}^2$$

We obtain

$$a = (0.36 \pm 0.18) \text{ nbarn.}$$

$$BR\ (\eta_c \rightarrow p\bar{p}).\ BR\ (\eta_c \rightarrow \gamma\gamma) = 4.3 \pm 3.6)\ 10^{-7}$$

Using MARK III results [4] BR ($\eta_c \rightarrow p\bar{p}$) = (1.8 ± 0.9) 10^{-3}

We deduce: BR ($\eta_c \rightarrow \gamma\gamma$) = (2.4 ± 2.0) 10^{-4}

and $\Gamma_{\gamma\gamma}$ = (2.7 ± 2.3) keV.

REFERENCES

[1] Particle data group PL 111B (1982).

[2] A. Peschard et al.; Stochastic cooling in the CERN-ISR during pbar-p colliding physics. CERN/LEP/ISR/RF/83-15.

[3] M. Macri; A clustered H_2 beam, in Physics with low energy cooled antiprotons, Plenum Publishing Corporation.

[4] W. Toki; SLAC-PUB-3262 (1983).

Inst. Phys. Conf. Ser. No. 73: Section 5
Paper presented at VII Eur. Symp. Antiproton Interactions, Durham 1984

QCD prediction for charmonium production in $\bar{p}p$ collisions

E.L. Berger[a)], P.H. Damgaard[b)] and K. Tsokos[c)]

[a)] Argonne National Laboratory, Argonne, IL 60439, USA
[b)] NORDITA, Blegdamsvej 17, DK-2100 Copenhagen Ø, Denmark
[c)] Department of Physics, University of Maryland, College Park MD 20742, USA

Abstract. Within the framework of perturbative QCD, the formalism of exclusive processes is used to give predictions for the production cross section of charmonium resonances.

1. Introduction

As is well known, a very detailed and impressive amount of information about the different charmonium states has already been obtained by experimental studies in e^+e^- colliders. Perhaps, then, the first question we should address in this talk is: why should we begin to search through the different charmonium channels again in $p\bar{p}$ collisions? There are several answers to this question. First, let us make a somewhat simplified (but basically correct) order-of-magnitude argument. The most obvious charmonium resonance is the ψ, which has quantum numbers $J^{PC} = 1^{--}$. This state is, of course, readily produced in e^+e^- collisions. In terms of Feynman diagrams the process can proceed through $e^+e^- \to \gamma^* \to \bar{c}c$. Only one intermediate off-shell photon is required. Thus, in the amplitude only one power of α_{EM} appears. In contrast, consider another charmonium state like the χ_2, which has quantum numbers $J^{PC} = 2^{++}$. Since it is charge-conjugate even, at least *two* intermediate photons are required: $e^+e^- \to 2\gamma^* \to \bar{c}c$. Therefore, even just at the amplitude level the process is (roughly) suppressed by a factor of 1/137. In the cross section this gives a suppression factor of $\sim 10^{-4}$ as compared to the $e^+e^- \to \psi$ cross section! Although this argument is rather sketchy (for one thing, it glosses over the differences between the wave functions of the ψ and the χ_2), the idea is right: there is no hope of producing the χ_2 state *directly* in e^+e^- collisions. (Of course, it can be seen in radiative decays, but that is a different story).

What is the difference if we start invoking the strong interactions? Consider the process $q\bar{q} \to \psi$. Just from the J^{PC} quantum numbers alone this process should be able to proceed via one intermediate gluon (just as we in the QED case had one intermediate photon), but that Feynman diagram obviously have a vanishing color factor. We are therefore forced to bring two

*Talk presented by P.H. Damgaard.

extra off-shell gluons into the game, i.e. $q\bar{q}\to 3g^*\to \bar{c}c$. This looks bad from the point of view of counting powers of the coupling constant: roughly speaking, a factor of α_s^3 is involved in this amplitude. However, in this case it turns out to be rather convenient that the strong interactions indeed are quite strong; at the charmonium resonances we can expect to find a QCD running coupling constant of about $\alpha_s \simeq 0.2$. Several powers of α_s are therefore quite harmless as far as cross section suppression is concerned. Moreover, in this case the production of the χ_2 resonance proceeds via two gluons: $q\bar{q}\to 2g^*\to \bar{c}c$. As compared to ψ production we would expect roughly the same order of magnitude for the cross section. Hence, if the ψ can be produced in $p\bar{p}$ collisions, then so should all other charmonium states which are allowed by the quantum numbers. This leads to the exciting experimental possibility that the whole of charmonium spectroscopy can be finished up in studies of $p\bar{p}$ collisions.

From the point of view of perturbative QCD, this turns out to be an exciting possibility as well. Assuming perturbative QCD to be applicable in this Q^2 range ($Q^2 \simeq 10$ GeV^2), a whole formalism has been developed to handle, among others, processes like $p\bar{p}\to$ heavy meson resonances. This is the formalism of exclusive processes in QCD, originally developed by Chernyak, Zhitnitsky and Serbo (1977), Efremov and Radyushkin (1980), Brodsky and Lepage (1979a,1979b) and by Duncan and Mueller (1980). The applicability of this formalism to processes involving baryons has been particularly stressed by Brodsky and Lepage. From this point of view, charmonium production in $p\bar{p}$ colliders is interesting because it gives us an idea of how well perturbative QCD works in the Q^2 region around 10 GeV^2. As we shall see, absolute predictions for the different cross sections can be computed, with only one other exclusive process involving baryons used as input. By simple scaling laws, the cross sections for $p\bar{p}\to b\bar{b}$ resonances or, in principle, $p\bar{p}\to t\bar{t}$ resonances follow as well.

A study of the $p\bar{p}\to\psi$ process (or rather its inverse: the decay of $\psi\to p\bar{p}$) had already been performed earlier (Brodsky and Lepage 1981), and instead we decided to look at processes involving two intermediate gluons, such as $p\bar{p}\to\chi_2$. While this study was underway, we learned of two similar calculations (Andrikopoulou 1984; Chernyak and Zhitnitsky 1983).

2. Computing the cross sections

Let us now give a short description of the method used in calculating these cross sections. The crucial point is that if Q^2 is sufficiently high, factorization of the annihilation amplitude into a short distance dominated 'hard' part and 'soft', in perturbation theory incalcuable, distribution amplitudes $\phi(x,Q^2)$ can be shown to hold. Having once demonstrated this factorization, the renormalization group takes care of the rest. The only other thing that is needed is the shape and absolute normalization of the distribution amplitudes $\phi(x,Q^2)$ at some fixed reference point $Q^2 = Q_0^2$. This can be found by a comparison with just one other exclusive process. As the distribution amplitudes are universal, it does not matter which process we choose, but of course one should pick a process for which the corresponding calculation is not beset by ambiguities, and for which good and reliable data exist. We chose the decay $\psi \rightarrow p\bar{p}$ as the 'reference process'. With that used as input, the production cross section for the other charmonium resonances can then be given.

The distribution amplitudes $\phi(x,Q^2)$ appear as follows. First choose a physical gauge such as the light cone gauge $A^+ = 0$. Assigning momenta $(P^+, P^-, 0_\perp)$ as in the infinite momentum frame, the constituent quarks of the proton are labelled by their relative longitudinal momenta $x_i = k_i^+/P^+$ and their transverse momenta $k_\perp^{(i)}$. The wave function $\psi(x,k_\perp)$ of the proton satisfies a bound state equation $\psi = SK\psi$, where S is a 3-particle propagator and K is a kernel built out of 2-particle irreducible (2PI) graphs, as illustrated in fig.1(a).

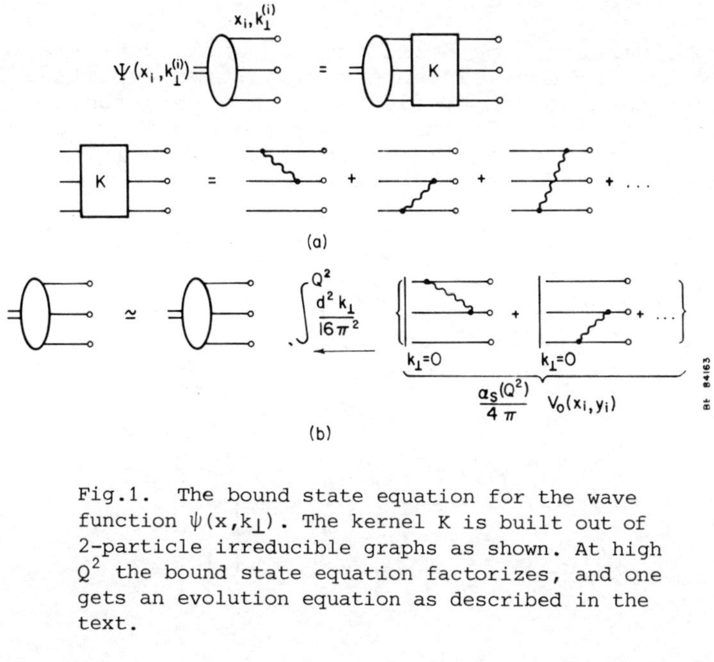

Fig.1. The bound state equation for the wave function $\psi(x,k_\perp)$. The kernel K is built out of 2-particle irreducible graphs as shown. At high Q^2 the bound state equation factorizes, and one gets an evolution equation as described in the text.

To lowest order in perturbation theory we now introduce the distribution amplitude $\phi(x,Q^2)$ by

$$\phi(x,Q^2) = (\ln(Q^2/\Lambda^2))^{-3\gamma_F/2\beta_0} \int^{Q} \left[\frac{d^2k_\perp}{16\pi^3}\right] \psi(x,k_\perp) \quad , \tag{2.1}$$

where the measure is

$$\left[\frac{d^2k_\perp}{16\pi^3}\right] = \left(\prod_{i=1}^{3} \frac{d^2k_\perp^{(i)}}{16\pi^3}\right) 16\pi^3 \delta^2\left(\sum_{j=1}^{3} k_\perp^{(j)}\right) . \tag{2.2}$$

The logarithmic factor in front of the integral is included in order to incorporate the effect of wave function renormalization of the quarks: γ_F equals the anomalous dimension of the quark fields, and $\beta_0 = 11 - 2n_f/3$, with n_f being the effective number of flavors.

At large Q^2 the bound state equation for $\psi(x,k_\perp)$ reduces to a differential equation for $\phi(x,Q^2)$, as illustrated in fig.1(b). In detail,

$$Q^2 \frac{\partial}{\partial Q^2} \phi(x,Q^2) = \frac{\alpha_s(Q^2)}{4\pi} \int_0^1 [dy] v(x,y) \phi(y,Q^2)/y_1 y_2 y_3 \ , \tag{2.3}$$

where $v(x,y)$ is a symmetric one-gluon exchange kernel, as shown in fig.1(b). At very large Q^2, the solution to eq.(2.3) is dominated by $\phi(x,Q^2) = C(Q^2)x_1x_2x_3$. At lower Q^2 we do not know what $\phi(x,Q^2)$ looks like. However, once known at some $Q^2 = Q_0^2$, the evolution to another Q^2 is dictated by (2.3).

These proton (antiproton) distribution amplitudes are to be convoluted with the short distance annihilation amplitude T_H, which for the χ_2 state we are considering has a perturbative expansion as illustrated in fig.2. The quark-antiquark pair form the bound state of the heavy meson, while the three parallel lines represent the colliding proton-antiproton pair.

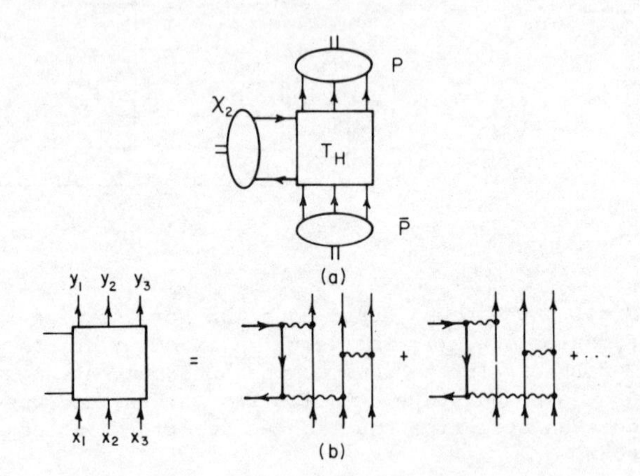

Fig.2. The short distance annihilation amplitude T_H.

Finally, the heavy quark-antiquark pair must be projected onto a J = 2 p-wave state of polarization $\varepsilon^{\mu\nu}$. The full annihilation amplitude can then be written

$$T = -iM^{-11/2}\sqrt{\frac{3}{8\pi}}\, R_p'(0)\, \varepsilon_\perp^{*\mu} g^{3\nu} \varepsilon_{\mu\nu} \int [dx][dy] \phi^*(y) \tilde{T}_H(x,y) \phi(x) \ , \tag{2.4}$$

where the polarization vector is $\epsilon_\perp^\mu = g^{1\mu} + ig^{2\mu}$, $\tilde{T}_H(x,y)$ is a normalized annihilation amplitude as shown in fig.2, and $R_p'(0)$ represents the derivative of the heavy meson wave function evaluated at the origin.

The overall normalization of the total cross section $p\bar{p}\to\chi_2$ can now be found by a comparison with the $\psi\to p\bar{p}$ decay rate. First a parametrization for $\phi(x)$ must be chosen. As $Q^2\to\infty$, we know that $\phi(x)\sim x_1x_2x_3$, and as a nonrelativistic guess one might imagine $\phi(x)\sim\delta(x_1-1/3)\,\delta(x_2-1/3)\,\delta(x_3-1/3)$. A convenient parametrization which interpolates smoothly between these two extremes is $\phi(x)\propto(x_1x_2x_3)^\eta$. At $\eta=1$ we get the asymptotic wave function, and as $\eta\to\infty$ we recover the δ-function parametrization. For reliable predictions one would hope for a relative insensitivity to η, and fortunately this appears to be the case. In fig.3 we show the ratio $R=T(\chi_2\to p\bar{p})/T(\psi\to p\bar{p})$ as a function of η. Despite almost ten order-of-magnitude differences in the numbers between $\eta=1$ and $\eta=4$, the *ratio* turns out to be very close to constant.

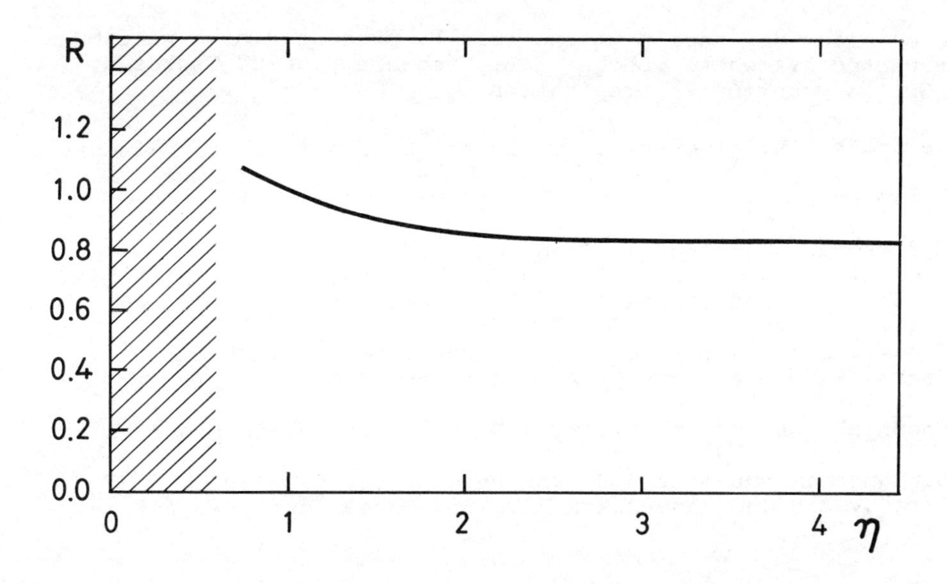

Fig.3. The ratio $R=T(\chi_2\to p\bar{p})/T(\psi\to p\bar{p})$ normalized to unity at the asymptotic wave function (i.e. η=1). The hatched area shows a region likely to be forbidden due to singular endpoint behavior.

To extract the production cross section we use the standard Breit-Wigner formula

$$\sigma = \frac{4\pi(2J+1)}{4M_\chi^2}\,\frac{\Gamma_{p\bar{p}}\Gamma_f}{(\sqrt{s}-M_\chi)^2+1/4\Gamma^2}\,, \tag{2.5}$$

where $\Gamma_f=\Gamma(\chi_2\to\text{observed final state})$, $\Gamma=\Gamma(\chi_2\to\text{all})$ and we have spin averaged over the initial p, $\bar{p}$ spins. For the asymptotic wave function ($\eta=1$) we get $\Gamma_{p\bar{p}}\simeq1.2\cdot10^{-4}\Gamma$, whereas for $\eta=4$ we get

$\Gamma_{p\bar{p}} \simeq 0.8 \cdot 10^{-4} \Gamma$. The full width of the χ_2 state is not well known, but an upper limit has been set at a few MeV. Thus, with $\Gamma = 2 \cdot 10^{-3}$ GeV and $M_\chi^2 \simeq 12.5$ GeV^2 all parameters in the cross section formula (2.5) are determined. This cross section can then be compared with experimental data.

Finally, we should make a comment about the production cross section for other charmed resonances. In the approximation in which we are working in this paper, the cross section for $p\bar{p} \to$ any $J = 0$ resonance (including the η_c) vanishes, simply by helicity conservation. The process $p\bar{p} \to \chi_1$, on the other hand, *is* allowed in this approximation. Unfortunately, for this process a collinear divergence appears in the total decay rate, signalling the breakdown of standard factorization for processes involving this state. What this means is that even though this process ($p\bar{p} \to \chi_1$) is allowed, and in fact proceeds through a two-gluon intermediate state, its branching ratio cannot be computed reliably.

A detailed discussion of the results presented here will be reported elsewhere (Berger E L, Damgaard P H and Tsokos K, NORDITA preprint in preparation).

We thank J.M. Alberty, B. Nizic and B. Söderberg for discussions.

References

Andrikopoulou A 1984 Z. Phys. C22 63
Brodsky S J and Lepage G P 1979a Phys. Lett. 87B 359
Brodsky S J and Lepage G P 1979b Phys. Rev. Lett. 43 545; (E) 43 1625
Brodsky S J and Lepage G P 1981 Phys. Rev. D24 2848
Chernyak V L, Zhitnitsky A R and Serbo V G 1977 JETP Lett. 26 594
Chernyak V L and Zhitnitsky A R 1983 Novosibirsk preprint 83-108
Duncan A and Mueller A H 1980 Phys. Lett. 93B 119
Efremov A V and Radyushkin A V 1980 Phys. Lett. 94B 245

Inst. Phys. Conf. Ser. No. 73: Section 5
Paper presented at VII Eur. Symp. Antiproton Interactions, Durham 1984

Experimental observations of isolated large transverse momentum leptons with associated jets in experiment UA1

D. DiBitonto

CERN - Geneva, Switzerland

Abstract. Events containing an electron or muon of large transverse momentum with two hadronic jets of large transverse energy are reported. These leptons are isolated with respect to the jets and together with the measured missing transverse energy give $(\ell\nu_T jj)$ masses clustering around the W mass. If interpreted as a novel decay of the W into $t\bar{b}$, where t is the sixth "top" quark of the Cabibbo current, then the mass of the top is bounded between 30 and 50 GeV/c^2.

1. Introduction

With the conversion of the CERN SPS into a $\bar{p}p$ Collider[1)] both the existence of the charged and neutral Intermediate Vector Bosons, $W^{\pm}$ and Z^0, and the dominant QCD phenomen of jets at high transverse energies have been demonstrated in a remarkably short period of time[2)]. The evidence reported for W production in both the electron and muon decay channels by the UA1 Collaboration has relied essentially on the identification of isolated high transverse momentum leptons accompanied by large missing transverse energy; for these events virtually no background was found as a result of the veto on jets opposite the lepton. In this study[3)], corresponding to a total integrated luminosity of 0.12 pb^{-1} for 1983, we are looking for leptons in the presence of jets possibly arising from new heavy quark decays of the W, namely into $t\bar{b}$ $(\bar{t}b)$ followed by a semi-leptonic decay of the top. The final state topology should contain two hadronic jets of large transverse energy identified with the b and $\bar{b}$ quarks, in addition to leptons from the top decay which should be isolated as a result of the large Q-value which is available in the reaction. The corresponding background from QCD jets faking leptons, however, is more severe.

2. Detection and Measurement of Leptons and Jets

The UA1 apparatus has already been described[4,5,6)]. Briefly it consists of a charged particle central detector (CD) followed by electromagnetic calorimeters of 27 radiation lengths and hadronic calorimeters of 8 absorbtion lengths, provided by the instrumented return yoke of the magnet. Drift chambers surrounding the outer hadron calorimeters provide muon identification.

Fast leptons emerging from the $\bar{p}p$ interaction region are identified as single minimum ionizing charged tracks in the central detector either depositing at least 12 GeV in two adjacent electromagnetic cells (electron trigger requirement) or penetrating the outer calorimetry and reconstructed in the muon drift chambers (muon trigger requirement). Electrons are further selected offline as a high transverse momentum track with good calorimetric matching in position, angle and energy, in addition to an electromagnetic profile requirement within the four segments of this calorimeter. A maximum energy requirement of less than 200 MeV in the outer hadron calorimeters following the track provides the final rejection capability against hadrons. The offline muon selection requires matching in position and angle between the central detector and muon chamber tracks and the minimum ionization loss in both calorimeters for isolated muons.

Jet finding is done in the calorimeters with the standard UA1 algorythm which associates both electromagnetic and hadronic cells in (η,ϕ) space with $\Delta R = (\Delta\eta^2 + \Delta\phi^2)^{\frac{1}{2}} < 1$. The initiators, which form the core of these jets, must have cells with $E_T \geq 1.5$ GeV, while only cells with $|\eta| < 2.5$ are used for jet finding. Jets are defined with $E_T > 8$ GeV (first jet) while other jets within the same event are defined with $E_T > 7$ GeV, uncorrected in both cases for energy and direction in the initial selection.

3. Event Selection

3.1 Electrons with at Least One Jet

From the electron trigger requirement of $E_T \geq 12$ GeV in two adjacent electromagnetic cells, neutral and charged electromagnetic clusters were first selected by requiring 90% isolation in a cone of $\Delta R < 0.7$. 2723 $(n)\pi^0$ + jet events were selected by the absence of a CD track pointing to the cluster, less than 200 MeV in the hadronic calorimeters immediately behind and final isolation cuts (to be defined), while electron/pion + jet events were first selected by the presence of a hard CD track ($p_T > 7$ GeV) matching the cluster 5σ in position.

49 $W \to e\nu$ events were then selected by requiring less than 600 MeV hadronic energy in the cluster and at least 15 GeV of missing transverse energy. This sample was essentially background free and used as a calibration sample for the candidate "electron" + jet events.

By requiring at least 1 GeV of hadronic energy behind the cluster with final isolation cuts, 169 $\pi^\pm$ + jet events were selected.

Finally, 152 "e" + jet events were selected by the requirement of less than 200 MeV hadronic energy. Two major sources of background dominated this sample, namely photon conversions in the beam pipe and detector walls and overlaps of non-isolated electrons. 43 conversions were removed from the sample of 152 events by scanning and program. To remove overlaps of non-isolated electrons the sample of 49 W events was used to define stricter cuts in energy-momentum matching between the CD and electromagnetic calorimeters ($|p^{-1} - E^{-1}| < 3\sigma$), electromagnetic shape in the four samplings and final isolation cuts:

$$\Sigma\, p_T < 1.0 \text{ GeV} \quad \text{(other tracks)}$$

$$\Sigma\, E_T < 1.0 \text{ GeV} \quad \text{(other cells)}$$

Table I
Electron + $\geq$ 2 Jet Events

Run/Event	$Q \cdot p_T^e$ (GeV)	N_J	E_T^{J1} (GeV)	E_T^{J2} (GeV)	E_T^{J3} (GeV)
A 5069/192	+15.0	3	19.5	14.1	13.0
B 6301/716	+19.5	2	14.2	13.9	
C 6899/804	-18.0	3	16.9	8.8	8.7
D 7443/509	-19.1	2	23.4	11.1	
E 8578/983	-18.3	2	22.6	15.5	

in a cone of $\Delta R < 0.4$ around the electron. This left a total of 19 events: 14 e + 1 jet events, 3 e + 2 jet events and 2 e + 3 jet events, where the first jet was defined with $E_T > 8$ GeV and other jets with $E_T > 7$ GeV (uncorrected). Events with at least two jets are summarized in Table I.

3.2 Muons with at Least One Jet

In the muon + jet selection events were selected containing at least one jet with $E_T > 8$ GeV (uncorrected) and a hard CD track with $p_T > 12$ GeV matching the muon chamber track. Isolation in a cone of $\Delta R < 0.4$ around the muon was required by demanding that the $\Sigma\ p_T$ of the remaining charged tracks be less than 10% of all tracks in this cone, and 20% for the corresponding E_T cells of the calorimeter. This left a total of 40 events. Events were then validated by scanning at the MEGATEK interactive facility after which fake high momentum tracks from $K \rightarrow \mu\nu$ decay were removed. The final requirement that there be no jet in a cone of $\Delta R < 1$ around the muon left 12 events: 7 μ + 1 jet events and 5 μ + $\geq$ 2 jet events, which are summarized in Table II. As in the case of the electrons, the muon track quality has been compared to and agrees well with the 14 $W \rightarrow \mu\nu$ events from the same data sample[2].

4. Background Estimates

4.1 Electron + Jet Events

The dominant background for electrons in this selection is from QCD jets faking electrons by random overlaps of (dominantly) charged and neutral pions. Two general methods are described to estimate this background: i) a global method in which the background shape is compared with QCD events, i.e, events containing at least two jets in addition to a cluster satisfying the electromagnetic trigger that is a mixture of charged and neutral pions and jets; ii) a direct method in which one measures the absolute flux of

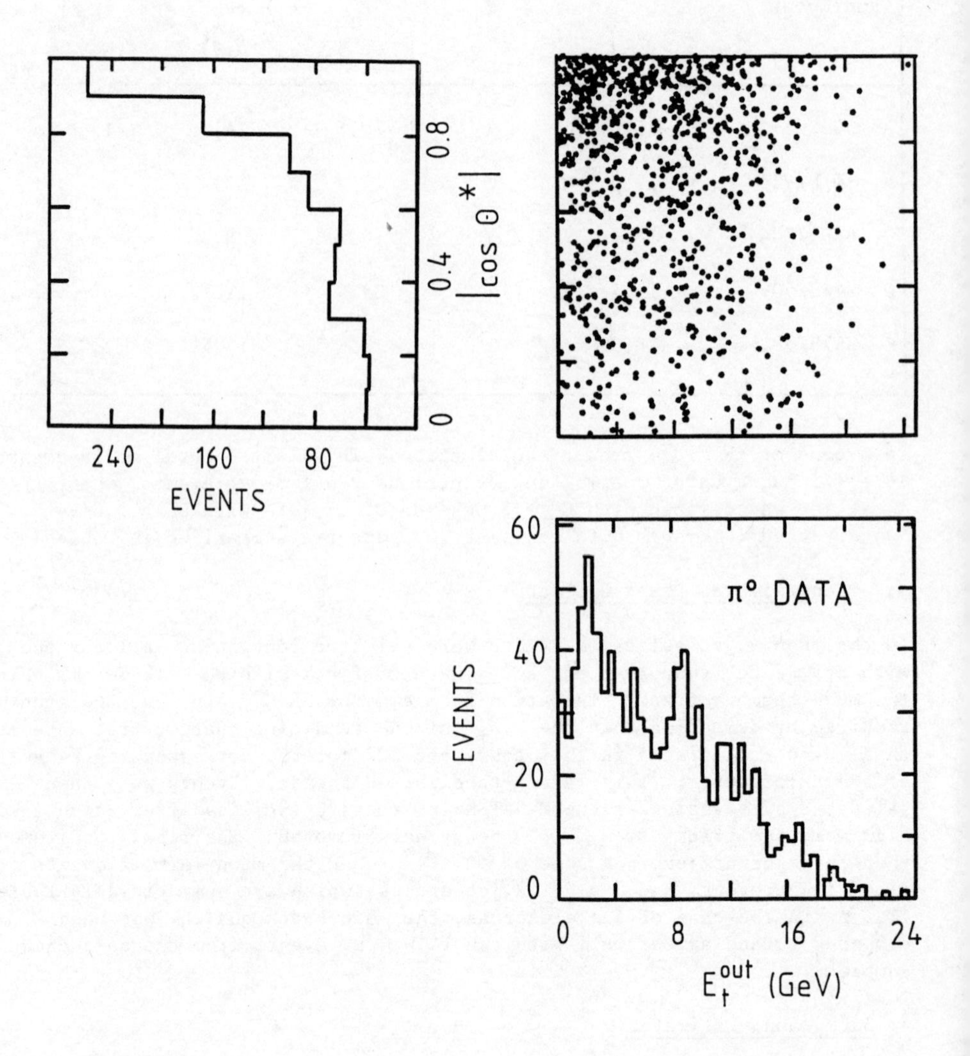

Fig. 1a. E_T^{out} versus $\cos\theta^*_{J2}$ for $\pi^0 + \geq 2$ jet events

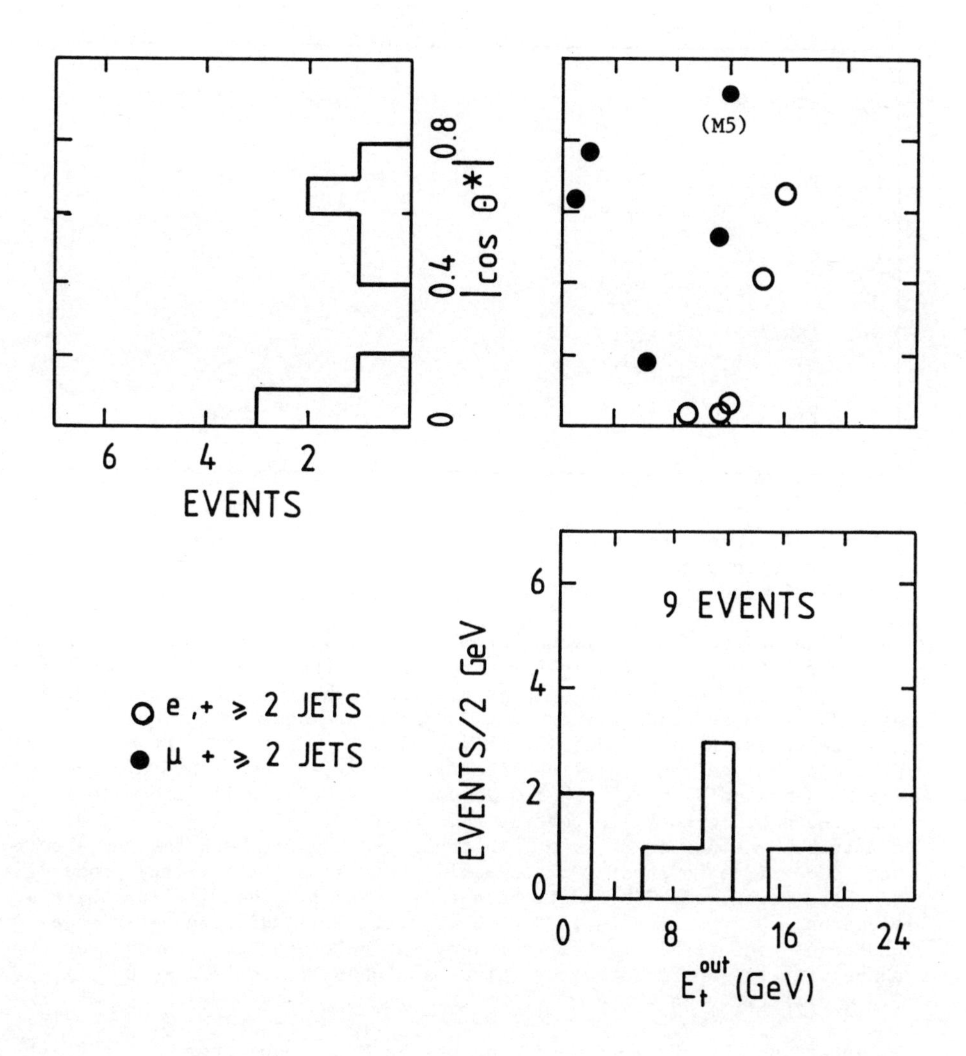

Fig. 1b. E_T^{out} versus $\cos\theta^*_{J2}$ for (e,μ) + $\geq$ 2 jet events

Table II
Muon + ≥ 2 Jet Events

Run/Event	$Q \cdot p_T^e$ (GeV/c)	N_J	E_T^{J1} (GeV)	E_T^{J2} (GeV)	E_T^{J3} (GeV)
M1 6639/118	+16.0 ±3.4	2	30.0	11.0	
M2 7700/487	-21.4 ±2.6	3	29.9	12.0	9.5
M3 7935/232	-16.2 ±3.3	2	22.3	11.3	
M4 7658/947	-12.5 ±2.7	2	17.8	14.4	
M5 7977/1182	+13.4 ±2.7	2	23.6	16.9	

$\pi^{\pm}(+n\pi^0)$ + jet events from the $\pi^{\pm}$ + jet selection and then estimates the probability that a pion so selected passes the same selection criteria as for the electrons.

The shape of the QCD background from π^0 + ≥ 2 jet events is shown in Figure 1 in which the transverse energy component of the isolated π^0 perpendicular to the plane formed by the $\bar{p}p$ axis and the highest E_T jet (J1) is plotted as a function of $\cos\theta^*_{J2}$, where θ^*_{J2} is the angle between the beam axis ($\bar{p}p$) and the lowest E_T jet (J2) in the (π^0,J1,J2) rest frame. From this scatter plot one can see that the five electron + ≥ 2 jet events are all bounded within a region RI = $\{E_T^{out} > 8$ GeV, $|\cos\theta^*_{J2}| < 0.7\}$, while the majority of background QCD events lie outside this region in RII = 1 - RI. Table III summarizes the number of events in these two regions both for the electron and π^0 + ≥ 2 jet events. On the basis of these statistics the probability that the background QCD events have a distribution identical to the five electron + ≥ 2 jet events is 8×10^{-5}, i.e., a 4 σ difference in shape. The choice of regions RI and RII, however, is arbitrary and a comparison of shape with the Kolomogorov test gives a probability of 5.2×10^{-4}, i.e., a 3.5 σ difference in shape.

To determine the absolute magnitude of the background from $\pi^{\pm}$ + ≥ 2 jet events, we estimate the probability that a $\pi^{\pm}$ satisfies the same selection criteria as for the electrons. Of the 169 $\pi^{\pm}$ + ≥ 2 jet events originally selected, 68 satisfy the electron trigger requirement of $E_T > 12$ GeV and from test beam data this probability is 1.5×10^{-3}, yielding a total of 0.1 background events. The corresponding number of background events in region RI is less than 0.05 events for comparison.

Finally the background from unseen conversions with one electron triggered with $p_T > 15$ GeV/c and the other with $p_T < 0.05$ GeV/c is less than 0.02 events, as determined from the measured π^0 flux in region RI.

Table III
Event distribution in E_T^{out} .vs. $\cos\theta^*_{J2}$

Region	π^0	e
π^0/electron + $\geq$ 2 jets (RI + RII)	933	5
RI = $\{E_T^{out} > 8$ GeV, $\lvert\cos\theta^*_{J2}\rvert < 0.7\}$	211	5
RII = 1 - RI	722	0

4.2 Muon + Jet Events

The dominant source of muon background for these events comes from pions and kaons decaying in the central detector drift volume. In the case of slow kaons, both parent and daughter tracks may form a "kink" in the track digitizations which is reconstructed as a fake high momentum track.

This decay background has been determined from a 5 nb^{-1} data sample triggered by jets with a low threshold (E_T > 15 GeV) and whose trigger requirement is also satisfied by the five muon + $\geq$ 2 jet events. We wish to determine the probability that decaying hadrons ($\pi, K \rightarrow \mu\nu$) from this data sample pass the same track quality cuts and are reconstructed with p_T > 12 GeV/c. The final background rate is then the convolution of this probability (per decaying hadron) with the measured p_T spectrum of the 5 nb^{-1} sample. For a mixture of 50% pions and 25% kaons at the Collider, this probability is typically 2×10^{-3} for a hadron with p_T of 13 GeV/c to be reconstructed with p_T > 12 GeV/c.

The results of this background calculation are summarized in Table IV. We find that the corresponding number of decay muon + $\geq$ 2 jet events is 0.4 events, and less than 0.08 events for $\lvert\cos\theta^*_{J2}\rvert < 0.8$. In Figure 1 the five muon + $\geq$ 2 jet events are plotted as a function of E_T^{out} and $\cos\theta^*_{J2}$ along with the e/π^0 + $\geq$ 2 jet events. Of the four muon + 2 jet events, event M5 is most likely a background event of QCD origin since jet J2 lies close to the beam axis with $\cos\theta^*_{J2}$ = 0.93.

5. Interpretation of Events

Weak production of the top via $W \rightarrow t\bar{b}$ is expected to show a clean experimental signature in the semi-leptonic decay channel of the top; for moderately large top quark masses two central hadronic jets of large transverse energy should be easily identified with the recoiling $\bar{b}$ jet and b jet from $t \rightarrow b\ell\nu$, while decay leptons should be isolated as a result of the large Q-value available in the reaction. Furthermore the cross section for this process is well known since its absolute value is normalized to the measured weak process $W \rightarrow (e,\mu)\nu$ cross section. Processes involving three jets in the final state from $p\bar{p} \rightarrow Wg$ are expected to be small with the initial state bremsstrahlung strongly peaked in the forward region.

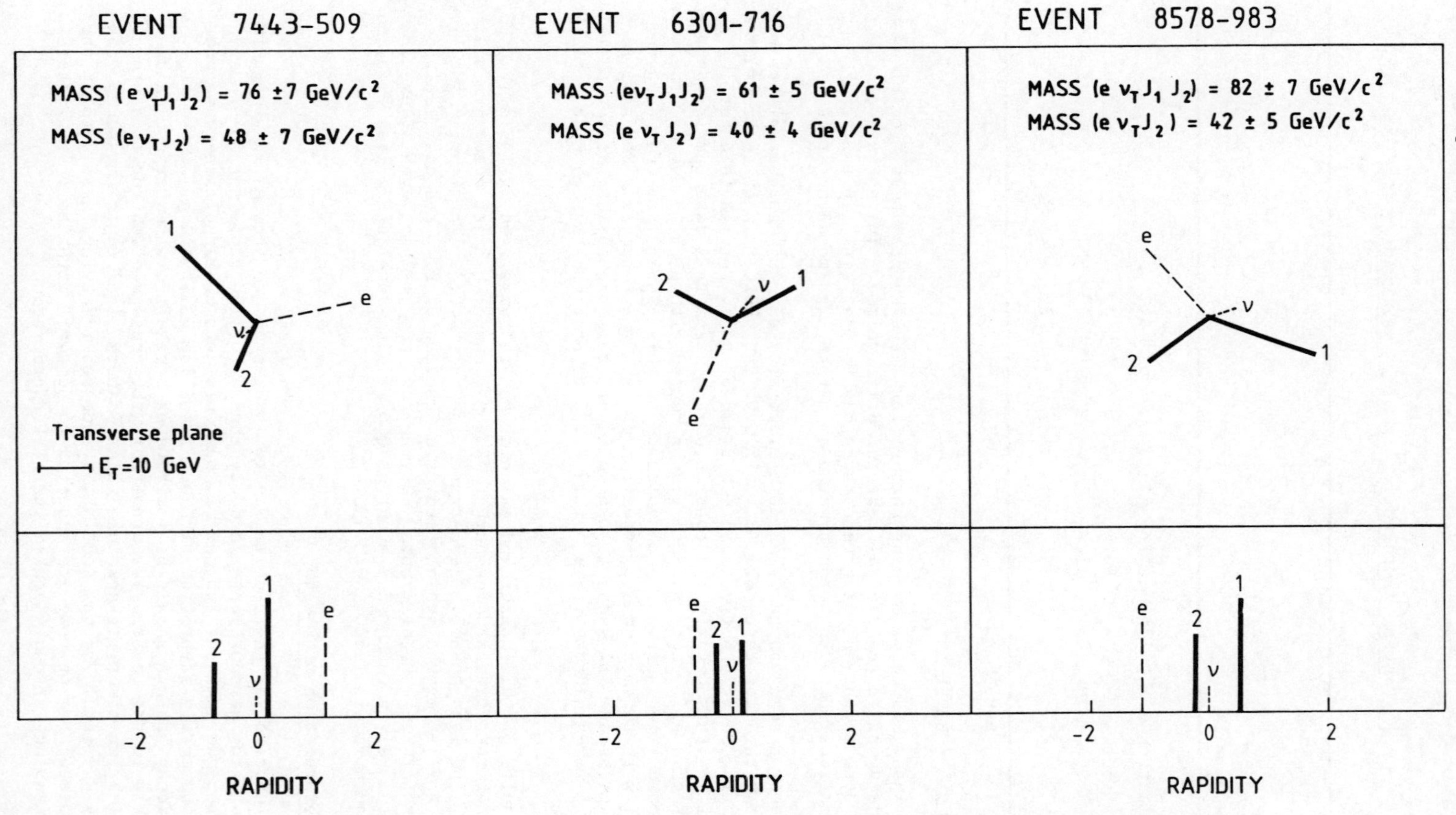

Fig. 2a. Electron + 2 jet events, view in transverse plane and rapidity.

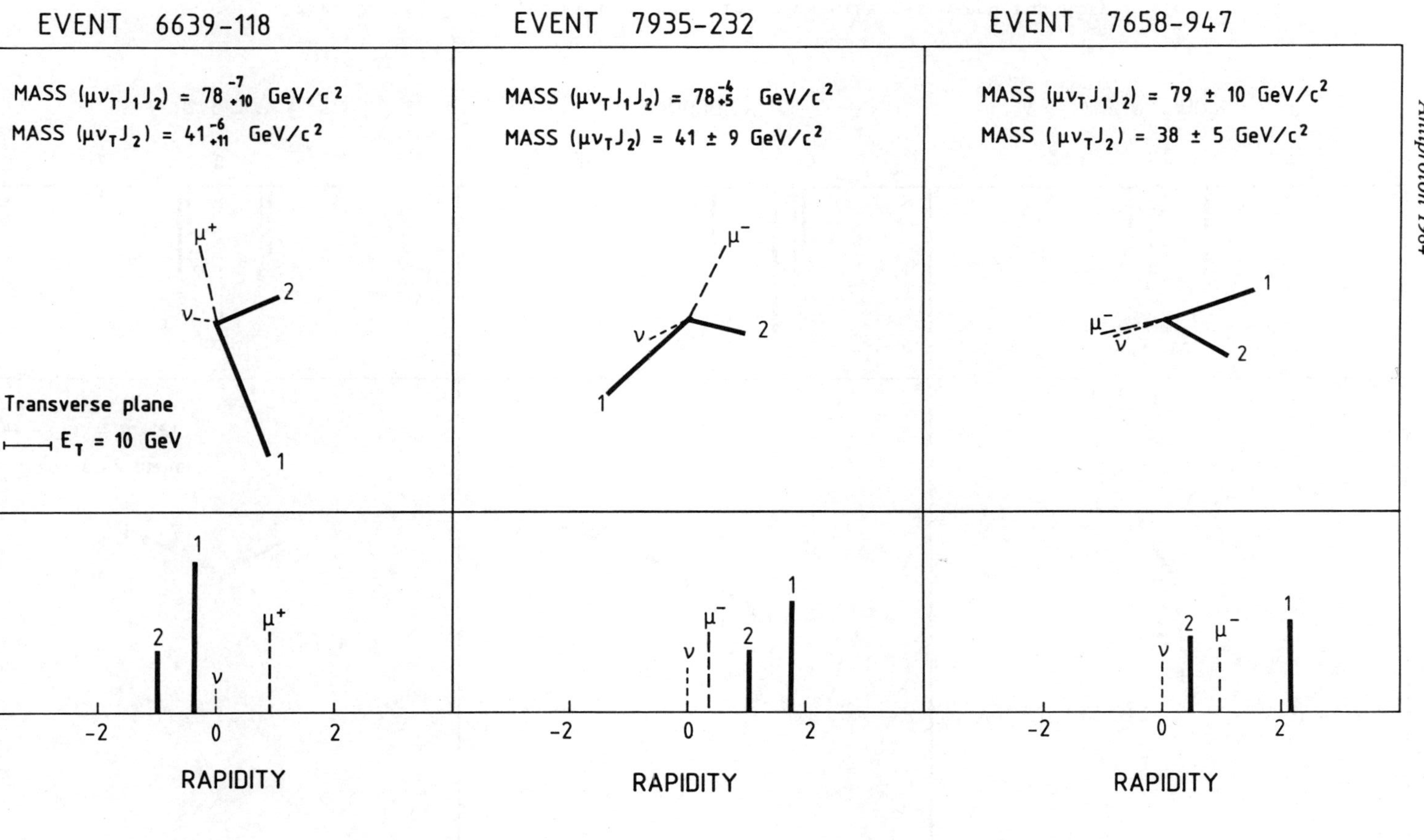

Fig. 2b. Muon + 2 jet events, view in transverse plane and rapidity.

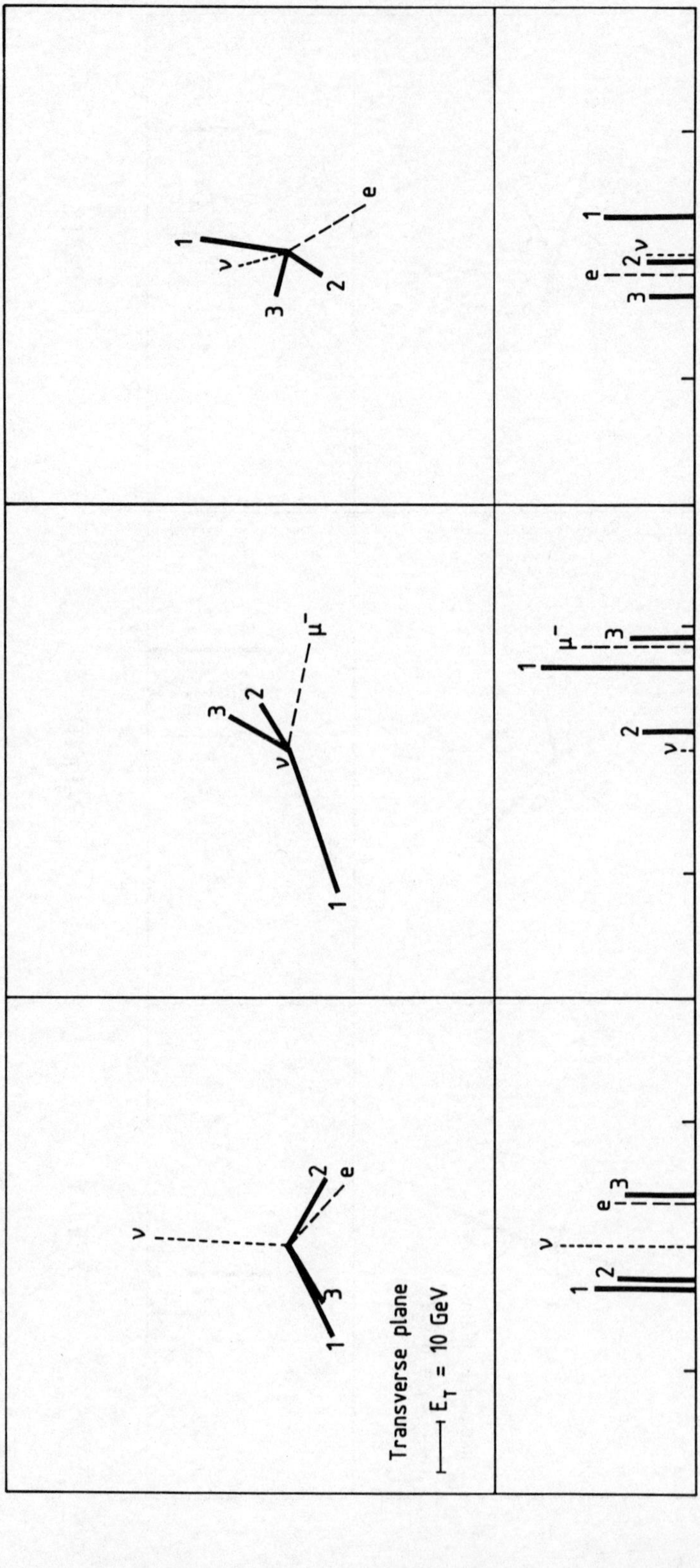

Fig. 3. (e,μ) + 3 jet events, view in transverse plane and rapidity.

Table IV
Decay Background Estimate for Muon Events

Topology	Muons	Background
μ + 1 jet	7	0.25
$\mu + \geq 2$ jets	5	0.40
$\mu + \geq 2$ jets $\lvert\cos\theta^*_{J2}\rvert < 0.8$	4	0.25
μ + 2 jets	4	0.1
μ + 2 jets $\lvert\cos\theta^*_{J2}\rvert < 0.8$	3	0.08

Strong production of the top by either gluon fusion, flavour excitation or single high mass diffraction dissociation is certainly not excluded. However, the cross sections for these processes are not well known, even though they may be much larger in the forward (diffractive) region. Furthermore the final state topology may involve anywhere from two to four jets with a large combinatorial background.[7)]

The six lepton + 2 jet events and the three lepton + 3 jet events are shown in Figures 2 and 3, respectively. (Event M5 with $\cos\theta^*_{J2} = 0.93$ has been omitted as a likely background event.) To test the $W \rightarrow t\bar{b}$ hypothesis we shall concentrate on the 2 jet events only (Figure 2).

The correlation between the ($\ell\nu_T$J1J2) mass and the ($\ell\nu_T$J2) mass is shown in Figure 4 for the six lepton + 2 jet events. Jets are corrected for energy and direction. The errors shown are dominated by the experimental resolution of the jet energy and momentum and may not necessarily reflect the systematic errors inherent in the determination of jet energy.

The strong correlation in the multijet mass near 80 GeV/c^2 suggests production and decay of the W Boson; in the corresponding ($\ell\nu_T$J2) mass a similar clustering near 40 GeV/c^2 suggests production and semi-leptonic decay of a new state (top), which in turn could be a final state decay of the W. According to this interpretation, Monte Carlo calculations[8)] at the parton level show that for top masses up to roughly 60 GeV/c^2 the recoiling $\bar{b}$ jet from W decay is always the highest E_T jet in 90% of all cases.

The clustering observed around the W mass for the six lepton + 2 jet events unfortunately is not a sufficient condition to prove the W decay hypothesis. In Figure 5 we show the same correlation plot, but this time for π^0 + 2 jet events of QCD origin. A similar clustering in the "top" invariant mass combination of ($\ell\nu_T$J2) is also observed near 40 GeV/c^2 and similarly near 80 GeV/c^2 in the multijet combination. When comparing the difference in shape between these spectra and those of the six lepton + 2 jet events, we

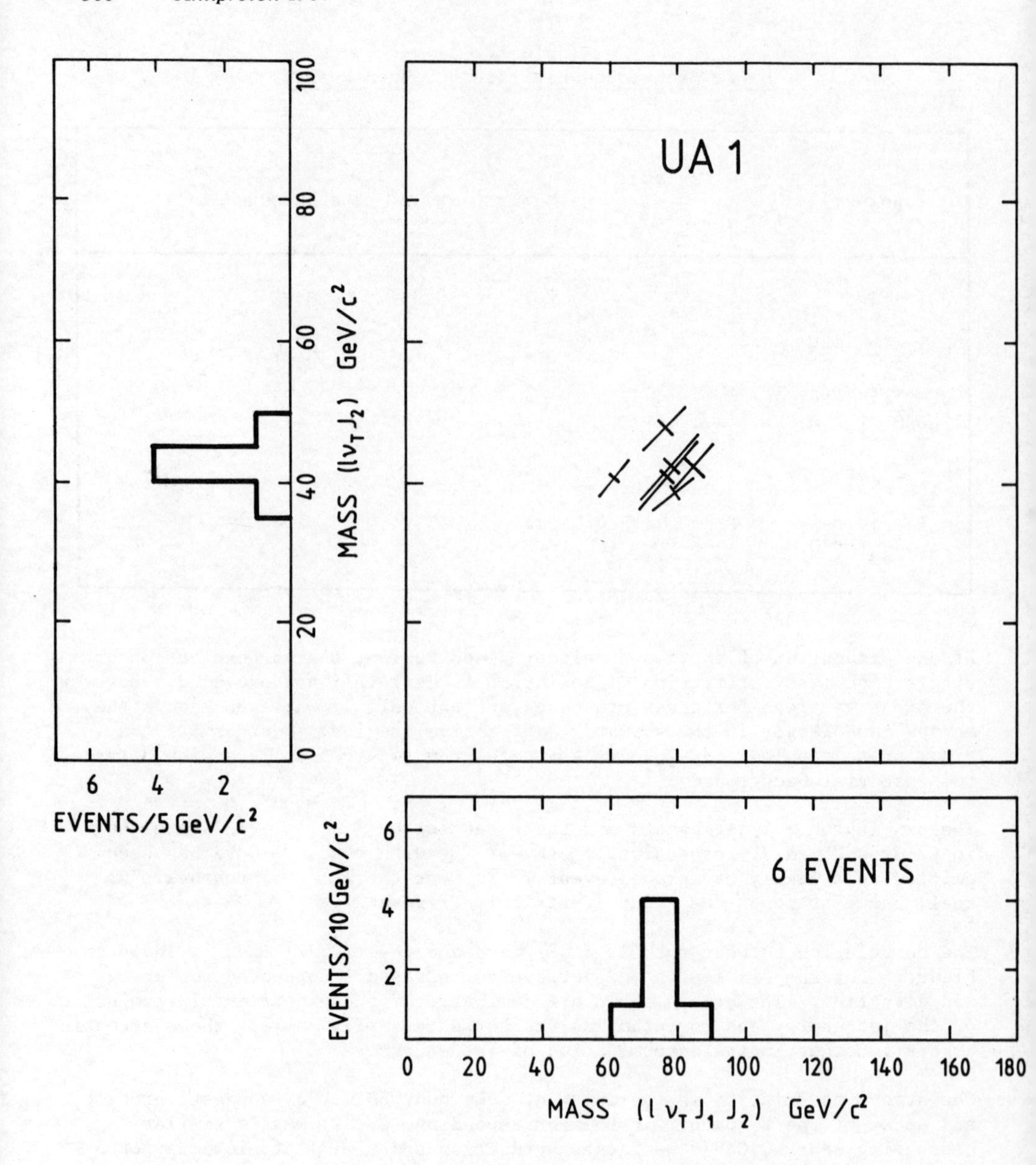

Fig. 4. $(\ell\nu_T J2)$ mass versus $(\ell\nu_T J1J2)$ mass for six lepton + 2 jet events.

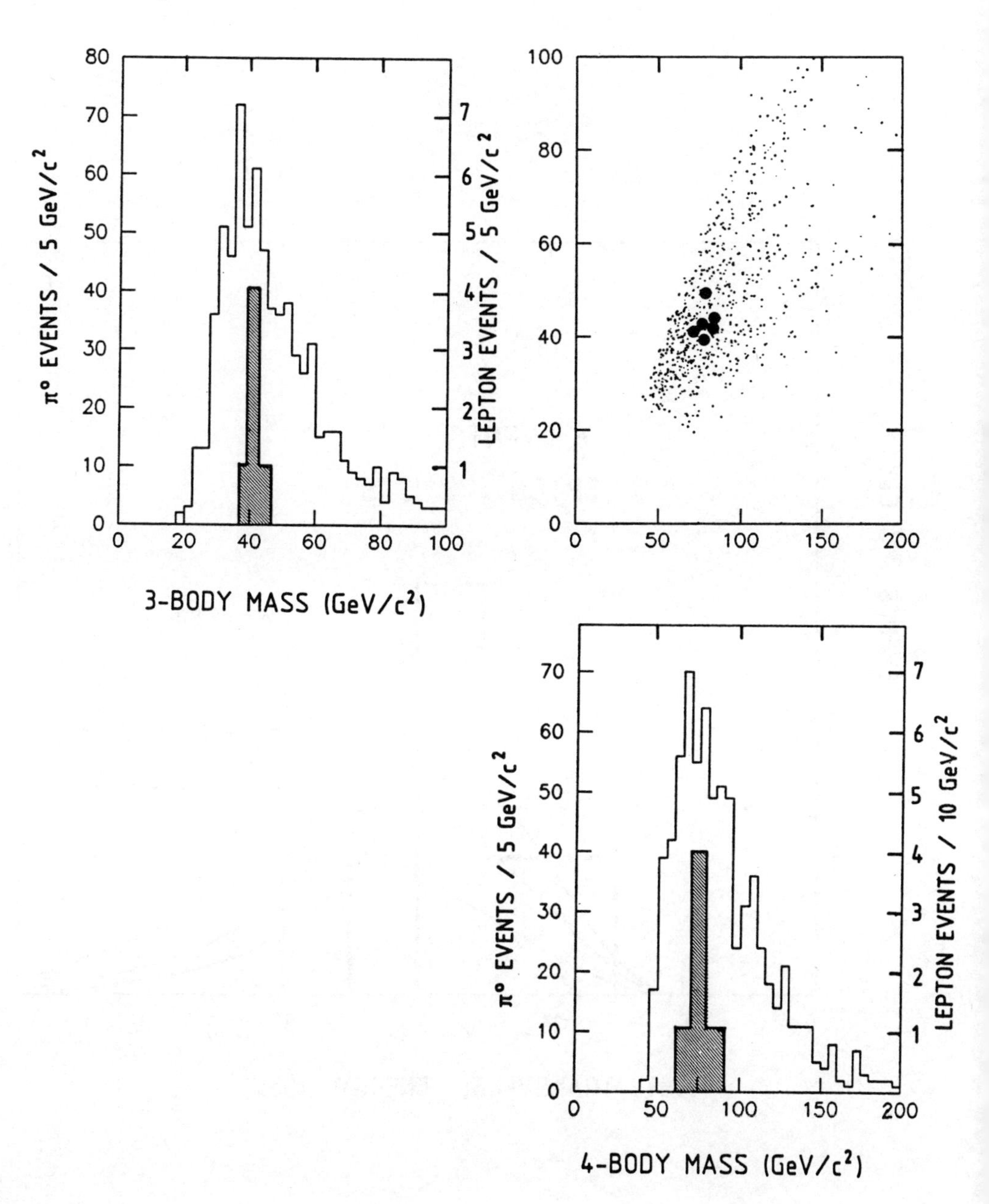

Fig. 5. ($\pi^0\nu_T$J2) mass versus ($\pi^0\nu_T$J1J2) mass for π^0 + 2 jet events.

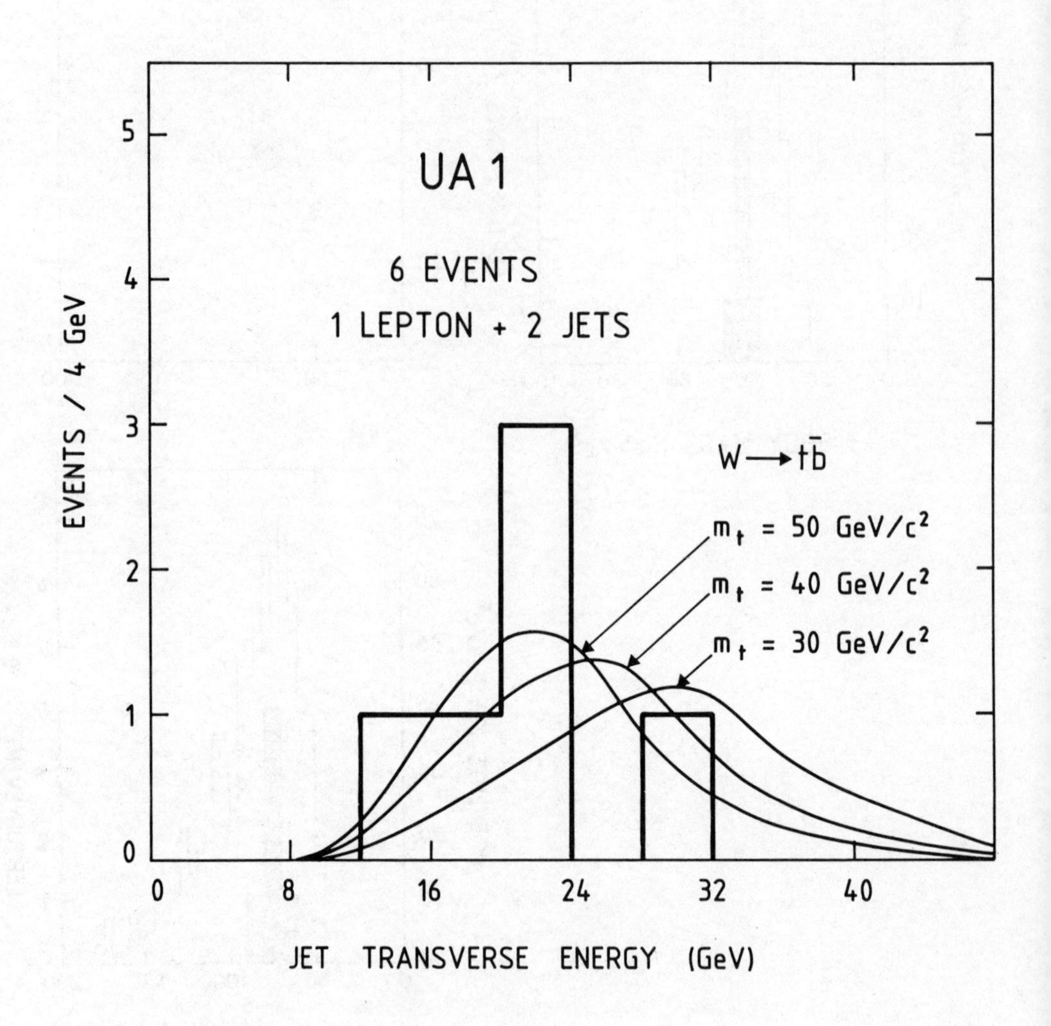

Fig. 6. Jet 1 (highest E_T jet) transverse energy distribution for lepton + 2 jet events.

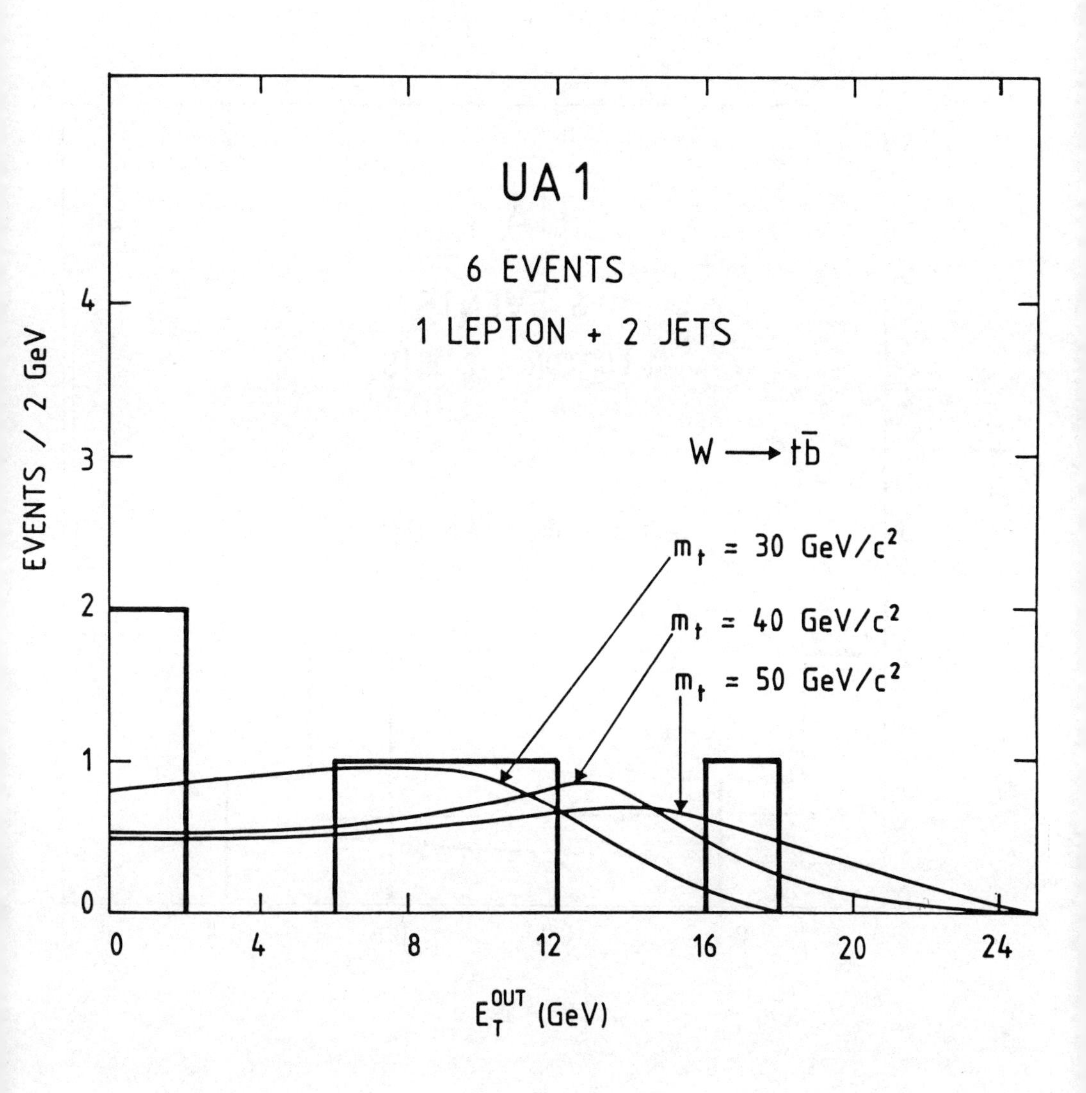

Fig. 7. E_T^{out} distribution for lepton + 2 jet events.

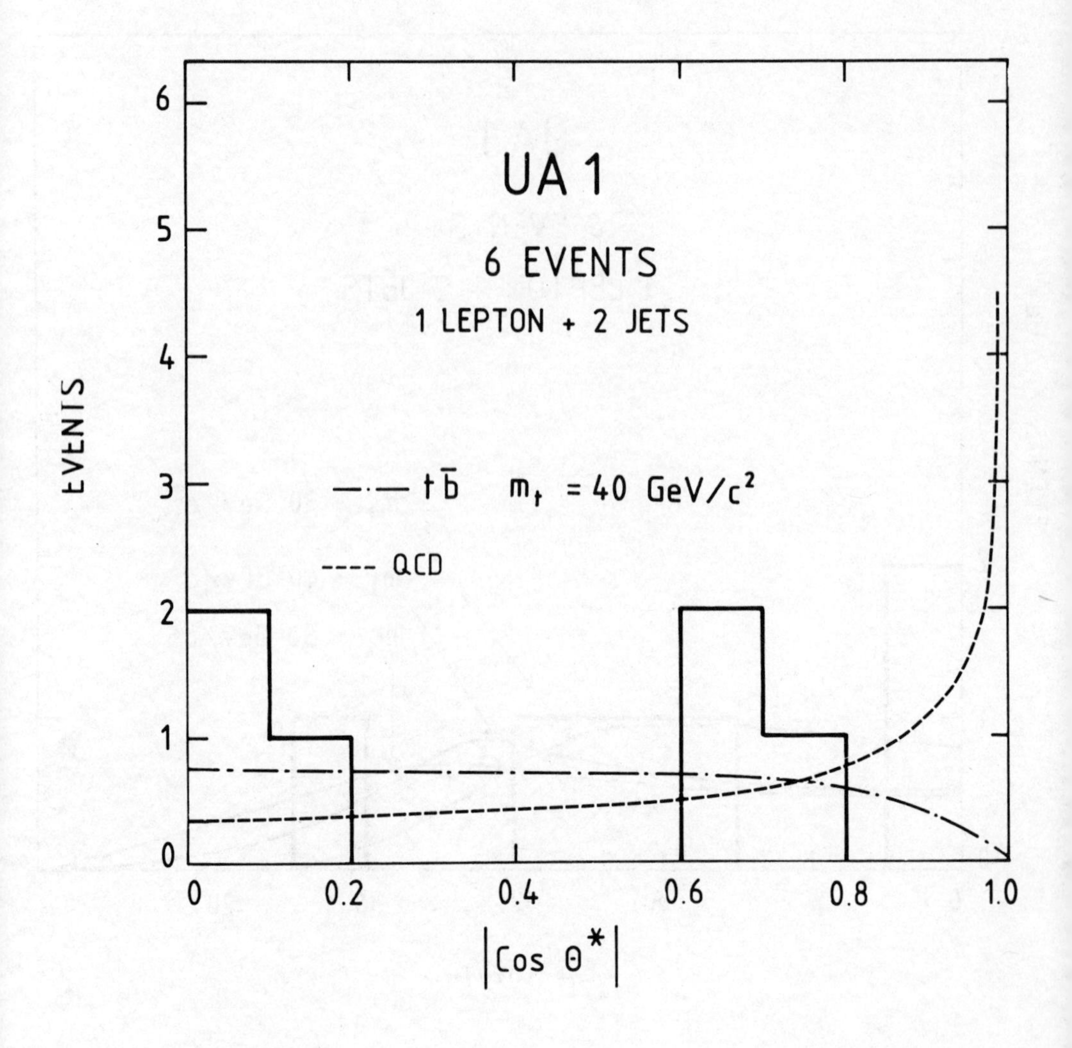

Fig. 8. $|\cos\theta^*_{J2}|$ distribution for lepton + 2 jet events shown with expectation from QCD processes.

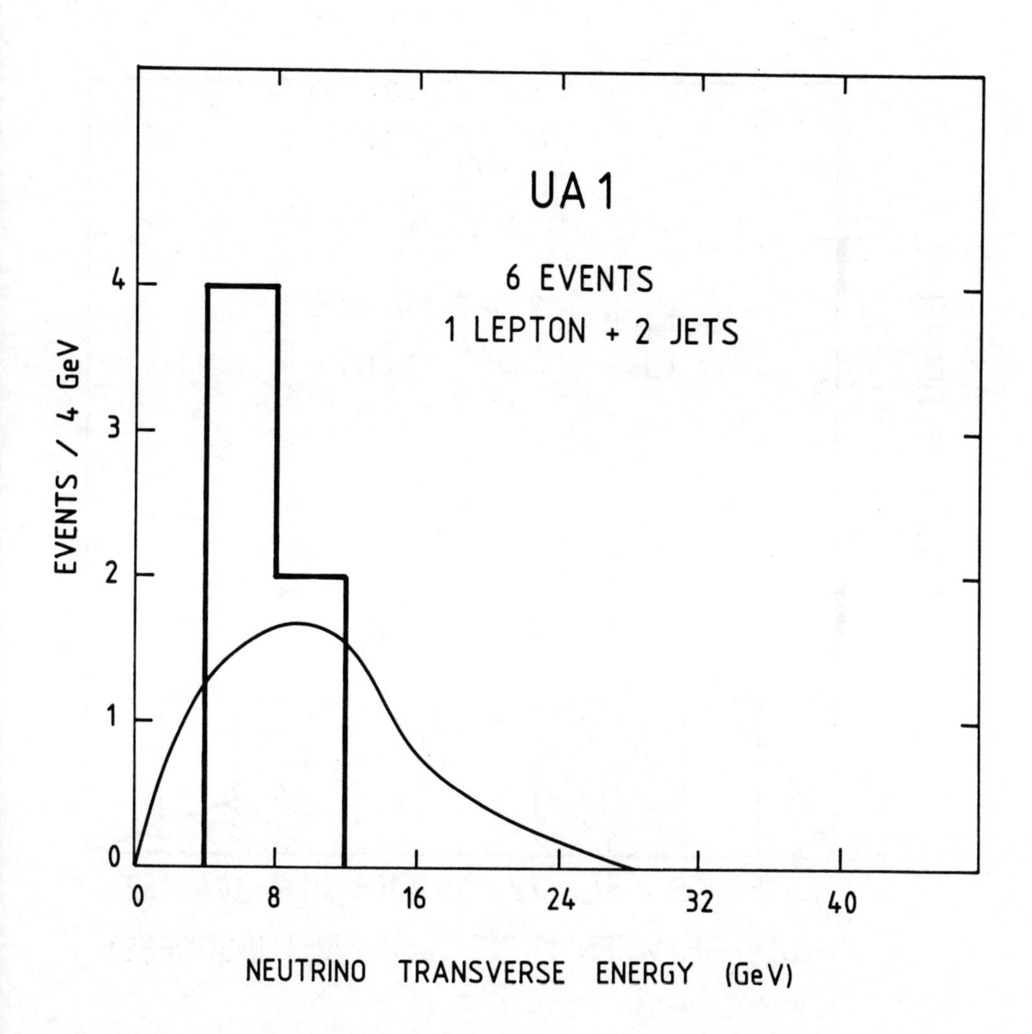

Fig. 9. Neutrino transverse energy distribution for lepton + 2 jet events, compared with expectation from $W \rightarrow t\bar{b}$.

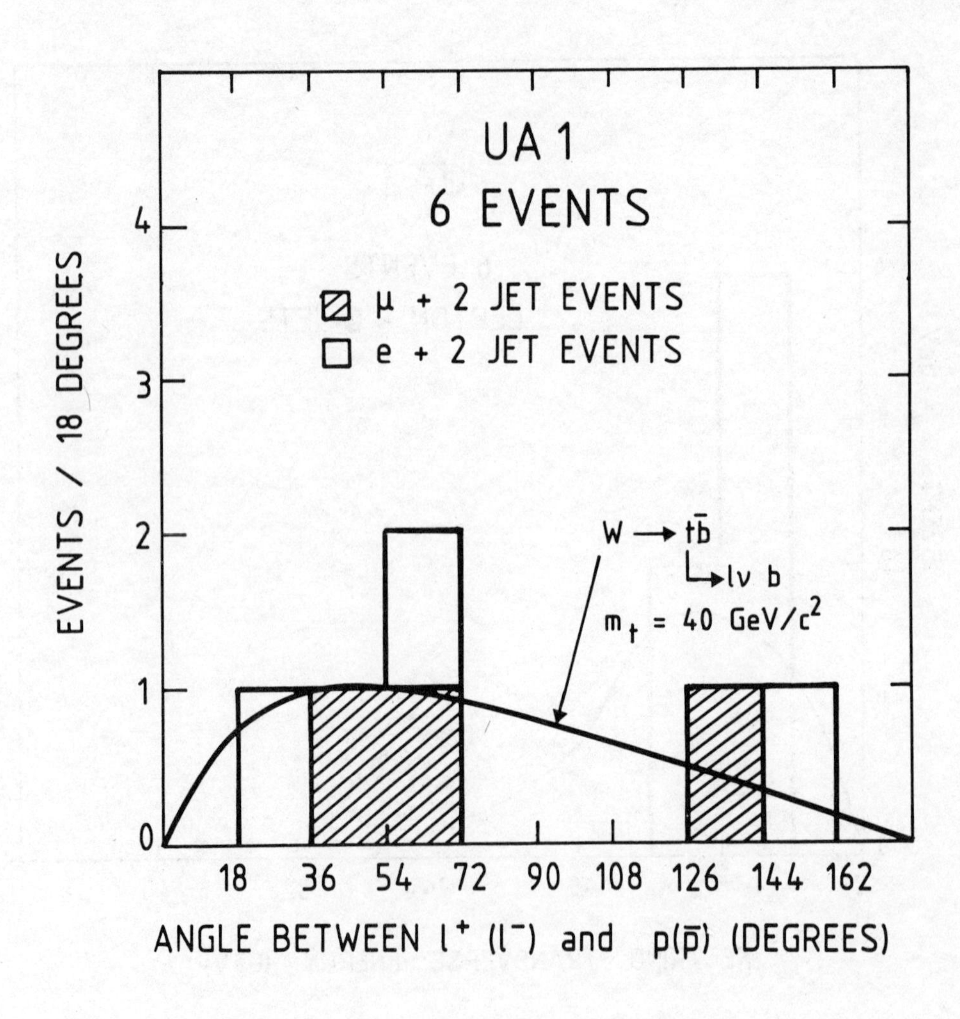

Fig. 10. Angle between $\ell^+(\ell^-)$ and $p(\bar{p})$ beam direction.

Table V
Rates and Efficiencies for $W \to t\bar{b}$

(m_t = 40 GeV/c^2, BR($t \to b\ell\nu$) = 10%, $\int$Ldt = 0.12 pb^{-1})

Cut	σ (pb)	Fraction Kept	No. of Events
No cuts	148.9	100%	18
$E_T^{\bar{b}} > 10$ GeV	136.5	91.7%	16.4
$E_T^{\ell} > 12$ GeV	58.3	39.2%	7.0
$E_T^{b} > 10$ GeV	22.7	15.2%	2.7

find only a confidence level of 1.3×10^{-1} (< 1 σ) for the ($\ell\nu_T$J2) combination and only 4.8×10^{-2} (∿ 2 σ) for the ($\ell\nu_T$J1J2) mass combination. It should be stressed, however, that the π^0 + 2 jet spectra are phase space spectra and not background spectra. The selection criteria are constrained in such a way that the phase space peaks near the same mass region in both selections.

In Figure 6 we plot the E_T spectrum of the highest E_T jet (J1) for these six events and compare these results with a Monte Carlo calculation of the process $W \to t\bar{b}$[8] which includes the experimental smearing. On the basis of this plot[9] a more likely top quark mass value of 50 GeV/c^2 is suggested by the same calculations. Experimental distributions for E_T^{out} of the leptons (Figure 7), $\cos\theta^*_{J2}$ (Figure 8), the transverse neutrino energy (Figure 9) and the angle between the lepton and beam axis (Figure 10) are also compared with the same calculations.

The rates and detection efficiencies for the process $W \to t\bar{b}$ for a top quark mass of 40 GeV/c^2 are summarized in Table V from the Monte Carlo calculations mentioned above[8]. Assuming a semi-leptonic branching ratio of 10% for the top, a total of 2.7 events per lepton channel is expected for a total integrated luminosity of 0.12 pb^{-1}. For an overall electron detection efficiency of 60% and 34% for muons, the number of (e + μ) + 2 jet events we should expect is then 2.6 events, to be compared to the 6 ± 2.4 events observed. If interpreted in terms of the top semi-leptonic branching ratio, then this branching ratio would be 23 ± 9 %, to be compared to 12.5% for b-quarks and 7 - 8% for lighter charmed quarks.

6. Conclusions

We observe a "non-trivial" signal in the channel of an isolated large transverse energy lepton + 2,3 associated jets. The 2 jet signal has an overall invariant mass clustering around the W mass, indicating a novel decay of the W. The rates and features of the 2 jet events do not satisfy the expectations for charmed and beauty decay. They are consistent, however, with the process $W \to t\bar{b}$, where t is the sixth "top" quark of the Cabibbo current. If indeed so, then the mass of the top is bounded between 30 and 50 GeV/c^2. We stress that the present uncertainty in the ($\ell\nu_T$j) mass is due to the determination of jet energy, and that more statistics is needed to establish on a firm basis these conclusions and the true nature of the effect observed.

Acknowledgements

I wish to thank the conference organizers for their splendid hospitality during my stay, and all members of the UA1 Collaboration who made these exciting results possible. In particular I wish to thank Professor Carlo Rubbia for the possibility to present these results on behalf of my colleagues.

References

1) C. Rubbia, P. McIntyre and D. Cline, Proc. Intern. Neutrino Conf. (Aachen, 1976)(Vieweg,Braunschweig,1977) p. 683;
Study Group, Design study of a proton-antiproton colliding beam facility, CERN/PS/AA 78-3 (1978), reprinted in Proc. Workshop on Producing high-luminosity, high-energy proton-antiproton collisions (Berkeley,1978) report LBL-7574, UC34, p. 189;
The staff of the CERN proton-antiproton project, Phys. Lett. 107B (1981) 306.

2) C. Rubbia, The physics of the proton-antiproton collider, Invited talk Intern. Europhysics Conf. on High-energy physics (Brighton,UK,1983) CERN-preprint EP/83-168 (1983);
C. Rubbia, Physics results of the UA1 Collaboration at the CERN proton-antiproton Collider, Invited talk 9th Hawaii Topical Conference in Particle Physics (Honolulu,Hawaii,1983) CERN-preprint EP/84-55 (1984).

3) C. Rubbia, talk at Neutrino 84 Conference (Dortmund,Germany,1984);
M. Della-Negra, Invited talk XV Symposium on Multiparticle Dynamics (Lund, Sweden,1984).

4) UA1 proposal, A 4π solid-angle detector for the SPS used as a proton-antiproton collider at centre-of-mass energy of 540 GeV, CERN/SPSC 78-06 (1978);
A. Astbury, Physica Scripta 23 (1981) 397;
M.J. Corden et al., Physica Scripta 25 (1981) 5,11;
B. Aubert et al., Nucl. Instrum. Methods 176 (1980) 195;
K. Eggert et al., Nucl. Instrum. Methods 176 (1980) 217,233.

5) M. Barranco Luque et al., Nucl. Instrum. Methods 176 (1980) 175;
M. Calvetti et al., Nucl. Instrum. Methods 176 (1980) 255;
M. Calvetti et al., The UA1 central detector, Proc. Intern. Conf. on Instrumentation for colliding beam physics (SLAC,Stanford,1982) (SLAC-250, Stanford,1982) p. 16.

6) The UA1 Collab., The UA1 Data Acquisition System, Proc. Intern. Conf. on Instrumentation for colliding beam physics (SLAC,Stanford,1982) (SLAC-250, 1982) p. 151.

7) D. DiBitonto, Heavy-Flavour Production in Experiment UA1, in: Proc. Third Moriond Workshop on Antiproton-proton Physics and the W Discovery (La Plagne,1983) (Editions Frontières, Gir-sur-Yvette,1983) p. 473;
E.L. Berger, D. DiBitonto, M. Jacob and W.J. Stirling, Phys. Lett. 140B (1984) 259.

8) P. Aurenche, R. Kinnunen and K. Mursula, LAPP preprint TH-106 (1984).

9) V. Barger, A.D. Martin and R.J.N. Phillips, Phys. Lett. 125B (1983) 343.

Inst. Phys. Conf. Ser. No. 73: Section 5
Paper presented at VII Eur. Symp. Antiproton Interactions, Durham 1984

How to search for new particles at p̄p colliders

V Barger

Physics Department, University of Wisconsin, Madison, Wisconsin 53706, USA

Abstract. Methods for identifying the t quark, a fourth generation of quarks and leptons, supersymmetric partners of the known particles, and the Higgs boson at $p\bar{p}$ colliders are discussed. The properties of events with single leptons, multileptons, and missing p_T along with hadronic jets are studied. The techniques include lepton isolation, jet broadness, Jacobian peaks, transverse masses and azimuthal correlations.

1. Search for the t Quark

1.1 Single lepton signals

The production of heavy quarks in $p\bar{p}$ collisions occurs via $W \to t\bar{b}$ decay and via $Q\bar{Q}$ hadroproduction ($Q = c,b,t$). The prediction for the $W \to t\bar{b}$ cross section is tied to the measured $W \to e\nu$ rate, and thus should be reliable. The hadroproduction cross section can be calculated from the lowest-order $q\bar{q},gg \to Q\bar{Q}$ fusion subprocesses. In addition, diffractive heavy-quark production may occur, but this contribution is more difficult to quantify. Figure 1 shows the fusion rates versus the heavy-quark mass; here a semi-empirical enhancement factor $K = 2$ has been assumed for all cases. For a t-quark mass greater than 35 GeV the $W \to t\bar{b}$ rate exceeds that from $t\bar{t}$ fusion.

Prompt leptons from semileptonic decays provide a signal for heavy quarks, as recognized in many studies of the t-quark signal at $p\bar{p}$ colliders (Abud et al 1978, Aurenche et al 1984, Ballocchi and Odorico 1984, Barger et al 1983abc 1984abcd, Berger et al 1984, Cabibbo and Maiani 1979, Chaichian et al 1983, Chau et al 1981, Collins and Spiller 1984ab, Desai and Lindfors 1983, Godbole et al 1983, Hagiwara and Long 1983, Halzen and Scott 1983, Horgan and Jacob 1981 1984, Lindfors and Roy 1984, Paige 1981 1983, Pakvasa et al 1979, Roy 1984, Sehgal and Zerwas 1984).

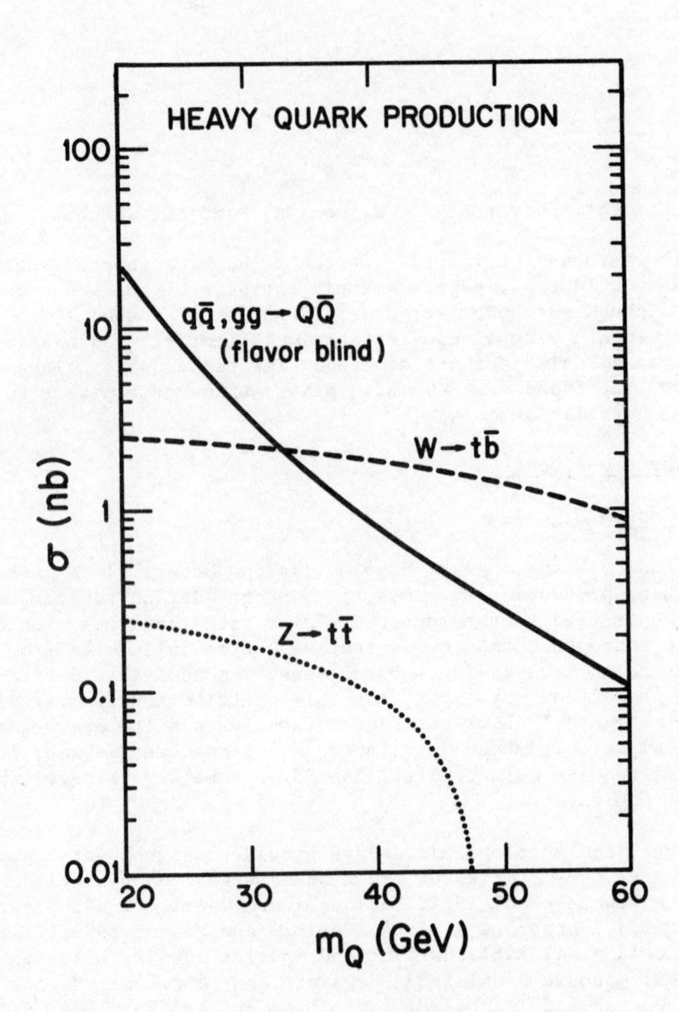

Fig. 1 Cross sections for heavy-quark production versus mass m_Q in $p\bar{p}$ collisions at $\sqrt{s}$ = 620 GeV versus the heavy-quark mass m_Q (Barger et al 1984e).

Characteristic topologies in the plane transverse to the beam axis of single-lepton (ℓ = e,μ) events are illustrated in Fig. 2.

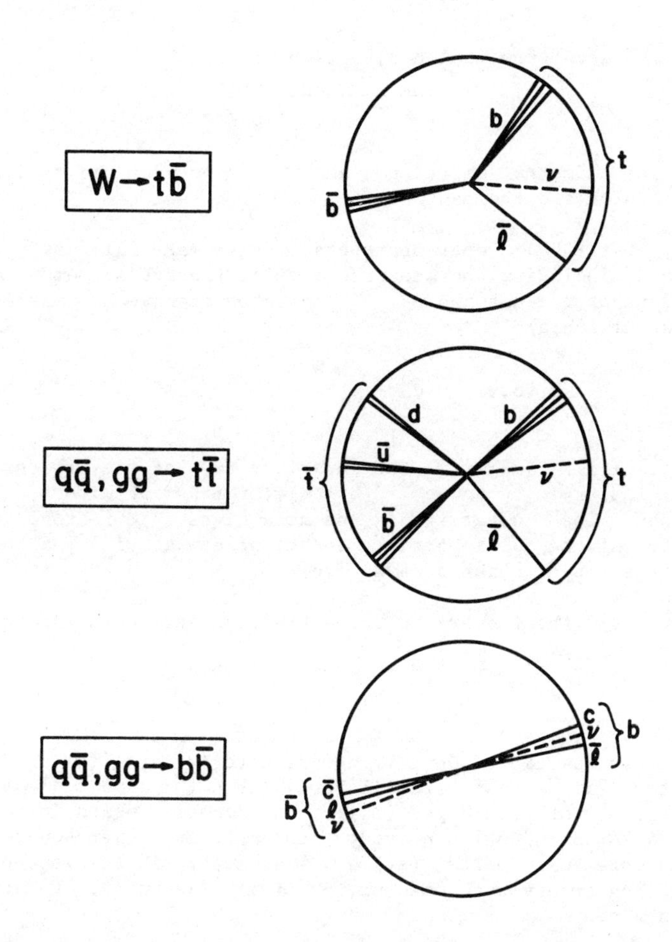

Fig. 2 Characteristic topologies of heavy-quark events in the transverse plane.

A lepton from t decay is generally "isolated" from the other decay fragments, whereas leptons from b or c decays are collimated within or near narrow hadronic jets. In events of $W \to t\bar{b}$ origin the $\bar{b}$ recoil jet opposite the t-decay fragments exhibits a Jacobian peak at $p_T(\bar{b}) \simeq (M_W^2 - m_t^2)/(2M_W)$. The topology of $t\bar{t}$ events is understandably more complicated. The background from $O(\alpha_s^2)$ $b\bar{b}$ production, which is larger than that of $c\bar{c}$, has back-to-back narrow jets. In order $O(\alpha_s^3)$ a more complicated three-jet topology also occurs in $b\bar{b}$ events (Kunszt et al 1980).

By imposing a lepton isolation cut, the $b,c \to \ell$ backgrounds can be suppressed, including the higher-order contributions, leaving the $t \to \ell$ signal (Barger et al 1984abc, Schmitt et al 1984). Typical isolation criteria used by the UA1 collaboration are

$$\sum_{\text{hadrons}} p_T < \text{few GeV}$$

within a "cone" about the lepton direction of

$$R = \sqrt{(\Delta\phi)^2 + (\Delta\eta)^2} \lesssim 0.7$$

where $\Delta\phi$ is the azimuthal difference and $\Delta\eta$ the pseudorapidity difference between the lepton and the hadron.

In the study of $t \to b\ell\nu$ events transverse masses make full use of the observed momenta, including the measured missing transverse momentum ν_T, and are independent of the unknown ν_L. The cluster transverse mass is defined by (Barger et al 1983a)

$$M_T^2(c,\nu) \equiv (c_T^o + \nu_T)^2 - (\vec{c}_T + \vec{\nu}_T)^2$$

where $c_T^o = (\vec{c}_T^{\,2} + m_c^2)^{\frac{1}{2}}$ and m_c is the invariant mass of the cluster. For $t \to b\ell\nu$ the cluster is $b\ell$ and $\vec{c}_T = (\vec{b} + \vec{\ell})_T$; for $W \to t\bar{b}$ with $t \to b\ell\nu$ the cluster is $\bar{b}b\ell$. The cluster transverse mass gives an effective two-body representation of the multi-particle decay, whose Jacobian peak occurs at the mass of the parent which decays to $c + \nu$.

The UA1 search for the t quark is currently focussed on the process

$$W \to t\bar{b}, \quad t \to b\ell\nu$$

which has several distinct identifying characteristics: (i) an isolated charged lepton, (ii) a fast $\bar{b}$ recoil jet, with $p_T(\bar{b})$ almost always larger than $p_T(b)$ for m_t not too large, (iii) joint Jacobian peaks in $p_T(\bar{b})$, $M_T(b\ell;\nu)$, and $M_T(\bar{b}b\ell;\nu)$ which provide both upper and lower bounds on m_t, as discussed more fully below, (iv) a 30% asymmetry in the lepton rapidity, (v) missing transverse momentum, and (vi) a known rate relative to $W \to e\nu$ for a given m_t.

To analyze two-jet events of $W \to t\bar{b}$ origin the following transverse variables are useful (Barger et al 1984a)

$$M_T^2(1) \equiv M_W^2 + m_b^2 - 2M_W[m_b^2 + \bar{b}_T^2]^{\frac{1}{2}}$$

$$M_T(2) \equiv M_T(b\ell;\nu)$$

$$M_T(3) \equiv M_T(\bar{b}b\ell;\nu) \ .$$

The variable $M_T(1)$ replaces $\bar{b}_T$; $M_T(2)$ and $M_T(3)$ are the cluster transverse masses of the t quark and W boson decay products respectively. Neglecting W_T and Γ_W, these variables have the kinematic bounds

$$m_t \le M_T(1) \ , \quad M_T(2) \le m_t \ , \quad M_T(3) \le M_W \ .$$

Thus the $M_T(1)$ distribution bounds m_t from above and the $M_T(2)$ distribution bounds m_t from below.

Figure 3 illustrates M_T distributions in lepton + two-jet events for m_t = 40 GeV including effects of W_T, Γ_W and measurement uncertainties on the missing neutrino momentum.

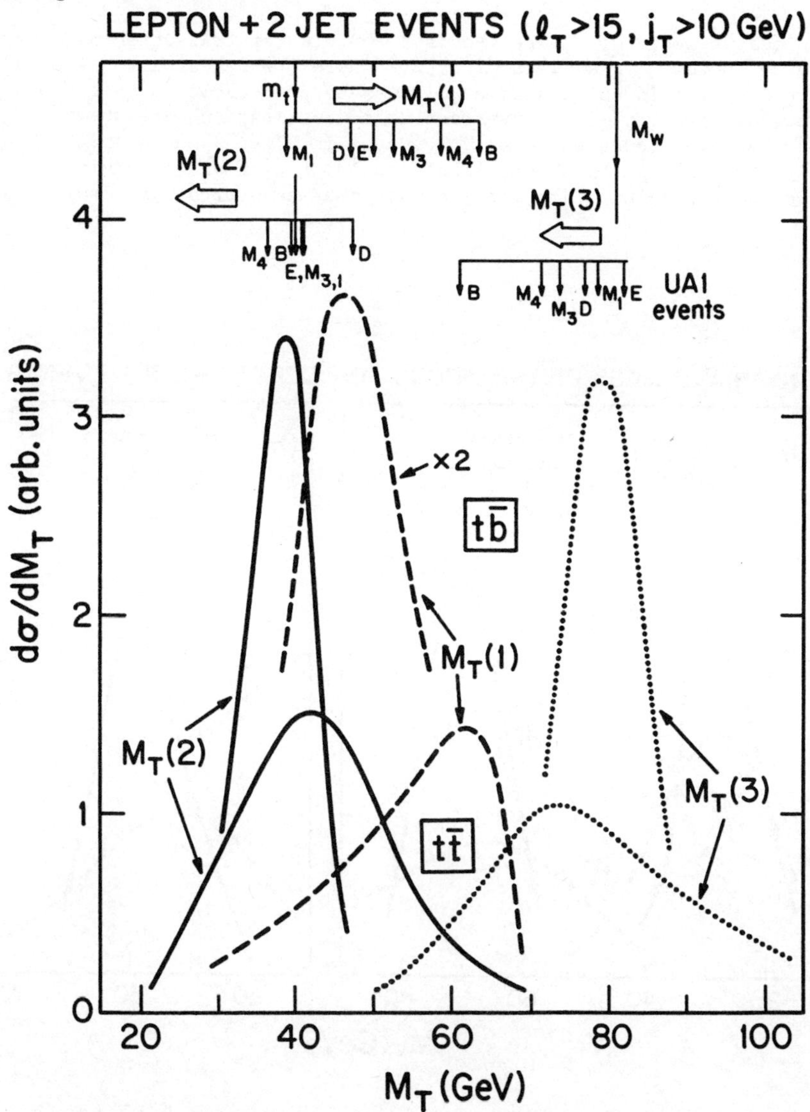

Fig. 3 Transverse mass distributions of lepton + 2-jet events calculated for $W \to t\bar{b}$ (upper curves) and $t\bar{t}$ fusion (lower curves) with $t \to b\ell\nu$ decay, taking m_t = 40 GeV (Barger et al 1984a). Acceptance cuts $p_{\ell T} \geq 15$ GeV and $p_{jT} > 10$ GeV are imposed. The central values of the UA1 $W \to t\bar{b}$ candidate events are represented by the vertical arrows at the top of the figure; these values have substantial uncertainties (see Rubbia 1984, DiBitonto 1984).

These calculations include isolation, acceptance and jet criteria similar to those used in the UA1 analysis. The vertical arrows at the top of the figure indicate the central data values for the three electron + 2-jet and three muon + 2-jet UA1 $W \to t\bar{b}$ candidate events. Also shown in Fig. 3 are the M_T distributions which result from $t\bar{t}$ hadroproduction via the fusion diagrams.

Accurate measurements of $\vec{\nu}_T$ are required for cluster transverse mass variables to be useful. In a class of events where the measured ν_T is consistent with zero within uncertainties, it may be preferable to test the m_c distributions directly. The invariant mass distributions are shown in Fig. 4. The comparisons with the UA1 events in Figs. 3 and 4 suggest that events B and M_4 are perhaps more likely to be of $t\bar{t}$ than $t\bar{b}$ origin, but this assessment does not take into account the measurement uncertainties.

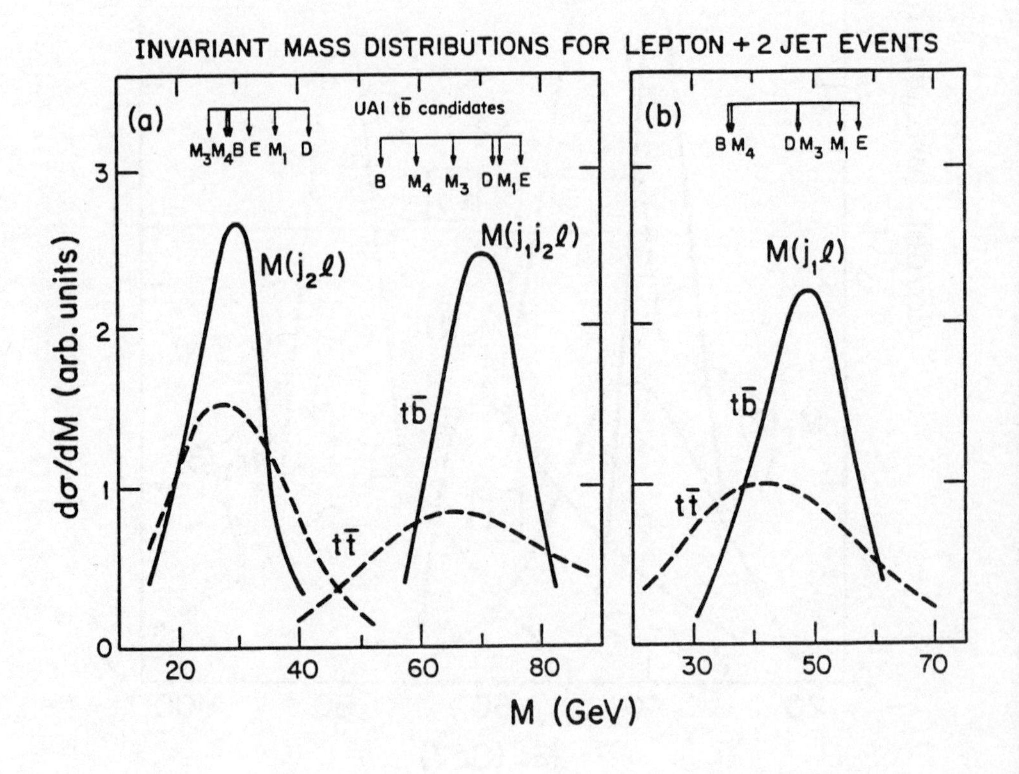

Fig. 4 Invariant mass distributions resulting from $W \to t\bar{b}$ (upper curves) and $t\bar{t}$ fusion (lower curves) with $t \to b\ell\nu$ decay for m_t = 40 GeV (Barger et al 1984a). The data are as in Fig. 3.

1.2 Dilepton signals

Double semileptonic decays of $c\bar{c}$, $b\bar{b}$, $t\bar{b}$ and $t\bar{t}$ heavy-quark pairs produced at $p\bar{p}$ colliders lead to dileptons which are another important signature for heavy quarks (Ali and Jarlskog 1984, Barger and Phillips 1984, Glover et al 1984, Halzen and Martin 1984). Figure 5 shows fusion rates for unlike-sign (US) and like-sign (LS) dilepton cross sections versus the minimum lepton transverse momentum p_T for which the leptons can be experimentally identified.

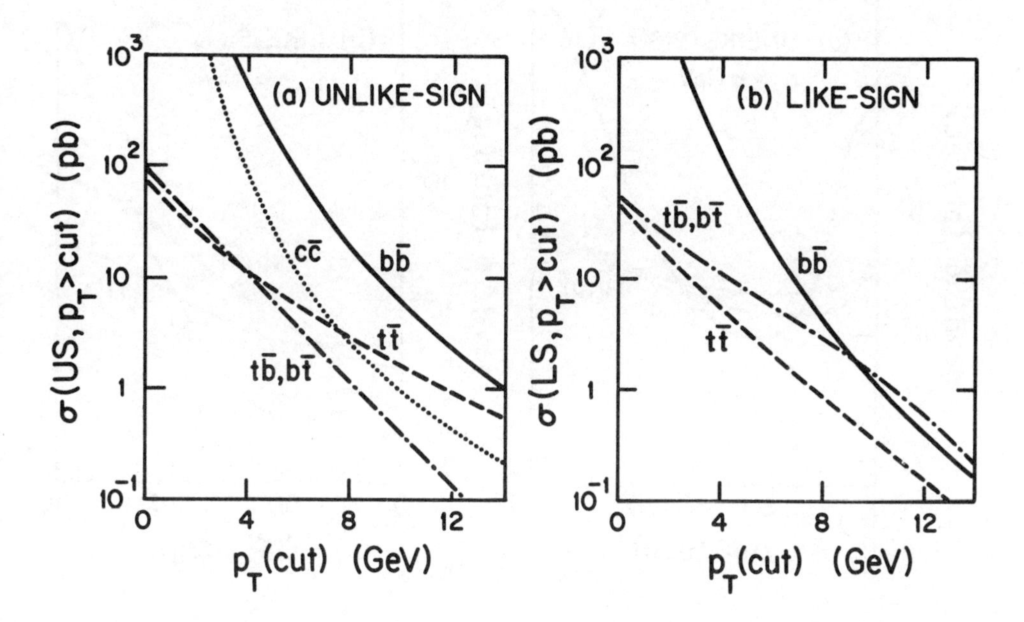

Fig. 5 Dilepton cross sections versus the p_T cut for $p\bar{p}$ collisions at $\sqrt{s}$ = 540 GeV with m_t = 35 GeV (Barger and Phillips 1984).

The various heavy-quark sources can be partially distinguished through the correlation of the leptons in relative azimuth $\Delta\phi$ about the beam axis (see Fig. 6).

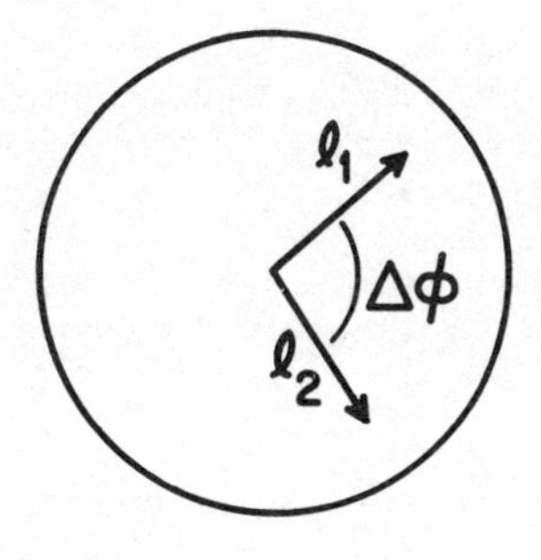

Fig. 6 Azimuthal angle $\Delta\phi$ between the transverse momenta of two leptons.

As shown in Fig. 7, the leptons from $b\bar{b}$ concentrate near $\Delta\phi \sim 180°$ for one lepton from each of b and $\bar{b}$ or near $\Delta\phi \sim 0°$ for both leptons from the same parent b. In contrast, t decay involves enough energy release that there is little azimuthal dilepton correlation.

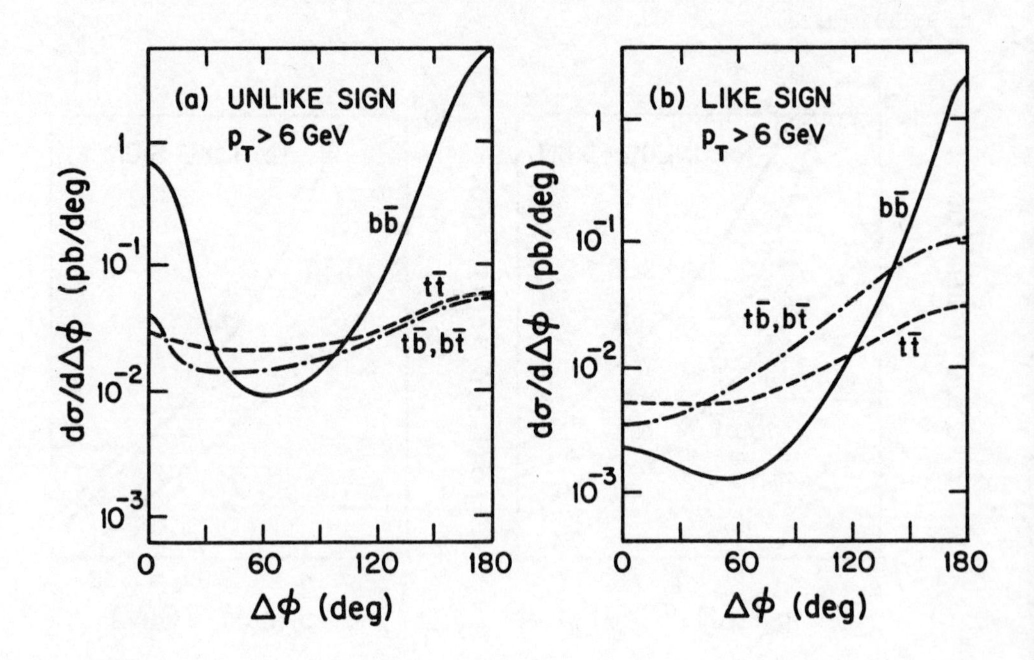

Fig. 7 Dilepton cross sections versus the azimuthal angle $\Delta\phi$ between the leptons for $\sqrt{s}$ = 540 GeV with m_t = 35 GeV (Barger and Phillips 1984).

An isolation cut on <u>one</u> lepton suppresses the $b\bar{b}$ and $c\bar{c}$ contributions over the range

$$40° < \Delta\phi < 120°$$

thus providing a selection for events of $t\bar{b}$ and $t\bar{t}$ origins. Figure 8 illustrates a "loose" isolation cut of this type, namely $\sum p_T < 4$ GeV from hadrons within a 30° cone about the lepton direction.

Without isolation cuts imposed, dilepton events can be used to detect B^o-$\bar{B}^o$ mixing. In the absence of mixing the $b\bar{b}$ sources of dimuons are as follows

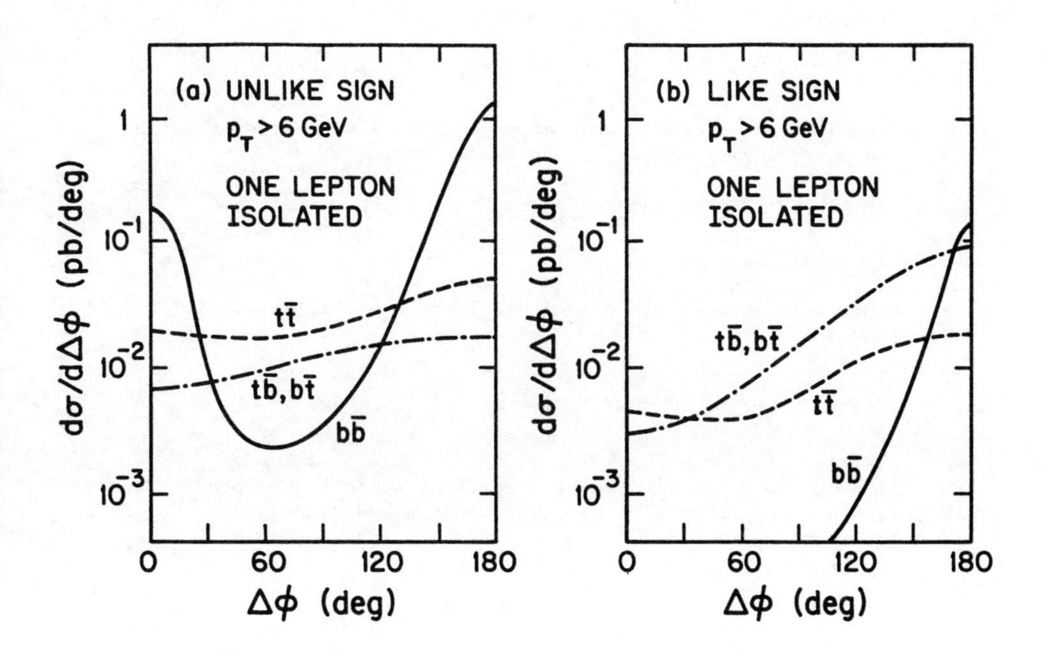

Fig. 8 Dilepton cross sections with at least one lepton loosely isolated (Barger and Phillips 1984).

$$\mu^- \leftarrow b \qquad \bar{b} \rightarrow \mu^+ \qquad \text{US}, \ \Delta\phi \sim 180°$$

$$\mu^+ \leftarrow c\mu^-\nu \leftarrow b \qquad \bar{b} \qquad \text{US}, \ \Delta\phi \sim 0°$$

$$\mu^+ \leftarrow c\bar{u}d \leftarrow b \qquad \bar{b} \rightarrow \mu^+ \qquad \text{LS}, \ \Delta\phi \sim 180° \ .$$

With mixing some of the US configurations become LS pairs, and vice versa; e.g.,

$$\mu^+ \leftarrow \bar{b} \swarrow b \qquad \bar{b} \rightarrow \mu^+$$

For dileptons in the $\Delta\phi \sim 180°$ peak the predicted LS to US cross-section ratio is

$$\frac{\sigma(LS)}{\sigma(US)} \simeq \begin{cases} \frac{1}{4} \text{ (no mixing)} \\ \frac{1}{2} \text{ (maximal } B_s^o - \overline{B}_s^o \text{ mixing)} \end{cases}$$

Future measurements of this ratio at $p\overline{p}$ colliders will determine the extent to which mixing occurs in the B^o-$\overline{B}^o$ system.

2. Search for the Fourth Generation

2.1 Heavy lepton

First consider a sequential left-handed fourth-generation doublet in the lepton sector

$$\begin{pmatrix} \nu_L \\ L \end{pmatrix}$$

with the neutrino ν_L assumed to be light. At $p\overline{p}$ colliders production occurs via $W \to L\nu_L$ (Barger et al 1983d, Cline and Rubbia 1983, Gottlieb and Weiler 1984) with $L \to \overline{\nu}_L$ + hadrons being the most promising decay signal: see Fig. 9.

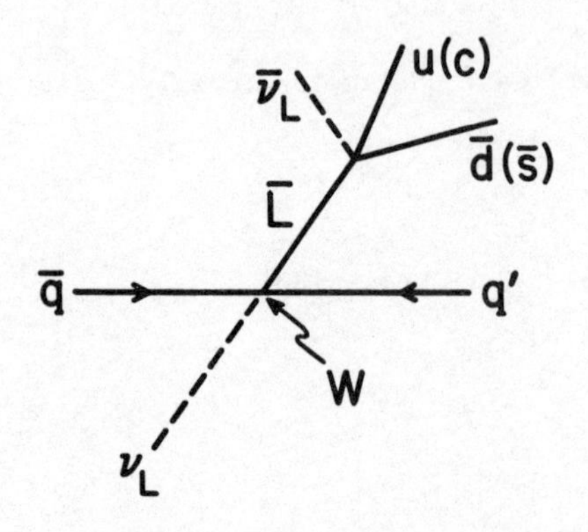

Fig. 9 Quark subprocess for heavy-lepton production and hadronic decay.

This gives events with unaccompanied large missing p_T balanced by two-quark jets. For $m_L \sim$ 30-50 GeV the signal relative to $W \to e\nu$ is substantial,

$$\frac{\sigma(p_{mT} > 20 \text{ GeV})}{\sigma(W \to e\nu)} \sim 10\% .$$

Figure 10 shows the distribution in missing transverse momentum (p_{mT}). The The τ background should be readily recognizable as a single, narrow energetic getic jet of low multiplicity and invariant mass.

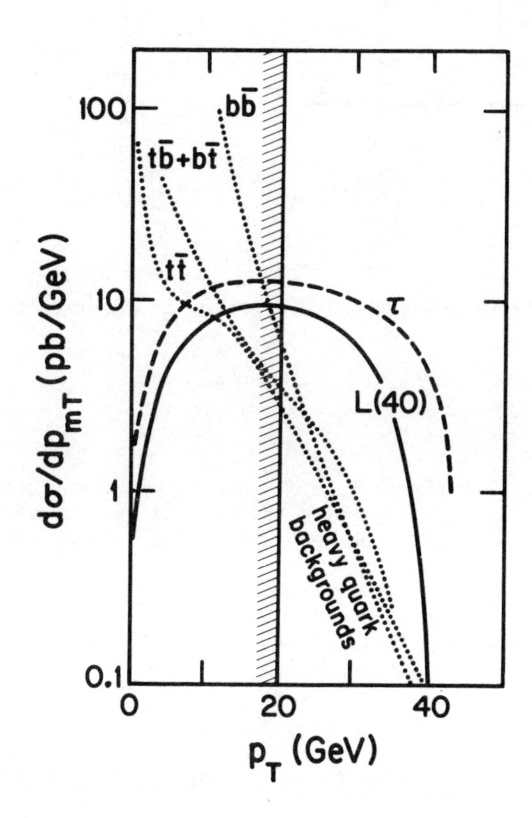

Fig. 10 Missing transverse momentum distribution from $W \to L\nu$ with $L \to \bar{\nu}_L q\bar{q}'$ for $m_L = 40$ GeV, compared with backgrounds from $W \to \tau\nu$ and from heavy-quark production with semileptonic decay (Barger et al 1984d).

2.2 Heavy unstable neutrino

The fourth-generation neutrino would be produced by $Z^o \to \nu_4\bar{\nu}_4$ decay. If ν_4 were heavy, it would decay via the charged weak current to channels like (Barger et al 1984f, Gronau 1983, Gronau and Rosner 1984, Thun 1984, Toussaint and Wilczek 1981)

$$\nu_4 \to e\bar{e}\nu_e \ , \ e\bar{\mu}\nu_\mu \ , \ eu\bar{d} \ , \ \ldots \ .$$

Such a heavy unstable neutrino could be detected at colliders through its decays in flight: see Fig. 11.

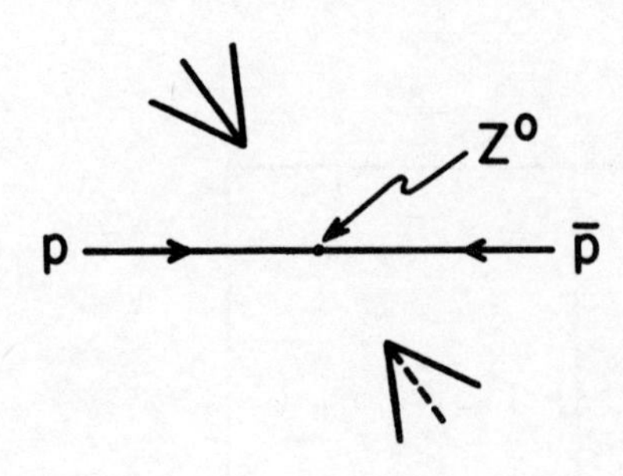

Fig. 11 Decays in flight of heavy unstable neutrinos produced in $p\bar{p}$ collisions via $Z^o \to \nu_4\bar{\nu}_4$.

The mean ν_L decay length for $m_\tau << m_4 << m_Z$ is (Barger et al 1984f)

$$\ell \simeq (40\ \text{cm})\left(\frac{15\ \text{GeV}}{m_4}\right)^6 \left(\frac{10^{-8}}{|U|^2}\right)$$

where $|U|^2 = |U_{4e}|^2 + |U_{4\mu}|^2 + |U_{4\tau}|^2$ and $U_{4\ell}$ is the neutrino mixing matrix element. Should a ν_4 state exist, collider experiments would be sensitive to its signal if the neutrino mixing is very small.

2.3 Heavy quarks

Next consider a possible fourth-generation quark doublet

$$\begin{pmatrix} a \\ v' \end{pmatrix} .$$

The a,v notation is suggested by the alphabetic labels of the other quarks

a b c d

s t u v

and the names amity and vitality have been adopted. A plausible generalization of the three-generation Kobayashi-Maskawa mixing matrix to the case of four generations is

$$U_{nm} = \begin{array}{c} \\ \\ \\ \\ \end{array}\begin{array}{cccc} d & s & b & v \end{array}$$

$$U_{nm} = \left(\begin{array}{ccc|c} 1- & \theta & \tfrac{1}{2}\theta^3 & \tfrac{1}{4}\theta^5 \\ \theta & 1- & \theta^2 & \tfrac{1}{2}\theta^4 \\ \tfrac{1}{2}\theta^3 & \theta^2 & 1- & \theta^3 \\ \hline \tfrac{1}{4}\theta^5 & \tfrac{1}{2}\theta^4 & \theta^3 & 1- \end{array}\right) \begin{array}{c} u \\ c \\ t \\ a \end{array}$$

where $\theta \simeq 0.23$.

For $m_v < m_t$ (overlapping generations) the v quark decays via a $v \to c$ transition which is highly inhibited by small mixing. As a consequence the v lifetime could be of order 10^{-13} sec. and measureable by microvertex detectors. Since hadroproduction is flavor-blind, $\sigma(v\bar{v}) > \sigma(t\bar{t})$; results for m_v = 30 GeV, m_t = 40 GeV are illustrated in Fig. 12.

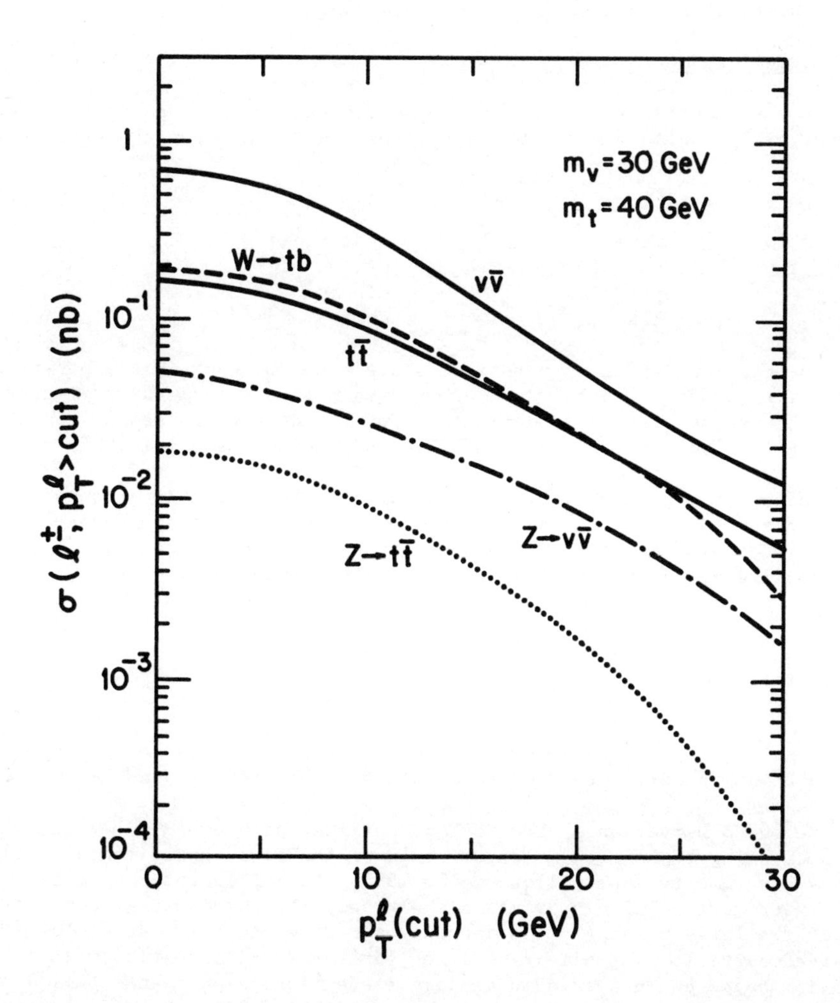

Fig. 12 Cross sections versus lepton p_T cut for primary leptons from decays of v and t quarks at $\sqrt{s}$ = 620 GeV, assuming m_v = 30 GeV and m_t = 40 GeV (from Barger et al 1984e).

Electroweak production differentiates v and t sources. The decay $W \to t\bar{b}$ is unlikely to have a v production counterpart since $W \to a\bar{v}$ is probably forbidden by kinematics and $W \to c\bar{v}, t\bar{v}$ are likely to be highly suppressed by mixing. If $m_v < m_t$, the v quark should also be discovered soon at the CERN $p\bar{p}$ collider.

If $m_v > m_t$, then the principal decay transition is $v \to t$ in which case v production becomes an additional source of t quarks. If $m_v > m_t + M_W$, the v quark decays via real W-boson emission.

3. Search for Supersymmetry Particles

In a supersymmetric theory the usual particles have superpartners that differ by 1/2 unit in spin but have other quantum numbers the same, e.g.

	spin		spin
photon (γ)	1	photino ($\tilde{\gamma}$)	½
gluon (g)	1	gluino ($\tilde{g}$)	½
lepton (ℓ)	½	slepton ($\tilde{\ell}$)	0
quark (q)	½	squark ($\tilde{q}$)	0

The interactions of the superpartners are fixed by gauge and supersymmetry, but their masses are unknown. For a given mass spectrum supersymmetry predictions are reasonably definite. For recent reviews see Dawson et al 1983, Ellis 1984, Haber and Kane 1984a and Nanopoulos et al 1984.

The photino may be the lightest superpartner and be stable or quasistable. It interacts feebly with hadrons and therefore would escape undetected in calorimeter experiments. Consequently the photino is a natural candidate for missing p_T in collider experiments. The decays of squarks

$$\tilde{q} \to q\tilde{\gamma}$$

or gluinos

$$\tilde{g} \to q\bar{q}\tilde{\gamma}$$

produced at $p\bar{p}$ colliders lead to events with missing p_T plus hadron jets.

The recent observation at the CERN $p\bar{p}$ collider of unexpected events with large missing transverse momentum could be the first experimental evidence for supersymmetry (UA1 collaboration 1984, UA2 collaboration 1984). The UA1 experiment found six events cleanly separated from backgrounds in which the large missing p_T is opposite a narrow hadronic jet of low charged-particle multiplicity. Three supersymmetry scenarios with a light photino have been proposed to explain these missing-p_T events (Allan et al 1984ab Barger et al 1984ghi, Ellis and Kowalski 1984ab, Haber and Kane 1984b, Reya and Roy 1984ab):

$$\text{(A)} \quad \tfrac{1}{2}m_{\tilde{g}} \gtrsim m_{\tilde{q}} \simeq 50 \text{ GeV}$$

$$\text{(B)} \quad \tfrac{1}{2}m_{\tilde{q}} \gtrsim m_{\tilde{g}} \simeq 50 \text{ GeV}$$

$$\text{(C)} \quad m_{\tilde{q}} \simeq 100 \text{ GeV}, \quad m_{\tilde{g}} \simeq 3 \text{ GeV} .$$

The relevant fusion subprocesses for explaining the monojet events in the three scenarios are shown in Fig. 13.

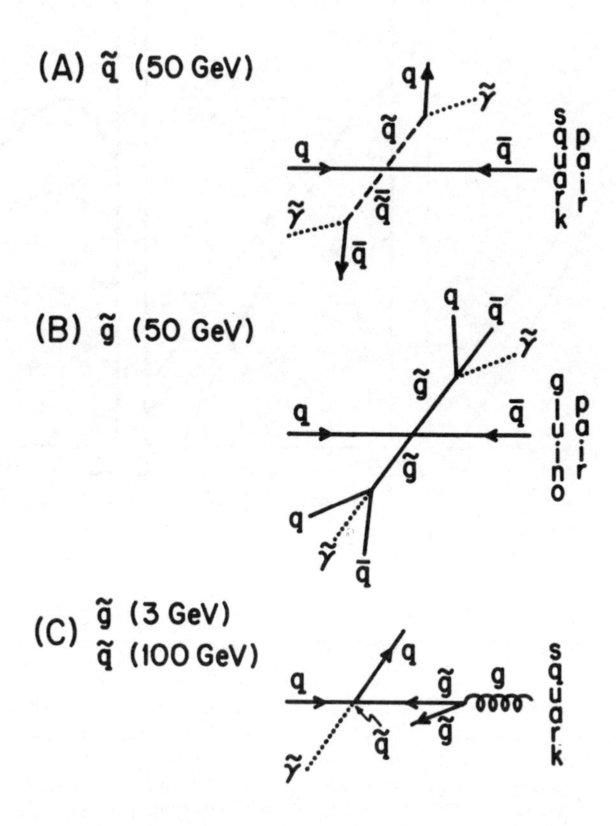

Fig. 13 Subprocesses for three supersymmetry scenarios proposed to explain the UA1 monojet events.

The three scenarios can be distinguished by their predictions for the number of jets and the E_T distribution of the monojets. Figure 14 shows the topological cross sections for one- to four-jet events with the large missing p_T and jet trigger conditions used in the UA1 event selection. The jets are defined by an algorithm similar to that used by UA1. Two parton momenta which satisfy $(\Delta\eta)^2 + (\Delta\phi)^2 < 1$ are coalesced; clusters form a jet if $E_T > 12$ GeV. Comparable numbers of monojet and two-jet events with missing p_T are predicted by both scenarios A and B, while C predicts monojet dominance.

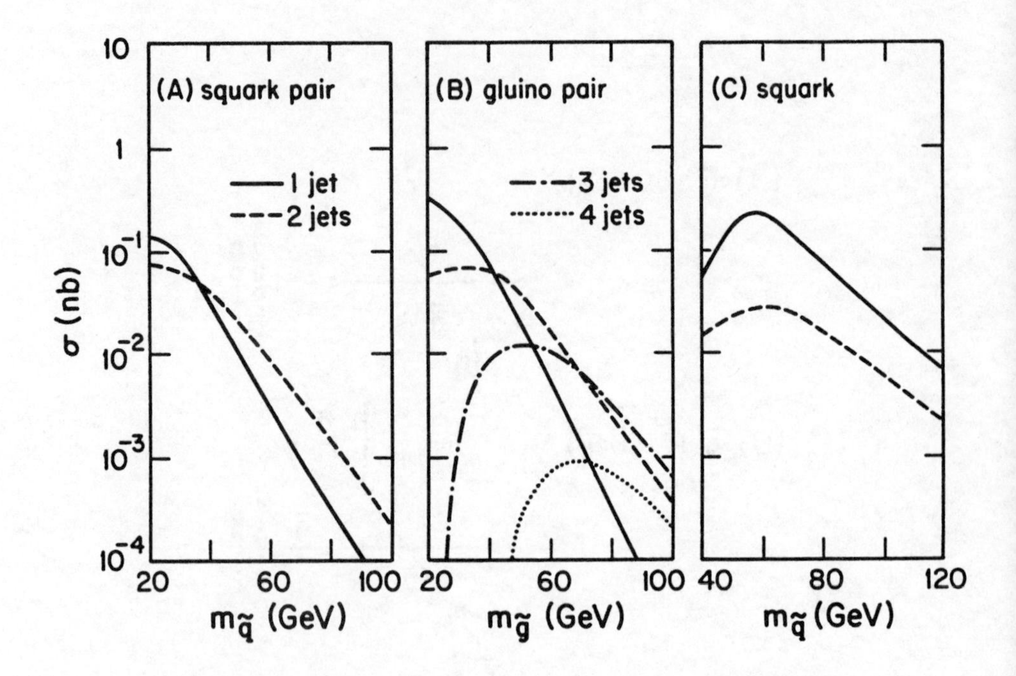

Fig. 14 Topological cross sections versus the mass of the decaying supersymmetric particle in the three scenarios (Barger et al 1984h).

Figure 15 shows the E_T distributions of the monojets. Scenario A gives a broad distribution, scenario B yields a soft distribution and scenario C predicts a Jacobian peak structure. The UA1 data values for the monojet events with large missing p_T are shown at the top of the figure. In scenarios A and B most of the 11 other observed monojet events with missing p_T between 15 and 30 GeV should be of supersymmetry origin rather than two-jet background.

A monojet can consist of multiparton contributions. This is generally the case in scenario B which consequently has the liability of predicting a large monojet invariant mass. In scenario A monojets of single-parton and multiparton origin are comparable in rate, while in C monojets are dominately single partons. Thus A and C have a better chance of explaining the observed monojets, since a single-parton jet should have a smaller invariant mass than a multiparton monojet.

The present UA1 data favor the description of scenario C. This scenario requires an approximate gluino-photino mass degeneracy, so that a gluino with mass of order a few GeV lives sufficiently long that it would not have been detected. Gluinos produced at the $p\bar{p}$ collider would then behave similarly to ordinary hadron jets in calorimeter experiments. Data from the next $p\bar{p}$ collider run should allow us to determine whether supersymmetry is the appropriate interpretation of the UA1 monojet events.

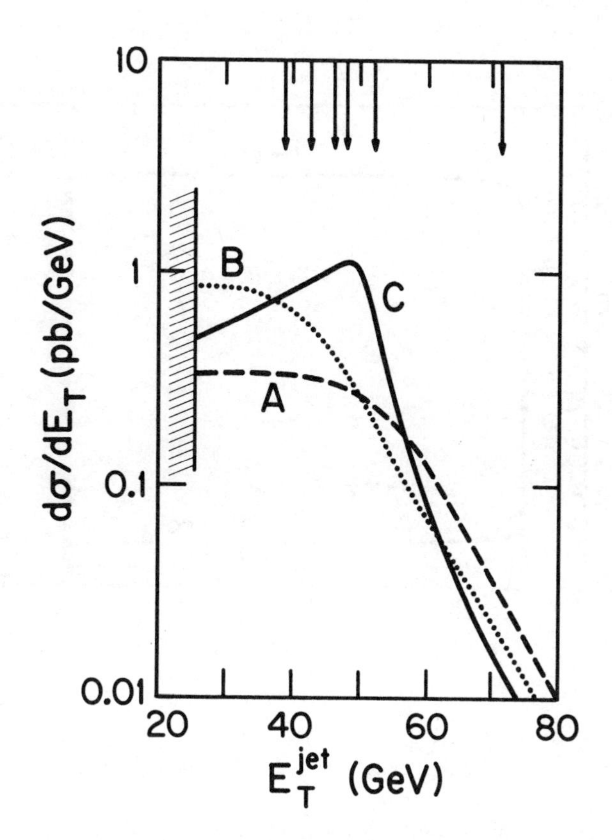

Fig. 15 Predicted monojet E_T distributions in the three scenarios: A (dashed curve), B (dotted curve), and C (solid curve).

4. Search for the Higgs Boson

The discovery of the Higgs boson is the final test of the standard electroweak gauge model. The most promising production mechanism for a light Higgs boson at $p\bar{p}$ colliders is radiation from Z^o or $W^{\pm}$ bosons (Keung 1982, Ioffe and Khoze 1976, Bjorken 1976, Glashow et al 1978), as shown in Fig. 16.

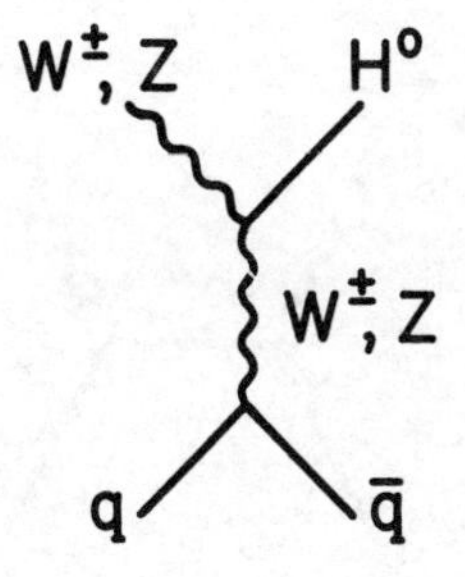

Fig. 16. Subprocess for Higgs boson production in $p\bar{p}$ collisions.

If the mass of the Higgs boson is between 10 GeV and $2m_t$, then $H^o \to b\bar{b}$ is the dominant decay mode; see Fig. 17. Microvertex detectors could be useful in identifying the $b\bar{b}$.

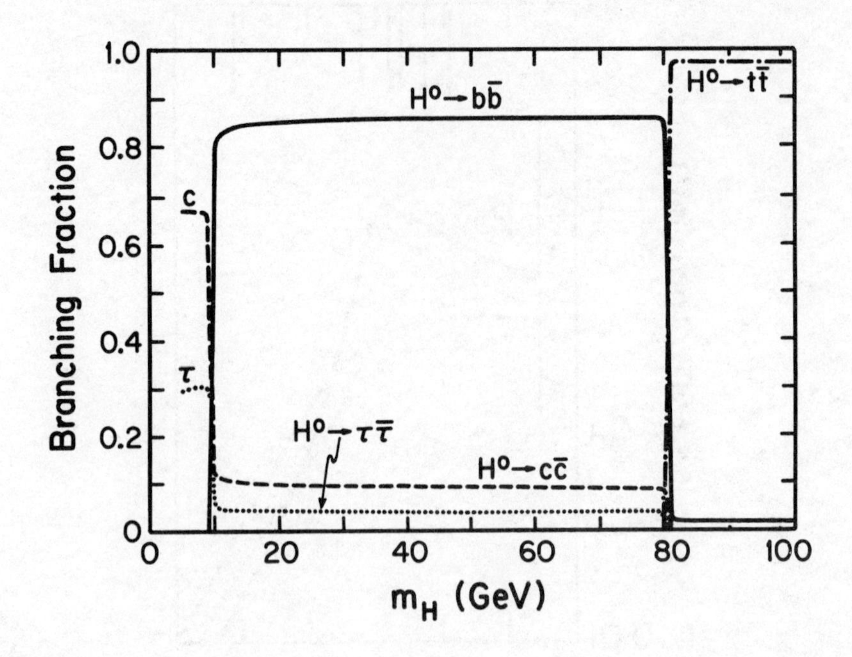

Fig. 17. Dominant branching fractions for Higgs boson decay.

There is an $O(\alpha_s^2)$ QCD background from the subprocess in Fig. 18 (Kunszt 1984), but estimates indicate that this background is well below the H^o signal. Unfortunately, the H^o signal is itself small, less than 10^{-2} of the $W \to e\nu$ or $Z \to e\bar{e}$ rates, making its detection extremely difficult.

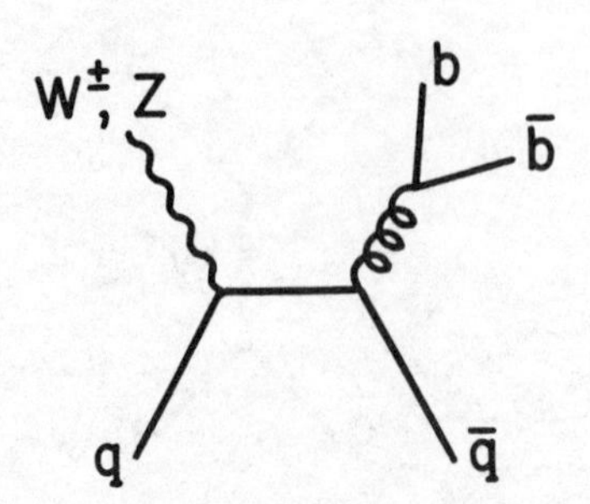

Fig. 18. QCD background to the Higgs production process in Fig. 15, with $H^o \to b\bar{b}$ decay.

5. Summary

Several techniques have been developed for new particle searches at $p\bar{p}$ colliders. Of particular importance are the following:

(i) lepton isolation: This suppresses $b,c \rightarrow \ell$ backgrounds and selects events of heavy-particle origin, such as the t quark.
(ii) Jacobian peaks in p_T and M_T: Two-body or effective two-body decay distributions provide a distinct signature. Examples are the $\bar{b}$ recoil jet of $W \rightarrow t\bar{b}$, the $M_T(be,\nu)$ cluster transverse mass of $t \rightarrow be\nu$ decay, and the $p_T(q)$ distribution of $\tilde{q} \rightarrow q\tilde{\gamma}$ decay.
(iii) azimuthal dilepton correlations: Dilepton events of $b\bar{b}$ origin have peaks at $\Delta\phi \sim 0°$ and $180°$ while those from $t\bar{b}$ and $t\bar{t}$ are more isotropic. The ratio of like-sign to unlike-sign events with $\Delta\phi \sim 180°$ measures B^o-$\bar{B}^o$ mixing.
(iv) missing p_T: This has proved to be a vital technique for new particle searches. Future applications include supersymmetric particle searches (decays to $\tilde{\gamma}$), detecting t semileptonic decays in $t\bar{b}$ and $t\bar{t}$ events, and searches for the heavy-lepton signal of a possible fourth generation.
(v) decay gaps: The detection of unstable particles through their decay gaps will be exploited in the future in searches for heavy neutrinos (ν_4) and for $H^o \rightarrow b\bar{b}$ decays. Microvertex detectors will allow heavy-flavor quantum numbers to be identified in $p\bar{p}$ collider experiments.

Much progress can be anticipated in the search for new particles at $p\bar{p}$ colliders. It will be exciting to see where the tantalizing results from the first $p\bar{p}$ collider experiments lead us.

Acknowledgments

I wish to thank M Pennington for his kind hospitality at this conference. I thank H Baer, N Glover, K Hagiwara, W-Y Keung, A D Martin, J Ohnemus, R J N Phillips and J Woodside for collaborations which led to the results presented here.

This research was supported in part by the University of Wisconsin Research Committee with funds granted by the Wisconsin Alumni Research Foundation, in part by the Department of Energy under contract DE-AC02-76ER00881 and in part by the SERC of Great Britain.

References

Abud M, Gatto R and Savoy C A 1978 Phys. Lett. 79B 435
Ali A and Jarlskog C 1984 CERN report TH.3896
Allan A R, Glover E W N and Martin A D 1984a University of Durham report DTP/84/20
Allan A R, Glover E W N and Grayson S L 1984b University of Durham report DTP/84/28
Aurenche P, Kinnunen R and Mursula K 1984 LAPP report TH.106
Ballocchi G and Odorico R 1984 Phys. Lett. 136B 126
Barger V, Martin A D and Phillips R J N 1983a Phys. Lett. 125B 339; 1983b Phys. Lett. 125B 343; 1983c Phys. Rev. D 28 145; 1984a CERN report TH.3972
Barger V, Baer H, Hagiwara K, Martin A D and Phillips R J N 1984b Phys. Rev. D 29 1923
Barger V, Baer H, Martin A D and Phillips R J N 1984c Phys. Rev. D 29 887

Barger V and Phillips R J N 1984 UW-Madison report PH/155
Barger V, Baer H, Martin A D, Glover E W N and Phillips R J N 1983d Phys. Lett. 133B 449; 1984d Phys. Rev. D 29 2020
Barger V, Baer H., Hagiwara K and Phillips R J N 1984e UW-Madison report PH/150
Barger V, Keung W-Y and Phillips R J N 1984f Phys. Lett. 141B 126
Barger V, Hagiwara K, Keung W-Y and Woodside J 1984g Phys. Rev. Lett. 53 641; 1984h UW-Madison report PH/197
Barger V, Hagiwara K and Keung W-Y 1984i UW-Madison report PH/183
Berger E L, DiBitonto D, Jacob M and Stirling W J 1984 Phys. Lett. 140B 259
Bjorken J D 1976 Proc. SLAC Summer Inst. SLAC-198
Cabibbo N and Maiani L 1979 Phys. Lett. 87B 366
Chaichian M, Hayashi M and Yamagishi K 1983 Marseille report CPT-83/PE.1539
Chau L L, Keung W-Y and Ting S C C 1981 Phys. Rev. D 24 2862
Cline D and Rubbia C 1983 Phys. Lett. 127B 277
Collins P D B and Spiller T 1984a Durham report DTP/84/4; 1984b Durham report DTP/222
Dawson S, Eichten E and Quigg C 1983 Fermilab report PUB-83/82-THY
Desai B R and Lindfors J 1983 Riverside report UCR-TH-83-2
DiBitonto D 1984 talk at VIIth European Symp. on Antiproton Interactions, Durham
Ellis J 1984 CERN report Ref.TH.3802
Ellis J and Kowalski H 1984a Phys. Lett. 142B 441; 1984b DESY report 84-045
Glashow S L, Nanopoulos D V and Yildiz A 1978 Phys. Rev. D 18 1724
Glover E W N, Halzen F and Martin A D 1984 Phys. Lett. 141B 429
Godbole R M, Pakvasa S and Roy D P 1983 Phys. Rev. Lett. 50 1539
Gottlieb S and Weiler T 1984 Phys. Rev. D 29 2005
Gronau M 1983 Phys. Rev. D 28 2762
Gronau M, Leung C N and Rosner J L 1984 Chicago report EFI 83/63
Haber H E and Kane G L 1984a Phys. Lett. 142B 212; 1984b Univ. of Michigan report UM-HE-TH-83-17
Hagiwara K and Long W F 1983 Phys. Lett. 132B 202
Halzen F and Scott D M 1983a Phys. Lett. 129B 341
Halzen F and Martin A D 1984 UW-Madison report PH/173
Horgan R and Jacob M 1981 Phys. Lett. 107B 395
Horgan R and Jacob M 1984 Nucl. Phys. B 238 221
Ioffe B L and Khoze V A 1976 Leningrad report 274
Keung W-Y 1982 AIP Conf. Proc. 85 186
Kunszt Z, Pietarinen E and Reya E 1980 Phys. Rev. D 21 733
Kunszt Z 1984 Bern report BUTP-84/10
Lindfors J and Roy D P 1984 Riverside report UCR-TH-83-5
Nanopoulos D V, Savoy-Navarro A and Tao Ch 1984 Phys. Rep. 105 1
Paige F 1981 AIP Conf. Proc. No. 85 p 168; 1983 talk at Gordon Conf.
Pakvasa S, Dechantsreiter M, Halzen F and Scott D M 1979 Phys. Rev. D 20 2862
Reya E and Roy D P 1984a Phys. Lett. 141B 442; 1984b Univ. Dortmund report 84/11
Roy D P 1984 Zeit. Phys. C 21 333
Rubbia C 1984 talk at Neutrino '84 conf., Dortmund
Schmitt I, Sehgal L M, Tholl H and Zerwas P M 1984 Phys. Lett. 139B 99
Sehgal L M and Zerwas P 1984 Nucl. Phys. B 234 61
Thun R 1984 Phys. Lett. 134B 459
Toussaint D and Wilczek F 1981 Nature 289 777
UA1 collaboration 1984 Phys. Lett. 139B 115
UA2 collaboration 1984 Phys. Lett. 139B 105

Discussion on Section 5

P.V. Landshoff (Cambridge): A question about your history table. In the old days, people used to put particles on straight line Regge trajectories - I think the ρ-trajectory now has six entries and is beautifully straight. Why do we never mention Regge trajectories these days?

Speaker, G. Karl: I agree with you: my history table is somewhat incomplete. It should not be interpreted as the only exciting discoveries in meson spectroscopy - certainly Regge trajectories belong there.

R.L. Jaffe (MIT): If you imagined turning off the second pair creation what does the final state look like? I believe experiment forces it to be a set of zero width resonances. Then it seems to me you are calculating the energy dependence of the width of narrow states into the $M\bar{M}$ channel.

Speaker, G. Karl: Yes, in principle the first set of quarks, with a linear potential would have a set of narrow states; nevertheless I believe that the rate of production at the first stage $GG \to q_1\bar{q}_1$ is smooth. The whole point of the calculation discussed is to check whether, in principle a continuous gluonic spectrum, with a constant rate of production of $q\bar{q}$ pairs (as a function of energy), could lead to mass bumps in the two meson final state. We find this can happen if pair creation is sequential.

H. Koch (KFK, Karlsruhe): Does your statement about mass peaks in the $q\bar{q} \to M_1M_2$ channel also hold for inclusive reactions like $\bar{p}p \to \gamma(\pi)X$?

Speaker, G. Karl: Yes, it does apply to inclusive final states, provided they are Zweig forbidden, i.e. the quarks in X do not come from $p\bar{p}$, say $p\bar{p} \to \pi\phi\phi$ would show the same behaviour as $\pi^-p \to \phi\phi N$ as a function of $\phi\phi$ mass.

W. Mitaroff (Vienna): The reaction $\pi^-p \to \phi\phi n$ should be suppressed by Zweig rule, but <u>is</u> observed, with the production cross-section enhanced by at least a factor of 10. That's independent of any final state mass bumps. How can you explain that?

Speaker, G. Karl: We have not yet looked at the rates expected and found in this reaction so I cannot comment in detail. But with the mechanism we discuss, the sequential fragmentation, the $\phi\phi$ production is less prohibited than the independent production of two single ϕ's or even a single ϕ production, which considerably restricts the energy and momenta of the two strange quarks first produced.

T. Kalogeropoulos (Syracuse): Are the Bessel functions responsible in your model for the peaks and their behaviour with energy?

Speaker, G. Karl: Yes, they give rise to near zeros of amplitudes.

P.V. Litchfield (Rutherford): Does your formula give bumps in a single J^P final state and is this the correct one for the claimed glueball candidates? Do the partial waves of your amplitude give loops in the Argand diagram?

Speaker, G. Karl: Yes, the formula applies to a given J^P channel for scalar mesons, or pseudoscalars, since the orbital angular momentum is also J. For vector mesons several couplings of the spins of the vectors are allowed and the formula we use is not able to discriminate between them. As far as I am aware, the formula we use doesn't give loops in an Argand diagram.

A. Lundby (CERN): In production experiments, only J/ψ charmonium has been seen. Macri will show this afternoon that all charmonium states were seen in $\bar{p}$'s on a hydrogen jet formation experiment at the ISR in a run of just a few weeks. Should one not do formation experiments on gluonium in the same way?

Speaker, G. Karl: In principle yes, but in practice the candidates at the moment would require $\phi\phi$ or $\eta\eta$ as beams which is unrealistic. It may be fun to search for $p\bar{p} \rightarrow \phi\phi$ at LEAR and look for the bumps of Lindenbaum.

U. Gastaldi (CERN): Of course, one cannot form states with mass below two proton masses in formation experiments with antiprotons. In the mass region below 1.8 GeV, $p\bar{p}$ annihilations at rest are very convenient and offer new possibilities. The convenience is that there are production experiments that, at LEAR energies, give one event per antiproton. The main new possibility is that the observation of broad new objects will become feasible by comparing mass spectra obtained with the same apparatus (so acceptances can be factorized out) requiring $p\bar{p}$ L X-rays in coincidence (P-wave annihilation) in one type of spectrum and K X-rays in coincidence (S-wave annihilation) in the second type of spectra.

J.D. Davies (Birmingham): Your data permits 20% S-wave annihilation. Would you not then see the structure observed by PS183?

Speaker, C. Amsler: In the S-wave, we would expect a 6σ effect with the branching ratio quoted by PS183 and hence about 1σ with 20% S-wave, if this structure is not produced from the P-wave.

M. Mandelkern (UC Irvine): Does the strong P-wave contribution imply anything about the radius of the annihilation potential?

Speaker, C. Amsler: The contributions of S and P-wave annihilations depend on the annihilation radius and on the strength of the Stark mixing, for which there are large uncertainties.

A.M. Green (Helsinki): Does your observation of P-wave dominance in $p\bar{p} \rightarrow \pi^+\pi^-$ imply that $q\bar{q}$ (^3P_o) dominates over $q\bar{q}$ ($^3S_1$) vertices?

Speaker, C. Amsler: No, the lower $KK/\pi\pi$ branching ratio in the P-wave compared to the S-wave implies that both annihilation and arrangement graphs contribute. These graphs have ^3P_o and $^3S_1$ vertices.

G.A. Smith (Penn State): What evidence do you have in the $K_S^oK_S^o$ channel for the angular momentum of the initial state?

Speaker, C. Amsler: As you know, K_SK_S is forbidden in $\bar{p}p$ S-wave annihilation. We expect to have 10-20 events in our P-wave sample. We have several candidates.

G.C. Phillips (Rice Univ.): Would you please explain the acceptance and the background of the measurements?

Speaker, D. Schultz: The acceptance of the spectrometer for this topology rises rapidly from 120 MeV/c, levels off around 200 MeV/c and then decreases slowly. The spectrum of annihilation pions is smooth in this region.

P.F. Dalpiaz (Ferrara): How much does the χ^2 of the whole spectrum change if you substantially alter the background shape?

Speaker, D. Schultz: If one fits the right and left hand sides of the spectra separately, one obtains a better χ^2. The observed effects do not, of course, go away.

M.C.S. Williams (Basel): You indicate that you have acceptance up to 500 MeV/c with this topology. Do you observe any sign of A_2 production.

Speaker, D. Schultz: Not in this sample of data.

R.L. Jaffe (MIT): Rumour has it that your spring run was even richer in structure than the 1983 run. Do you have any remarks to make on these data?

Speaker, D. Schultz: There is a lot of structure in our data from the recent run in April. The analysis of the data is still being refined. In addition we are doing Monte Carlo studies of the spectrometer. It may be some time before we fully understand the spectrum.

D. Garreta (Saclay): In your April run, does your 200 MeV/c line survive with the same yield?

Speaker, D. Schultz: The line is still there. It is a little early to quote a yield, but it looks about the same.

J. Vandermeulen (Liege): Have you determined the temperature of the K spectrum?

Speaker, T. Kalogeropoulos: No. The spectrometer has a long path and most $K^\pm$'s decay. The temperature for kaons was determined in K^o bubble chamber data. Roy reported, at the Antinucleon Symposium in Syracuse, temperatures of 124, 84 and 54 MeV for $\pi^\pm$, K^o, Λ, respectively. In the present work a $\pi^\pm$ temperature of 91 MeV is obtained. The difference in $\pi^\pm$ temperatures may reflect differences in resolution. In this case, the K temperature may be lower than 84 MeV.

T. Kalogeropoulos (Syracuse): Comment to L. Pinsky. Your peak coincides with the 1897 MeV seen in $\bar{p}d \to pX$. The magnitude of your effect may be difficult to reconcile with the observations on the 1897 MeV which requires recoil "spectators" > 175 MeV/c.

T. Bressani (Torino): How do you plan to evaluate the efficiency of the $\bar{n}$ calorimeter as a function of $\bar{n}$ energy, and what could be the systematic error in the final data because of this effect?

Speaker, L. Pinsky: We have exposed the calorimeter to a $\bar{p}$ beam. This, as you know, has the problem that a charged particle must enter the detectors and that a different inherently charged-prong multiplicity exists for $\bar{n}$'s and $\bar{p}$'s. We will have to do this calculation very carefully, including a detailed Monte Carlo of the calorimeter. I am not prepared at this time to quote an absolute estimate of the systematic error in this calibration.

P.J. Litchfield (Rutherford): Is the structure in A_o consistent with that in A_2 and A_4 on the assumption of a 2^{++} resonance?

Speaker, K. Nakamura: A_2 and A_4 show a wider structure than A_o, which is proportional to $\sigma_{K^+K^-}$. But, considering the rather poor statistical accuracy at the lowest momentum (390 MeV/c), these structures in A_o, A_2 and A_4 are not inconsistent with the assumption of a 2^{++} resonance.

C.M. Laloum (Coll. de France): Long ago in elastic scattering a 4^{++} effect was observed. Can you exclude such an effect in your data rather than your 2^{++} in K^+K^-?

Speaker, K. Nakamura: We have fitted the K^+K^- data with Legendre polynomials up to 6th order, because we found no significant improvement when higher terms were included. So, a spin assignment higher than 2 is unlikely.

T. Kalogeropoulos (Syracuse): Two questions, do you have data on $K^o\bar{K}^o$ and does the $\pi^+\pi^-$ cross-section scale with the total cross-section?

Speaker, K. Nakamura: Firstly, we did not measure the $K^o\bar{K}^o$ channel and secondly, the $\pi^+\pi^-$ cross-section seems to have a (A + B/p) behaviour, but we have not yet performed this type of fit.

U. Gastaldi (CERN): How many J/ψ events did you collect and what is the energy resolution of the experiment at the J/ψ mass?

Speaker, M. Macri: We have collected a sample of 200 J/ψ's. The mass resolution from the machine measurements of $\bar{p}$ momentum from the whole sample is 150 keV.

P.J. Litchfield (Rutherford): What fraction of your η_c events are background?

Speaker, M. Macri: The analysis is preliminary. We have to complete the $\pi^o\pi^o$ study and the actual situation is that the background is 15%

P.F. Dalpiaz (Ferrara): Do you take data on the η_c'?

Speaker, M. Macri: No.

M. Sakitt (Brookhaven): What fraction of your total expected yield does the result on the χ states represent?

Speaker, M. Macri: Of the total integrated luminosity, the fraction devoted to χ_1, χ_2 data taking represents 40%.

T. Kalogeropoulos (Syracuse): Can you do the charmonium $\rightarrow J/\psi(e^+e^-) + \gamma$ without the Cherenkhov, using only the calorimeter.

Speaker, M. Macri: No.

R.L. Jaffe (MIT): (1) Is there any theoretical reason to prefer a wavefunction of the form $(x_1x_2x_3)^\eta$ at less than asymptotic Q^2.
(2) Given the answer to (1) have you studied the sensitivity of your results to other choices?

Speaker, P. Damgaard: The ansatz $\phi(x) \propto (x_1x_2x_3)^\eta$ at intermediate Q^2 is just a convenient parametrization, which scans a large region of the allowed "wavefunction space". Furthermore, we know that at large Q^2 $\phi(x) \propto x_1x_2x_3$, so it seems to be a sensible parametrization. The answer to (2) is no.

G. Karl (Guelph/Oxford): The formalism of Brodsky and Lepage has been criticized by Isgur and Llewellyn Smith for being inapplicable for elastic formfactors at presently available Q^2. Why do you think it will work for $p\bar{p} \rightarrow \psi$ or χ?

Speaker, P. Damgaard: I think almost everybody agrees that the nucleon formfactor calculations are not very reliable. There are several reasons for this, and some of them are rather technical. Well, first of all, there is a sign disagreement between the two existing calculations of nucleon formfactors (one by Brosky and Lepage, the other by Chernyak and Zhitnitsky), and we don't know who is right. One of the more technical reasons why the lowest order formfactor calculations are unreliable is that the proton formfactor actually vanishes at the asymptotic wavefunction $\phi(x) \propto x_1x_2x_3$. This is presumably just an accident of the lowest order calculation, and it is very unlikely to persist in higher orders. For this, and other reasons, we decided not to use the formfactor calculations to normalize our cross-sections. Note that had we done that, we would have been able to predict both the $p\bar{p} \rightarrow \psi$ and the $p\bar{p} \rightarrow \chi_2$ cross-sections simultaneously.

D. Geesaman (Argonne): Why do you think at such relatively low Q^2 at each quark vertex perturbative QCD applies. This is, of course, related to the discussion of Isgur and Llewellyn Smith, which suggest these arguments are only valid at much higher Q^2.

Speaker, P. Damgaard: Clearly, the charmonium states are just on the borderline of where we want to work. At this Q^2, other effects could be important. However, as far as QCD perturbation theory goes, I don't think there is a problem. You should keep in mind that the QCD Λ "is not as big as it used to be", and we now think that it may be as small as $\sim$ 150 MeV. If this is true then there should be no problem applying perturbative QCD to the region of the charmonium states.

M. Macri (Genova): Can you comment on how knowledge of the angular distribution of χ_1, χ_2 can influence the massless quark model?

Speaker, P. Damgaard: I didn't have time to mention predictions of angular distributions in my talk. In the limit of massless u and d quarks, it is straightforward to derive a whole list of predictions for angular distributions. It should be kept in mind though, that these predictions only really check whether we have vector-like gluons. Scalar gluons have been ruled out countless times by now, but here should be yet another opportunity.

R. Cester (Torino): How would you explain the $\eta_c \to \bar{p}p$ branching ratio, which is comparable to that for $J/\psi \to \bar{p}p$?

Speaker, P. Damgaard: When I said that the process $p\bar{p} \to \eta_c$ is not allowed in QCD, I made the approximation that the proton valence quarks, the u and d quarks, were massless. In this approximation it is certainly true that you cannot produce a J = 0 state, like the η_c, in $\bar{p}p$ collisions, simply by helicity conservation. If you put in a quark mass, m, the amplitude for producing the η_c is down by $\sim m^2/Q^2$, which is why I said that in the large Q^2 region such a process is forbidden. Of course, at the η_c $Q^2 \simeq 10$ GeV^2, so who knows? The fact that experimentally the branching ratios for $\eta_c \to p\bar{p}$ and $\psi \to p\bar{p}$ appear to be of almost comparable size is, in my opinion, somewhat surprising. It would be very nice if the predicted suppression could be checked at higher Q^2 in, say, $b\bar{b}$ spectroscopy.

P.V. Landshoff (Cambridge): You probably said it, but what did you take for the J/ψ and χ wavefunctions?

Speaker, P. Damgaard: No, I didn't have time to mention this in my talk. For the ψ we get the wavefunction at the origin, for the χ_2 we get the first derivative. When we take the branching ratios, these factors cancel out.

C. Amsler (Zurich): In 2 and even more in 3 jet events, there is a combinatorial problem. Where are the "wrong" combinations in your plots?

Speaker, D. Dibitonto: For the 2 jet events, we assume that the smallest E_T jet is coming from top decay from $W \to t\bar{b}$. For top masses up to $\sim$50 GeV/c^2, this assignment is correct 90% of the time according to our Monte Carlo.

A. Donnachie (Manchester): How sensitive are the results to the definition of jets, i.e. $E_T \geqslant 8$ GeV for the first jet and $\geqslant 7$ GeV for the second?

Speaker, D. Dibitonto: In most of the 2 jet events, the number of jets is rigorously stable when the jet E_T threshold in the algorithm goes to zero. However, the lepton + 1 jet events are still under study.

T. Akesson (CERN): How do you define the jet-energy that goes into the invariant mass? Do you correct for the tail of the jet fragmentation that is lost by the jet finding algorithm?

Speaker, D. Dibitonto: The jet energy and momentum are defined as the scalar and vector sum, respectively, of both electromagnetic and hadronic energy cells within a cone of $\Delta R = (\Delta\eta^2 + \Delta\phi^2)^{\frac{1}{2}} \leqslant 1$. The systematic energy-momentum corrections to the uncorrected jets are determined from a full detector simulation of $W \rightarrow$ jet-jet from Isajet over the entire apparatus, which for uncorrected jets gives a reconstructed W mass of 63 GeV/c^2.

B. Andersson (Lund): Do you have any more precise comments about the fact that your (π^0 + 2 or more jets) sample also tend to cluster around the same masses?

Speaker, D. Dibitonto: I believe that the event selection is so constrained kinematically (lepton isolation, $\geqslant 2$ jets) that the phase space naturally peaks near the W mass region in the ($\ell\nu j_1 j_2$) mass, i.e., the 6(ℓ + 2 jet) events are consistent with $W \rightarrow t\bar{b}$ decay, but not proven to be so by this scatter plot alone. What we do claim is that experimentally these leptons are prompt, i.e. that they cannot be due to backgrounds of QCD origin, and that they are consistent with the $W \rightarrow t\bar{b}$ interpretation.

S. Loucatos (Saclay): Do you have results on diffractive top production?

Speaker, D. Dibitonto: Work is still in progress and I hope to have preliminary results soon.

M. Macri (Genova): What is the theoretical need for a fourth generation?

Speaker, V. Barger: Some models require heavy fermions. For example, if renormalization group equations describing the evolution of couplings are to have infrared fixed points, then a fourth generation of quarks and leptons is predicted. A heavy sequential fermion is also necessary in supersymmetry models broken by radiative corrections.

Inst. Phys. Conf. Ser. No. 73: Section 6
Paper presented at VII Eur. Symp. Antiproton Interactions, Durham 1984

Proton–proton and antiproton–proton elastic scattering at ISR energies in the t interval $0.05 < |t| \leq 3.0$ GeV^2

F. Rimondi

Dipartimento di Fisica dell'Universita' and INFN, Bologna, Italy.

Abstract. Final results of the measurement of the pp and $\bar{p}p$ elastic differential cross sections in the interval $0.05 < |t| < 0.85$ GeV^2 at $\sqrt{s}$ = 31, 53 and 63 GeV are reported. Preliminary results, at $\sqrt{s}$ = 53 GeV, on the $\bar{p}p$ elastic t distribution in the interval $0.5 < |t| \leqslant 3.0$ GeV^2 show a possible difference between the pp and the $\bar{p}p$ t distribution in the dip region.

1. Introduction

The differential cross sections for pp elastic scattering have been extensively measured at all available energies and, in particular, at the CERN-ISR [1-6]. New energy regions in which to explore $\bar{p}p$ elastic scattering have been provided by the recent injection of antiprotons into one ring of the CERN-ISR and into the CERN-SPS collider [7-10].
The successful operation of the ISR with antiprotons has made possible, in particular, the comparison of pp and $\bar{p}p$ elastic cross sections measured in the energy range of that machine with the same detector.

This comparison was one of the tasks of the experiment R420. Preliminary results from this experiment have been already reported at other conferences [11].

Final results on pp and $\bar{p}p$ elastic differential cross sections in the t range $0.05 < |t| < 0.85$ GeV^2 at three energies, that is at $\sqrt{s}$ = 31, 53 and 63 GeV, and preliminary results on $\bar{p}p$ elastic t distribution at $\sqrt{s}$ = 53 GeV for $0.5 < |t| \leqslant 3.0$ GeV^2 will be presented.

2. Apparatus and Trigger

The experiment was performed at the CERN-ISR using the Split Field Magnet (SFM) detector [12].
The SFM was used with a two stage trigger: i) a fast trigger which required at least one hit in each of two scintillator banks surrounding the outgoing beams and no signal from proportional wire chambers in the central region, and ii) a slow trigger which required two track candidates, one on each side of the detector. Data were also taken with only the fast trigger in order

to evaluate the slow trigger efficiency.

3. Low t Region ($0.05 < |t| < 0.85$ GeV^2)

Data were collected at $\sqrt{s}$ = 31, 53 and 63 GeV. In the $\bar{p}p$ runs, the stored antiproton current was 2-6 mA and the proton current was about 10 A, resulting in luminosities of approximately 10^{27} cm^{-2}sec^{-1}. The pp runs were performed with luminosities at least 10^3 times larger.

The raw data were processed through the standard SFM off-line program for track finding and a special off-line program for track and vertex fitting. The colinearity of the tracks in the centre-of-mass system was used as a constraint on the fit. The background events were efficiently rejected on the basis of the quality of the fit.

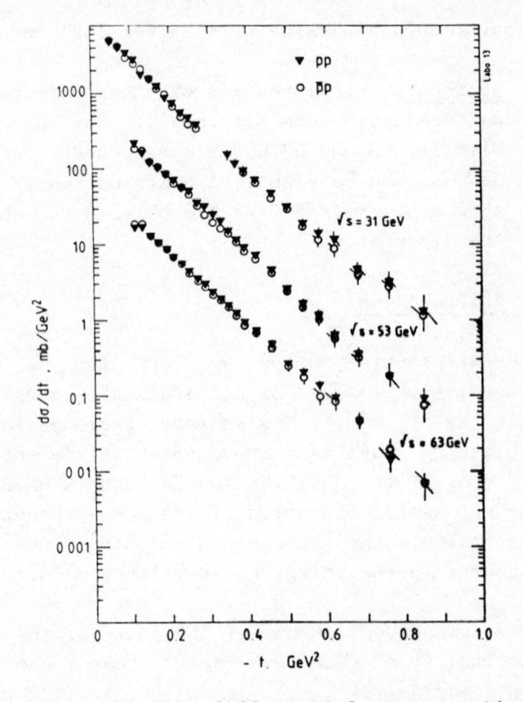

Fig.1. Elastic differential cross sections at $\sqrt{s}$ = 31,53 and 63 GeV. Only t dependent errors are shown. The systematic scale error is estimated at 10% for pp data and 15% for $\bar{p}p$ data. The 31 (53) GeV data have been scaled by a factor 100 (10) for clarity.

The acceptance of the apparatus as a function of the square of the four-momentum transfer, t, and of the azimuthal angle, ϕ, was determined by processing simulated events. In order to minimize problems arising from the trigger and from the SFM geometry, only those events, for a given bin in t, laying inside a limited ϕ range, chosen to have a high acceptance, were allowed. After all cuts we had between 50000 and 180000 events at each energy.

Using simulated events, we calculated corrections to the data for the following effects: i) nuclear absorption (about 15%) ,ii) Coulomb multiple scattering (negligible in our t range), and iii) slow trigger efficiency (1-4%).

The normalization for the pp cross section at each energy was obtained by comparison to previous measurements [1-6] . The normalization for the $\bar{p}p$ data at each energy was then obtained by assuming equality with the pp cross section at the same energy in the region near $|t| = 0.3$ GeV^2. Because of the very small slope difference between the pp and $\bar{p}p$ cross sections, the normalization is not very sensitive to the t value chosen for this procedure. This method gives results that agree with other ISR $\bar{p}p$ measurements [8] and with our own luminosity measurements, which are accurate to 10% for pp measurements and to 15% for $\bar{p}p$ measurements.

Figure 1 gives the differential cross sections for pp and $\bar{p}p$ elastic scattering at $\sqrt{s}$ = 31, 53 and 63 GeV. The errors shown reflect only the t dependent uncertainties, which include uncertainties due to acceptance as well as statistical uncertainties.

For $0.17 < |t| < 0.85$ GeV^2 both pp and $\bar{p}p$ data may be represented by simple exponential form

$$(1) \qquad d\sigma/dt = a \exp(bt)$$

The $\sqrt{s}$ = 31 GeV cross sections, which have been measured over a larger t range, $0.05 < |t| < 0.85$ GeV^2, give better fits with either two slopes or a quadratic exponential of the form

$$(2) \qquad d\sigma/dt = A \exp(Bt+Ct^2)$$

The fits to the data at $\sqrt{s}$ =53 and 63 GeV show no significant improvement when the quadratic term is added.
From the quadratic exponential fits one may compute the local slope at certain t value via the relation $b = B+2C|t|$.
The slopes determined by the exponential fits correspond to the local slopes at a mean value of 0.4 GeV^2.
The comparison of the b values at $|t| = 0.4$ GeV^2 from the quadratic exponential fits and from the simple exponential fits shows a good agreement between the two determinations.

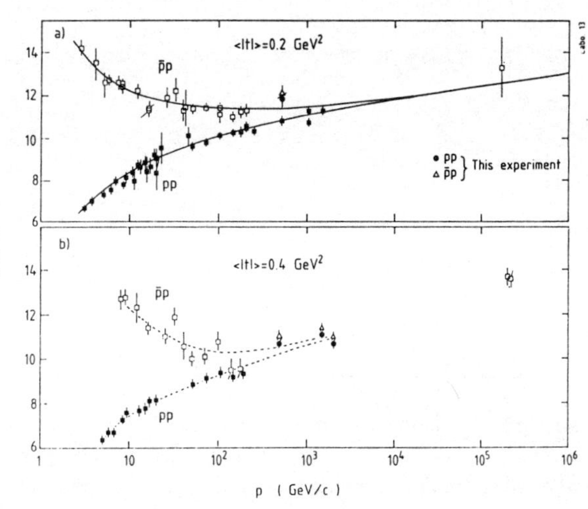

Fig.2. Compilation of local slopes at $|t|$=0.2 and 0.4 GeV^2 as functions of equivalent laboratory beam momentum. The solid lines in fig. 2a are the fits of reference 13, the dashed lines in fig. 2b are included only to show the trend of the data.

A compilation of local slopes at $|t| = 0.2$ and $|t| = 0.4$ GeV^2 as functions of equivalent laboratory beam momentum is shown in figure 2. Our local b values for pp scattering agree within errors with previous measurements.

The b values for $\bar{p}p$ scattering agree with and join smoothly to measurements at lower and higher energies [7-10].

From figures 1 and 2 it is seen that there may be a slight difference in the slopes of the pp and $\bar{p}p$ differential cross sections. We have fitted the ratios of the differential cross sections for $|t| > 0.17$ GeV^2 to an exponential dependence

$$(3) \qquad [d\sigma/dt]_{\bar{p}p} \, / \, [d\sigma/dt]_{pp} = A' \exp(\Delta b \; t)$$

The results of the fits to eq. 3 are given in table 1. There may be a difference in slope, but because of the size of the errors it is not possible to determine if there is an energy dependence. Note that because of our normalization procedure, the fit results for A' are not significant.

Table 1. Fits to cross section ratios

$\sqrt{s}$ GeV	$\|t\|$ GeV^2	Δb GeV^{-2}
31	0.17-0.85	0.28 ± 0.23
53	0.17-0.85	0.37 ± 0.18
63	0.17-0.85	0.29 ± 0.17

4. High t Region ($0.5 < |t| < 3.0$ GeV^2)

The $\bar{p}p$ data were taken at $\sqrt{s}$ = 53 GeV during the last ISR antiproton run at the end of 1983. The luminosity was of the order of 6×10^{27} cm^{-2} sec^{-1}. Some data were collected during the previous runs at the same energy. In total about four million triggers were recorded, corresponding to an integrated luminosity of about 4.5×10^{33} cm^{-2}.

These raw data were filtered to avoid full reconstruction of low t elastic events, then processed with a slightly modified version of the standard SFM off-line chain.

The final sample of full reconstructed events consists of about 15000 $\bar{p}p$ elastic events, 7879 having $|t| > 0.5$ GeV^2. Proton-proton comparison data, taken in the same conditions, are available, but not yet completely reconstructed.

The uncorrected t distribution of the antiproton elastic events is shown in figure 3. For comparison, the t distribution for a subsample of the pp data is also shown.

The errors are statistical only and, as already said, no corrections are applied to the data. The relative normalization is obtained requiring that the two distributions subtend the same area.

Estimates of the acceptance of the apparatus, computed with a Monte-Carlo, as a function of t and ϕ, appear almost flat for $|t| \gtrsim 1$ GeV^2. The acceptance corrections, averaging around 60%, are the largest ones, while nuclear absorption, essentially t independent, is about 15%, as for the low t region.

The comparison of the pp distribution (fig.3) with results from previous measurements [14] shows a good agreement of the shapes within the errors.

The $\bar{p}p$ distribution at $|t| \simeq 1.4$ GeV^2 indicates a possible different behaviour with respect to the pp distribution and seems to exhibit a shoulder rather than a clear dip.

In order to arrange this result into the frame of the existing data it may be worth having a brief survey of pp and $\bar{p}p$ elastic differential cross sections as functions of the energy.

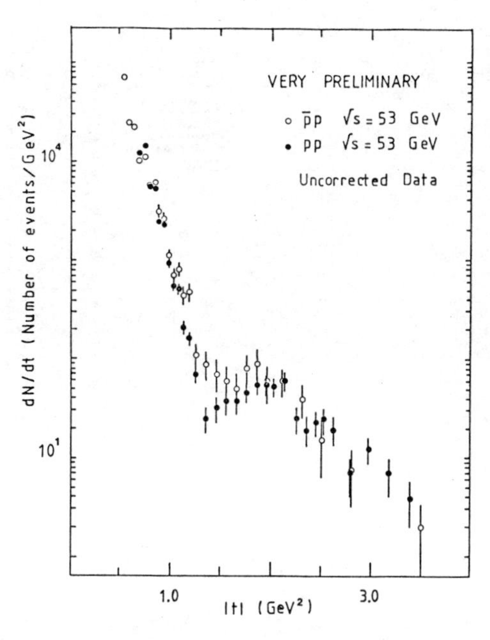

Fig.3. Uncorrected pp and $\bar{p}p$ elastic scattering t distributions at $\sqrt{s}$ = 53 GeV.

Figure 4a shows pp elastic differential cross sections at various incident momenta ranging from 24 to 1064 GeV/c [14–18].

In the pp 24 GeV/c data from Allaby et al. [15] a change in slope appears around $|t| \simeq 1.3$ GeV^2, followed by a less pronounced shoulder. This structure is more evident at 50 GeV/c where a prominent change in slope is observed at $|t| \simeq 1.4$ GeV^2. It then develops, with increasing energy, into a dip at 150 and 200 GeV/c, and the dip is seen well pronounced in the pp ISR data at $\sqrt{s}$ = 53 GeV.

The $\bar{p}p$ elastic differential cross sections are shown in figure 4b at three energies ranging from 10.1 GeV/c incident momentum to $\sqrt{s}$ = 540 GeV [19–21].

At 10.1 GeV/c the $\bar{p}p$ differential cross sections show two structures: a change in slope at $|t| \simeq 0.5$ GeV^2 and a broad flattening at $|t| \simeq 2$ GeV^2. When the incident momentum reaches 50 GeV/c the data show a dip at $|t|$ around

1.5 GeV^2.

This dip is already observed at 30 GeV/c [22] and still present at 100 and 200 GeV/c [23] (data not reported in the figure), but is not exhibited by the data at $\sqrt{s}$ = 540 GeV.

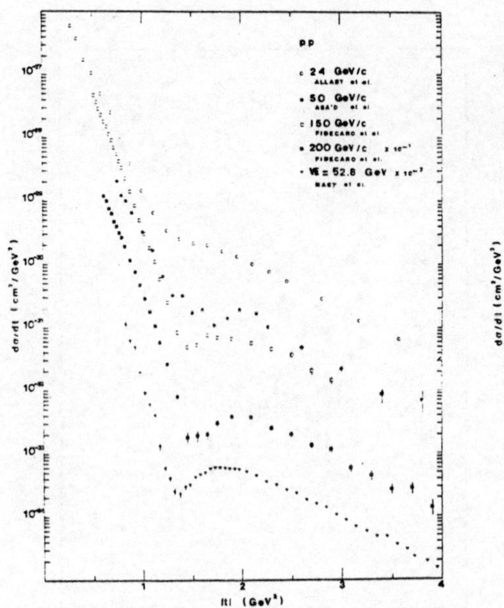

Fig.4a. pp elastic differential cross sections at 24 GeV/c [15], 50 GeV/c [16], 150 GeV/c [17], 200 GeV/c [18] ($x10^{-1}$) and at $\sqrt{s}$ = 53 GeV [14] ($x10^{-2}$).

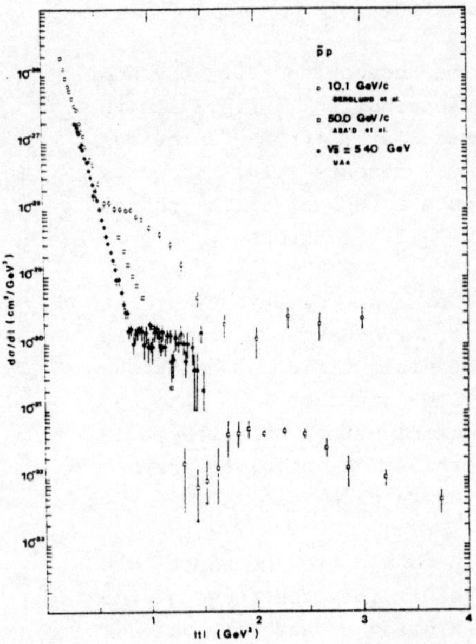

Fig.4b. $\bar{p}p$ elastic differential cross sections at 10.1 GeV/c [19] , 50 GeV/c [20] and at $\sqrt{s}$ = 540 GeV [21].

Our measurement of the $\bar{p}p$ elastic cross section falls in the large gap between a hundred of GeV/c of incident momentum and the SPS collider energies.

The shape of the distribution, if confirmed by the final analysis,indicates the filling up of the dip at the ISR energies.

5. Conclusions

We have measured the pp and $\bar{p}p$ elastic differential cross sections at three energies, i.e. $\sqrt{s}$ = 31, 53 and 63 GeV, in the t interval $0.05 < |t| < 0.85$ GeV^2.

Both pp and $\bar{p}p$ distributions have simple exponential shapes, for

$0.17<|t|<0.85$ GeV^2, with similar slopes.
At $\sqrt{s}$ = 31 GeV the data for $0.05<|t|<0.85$ GeV^2 are consistent with a change in slope near $|t|\simeq 0.15$ GeV^2.

Differences in slope between pp and $\bar{p}p$ are seen, but because of the size of the errors it is not possible to conclude if there is any energy dependence.

At $\sqrt{s}$ = 53 GeV the measurement has been extended to $|t|$ around 3 GeV^2. Preliminary results show a possible difference between pp and $\bar{p}p$ distributions in the dip region.

Acknowledgments

I would like to thank all members of the ABCDHW Collaboration. I am indebted to my colleagues for useful discussions and for their help in the correction of the manuscript.

References

(1) M. Holder et al., Phys. Lett. 35B (1971) 355.
(2) G. Barbiellini et al., Phys. Lett. 36B (1971) 400; 39B (1972) 663.
(3) L. Baksay et al., Nucl. Phys. B141 (1978) 1.
(4) N. Kwak et al., Phys. Lett. 58B (1975) 233.
(5) U. Amaldi et al., Phys. Lett. 36B (1971) 504; 66B (1977) 390.
(6) H. Schopper, Editor Landolt-Bornstein New Series, Vol. 9 (1980).
(7) M. Ambrosio et al., Phys. Lett. 115B (1982) 495.
(8) D. Favart et al., Phys. Rev. Lett. 47 (1981) 1191.
N. Amos et al., Phys. Lett. 120B (1983) 460; 128B (1983) 343.
(9) R. Battiston et al., Phys. Lett. 115B (1982) 333; 117B (1982) 126 ; 127B (1983) 472.
(10) G. Arnison et al., Phys. Lett. 121B (1983) 77; 128B (1983) 336.
(11) F. Fabbri, XVIIIth Rencontre de Moriond, 23-29 Jan. 1983.
A. Breakstone et al., Contribution to the International Europhysics Conference on High Energy Physics, Brighton (UK) 1983.
(12) W. Bell et al., Nucl. Instr. & Meth. 124 (1975) 437; 156 (1978) 111.
(13) J. P. Burq et al., Nucl. Phys. B217 (1983) 285.
(14) E. Nagy et al., Nucl. Phys. B150 (1979) 221.
(15) J. V. Allaby et al., Nucl. Phys. B52 (1973) 316.
(16) Z. Asa'd et al., Phys. Lett. 128B (1983) 124.
(17) G. Fidecaro et al., Nucl. Phys. B173 (1980) 513.
(18) G. Fidecaro et al., Phys. Lett. 105B (1981) 309.
(19) A. Berglund et al., Nucl. Phys. B176 (1980) 346.
(20) Z. Asa'd et al., Phys. Lett. 108B (1982) 51.
(21) G. Matthiae et al., Contribution to the International Europhysics Conference on High Energy Physics, Brighton (UK) 1983.
(22) Z. Asa'd et al., Phys. Lett. 130B (1983) 335.
(23) D. H. Kaplan et al., Phys. Rev. D26 (1982) 723.

Inst. Phys. Conf. Ser. No. 73: Section 6
Paper presented at VII Eur. Symp. Antiproton Interactions, Durham 1984

Odderon effects at the Collider and beyond

Basarab Nicolescu

Division de Physique Théorique*, Institut de Physique Nucléaire, 91406 Orsay, France and LPTPE, Université Pierre et Marie Curie, Paris, France

Abstract. We review the theoretical and experimental indications for possible differences between antiproton-proton and proton-proton interactions at high energies, with special emphasis on the TeV energy range.

1. Introduction

The new high-energy (ISR and collider) data has renewed the interest in soft hadronic physics. The advent of accelerators exploring the TeV energy range offers an unique opportunity for the understanding of non-perturbative aspects of QCD and, therefore, for explaining 99.9% of the events.

Of course, the present status of the theory is such that one can not perform detailed computations for low p_T physics without introducing extra dynamical assumptions. However, even at the present moment, we have powerful tools for studying soft physics. On one side, we have interesting models, which are able to adapt many of the features of the data. On other side, we have asymptotic theorems [1,2] , which are rigorous results derived from general principles and which induce fascinating finite-energy effects, even without invoking the assumption of early asymptoticity.

An interesting question in the framework of collider and TeV physics is the following : Is there a difference between $\bar{p}p$ and pp scattering at high energies ?

Conventional wisdom, based upon theoretical ideas and experimental data of the 1970's gives a negative answer to this question. However, there is no fundamental theoretical understanding of why this should be so.

In this talk I will present a point of view, based upon asymptotic theorems, showing that it is possible to have important differences between antihadron-hadron and hadron-hadron interactions at high energies. Such differences are induced by the odderon (not to be confused with "odoronium" !) - a name which designates a singularity located at $\ell=1$ in the complex-ℓ plane and contributing to the odd-under-crossing amplitude [2]. The recent preliminary R420 data, presented at this conference [3], seem to offer the first clean experimental support in favor of such a point of view.

---

* Laboratoire Associé au C.N.R.S.

2. Brief history of the odderon idea

We introduced the odderon idea in 1973 [4], just after the discovery of increasing pp total cross-sections. Our starting point was a reformulation of the principle of maximal strength of strong interactions [5] in the presence of increasing σ_T. Namely, we considered that the strong interactions can be such that one has simultaneously the maximal behaviour allowed by general principles both for the total cross-section ($\sigma_T \propto \ln^2 s$ [6]) and for the difference of antihadron-hadron and hadron-hadron total cross-sections ($\Delta\sigma \propto \ln s$ [7]). Such a situation corresponds to a froissaron-odderon dominated asymptotic forward (t=0) physics : the exchange of a triple-pole at $\ell=1$ in the even-under-crossing amplitude F_+ (the froissaron) and of a double-pole at $\ell=1$ in the odd-under-crossing amplitude F_- (the "maximal" odderon).

The crossing-symmetrical asymptotic form of the F_+ and F_- amplitudes is therefore [4] :

$$\frac{F_+(s,t=0)}{is} \xrightarrow[s \to \infty]{} C_+\{\ln[(\frac{s}{s_+})\ e^{-i\pi/2}]\}^2 \tag{1}$$

$$\frac{F_-(s,t=0)}{is} \xrightarrow[s \to \infty]{} iC_-\{\ln[(\frac{s}{s_-})\ e^{-i\pi/2}]\}^2 \quad , \tag{2}$$

where C_+, C_-, s_+ and s_- are real constants. In terms of F_+ and F_- one has

$$F_{AB} = F_+ + F_- \quad , \qquad F_{\bar{A}B} = F_+ - F_- \tag{3}$$

It is important to note that our formulation involves the simultaneous saturation of the axiomatic bounds for σ_T and $\Delta\sigma$ only in the sense of the analytical forms. For example, the constant C_+ can very well be different from its axiomatic bound [8] $\pi/m_\pi^2 \simeq 60$ mb. The toy model of Ref. 4 shows how this can be dynamically realised.

It is also important to note that, by considering the usual Regge exchanges in addition to the froissaron and the odderon contributions [see, e.g., Refs. 2, 9-12] we were lead to the conclusion that, in spite of the extreme remoteness of the asymptotic regime, asymptopia nevertheless manifests itself via interesting finite-energy effects. The use of derivative relations [9] is important in justifying such finite-energy effects. The physical picture which emerged [9-12] is the following : the low-energy ($5 < \sqrt{s} < 100$ GeV) phenomena are described by Regge physics (Regge poles and cuts), while the asymptotic regime is described by ln s physics. The asymptotic singularities (the froissaron and the odderon) appear at low energies as corrections to the usual Regge exchanges. In the intermediate energy range (i.e. at collider and TeV energies) we expect new effects to occur as a result of the gradual emergence of the ln s physics.

Finally, a remark concerning the terminology is necessary. We first introduced the term "odderon" [10] to describe a possible pole in F_- at $\ell=1$, corresponding to

$$F_-(s, t=0) = - D_- s \tag{4}$$

The term was chosen because it describes the analog of the Pomeron in the odd-under-crossing amplitude. However, in the current literature the term has been generalized to refer to any $\ell=1$ singularity in F_-. For precision, we shall call the behaviour (2) "maximal odderon" asymptotic behaviour.

When it was first formulated, the odderon idea was considered by many, in a period over-dominated by the Regge approach, as being strange and even heretical. The odd-under-crossing amplitude F_- was believed as being dominated by Regge poles located near $\ell=1/2$ and therefore the total and differential cross-sections for antihadron-hadron and hadron-hadron scattering were believed to become practically identical at ISR energies and beyond.

Nevertheless detailed phenomenological studies were performed by us [e.g. 9, 10] and also by other groups [13], by using experimental data available in that period. I quote just a few of the finite-energy effects which were studied in the past :

1) minima or zeros in $\Delta\sigma$ [2, 9] ;

2) minima in the forward differential cross-sections [10] ;

3) zeros in $\rho(s) = \text{Re } F(s, t=0)/\text{Im } F(s, t=0)$ [9] ;

4) dramatic effects in the pion-nucleon charge-exchange polarization [10].

The experimental situation before 1982, concerning the odderon problem, was highly ambiguous. The data were not in contradiction with the odderon approach, however no clear experimental evidence was found for it. Even if we think, as Dyson, that "absence of evidence is not the same thing as evidence of absence" [14], the odderon situation, from the experimental point of view, was quite unsatisfactory. We had to wait for new data.

The current revival of interest in the odderon idea is connected, I think, with four facts :

1) the recent high-energy data, some of them, like the R420 results (see next section), pointing towards the existence of the odderon ;

2) the possibility of having both $\bar{p}p$ and pp TeV accelerators in a not too distant future ;

3) the recognition of the role of the odderon in reggeon field theory [15, 16] ;

4) the phenomenological success of the Donnachie-Landshoff model [17, 18], which includes in addition to Regge exchanges, at large t, a 3-gluon exchange contribution, which is, in fact, an odderon contribution, similar to that of eq. (4). This 3-gluon-odderon gives a remarkable dynamical support for the odderon idea, even if its precise form at small t is not yet known.

3. Experimental hints for maximality and the existence of the odderon

The recent collider UA4 data [19] for $(\sigma_T)_{\bar{p}p}$ at $\sqrt{s} = 0.546$ TeV as well as the cosmic-ray data [20] for $(\sigma_T)_{pp}$ at $\sqrt{s} = 30$ TeV are nicely compatible with the maximal (froissaron) $\ln^2 s$ behaviour of σ_T [21].

The situation concerning the forward slope is somewhat more ambiguous, both on the theoretical and the experimental side. The AKM asymptotic theorem [22] implies a non-uniform behaviour of the slope : namely, if $\sigma_T \propto \ln^2 s$ at $s \to \infty$, then

$$b(s,t)\Big|_{s \to \infty} \begin{cases} \propto \ln^2 s, \text{ at } t=0 \\ \lesssim \ln s, \text{ at fixed } t \neq 0 \end{cases} \tag{5}$$

At finite s, the transition between the two regimes has to occur at some small t, but the corresponding precise t-value is unknown. Experimentally, the UA4 data [19] for small t are compatible with only a small $\ln^2 s$ increase of the slope between ISR and collider energies [21]. The Regge-pole model, which gives an effective ln s behaviour of the slope, is also marginally compatible with these data [21]. However, one has to note that the neglect of the $\ln^2 s$ term in reproducing $t \neq 0$ data corresponds to a very ugly situation : it requires a big coefficient of t in the exponent of the froissaron in order to cut its t-dependence, and induces a huge variation of the slope in the immediate neighbourhood of t = 0 [21].

Of course, these hints for maximality of F_+ (corroborated by the behaviour of ρ [21]) do not imply a situation of early asymptoticity. In fact, we have shown that the overall forward data for $\bar{p}p$ and pp scattering indicate the extreme remoteness of asymptotic regime [21]. The new UA4 data for σ_{el}/σ_T [19], showing an increase of this ratio between the ISR and collider energies, are particularly significant in this context [21].

The corresponding experimental hints for the odderon exchange come from two different sources : the polarization in $\pi^- p \to \pi^\circ n$ at Serpukhov energies and the difference

$$\Delta\left(\frac{d\sigma}{dt}\right) \equiv \left(\frac{d\sigma}{dt}\right)_{\bar{p}p} - \left(\frac{d\sigma}{dt}\right)_{pp} \tag{6}$$

at ISR energies.

The polarization in $\pi^- p \to \pi^\circ n$ is an ideal case for the study of odderon effects [10], even at quite low energies, because : a) only F_- contributes to $\pi^- p \to \pi^\circ n$; b) the polarization is, by definition, sensitive to small new contributions added to conventional exchanges (like the ρ and ρ' Regge poles). If the odderon is present, we expect a new dynamical zero to appear in the polarization at fixed s, as a result of a cancellation between the $\rho \otimes \rho'$ and $\rho \otimes$ odderon contributions [10, 23]. This is precisely what was observed recently in the Serpukhov data at p_L = 40 GeV ($\sqrt{s}$ = 0.009 TeV) [24], which show a zero in P at $|t| \simeq 0.3 - 0.4\ (\text{GeV/c})^2$ with an accompanying minimum at $|t| = 0.5\ (\text{GeV/c})^2$.

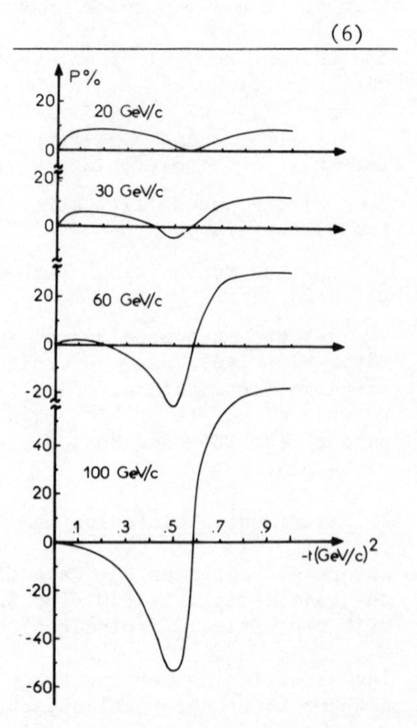

Fig. 1 Theoretical predictions for the polarization in $\pi^- p \to \pi^\circ n$, based on the maximal odderon asymptotic behaviour [23].

The position of the new zero will vary with energy, approaching $t = 0$ as the energy increases (Fig. 1). At the same time, the polarization becomes more and more negative at small t (in contrast with the positive polarization observed experimentally at low energies), as a result of the increasing dominance of the $\rho \otimes$ odderon contribution. At Fermilab energies we predict a gorgeous polarization (Fig. 1), negative everywhere in the small-t region, $0 < |t| < 0.6$ (GeV/c)^2. This is true for both maximal odderon and odderon-pole cases [23].

The cleanest indication of an unusual behaviour of the odd-under-crossing amplitude F_- comes from the recent preliminary R420 data for $\Delta(d\sigma/dt)$ at "large" t [3] ($|t| \gtrsim 1$ (GeV/c)^2). Let us see first what is generally expected for this difference if the odderon is present.

Of course,

$$\Delta(\frac{d\sigma}{dt}) \propto \text{Re } F_+ \cdot \text{Re } F_- + \text{Im } F_+ \cdot \text{Im } F_- \tag{7}$$

If the dominant contribution Im F_+ goes through a zero, then a dip will be induced in at least one of $d\sigma/dt$. However, $\Delta(d\sigma/dt) \neq 0$ if $F_- \neq 0$ (eq. 7). If F_- is controlled by Regge exchanges, one expects that $F_- \simeq 0$ at ISR at collider energies. Therefore, if $\Delta(d\sigma/dt) \neq 0$ we need a new contribution in F_-.

Moreover, $F_- \neq 0$ will contribute (because of the crossing principle) with opposite signs to $\bar{p}p$ and pp. Such a $F_- \neq 0$ will fill the dip in $\bar{p}p$, producing a shoulder, while maintaining the dip in pp.

Finally, if F_- goes through a zero in t at fixed s or if the phases are such that the combination (7) is zero, we get a zero in $\Delta(d\sigma/dt)$. In other words, we expect also cross-overs at high t.

The general picture of the odderon effects at ISR and collider energies [25] is compatible with the preliminary R420 $\bar{p}p$ data [3] at $\sqrt{s} = 0.053$ TeV (Fig. 2), which seem to show, when compared with the pp data of Ref. 26, a "dip in pp-shoulder in $\bar{p}p$" effect.

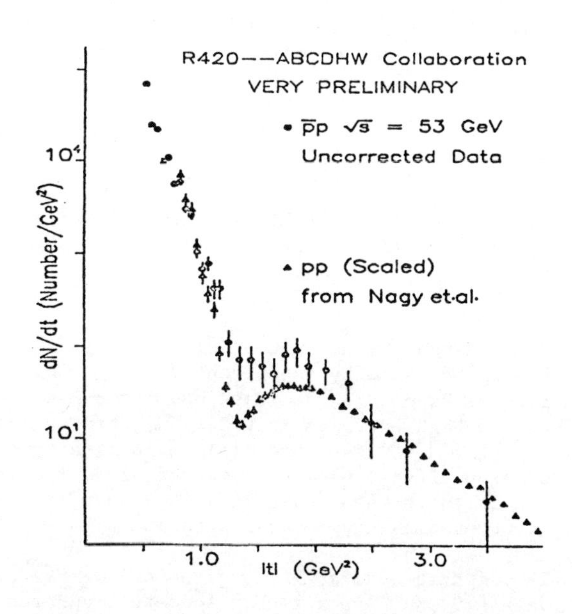

Fig. 2 The new "dip in pp-shoulder in $\bar{p}p$" experimental effect [3, 26].

The R420 experiment is a very significant one and I have to say that it is very silly that no more data were taken before closing the ISR. It has to be realized that, in spite of the present experimental uncertainties concerning the absolute magnitude of $\Delta(d\sigma/dt)$, the new effect is huge when compared with what is expected from the usual Regge exchanges. Most of the existing models are incompatible with the R420 effect. It is not very meaningful to think that this effect represents a small perturbation, by remarking that $d\sigma/dt$ drops by several orders of magnitude from the forward direction to the t-region where the new effect is observed : it is like saying that a dip or a shoulder is the same thing ! The challenge is to build a model involving a reasonable F_- contribution, which model can explain both the R420 effect and the somewhat unexpected big $d\sigma/dt$ for $\bar{p}p$ at large t at collider energies (preliminary UA4 data [27]).

One can ask if this new R420 effect will persist at higher energies. An asymptotic theorem [28] tells us that, under some general and reasonable assumptions (e.g. in the absence of oscillations) one expects $\Delta(d\sigma/dt)$ to go towards zero as $s \to \infty$. However, even if that would be the ultimate situation, given the probable extreme remoteness of the asymptotic regime one can wonder when this should happen. In any case, the "dip in pp-shoulder in $\bar{p}p$" is at least a significant finite-energy effect. We expect such an effect to persist at collider and TeV energies.

One has to note that the 3-gluon-odderon of Donnachie and Landshoff [17,18] can induce the R420 effect. However, one can wonder why there is, in this model, one order of magnitude discrepancy with the UA4 data at large t [27]. One possibility is that some stronger odderon (like the maximal one) is needed. To clarify the situation we have to wait for final R420 and UA4 data.

Before concluding, I would like to say few words about other experimental quantities which could be able to detect odderon effects at collider and TeV energies.

One of them is

$$\Delta\sigma \equiv (\sigma_T)_{\bar{p}p} - (\sigma_T)_{pp} \tag{8}$$

The odderon-pole (4) gives no contribution to $\Delta\sigma$, allowing it to be Regge-behaved at high energies. However, the maximal odderon gives a contribution

$$\Delta\sigma = -\,2\pi\, C_-\, \ln s \tag{9}$$

The constant C_- has not a definite sign from general principles (this sign has to be fixed by experiment). If $C_- > 0$, then $\Delta\sigma \to -\infty$ at $s \to \infty$ and therefore we expect [2] at least one cross-over between $(\sigma_T)_{\bar{p}p}$ and $(\sigma_T)_{pp}$ at finite high-energies ; if $C_- < 0$, then $\Delta\sigma \to +\infty$ at $s \to \infty$, and therefore we expect [2] at least one minimum in $\Delta\sigma$ at finite high-energies. The maximal odderon effects will be detected via a (positive or negative) curvature in the $\Delta\sigma$ curve when drawn in a ln - ln plot (the straight line being the corresponding Regge-model prediction).

The highest available energies for $\Delta\sigma$ are the ISR energies. The recent ISR data [29, 30] are not able to detect odderon contributions which are [11, 21, 31] (and have to be) small at ISR energies in the forward direction ; they are at most comparable with the forward Regge contributions to F_- at these energies [11]. (As I already said, the situation is dramatically different at sufficiently large t.) The low-energy data can still be used in order to extract bounds on the maximal odderon effects at higher energies (from compatibility with the existing data). By using the approach of Refs. 11 and 21 we obtain the extrapolation of $\Delta\sigma$ at collider and TeV energies shown in Fig. 3. As is seen from Fig. 3, we expect a minimum in $\Delta\sigma$ at $\sqrt{s} \simeq 0.1$ TeV. At $\sqrt{s} = 10$ TeV, the positive difference $\Delta\sigma$ can reach

2-4 mb, $(\sigma_T)_{pp}$ being there ≃ 110 mb. It has to be noted that, at $\sqrt{s}$ = 10 TeV, the froissaron contribution is only two times bigger than the Pomeron-pole contribution, and therefore this energy is still a highly non-asymptotic one [21].

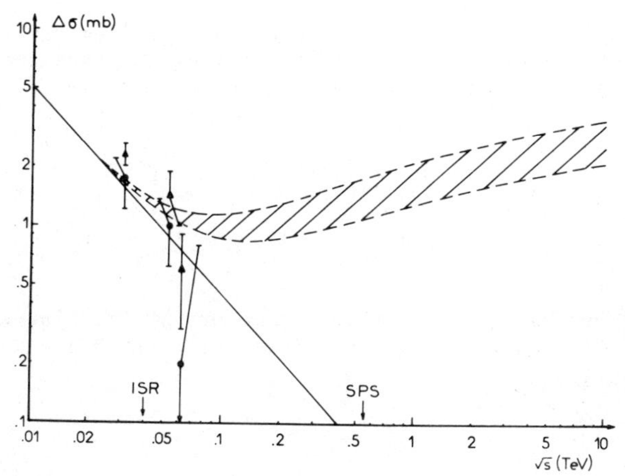

Fig. 3 Odderon effects in $\Delta\sigma$ at collider energies and beyond [11, 21].

Another interesting quantity is

$$\Delta\rho \equiv \rho_{\bar{p}p} - \rho_{pp} \tag{10}$$

If the odderon is absent, $\rho_{\bar{p}p}$ becomes practically the same as ρ_{pp} for $\sqrt{s}$ > 0.1 TeV (see the solid curves in Fig. 4). An odderon-pole leads to values of $\Delta\rho$ at TeV energies which are still very small. However, the maximal odderon induces here also interesting effects. We expect [11] a zero in $\Delta\rho$ at $\sqrt{s}$ ≃ 0.1 TeV. For $\sqrt{s}$ > 0.1 TeV, $\Delta\rho$ becomes negative reaching a value of -6% to -10% at $\sqrt{s}$ = 10 TeV (Fig. 4).

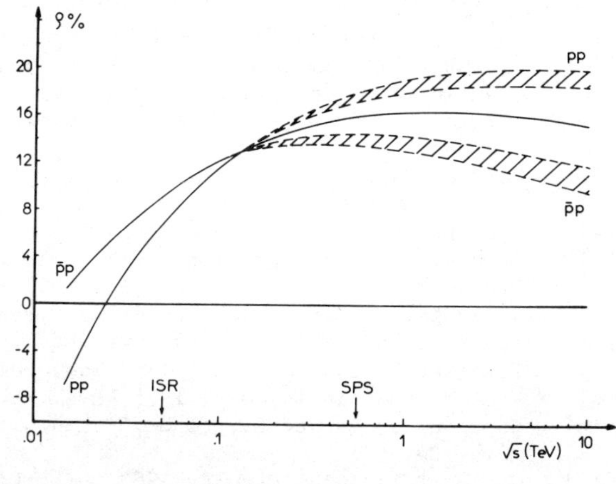

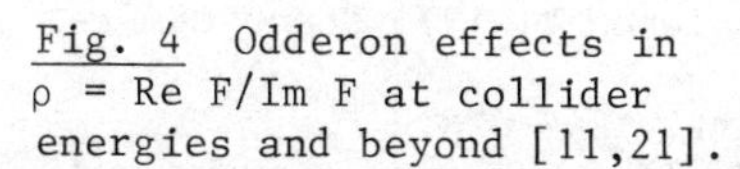
Fig. 4 Odderon effects in ρ = Re F/Im F at collider energies and beyond [11,21].

It is interesting to note that a high-precision experiment performed in $\bar{p}p$ at collider energies can lead to interesting conclusions, even if a pp collider would be not simultaneously available. If the $\rho_{\bar{p}p}$ value will be significantly different from what is predicted from dispersion relations with only Regge contributions in F_-, an indication for the presence of the odderon can still be obtained. Measurements of the real parts at collider energies would be therefore welcome.

4. Conclusions

Even if I can not go as far as Alan White goes when he says that "future generations will believe it vital to test" if there is or there is not a difference between pp and $\bar{p}p$ scattering at high energies [32], I still think that this test is important, especially in connection with non-perturbative aspects of QCD. Experiments made <u>both</u> in pp and $\bar{p}p$ scattering are therefore of great interest. It is possible that the TeV energy range will reveal a new "scale" of strong interactions with a host of accompanying theoretical and experimental surprises.

⁂

I dedicate this talk to the memory of Dr. Claude Broll, who died in an unfortunate accident few days after participating at the Durham Conference.

REFERENCES

[1] Eden R.J. 1971 Rev. Mod. Phys. 43 15 ; Roy S.M. 1972 Phys. Rep. 5C 125 ; Fischer J. 1981 Phys. Rep. 76 157 ; Martin A. 1982 Z. Phys. C15 185.

[2] For a previous review of the odderon problem and asymptotic theorems, see Nicolescu B. 1982 Proc. 6th European Symp. on Nucleon-antinucleon and quark-antiquark interactions, Santiago de Compostela, ed. C. Pajares (Madrid : Sociedad Española de Fisica) pp. 273-8.

[3] R420 collab. 1984, Breakstone A. et al., results presented by Rimondi F. at this conference.

[4] Lukaszuk L. and Nicolescu B. 1973 Lett. Nuovo Cimento 8 405.

[5] Chew G.F. and Frautschi S.C. 1961 Phys. Rev. Lett. 7 394.

[6] Froissart M. 1961 Phys. Rev. 123 1053 ; Martin A. 1963 Phys. Rev. 129 1432 ; Martin A. 1966 Nuovo Cimento 42A 930.

[7] Roy S.M. and Singh V. 1970 Phys. Lett. 32B 50.

[8] Lukaszuk L. and Martin A. 1967 Nuovo Cimento 52A 122.

[9] Kang K. and Nicolescu B. 1975 Phys. Rev. D11 2461.

[10] Joynson D., Leader E., Lopez C. and Nicolescu B. 1975 Nuovo Cimento 30A 345.

[11] Gauron P. and Nicolescu B. 1983 Phys. Lett. 124B 429.

[12] Gauron P. and Nicolescu B. 1983 Proc. XVIIIth Rencontre de Moriond, ed. J. Tran Thanh Van (Dreux : Editions Frontières) pp. 107-14.

[13] See, e.g., Diu B. and Ferraz de Camargo A. 1977 Lett. Nuovo Cimento 20 609 ; Bouquet A. and Diu B. 1978 Nuovo Cimento 43A 53.

[14] Dyson F.J. 1983, "Unfashionable Pursuits", in The Mathematical Intelligencer 5 n°3 47.

[15] Bartels J. 1979, Lectures on "High Energy Behaviour of Nonabelian Gauge Theories", DESY preprint 79/68.

[16] White A.R. 1983 Proc. XIIIth Int. Symp. on Multiparticle Dynamics Volendam, ed. W. Kittel, W. Metzger and A. Stergiou (Singapore : World Scientific) pp. 799-818.

[17] Donnachie A. and Landshoff P.V. 1984 Nucl. Phys. B231 189 ; Donnachie A. and Landshoff P.V. 1984 Cambridge preprint DAMTP 84/6.

[18] Donnachie A., talk at this conference.

[19] UA4 collab. 1984, Bozzo M. et al., results presented by Braccini P.L. at this conference, CERN/EP preprints 84-90 and 84-91.
[20] Baltrusaitis R.M. et al. 1984 Phys. Rev. Lett. 52 1380.
[21] Gauron P. and Nicolescu B. 1984 Phys. Lett. 143B 253.
[22] Auberson G., Kinoshita T. and Martin A. 1971 Phys. Rev. D3 3185.
[23] Gauron P.,Leader E. and Nicolescu B. 1984 Phys. Rev. Lett. 52 1952.
[24] Apokin V.D. et al. 1982 Z. Phys. C15 293.
[25] Gauron P., Leader E. and Nicolescu B., to be published ; in this work we consider the froissaron as described by a 3/2 cut in the complex ℓ-plane collapsing to a triple pole at t=0, while the odderon corresponds to two complex conjugate poles collapsing to a double pole at t=0.
[26] Nagy E. et al. 1979 Nucl. Phys. B150 221.
[27] UA4 collab. 1984, results presented by Braccini P.L. at this conference.
[28] Cornille H. and Martin A. 1972 Phys. Lett. 40B 671.
[29] R210 collab. 1982, Ambrosio M. et al. Phys. Lett. 113B 87 and 115B 495.
[30] R211 collab. 1983, Amos N. et al. Phys. Lett. 120B 460 and 128B 343.
[31] Block M.M. and Cahn R.N. 1983 Phys. Lett. 124B 229.
[32] White A.R. 1984 Argonne preprint ANL-HEP-CP-84-19, presented at the workshop "$\bar{p}$-p Options for the Super-Collider", organized by Argonne National Laboratory and the University of Chicago, Feb. 13-17, 1984.

Inst. Phys. Conf. Ser. No. 73: Section 6
Paper presented at VII Eur. Symp. Antiproton Interactions, Durham 1984

Results from the UA5 experiment at the CERN p̄p Collider

UA5 Collaboration (Bonn-Brussels-Cambridge-CERN-Stockholm)

Presented by Ch. Geich-Gimbel

Physikalisches Institut der Universität Bonn, Germany

Abstract. Results from the UA5 experiment at $\sqrt{s}$ = 54o GeV at the CERN SPS collider are presented. Production rates and differential cross sections are given for K^o_s , Λ and Ξ^-. Forward-backward multiplicity correlations and two-particle rapidity correlations are investigated. In terms of a simple cluster approach for the effective cluster size a value of k_{eff} $\sim$ 2.5 (charged particles) is obtained, with a decay width $\delta \sim$ 0.7 in units of pseudorapidity. Finally the occurrence of events with unusually large track density in small rapidity intervals is discussed.

1. Introduction

The UA5 detector, consisting of two large streamer chambers (6m x 1.25m x 0.5m visible volume) was used in two SPS collider runs at $\sqrt{s}$ = 54o GeV in October 1981 and September 1982. In the second run, from which about 7ooo non-single diffractive events were completely measured and analyzed, a beryllium beam pipe was introduced to reduce background from conversions. The experimental setup allows good reconstruction of neutral strange particles (K^o_s and $\Lambda(\bar{\Lambda})$) by their decay into two charged particles, as the fiducial volume starts at a distance of only 6cm from the beam axis. For details of the apparatus see {1}.

Results for the production of K^o_s and Λ from the 1981 data have been published {2} and were based on 21oo fully measured non-single diffractive events. From the 1982 data with 7ooo events fully measured so far about four times the statistics in V^o-production is available. We also include preliminary results on Ξ^--production from the 1982 data {4}.

Additionally, with 1982 data multiplicity distributions and correlations in particle production have been studied in more detail. Deviations from KNO-scaling, favouring high multiplicity events have been published {3}. Here two-particle rapidity correlations and fluctuations of the particle density in small rapidity intervals will be discussed.

2. Strange particle production

2.1 Inclusive K^o_s and Λ production

The inclusive rates for K^o_s and $\Lambda(\bar{\Lambda})$ production obtained from the 1982 data, being in good agreement to the published 1981 data {2}, are shown in table 1.

		UA5 1982 data, 7138 events $\|\eta\|<3.5$	$\|\eta\| < 3$	UA5 1981 data {2} 21oo events, $\|\eta\| < 3$
Accepted	K^o_s	366	34o	92
	$\Lambda(\bar{\Lambda})$	28o	268	46
Corrected number	K^o_s	1.1±0.1	1.0±0.1	1.0±0.2
per event	$\Lambda(\bar{\Lambda})$	0.48±0.08	0.43±0.08	0.35±0.10

Table 1. Inclusive K^o_s and Λ rates. Errors are statistical only.

2.2 Differential distributions for K^o_s and Λ production

In fig.1 the p_T-distribution for K^o_s is shown both for 1981 and the 1982 data - and also for $K^{\pm}$ production as published by the UA2-Collaboration {5} for $|\eta| < 0.8$. As can be seen from figure, a purely exponential fit of the form $dN/dp_T^2 = A \exp(-Bp_T)$, giving B=4.2, hence p_T = 48o MeV/c, is likely to underestimate the cross section at larger p_T. On the other hand, a power law fit of the form $dN/dp_T2 = A\{p_{T_o} / (p_T + p_{T_o})\}^B$ {6} would overestimate the cross section at small p_T {7}, again giving too small a value for $\langle p_T \rangle$.

Therefore a combined fit, i.e. applying a power law fit above p_T= 0.4 GeV/c, and exptrapolating to lower p_T by an exponential, was used, leading to a preliminary value of $\langle p_T \rangle$= 56o ± 6o MeV/c for K^o_s.

In the case of $\Lambda(\bar{\Lambda})$ -production the p_T-distribution, see fig.2, can satisfactorily be described by a pure exponential, giving $\langle p_T \rangle$ = 68o ± 6o MeV/c. Comparing the p_T-distributions for K^o_s and Λ it might be noted that the onset of the deviation from an exponential dependence seems to be shifted towards higher p_T for the heavier particles. Data for inclusive p,$\bar{p}$-production from UA2 {5} are included in fig.2, which roughly exhibit the same behaviour as the Λ's.

Next it was investigated whether there are indications for multiplicity dependent effects on the strange particle production, which could signal the occurrence of new phenomena in high multiplicity events, e.g. phase transitions, quark gluon plasma {8}. Fig.3 shows the inclusive p_T-distributions in two multiplicity bins, namely $n_{ch} \gtrless 3o$, not differing significantly, the average number itself of clean K^o_s and coplanar V^o's (K^o_s and Λ, $\bar{\Lambda}$) versus the observed multiplicity is plotted in fig.4. As the number of strange particles rises almost linearly with the charged multiplicity, the frequency of neutral strange particles per charged track is consistent with being flat over the entire multiplicity range.

Finally, $d\sigma/d\eta$ and $d\sigma/dy$, i.e. the pseudorapidity and rapidity distributions for K^o_s as shown in fig.5 and 6, are more or less flat in the central region out to y or $\eta \sim 2.5$.

An interesting conclusion might be drawn from fig.7. There the mean transverse momentum for three kinds of particles, i.e. pions, kaons and baryons (protons, antiprotons or Lambdas where available) is plotted versus the C.M. energy from PS-, FNAL-, ISR- and SPS Collider-experiments. For the lightest particles considered, pions,the increase of $\langle p_T \rangle$with rising energy is weakest and apparently smooth, whilst the increase is steeper the heavier the particle considered - or the more primordial the particle might be in a fragmentation (resonance decay?) chain. For a discussion of the rôle the proposed UA5/2 experiment {1o} could play in this investigation, the reader is referred to D.R.Ward's presentation at this conference {9}.

2.3 Observation of Ξ^--production {4}

One example of a streamer chamber picture with an event, where a Ξ^- *) decays into $\Lambda\pi^-$ with a subsequent decay $\Lambda \to p\pi^-$, can be seen in fig.8. In a systematic search in about 7,ooo fully measured minimum bias events 17 Ξ^--candidates, where a V^o is pointing to a 'kink', were found. All 17 candidates give good kinematical fits to the $\Xi^- \to \Lambda\pi^-$, $\Lambda \to p\pi^-$ hypothesis, one of which has a higher probability for $\Omega^- \to \Lambda K^-$, but assigning 5.9 lifetimes to the Ω^- instead of 2.5 if considered as a Ξ^-.

Cuts against regions, where the scanning efficiency is low (distance of kink vertex to chamber base > 3cm, kink angle, V^o-opening angle both > 3°, Λ-path 3cm and $|\eta| < 3.5$) exclude one candidate. Thus we are left with 16 events with a Ξ^-.

Possible background sources studied were:

- interactions with the chamber gas nuclei of the type
  $\pi^\pm(K^\pm)\ N \to K^o_s/\Lambda$ + 1 charged track
  $\pi^\pm N \to \pi^o$ + 1 charged, $\pi^o \to \gamma\gamma$, one $\gamma \to e^+e^-$
- decays of charged strange particles
  $\Sigma^+(K^\pm) \to p(\pi^\pm)\pi^o$, $\pi^o \to \gamma\gamma$, one $\gamma \to e^+e^-$
- random association of a V^o to a kink.

These backgrounds were estimated to cause less than 1.5 fake Ξ^--signatures in total.

Also - a Monte Carlo simulation of 1o,ooo events, where Ξ^-production is explicitly excluded, does not give a possible Ξ^--signature. Therefore the observed Ξ^--signal is genuine.

Some kinematical distributions for the Ξ^-'s are shown in figs.9-15 and explained in the figure captions. The curves represent Monte Carlo results, assuming a $\langle p_T \rangle$ of 1.1 GeV/c for Ξ^- production, see below. The data are compatible with the simulations.

In order to obtain $\langle p_T \rangle$for Ξ^- production a maximum likelihood fit was made to the p_T-distribution. Assuming a form $d\sigma/dp_T^2 = A\ \exp(-Bp_T)$ one obtains $\langle p_T \rangle = 1.1 + 0.3/-0.2$ GeV/c. As this value is determined only from events with 0.7 GeV/c $< p_T(\Xi^-) <$ 3.7 GeV/c a deviation from the exponential form chosen outside this p_T-range could lead to a bias in the $\langle p_T \rangle$ calculated this way.

The inclusive Ξ^--production rate is very sensitive to the mean transverse momentum, as can be seen from fig.16. For the full p_T-range, the best value is 0.08 +0.03/-0.02 (Statistical error only) Ξ^- per event. At $\sqrt{s}$ = 34 GeV in an e^+e^--experiment at TASSO {11} the inclusive Ξ^--rate was found to be 0.026 ± 0.008 ± 0.009 per hadronic event. Bearing in mind the uncertainties in extrapolating the observed Ξ^--signal to p_T= 0 one nevertheless would

*) In our non-magnetic detector Ξ^- cannot be distinguished from $\overline{\Xi^-}$ (nor Λ from $\bar{\Lambda}$). We denote either of them by Ξ^- and Λ.

conclude that S = − 2 content of the final state indicated by Ξ^- production rises with increasing CM energy.

Trying to follow simple factorization arguments, i.e. suspecting the ratio $\bar{\Lambda}(\bar{s}\bar{u}\bar{d})$ / $\bar{p}(\bar{u}\bar{u}\bar{d})$ to be similar $\bar{\Xi}(ssd)$ / $\Lambda(sud)$, leads to somewhat contradictory results. Naively the $\bar{\Lambda}/\bar{p}$-ratio represents the so-called strangeness suppression factor, which is generally believed to be about 0.3 (measured to be 0.38 ± 0.07 in our experiment {13}). For the $\Xi^-/\bar{\Lambda}$-ratio only 0.06 ± 0.02 is obtained at $\sqrt{s}$ = 63 GeV at the ISR {12} for p_T between 1.2 and 2.4 GeV/c and $|y| < .2$, similarly at TASSO 0.09 ± 0.03 ± 0.03 (all p_T) {11}, but we find Ξ^- 0.35 ± 0.11 for p_T > 1.0 GeV/c (Λ's possibly originating from Ξ^- decay not subtracted).

The latter value is problematic in the sense that due to the much different $\langle p_T \rangle$ for Λ and Ξ^- the p_T > 1.0 GeV/c condition implies a severe overestimation of the Ξ^-/Λ-ratio. Attempting to cure this bias the said exptrapolation to p_T= 0 would give an "all p_T" Ξ^-/Λ ratio of 0.17 ± 0.06, much closer to the ISR and PETRA values.

Again, future runs at $\sqrt{s}$ = 200 GeV and 900 GeV {9,10} with the same detector and analysis chain could tell, whether the simple factorization arguments become valid, if the available energy is sufficiently large.

3. Study of correlations in particle production

In the past various analyses at the ISR {14-18} have proved the existence of both short range and long range correlations in particle production and discussed in terms of cluster models. Within specific cluster models their relevant parameters, the decay multiplicity and width, are accessible by the investigation of two particle rapidity correlations {19,20}. The formalism will be briefly described below.

3.1 Inclusive two particle rapidity correlations

For two particle rapidity correlations one usually defines an inclusive correlation function as

$$C(\eta_1,\eta_2) = \rho^{(2)}(\eta_1, \eta_2) - \rho^{1}(\eta_1)\rho^{(1)}(\eta_2),$$

where $\rho^{(1)}$ denotes the charged particle rapidity density $\rho^{(1)} = 1/\sigma\ d\sigma/d\eta$ and $\rho^{(2)} = 1/\sigma\ d^2\sigma/d\eta_1 d\eta_2$ the two-particle density. In the case of uncorrelated particle production the correlation function $C(\eta_1,\eta_2)$ vanishes. As the correlation is strongest for $\eta_1 = \eta_2$, decreasing for increasing difference in rapidity it is convenient to express this function only in terms of the rapidity difference $\Delta\eta = \eta_1 - \eta_2$, integrating over a plateau region of $|\eta_1 + \eta_2| = 2$, see fig.17.

Originally, in the context of two-component models, the correlation function was broken down in terms for intrinsic correlations within each component and a crossed term describing long range correlations {21}. When summing over different multiplicities n the correlation function analogously contains a short range correlation term

$$C_S = \sum_n \frac{\sigma_n}{\sigma} C_n(\eta_1, \eta_2)$$

and a long range correlation term

$$C_L = \sum_n \frac{\sigma_n}{\sigma} \{\rho^{(1)}(\eta_1) - \rho_n^{(1)}\eta_1)\}\{\rho^{(1)}(\eta_2) - \rho_n^{(1)}(\eta_2)\},$$

where the index n denotes a fixed multiplicity n in the semi-inclusive cross section and the corresponding particle densities. The long range contribution, also shown in fig.17, sums the products of differences between the inclusive and semi-inclusive particle densities and depends on the shape of the multiplicity distribution. This term differs from zero even in absence of "true" correlations between the charged particles produced. Thus, the dynamical correlations can be present only in the first term, C_S, containing two-particle densities {15}.

The determination of the short range correlation component implies the measurement of semi-inclusive rapidity densities to obtain $C(\Delta\eta) - C_L(\Delta\eta)$ as plotted in fig.17 {24}.

3.2 Semi-inclusive charged particle rapidity correlations

The short range component is usually fitted by the sum of a Gaussian function and a background term proportional to the product of single particle densities according to

$$C_S(\Delta\eta) = \frac{\langle k(k-1)\rangle}{\langle k\rangle} \frac{\rho^{(1)}(\eta=0)}{2\sqrt{\pi}\delta} \exp\{-\Delta\eta^2/4\delta^2\} - A\,\rho^{(1)}(\eta_1)\,\rho^{(1)}(\eta_2),$$

where, within the cluster model, k stands for the decay multiplicity of the clusters and δ for their decay width in units of pseudorapidity {19}.

The Gaussian and the background fitted are shown for a fixed multiplicity of observed charged particles, here for $2o < n_{obs} < 25$, in fig.18. The quantities $\langle k(k-1)\rangle/\langle k\rangle_n$ and δ_n for different observed multiplicities are given in fig. 19 and 2o, as function of the corrected normalized multiplicity $z = n/\langle n\rangle$. For comparison, two ISR-results {15,18} have been added to these figures. Neither the quantity $\langle k(k-1)\rangle/\langle k\rangle$, related to the decay multiplicity of the clusters produced nor δ, the decay width do seem to have significantly increased from ISR to Collider energies.

In a study of forward-backward multiplicity correlations {22,23} the number of charged particles emitted into one hemisphere, for instance into the backward one, n_B was shown to be related to the number of forward going particles n_F. In the ananlysis it was averaged over n_B at fixed n_F, and a linear relationship $\langle n_B(n_F)\rangle = a + bn_F$ was found. Demanding a central gap of size 2 in units of η to protect against the influence of resonances emitting their decay products into either region for the slope parameter b a value of 0.41 ± 0.01 was obtained {22} from our 1981 data. The same analysis on the new data from 1982 gave consistent results, namely b = 0.42 ± 0.02 for this gap size {24}.

Detailed studies revealed that the slope parameter as well as its dependence on the gap size chosen can well be reproduced {24} by the UA5 cluster Monte Carlo, agreeing with the main features of multiparticle production at $\sqrt{s}$ = 54o GeV. The effective cluster size input into this Monte Carlo is $k_{eff} = \langle k\rangle + D_k^2/\langle k\rangle \sim 2.6$ (charged particles), D_k being the dispersion of the cluster decay multiplicity.

This figure turns out to be consistent with $\langle k(k-1)\rangle/\langle k\rangle \sim 1.4$, obtained from the analysis of two-particle rapidity correlations, as reported here, when rewriting $\langle k(k-1)\rangle/\langle k\rangle = \langle k\rangle + D_K^2/\langle k\rangle - 1 = k_{eff} - 1$.

4. Fluctuations in particle density

As scaling deviations favouring high multiplicity events were observed in our data {2}, it became interesting to investigate the multiplicity distribution for limited (central) regions with $|\eta|$ smaller than some η_{cut}. In fig.21 four examples for different η_{cut} are given. One observes the fluctuations in multiplicity to grow the smaller the rapidity interval chosen becomes.

Next it was searched whether there exist events which have unusually high particle densities in narrow rapidity intervals, which could hint at some new physics {25}. In principle this might work as a jet finding procedure. In practice a window of $\Delta\eta$ = 0. 5 was slit over the multiplicity distributions, rendering "spike" events, three of which are shown in fig.22. There the particle density $dn/d\eta$ reaches values of up to about 3o, in contrast to $\langle dn/d\eta \rangle \sim 3$ in the central plateau region.

Remarkably enough these events do not exhibit any jet structure, as can be seen from fig.23. Secondly, such events also occur by the UA5 cluster Monte Carlo {26}, which allows for randomly emitted clusters to be superimposed in rapidity.

A more quantitative analysis, using an arbitrary cut (N_{max} (per $\Delta\eta$=0.5) = 0.1 x n_{ch} + 7), see fig.24, to define a "spike"sample both in data and Monte Carlo, gave the following results:Frequency of spike events 0.7 % (data), 0.5 % (MC), $\langle n_{ch} \rangle$ = 47 (data), 41 from Monte Carlo and $\langle N_{max}(\Delta\eta=0.5)\rangle$ = 12.9 for data, respectively 12.2 for MC.

In conclusion one might not suspect the occurrence of such events with extraordinarily high particle density in small rapidity intervals to originate from new or spectacular physics.

5. Conclusions

At $\sqrt{s}$ = 54o GeV the production of K^o_S with $\langle p_T \rangle$ = 56o ± 6o MeV/c and a rate of 1.1 ± 0.1 per non-single diffractive event has been observed. The corresponding figures for $\Lambda(\bar{\Lambda})$-production are $\langle p_T \rangle$= 68o ± 6o MeV/c at a rate of 0.48 ± 0.08 per event.

Furthermore, there is sizeable $\Xi^-(\bar{\Xi}^-)$ production present, namely 0.08+0.03/-0.02 per event, with $\langle p_T \rangle \sim 1.1$ GeV/c. For the Ξ^-/Λ-ratio, which could serve as a measure for the strangeness suppression factor λ , a preliminary value of 0.17 ± 0.06 is obtained - somewhat higher than at ISR or PETRA energies {11,12}.

The mean transverse momentum seems to increase much stronglier for heavy particles than for pions, when comparing results from PS, FNAL, ISR and the CERN $\bar{p}p$-Collider.

An analysis of short range two-particle rapidity correlations confirmed the result from studies of forward-backward multiplicity correlations {22}, assigning a value of $\sim$ 2.5 for the effective size (charged particles) of clusters, with a decay width of $\sim$ 0.7 in units of pseudorapidity. These figures agree quite well with earlier ISR results {14-18}.

In small rapidity intervals particle densities of up to about ten times the average central particle density were found. Such events can well be produced by the standard UA5 cluster Monte Carlo {26} and are not believed to indicate new physics.

REFERENCES

{ 1} K.Alpgard et al., Phys.Lett, 1o7B(1981) 31o;
Phys.Scr.23 (1981) 642.
{ 2} K.Alpgard et al., Phys.Lett, 115B(1982) 65.
{ 3} G.J.Alner et al., Phys.Lett, 138B (1984) 3o4.
{ 4} G.J.Alner et al., UA5-Coll., "Observation of $\bar{\Xi}^-$ production in $p\bar{p}$ interactions at $\sqrt{s}$ = 54o GeV", to appear.
{ 5} M.Banner et al., Phys.Lett, 122B (1983) 322.
{ 6} G.Arnison et al., Phys.Lett, 118B (1982) 167.
{ 7} P.C.Carlson, "New Results from UA5", invited talk at the IVth Int. Workshop on $p\bar{p}$ Collider Physics, Bern, 5-8th March 1984
{ 8} L.Van Hove, CERN-TH.3924, and references quoted therein.
{ 9} D.R.Ward, "Future plans for the UA5 Experiment - $p\bar{p}$ Interactions at 9oo GeV C.M. energy", paper to this conference.
{1o} UA5-Coll., CERN SPS Proposal 184, CERN/SPSC 82-75
{11} M.Althoff et al., Phys.Lett. 13oB (1983) 34o
{12} T.Åkesson et al., CERN-EP/84-26, March 1984
{13} K.Böckmann, UA5-Coll., "Particle production in $p\bar{p}$ interactions at 54o GeV and strange quark suppression", Proceedings of the VI Warsaw Symposium on Elementary Particle Physics, Kazimierz, May 1983, and Th.Müller: "Strangeness suppression at Collider energy", Proceedings of the XV International Symposium on Multiparticle Dynamics, Lake Tahoe, June 1983.
{14} K.Eggert et al., Nucl.Phys. B86 (1975) 2o1
{15} S.R.Amendiola et al., Nuovo Cimento 31A (1976) 17
{16} D.Drijard et al., Nucl.Phys. B155 (1979) 269
{17} S.Uhlig et al., Nucl.Phys. B132 (1978) 15
{18} W.Bell et al., Z.Phys. C22 (1984) 1o9
{19} E.L.Berger, Nucl.Phys.85B (1975) 61
{2o} J.Benecke and J.Kühn, Nucl.Phys. B14o (1978) 179
{21} A.Bialas, Proceedings of the IVth Int.Symp. on Multiparticle Hadrodynamics, Pavia 1973, p.93
{22} K.Alpgard et al., Phys.Lett 123B (1983) 1o8
{23} G.Ekspong, Proceedings of the 3rd Topical Workshop on Proton Antiproton Collider Physics, CERN, Yellow Report 83-o4, p.112
{24} K.Böckmann and B.Eckart, "Short and long range Correlation of Hadrons in $p\bar{p}$ Collisions at 54o GeV", presented by K.Böckmann for the UA5 collaboration at XV.Int.Symp. on Multiparticle Dynamics, Lund, 1o-16 June 1984
{25} J.G.Rushbrooke, "Inelastic hadron-hadron processes at C.M.energies up to 4o TeV", CERN/EP/84-34 to appear in Proceedings of the workshop on pp options for the supercollider, 13-17 February, 1984, University of Chicago.
{26} B.Eckart, UA5-Coll., private communication.

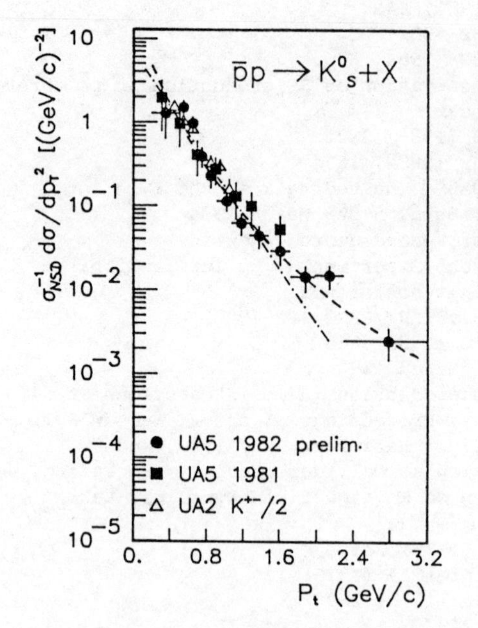

Fig.1 p_T-distribution for K^O_s. UA5-1981 data are from (2), UA2 $K^{\pm}$ from (5).

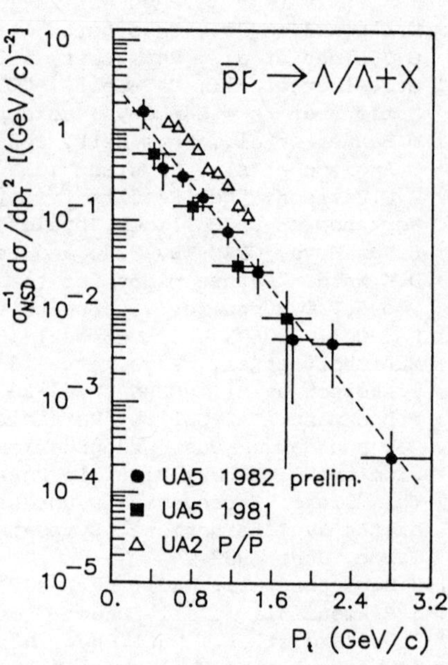

Fig.2 p_T-distribution for $\Lambda(\bar{\Lambda})$. UA5-1981 data are from (2), UA2 $p,\bar{p}$ - data from (5).

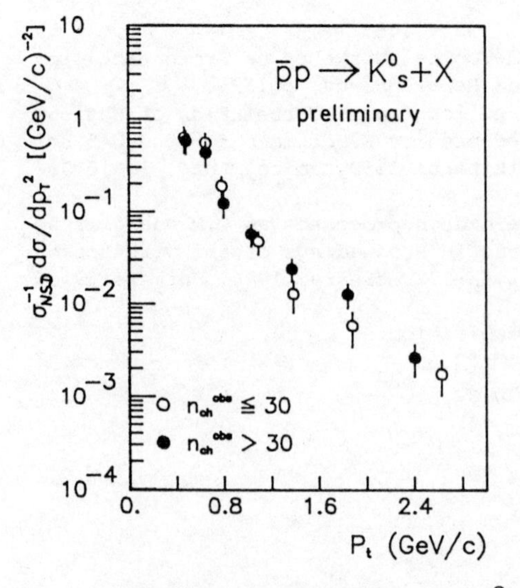

Fig.3 p_T-distribution for K^O_s for two different bins of observed charged multiplicity.

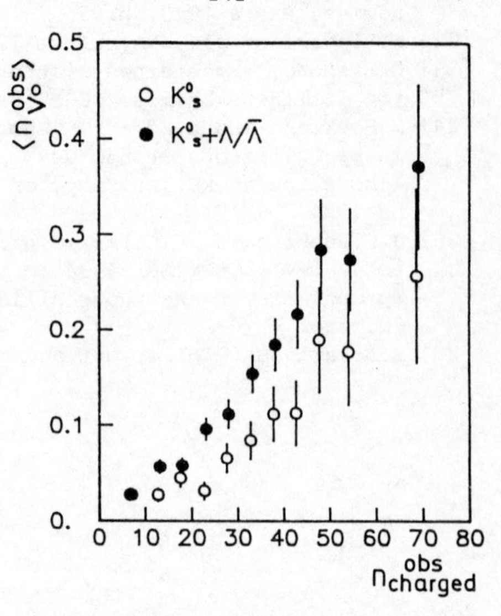

Fig.4 Observed K^O_s and coplanar-V^O multiplicity versus observed charged multiplicity.

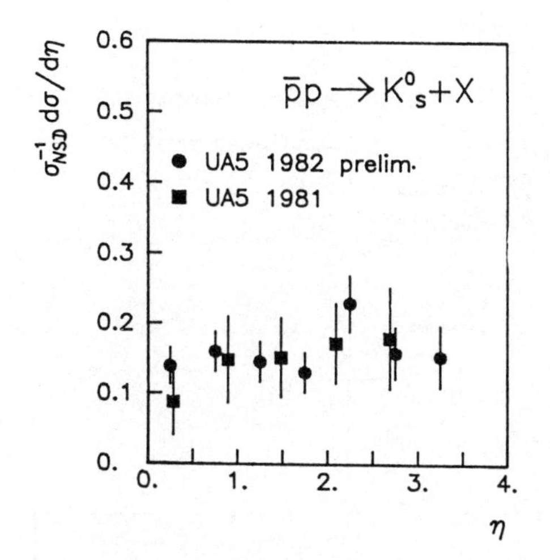

Fig.5 Pseudo-rapidity distribution for K^0_s.

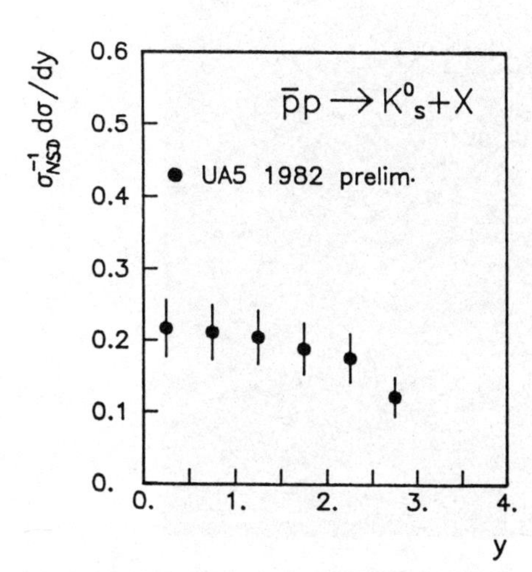

Fig.6 Rapidity distribution for K^0_s.

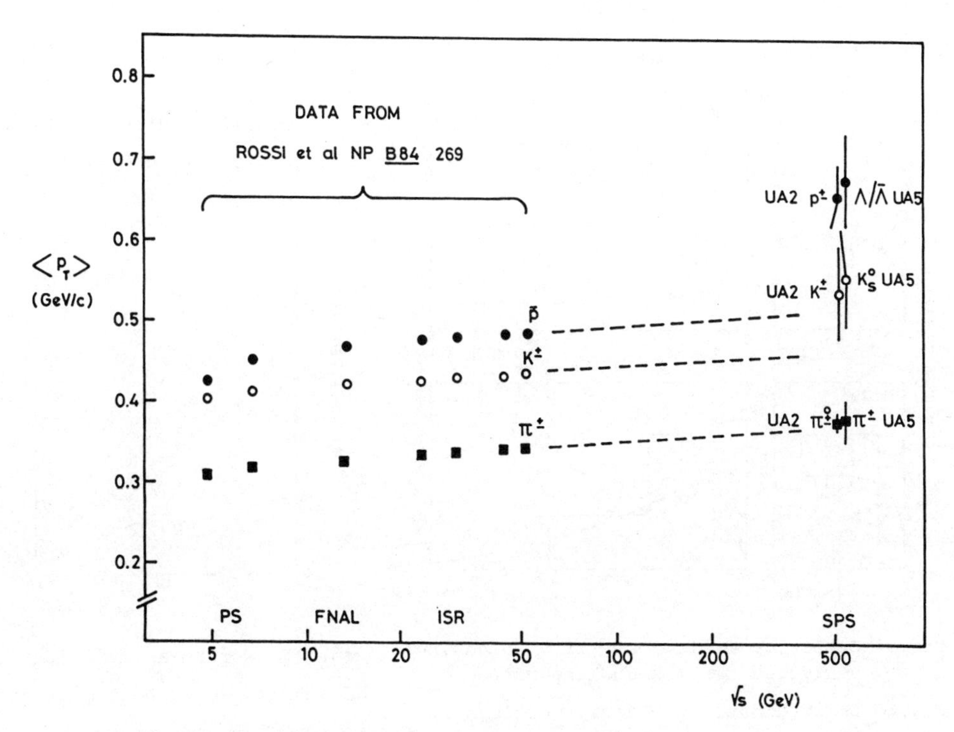

Fig.7 Mean transverse momentum for various particles at PS, FNAL, ISR Experiments and the $p\bar{p}$ Collider. UA2 data are from (5).

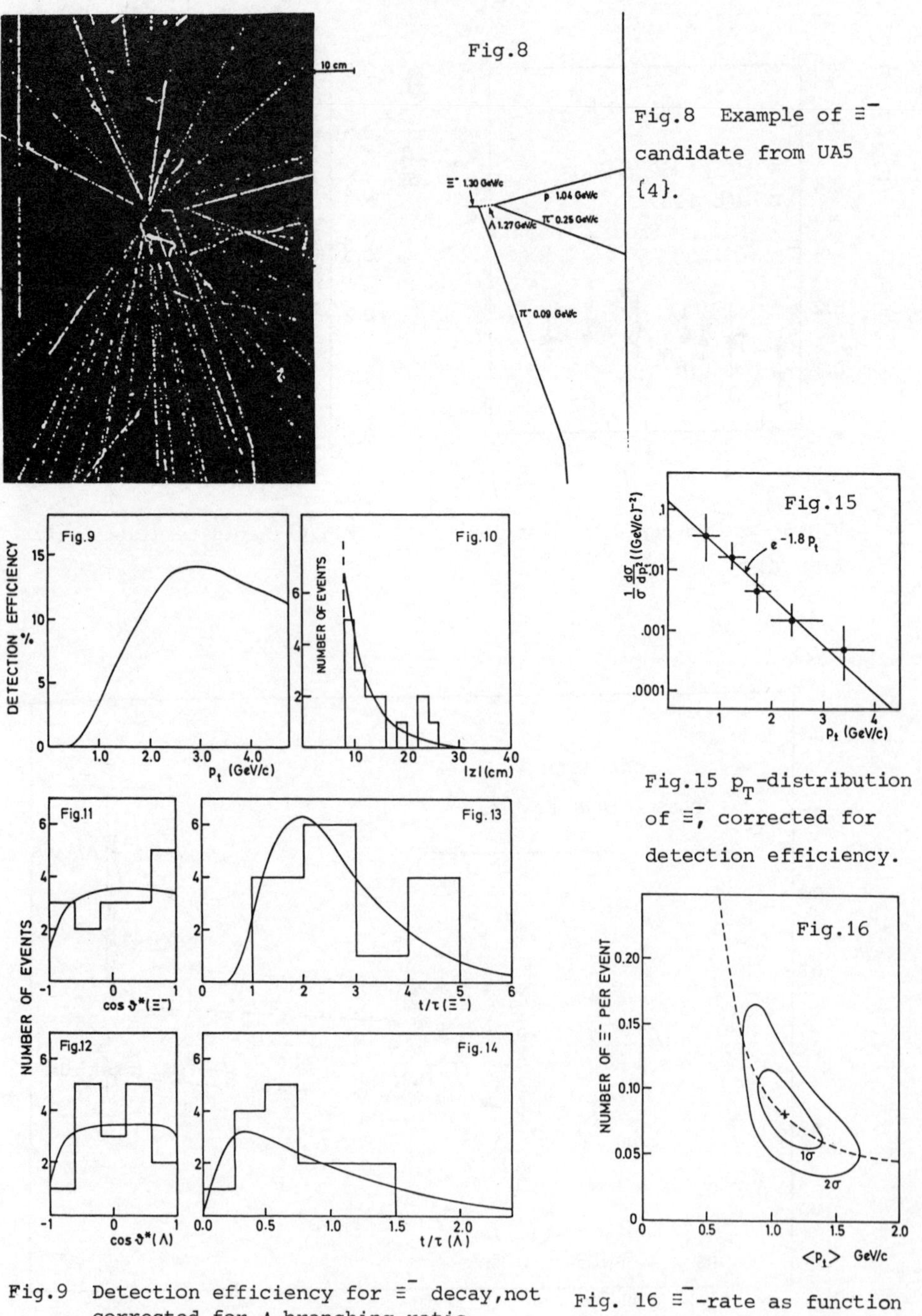

Fig.8 Example of Ξ^- candidate from UA5 {4}.

Fig.15 p_T-distribution of Ξ^-, corrected for detection efficiency.

Fig.9 Detection efficiency for Ξ^- decay, not corrected for Λ branching ratio

Fig.10 z-coordinate of Ξ-decay vertices

Fig.11,12 $\cos\Theta^*$-distribution for $\Xi^- \to \Lambda$ decays and $\Lambda \to p$ decays

Fig.13,14 lifetime distributions for Ξ^- and Λ.

Fig. 16 Ξ^--rate as function of assumed $\langle p_T \rangle$. Best estimate indicated by the cross, surrounded by 1σ and 2σ confidence level contours.

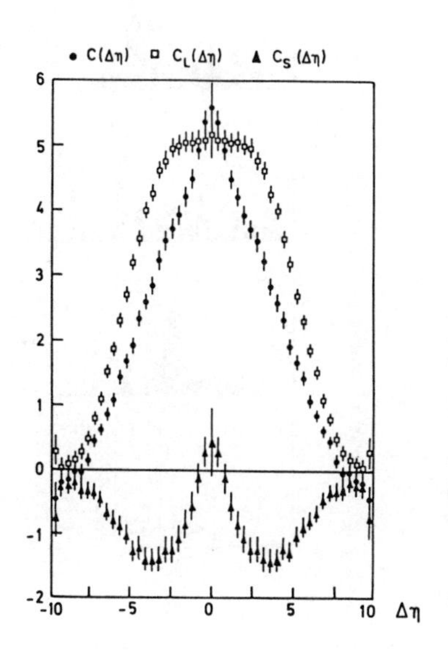

Fig.17 Inclusive charged particle correlation function $C(\Delta\eta)$ and its long and short range components, $C_L(\Delta\eta)$ and $C_S(\Delta\eta)$

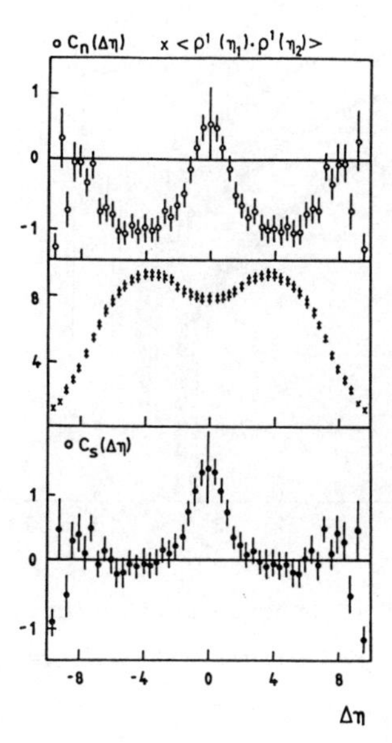

Fig. 18 Correlation function for observed multiplicity between 20 and 25, background term and short range term after back-ground subtraction, see text.

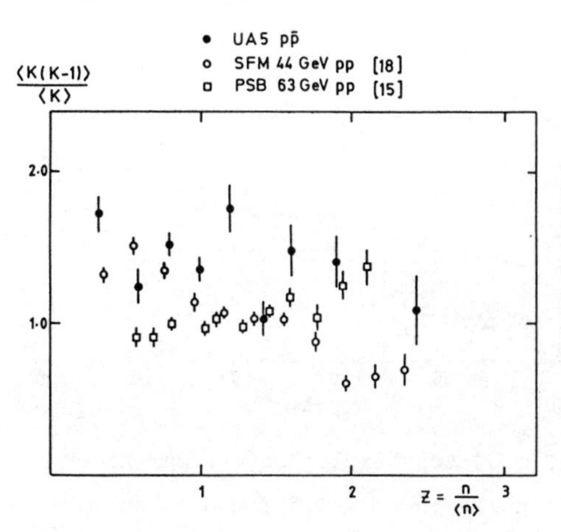

Fig. 19 Multiplicity dependence of $k_{eff} - 1$ (charged cluster decay size)

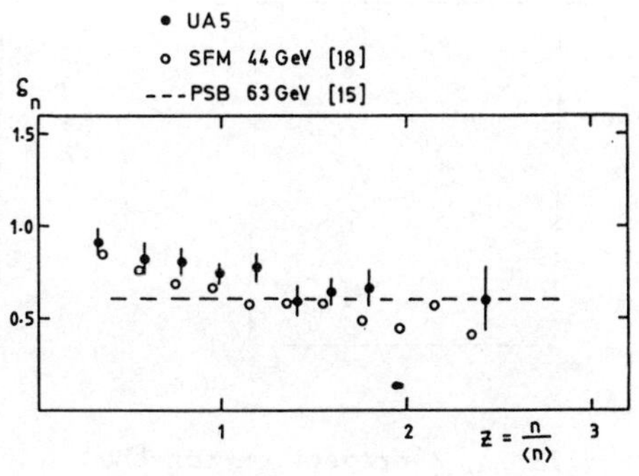

Fig. 20 Multiplicity dependence of the cluster decay width δ.

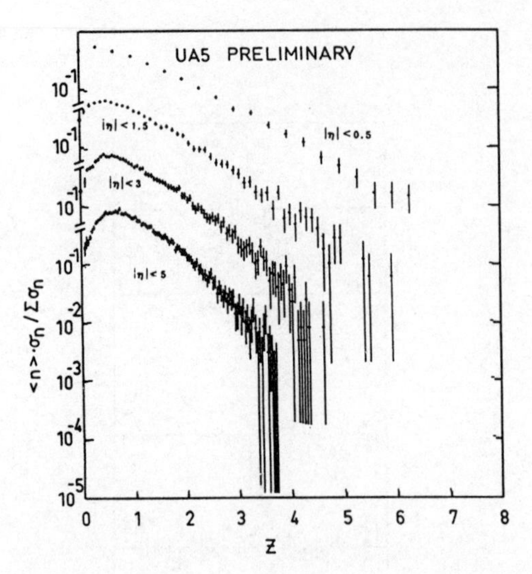

Fig.21 Corrected multiplicity distributions (non-single diffractive events), plotted as a function of $z = n_{ch}/<n_{ch}>$ for different pseudorapidity regions.

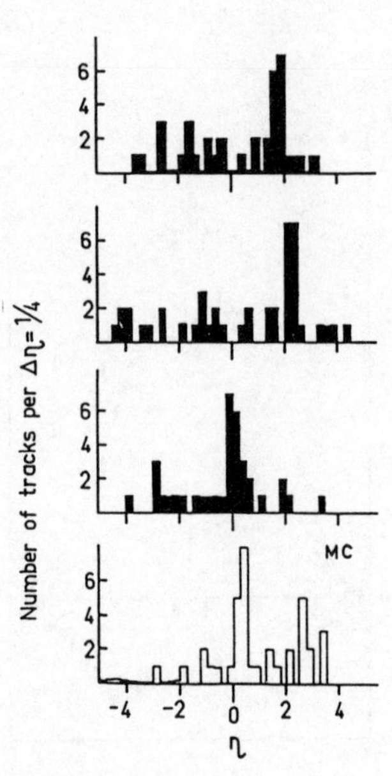

Fig.22 Track profile for three events (data), and one MC-event, bottom.

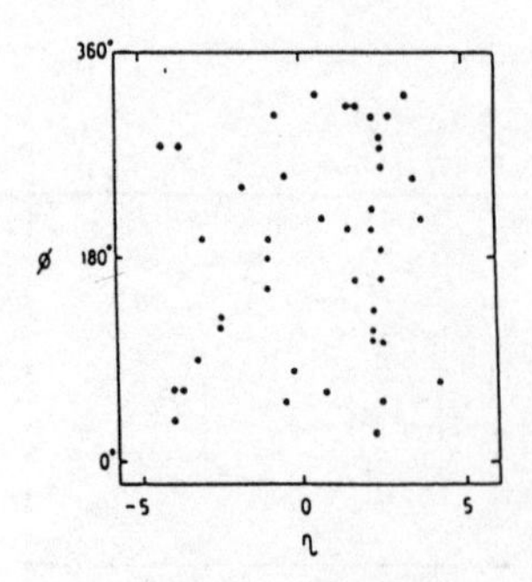

Fig.23 η, ϕ projection for the top event of fig. 22. ϕ is azimuthal angle around beam axis.

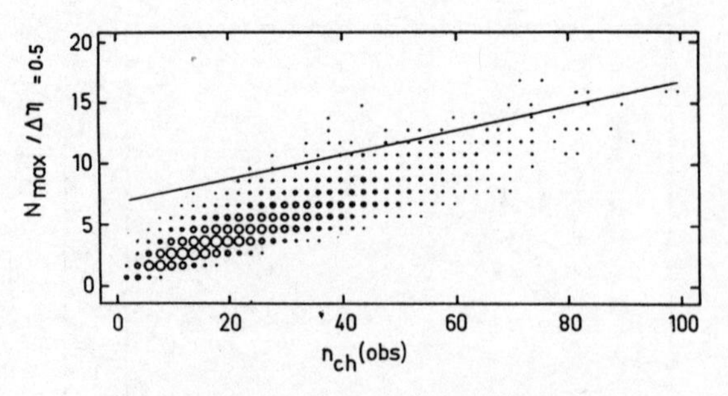

Fig.24 Scatterplot of maximum track density in Δη = 0.5 found in an event versus its total charged multiplicity. For the straight line, by which the "spike" sample is defined, see text.

Inst. Phys. Conf. Ser. No. 73: Section 6
Paper presented at VII Eur. Symp. Antiproton Interactions, Durham 1984

Proton–proton and proton–antiproton elastic scattering at high energy

A. Donnachie

Department of Physics, University of Manchester

1. Total Cross Sections and Small |t|

General theorems impose constraints on amplitudes and cross sections at asymptotically large energies, the best known of which is the Froissart bound

$$\sigma_T < C \log^2(s/s_o) \qquad (1.1)$$

where C and s_o are constants.

Because of this, σ_T is frequently parametrised by a sum of terms, none of which rises more rapidly than $\log^2 s$. In this type of parametrisation, it is assumed that the Froissart bound is already saturated, but this is not necessarily the case. Other possibilities must be considered and it is essential to look carefully at all aspects of the data before being tempted into simplifying (and possibly erroneous) assumptions. There are three possibilities to be considered.

(i) The Froissart bound is not saturated at all i.e.

$$\lim_{s\to\infty}(\sigma_T/\log^2 s) = 0 \qquad (1.2)$$

An interesting special case for this is the "critical Pomeron" (Moshe 1978; Baumel, Feingold and Moshe 1982) for which

$$\sigma_T \sim (\log s)^{0.3}$$
$$\sigma_{El}/\sigma_T \text{ decreases with increasing s} \qquad (1.3)$$

This is certainly disfavoured by present data. The cross section is increasing too fast with increasing s and σ_{El}/σ_T is also increasing, from a value of 0.185 ± 0.005 at ISR energies to 0.213 ± 0.008 at Collider energies (Braccini 1984). These results appear to put the "critical Pomeron" in a rather untenable situation.

(ii) The Froissart bound is saturated and we are already in the asymptotic regime, i.e.

$$\sigma_T \sim \log^2 s$$

$$\sigma_{El}/\sigma_T = \text{constant} \neq 0 \qquad (1.4)$$

$$\sigma_T/b = \text{constant}$$

where b is the slope of the diffraction peak at t = 0.

A particular example of this is provided by the geometrical scaling model (Dias de Deus and Kroll 1978, 1983). Neglecting crossing odd contributions the diffraction amplitude in this model may be written as

$$\text{Im } F(s,t) = \text{Im } F(s,0)\ \phi\ (\tau) \qquad (1.5)$$

$$\text{Re } F(s,t) = \text{Re } F(s,0)\, d(\tau\phi)/d\tau$$

where $\tau = -t\sigma_T$ and $\phi(\tau)$ is a universal scaling function. The differential cross section then has the form

$$\frac{d\sigma}{dt}(s,t) = \frac{\sigma_T^2(s)}{16\pi}\ \{\phi^2(\tau) + \rho^2(s)\}\ \left[\frac{d(\tau\phi)}{d\tau}\right]^2 \qquad (1.6)$$

where $\rho(s) = \text{Re } F(s,0)/\text{Im } F(s,0)$. Since ρ is known to be small eqns. (1.5) and (1.6) immediately give the results contained in eqns. (1.4b) and (1.4c). Since σ_{El}/σ_T is almost constant in the ISR range, it was assumed that the asymptotic regime is reached there and ϕ obtained by numerically fitting the data. This then gives a prediction at Collider energies.

The model can accommodate the change in σ_T between ISR and Collider energies, but as we have already seen σ_{El}/σ_T is increasing and, in addition, b shows no obvious $\log^2 s$ dependence in this energy range.

(iii) The Froissart bound is eventually saturated, but we are still far from the asymptotic regime.

This arises naturally in Regge theory. The cross section is dominated by Pomeron exchange, with the Pomeron trajectory $\alpha_P(t)$ having an intercept $\alpha_P(0)>1$. Double-Pomeron exchange becomes increasingly relevant in the ISR to Collider energy range, and at Collider energies the interference between single-Pomeron and double-Pomeron exchange is significant. There is no disagreement in this approach with any aspect of the data on σ_T or on $d\sigma/dt$ at small $|t|$ from ISR to Collider energies (Donnachie and Landshoff 1984 a,b).

Theoretical support for $\alpha_P(0) > 1$ is provided by the old calculation of Cheng and Wu (1970) who considered the scattering of quarks by the exchange of infinite "towers" in Abelian theory with massive gluons. In this model, single tower exchange yields fractional power behaviour for the total cross section. Summing s-channel iterations of single tower exchange yields the following three results. Firstly, at energies not too high the amplitude is well approximated by fractional power single tower exchange. Secondly, as the energy is increased, successively more and more towers become important. Thirdly, at sufficiently high energy, the Froissart bound is eventually obeyed (Cardy 1971,1974).

This suggests that at presently accessible energies, the amplitude should be expanded in the form

$$T = T_P + T_{PP} + T_{PPP} + \ldots\ldots \tag{1.7}$$

or more generally, including other exchanges $R(= \rho, \omega, f, A_2)$

$$T = T_P + T_R + T_{PP} + T_{PR} + T_{PPP} + \ldots \tag{1.8}$$

and the series should be truncated beyond the exchange of a suitable number of Pomeron/Reggeon exchanges.

2. Differential Cross Sections: $|t| < 1\ (\mathrm{GeV/c})^2$

Cheng and Wu (1970) found that the quark-quark scattering amplitude has the eikonal form

$$T(s,t) = is \int d^2b \exp(i\underset{\sim}{q}.\underset{\sim}{b}) \left[1 - \exp(i\chi(s,b))\right] \tag{2.1}$$

where $q^2 = -t$.

The possible role of non-Abelian theories in describing high energy hadron-hadron scattering was initally discussed by Low (1975) and Nussinov (1975, 1978). This is a dynamical model, in which the forward amplitude is dominated by the exchange of two coloured gluons. The amplitude for one gluon exchange is zero for colour-singlet hadrons and the amplitude for two gluon exchange is purely imaginary in the $s \to \infty$ fixed $|t|$ limit. Two gluon exchange thus provides a QCD model for the bare Pomeron. Gunion and Soper (1977) extended this model to describe hadron scattering in the eikonal approximation.

Thus although there is no formal justification, it is tempting to follow Chous and Yang (1979) and assume that the eikonal form is applicable to the pp amplitude. If this is done, and the second exponential in eqn. (1.9) expanded in powers of χ , then the first term can be equated with single-Pomeron exchange (thus specifying χ), the second term can be equated with double-Pomeron exchange and so on. The resulting double and triple-Pomeron terms appear to have the correct s- and t- dependence, but the normalisation is not that required by the data, and has to be reduced (Donnachie and Landshoff 1984b).

A comparatively simple model on these lines has been developed by Donnachie and Landshoff (1984a,b). They note first of all that quark additivity for total cross sections plus the dominance of helicity non-flip in forward elastic scattering, implies that the Pomeron couples primarily to a single valence quark with a more-or-less constant γ^μ coupling (Landshoff and Polkinghorne 1971, Jaroskiewicz and Landshoff 1974) but with the addition of a Regge signature factor which gives it even C-parity. When single Pomeron exchange is dominant, it gives for pp or $\bar{p}p$ elastic scattering,

$$\frac{d\sigma}{dt} = \frac{[3\beta F_1(t)]^4}{4\pi} \left(\frac{s}{m^2}\right)^{2\alpha_P(t)-2} \tag{2.2}$$

where β is the strength of the coupling of the Pomeron to a quark, and $F_1(t)$ is the Dirac form factor of the proton:

$$F_1(t) = \frac{(4m^2 - 2.79t)}{(4m^2 - t)} \frac{1}{(1 - t/0.71)^2} \tag{2.3}$$

Except that the Dirac form factor $F_1(t)$ enters rather than $G_M(t)$, this formula bears some resemblance to that of Chou and Yang (1979, 1983), but unlike them Donnachie and Landshoff (1984a,b) include in the amplitude the explicit energy dependence and phase required by Regge theory. While the latter is not relevant for the simple formula (2.2) it is of significance in the evaluation of the double-Pomeron amplitude and the interference between the single-Pomeron and double-Pomeron amplitude and between them and other contributions to the cross section. We will amplify this point in the next section.

Returning for the moment to single-Pomeron exchange, note that eqn. (2.2) can be modified to describe any Pomeron-dominated hadron-hadron elastic scattering. For πp, we have

$$\frac{d\sigma}{dt} = \frac{[3\beta F_1(t)]^2 [2\beta F_\pi(t)]^2}{4\pi} \left(\frac{s}{m^2}\right)^{2\alpha_P(t)-2} \tag{2.4}$$

and for pd

$$\frac{d\sigma}{dt} = \frac{[3\beta F_1(t)]^2 [6\beta A(t)]^2}{4\pi} \left(\frac{s}{m^2}\right)^{2\alpha_P(t)-2} \tag{2.5}$$

At small angles, the deuteron effectively has only one electromagnetic form factor A(t), which has been measured (Elias et al 1969). The pion form factor $F_\pi(t)$ is not well known, but it is traditional to assume that

$$F_\pi(t) = \frac{1}{(1-t/0.71)} \tag{2.6}$$

Fixing the Pomeron trajectory

$$\alpha_P(t) = 1 + \xi + \alpha_P' t \tag{2.7}$$

and β from pp data gives $\varepsilon = 0.08$, $\alpha_P' = 0.25$ and $\beta^2 = 3.21\ \mathrm{GeV}^{-2}$ in the single-Pomeron exchange approximation. Equations (2.4) and (2.5) then give an absolute prediction (shape and normalisation) of the πp and pd differential cross sections. The results are shown in Figs. 1 and 2. In Fig. 2, the predictions are the black points: the error bars on these points reflect the experimental errors in A(t).

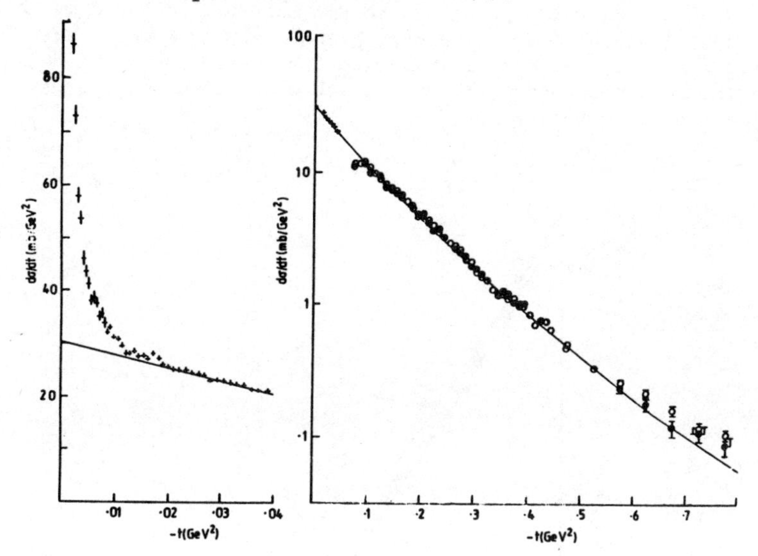

Figure 1. πp elastic scattering. The data are from Akerlof et al (1976) and Burq et al (1983).

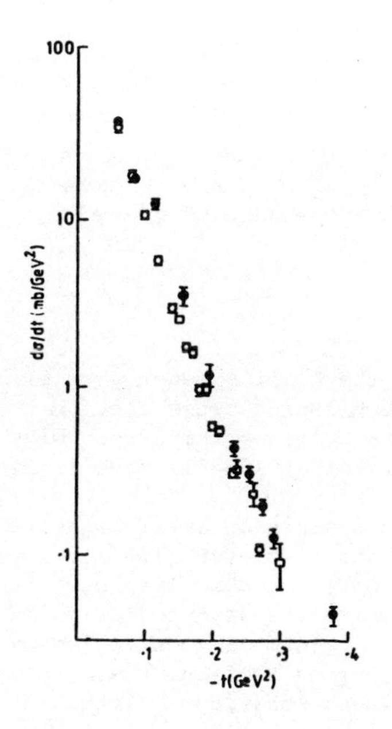

Figure 2. pd elastic scattering. The data (open symbols) are from Armitage et al (1978) and Goggi et al (1979).

For pp scattering it is necessary to include double-Pomeron exchange even at quite small t and particularly at the Collider energy. The double-Pomeron term is conventionally normalised by using it to cancel the imaginary part of the single-Pomeron term in the dip region at ISR energies. It then gives a negative contribution to the total cross section of about 7% at ISR energies, so β^2 must be increased by this amount, i.e. $\beta^2 = 3.43\ \text{GeV}^{-2}$. Figure 3 shows single-Pomeron exchange (broken line) and the effect of including double-Pomeron exchange (solid line) at the Collider energy, together with elastic scattering data from UA1 (Arnison et al 1983).

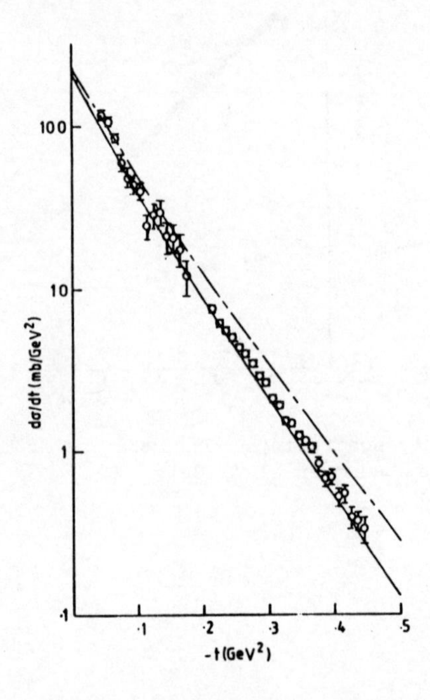

Figure 3. $\bar{p}p$ elastic scattering at the Collider (UA1, Arnison et al al (1983)). The broken curve is single-Pomeron exchange. The solid curve includes double-Pomeron exchange also.

This model predicts a total cross section at the Collider energy of 62.5 mb and $\rho = 0.12$, in excellent agreement with the measured cross section (Braccini 1984) of 61.9 ± 1.5 mb, assuming $\rho = 0.15$. A comparison with the data from UA4 (Battiston et al 1983, Bozzo et al 1984)is made in Fig. 4.

Of course this approach is not unique, and an example of an alternative viewpoint is provided by the model of Gauron and Nicolescu (1984). They assume that low energy phenomena ($5 < \sqrt{s} < 100$ GeV) are described by conventional Regge physics (poles and cuts) while the asymptotic regime is described by log s physics. This asymptotic regime is controlled by new entities, and in particular for the total cross section, they introduce the Froissaron which has specific $\log^2 s$ behaviour. Thus

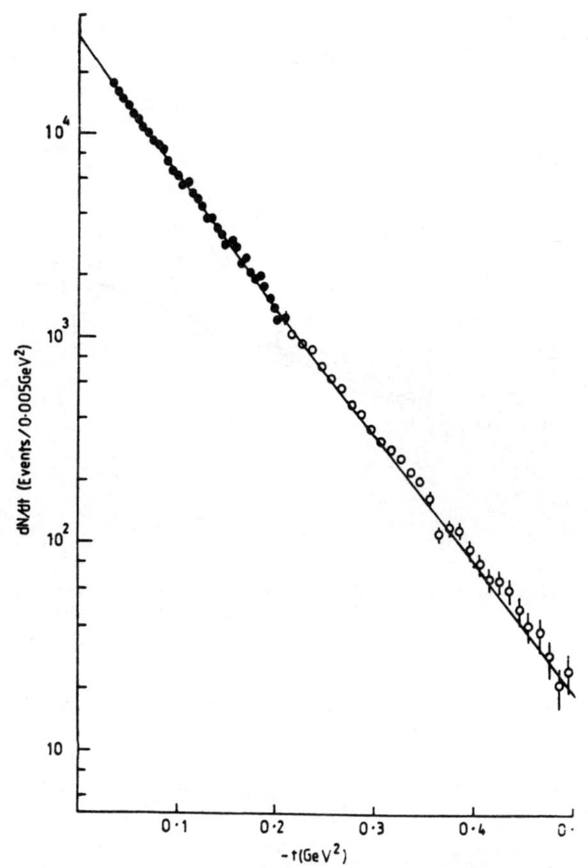

Figure 4. $\bar{p}p$ elastic scattering at the Collider (UA4, Battiston et al 1983, Bozzo et al 1984). The line is the calculation of Donnachie and Landshoff (1984b).

their amplitude, in contrast to a series of purely power terms as in eqns. (1.7) and (1.8) has the general form

$$T = T_P + T_R + C \log^2 s \qquad (2.8)$$

This model fits the t = 0 data rather well, and a typical example of their results are shown in Fig. 5. They do require the inclusion of the Froissaron to explain the data, but this is because they have omitted any contribution from the double-Pomeron. We know that this must be there because of the dip structure in pp scattering and as we have outlined above, the double-Pomeron contribution required for this is precisely that required to give the correct energy dependence of the t = 0 data.

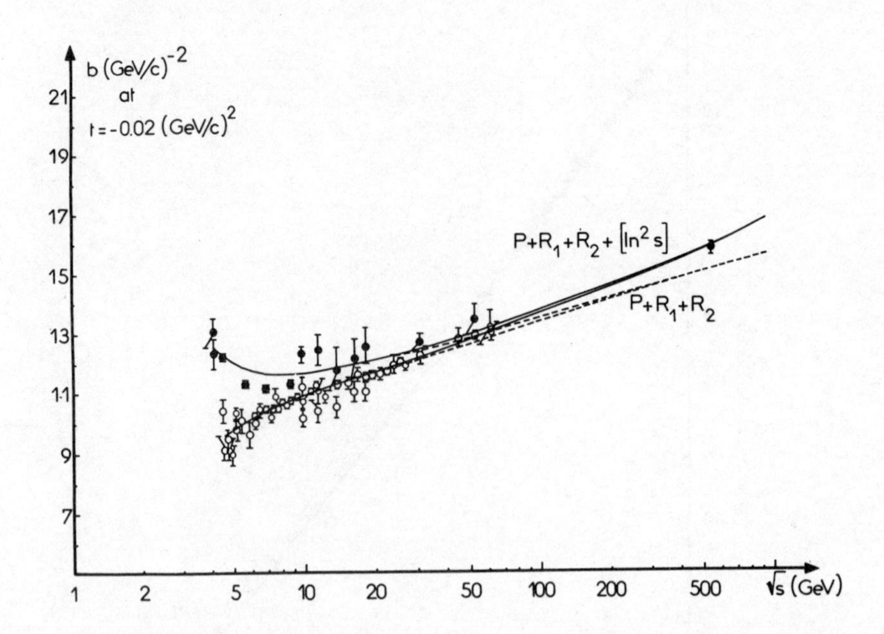

Figure 5. $\bar{p}p$ and pp slopes at $t = -0.02\ (GeV/c)^2$. The curves are the calculation of Gauron and Nicolescu (1984).

3. Differential Cross Sections: the Dip

The pronounced dip observed in pp elastic scattering at ISR energies arises naturally in the picture of single-Pomeron plus double-Pomeron interference. Both these amplitudes are predominantly imaginary and their imaginary parts have the opposite sign, so that the normalisation of the double-Pomeron term relative to the single-Pomeron can be specified by requiring their imaginary parts to cancel at the dip. Since the double-Pomeron term has less t-dependence but more energy dependence than the single-Pomeron, a consequence of this mechanism is that the dip position moves to smaller $|t|$ as the energy increases in agreement with experimental observation (Fig. 6). The curves are from the model of Donnachie and Landshoff (1984a).

Of course this simple picture is valid only at sufficiently high energy; at low energies the situation is much more complicated. We will not explore this in detail, but simply note that the behaviour of the pp differential cross section can be fully explained within the Regge pole plus Regge cut picture (as indeed can "low energy" $\bar{p}p$, πp and Kp data), by means of complicated interferences among the different terms. The results from a recent calculation (Collins and Kearney 1984) for pp and $\bar{p}p$ are shown in Figs. 7a and 7b. This complexity dies out with increasing energy and by ISR energies the dominance of single-Pomeron plus double-Pomeron up to and just beyond the dip in pp scattering is automatic.

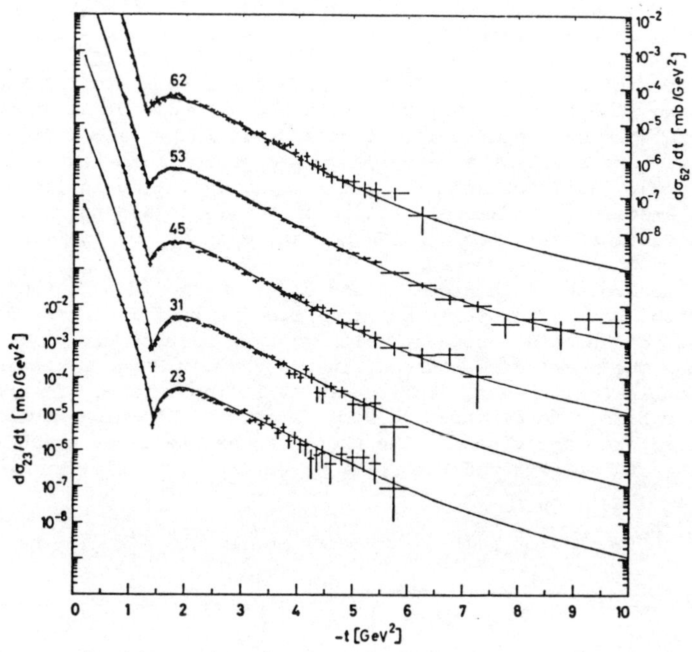

Figure 6. pp elastic scattering at ISR energies. The data are from Nagy (1979) and the curves are the calculation of Donnachie and Landshoff (1984a).

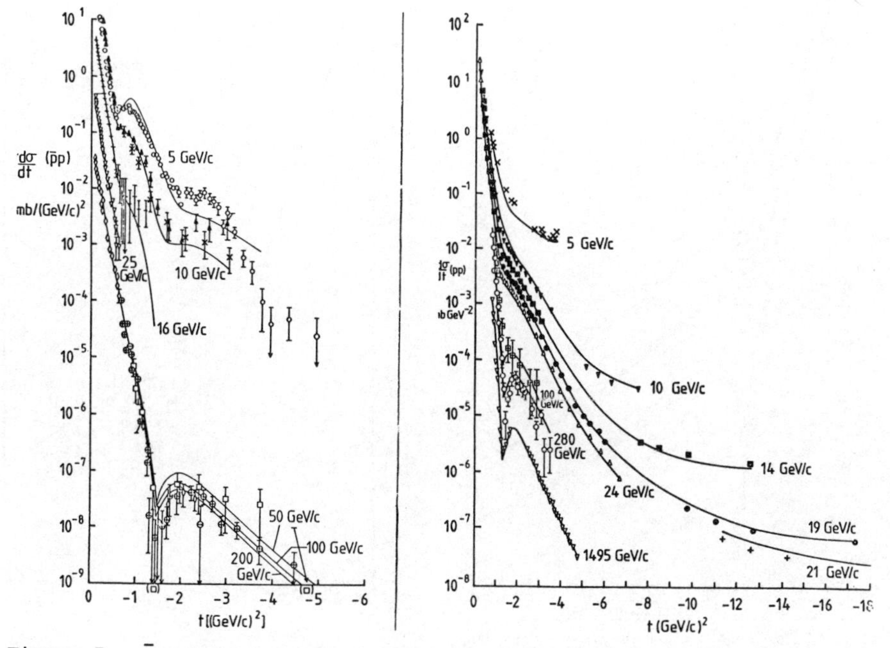

Figure 7. $\bar{p}p$ and pp elastic scattering from PS to SPS and ISR energies. The curves are from the calculation of Collins and Kearney (1984).

While the ISR pp data show this pronounced dip, the Collider $\bar{p}p$ data have no dip, but a shoulder instead. Does this invalidate our simple picture?

By itself it need not, as is illustrated by the calculation of Bourrely, Soffer and Wu (1984). They worked within the eikonal formalism of eqn. (2.1) and related the leading term in the expansion to a modified nucleon form factor: essentially the electromagnetic proton form factor multiplied by a slowly varying function. The eikonal was evaluated in its entirety, and the parameters in the model (six in all) were chosen to give the best possible simultaneous fit to the ISR pp data and the Collider $\bar{p}p$ data.

Their results are shown in Figs. 8a and 8b. The shoulder in the Collider data is reproduced rather well, but the price is to fill in the dip in the ISR pp data too quickly. However, although the results are qualitative and although the model does not contain the correct Regge phase in the Pomeron, double-Pomeron etc. amplitudes, it is indicative. It appears that it is possible to construct a Regge picture of Pomeron, double-Pomeron, etc. exchange based on the photon-Pomeron analogy which will describe the ISR pp data and the Collider $\bar{p}p$ data in isolation. Is this sufficient?

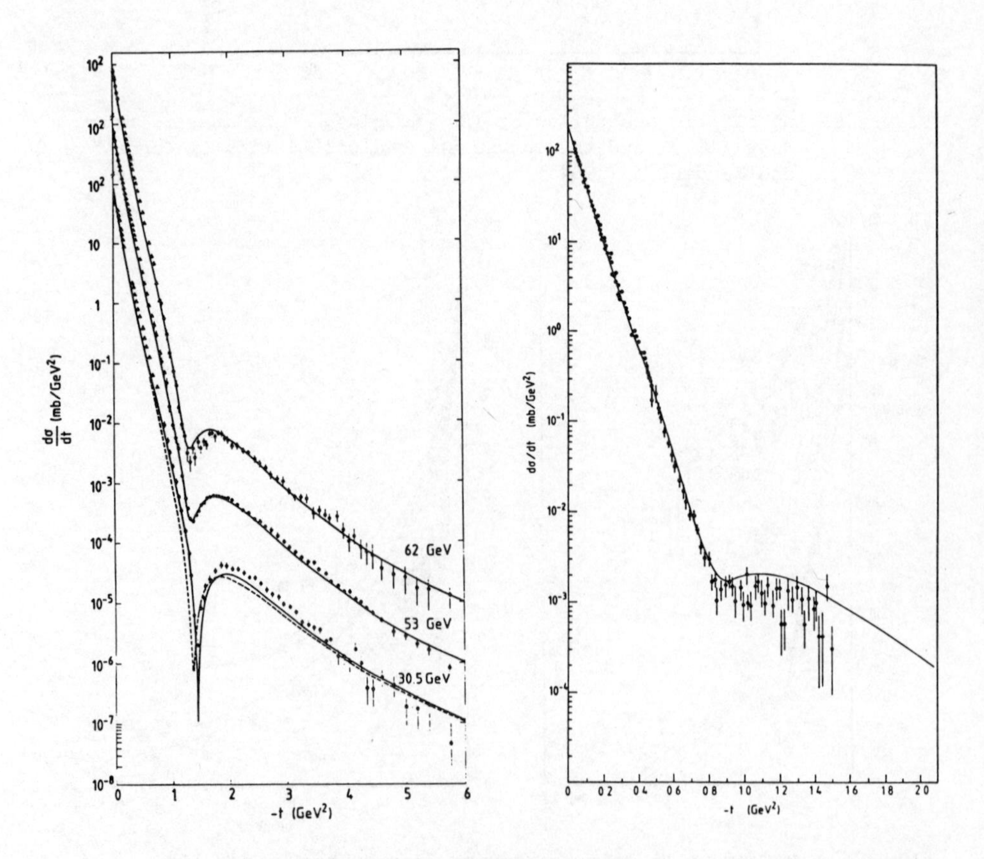

Figure 8. The calculation of Bourrely, Soffer and Wu (1984) compared with ISR pp data and Collider $\bar{p}p$ data.

The answer to this question lies in the behaviour of the $\bar{p}p$ cross section at ISR energies. The models discussed so far contain explicitly the requirement that at sufficiently high energy (i.e. top ISR energies and above)

$$\frac{d\sigma}{dt}(pp) = \frac{d\sigma}{dt}(\bar{p}p) \tag{3.1}$$

since the Pomeron (and hence the double-Pomeron etc) has even C-parity i.e. all the terms are unchanged when going from pp to $\bar{p}p$.

A crucial test of this is to compare pp and $\bar{p}p$ cross sections, at the same energy, to check for the presence of C = -1 exchange. A general discussion of "odderons" is given in the paper by Nicolescu (1984) and the discussion here will be restricted to the particular model of Donnachie and Landshoff (1984a).

4. Differential Cross Sections: Large |t|

At PS energies, the dimensional counting rule seems to apply to pp scattering. This rule predicts that the differential cross section for the reaction $AB \to CD$ at large $|t|$ is of the form

$$\frac{d\sigma}{dt} \sim s^{-n} f(\Theta)$$
$$n = n_A + n_B + n_C + n_D - 2 \tag{4.1}$$

where n_A, n_B, n_C, n_D are the total number of valence quarks in the four hadrons. For pp, n = 10 and this agrees well (Landshoff and Polkinghorne 1973) with pp elastic scattering data in the PS energy range.

The form of $f(\Theta)$ in equation (4.1) is such that at fixed t the differential cross section falls sharply with increasing energy. This fall slows abruptly through the SPS energy range and stops completely through the ISR energy range, where at fixed large $|t|$ the cross section is independent of energy (Fig. 9).

This rather sudden change suggests that a new mechanism begins to dominate for $t > 3\text{-}4\,(\mathrm{GeV}/c)^2$ and Donnachie and Landshoff (1979) have suggested that this is due to a triple gluon exchange mechanism. This model predicts that at large $|t|$

$$\frac{d\sigma}{dt} \text{ is energy independent} (\;|t| \text{ fixed})$$

$$\frac{d\sigma}{dt} \sim |t|^{-8} \quad \text{(energy fixed)}$$

We have seen in Fig. 9 that the first of these predictions is correct, and the second is also in remarkably good agreement with the data (Fig. 10) for $t > 3\text{-}4\,(\mathrm{Gev}/c)^2$.

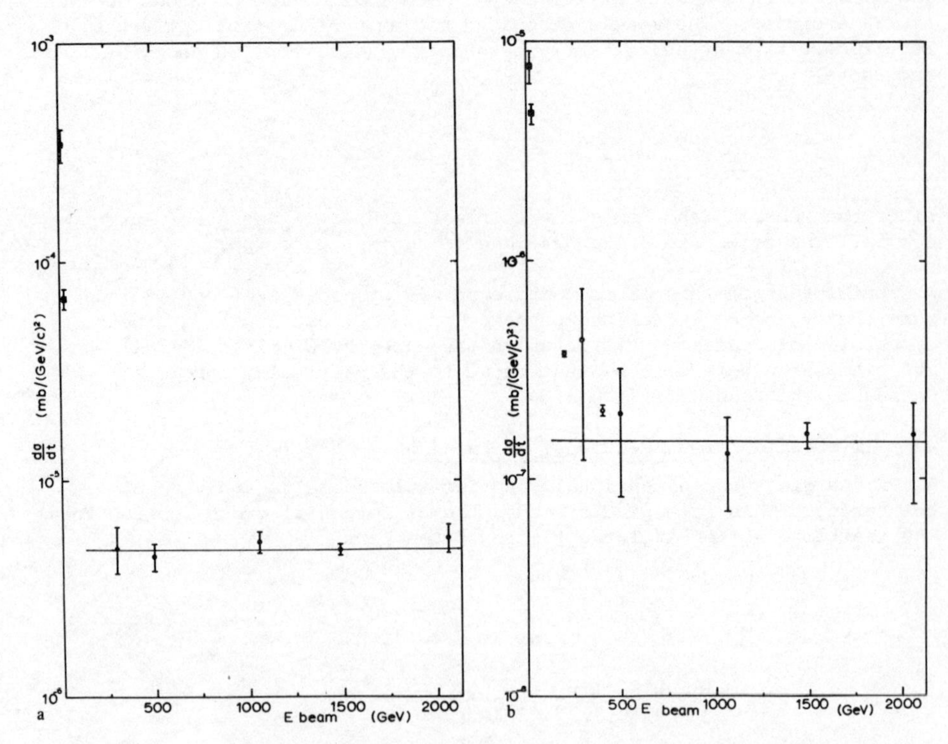

Figure 9. Energy dependence of the differential cross section for pp elastic scattering at $t = -3.5(GeV/c)^2$ and $t = -5.4 (GeV/c)^2$. The data are from Hartmann et al (1977), Conetti et al (1978), Nagy et al (1979) and Allaby et al (1973).

The triple gluon exchange is damped at small $|t|$, and the damping occurs over a distance comparable to the radius of the 3-quark system (Donnachie and Landshoff 1984a). This contribution is purely real and its sign in pp differs from that of the combined Pomeron plus double-Pomeron terms, partially cancelling them and helping to enhance the dip. However, since triple gluon exchange has C-parity = -1, the real parts add in $\bar{p}p$ with the result that there should be no dip in $\bar{p}p$ scattering at ISR energies.

The model naturally predicts that there should be no dip in $\bar{p}p$ scattering at Collider energies, in qualitative agreement with observation. However, the magnitude of the predicted cross section at the shoulder is much too low and it is likely that a significant part of this effect is due to the emergence of the triple-Pomeron. Since the triple-Pomeron is negligible at ISR energies, a crucial test of the C = -1 contribution in the dip region is a direct comparison of $\bar{p}p$ data there. Despite their limited statistics, the two ISR experiments with $\bar{p}p$ elastic data in this region can hopefully clarify this point.

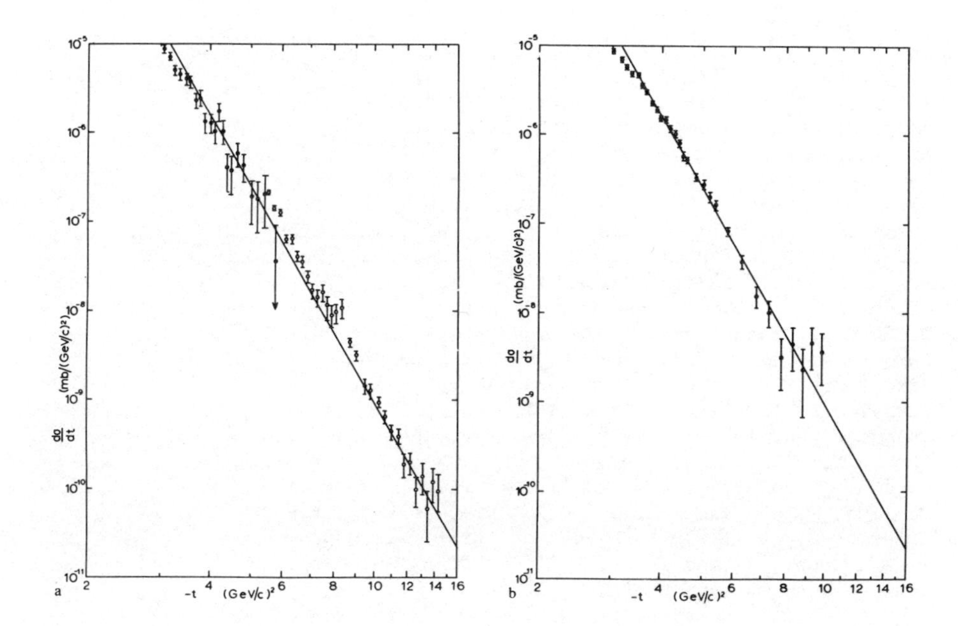

Figure 10. Differential cross section for pp elastic scattering at (a) 400 GeV (Conetti et al (1978), ⊙) and 494 GeV (Nagy et al (1979), ●) and at (b) 1482 GeV (Nagy et al (1979), ●). The line is $d\sigma/dt = 0.09\ t^{-8}$.

5. Conclusions

Almost all aspects of high energy pp and $\bar{p}p$ elastic scattering can be readily understood in terms of the conventional picture of Pomeron dominance, with double-Pomeron, triple-Pomeron etc. becoming increasingly important with increasing energy. The dip in the pp data at ISR energies is certainly due to the interference between the single Pomeron and double-Pomeron terms, and the shoulder in the $\bar{p}p$ data at Collider emergies can probably be most easily explained by the emergence of the triple-Pomeron. The forthcoming measurement of ρ by the UA4 Group at the Collider will provide an important additional constraint. There is certainly no need to invoke any new exchanges to explain the present data for $|t| > 2 - 3(\text{GeV/c})^2$.

The only additional requirement is for t > 3/4 $(\text{GeV/c})^2$, where the simplest explanation of the data is in terms of triple-gluon exchange. An open question is how this C = -1 exchange can be extrapolated to lower values of $|t|$ and how its intereference with the dominant C = +1 exchanges is manifested. How different is $\bar{p}p$ scattering from pp scattering in the dip region at ISR energies? The results of the ISR experiments are eagerly awaited.

References

Akerlof, C.W. et al 1976, Phys. Rev. D14, 2864.
Allaby, J. et al 1973, Nucl. Phys. B52, 316.
Amos, N. et al 1983a, Phys. Lett. 120B, 460.
Amos, N. et al 1983b, Phys. Lett. 128B, 343.
Armitage, J.C.M. et al 1978, Nucl. Phys. B132, 365.
Arnison, G. et al 1983, Phys. Lett. 128B, 336.
Battison, R. et al 1983, Phys. Lett. 127B, 472.
Baumel, J. Feingold, M. and Moshe, M. 1982, Nucl. Phys. B198, 13.
Bourrely, C. Soffer, J. and Wu, T.T. 1984, Nucl. Phys. (in press).
Bozzo, M. et al 1984, CERN/EP 84-90.
Braccini, P.L. 1984 These proceedings.
Brodsky, S.J. and Farrar, G.R. 1973, Phys. Lett. 31, 1153.
Burq, J.P. et al 1983, Nucl. Phys. B217, 285.
Cardy, J.L. 1971, Nucl. Phys. B28, 455.
Cardy, J.L. 1974, Nucl. Phys. B75, 413.
Cheng, H. and Wu, T.T. 1970, Phys. Rev. Lett. 24, 1456.
Chou, T.T. and Yang, C.N. 1979, Phys. Rev. D19, 3268.
Chou, T.T. and Yang, C.N. 1983, Phys. Lett. 123B, 457.
Collins, P.D.B. and Kearney, P.J. 1984, Zeits, fur Phys. C22, 277.
Conetti, S. et al 1978, Phys. Rev. Lett. 41, 924.
Dias de Deus, J. and Kroll, P. 1978, Acta Phys. Pol. B9, 159.
Dias de Deus, J. and Kroll, P. 1983, J. Phys. G9, L81.
Donnachie, A. and Landshoff, P.V. 1979, Zeits. fur Phys. C2, 55.
Donnachie, A. and Landshoff, P.V. 1984a, Nucl. Phys. B231, 189.
Donnachie, A. and Landshoff, P.V. 1984b, Nucl. Phys. (in press).
Elias, J.E. et al 1969, Phys. Rev. 177, 2083.
Gauron, P. and Nicolescu, B. 1984, Phys. Lett. 143B, 253.
Goggi, G. et al 1979, Nucl. Phys. B149, 387.
Gunion, J.F. and Soper, D.E. 1977, Phys. Rev. D15, 2617.
Hartmann, J.L. et al 1977, Phys. Rev. Lett. 39, 975.
Jaroskiewicz, G.A. and Landshoff, P.V. 1974, Phys. Rev. D10, 170.
Landshoff, P.V. and Polkinghorne, J.C. 1971, Nucl. Phys. B32, 541.
Landshoff, P.V. and Polkinghorne, J.C. 1973, Phys. Lett. 44B, 293.
Low, F.E. 1975, Phys. Rev. D12, 163.
Matveev, V.A. Muradyan, R.M. and Tavkelidze, A.N. 1973, Lett. Nuovo Cimento, 7, 719.
Moshe, M. 1978, Phys. Rep. 37C, 255.
Nagy, E. et al (1979), Nucl. Phys. B150, 221.
Nicolescu, B. 1984, These proceedings
Nussinov, S. 1975, Phys. Rev. Lett. 34, 1286.
Nussinov, S. 1978, Phys. Rev. D14, 246.
Sivers, D.I. Brodsky, S.J. and Blankenbecler, R. 1978, Phys. Rep. 23C, 1.

Inst. Phys. Conf. Ser. No. 73: Section 6
Paper presented at VII Eur. Symp. Antiproton Interactions, Durham 1984

The Lund string model

Bo Andersson

Department of Theoretical Physics
University of Lund, Sweden

Abstract: I describe some of the main features of the Lund String Fragmentation and touch at some of the developments of the model, in particular the possibility to derive a nonperturbative probability for the production of multijets in hard events.

1. Introduction

Ten or even five years ago almost everybody in high energy physics believed that we only needed to go to sufficiently large energies to be able to disentangle partonic features. We must now, however, agree upon that the hadronization process is a major disturbance in connection with all tests of basic QCD structures.

An example of this is that despite the very difficult and hard work performed over several years at the e^+e^- annihilation machines at DESY and SLAC, we only know the basic coupling-constant α_s to be between say 0.14 and 0.24. The error-bars on the experimentally determined distributions are much smaller, of the order of 5-10 % in general, and there are lots of tests. Due to strongly model-dependent features of the comparisons between the theoretical calculations and the experimental observations, it is fundamentally impossible to do better at least if you accept all presently proposed models.

Another example is the observation of the large x-sections for very high transverse energy jets at the S$p\bar{p}$S-Collider at CERN. Theorists can deduce the cross-section qualitatively, i.e. to within a factor of two to three theoretically calculated curves are in agreement with the observations. My own feeling is that until we have mastered the problem of understanding the transition between the low p_T (major) event structures and the supposedly mainly gluonic jets at the collider, we will hardly be able to do much better. Both in this case and in e.g. the calculation of the properties of T-decay into hadrons there is a fundamental difficulty again. Now it is related to whether a partonic state is characterized solely by the energy momenta and quantum numbers of the partons or whether there are additional topological quantum-numbers, i.e. how the partons are "connected" by the colour force-fields. I will come back to that at the end of the lecture.

I feel that there are at the moment essentially only two reasonably consistent schemes for hadronization. I am however perfectly willing to admit that also these schemes need much more understanding before they can be taken as more than phenomenological rules of the thumb. One of them was presented by Bryan Webber the other day and is based upon a clever use of perturbative QCD jet-calculus. The other one I will talk about from now on [1].

The main assumption of the Lund Model is that the final state hadrons in a "hard" event stem from the breakups of the force-fields between the partons. The model also includes low p_T hadronic reactions but in that case the dynamics at present has the character of a cook-book recipe and I will not dwell upon the details.

As a model for the force-fields the Lund model uses the massless relativistic string which for many purposes can be taken as a good old rubber band. Its relativistic and causal properties make it into a theoretical laboratory which is often full of surprises however. For simplicity I will in general assume that the string state is produced instantaneously and stretches out in its "cold" transverse groundstate, but I will comment upon these assumptions at the end. The energy-density in that field or the stringtension κ will be taken as $\kappa \simeq 1$ GeV/fm, which is in accordance with bag-models, Regge-slopes and p_T-width etc. This is commonly the main energy-scale of hadronization.

From a general point of view the scheme discussed by Bryan Webber and the one I am talking about are probably only two different ways of implementing the same dynamics and it seems presently that almost all features (maybe with the exception of baryon production and polarization) that come out of one of them also is around in the other.

A general word of warning may, however, be in place at this point. As long as only the properties of "all particles", "all charged particles" or "all pions" in hadronic events are considered, it is seldom necessary to go outside the time-honoured incoherent longitudinal phase-space dynamics to describe the data. Features of the fragmentation dynamics which contain more information and necessitate a more sophisticated description are e.g.

1. Strangeness-antistrangeness, baryon-antibaryon and heavy quark production (in particular correlation studies). Because strange particles and baryons are more rare, such pairs are often correlated and can exhibit the local properties of the production process. Also the larger masses imply that the particles carry more information on the force fields between the partons.
2. Vector meson production studies, which more directly exhibit the quantum number and energy-momentum flow of the original partons than their final decay products.
3. Transverse momentum correlations, in particular in connection with heavy particles, which should exhibit degrees of freedom in the force fields like soft gluon emission etc.
4. Polarization properties, which exhibit further degrees of freedom than the purely kinematical energy momentum flows.

2. A Simple Picture in Space-time of the Structure of Hadronization

The Lund model has provided a simple space-time structure for the hadronization process [2]. In order to understand this structure we consider fig. 1, which describes the motion of a massless $q\bar{q}$-pair in space-time when the force between them is a constant. In the cms (fig. 1a), the pair starts to move apart along opposite light-cones, thereby building up a force-field (the hatched area) in between them. When all their energy-momentum is used up they turn around and the motion continues in a periodic way which has been termed "the yo-yo-mode". In a moving frame the corresponding motion is again along opposite lightcones (fig. 1b), but as they no longer have the same energy-momentum, they are stopped and turned around at different times. In that way the whole system moves in space-time.

The system will exhibit Lorentz-contraction in space-size and time-dilation of the period of motion in such a way that the space-time surface spanned during one period is invariant. It is easily seen that this surface is proportional to the mass-square of the system. In case the $q\bar{q}$-pair is massive, the corresponding motion will be along hyperbolas with the same lightcones as asymptotes.

In fig. 2 the motion of a very energetic pair is depicted. Then a large amount of energy is stored in the force-field between the pair and new $q\bar{q}$-pairs can be produced from this energy along the field. We note that the field vanishes between the produced pairs as they are dragged apart. Later on new pairs are produced and the whole final state appears as a set of small yo-yo-mesons going apart in space-time. If we look upon an enlargement of the process in figs. 3, we note that while in one system (fig. 3a) the vertex number 1 seems to be the first in time and the vertex 2 comes later, this time-ordering will change in another frame (fig. 3b). We conclude that it is always the slowest yo-yo-mesons which are the first to disentangle in any frame and that all vertices are equivalent. You may start wherever you like and "peel off" one meson at a time either along the positive or along the negative lightcone.

There are evidently many different ways "to order" such a process. Although timeordering is frame-dependent as noted above, one may of course choose one particular frame more or less arbitrarily. A useful way is to order the vertices along the lightcone so that the first (rank) meson contains the original q-particle q_o and the $\bar{q}_1$, from "the last vertex" along the lightcone, the second rank meson is composed of $q_1\bar{q}_2$ etc. It is then useful to define the fractional energy-momentum Z_1 of the first rank meson with respect to the original q_o-energy-momentum

$$Z_1 = \frac{(E+p)_{meson}}{(E+p)_{q_0}} \tag{1}$$

which is a Lorentz-invariant. Noting that q_o loses energy and momentum to the string at the rate K in any frame it means that q_o will be able to go the distance ℓ along the lightcone (fig. 4) such that

$$(E+p)_{q_0} = K \cdot \ell \tag{2}$$

Then the first rank meson is produced out of a fraction $Z\ell$ of the force-field. The corresponding (E-p)-component is given by the size of the meson (fig. 4)

$$(E-p)_{meson} = \frac{m^2}{Z_1 \cdot K \cdot \ell} \tag{3}$$

Consequently it is sufficient to give the distribution in Z_1, Z_2 etc to describe the whole process in this one space - one time dimensional model.

Evidently we could just as well focus on the antiquark "end" of the jet and define Z'_1, Z'_2 etc for the mesons produced over there.

An immediate question is whether there is any classical stochastical process such that the distribution in Z_1, Z_2 etc is compatible with the corresponding distribution Z'_1, Z'_2 etc. It turns out that there is only a unique process which is consistent and the fragmentation function [3] is for each "step"

$$f(z)dz = N \frac{dz}{z}(1-z)^{\alpha} \exp -\left(\beta \frac{m^2}{z}\right) \tag{4}$$

It is worthwhile noting that

1. The distribution depends upon the mesonmass m in such a way that heavier mesons have a strong peaking towards large Z-values (the mean Z-value $\langle Z \rangle$ is $\langle Z \rangle \sim 1 - \frac{\alpha+1}{\beta m^2}$).
2. The parameters α and β are related to the total meson multiplicity and the correlation length in rapidity between neighbouring mesons. Theoretically one can give an interpretation of α in terms of Reggeslopes and β in terms of the Wilson area law in QCD [3].

Although we have up to now considered only a single-space-dimension the process as such can be extended to several dimensions as a process on the string space-time history. Recently we have found that the process not only is able to give a description of the breaking into hadrons of a given string-state, i.e. a given configuration of colour-connected partons. It can also be extended to give the relative probability for different partonic states. The result is that if we define the sum of the probabilities for all hadronic states that can be produced from the given string state weighted with the phasespace then we obtain a result which agrees with the perturbative QCD-results wherever perturbative QCD is valid. But this result goes much further, because we have in that way a formula which gives a result valid everywhere, i.e. in a unique way it interpolates between the regions of phasespace where perturbative QCD is valid.

3. A model for the production matrix-element - the tunnelling process

In connection with the production of massive $q\bar{q}$-pairs (mass μ) we note that although the motion in the yo-yo states has a similar classical interpretation as for massless pairs, the pair cannot in a constant force-field be produced in a single space-time point. Classically they are produced a distance 2ℓ apart such that

$$2\mu = 2\kappa \cdot \ell \tag{5}$$

There is actually a simple time-honoured dynamical process available [4,5,6]. It turns out that in the presence of a constant electro-(magnetic) field the no-particle state is unstable in QED and its rate of decay may be interpreted as the production rate of particle-antiparticle states which are coupled to the field.

A simpler but equivalent formulation which is more useful for our purposes is to consider the production as a tunnelling process in a linear potential. With a knowledge of the wavefunction ψ_c of the particle at the classical turningpoint ($x=+\ell$) (and the wavefunction $\bar{\psi}_c$ for antiparticle similarly at ($x=-\ell$) we may interpolate the wavefunctions into the classically forbidden regions $|x|<\ell$, e.g. by WKB-methods. More precisely, with a knowledge of the wavefunctions along the classical space-time orbits in terms of given wave-packages it is possible to compute the overlap-integral of the wavefunctions [7]. We note in particular that the main contribution stems from a small surrounding of the origin ("the production vertex of the virtual pair") [7] and that we have in a linear potential $V=\kappa x$ with $E=0$:

$$\psi(0) \simeq \psi_c \exp i \int_{x_c}^{0} p\,dx = \psi_c \exp - \int_0^{x_c} dx \sqrt{\mu^2-(\kappa x)^2} = \psi_c \exp - \frac{\mu^2 \pi}{4\kappa} \tag{6}$$

(Note that in a linear potential the energy-eigenvalue is related to the gauge-choice, i.e. the zero in the potential.) This result is inherently non-perturbative - an expansion in κ will have vanishing coefficients to each order. The production probability of the pair is then

$$|\psi(0)\bar{\psi}(0)|^2 \propto \exp - \frac{\mu^2 \pi}{\kappa} \tag{7}$$

and we note that with the value of κ determined above we obtain a relative rate of

$$u:d:s:c \simeq 1:1:1/3:10^{-11} \tag{8}$$

Consequently charm-pairs are never produced in a <u>soft</u> hadronization process, while strangeness will occur on the level expected from ordinary SU(3)-breaking.

In the model in 3 space-dimensions we obtain in a force-field without transverse excitations ("uniform") that transverse momentum is locally conserved and that the tunnelling mechanism may again be used [7]. If the quark and antiquark each takes $\vec{k}_T$ and $-\vec{k}_T$ respectively, we obtain the production probability

$$dP \propto d^2k_\perp \exp - \left(\frac{\mu_\perp^2 \pi}{\kappa}\right)$$
$$\mu_\perp = \sqrt{\mu^2 + k_\perp^2} \tag{9}$$

in such a way that if this is the only source of transverse momentum the final state mesons should have a gaussian p_T-spectrum with

$$\langle p_\perp^2 \rangle \simeq (0.35)^2 \, (\mathrm{GeV}/c)^2 \tag{10}$$

The tunnelling process applied to spin ½-particles has some very interesting (and observable) consequencies for polarization. To see the main properties we again consider a semi-classical picture (fig. 5) [8]. In case the field vanishes in between the pair during the production process, then we note that the produced pair has obtained an orbital angular momentum $\vec{L}$ along the direction

$$\vec{L} \simeq \vec{e} \times \vec{k}_{\perp} \tag{11}$$

(with $\vec{e}$ the direction of the force-field, i.e. the direction along which the particles are dragged). The size of $|\vec{L}|$ can be estimated using equation (5) to be

$$|\vec{L}| \simeq \frac{2\mu_{\perp} k_{\perp}}{\kappa} \simeq 1 \tag{12}$$

for $k_T \simeq 0.3$ GeV/c. In a force-field without local concentrations of in particular angular momentum the total angular momentum should be conserved

$$\vec{J} = \vec{L} + \vec{S} \tag{13}$$

We therefore expect the produced particles to be polarized along

$$\vec{k}_{\perp} \times \vec{e} \tag{14}$$

as soon as $|\vec{k}_T|$ is of the order of the mean $\langle k_T \rangle$. The key features in these considerations are the uniformity of the field and its vanishing in between the produced pair.

It is an interesting and very helpful thing that the Λ-particle exhibits its polarization in its decay distribution and that an eventual Λ-polarization is carried entirely by the s-quark. In SU(6)-language Λ is a $(ud)_o s$ state with $(ud)_o$ a diquark (colourantitriplet) with spin and isospin 0. Our considerations above predict both the size and the sign of the Λ-polarization for large X_F-values in a proton fragmentation region where the main Λ-production mechanism is that an ss-pair is produced and the s-quark combines with the already existing $(ud)_o$-diquark. Actually the results agree almost exactly with the observed experimental Λ-polarization [8] for the ISR experiments with large X_F [9]. For the Σ^o and Σ^+ polarizations we would obtain the opposite sign and roughly about 1/3 of the size. We note that for smaller X_F-values there are also other possible production mechanisms for the Λ. In particular many such Λ-particles should stem from Y*-decays. Therefore the Λ-polarization should be essentially smaller for smaller X_F-values (as observed by the NAL hyperon beam experiments).

4. A Model for Gluon Emission and Some Dynamical Consequences

Up to now we have only considered the stretching and breaking-up of a forcefield between triplet(3) and antitriplet($\bar{3}$) charges. A generalization of the notion of a constant forcefield of the kind we have used is the relativistic massless string (RMS) with a 3 and $\bar{3}$ at the end-points. Just as for an ordinary classical rubberband it is possible to consider local excitations on the RMS and the Lund group has in refs [10] proposed such "kink"-solutions of the string equations as models for gluons. The gluon kink moves with the velocity of light like a massless particle.

In fig. 6 we describe the motion of a ($q\bar{q}g$) state starting at the space-time point 0 and going apart thereby stretching the string from the 3 via the 8 to the $\bar{3}$. This means in particular that the gluon is taken as an excitation of the force-field in such a way that the force-field becomes "bent" by the gluon motion. ("A confined gluon does not get away." "There is no such thing as a free gluon.")

This gluon model has the particular property that one may use the same rules as we have used before for the breakup of each string-piece in its own restframe.

In fig. 6 we depict the development of the $q\bar{q}g$-state, how the energy is transferred to the outward-moving string pieces and how the string breaks up by repeated ($q\bar{q}$) production and yo-yo-meson-breakoffs. We note as particular features of the process that

1. most of the meson states are produced in motion along one of the three directions of the original partons (note the Lorentz contraction of the string size along the direction of motion);
2. there are final state meson states produced only in the angular segments between the $\bar{q}g$- and the gq-directions but as there is no string connecting the $q\bar{q}$ directly, no states are produced in that angular segment. (For the observable consequences of this feature, cf below.)

At this point it is useful to consider "the gluon disturbance length" in rapidity, i.e. inside which rapidity-range (due to the relation pseudo-rapidity-rapidity this also means a corresponding angular range) the properties of a gluon emission are noticeable. We note that the force-field is bent and dragged outwards by the gluon in such a way that the gluon momentum transverse to the general $q\bar{q}$-direction is transferred to the moving string. In particular the amount of this momentum which is carried by a particular string-piece is proportional to its size [10]. A meson moving with an energy E will appear along the direction transverse to the gluon emission as Lorentz contracted, i.e. with a size proportional to $\frac{m}{E}$ where m is the meson mass. Consequently the amount of gluon momentum transferred to this meson will be proportional to

$$\ell_0 \frac{m}{E} = \ell_0 \frac{1}{\cosh(y-y_g)} \tag{15}$$

where $y-y_g$ is the (pseudo-)rapidity difference between the meson and the gluon emission direction and ℓ_0 is the rest-size of the meson yo-yo.

We conclude from these simple considerations, i.e. basically relativistic kinematics that the gluon will be noticeable inside a (pseudo-)rapidity range of about ± 1 unit in rapidity around the gluon emission direction. This range is <u>independent</u> of the gluon energy. The larger the gluon energy the more particles are produced however and the more energy and momentum given to each particle with a peak along the gluon direction.

For a very large gluon energy it is possible to construct a gluon fragmentation distribution into mesons. We note that at present e^+e^--annihilation energies there is little sensitivity to fractional z-values smaller than 0.1-0.2, i.e. the large "central" rise of the gluon fragmentation is not noticeable.

There is in the model a natural partition into hard gluon emission and soft and collinear gluon emission. Hard gluon emission corresponds to situations such that the invariant mass available in the string-fields around the gluon is sufficiently close to e.g. the quark, then the force-field in between them is too small (i.e. contains too little energy) to produce particles. In that case (collinear gluons) the "first" break may occur on "the other side" of the gluon and both partons end up in the same final state meson (which therefore obtains energies larger than each of the partons). This provides in the Lund model a natural cutoff for collinear gluons. It turns out that the mass-scale M_o necessary for noticeable gluon-effects is larger than 2 GeV. Actually, choices of M_o between 2 and 4 GeV do not influence the observable final state hadron distributions. It only corresponds to a theoretical reshuffling between "2-jet events" and "3-jet events".

The emission of a gluon with some momentum transverse to the general $q\bar{q}$-direction evidently implies a recoil-effect for the q and $\bar{q}$. This recoil effect will be distributed among the fast final state particles along the q (and $\bar{q}$) directions in accordance with their fractional energy-momentum. The gluon transverse momentum is distributed in accordance with eq. (15) and it is then easy to see that (sufficiently) collinear gluons do not give any transverse momentum effects. Actually, a close study has shown [11] that gluon emission inside "the quark fragmentation region" (remember the 2 units in (pseudo-)rapidity) does not have any noticeable p_T-effects. Gluons emitted outside that region do have p_T-effects even if the gluon transverse momentum is smaller than the cutoff mass M_o, discussed above and we term such gluons "soft gluons".

We feel that this property of "infrared stability" which implies a "soft" turning-over between 3-jet and 2-jet events, 4-jet and 3-jet events etc is an essential and necessary feature of any hadronization model. It should be remembered that while in QED soft photon-emission is in principle always detectable (by a sufficiently fine-grained and distant detector) a confined QCD-gluon detector must always be related to the final state hadron-observables. As the density of such hadrons in phase-space in general is low and there is a nonvanishing mass-scale related to it we feel that treatment of gluon-emission should be rather different in the two theories. The gluon emission "degrees" of freedom in the Lund model have a set of observable features:

A. Hard gluon emission evidently distorts the whole event and may be experimentally disentangled e.g. by thrust-cuts or other event-topological measures. It occurs in principle on the scale of α_s which for Petra energies implies the order of 10-20 %.
B. Collinear gluon emission has no observable transverse momentum implications but "disturbs" the emission of fast (large z-)particles in the iterative process. The reason is that in the Lund model large z-particles are produced by a fast breaking of the force-field. The emission of a collinear gluon implies some concentration of energy in the gluon, which until it is delivered to the field is not available for particle production. Calculation of this effect leads to results similar to the ones obtained in jet-calculus, i.e. P is modified from

the constant 1 to

$$P \approx (1+\beta)(1-z)^{\beta} \tag{16}$$

with β a number determined by α_s and the particle masses.

C. Soft central gluon emission will on the one hand lead to small "p_T-bends" in the event-structure for centrally produced particles, on the other hand to recoil-effects in the q- and ($\bar{q}$-) fragmentation regions. These effects can be calculated [11] in case the gluon emission process can be taken as an independent stochastical (Poissonian) process [12]. The result is process-dependent and as examples I would like to mention:

C1: in e^+e^--annihilation, the handling of the events tends to minimize the effects because one always tries to define a thrust-axis with regard to which particles have as small a p_T as possible. In this case there is a general p_T-broadening such that the mean $\langle p_T \rangle$ from eq. (10) is increased to

$$\langle p_\perp^2 \rangle \approx [0.43 - 0.45]^2 (\text{GeV/c})^2 \tag{17}$$

C2: in leptoproduction, there is a given axis and one of the q or $\bar{q}$-constituents is marked out by the hard process in such a way that gluon emission can be related to this parton. Then there is both a p_T-recoil and an expected (at present energies) noticeable p_T-compensation in the center.

C3. Drell-Yan, where the spectators are expected to be unaffected by the process, the emission of both hard and soft gluons should be rather central.

I would finally like to mention that the features mentioned under 2 above, i.e. that there should be particles emitted in the $\bar{q}$g- and gq-angular segments but not in the remaining q$\bar{q}$-segment have observational consequencies. Due to the distorted configurations with an, in general, softer gluon jet it takes some hard work to disentangle the effect. We note, in particular, that a pion produced along the (moving) string-segments will due to the string-motion obtain a p_T of the order of only a few hundred MeV/c and this is partly washed out by the fragmentation p_T from the tunnelling processes. In the figs. 7 we exhibit a momentum space distribution of the hadrons in connection with (fig. 7a) hard gluon emission, (fig. 7b) collinear gluon emission and (fig. 7c) soft central gluon emission. The particles are generally produced along momentum space hyperbolas but due to the fragmentation p_T there is a smearing-effect (the hatched area). The predicted effect has, nevertheless, been exhibited by the JADE-group at Petra [13]. It is an interesting feature that this so-called "string-effect" is mirrored by a correct treatment of higher order perturbative QCD in the jet-calculus discussed by Bryan Webber.

I have in this section presented a few of the dynamical consequences of a causal and relativistically covariant confined gluon model. The model has some further properties, to be discussed in the next section, which may have farreaching consequences for the treatment of many-gluon processes.

5. Is There a Field-topological Quantum Number Necessary to Label the Final States?

The title of this section could be even more provocative and be formulated as "Do the force-fields have an independent quantum-mechanical state-property?" There is also a third way to say the same thing: "Even if different colours cannot be distinguished, the direction of the colour field (colour flow) between the charges may be observable".

In a model like the Lund model, where the final state particles stem from the force-fields between the initial partons, it is evidently necessary to consider the topological features of the confined field, i.e. how to span the force-fields. In the cases considered up to now there is only one way to do that, but for e.g. 4-jet events in e^+e^--annihilation and leptoproduction, for 3-gluon decays of heavy onia and for e.g. large p_T gluon-gluon scattering there are in general several topologically different configurations.

In fig. 8 the two different possible ways to span the force-field for a state containing a $q\bar{q}gg$ is shown, In fig. 9 we show the expected configurations for a 3-gluon decay of heavy onia in the Lund model [14] (a triangular string with the gluons in the corner). We note that in this case the colour flow, defined e.g. from colour to anticolour, will have two different directions. This means that e.g. an $s\bar{s}$-pair produced along the force-field will be dragged by the force-field either with the s-quark towards the gluon a, and $\bar{s}$ towards c or else in the opposite way. This feature is different from ordinary flavour-compensation. Finally in the figs. 10 we exhibit the six different configurations possible for gluon-gluon scattering in hadron-hadron-interactions. The emission of a gluon 1 and gluon 2 from the initial hadrons will leave these objects as 8-states $\bar{1}$ and $\bar{2}$. After the process 1+2→3+4 the final state gluons 3 and 4 must be connected via colour force lines to $\bar{1}$ and $\bar{2}$ and as shown in the figs. 10 this can be done in several topologically distinct ways, both with regard to the colour flow direction and with regard to field-topologies.

If we assume that the different field-configurations correspond to different states, then we may partition the matrix-elements in a gauge-independent way into different configurations. In order to compute the cross-section, however, one in general squares the matrix-elements and sums over the different possible colour configurations and this will leave interference terms in the ordinary treatment. The relative size of the interference term is, in general, small of the order of $\frac{1}{N^2}$ with N the number of colours in the gauge group. This is the fundamental difficulty mentioned above because there are no interference terms if the configurations correspond to different states. It is not known at present how to circumvent this difficulty.

Acknowledgement: I would like to thank the best collaborators any man can have, i.e. Gösta Gustafson, Gunnar Ingelman, Torbjörn Sjöstrand, Hans-Uno Bengtsson, Olle Månsson, Bo Söderberg, and the other members of the Lund group, and Lena Brännström for her help with this manuscript.

References

1. B. Andersson, G. Gustafson, G. Ingelman, T. Sjöstrand, Physics Reports 1983, 97, 31.

2. B. Andersson, G. Gustafson, C. Peterson, Z. Physik C1 (1979) 105.

3. B. Andersson, G. Gustafson, B. Söderberg, Z. Physik C20 (1983) 317.

4. W. Heissenberg, H. Euler, Z. Physik 98 (1936) 714.

5. J. Schwinger, Phys. Rev. 82 (1951) 664.
E. Brezin, C. Itzykson, Phys. Rev. D2 (1970) 1191.

6. A. Casher, H. Neuberger, S. Nussinov, Phys. Rev. D20 (1979) 179.

7. B. Andersson, G. Gustafson, T. Sjöstrand, Z. Physik C6 (1980) 235.

8. B. Andersson, G. Gustafson, G. Ingelman, Phys. Lett. 85B (1979) 417.

9. S. Erhan et al, Phys. Lett. 82B (1979) 301.

10. B. Andersson, G. Gustafson, Z. Physik C3 (1980) 223.
B. Andersson, G. Gustafson, T. Sjöstrand, Z. Physik C6 (1980) 235.
B. Andersson, G. Gustafson, T. Sjöstrand, Phys. Lett. 94B (1980) 211.

11. B. Andersson, G. Gustafson, T. Sjöstrand, Z. Physik C12 (1982) 49.

12. G. Parisi, R. Petronzio, Nucl. Phys. B154 (1979) 427.

13. JADE-Collaboration, A. Bartel et al, Phys. Lett. 101B (1981) 129.

14. B. Andersson, G. Gustafson, Z. Physik C3 (1980) 223.

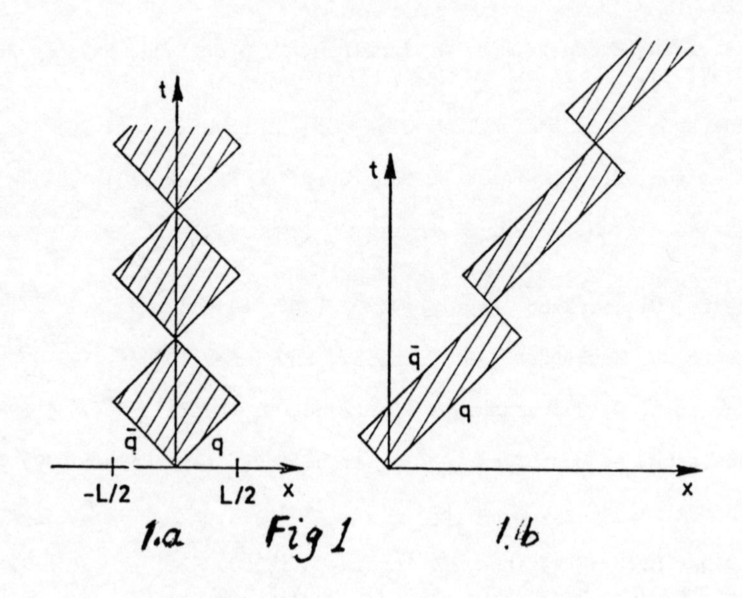

1.a Fig 1 1.b

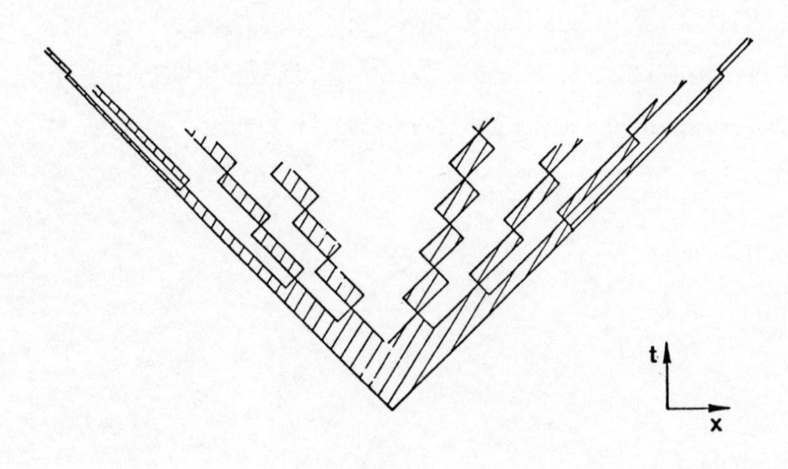

Fig 2

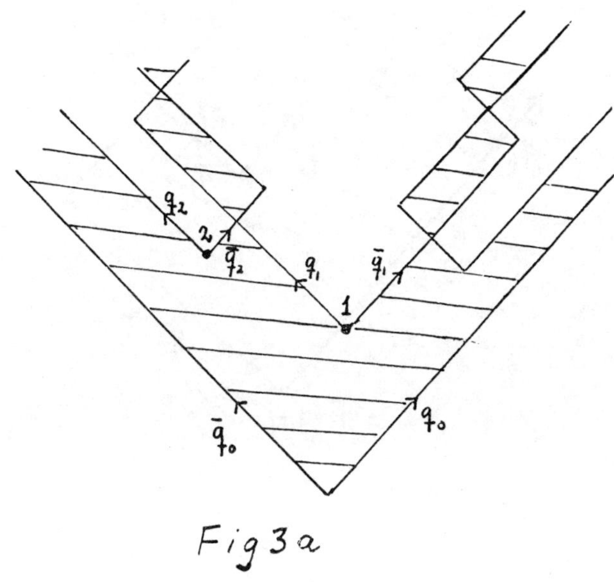

Fig 3a

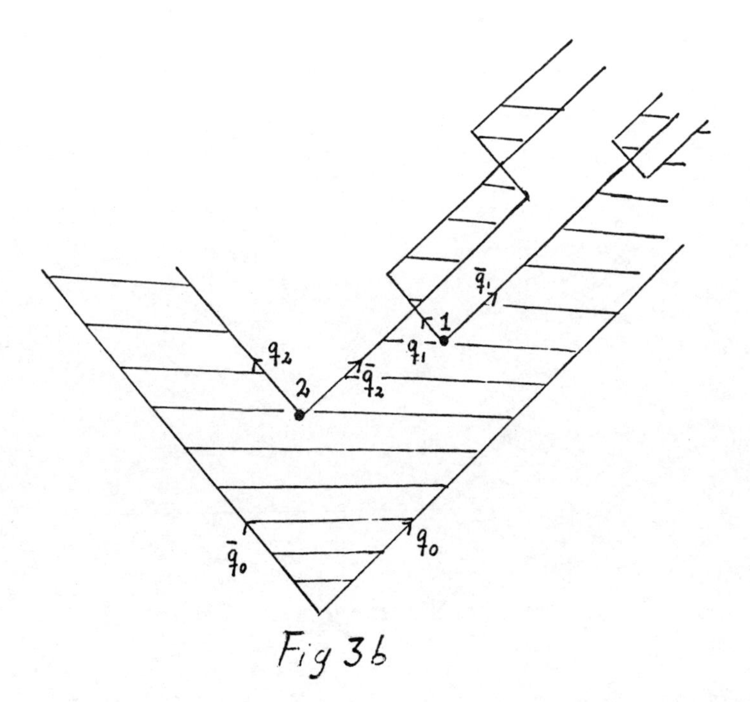

Fig 3b

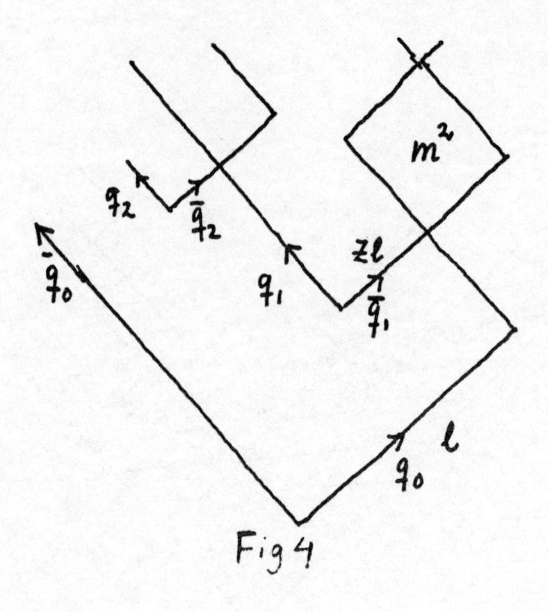

Fig 4

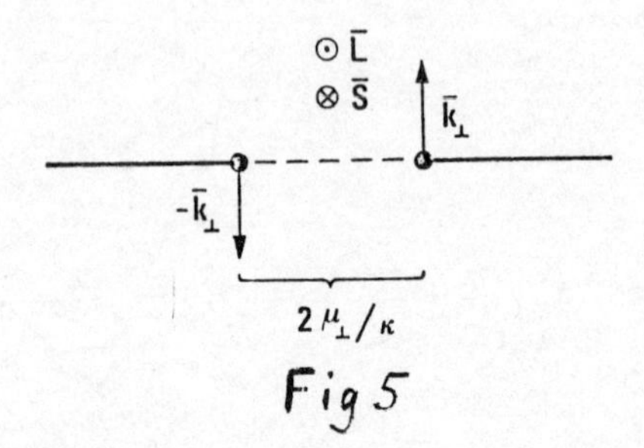

Fig 5

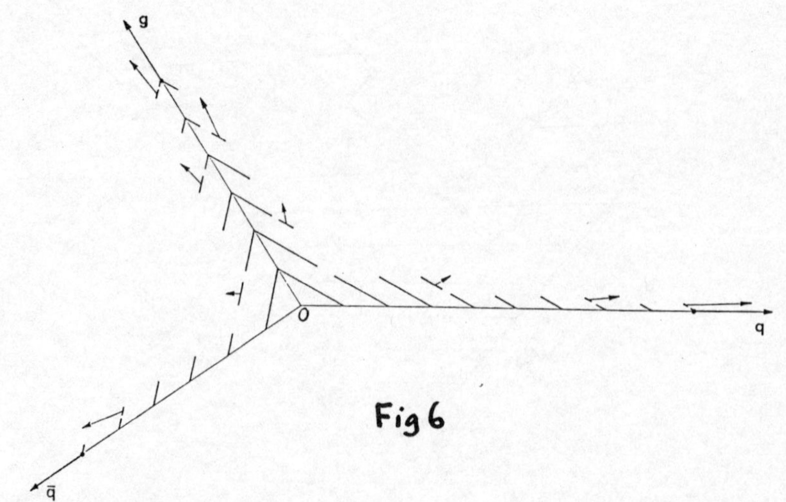

Fig 6

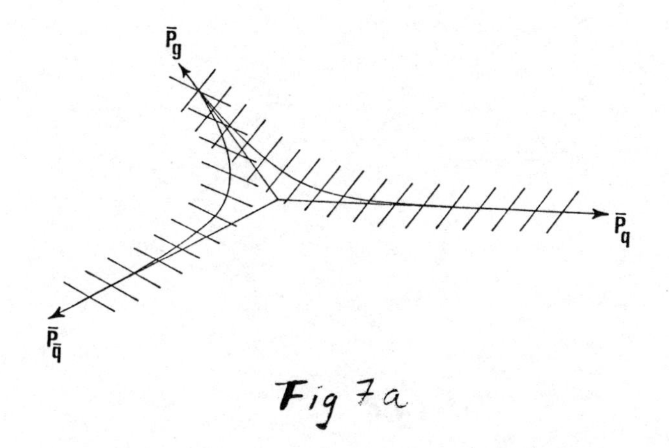

Fig 7a

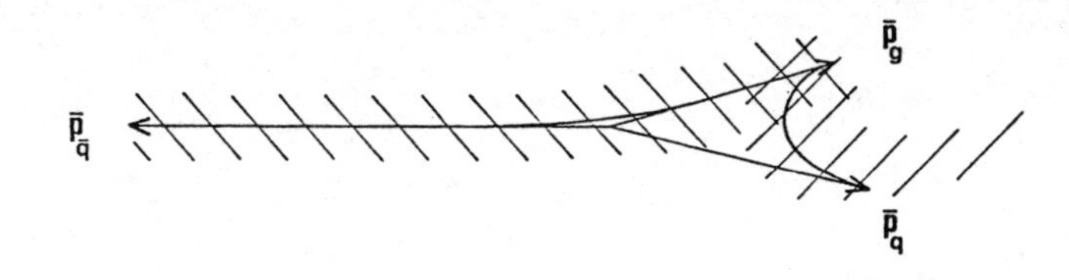

Fig 7b

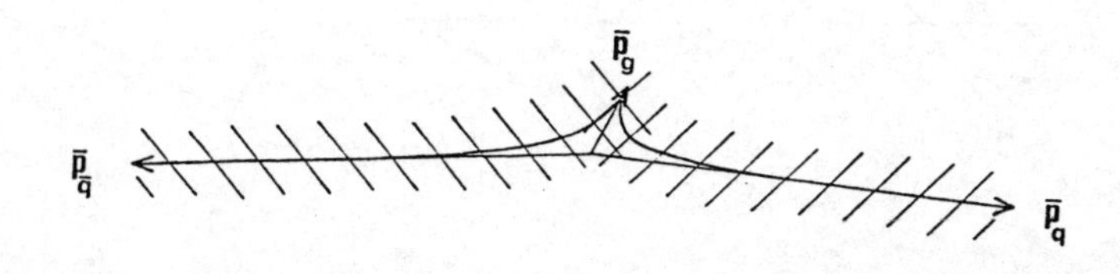

Fig 7c

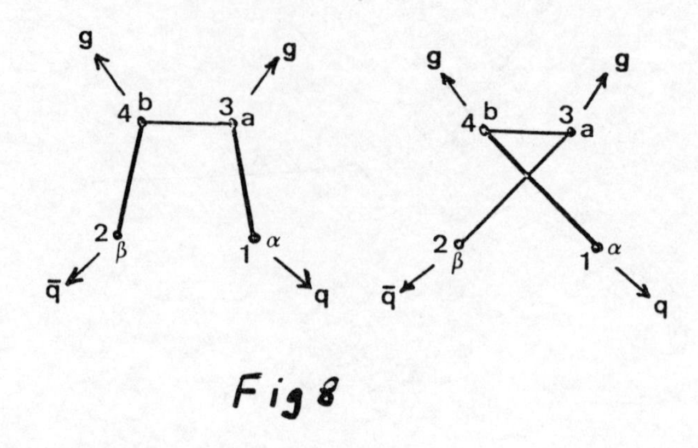

Fig 8

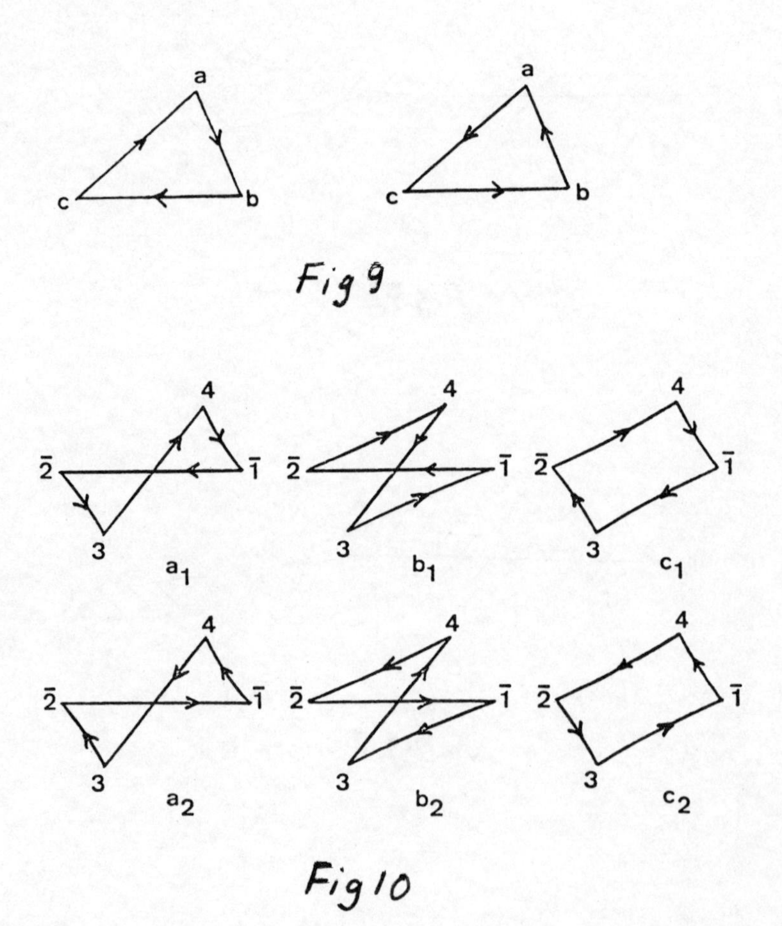

Fig 9

Fig 10

Discussion on Section 6

P.V. Landshoff (Cambridge): Have you considered the measurement of σ_{tot} by the Fly's Eye experiment at 25 TeV?

Speaker, B. Nicolescu: Yes, our extrapolation is compatible with this experimental result (see figure). In other words, all existing experimental data support the $\ell n^2 s$ behaviour of σ_{tot}. However, the coefficient in front of $\ell n^2 s$ is smaller than was previously thought.

C. Broll (Annecy): Can you tell us more about odderon effects in the πN charge exchange reaction?

Speaker, B. Nicolescu: The general effect is that the polarization becomes more and more negative at small t, as a result of the increasing dominance of the $\rho \otimes$ Odderon contribution. At Fermilab energies, the polarization is predicted to be negative everywhere in the small t region, $0 < |t| \lesssim 0.6\ (GeV/c)^2$.

S. Loucatos (Saclay): Is the longitudinal phase space Monte Carlo inconsistent with your data on multiplicity correlations?

Speaker, C. Geich-Gimbel: Yes, the cluster model Monte Carlo reproduces the bulk of data much better.

T. Akesson (CERN): What are the assumptions in the Monte Carlo that you compared your data with?

Speaker, C. Geich-Gimbel: The most important of these are independent emission of small mass clusters with a truncated Poissonian multiplicity distributions and particle ratios.

P.V. Landshoff (Cambridge): Can the clusters be ordinary resonances?

Speaker, C. Geich-Gimbel: Part can perhaps be, but not all. Their decay wouldn't give $\langle K \rangle \approx 2$ charged particles.

B. Andersson (Lund): Your cluster multiplicity is essentially larger than you would find if you take all known PDG-resonances in "reasonable" ratios.

Speaker, C. Geich-Gimbel: Answer as for previous question.

W. Mitaroff (Vienna): Which are the basic parameters of your calculation and what are their "best" values?

Speaker, B. Andersson: They are the intrinsic p_T of the quarks, which is ~ 0.35 GeV/c (though adding soft photons you get different results for different processes, e.g. 0.44 GeV/c for e^+e^- annihilation), the ratio $s : u : d \simeq 0.3 : 1 : 1$ and the vector to pseudoscalar ratio, which is about one. The Petra groups report $\alpha_s \simeq 0.17$, which corresponds to

$\Lambda_{QCD} \simeq 400$ MeV - I believe myself that Λ is close to the mean hadronic mass-scale.

A. Donnachie (Manchester): In your talk you referred to that of Bryan Webber and said that you and he are calculating the same thing by different techniques. Is this true for heavy quark production, since naively it would appear that the cluster model can more easily give heavy quarks than the string model, since once a high mass cluster is formed the only inhibition to heavy quark production is phase space. This may be particularly relevant for charm.

Speaker, B. Andersson: No, Bryan Webber actually produces charm in general not from his "partonic clusters", but from the (small) probability that a (hard) gluon decays into a $c\bar{c}$ pair. His mass-clusters strongly peak at low masses.

S. Loucatos (Saclay): Do you intend to use your string model to reproduce initial state radiation in hadron collisions?

Speaker, B. Andersson: Yes, that is our intention, but this subject is far too big to cover now.

Future plans for the UA5 experiment—$\bar{p}p$ interactions at 900 GeV CM energy

UA5 Collaboration (Bonn-Brussels-Cambridge-CERN-Stockholm)

Presented by D.R. Ward

Cavendish Laboratory, Madingley Road, Cambridge.

Abstract. The next run of the UA5 experiment will be at $\sqrt{s}$ = 900 GeV, using the CERN Sp$\bar{p}$S collider in a new ramped mode. The motivation for the experiment will be discussed, and the present status summarized.

1. Introduction

The idea of raising the energy of the Sp$\bar{p}$S collider to $\sqrt{s}$ = 900 GeV was suggested in 1982 by Rushbrooke [1]. The SPS machine cannot sustain continuous 450 GeV beams because the power dissipation in the magnets would be too great. It was therefore proposed continuously to cycle the energy of the machine between 450 GeV and 100 GeV per beam, as indicated schematically in fig.1. The lengths of flat top and bottom in such a cycle

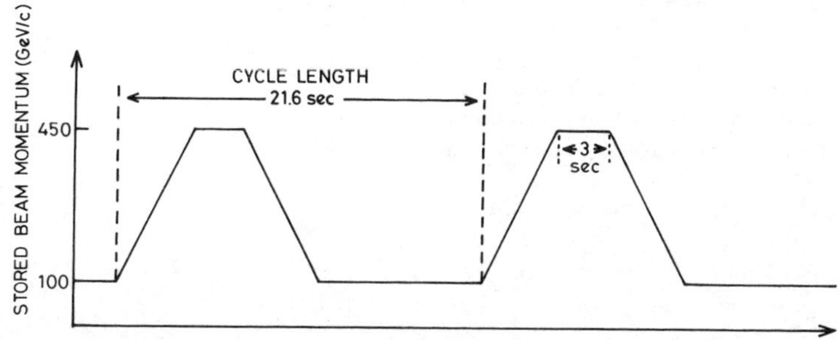

Fig.1 Ramped collider scheme - showing the variation of beam momentum with time (not to scale).

would be chosen so as to make the average power dissipation over the cycle equal to that in 270+270 GeV continuous operation, or fixed target operation.

Late in 1982 the UA5 collaboration proposed [2] to make an exploratory investigation of $p\bar{p}$ interactions at $\sqrt{s}$ = 900 GeV using the Sp$\bar{p}$S collider in such a ramped mode. This experiment would provide an overview of the main features of hadron interactions over a very large range of c.m. energies, using the new data at $\sqrt{s}$ = 900 and 200 GeV together with existing data from the same detector at $\sqrt{s}$ = 540 GeV and 53 GeV.

This experiment was approved (as UA5/2) in 1983, subject to machine tests confirming the feasibility of the ramped collider scheme. In §2 the motivation of the experiment will be further discussed, and in §3 the results of the machine tests so far and the current status of the experiment will be outlined.

2. Motivation

There were two main reasons for proposing the UA5 900 GeV run. The primary motivation was to search for the unusual events observed in cosmic ray emulsion chamber experiments at these energies. The best known events of this type are Centauro events [3] (high multiplicity, high p_T events with no photons produced), while other exotic phenomena are Mini-Centauros [3] (like Centauros but of lower multiplicity), Geminions [3] (two very high p_T hadrons or jets) and Chirons [4] (very high p_T events in which apparently very tight jets of particles, including hadrons, are seen with relative $p_T \sim$ tens of MeV).

None of these phenomena has been seen at the $Sp\bar{p}S$ collider at 540 GeV [5,6]. There is, however, evidence for a threshold just above the present collider energy. Fig.2 (taken from ref.[7]) shows an estimate of the proportion of exotic events in the Mt.Chacaltaya experiment as a function of energy. The cosmic ray data suggest a significant energy dependence, and are quite consistent with the non-observation of Centauros at the present collider energy.

If, therefore, the cosmic ray data are to be interpreted as resulting from proton-nucleon collisions we may expect to see unusual events at the $\sim$ 5-10% level at $\sqrt{s}$ = 900 GeV. Clearly such a finding would be of immense importance for particle physics, whilst even a negative result could have important cosmological implications, indicating that the cosmic ray data could not be explained in terms of proton interactions, and thus implying a non-protonic component in the cosmic ray spectrum.

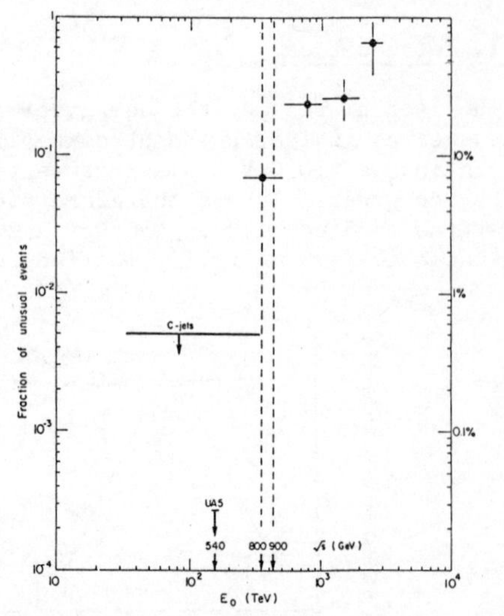

Fig.2 From ref.[7]. Fraction of unusual events vs. energy.

The second motivation for UA5/2 was to continue our studies of "normal" hadronic interactions. The acquisition of data at both 200 and 900 GeV will of great value in determining the s-dependence of the phenomena studied. A few examples will suffice to illustrate this point:

i) We have seen clear violations of KNO scaling in the multiplicity distribution at 540 GeV [8]. As shown in fig. 3 this is manifested as a growing proportion of high multiplicity events. What is not yet clear is whether this represents a new phenomenon or whether the apparent scaling seen up to ISR energies was merely a result of the small energy range studied. Evidently data at other energies will help to resolve this.

ii) Collider data on kaon and baryon production [9-11] seem to indicate an increase in $\langle p_T \rangle$ for those particle types compared to ISR energies (see fig.4). Clearly data at 200 GeV and 900 GeV will help to clarify whether this is a smooth or an abrupt change with c.m. energy.

(iii) The Ξ^-/Λ ratio measured by UA5 [11,12] is substantially larger than at lower energies. How does this depend on energy?

3. Machine tests and current status

Machine tests for the ramped collider scheme were carried out in May/June 1984. The cycle used had a total period of 21.6 sec, including a 3 sec. flat top and a 9.6 sec. flat bottom. The SPS supplied timing pulses to the experimental area at the start and end of the flat regions. The tests used single bunches of $\sim 10^{10}$ protons.

The key quantities to measure were the beam lifetime and the background levels in the interaction region. By the end of the machine development period the beam lifetime was typically $\sim$ 1 hour for the first half hour after a fill, levelling out to a lifetime of $\sim$ 3-5 hours thereafter. This corresponds to a fractional beam loss of $<10^{-3}$/cycle. The Q-value was stable to .02 over the whole cycle, and the SPS experts now believe they can reduce this to .005, corresponding to a beam lifetime of $\sim$ 15 hours. The final coasts were sustained for $\sim$ 2 hours. Of course the luminosity lifetime depends on the lifetime of the $\bar{p}$ bunch as well, and effects of beam-beam interaction could become

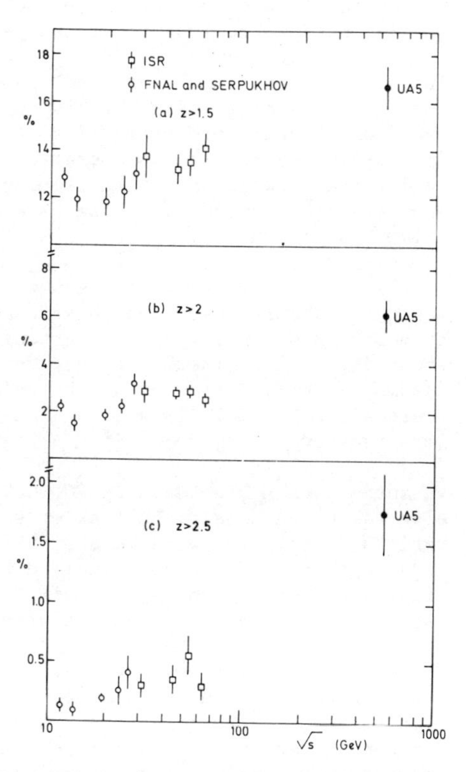

Fig.3 Fraction of high multiplicity events in non single-diffractive events, vs.$\sqrt{s}$.

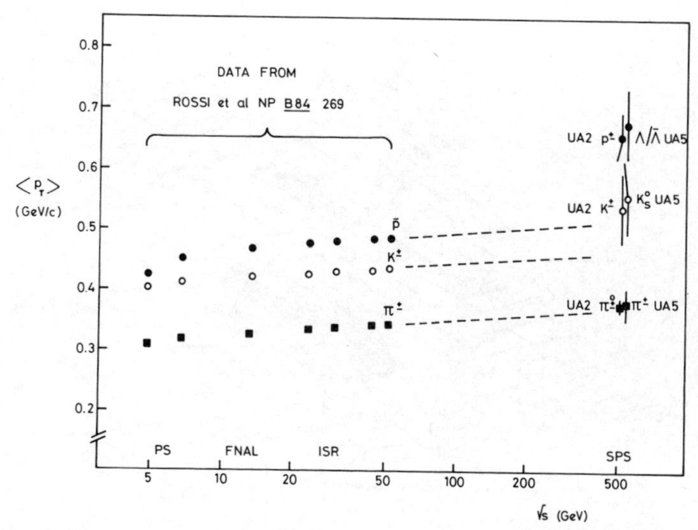

Fig.4 Average p_T for different particle types vs. $\sqrt{s}$.

important. Nonetheless the present results are extremely encouraging. The present status of the ramped collider is comparable to that of the continuous Sp$\bar{p}$S collider just before the first runs in 1981.

The level of in-time background was measured in the experimental area during the machine tests. We found the level of background throughout the cycle was roughly proportional to the beam momentum, and was no worse than in 270+270 GeV continuous operation at comparable beam currents. In particular the background during ramping was no worse than in the flat regions. We therefore conclude that the background levels should be low enough to allow UA5 to run.

In the light of these tests we now envisage the following scenario for data taking. Assuming usual bunch sizes, $\sim 10^{11}$ p, and $\sim 5\ 10^{9}$ $\bar{p}$. and running at normal β we anticipate a starting luminosity of $\sim 3\ 10^{26}$ cm^{-2} s^{-1}. This would give an inelastic trigger rate of $\sim$ 15 Hz. If we conservatively assume a beam lifetime of $\sim$ 3 hours we would be able to take data for $\sim$5-6 hours after a fill. This is an acceptable prospect for data taking for a first survey.

We anticipate taking data at 200 GeV and 900 GeV in parallel, and possibly also taking data near the tops of the ramps, at $\sqrt{s} \sim$ 700-900 GeV. A variety of triggers will be available, a two arm ("minimum bias") trigger to accept most of the non single-diffractive inelastic cross-section, a single arm trigger to capture single-diffractive events, and a transverse energy trigger using our 90° calorimeter.

The detector (fig.5) will be basically unchanged for the 900 GeV run. We shall therefore be using a proven and well understood detector. One significant change will be the introduction for part of the run of a "photon converter", a thin plate placed just beneath the upper streamer chamber to enhance photon detection. This will assist the search for Centauro-like events.

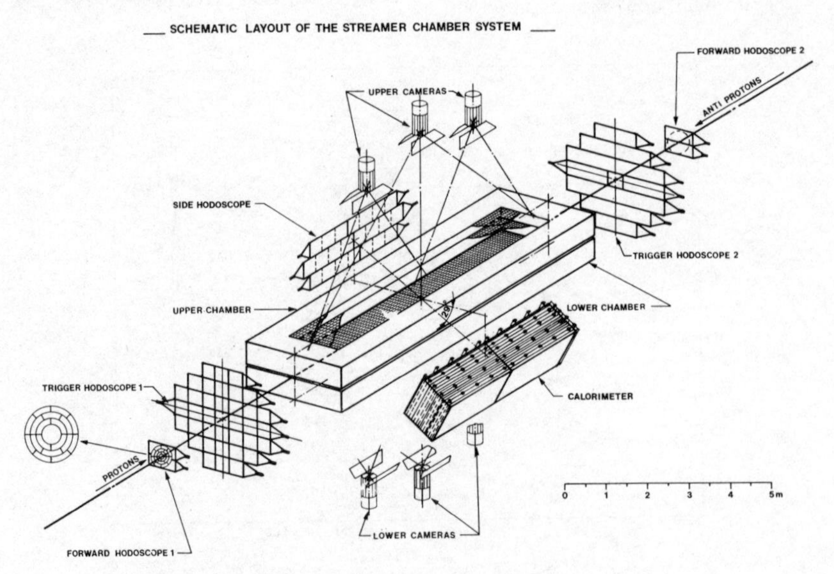

Fig.5 Schematic layout of the UA5 detector.

UA5/2 is tentatively scheduled to run in period 1 (∿ March) of 1985. From the first runs of UA5 results were available within a month of taking data, so we expect to have our first results at $\sqrt{s}$ = 900 GeV by spring or early summer of 1985.

Acknowledgements

The UA5 Collaboration gratefully acknowledge the dedicated efforts of the staff of the SPS division who have contributed to the ramped collider project.

References

[1] J.G. Rushbrooke, CERN report EP/82-6.
[2] UA5 Collab., CERN SPS Proposal 184, CERN/SPSC 82-75.
[3] C.M.G. Lattes et al., Phys. Rep. C65 (1980) 151.
[4] J. Chinellato et al.,"Chirons" paper to Paris Conf. (1982).
[5] K. Alpgård et al., Phys. Lett. 115B (1982) 71.
[6] G. Arnison et al., Phys. Lett. 122B (1983) 189.
[7] J.G. Rushbrooke, "Comparison of cosmic ray physics with latest accelerator data", paper to Paris Conf. (1982).
[8] G. Alner et al., Phys. Lett. 138B (1984) 304.
[9] M. Banner et al., Phys. Lett. 122B (1983) 322.
[10] K. Alpgård et al., Phys. Lett. 115B (1982) 65.
[11] Ch. Geich-Gimbel, "Results from UA5", paper to this conference.
[12] G. Alner et al., 'Observation of Ξ^- production in $p\bar{p}$ interactions at $\sqrt{s}$ = 540 GeV", submitted to Phys. Lett. B.

Inst. Phys. Conf. Ser. No. 73: Section 7
Paper presented at VII Eur. Symp. Antiproton Interactions, Durham 1984

Physics from future colliders

G. L. Kane

Randall Laboratory of Physics, University of Michigan, Ann Arbor, MI 48109, U.S.A.

Some of the physics that provides motivation for building future (mainly hadron) colliders is discussed. The main topics are ways to study the origin of SU(2) breaking, including production of heavy Higgs bosons and techniques for production and study of W pairs from standard or new physics; production of new heavy particles, especially some suggested by supersymmetry with large missing P_T as a signature; and rare decays and interactions of τ, c, b, t, $W^{\pm}$. Some comments on comparisons of energy, luminosity, and beams ($\bar{p}p$, pp, e^+e^-) are given.

Introduction

In a talk about future experiments it is important to define the context in which one expects the experiments to occur (1,2,3,4). We now have the remarkably successful Standard Model (SM). It has been so well tested that at the level of 100 GeV physics we do not expect anything to be wrong with the Standard Model, at least as an effective field theory. That means we should not expect any large deviations from SM predictions, and that we can use SM techniques to calculate production and decay of any particle that has electric charge, color, or weak isospin quantum numbers.

But there are many ways in which the SM is incomplete. These are well known: we do not understand the origins of mass, although we can parameterize it. We do not understand why or how SU(2) is broken. We do not understand why parity or CP conservation is violated in weak interactions. We do not know if quarks and leptons can be unified. And so on. A new interaction may be responsible for the origin of mass, or for understanding the form the SM takes. Consequently, it is expected that experimental clues to a new interaction will appear, perhaps as small deviations from SM predictions, or perhaps in the form of new particles.

Some things will be settled soon. If the SM is a valid, effective field theory, we know from indirect evidence on b-quark decay and interactions that a t-quark exists, but we would still like to know its mass and particularly its decay modes, which may not take too long. Apparently unexpected events have appeared at CERN and at DESY recently--soon we will know if they are indeed new physics.

In the next few years we will have data from $\bar{p}p$ colliders, first in 1984 at $\sqrt{s} \simeq 620$ GeV and luminosity $\mathcal{L} \gtrsim 3 \times 10^{29}$ cm^{-2} sec^{-1}, then in 1986-1987 from CERN at about an order of magnitude higher luminosity and from FNAL at 2 TeV and $\mathcal{L}=10^{30}$-10^{31} cm^{-2} sec^{-1}. About 1989-90 HERA will give us ep collisions at $\sqrt{s} = 320$ GeV. And somewhere in the 1990's we will have the U.S. SSC at 40 TeV, $\mathcal{L} \simeq 10^{33}$ cm^{-2} sec^{-1}, and/or perhaps the European LHC at similar $\mathcal{L}$, and with $\sqrt{s}$ from 10-18 TeV, probably both pp. There are no clear plans for e^+e^-machines beyond TRISTAN (1986), SLC (1986-7), and LEP (1989).

What are the windows for new physics at Super Colliders? We can look for heavy particles or for new interactions. Many approaches to thinking about what the new physics might be suggest new particles, such as heavy Higgs bosons, supersymmetric states, technicolor pseudo-Goldstone bosons, mirror fermions, new families, colored Higgs, etc. If the new physics is going to explain why the SM takes the form it does, around 100 GeV, the new physics will have to show up not far above that scale. The effects could appear through new interactions as well, such as lepton number violating, flavor changing, neutral currents, or non-SM CP violation, or right-handed currents. All of these can be studied at hadron Super Colliders.

Some talks along these lines mainly have shown cross sections for production of heavier states. Rather than just do that, I will follow the procedure (suggested to me at the recent ICFA meeting) of discussing "Some Promising Experiments of the Future".

Before I begin with the explicit experiments, I want to briefly discuss whether experiments can be done at future high energy, high luminosity, colliders. There has already been considerable study at workshops about these questions (1,2,3,6). On the whole the answers have been guardedly optimistic, with some dissent; most efforts have concluded that the limiting factor may be expense, at least if existing technologies are extrapolated. So far little consideration has been given to innovative detectors, such as a high field solenoid that bends around and captures all particles with P_T below some amount, say 1 GeV, combined with simpler devices outside that can handle the limited event rate from larger P_T particles. When detectors have run at CERN with $\mathcal{L} \sim 10^{31}$ and at the FNAL collider, the combination of the workshop study and the experience gained can reasonably be expected to give rise to workable detectors for future super colliders.

The spirit of the following discussion is meant to be speculative and to raise questions--what kind of machines, detectors, and ideas do we need to proceed effectively?

<u>Experiment(s) 1--How can we study SU(2) breaking experimentally?</u>

The best way to proceed would be to detect a Higgs-boson, and study its mass, parity, and interactions. When can we hope to do that? Fig. 1 shows various possibilities for a minimal SM H° (i.e. a theory with a single Higgs doublet); if one includes (7) theories with 2 or more Higgs doublets, then some branching ratios are larger, some new decays are allowed so that a few additional possibilities for detection occur, and the spectrum includes 3 neutral Higgs and a charged pair $H^{\pm}$. If t is observed decaying semileptonically, one can show $M(H\pm) \gtrsim 60$ GeV.

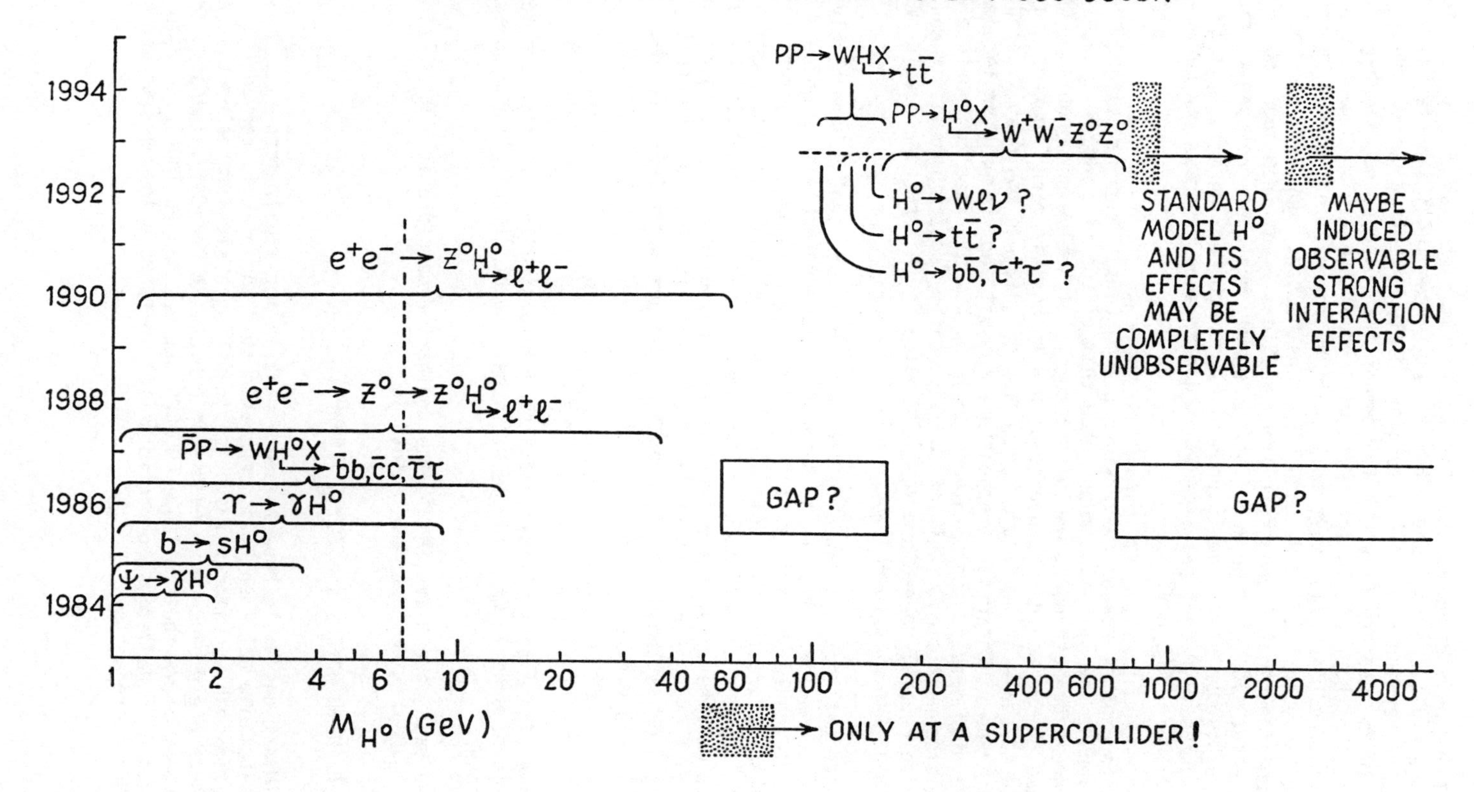

OBSERVABILITY OF A STANDARD MODEL HIGGS BOSON
1994
1992
1990
1988
1986
1984
1
2
4
6
10
20
40
60
100
200
400
600
1000
2000
4000
M_{H^o} (GeV)
$e^+e^- \to Z^oH^o \to \ell^+\ell^-$
$e^+e^- \to Z^o \to Z^oH^o \to \ell^+\ell^-$
$\bar{P}P \to WH^oX \to \bar{b}b, \bar{c}c, \bar{\tau}\tau$
$\Upsilon \to \gamma H^o$
$b \to sH^o$
$\Psi \to \gamma H^o$
$PP \to WHX \to t\bar{t}$
$PP \to H^oX \to W^+W^-, Z^oZ^o$
$H^o \to W\ell\nu$?
$H^o \to t\bar{t}$?
$H^o \to b\bar{b}, \tau^+\tau^-$?
STANDARD MODEL H^o AND ITS EFFECTS MAY BE COMPLETELY UNOBSERVABLE
MAYBE INDUCED OBSERVABLE STRONG INTERACTION EFFECTS
GAP ?
GAP ?
ONLY AT A SUPERCOLLIDER !

FIGURE 1.

Must there be a Higgs Boson? As is well known, the answer is yes, to keep the amplitude for (say) $f\bar{f}\rightarrow W_L W_L$ from exceeding its unitarity bound (f is any fermion and W_L is a longitudinally polarized guage boson--by W we always mean any of $W^{\pm}$, Z°). Rather than discuss details here of the usual methods of finding H°, I will emphasize (8) what happens when H° gets very heavy, and ways to look in WW interactions even if an explicit way to find an H° cannot be constructed. However, it should be emphasized that there are regions in Fig 1 (e.g. from about 60 GeV up to about 175 GeV) where no convincing way has been proposed to find a Higgs boson even if one is present, and much more work is needed on this subject. Recent analysis (9) suggests some progress can be made for 100 GeV $\lesssim$ 160 GeV.

There is another reason to consider heavy Higgs. Suppose a light H° is discovered, say with mass much below that of the Z° . While it would be a great breakthrough, would that completely solve our problem? No, because (1) we would not understand how to keep it much lighter than the basic scale of electroweak interactions, (2) we would still need to look for new interactions responsible for setting the weak scale and for fermion masses, and (3) there could be a whole spectrum up to a TeV of related particles. While it would greatly focus thinking, it would still be crucial to explore the TeV region for effects that could determine the electroweak scale.

It turns out that the rates to produce a heavy H° are not impossibly small--see below. But first, suppose it were possible to produce a heavy H°--could we observe it (4)? The answer appears to be that once its mass is greater than some number around 600-800 GeV (the exact number depends on what cuts can be made, on detector capabilities, and on how many heavy quarks exist), we can no longer directly observe an H°. That is a prediction of the SM, for H° and also for some heavy quarks Q or leptons L; it is a problem for any particle F that decays by emitting a W,

$$F\rightarrow W+f.$$

Because of the possibility of emitting a longitudinal W, the rate for this is of order $G_F M_F{}^3$ and the width gets to be of order the mass as M_F increases.

In the case of a heavy SM H°, the dominant decays are to W^+W^-, $Z^{\circ}Z^{\circ}$ once $M_H>2M_Z$. The tree level width is

$$\Gamma\ (H^{\circ}\rightarrow W^+W^-+Z^{\circ}Z^{\circ}) = 3G_F M_H^3\ /16\pi\sqrt{2}\ .$$

If $\Gamma/M \ll 1$ one would hope to see a peak in the W pair mass distribution, but when Γ is comparable to M the peak is spread out. Some reasonable criterion, such as requiring a signal greater than $\sqrt{N}$ if there are N background events in a bin of given ΔM_{WW}, leads to the statement above that for $M_H \gtrsim$ 600-800 GeV one can no longer hope to observe H°. Radiative corrections (10) to Γ_H make the results worse, as they increase Γ_H.

It should be noted that although the above effect is due to M_H getting large and to emitting a longitudinal W, reminiscent of the unitanty violations without a Higgs in the SM, there is no unitarity violation here but simply a prediction of the SM. And the effect occurs equally well for $Q \to qW$; it does not depend on M_H getting larger, but is due in a sense to large Yukawa couplings.

How can we get around this to study SU(2) breaking? We don't know that we can. What we can try is to study WW scattering in general, and look for any kind of deviations from the SM predictions. That would include a mass plot bump, but there are other possibilities too.

The SM, without a Higgs contribution, predicts that W pairs (W^+W^-, $Z°Z°$, $W^+Z°$) will be produced from the diagrams

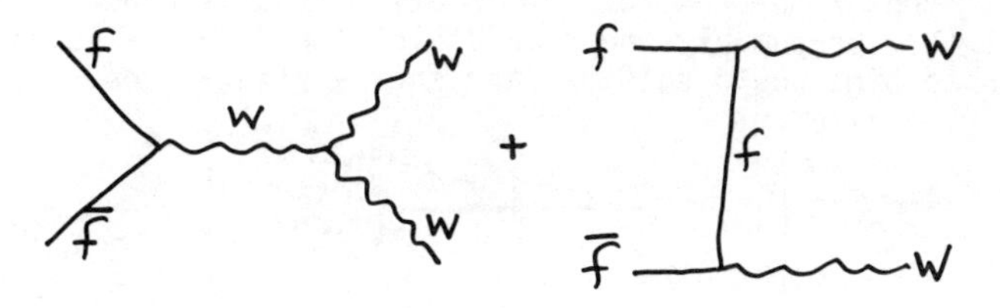

Fig 2

One can proceed by calculating these contributions and comparing the resulting predictions with experiment. Any deviation is a signal of new physics. New physics could arise from a Higgs contribution, from a modification of the non-Abelian WWW vertex shown above, or from new sources of W pairs.

There are several possible signatures of deviations suggested so far:
(1) The SM W pairs peak in the forward and backward directions (11),

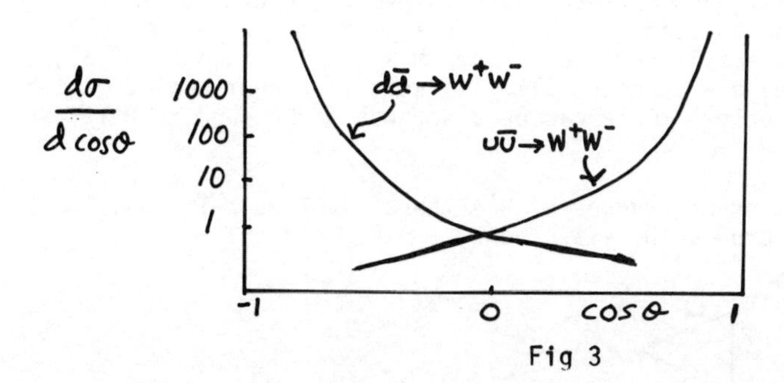

Fig 3

with cosΘ in each case the angle between the q (not the $\bar{q}$) and the W^+. Other sources could introduce extra W's at 90° or a forward/backward asymmetry. For example, a heavy Higgs will produce W's essentially isotropically, so we should look for extra W's at 90°.

(2) Essentially all (11) of the W's produced as in fig 2 are transversely polarized. New sources of W's may produce mainly

longitudinal ones--e.g. the H° decays mainly to $W_L^+W_L^-$ and $Z_L^\circ Z_L^\circ$. It is very important to learn how to observe the polarizations of W's. Probably this can be done by observing the decay into $q\bar{q}$ and determining the plane in which the $q\bar{q}$ lie. Then the distribution of that plane with respect to the production plane or the other W plane is different for W_L and W_T. It will be easier to do that for $Z^\circ \to \ell^+\ell^-$, but but to get reasonable statistics it is necessary to learn to do it for the jet-jet decays.

(3) The processes of fig 2 give a ratio (1) of Z°Z° to W^+W^- of about 1/7, while other sources will give different results; e.g. from H° decay $Z^\circ Z^\circ/W^+W^-=1/2$. Here probably the leptonic modes are required to tell W± from Z°, but perhaps the statistics will be good enough for that. Maybe it will be possible to plot M_{jj} from the jj decay and to separate Z° from $W^\pm$ because of the 12 GeV mass difference. Combining with point (1), presumably one should plot the $Z^\circ/W^\pm$ ratio vs cosθ; perhaps three bins would suffice, and show a rise at cosθ ~0.

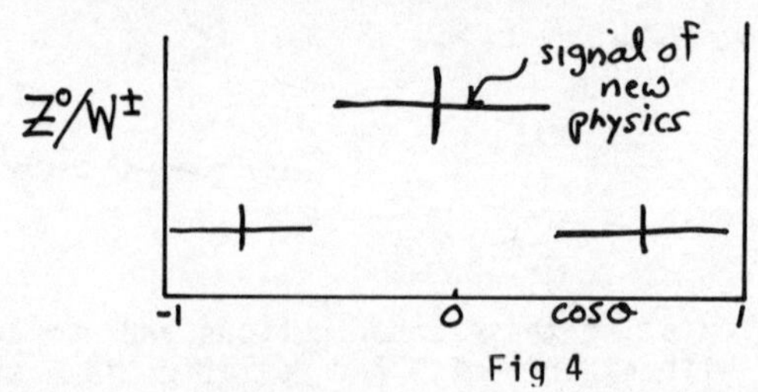

Fig 4

(4) If light Higgs do not exist and the WW interaction gets stronger, perhaps there will be extra production of W's. This has been discussed (12) by Chanowitz and Gaillard, who speculate that the rate for producing additional W's beyond those predicted by the SM may be large enough to observe at a high £ hadron collider. This would be an exciting possibility, and further study is needed to consider backgrounds and how to detect signals. For example, in the SM all quark jets and W's easily radiate further W's, so the chances of a jet pair each radiating a W is not small. If we also cannot easily tell a W from a jet (see below) there can be a serious background to multi-W states.

Next consider what other sources of W's there might be. The usual source (13) is to produce H° via gluons,

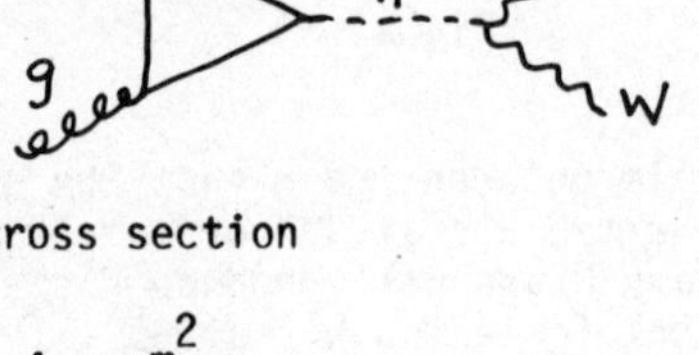

which gives a cross section

$$\sigma_{gg} \sim \left(\frac{m_Q}{m_H}\right)^4 \left(\ln\frac{m_Q^2}{m_H^2}\right)^2$$

as M_Q or M_H are varied at fixed s--note that σ_{gg} falls rapidly with M_H. Recently it has been pointed out (4,14) that at energies where

$$s >> \hat{s} >> m_{H^\circ}^2 >> m_W^2 ,$$

another production mechanism is important; for example,

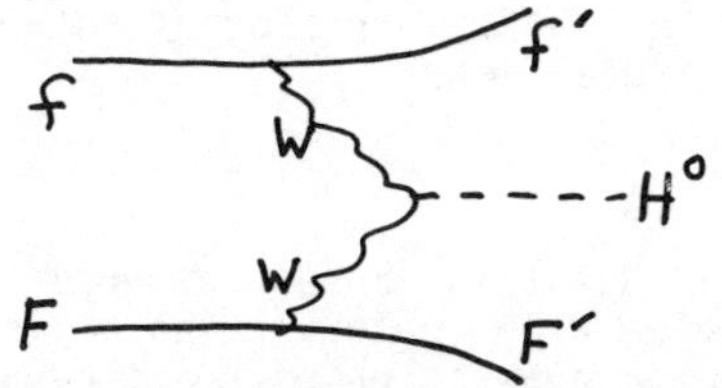

where F,F',f,f' are any appropriate fermions (leptons or quarks). One important consequence is that with this mechanism one can produce heavy H° at e^+e^-colliders well as at hadron colliders. In fact, it is difficult to imagine a decision to build a very high energy e^+e^- collider if there were not a mechanism such as this to allow production and study of a heavy H° with M $\gtrsim$ 1 TeV.

One can generalize producing W pairs (or multiple W's) via any model for strong WW interactions,

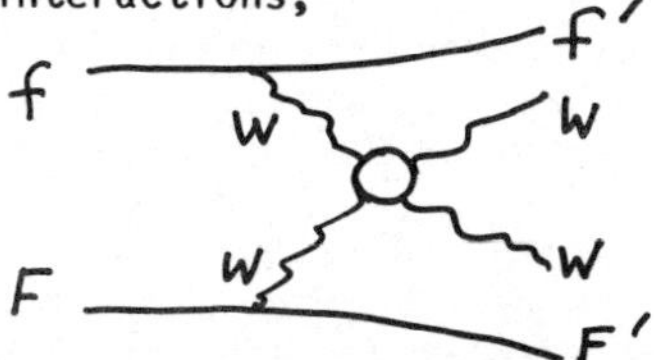

The structure functions that allow us to work in the "effective W" approximation have been given for W_L and W_T (15,16). It is not as extreme as one might imagine to treat W's as essentially massless--if M_H =300 GeV, $\sqrt{s}$=2 TeV, one can essentially satisfy $s >> \hat{s} >> M_H^2 >> M_W^2$. The $W^\pm$ structure functions are, making all simplifying approximations,

$$F_T (X) \simeq \frac{\alpha_2}{8\pi} \frac{1+ (1-x)^2}{x} \ln \frac{P_{T,\ max}^2}{m_W^2}$$

$$F_L (x) \simeq \frac{\alpha_2}{4\pi} \frac{1-x}{x} .$$

Because of the ln (P_T^2, max/m_W^2) factor, analogous to that for photons, F_T will dominate if the longitudinal and transverse couplings are the same. However, as discussed above, often when heavy objects (X = H°, Q, L) are involved, the coupling to W_L will dominate by M_X^2/M_W^2, and the longitudinal contribution will dominate. In any case, with these structure functions one can quickly do a parton model calculation of any cross section one might want to study at a future collider.

For $q\overline{q} \rightarrow H^\circ X$ one obtains from this calculation,

$$\sigma_{WW} \simeq \frac{\sigma_2^3}{16\, m_W^2} \ln \frac{M_H^2}{m_W^2}$$

Since this does not fall with M_H, as M_H increases at some value of M_H σ_{WW} must dominate H° production via gluons,

$$\sigma_{WW} > \sigma_{gg}.$$

For three families with $M_t \simeq 40$ GeV this happens at about 500 GeV, unfortunately about in the region where observing an H° is difficult or impossible. If there is a fourth generation of heavier quarks, the crossover does not occur until $M_H > 1$ GeV. So σ_{WW} may never dominate in a region where it is measurable in hadron reactions; but we can hope that it may produce sufficient additional W's to allow some effect to be seen at larger M_{WW}.

Can such studies be done (1,2,3,17)? What about the rates and backgrounds? They are given in several forms in recent studies, from which we can examine a few numbers.

As one goes from 10 TeV to 40 TeV, the cross section for $q\overline{q} \rightarrow WW$ increases by about a factor of 30, and the window in WW mass that can be studied increases by about a factor of 2, from about M(WW) = 1.5 TeV to about M(WW) = 3 TeV for $\mathcal{L} = 10^{33}/\text{cm}^2\text{sec}$. At 40 TeV, $\sigma(gg \rightarrow H^\circ) \approx 10^{-38}$ cm^2, $\sigma(ww \rightarrow H^\circ) \simeq 10^{-37}$ cm^2, while $\sigma(q\overline{q} \rightarrow ww) \simeq 10^{-35}$ cm^2. The SM The SM background clearly dominates over WW from a Higgs. If one insists on a very clean signature such as ZZ with both Z's giving a lepton pair, both background and signal run out by M(WW) $\lesssim$ 700 GeV, so it will be necessary to use jj modes for one or both W's.

Another serious problem, first pointed out by Paige (18), is that an energetic W looks like a jet. If we want to study WW scattering at 1 TeV, normally we have to identify 500 GeV W's. Typically $W \rightarrow 2j$, while a real 500 GeV jet can radiate and also have two cores, making separation difficult. One way to separate is to make a cut on jet effective mass. One can also cut on multiplicity, on having exactly two cores, and other distributions. Present arguments (18) suggest it is possible to get a factor of about 30 suppression of the j background.

It is not clear that a factor 30 is sufficient. In the SM the entire jj cross section integrated for 500 GeV $< P_T < \infty$ is quite large, over 10^{-32} cm^2. One can eliminate gluon jets by requiring one $W \rightarrow \ell\nu$. But the SM cross section for W+j integrated from 500 GeV $< P_T < \infty$ is larger than 10^{-35} cm^2. Another factor of 30 reduction puts us in a position where the number of events of (1) W+j that fake WW, (2) WW from $\overline{q}q$, and (3) expected WW from a heavy H°, are all about equal. With further study, perhaps one will become optimistic that any new sources of W pairs or WW interactions can be discovered at an appropriate future collider.

Experiment(s) 2 -- Search for New, Rare, Heavy Particles

One reason, usually the main one mentioned as justification, to build a new machine at higher energies, is to pass the threshold in energy to find a few events of the production of a new particle. Since the cross section to make a new state of mass M falls as $1/M^2$, as M increases one must go to a higher luminosity machine to expect to see the same number of events as at a lower M; put differently, for this class of physics there is also a threshold in luminosity and one must take care to be above it as well as above the threshold in energy. Once above the threshold, the event rate increases more rapidly with than with $\sqrt{s}$, and careful study is needed to determine the best mix of and $\sqrt{s}$ to maximize signal/noise.

Many signatures are possible for new, rare events , of course. We will go through the example of particles suggested by supersymmetry, because it is of great current physics interest, and because it introduces several new features of the kind that can arise elsewhere. The experiences of the past few months with UA1 and UA2 have affected this analysis as is clear from the UA1, UA2 talks at this conference. What would we observe if we started to produce supersymmetric partners at a collider?

To answer that question, first we briefly review the theory (19). In supersymmetry, every particle has associated with it a partner which is identical in every way except that it differs by half a unit of spin. With the photon (γ) there is a spin-1/2 photino ($\tilde{\gamma}$), with the gluon (g) a spin-1/2 gluino ($\tilde{g}$), also a color octet, with the $W^{\pm}$ a spin-1/2 wino ($\tilde{w}^{\pm}$), with the neutrino (ν) a spin-zero scalar-neutrino ($\tilde{\nu}$), etc. If these superpartners had the same mass as their associated particles, they would surely have been observed, so we assume the symmetry is broken and they have acquired extra mass. At present there is no commonly accepted pattern of supersymmetry breaking, so one must study all possible masses experimentally.

Fortunately, however, the most important couplings are related by the supersymmetry to gauge couplings, so they are known and can be used to compute rates and constrain the theory if effects are not observed. To a good approximation, one can generate all vertices of a supersymmetric theory by taking vertices of the SM and replacing particles by their partners in pairs. Thus

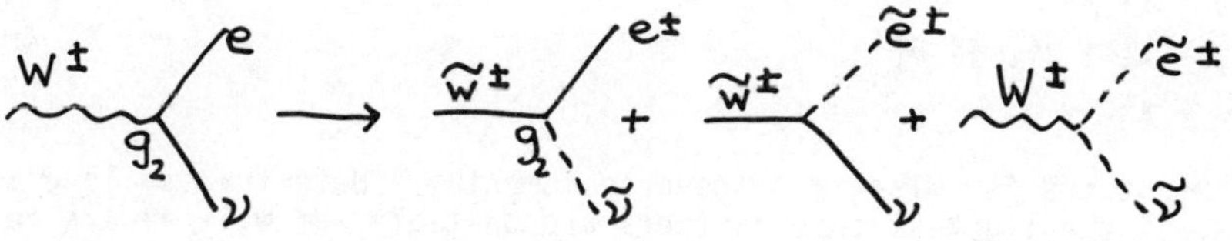

and all vertices have strength g_2. The space-time coupling is the minimal one. By combining the vertices one can produce all supersymmetric predictions.

In general, the lightest superpartner will be stable, and all superpartners will decay into it. It can be $\tilde{\gamma}$ or $\tilde{\nu}$ --- it does not matter for our purposes, since both interact with cross sections that are of order ν cross sections, and thus they escape collider detectors.

If $\tilde{\gamma}$ is lightest, it will not be precisely the partner of a photon, but a mixture with other neutral gauge boson partners. We will speak as if the lightest state is a photino, but the reader may wish to keep in mind that the actual state may be somewhat different.

Then the other states decay:

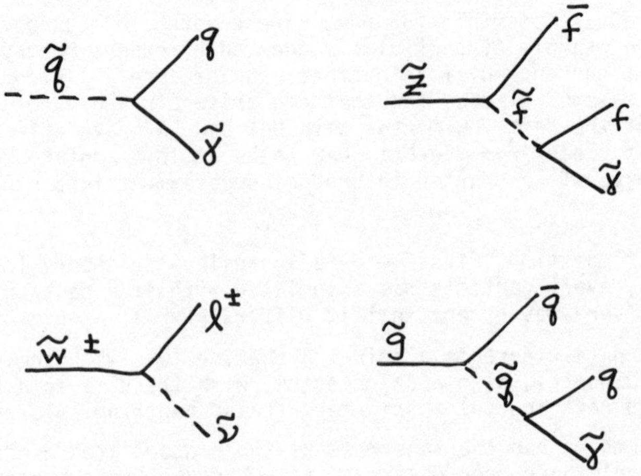

where f is any lepton or quark. At a collider, quarks appear as jets, $\ell^{\pm}$ as e or μ, $\tau^{\pm}$ as low multiplicity, narrow hardon jets, and $\tilde{\nu}$, $\tilde{\gamma}$ escape. Thus the individual signatures are

$$\tilde{\ell}^{\pm} \to \ell^{\pm}\, \not{p}_T$$
$$\tilde{q} \to j\, \not{p}_T$$
$$\tilde{g} \to jj\, \not{p}_T$$
$$\tilde{z} \to \ell^{+}\ell^{-}\, \not{p}_T,\ jj\, \not{p}_T$$
$$\tilde{w} \to \ell^{\pm}\, \not{p}_T$$

where $\not{p}_T$ stands for missing transverse momentum. Note that so long as all except the lightest superpartners are unstable, $\not{p}_T$ will always be present and no mass peaks (in jj or $\ell^{\pm}$j) will ever occur as a consequence of the production of supersymmetric partners . No hard photons will ever arise from decay of superpartners. $W^{\pm}$, Z° could arise in decays of sufficiently heavy superpartners, e.g. if $\tilde{m}_W$, $\tilde{m}_Z$, $\tilde{m}_g$, $\tilde{m}_q$ or $\tilde{m}_{\ell}$ were greater than m_W.

Presumably the content of $\tilde{q}$, $\tilde{g}$ in hadrons is sufficiently small that superpartners will be produced in pairs starting from $q\bar{q}$, qg, gg; if any superpartner were light enough to carry an appreciable fraction of the proton's momentum, it would have produced very copiously already. Then one expects constituent processes such as

$$q\bar{q} \to \tilde{w}\tilde{g},\ \tilde{z}\tilde{\gamma},\ \tilde{g}\tilde{\gamma}$$

$$qg \to \tilde{q}\tilde{g}$$

$$gg,\ q\bar{q} \to \tilde{g}\tilde{g},\ \tilde{q}\bar{\tilde{q}}.$$

In the transverse plane, a final state such as $\tilde{w}\tilde{g}$ gives, for example

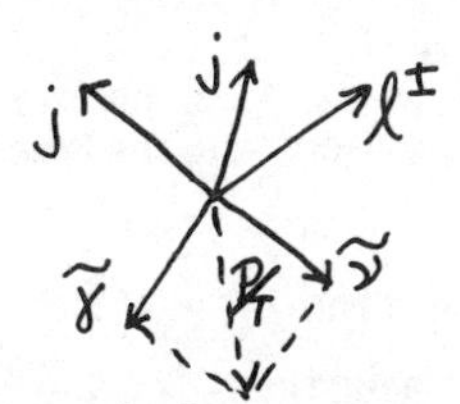

so the signature is $jj\ell^{\pm}\not\!P_T$. Similarly, $\tilde{q}\tilde{g}$ gives

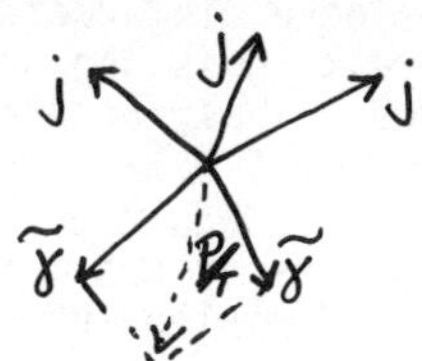

leading to $jjj\not\!P_T$. Some of the other processes listed above give signatures $jj\not\!P_T$, $jjjj\not\!P_T$, $j\not\!P_T$, $\ell^{\pm}j\not\!P_T$, $\ell^{+}\ell^{-}\not\!P_T$. Since the cross sections can be calculated in a supersymmetric theory, both the signatures and the rates must be correct to interpret the results as the production of supersymmetric partners --- that events with such signatures should be produced, at the right rate, is a prediction of supersymmetry.

Some important demands on detectors arise here. First, the above signatures occur in an ideal world. In practice, sometimes jets overlap, sometimes jets radiate gluons and spread out, etc. Thus, it is not straightforward to compare theory and experiment when jets are involved. This difficulty can be minimized if detectors are designed to detect jets and to separate multiple core events as well as possible. Similarly, the $\not\!P_T$ is a crucial part of the signature and should be as well measured as possible.

We have seen that supersymmetry always gives some events with $\not\!P_T$. There are also events with no missing P_T but which still have missing energy. They arise, for example, when one interchanges $\tilde{\gamma}$ and a j in the above events, so the $\tilde{\gamma}$ and $\tilde{\nu}$ (or two γ) are approximately back-to-back and there is little net missing P_T. There is, of course, a continuum from large $\not\!P_T$ to no $\not\!P_T$. Some of these events are quite spectacular in that they appear to violate lepton number conservation, since one has a hard $e^{\pm}$ or $\mu^{\pm}$ but no missing momentum. If it could be established that such events did not occur it would give another way of putting strong limits on supersymmetry. To improve limits would require better jet/electron separation or jet/muon separation in future detectors. The presence of events with E but no $\not\!P_T$ suggests more thought should be given to constructing detectors sensitive to missing central region energy.

Further, in all of the supersymmetric events, the missing P_T is not due to a single particle. If K_μ is the 4-vector of the missing momentum, then if the momentum is carried off by a single particle (e.g. ν), $m^2=K^2=0$, while if two or more particles carry away momentum, $K^2 > 0$. While it is obviously difficult to measure K^2, it may be possible for some events to establish that $K^2 = \not{E}^2 - \not{P}_T^2 - \not{P}_L^2 > 0$.

One of the ways to produce supersymmetric partners--associated production (19,20) of "neutralinos", the partners of gauge bosons--should be emphasized because (a) it is a good way to look, and (b) if superpartners are not found, it will probably eventually provide the strongest constraint on the idea that supersymmetry is relevant to understanding the weak scale.

Basically, we want to look at production of $\tilde{\gamma} + \tilde{z}$. However, in any real theory these are four neutral gauginos, $\tilde{\gamma}$, $\tilde{z}$, $\tilde{h}_1^0$, $\tilde{h}_2^0$. They couple through interactions with Higgs bosons, so they necessarily mix with one another, and the mass eigenstates never interact like the weak eigenstates. If the supersymmetry breaking is the same order as the weak scale, the mixing is always large. Thus, we should speak of producing new eigenstates; for simplicity I will refer to them as $\tilde{\gamma}'$, $\tilde{z}'$. (Readers who wish to pursue this should be careful of the existing literature, as a number of papers do not take the mixing into account and therefore give incorrect and misleading results.)

One reason this reaction will be important is that a nice effect occurs, and the determinant of the mass matrix contains a factor $\sin^2\theta_W$. Since the determinant of the mass matrix is the product of the eigenvalves, some of the masses must be small compared to the typical entries. Numerically one gets a result as shown in the figure, where the sum of masses of the two lightest neutralinos is shown as a function of the two important supersymmetry breaking parameters M,μ. Up to quite large M,μ one has rather light neutralinos, so they will be kinematically accessible. At lower energy and higher luminosity machines the lighter two neutralinos will be produced, while at higher energy machines, one will produce the two most strongly coupled neutralinos.

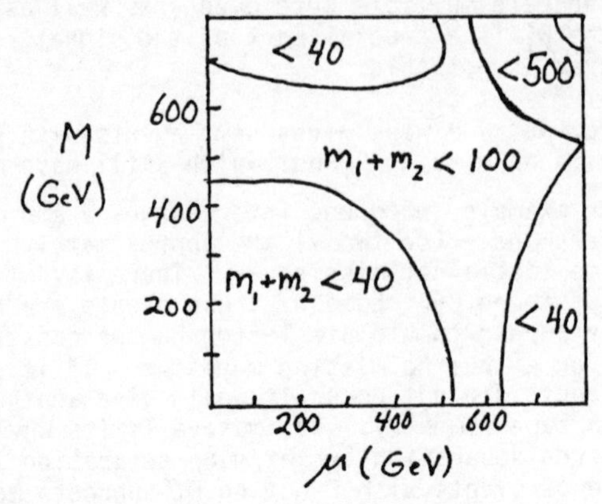

The cross section is somewhat smaller than might be expected because of the mixing. The production is via

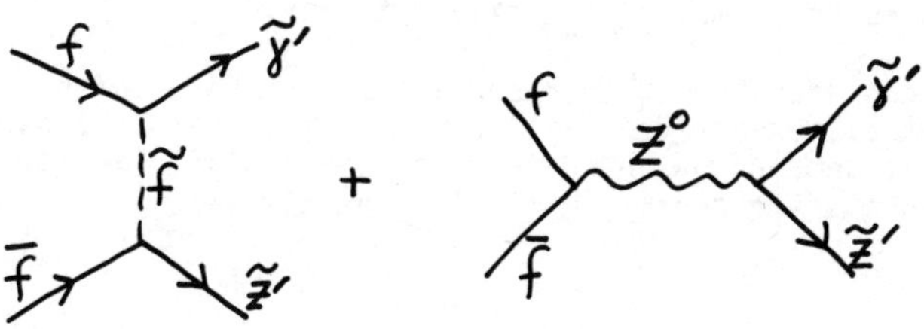

where f is any fermion, either a quark or an electron. Because of the mixing of higgsino into $\tilde{\gamma}'$, $\tilde{z}'$, their coupling to fermions is reduced. The off-diagonal coupling to Z^o is not zero so long as $\tilde{\gamma}$ and $\tilde{z}$ are not pure Marjorana or Dirac fermions, and usually it is a significant coupling. Then, depending somewhat on masses, one finds $\sigma(f\bar{f} \rightarrow \tilde{\gamma}' \tilde{z}') \simeq 0.1$ pb. This is large enough to observe at some present and future e^+e^- and hadron colliders.

The other nice feature of neutralino associated production is that it has a very nice signature. The lightest neutralino, $\tilde{\gamma}'$, is expected to be stable, and will escape a collider detector (as discussed for the photino above). The heavier neutralino will decay,

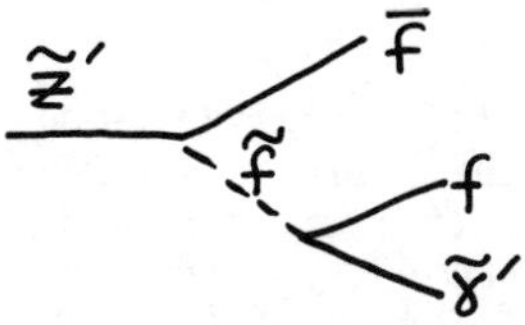

where f is any lepton or quark. If f is e or μ, the signature is an event with an e^+e^- or $\mu^+\mu^-$ pair on one side, opposite nothing,

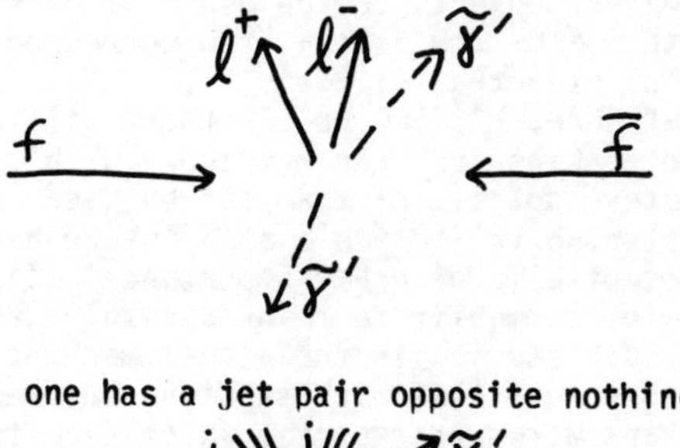

If f is a quark, one has a jet pair opposite nothing,

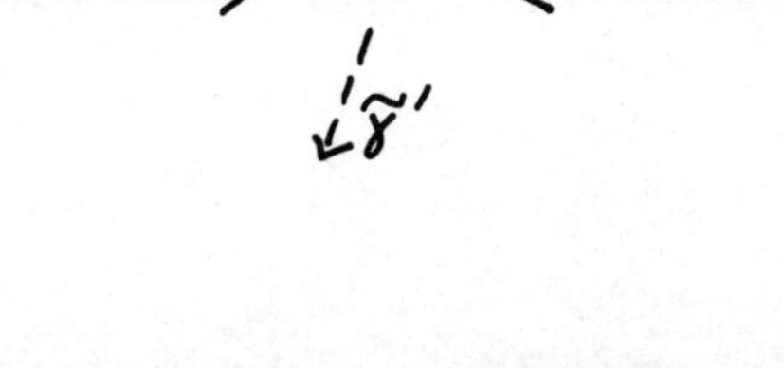

In either case, there is a clear one-sided event with $\ell^+\ell^-$or jj, a signature which is not expected in the SM. Background from $\gamma\gamma$ events at e^+e^-colliders or from $\tau^+\tau^-$is easily cut away.

We have concentrated in this section on new kinds of particles due to supersymmetry. Other ideas will produce other kinds of particles, but the general demands on new detectors, particularly involving missing P_T and missing energy, will often be similar.

Experiment(s) 3 -- Rare Effects at SuperColliders

There are two reasons (4) for building a high luminosity Super Collider,

(1) to probe higher mass scales, as we have discussed above, and

(2) heavier particles such as $\tau^{\pm}$, c, b, t, $W^{\pm}$ can be produced in large quantities and studied, particularly for rare decays and for CP violation.

The numbers are large; at any hadron collider with $\sqrt{s} \gtrsim$ 8-10 TeV, and $=10^{33}$ $cm^{-2}sec^{-1}$, in 10^7 sec the yields are > 10^{12} c's, ~10^{12} b's, ~10^{11} t's >10^{10} τ's, ~10^9 W's. If a way can be found to study these particles, their patterns of rare decays, the possibility of finding decays forbidden by the SM, and the possibility of studying CP violation in systems other than Kaons, offer new and perhaps crucial ways to address basic particle physics questions.

The study of these heavier objects can be begun at existing colliders. At the CERN SppS, with $=10^{30}$ $cm^{-2}sec^{-1}$, the corresponding numbers are about 10^9 c's, 10^7 b's, ~10^3 t's, ~10^6 τ's, ~10^4 W's. A further increase will occur at FNAL. At SLC or LEP there will be ~10^6 c and b, and over 10^5 τ's, and perhaps CESR can produce ~10^6 b's. Although these are sizeable rates, not all of them can be used, even at e^+e^-colliders. The step up to ~10^{12} b's at a future hadron collider is a qualitative one, potentially of great importance. It is important for the particle physics community to think carefully about the feasibility of doing such studies in the actual machine environment, how important it is to ensure that such studies can eventually be done. There are three relevant workshop reports, which seem to be guardedly optimistic that some utilization can be made of the potential event rates to study interesting physics, but so far no thorough study has been done. Clearly some vertex detection capability must be introduced to "trigger" on the existence of long-lived particles in the events, in order to obtain a sample sufficiently enriched in the heavier particles. Everyone is familiar with the situation at the ISR and at fixed target hadron machines, where charm is copiously produced but not well studied; if new detector techniques cannot beat that situation, then the large quantities of b-quarks, etc., are of no value.

We can briefly explore some of the constraints and the physics possibilities. The heavier particles c, b, t, W will be produced via perturbative QCD contributions, in the central region, with an approximately flat rapidity distribution. That means that in angle they are sharply peaked forward, with a large fraction going within a degree or two of the beam. So experimentation could either be considered for the central region, in which case there is an immediate cost of a factor of 20 or so in rate, or for a position near the beam pipe, in which case backgrounds may be prohibitive. In P_T probably the distributions peak for b, t at $P_T \sim 5$-10 GeV, so requiring $P_T \gtrsim 5$ GeV only costs a factor of 3-4 in rate. In the central region tracking itself may not be a problem since the number of charged particles per unit rapidity per radian in ϕ is only about one.

There are four categories of physics to consider. I will concentrate on b-quark physics; similar remarks can be made for the other particles. It is very important to keep in mind while thinking about rare decays or mixing or CP violation that the physics of these interactions is not understood. Often there is a SM prediction for some observable, but (differently from most situations) here we need not expect it to be right, because other effects could be present, giving much larger or very different effects.

(1) Look for rare decays which are forbidden in the SM. Many possibilities come to mind, such as $B \to \tau\mu$, $B_0 \to \mu e$, $T \to \tau\mu$, μe, $D \to \mu e$, $B \to \pi\mu e$, etc. Consider as an example, $B \to \mu e$. Suppose one can look in the central region and somehow learn that a sample of events was enriched in long lived particles, e.g. by finding μ or e with fairly large P_T and somewhat isolated, or by a "lifetime trigger". Then one can look opposite for an isolated μe pair whose effective mass is M_B (or M_D). A detector should be able to provide reasonable resolution in M_B. No background is expected at all for isolated μ + e, and requiring a few events at a given $M(\mu e)$ would surely eliminate any possible background. With 10^{12} b's or c's to start with, one could perhaps hope to detect a signal if it were as little as 10^{-8} in branching ratio.

Is that an interesting sensitivity? One might imagine that such interactions would arise through whatever physics was responsible for the origin of fermion masses. If some new interaction were involved, it would be unlikely to be diagonal for the mass eigenstates. So, perhaps one can speculate that such forbidden decays will occur with a factor M_f at each vertex. Further, since b decays are suppressed by a mixing angle factor U_{bc}^2, it seems a reasonable guess that

$$\frac{BR(B^0 \to \mu e)}{BR(K_L \to \mu e)} \sim \left(\frac{m_b}{m_s}\right)^2 \frac{U_{su}^2}{U_{bc}^2} \simeq 2\times 10^4$$

Thus, a sensitivity of 10^{-8} in $B \to \mu e$ may be comparable with 10^{-12} in $K_L \to \mu e$.

(2) There are a number of interesting B decays that are predicted to be rare but non-zero in the SM.

(a) The decay $B \to \tau\nu$ is one of two possible and difficult ways to measure the fundamental mixing angle U_{bu}, since the annihilation occurs via

[The other way is $\bar{\nu}N \to bX$ with production off a u-quark; to separate from production off charm requires events at large x. It may be very difficult to find candidate events or to determine that they came from $u \to b$.] The expected branching ratio is known (22) to be below 10^{-4} with existing limits, and could be much smaller. The theoretical rate depends on the wave function factor f_B, which is only known to about a factor of two so far, but a determination of U_{bu} to a factor of two or three would be very worthwhile.

(b) An interesting decay in the SM is $B \to k\ell^+\ell^-$, $K^*\ell^+\ell^-$. It appears to be a flavor-changing-neutral-current decay, but it is actually induced at one loop in the SM as a radiative correction. At the quark level it is completely calculable when M_t is measured, although the hadronic

wave functions required to put b in B and s in K or K* introduce a limited uncertainty in the predicted rate. Deshpande and I (23) estimate the branching ratio to be about 10^{-5}.

(c) A decay with a nice window on unknown contributions is $B \to \tau^+\tau^-$. It is analagous to $K_L \to \mu^+\mu^-$, but is enhanced by $(M_\tau/M_\mu)^2$ since the rate is helicity suppressed. The branching ratio is enhanced by $U_{su}^2/U_{bc}^2 \simeq 20$, so the net enhancement over $K_L \to \mu\mu$ is about 6000, giving an expected BR of about 6×10^{-5}.

(3) Mixing of B° and $\overline{B}^\circ$ can occur. In the SM the mixing effect is large, of order 10%. The simplest way to look is to tag one B (charged or neutral) to be definitely a B (or a $\overline{B}$), e.g. by its semileptonic decay, or any other method. Then, look for the semileptonic decay of the opposite B. If $\overline{B}^\circ \leftrightarrow B^\circ$ one would find like-sign dileptons rather than opposite sign ones. Care is needed to only use prompt leptons, which might be distinguished by their large P_T, as the secondary leptons from $b \to c \to \ell$ will mimic the like-sign effect.

(4) Finally, one can look for CP violating effects. These are predicted to be small in the SM, but there is no evidence that they are not large ($\gtrsim$ 10%) in some modes, and there are other models of CP violation where they can be large. Measurements are difficult, but perhaps possible. Comparisons can be made of opposite sign channels, e.g. $\ell^+\ell^+$ vs $\ell^-\ell^-$events if mixing is present, or with exclusive channels (compare ψK_S and ψ K_L, or $D^+\pi^-$and $D^-\pi^+$). Following one B on its lifetime curve has been considered (24). Detailed analyses of $B \to D\pi\ell\nu$ can be attempted (25) -- there will be about 10^{10} events/year of this, so if 10^5 could be tagged a 1% effect could be found. The decay $B_L^\circ \to \pi^\circ\ell^+\ell^-$ is forbidden by CP invariance; perhaps in some models it could be very large.

Acknowledgements

I have benefitted in preparing a talk such as this from many discussions with colleagues at workshops over the past two years. I am particularly indebted to Frank Paige for frequent interesting interactions.

References

1. Proc. of the 1982 DPF Summer Study on Elementary particle Physics and Future Facilities (Snowmass CO, 1982), ed. by R. Donaldson, R. Gustafson, and F. Paige.

2. Summary Report of the PSSC Discussion Group Meetings, June 1984, ed. by Phyllis Hale and Bruce Winstein, FNAL.

3. $\overline{p}p$ Options for the Supercollider, Proceedings of a DPF Workshop, Feb. 1984, ed. by J.E. Pilcher and A.R. White.

4. "Windows for New Physics at Super Colliders", Invited Talk at the Conference on Physics of the XXI Century, Tucson, AZ, Dec. 1983.

5. E. Eichten, I. Hinchliffe, K. Lane, and C. Quigg, Fermilab-Pub-84/17-T (1984).

6. Proceedings of the 1983 DPF Workshop on Collider Detectors, LBL, Feb. 1983, ed. by S. Loken and P. Nemethy.

7. H. E. Haber, G.L. Kane, and T. Sterling, Nucl. Phys. <u>B161</u> (1979) 493.

8. See also M.K. Gaillard in ref 3, and F. Paige in ref 2,3.

9. J. Gunion et al., private communication.

10. J. Fleisher and F. Jegerlehner, Phys. Rev. D23 2001 (1981).

11. See, for example, recent calculations by C.L. Bilchak, R.W. Brown, and J.D. Stroughair, Phys. Rev D29 375 (1984).

12. Michael S. Chanowitz and Mary K. Gaillard, Phys. Lett. 142 (1984) 85.

13. H. Georgi, S.L. Glashow, M. Machacek, and D.V. Nanopoulos, Phys. Rev. Lett. 40 (1978) 692.

14. R.N. Cahn and Sally Dawson, Phys. Lett. B136 196 (1984).

15. "The Effective $W^{\pm}$, Z° Approximation for High Energy Collisions", G.L. Kane, W.W. Repko, and W.B. Rolnick, Michigan preprint, June 1984, to be published in Phys. Lett.

16. Sally Dawson, preprint LBL 17947.

17. In addition to results given in refs. 1,2,3, I use some of my own calculations and some unpublished results from calculations of F. Paige, particularly with ISAJET.

18. F. Paige, ref. 3 and private communication.

19. A review along these lines is H.E. Haber and G.L. Kane, Michigan preprint UM TH 83-17 (1984), to be published in Phys. Rep. C.

20. J.-M. Frere and G.L. Kane, Nucl. Phys. B223 (1983) 331; J. Ellis, J.-M. Frere, J.S. Hagelin, G.L. Kane, and S.T. Petcov, Phys. Lett. 132B (1983) 436.

21. L.L. Chau et al., ref. 1; D. Loveless et al., ref. 3; G. Trilling et al., Snowmass II.

22. Ling-Lie Chau, Workshop on Search for Heavy Flavors, Como, Italy, Sept. 1983.

23. N. Deshpande and G.L. Kane, to be published.

24. See the calculations of M Machacek in the Snowmass II proceedings, and references there to earlier studies.

25. This proceeds exactly as the similar analysis for $D \rightarrow k\pi\ell\nu$; see G.L. Kane, K. Stowe, and W.B. Rolnick, Nucl. Phys. B152 (1979) 390.

Discussion on Section 7

J.D. Davies (Birmingham): Will you consume all the PS protons?

Speaker, D.R. Ward: I should think it unlikely that we will need all the protons for the whole period.

S. Loucatos (Saclay): Would UA1 take data simultaneously?

Speaker, D.R. Ward: I think there is a problem with their new vacuum chamber, because the beam size blows up at low momenta. But you should ask UA1!

D. Dibitonto (CERN): UA1 comment - no decision has been taken on this yet. It depends on the schedule for normal collider running, which begins soon afterwards.

B. Nicolescu (IPN, Orsay): Don't you think that low p_T physics is also "physics at future machines"?

Speaker, G.L. Kane: Yes, it is physics too. But presumably it is part of the Standard Model, which will eventually emerge from a full understanding of non-perturbative QCD. I have concentrated here on some of the questions which seem to motivate the particle physics community most strongly to build new machines, those questions which may focus our attention on physics beyond the Standard Model.

Relative impulse approximation for antiproton scattering

A T M Aerts

CERN, CH 1211 Geneva 23, Switzerland

Abstract. A comparison is made between the relativistic and non-relativistic impulse approximation predictions for the differential cross-section and polarization observables for antiproton scattering off ^{40}Ca at T_{lab} = 180 MeV. It is found that for the model considered here a clear distinction between these two approximations can only be made on the basis of polarization measurements.

1. Introduction

In the past five years, the description of elastic proton-nucleus scattering by means of a phenomenological Dirac optical potential has proved to be a success (Arnold et al. 1979, 1981, 1982). Not only has one been able to fit the differential cross-section data for scattering off J = 0 nuclei such as ^{40}Ca and ^{208}Pb, at a variety of momenta, but also the polariza-tion and spin rotation observables were reproduced quite well. Certainly in this respect the Dirac approach represents an improvement (Clark et al. 1982) to the non-relativistic approach.

It has also been possible to understand the parameters of the Dirac potentials in terms of more fundamental quantities. The Dirac potentials for low projectile energy have been shown consistent with those found in relativistic models for nuclear matter (Jaminon et al. 1980, Anastasio et al. 1983). At higher energies, the impulse approximation (McNeil et al. 1983) to the Dirac potential has been found to be in remarkably good agreement with the phenomenological model (Shepard et al. 1983). This approximation allows a direct calculation of the Dirac potential from the empirical NN scattering amplitudes and the measured nuclear densities in a parameter-free way.

The start of a $\bar{p}$-nucleus scattering program at the LEAR facility at CERN (Garreta 1984a) has brought within reach the possibility of testing the Dirac approach also for antiproton scattering. Regarding the antiproton as just another spin ½ particle, one may expect a similar amount of success for antiprotons as one has had for proton elastic scattering. This, however, is by no means guaranteed.

An important difference with proton-nucleus scattering is that no empirical $\bar{N}N$ amplitudes exist. Instead, one has to use as input amplitudes generated with models for the fundamental $\bar{N}N$ interaction (Dover and Richard 1980, Côté et al. 1982, Timmers et al. 1984). Since these models have, in view of the limited data available on $\bar{N}N$ scattering, not been constrained as fully as one would have liked, one cannot expect that the computed amplitudes give as complete a description of the $\bar{N}N$ interaction as the empirical amplitudes would do. The Dirac approach then may provide us with additional constraints for the fundamental interaction obtained by systematically studying $\bar{p}$-nucleus scattering.

Here we report on the initial stage of such a program (Aerts 1984). We briefly recall how a Dirac optical potential is constructed in Section 2. In Section 3, we present a comparison between the results of a relativistic and a non-relativistic calculation of the scattering observables.

2. Relativistic Impulse Approximation

The elastic scattering of an antiproton off a J = 0 nucleus is assumed to be described by the following co-ordinate space Dirac equation:

$$(\not{p} - m - U(E))\psi_{\vec{k},s}(t,\vec{r}) = 0 \quad (1)$$

In Eq. (1), $\vec{k}$ is the asymptotic momentum of the antiproton, related to its energy by $E = (\vec{k}^2+m^2)^{\frac{1}{2}}$; s is its spin orientation; U(E) is the energy dependent optical potential representing the interaction of the projectile with the ground state nucleus.

In order to construct the optical potential, we will use a single scattering, relativistic impulse approximation (McNeil et al. 1983), which gives U directly in terms of the nuclear densities ρ_i and $\bar{N}N$ invariant amplitudes F_i:

$$\langle\vec{k}'|U(E)|\vec{k}\rangle = -(2\pi/m)\left[F_s(\vec{q})\rho_s(\vec{q}) + \gamma^\circ F_v(\vec{q})\rho_v(\vec{q})\right] \quad (2)$$

with $\vec{q} = \vec{k}-\vec{k}'$. The only invariant amplitudes contributing to the potential for J = 0 nuclei are the Lorentz scalar (s), the time component (γ°) of the four-vector (v), and a tensor amplitude, which has been neglected in Eq. (2) (its relative strength is ~5%).

The optical potential in co-ordinate space, to be used in Eq. (1), is obtained by Fourier transformation of the expression of Eq. (2). This amounts to a folding of the amplitude with the nuclear densities.

3. Results and Discussions

The impulse approximation sketched in the previous section works best when nuclear medium corrections to the elementary interactions are unimportant. This has restricted its successful application for protons to processes where the projectile has a momentum of the order of its mass or larger. The highest projectile momentum, however, for which final antiproton-nucleus data will become available in the near future is P_{lab} = 600 MeV/c (Garreta 1984a). When we consider that the antiproton is much more strongly absorbed than the proton, we may anticipate that nuclear medium effects may not be as important for the antiprotons as they are for the protons. The impulse approximation, then, may already work well for the antiproton at P_{lab} = 600 MeV/c.

We will take as a specific example the scattering of $\bar{p}$'s off ^{40}Ca at P_{lab} = 600 MeV/c, or T_{lab} = 180 MeV, and solve Eq. (1) in the eikonal approximation (Amado et al. 1983, Wallace and Friar 1984). For simplicity, we will take $\rho_s = \rho_v$ in Eq. (2) and we will make the forward scattering approximation $F_i(\vec{q}) \to F_i(\vec{0})$. Writing $U(r) = S(r)+\gamma^\circ V(r)$, we have

$$S(r) = S_0(1+e^{(r-a)/b})^{-1} \; ; \qquad V(r) = V_0(1+e^{(r-a)/b})^{-1} \quad (3)$$

where we have parametrized the densities by simple two-parameter Fermi-distributions. For ^{40}Ca, a = 3.55 fm and b = 0.64 fm are reasonable values for the parameters. From the $\bar{N}N$ model amplitudes of Timmers et al., we deduce that S_0 = (-100.0-i422.2) MeV and V_0 = (94.4+i286.8) MeV. We have considered here the case without Coulomb interaction.

The results of the calculation for the differential cross-section and the polarization observables are displayed in Figs. 1(a-c), labelled with "Dirac". For comparison, we have included the results of a non-relativistic impulse approximation calculation (see Shepard et al. 1983). The term "non-relativistic" here refers to the transformation properties of the potential, as opposed to that of the Dirac potential, not to the kinematics which is relativistic. These curves are labelled "Schrödinger".

From Fig. 1a we see that we cannot distinguish between the Dirac and Schrödinger curves. The closeness of these curves is due to the fact that the Schrödinger and the Dirac potential in its Schrödinger equivalent form are equal up to terms of second and higher order in $(S_0-V_0)\rho(r)/2m$, which is usually small compared with unity. From this we see that the $\bar{N}N$ interaction helps to suppress terms of higher order than linear in the density. This we take as an indication that nuclear medium effects may indeed still be comparatively weak at this energy. The model dependence of this observation is currently under investigation (Aerts 1984). The fact that variations of a few per cent in a and b do not lead to noticeable differences between the predictions of the relativistic and non-relativistic impulse approximations, leads one to suspect that the two curves will still be close when we use the full potential of Eq. (2). A comparison with the preliminary experimental data (Garreta 1984b) shows that our crude model already reproduces the general features for $d\sigma/d\Omega$, but predicts the

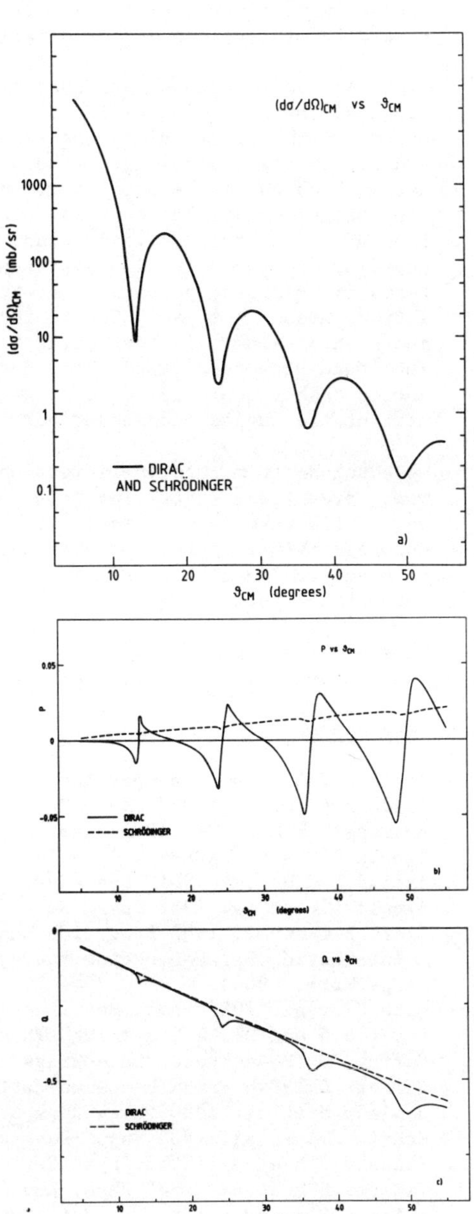

Fig. 1. Differential cross-section (a), polarization (b) and spin rotation (c) for elastic $\bar{p}$-^{40}Ca at T_{lab} = 180 MeV.

position of the diffraction minima at angles θ which are 1°-3° too small. Further refinements such as the use of more realistic densities and the inclusion of eikonal correction terms and the final experimental data are needed before one can make a detailed comparison.

Figs. 1b and 1c demonstrate that the main difference between the predictions of the non-relativistic and the relativistic impulse approximations occurs in the polarization observables P and Q. This is well known from elastic proton scattering. The distinction is most obvious in the polarization P(θ) where the Dirac curve shows much more structure than the Schrödinger curve. The same is to a lesser extent true for the spin rotation Q(θ). For both approximations the phase difference between the spin non-flip and the spin flip amplitude is about -90°. The non-linear density terms in the Dirac potential show up here by generating small phase oscillations around θ = -90°. The Schrödinger phase is almost stationary, just above this value. In both cases the relative increase of the spin flip amplitude with respect to the non-spin flip amplitude is clear. We also notice that the predicted polarizations are rather weak, less than 5% for most of the angles considered here!

We conclude from this investigation that nuclear medium effects are rather weak, even for $\bar{p}$ scattering at a low T_{lab}= 180 MeV. Experimental data on P and Q will make it much easier to distinguish between the relativistic and non-relativistic impulse approximations. It will certainly be quite useful to have such data when one wants to test the various model predictions for the optical potential.

We would like to thank J.J. de Swart and P.H. Timmers for providing us with the $\bar{N}N$ amplitudes in numerical form.

References

Aerts A T M 1984 in preparation
Amado R D et al. 1983 Phys. Rev. C28 1663
Anastasio M R et al. 1983 Phys. Rep. 100 327
Arnold L G and Clark B C 1979 Phys. Lett. 84B 46
Arnold L G et al. 1981 Phys. Rev. C23 1949
Arnold L G et al. 1982 Phys. Rev. C25 936 and references therein
Clark B C et al. 1982 Proc. Indiana Univ. Cyclotron Facility Workshop: The Interaction Between Medium Energy Nucleon in Nuclei, ed. H O Meyer (A I P New York, 1983)
Côté J et al. 1982 Phys. Rev. Lett. 48 1319
Dover C B and Richard J M 1980 Phys.Rev. C21 1466
Garreta D 1984a These Proceedings
Garreta D 1984b Private communication
Jaminon M et al. 1980 Phys. Rev. C22 2027
McNeil J A et al. 1983 Phys. Rev. Lett. 50 1439
Shepard J R et al. 1983 Phys. Rev. Lett. 50 1443
Timmers P H et al. 1984 Phys. Rev. D29 1928
Wallace S J and Friar J L 1984 Phys. Rev. C29 956

Low energy nucleon–antinucleon interactions and a hybrid model

F. Ando

Iryo-Gijutsu Tandai, Shinshu University, Matsumoto, Japan

Abstract. A reaction matrix formalism is applied to study of the nucleon antinucleon interactions. The internal and outer parts are described by the 3 quark-3 antiquark and the conventional hadronic wave functions, respectively. The phenomenological transition potentials connecting the two regions are then studied microscopically with the one-gluon exchange interaction previously considered.

1. Introduction

Jaffe and Low (1979) proposed a reaction matrix method in the analysis of low energy hadron-hadron scattering to reveal the quarkgluon eigenstates. The hadron-hadron scattering experiments in the external region can predict the P-matrix poles which in turn are eigenstates of the confined 'bag' states. A number of similar attempts (Jaffe 1977, Kim and Orlowski 1983, Simonov 1981, 1982, Lomon 1983, Henley et al 1983) have been made for NN, meson-meson interactions with reasonable success. In general, the short range hadron-hadron interaction should be described with quark and gluon degrees of freedom, whereas the long range component is well described with the phenomenological hadronic degrees of freedom. A natural question to ask is whether the methods may be extended to study the nucleon antinucleon interaction at low energies.

The reaction at low energies have been extensively studied in quark language (Faessler et al 1982, Green and Niskanen 1984, Maruyama and Ueda 1981, 1983, Maruyama 1983, Ando 1983), either in terms of one-gluon exchange interaction or quark rearrangement or both. Very recently Maruyama and Ueda (1984) have proposed to analyse the $N\bar{N}$ potential with quark rearrangement and one-meson exchange, with a qualitative sucess if a repulsion at short distance is introduced by hand. Alternatively, one may try a posibility of applying a reaction matrix theory to $N\bar{N}$ problem.

Perhaps the earliest attempt to understand the nucleon-antinucleon annihilation in terms of the reaction between bags was by Friedman, Hwang and Wilets (1981), who assumed that a nucleon and a antinucleon annihilate when the bags overlap, and that the reaction takes place through doorway states consisting of quarks and one gluon, which results in the imaginary part of the annihilation potential, whereas the real part is taken over from the phenomenological potential. In the advent of LEAR operation, a closer look at the interaction in the framework of the reaction matrix theory seems to be profitable.

2. Model

At this preliminary stage, a rigorous formulation of $N\bar{N}$ reaction theory is beyound our scope. Perhaps the simplest thing to do is just to extend one of the NN scattering formalisms minimally to include inelastic as well as annihilation channels, causing some complications. Note also that, unlike NN channels, no resonances nor bound states have been established here yet. Bearing these in mind, we go on to describe our reaction matrix model.

There are two regions of interactions: the external where separated hadrons interact with exchange of mesons, describable with the conventional phenomenological potentials, and the internal region where the $N\bar{N}$ becomes compound bag states. Therefore the wave function consists of two components: the hadronic and quark sectors,

$$\psi_h = \mathcal{A}\, \varphi(\underline{r})\, \phi_A(\underline{r}_1,\underline{r}_2,\underline{r}_3)\, \phi_B(\underline{r}_4,\underline{r}_5,\underline{r}_6) \quad r \geqslant R \ ,$$

$$\psi_q = \sum \alpha_\nu\, \phi_\nu^{3q3\bar{q}}(\underline{r}_1,\cdots,\underline{r}_6) \qquad r \leqslant R \ ,$$

where r is the relative coordinate of the two hadronic clusters A and B. Note here that R is comparable to but not the same as the bag radius. The two regions are coupled via transition interactions, thus the hamiltonian is written as

$$H = H_o + V \quad ,$$

where H_o represents the conventional hadronic interaction describing $N\bar{N}$, two- and three-meson states for $r \geqslant R$, and the bag hamiltonian for $r \leqslant R$. With the help of the projection operators for the two regions, we have

$$(E - H_{qq})\, \phi_\nu = 0 \quad ,$$

with suitable boundary conditions. The ϕ_ν are eigenstates of H_{qq} consisting of $3q3\bar{q}$ bag ground and excited states, with one-gluon exchange (t-channel) and self-energy. However the annihilation interactions are not included.

The transition between quark compound states and hadronic states are represented by QHP and PHQ. The simplest of such interaction within QCD is to take the annihilation via one-gluon exchange. Thus the transitions

$$(3q3\bar{q}) \rightarrow \begin{cases} N\bar{N} \\ 2 \text{ mesons} \\ 3 \text{ mesons} \end{cases}$$

take place through V.

The quark wave function ψ_q may be projected out, and ψ_h satisfies

$$(E - H_{hh})\,\psi_h = V_{hqh}\,\psi_h \ ,$$

where the interaction V_{hqh} contains the effect of the internal interaction. Phenomenologically it is convenient to restrict the form of V_{hqh}, e.g., being non-zero only at the surface R (Simonov 1981). This is justified since the quantity, which is essentially a product of the penetration factor and the reduced width in the language of compound nucleus, is a function of R.

Its energy dependence is determined from the boundary condition chosen at the surface in general. If viewed from the external region, the transition takes place only at the boundary.

The interaction such as rearrangement and annihilation into vacuum are not quite fit in this schme,perhaps to be included with phenomenological coupling parameters. Alternatively, such processes may take place through the multi-quark exchange (hadron exchange), so that the ansatz of sharp boundary may not be suitable, instead an extra transitional region may have to be added.

The bag compound states may be compared to the nuclear compound states which decay into various reaction modes. Different boundary conditions on the compound state lead to equivalent but apparently different formulations of the nuclear reaction theory. In particular, the transition (decay) amplitudes are directly related to the various boundary conditions at the channel radius. Thus in the spirit of reaction theory, it may make little sense to evaluate the amplitudes with microscopic interactions. On the other hand, there are a number of successful microscopic calculations on $N\bar{N}$ reaction crosssections, e.g., the branching ratios of various decay channels with specific wave functions. As a first step toward connecting the two approaches which are far apart at present, we are attempting to evaluate the quark gluon interactions over the 'bag' volume.

In a study of $N\bar{N}$ annihilation (Ando 1983), it was argued that the processes with two-meson final states must be included along with the three-meson states, mainly from a consideration of color structure. This should be true here,with quarks being confined in a bag. The quark-gluon vertex is given, with obvious notations, by

$$-ig\,\bar{\psi}_l(x)\left(\tfrac{1}{2}\lambda^a\right)_{lm}\gamma_\mu\psi_m(x)\,A^a_\mu(x)\,.$$

The annihilation of $N\bar{N}$ is assumed to take place through gluon exchange between a quark and an antiquark in bag compound states of the internal region. As illustrated in the figures, there are two types of interactions leading to three-meson final states (a) and (b), while (c) refers to a two-meson state. The transition matrix elements of Fig. are evaluated e.g., with the usual bag wave functions. The relevant integral should be done over the internal region. The result expressed as a function of R is essentially $V_{hqh}(r,r')$, where r(r') is the relative distance between the hadronic clusters in the external region.

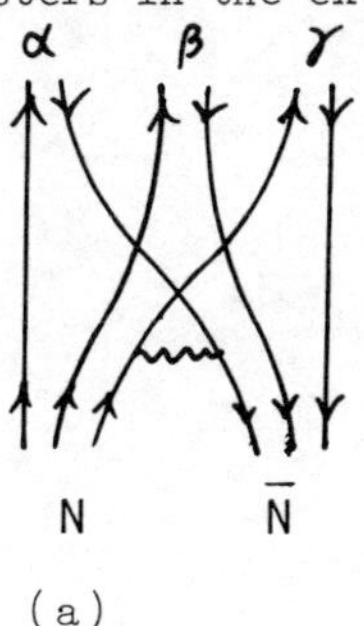

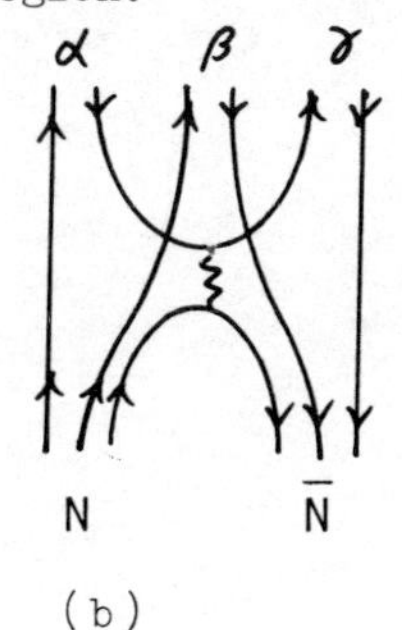

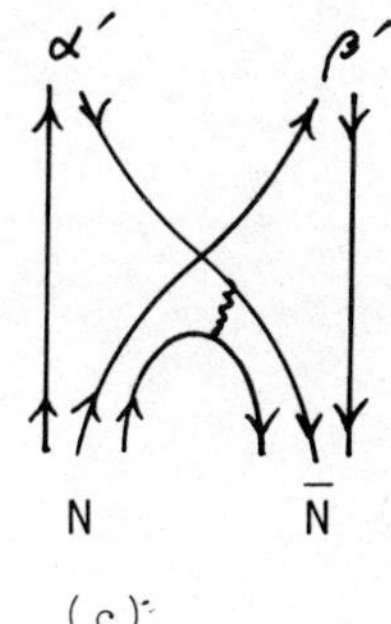

Fig. $N\bar{N}$ transitions into three and two mesons (a) exchange, (b) and (c) annihilation interactions.

3. Discussions and outlook

In this preliminary report, we are trying to convey an important role of nuclear reaction theory in the study of hadronic interactions, especially the nucleon-antinucleon reaction. In particular, we emphasize the concept of compound nuclear states should be the first to be scrutinized. A possible application of our model is to estimate the decay branching ratios and energydependences of low energy $\bar{N}N$ transition into mesons.

As mentioned before, the rearrangement interaction, which is favored in data fitting over the one-gluon exchange interaction, does not fit in our scheme. In fitting data, however, one usually use adjustable parameters which obscure different roles among elementary processes. Secondly, the process takes place at an intermediate range, which could be treated in terms of the direct interaction.

We should say that a detailed formulation is to be worked out now so that the compound reactions are clearly separated from other kinds of reactions.

References

Ando F 1983 Proc.Intntl. Summer School on NN and Nucl. Many-Body Problems (Chang Chun)
Faessler A et al 1982 Phys.Rev. D26,3280
Friedman R A et al 1981 D23, 1103 Phys. Rev.
Green A M and Niskanen 1984 Nucl. Phys. A412, 448
Henley E M et al 1983 Phys. Rev. C28, 1277
Jaffe R L 1977 Phys. Rev. D15, 267, 281
Jaffe R L and Low F E 1979 Phys. Rev. D19, 2105
Kim Y E and Orlowski 1983 Purdue Univ. Prep. 5
Lomon E 1983 C.T.P.#1130 and ref. therein
Maruyama M 1983 Prog.Theor. Phys. 69,937
Maruyama M and Ueda T 1981 Nucl. Phys. A364, 297
——————— 1983 Phys. Lett. 124B, 121
——————— 1984 OUAM-3-1
Simonov Yu.A 1981 Phys. Lett. 107B, 1
——————— 1982 Sov. J. Nucl. Phys. 36, 422

Inst. Phys. Conf. Ser. No. 73: Section 8
Paper presented at VII Eur. Symp. Antiproton Interactions, Durham 1984

A model of antiproton annihilation in nuclei implying two nucleons

J CUGNON and J VANDERMEULEN

Université de Liège au Sart Tilman, Bâtiment B.5, B-4000 LIEGE 1, Belgium

Abstract. We examine the possibility of an annihilation mechanism in nuclei implying in the primary stage more than one nucleon. If this results in the formation of an extended quark-gluonic blob, the production of strange particles is enhanced[1,2].
We first present arguments showing that annihilation on two nucleons is far from negligible. Second, we examine properties of such an annihilation by means of a phase space model. We show that considering a conventional hadronic phase only also predicts some strangeness enhancement.

1. Probable existence of b=0 annihilations.

We assume that, inside a nucleus, the collision of an incoming $\bar{p}$ with a nucleon leads in a first stage to the formation (with cross section σ) of a hadronic fireball with baryon number b=0. The fireball has a finite lifetime τ and a nonvanishing velocity β_f. Therefore, it may collide with another nucleon inside the nucleus and coalesce with it, forming a b=1 fireball (with cross section σ') which eventually decays into a number of hadrons inside the nucleus.

A calculation assuming a homogeneous medium gives, for the fraction of (b=1) annihilations : $R = \frac{P(b=1)}{P(b=1)+P(b=0)}$, the values in the table.

	p_{lab} (GeV/c)	0.6	2.0
$\sigma' = \sigma$	τ = 0.3 fm/c	0.18	0.22
	τ = 1 fm/c	0.43	0.49
$\sigma' = \frac{\sigma}{3}$	τ = 0.3 fm/c	0.07	0.09
	τ = 1 fm/c	0.20	0.24

The probability of forming a b=1 fireball is sizeable, even for a small lifetime of the b=0 fireball. These figures would be somewhat lowered if short range correlations were taken into account, because it is then unprobable to find two nucleons very close to each other but this effect is counterbalanced by the range for the annihilation process.

2. The decay of the fireballs.

We assume that both the b=0 and b=1 fireballs decay statistically into stable hadrons. Our treatment is based on the idea that, due to the finite fireball mass, the constraints of energy-momentum conservation must be significant. The rate for a given multiplicity is written :

$$f_n(\sqrt{s};m_1,\ldots,m_n) = C^n R_n(\sqrt{s};m_1,\ldots,m_n) \quad ,$$

where R_n is the invariant momentum space integral. We take

$$C = \lambda C_0 \quad ; \quad C_0 = \frac{\pi r^2}{h^2} = \frac{1}{4\pi m^2}$$

where h is the Planck constant ; r is the Compton wave-length associated to a characteristic hadronic mass m , which we choose equal to the pion mass.

Applied to $\bar{p}p$ annihilation at rest (b=0 fireball), the model with $\lambda = 1.60$ gives the correct mean pion multiplicity and a reasonable breakdown between the various pion multiplicities. At the same time, the predictions for kaonic channels are good.

Channel type	Percentage		$\langle n_\pi \rangle$	
	experiment.[3]	model	experiment.	model
π's	95.4 ± 1.6	95.5	5.01 ± 0.23[4]	5.05
$K\bar{K}\pi$'s	4.6 ± 1.6	4.5	$\sim$ 2	1.6

The inclusive pion spectrum can also be computed in the assumption of uniform phase space distributions. The predictions turn our rather close to the observed spectrum. It is also reasonably akin to a Boltzmann spectrum with T = 105 MeV.

3. The b=1 fireball.

The possible final states contain, in addition to a baryon in one of the various possible states (N,Λ,Σ or Ξ), a system of mesons of opposite strangeness. On the basis of a result obtained in the frame of the statistical bootstrap model[5], that the proper volume of a particle (or cluster) must be proportional to its mass, we have put

$$\frac{C(b=1)}{C(b=0)} = (1.5)^{2/3} \quad .$$

We note an enhancement of strange channels. The fraction of channels containing a $K\bar{K}$ pair is somewhat reduced compared to the b=0 case, but the channels with one hyperon take up a substantial fraction of the total rate. We predict that 8.8 % of the b=1 annihilations will produce a hyperon. The average number of kaon (S=+1) per annihilation is predicted to be 0.12 compared to 0.05 in the case b=0. (It is still larger with a b-independent C.)

PREDICTED BRANCHING RATIOS FOR (b=1) ANNIHILATION ($\sqrt{s}=3M_N$)

Channel type	Percentage	$<n_\pi>$
$N\pi$'s $N\pi$	88.5 5.2×10^{-2}	4.73
$NK\bar{K}\pi$'s	2.7	1.16
$\Lambda K\pi$'s	2.9	2.51
$\Sigma K\pi$'s	5.5	2.32
$\Xi KK\pi$'s	0.4	0.39
all	100.	4.42

4. Discussion.

We have given indications which tend to show that annihilations inside nuclei implying two (and perhaps more) nuclei are likely.

The model that we consider here is based on kinematical constraints operating in a pure hadronic phase. The most important consequence is the prediction that strangeness production is favoured in $b>0$ compared to $b=0$ annihilations. This is essentially due to the fact that the transformation of a nucleon into a hyperon requires only 35 to 50 % of the energy necessary to create a kaon. It is interesting to compare our results with those obtained in refs.[2,6] where the authors assumed that the annihilation gives rise to the formation of a quark-gluon blob. We consider the ratio $\bar{R}$ of the antistrange quark to the total antiquark content of the produced particles[7]. In the quark-gluon phase, one has

$$\bar{R} = \frac{\bar{s}}{\bar{q}} = \frac{1}{4}\sqrt{\frac{\pi}{2}}\left(\frac{m_s}{T}\right)^{3/2} \exp\left(\frac{\mu-m_s}{T}\right) ,$$

where m_s is the mass of the strange quark. For zero chemical potential and $m_s/T \approx 1$ to 2 , one has $\bar{R} \approx 0.12$. For a non-zero baryon density ($\mu \neq 0$), this ratio is somewhat higher. Experimentally, annihilation at rest gives (by using pion and kaon multiplicities) $\bar{R} = 0.009$. This result is accounted for in our model. The small value reflects the large difference between pion and kaon masses. If, on the other hand, a quark-gluon model is applied to fit the $\bar{N}N$ annihilation data, it is then necessary to assume that the strange quark density is not saturated, by a fraction $F \approx 0.1\text{-}0.2$ [6].

When we consider $b=1$ annihilation in our model, we find a ratio

$$\bar{R} = \frac{f(NK\bar{K}) + f(\Lambda K) + f(\Sigma K) + 2f(\Xi KK)}{<n_\pi> + f(NK\bar{K})} = 0.027 .$$

This is a substantial increase from the $b=0$ case. For $b=2$, we find $\bar{R} = 0.037$.

We thus arrive qualitatively at the same conclusion as the one obtained in ref.[1] assuming the formation of a quark-gluon blob. But we stress that our result only comes from kinematical constraints inside the hadron phase.

The (S=+1) kaons only scatter elastically with the nucleons ; therefore the (S=+1) content is preserved by the intranuclear cascade. (At the energies covered by LEAR, the energy of the annihilation pions is not sufficient to produce kaons with significant rate.)

Another signal is the presence of a high energy tail in the proton spectrum. Indeed, b=1 annihilations produce protons with a temperature of $T \approx 110$ MeV . The nucleus is not completely transparent to such protons, but $\sim$ 30-50 % of them would emerge out of medium heavy targets like Ca or Ag. These protons would then be observable, since the so-called "participant" protons ejected during the cascade will have a rather lower temperature[8].

References.

1) Rafelski J 1980 Phys. Lett.91B 281 ; Workshop on Physics at LEAR Erice 1982
2) Rafelski J and Müller B 1982 Phys. Rev. Lett.48 1066
3) Baltay C et al. 1966 Phys. Rev.145 1103
4) Ghesquiere C 1974 in Proceedings of the Symposium on Antinucleon-Nucleon Interactions Liblice ed L Montanet (Report CERN 74-18 p.436)
5) Hagedorn R, Montvay I and Rafelski J 1980 in Hadronic Matter at Extreme Energy Density eds N. Cabibbo and L Sertorio (New York:Plenum) p.49
6) Phatak S C and Sarma N 1984 Bhabha Institute preprint
7) Müller B 1983 Preprint UFTP 125/83
8) Cahay M, Cugnon J and Vandermeulen J 1983 Nucl. Phys.A393 237

Paper presented at VII Eur. Symp. Antiproton Interactions, Durham 1984

Parametrization of low energy $\bar{p}p$ scattering data

W. Derks, P.H. Timmers, and J.J. de Swart

Institute for Theoretical Physics, Nijmegen, The Netherlands

Abstract. Several parametrizations of $\bar{p}p$ differential cross sections, from which σ_T and ρ are determined, were tested by simulating experiments with the Nijmegen $\bar{N}N$ model. Special attention is given to spin dependence and Coulomb effects. It is found that experimental results on the total cross section and the rho parameter are questionable.

1. Introduction

The aim of many $\bar{p}p$ scattering experiments is to find the total cross section σ_T or the real-to-imaginary ratio ρ of the forward amplitude. This is done by parametrizing the elastic differential cross section $d\sigma/dt$ using typical high energy approximations. However, the same parametrization is used in the low energy region.

We tested the validity of this method by generating differential cross sections with the Nijmegen coupled channel $\bar{N}N$-model (Timmers, v.d. Sanden, and de Swart, 1984) and analyzed them in the same way as the experimentalists do (for references to experimental papers see also Timmers, v.d. Sanden and de Swart, 1984). In this way we can compare the results of this analysis with the exact results as calculated for the model.

2. Parametrization of the differential cross section

In the forward direction the nuclear amplitude (in the absence of Coulomb interaction) is often parametrized as

$$f(t) = \frac{1}{4\sqrt{\pi}}\,\sigma_T\,(i+\rho)\,e^{-bt/2} \tag{1}$$

where it is assumed that f falls off exponentially with the same slope for the real and imaginary part and that the phase of the amplitude is approximately constant for small t (here t is the absolute value of the momentum transfer squared). Moreover it is assumed that there is no spin dependence.

If the Coulomb potential is present, one has to add the Coulomb amplitude and, following a calculation of Locher (1967), due to the Coulomb-nuclear final state effects and form factor effects one has to multiply f with a phase factor $\exp[-i\,\delta(t)]$ where $\delta(t)$ is given by

$$\delta(t) = \delta_L(t) - \delta_C(t) = \frac{\alpha}{\beta}\,[\ln R^2 t + \gamma] \tag{2}$$

Here $\delta_C(t)$ is the phase of the Coulomb amplitude, α is the fine structure constant, β is the laboratory velocity of the incident $\bar{p}$, $R^2 = 9.5\ \text{GeV}^{-2}$

and γ is Eulers constant (0.5772...). Therefore if one defines an "electromagnetic" amplitude:

$$f_L(t) = \frac{2\sqrt{\pi}\alpha}{\beta t} F^2(t)\, e^{i\delta_L(t)} \tag{3}$$

with F(t) the proton electromagnetic form factor:

$$F(t) = [1 + \frac{t}{\Lambda^2}]^{-2} \tag{4}$$

where $\Lambda^2 = 0.71\ GeV^2$, the differential cross section can be found by

$$\frac{d\sigma}{dt} = |f(t) + f_L(t)|^2 \quad . \tag{5}$$

The result of Lochers calculation that the Coulomb interaction changes f only by a phase factor is certainly not valid in the low energy region because there the Coulomb attraction also produces an enhancement of the absolute value of the nuclear amplitude.

Equation (5) leads to a splitting of $d\sigma/dt$ into three parts: $d\sigma_N/dt = |f|^2$, $d\sigma_C/dt = |f_L|^2$ and $d\sigma_I/dt = 2\ \mathrm{Re}\ f^* f_L$. For the nuclear part we find:

$$\frac{d\sigma_N}{dt} = \frac{1}{\pi} \left(\frac{\sigma_T}{4}\right)^2 (1+\rho^2)\, e^{-bt} \quad . \tag{6}$$

Note that if Lochers calculation is valid here, σ_T in equation (6) is the total cross section for scattering without Coulomb interaction.

It has been shown by Lacombe et al. (1983) that if spin dependence is important equation (6) should be replaced by

$$\frac{d\sigma_N}{dt} = \frac{1}{\pi} \left(\frac{\sigma_T}{4}\right)^2 (1+\rho^2)(1+\eta^2)\, e^{-bt} \tag{7}$$

where η is a function of the various amplitudes at t = 0.

With the Nijmegen $\bar{N}N$-model we tested these parametrizations by generating differential cross sections and analyzing them with equations (3), (6) and (7). Then we compared the fitted values of σ_T and ρ with the exact model values with and without Coulomb interaction.

3. Results

One of the high energy approximations that breaks down in the low energy region is the assumption of exponential fall-off with the same exponent for all helicity amplitudes. Fortunately this has little effect on the fitted values of σ_T and ρ.

The Coulomb effects do have influence, which is at least of the same order of magnitude as the current experimental errors. The model cross section with Coulomb σ_T^C is closer to the σ_T that is found from fitting the differential cross sections with formula (7) than the model cross section without Coulomb σ_T^{NC} (see figure 1, see next page). This is contrary to what one would expect from Lochers calculation.

We have parametrized the ratio $r_1 = \sigma_T^C/\sigma_T(\mathrm{fit})$ by

$$r_1 = \sum_{n=0}^{3} a_n\, p_{lab}^{-n} \tag{8}$$

The values of the a_n's are given in table 1. Spin dependence is not negligible. Fitting the differential cross sections with $\eta = 0$ leads to wrong values for σ_T and ρ (see figs 2 and 3, next page), although the χ^2 of the fit is roughly the same. We also parametrized the ratio $r_2 = \sigma_T(\eta = 0)/\sigma_T(\eta = \text{model value})$ with the same form:

$$r_2 = \sum_{n=0}^{3} b_n \, p_{lab}^{-n} \qquad (9)$$

The same was done for the difference $d = \rho(\eta = 0) - \rho(\eta = \text{model value})$:

$$d = \sum_{n=0}^{3} c_n \, p_{lab}^{-n} \qquad (10)$$

The values of the b_n's and c_n's can again be found in table 1.

Unfortunately one cannot fit ρ, η and σ_T at the same time because they are strongly correlated. As one can see in the figures, fitting with $\eta = 0$ can lead to errors of 10% or more in σ_T. The same can happen if one or two of the parameters are not fitted but taken from other sources that often give wrong values. It is possible that the effect of spin dependence is even larger in the Paris model (Coté et al., 1982), because there η is not only important in the very low energy region.

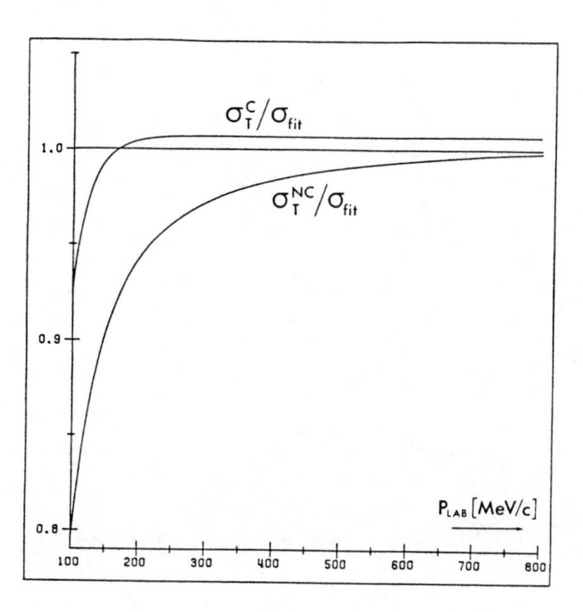

Fig. 1: model values with and without Coulomb divided by the fitted value of σ_T.

4. Conclusions

In fitting the differential cross sections spin dependence cannot be neglected. If it is properly taken into account and formula (7) is used to fit the data one finds a σ_T that corresponds to the total cross section for an interaction that includes also the Coulomb potential. This interpretation of the parameters in (7) is important if one wants to compare model results with experimental data. The results suggest that many experimental values for ρ and σ_T are quite questionable. Perhaps it is more useful if experimentalists give not only values for σ_T and ρ but also the values of $d\sigma/dt$ from which they extracted these parameters. Another possibility is fitting σ_T and ρ with $\eta = 0$ and then using formula's (9) and (10) to correct them.

n	a_n	b_n	c_n
0	1.00921	0.95547	$-0.12058 \cdot 10^{-1}$
1	$-0.36229 \cdot 10^{-2}$	$0.43827 \cdot 10^{-1}$	$0.70986 \cdot 10^{-2}$
2	$0.15647 \cdot 10^{-2}$	$-0.43771 \cdot 10^{-2}$	$0.16871 \cdot 10^{-2}$
3	$-0.20468 \cdot 10^{-3}$	$0.17797 \cdot 10^{-3}$	$-0.79980 \cdot 10^{-4}$

Table 1: values of a_n, b_n and c_n in GeV^n. The parametrization is <u>only</u> valid for 100 MeV/c < p_{lab} < 800 MeV/c.

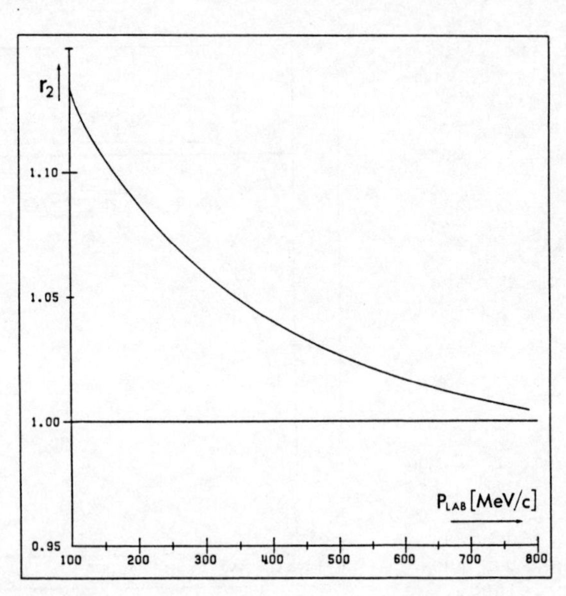

Fig. 2: value of σ_T from a fit with η = 0 divided by the result if η is taken equal to the model value.

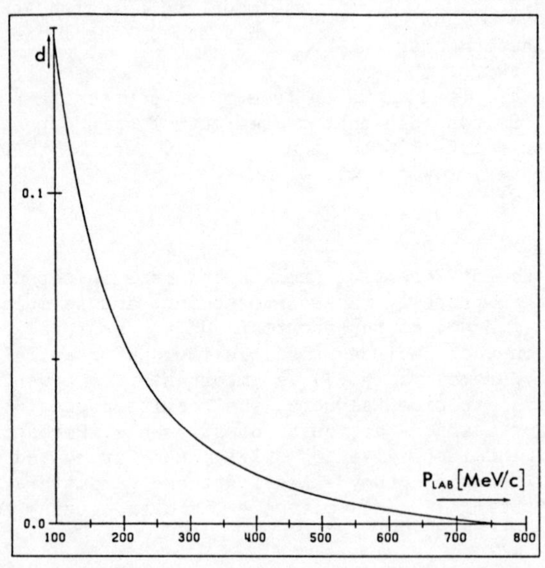

Fig. 3: difference of values of ρ fitted with η = 0 and fitted with the model value for η.

<u>References</u>

Coté J. et al. 1982 Phys.Rev.Lett. <u>48</u> 1319.
Lacombe M. et al. 1983 Phys.Lett. <u>124B</u> 443.
Locher M. 1967 Nucl.Phys. <u>B2</u> 525.
Timmers P.H., van der Sanden W.A. and de Swart J.J. 1984 Phys.Rev. <u>D29</u> 1928

The antiprotonic x-ray cascade in Mo isotopes (PS 186)

F J Hartmann[a], H Daniel[a], W Kanert[a], E Moser[a], G Schmidt[a], T von Egidy[a], J J Reidy[b], M Nicholas[b], M Leon[c], H Poth[d], G Büche[d], A D Hancock[d], H Koch[d], Th Köhler[d], A Kreissl[d], U Raich[d], D Rohmann[d], M Chardalas[e], S Dedoussis[e], M Suffert[f], A Nilsson[g]

a Technische Universität München, D-8046 Garching, Germany
b University of Mississippi, University, Mississippi 38677, USA
c Los Alamos National Laboratory, Los Alamos, New Mexico 87545, USA
d KFK und Universität Karlsruhe, D-7500 Karlsruhe, Germany
e University of Thessaloniki, Thessaloniki, Greece
f CRN et ULP, Strasbourg, France
g Research Institute for Physics, Stockholm, Sweden

Abstract. The antiprotonic x-ray cascade has been measured for the four molybdenum isotopes ^{92}Mo, ^{94}Mo, ^{98}Mo, and ^{100}Mo. Isotope effects in the cascade were not observed. The x-ray intensities were found to be well reproduced by cascade calculations starting at n=20 with a modified statistical or a linear distribution. In these calculations K and L electron depletion and refilling were taken into account. Nuclear absorption widths derived from a recent set of optical potential parameters were used.

1. Introduction

The atomic cascade of exotic particles certainly is interesting in itself, but it also deserves attention because it may influence other phenomena of the interaction of stopped exotic particles with matter (nuclear capture or annihilation, exotic atom level energies etc.). The kaonic, muonic and also pionic cascade has already been studied in detail (Godfrey and Wiegand 1975, Vogel 1980, Hartmann et al 1982, Hartmann et al 1984). Now the high flux $\bar{p}$ beams at LEAR make investigations of the antiprotonic cascade easily possible. We measured the antiprotonic x-ray cascade in ^{92}Mo, ^{94}Mo, ^{98}Mo, and ^{100}Mo.

2. The atomic cascade

When antiprotons enter a target they are slowed down and either annihilated in flight or captured at kinetic energies below 1 keV into highly excited bound states. The particle then cascades down the antiprotonic-atom level scheme, first by Auger effect and later by radiative transitions. When it reaches states with low angular momentum quantum number ℓ it annihilates. In the course of the cascade the host atom is deprived of electrons. The measurement of the antiprotonic x-ray cascade shall help to answer three questions:

a) what are the first bound states in the antiprotonic atom (characterized e.g. by the principal quantum number n and by ℓ),

b) what role is played by annihilation, and
c) how is the electron shell of the antiprotonic atom affected during the cascade.

This end is best attained by a comparison of the experimental results with model cascade calculations.

3. Experiments

The experiments were performed at LEAR. Antiprotons of 300 MeV/c were stopped in Mo targets of 2.5*3.0 cm² area and 100 mg/cm² thickness. Six semiconductor detectors registered antiprotonic and electronic x rays as well as nuclear γ rays. For the measurement of the antiprotonic x rays two pure Ge detectors were most suited: one with 5cm² area and 1.3cm thickness, the other with 2cm² area and 0.7 cm thickness. A typical spectrum from the smaller detector is shown in Fig. 1.

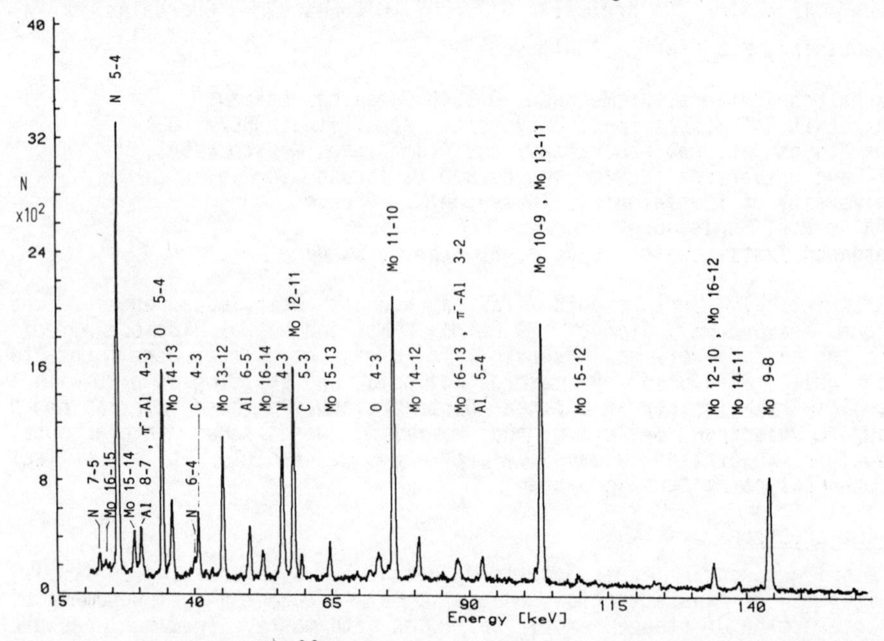

Fig. 1. Spectrum from ^{98}Mo, as taken with the smaller Ge detector. One channel corresponds to 0.082 keV.

The experimental configuration is described in more detail in the contribution of PS176 to this conference (Koch et al 1984). The line intensities were corrected for target absorption and detector efficiency and normalized to the Mo(11-10) intensity. Thirty antiprotonic x-ray lines have been evaluated up to now for each of the isotopes ^{94}Mo, ^{98}Mo and ^{100}Mo.

4. Results

The experimental intensities for ^{98}Mo are shown in Fig. 2 together with the results of cascade calculations to be described in section 5. The small Mo(7-6) intensity is mostly due to the nuclear resonance effect (von Egidy 1984). To show that there is no isotope effect in the cascade, the intensities of the strongest x-ray lines are compared in Table 1.

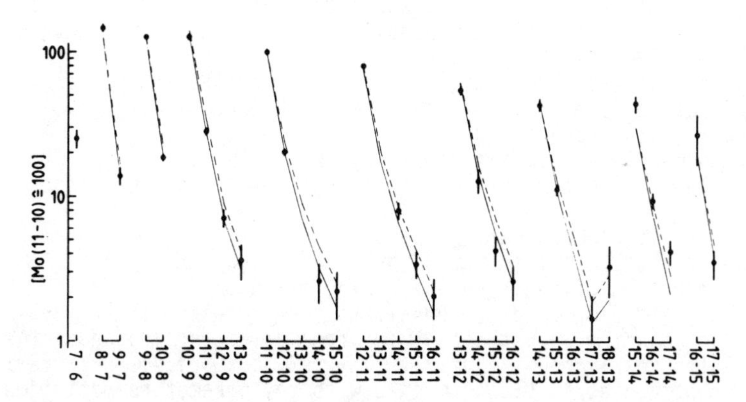

Fig. 2. Experimental and calculated x-ray intensities in Mo. Dashed line: statistical distribution; full line: modified statistical distribution. The (10-9) intensity comprises the (13-11) intensity and the (11-9) intensity comprises the (13-10) intensity

Table 1. Intensities of strong x-ray lines in ^{94}Mo, ^{98}Mo and ^{100}Mo

Transition	Intensity [Mo(11-10)= 100]		
	^{94}Mo	^{98}Mo	^{100}Mo
Mo(14-13)	42 ± 3	43 ± 4	46 ± 4
Mo(17-15)	3.8 ± 0.7	3.5 ± 0.9	4.2 ± 0.6
Mo(13-12)	51 ± 3	54 ± 4	55 ± 3
Mo(16-14)	10.1 ± 0.5	9.4 ± 1.1	10.9 ± 0.7
Mo(12-11)	79.6 ± 1.8	81.2 ± 2.4	82.1 ± 2.0
Mo(15-13)	10.7 ± 0.4	11.2 ± 1.0	11.0 ± 0.6
Mo(11-10)	100.0 ± 2.2	100 ± 3	100.0 ± 2.3
Mo(14-12)	16.9 ± 0.5	15.6 ± 1.0	15.7 ± 0.9
Mo(10-9+13-11)	137 ± 3	134 ± 5	135 ± 3
Mo(12-10)	22.6 ± 0.7	20.2 ± 1.0	22.8 ± 0.8
Mo(9-8)	127 ± 3	126 ± 3	125 ± 3
Mo(11-9+13-10)	28.5 ± 1.2	28.2 ± 1.6	26.3 ± 0.9
Mo(8-7)	121 ± 3	121 ± 5	5.2 ± 0.7[a]
Mo(10-8)	20.0 ± 0.7	13.3 ± 1.1	18.6 ± 1.1
Mo(9-7)	15.0 ± 0.8	13.9 ± 2.1	16.8 ± 1.6

[a] the low (8-7) intensity in ^{100}Mo is due to the E2 resonance effect

5. Cascade calculations

To simulate the antiprotonic x-ray cascade a modified Hüfner cascade code (Hüfner 1966, Bergmann 1979, Hartmann 1983) was used. In the calculation the cascade is assumed to start at n=20 (which should be well below the first bound state in the $\bar{p}$ atom) with an ℓ distribution to be adjusted. For the electronic K shell the depletion by Auger effect and refilling from higher shells is allowed for. In calculating the K refilling rate the momentary number of L and M electrons was taken into account. L shell

depletion by Auger effect and K electron refilling from the L shell was also calculated. The L shell was assumed to be refilled with a constant rate determined by relaxation time and density of the conduction electrons (Ashcroft and Mermin 1981) in Mo.
The $\bar{p}$ annihilation rate was calculated for circular orbits using a fit value for the effective scattering length recently reported (Batty 1981)

$$a = (1.53 + 2.50i)\ \text{fm}.$$

It was scaled to other levels in the usual way (West 1958).
Three different types of initial distributions were used in our cascade calculations:

a) the statistical distribution $p(\ell) \propto 2\ell+1$,
b) the modified statistical distribution $p(\ell) \propto (2\ell+1)\exp(\alpha\ell)$, and
c) a linear distribution of the form $p(\ell) \propto c+\ell$, $p(\ell) \geqslant 0$.

The statistical distribution leads to unsatisfactory results (χ^2_ν = 3.5). Distributions b and c after adjustment of one free parameter reproduce the experimental results nearly equally well (b:χ^2_ν= 1.3, c:χ^2_ν= 1.4). Figure 3 shows the different ℓ distributions in n=20 with the parameters optimized where possible.
In conclusion one can say that the experimental data may be well reproduced by cascade calculations.

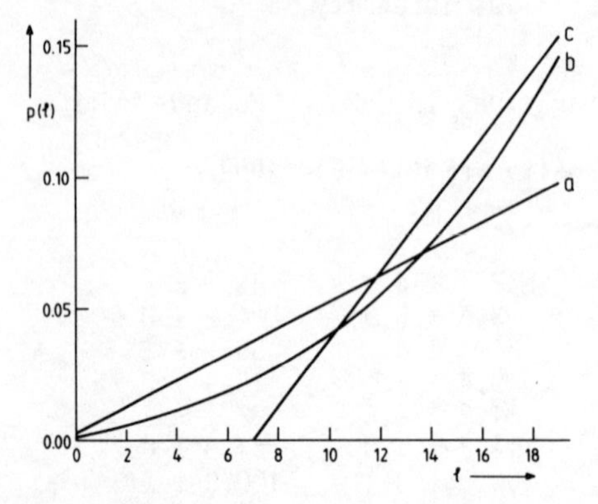

Fig. 3. Initial distributions for the cascade calculations.
a) statistical distribution,
b) modified statistical distribution, α =0.075,
c) linear distribution
$p(\ell) \propto -7+\ell$, $\ell \geqslant 8$,
$p(\ell) = 0$, $\ell < 8$.

Acknowledgments

We wish to thank H. Angerer, H. Hagn and P. Stoeckel for technical help and the LEAR staff for providing us with the intense $\bar{p}$ beam. Support of the German BMFT is gratefully acknowledged.

References

Ashcroft N W and Mermin N D 1981 Solid State Physics (Philadelphia)
Batty C J 1981 Nucl. Physics A372 433
Bergmann R 1979 Thesis Technische Universität München (Munich)
Godfrey G L and Wiegand C E 1975 Physics Letters 56B 255
Hartmann F J et al 1982 Zeitschrift Physik A305 189
Hartmann F J 1983 Computer program CASH
Hartmann F J et al 1984 Verh. DPG(VI) 19 940
Hüfner J 1966 Zeitschrift Physik 195 365
Koch H et al 1984 Contribution to these proceedings
Vogel P 1980 Phys. Review A22 1600
von Egidy T et al 1984 Contribution to these proceedings
West D 1958 Report on Progress in Physics 21 271

Inst. Phys. Conf. Ser. No. 73: Section 8
Paper presented at VII Eur. Symp. Antiproton Interactions, Durham 1984

Calculation of parton fragmentation functions from jet calculus: gluon applications

K. E. Lassila and Alexander Ng

Department of Physics and Ames Laboratory-DOE
Iowa State University, Ames, Iowa

Abstract. We present a method for calculation of general parton fragmentation functions based on jet calculus plus meson and baryon wave functions. Results for gluon fragmentation into mesons and baryons are discussed and related to recent information on upsilon decay into gluons. The expressions derived can be used directly in e^+e^- cross section predictions and will need to be folded in with baryon parton distribution functions when used in $p\bar{p}$ collisions.

The actual existence of jets strongly confirms QCD expectations about the behavior of quarks and gluons; the parton jet characteristics reflect the confinement mechanism. In this framework, a parton in some interaction is given a large Q^2 off-shell mass value and it evolves into a jet of color singlet hadrons. The quantity in this break-up which is measured experimentally is called the parton fragmentation function, denoted by

$$D_{i\to h_i\ldots h_m}(\bar{x}_1\ldots.\bar{x}_m;Q^2,Q_0^2).$$

The parton i, which we shall take as a gluon in the present applications, breaks up inclusively into hadrons $h_1\ldots h_m$, which have longitudinal momentum fractions $\bar{x}_1\ldots.\bar{x}_m$ of that carried by i. The initial Q^2 value given parton i by the hard interaction is degraded by the various possible QCD branching processes down to Q_0^2 where the final state hadrons appear. Our basic expression for the fragmentation function is:

$$D_{i\to h_i\ldots h_m}(\bar{x}_1\ldots.\bar{x}_m;Q^2,Q_0^2) = \int_0^1 \prod_{j=1}^{n} \frac{dx_j}{x_j} F_{i\to a_1\ldots a_n}(x_1\ldots x_n;Q^2,Q_0^2)$$

$$\times R_{a_1\ldots a_n}^{h_1\ldots h_m}(x_1\ldots x_n;\bar{x}_1\ldots\bar{x}_m). \tag{1}$$

This expression is constructed by multiplying two probabilities

$F_{i\to a_1a_2\ldots a_n}$ and $R_{a_1a_2\ldots a_n}^{h_1h_2\ldots h_m}$. The former, F, is the probability for parton i to breakup inclusively into partons $a_1a_2\ldots a_n$ with corresponding momentum fractions $x_1x_2\ldots x_n$. It is constructed according to the rules of jet calculus (Konishi et al.) for finding n quarks and antiquarks ($a_1 a_2 \ldots a_n$) which will appear in the final state as the bound state constituents of mesons and baryons. The jet calculus diagrams for finding 2, 3 and 4 (inclusive) partons in the final state of the QCD jet are shown in Fig. 1. The second quantity R gives the probability that quarks with antiquarks make mesons and sets of three quarks make baryons. It is constructed from the form factor analysis of bound states (Lepage and Brodsky). A recent analysis (Grössing et al.) strongly suggests that the main contribution comes from the edges of phase space. Thus, for meson M and baryon B production, $R^M \sim x_1 x_2$ and $R^B \sim x_1 x_2 x_3$, though simple, is justified. With the jet calculus expressions for each diagram, we can calculate fragmentation functions, e.g., $D_{i\to M}, D_{i\to B}, D_{i\to M_1 M_2}$.

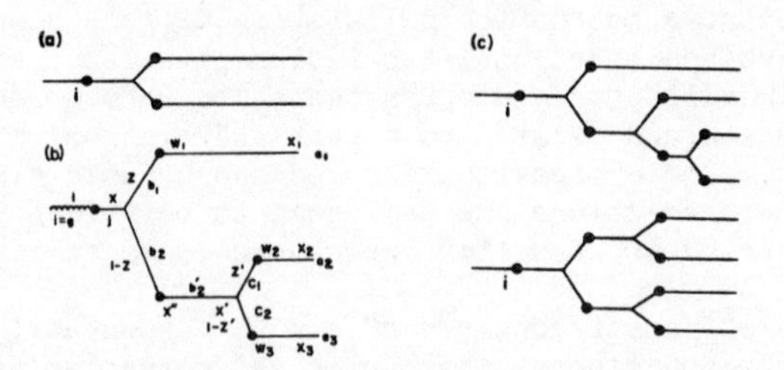

Fig. 1. Jet calculus diagrams giving probability for production of (a) two, (b) three, and (c) four partons. The big dots on the lines are the KUV propagators. The momentum fractions given by w, x, z labels are shown on the partons lines labelled a, b, c in (b) and enter as labelled into Eq. (2).

The calculations of $D_{i\to M}$ and applications are given in (Jones et al.) and (Chang and Hwa). These results were extended to quark (i≡g) fragmentation into baryons by (Migneron et al.) and (Eilam and Zahir). Here we emphasize gluon (i≡g) fragmentation into baryons because of the SFM experiment (Breakstone et al.) studying gluon fragmentation in $p\bar{p}$ experiments and upsilon decay into gluons (Giles). It is useful to begin by writing $F_g \equiv F_{g\to a_1 a_2 a_3}$, as depicted in Fig. 1(b), explicitly,

$$F_g = \Sigma \int_{Y_o}^{Y} dy \int_{Y_o}^{y} dy' \int_0^1 dx dx' dx'' dz dz' dw_1 dw_2 dw_3 D_{a_1 b_1}(w_1, y-Y_o)$$

$$\times D_{a_2 c_1}(w_2, y'-Y_o) D_{a_3 c_2}(w_3, y'-Y_o) P_{b_2\to c_1 c_2}(Z') D_{b_2 b_2}(x'', y-y') P_{j\to b_1 b_2}(z)$$

$$\times D_{jg}(x, Y-y)\delta(x_1 - w_1 z x)\delta(x' - x''(1-z)x)\delta(x_2 - w_2 z' x')\delta(x_3 - w_3(1-z')x'). \quad (2)$$

The $D_{\alpha\beta}$, with α,β=a,b,c as on Fig. 1b, are the KUV parton propagators the moments of which satisfy Altarelli-Parisi (AP) evolution equations in terms of the variable $y \sim \ln \ln Q^2/\Lambda^2$ with Y its value at the initial Q^2 and Y_o at the final degraded of $Q_o{}^2$ value. The QCD scale parameter is Λ and the P's are the AP splitting functions for QCD vertices. The delta functions give momentum conservation and the sums and integrals are done over unobserved parton quantum numbers. When one or more of the final partons a_1, a_2, a_3 are gluons, they are converted to $q\bar{q}$ pairs by inserting an appropriate number of $P_{g_j \to q_j \bar{q}_j}$ into expression (2) so the converted quark or antiquark is used in Eq. (1) to make the B or M in momentum conserving fashion. Thus, $D_{g\to B}$ can be split into parts labelled by a superscript for the number of gluons among $a_1 a_2 a_3$,

$$D_{g\to B}(x,Q^2) = D^{(o)}_{g\to B}(x,Q^2)+D^{(1)}_{g\to B}(x,Q^2)+D^{(2)}_{g\to B}(x,Q^2)+D^{(3)}_{g\to B}(x,Q^2). \quad (3)$$

These distributions calculated at Q^2=1089 GeV^2 for proton production in both a u quark and a gluon jet are shown as functions of the momentum fraction x(of B=proton) in Fig. 2. This calculation was done by transforming Eq. (1) into moment space, calculating all moments possible on a VAX computer and converting back to x space. The partial contributions of each term in Eq. (3) and the sum are shown labelled by the number of gluons except for the $D^{(o)}_{g\to B}$ term which is labelled "qqq" on Fig. (2a) and is too small to be shown in Fig. (2b). The leading quark effect is automatically contained in the jet calculus and is evident clearly in u→P (especially in the ggq term).

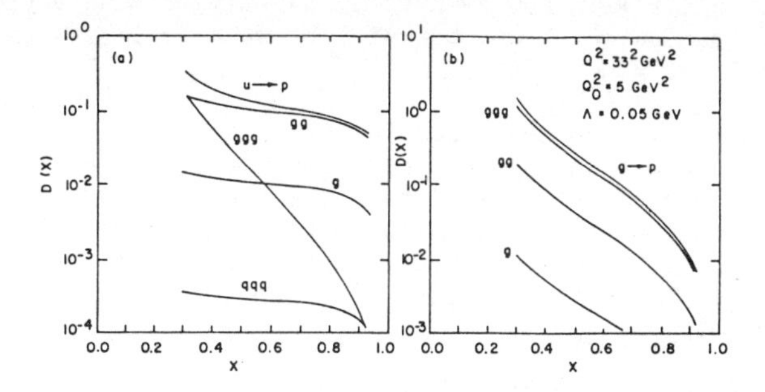

Fig. 2. Graph showing the fragmentation function (a) for u→p and (b) for g→p, where p=proton. The partial contributions from different numbers of gluons converted to $q\bar{q}$ pairs are also shown, see Eq. (3).

The x dependence of gluon and quark fragmentation is clearly different, a result agreeing with Gile's statement on the CLEO upsilon measurements: The assumption that gluons fragment like quarks does not describe the data at all. An interesting comparison can be made by plotting the summed contributions on the same graph, Fig. 3, where (3a) shows the fragmentation of a u quark and a gluon into protons. Despite the limited x range (strictly due to limitations of the VAX computer in its ability to calculate moments), one can see that gluon breakup to protons is actually more probable overall than u quark breakup into protons. The same statement applies for Λ baryons, the curves for which are not shown here. The net effect is that gluons can be identified by the baryon content of their jet. In fact, data now (Giles) indicates that in e^+e^- annihilation, the Υ region has an unexpectedly large number of baryons which our description attributes to the behavior of gluon jets quite naturally, since Υ decays into gluons.

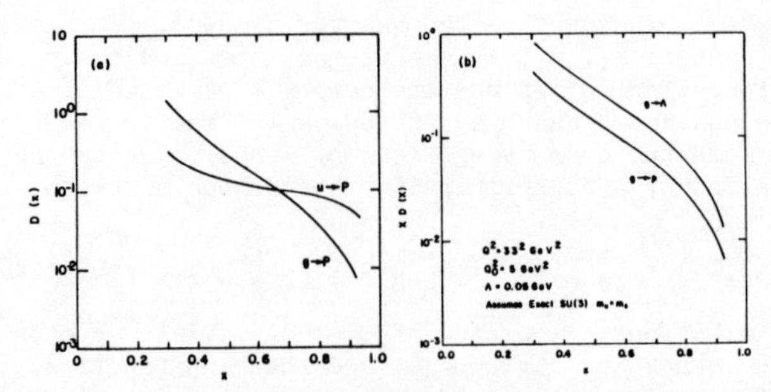

Fig. 3. Comparison of (a) u quark and gluon fragmentation into protons and (b) gluon fragmentation into protons and lambdas, both as functions of momentum fraction x.

As noted in Fig. (3b) our jet calculus is assumed exactly symmetric between u, d, and s quarks. However, because the s quark is believed heavier than u or d, its generation in jet calculus should be suppressed by a factor comparable to 1/3.4 found by (Malhotra and Orava) in their phenomenological analysis of quark suppression. Thus, Λ production at all x will not be exactly two bigger than p production as comes out of our calculation, but will be down by 2/3.4 for all x (a characteristic of gluon jets).

A final observation we can draw from Eq. (1) and Fig. (1) is that the Q^2 dependence in this model becomes much stronger the more KUV propagators are required to obtain the requisite number of final state partons. Thus, baryon production will vary much more rapidly than meson production, with 2 meson correlations having a yet stronger Q^2 dependence. This should be an aspect that can be explored with future experimental data.

In conclusion, we have presented an approach to the direct calculation of fragmentation functions independent of Monte Carlo methods. The results seem quite promising, especially in relation to new gluon fragmentation data discussed above and other applications, as our calculations of q,g→two mesons, will have to be pursued elsewhere.

References

A. BREAKSTONE et al., Phys. Lett. 135B, 510 (1984).
V. CHANG and R. C. HWA, Phys. Rev. D23, 728 (1981).
G. EILAM and M. S. ZAHIR, Phys. Rev. D26, 2991 (1982).
R. GILES "Hadron Production in the T Region", XXII International Conference on High Energy Physics (to be published, 1984).
G. GROSSING, A. NG and K. E. LASSILA, Acta Phys. Austriaca 55, 233 (1984).
L. M. JONES, K. E. LASSILA, U. SUKHATME and D. WILLEN, Phys. Rev. D 23, 717 (1981).
K. KONISHI, A. UKAWA and G. VENEZIANO, Nucl. Phys. B157, 45 (1979), KUV in the text.
G. P. LEPAGE and S. J. BRODSKY, Phys. Rev. Lett. 43, 545, 1625 (1979).
P. K. MALHOTRA and R. ORAVA "Measurement of Strange Quark Suppression in Hadronic Vacuum", FERMILAB-Pub-82/79.
R. MIGNERON, L. M. JONES and K. E. LASSILA, Phys. Lett. 114B, 184 and Phys. Rev. D 26, 2235 (1982).

Inst. Phys. Conf. Ser. No. 73: Section 8
Paper presented at VII Eur. Symp. Antiproton Interactions, Durham 1984

Antineutron production from the p̄–carbon interaction at 590 MeV/*c*

K Nakamura†¶, T Fujii†, T Kageyama†¶, F Sai†, S Sakamoto†$, S Sato†#, Y Takada§, T Takeda†, T Takahashi†, T Tanimori† and S S Yamamoto†

†Department of Physics, University of Tokyo, Tokyo 113, Japan
§Institute of Applied Physics, University of Tsukuba, Sakura, Ibaraki 305, Japan

Abstract. The antineutron angular distribution in the reaction $\bar{p}C\to\bar{n}X$ has been measured at 590 MeV/c. The shape of the distribution is found to be similar to that of the charge exchange reaction $\bar{p}p\to\bar{n}n$. This fact indicates that the quasi-free process is the dominant mechanism for $\bar{p}C\to\bar{n}X$. The ratio $\sigma(\bar{p}C\to\bar{n}X)/\sigma(\bar{p}p\to\bar{n}n)$ is found to be 0.86±0.05.

1. Introduction

The antiproton ($\bar{p}$) is a unique probe for studying the complex nucleus because of its extremely short mean free path in nuclear matter and the large probability of annihilation. However, current experimental information on the low-energy antiproton-nucleus ($\bar{p}A$) interactions is quite limited. It is only recently that $\bar{p}A$ elastic-scattering data have become available (Nakamura et al 1984a, Garreta et al 1984). To obtain comprehensive understanding of the $\bar{p}A$ interactions, a variety of $\bar{p}A$ reactions have yet to be studied.

The antineutron ($\bar{n}$) production by $\bar{p}$'s from complex nuclei below one-pion-production threshold is considered to be a quasi-free process corresponding to an elementary process of $\bar{p}p\to\bar{n}n$. Therefore, we have investigated the reaction $\bar{p}C\to\bar{n}X$ for the purpose of studying the incoherent production mechanism of the $\bar{p}A$ interactions. Since we measured (Nakamura et al 1984b) the $\bar{p}p\to\bar{n}n$ differential cross section at the same beam momentum with the same apparatus, it is possible to directly compare the $\bar{n}$ angular distribution obtained for the reaction $\bar{p}C\to\bar{n}X$ with that of the elementary process.

2. Experimental Details

The experiment was carried out with a low-momentum separated beam (K3) from the 12 GeV proton synchrotron at KEK. The experimental arrangement shown in Fig. 1 was used for measurement of the differential cross sections for the reactions $\bar{p}p\to\bar{p}p$, $\bar{n}n$, $\pi^+\pi^-$ and K^+K^- (Nakamura and Tanimori 1984). For the present measurement, the liquid-hydrogen target in Fig. 1 was replaced

¶Present address: National Laboratory for High Energy Physics (KEK), Oho, Ibaraki 305, Japan.
$Present address: Rutherford-Appleton Laboratory, Chilton, Didcot, Oxon, UK.
#Present address: Computer Engineering Division, NEC Corporation, Fuchu-shi, Tokyo 183, Japan.

by a 1.806±0.006 g/cm^2 thick carbon target. The target was surrounded by a five-layer cylindrical drift chamber (CDC) which was placed in a dipole magnetic field of 2.5 kG at the center. The $\bar{p}$ beam of 590 MeV/c was defined by the counters C1 (not shown in Fig. 1), C2 and C3, and its trajectory was determined by two multiwire proportional chambers with bidimensional readout and CDC. Outside of the magnetic field were located planar drift chambers DC1-DC4. On the surface of the magnet pole pieces were attached scintillation counters (called pole-face counters, not shown in Fig. 1) in order to detect charged annihilation products. In the forward and backward directions, two walls of TOF counters TF and TB were located. Another TOF counter plane TX was placed to cover the rather large gap between DC1 and DC4.

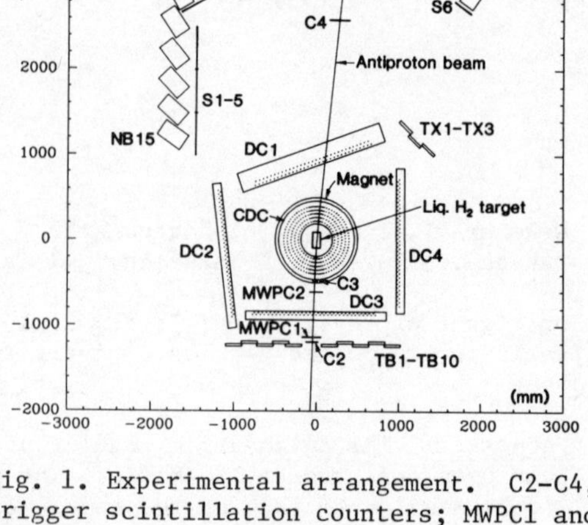

Fig. 1. Experimental arrangement. C2-C4, trigger scintillation counters; MWPC1 and MWPC2, multiwire proportional chambers with bidimensional readout; CDC, five-layer cylindrical drift chamber; DC1-DC4, four-layer drift chambers; TF1-TF11, TB1-TB10 and TX1-TX3, time-of-flight scintillation counters; NB1-NB16, iron-scintillator sandwich counters; S1-S6, scintillation counters.

The $\bar{n}$'s were detected by iron-scintillator sandwich counters NB. An NB module was 20-cm wide and 80.6-cm high, and consisted of 7 layers of 1-cm thick scintillator interleaved between iron plates with a total thickness of 24.8 cm. Fifteen NB modules were arranged on the left with respect to the beam to measure the $\bar{n}$ angular distribution. On the opposite side, another NB module (NB16) was placed in order to study the effect of the neutron background using the $\bar{p}p \to \bar{n}n$ events in which the $\bar{n}$ was detected by NB16, and consequently the neutron hit one of the NB modules on the left according to the two-body kinematics. All NB modules were placed behind scintillation counters (TF or S1-S6).

The $\bar{n}$ detection efficiency of the NB module was estimated to be $1.67p^2+1.5p+72.2\pm4\%$ (p is the $\bar{n}$ momentum in GeV/c) from the measured detection efficiency for $\bar{p}$'s and a Monte Carlo simulation of the energy deposit of $\bar{p}$'s and $\bar{n}$'s annihilating in the NB module.

The identification of $\bar{n}$'s were performed on the basis of the pulse-height and TOF information from the NB modules. Additionally, we imposed the following selection criteria: (a) No charged-particle trajectory other than the incident $\bar{p}$ should exist in the CDC; (b) There should be no TF, TB and/or S hits with timing compatible with γ's from π°'s or fast $\pi^\pm$'s (missed by the CDC due to inefficiency) produced by $\bar{p}p$ annihilation in the target; and (c) There should be no pole-face counter hits. Further details of the $\bar{n}$ identification and the applied corrections are the same as those for the analysis of the $\bar{p}p \to \bar{n}n$ reaction, and are described in Nakamura et al (1984b).

3. <u>Results and Discussion</u>

In this experiment, the $\bar{n}$ energy could not be measured, and therefore we only present the angular distribution. Generally, the $\bar{n}$ angular distribution from the $\bar{p}A$ interaction is given by $d\sigma/d\Omega=\int dE\, d^2\sigma/d\Omega dE$, where E is the $\bar{n}$ energy. Since the $\bar{n}$ detection efficiencies of the NB modules are nearly energy independent, the observed $\bar{n}$ angular distribution is considered to directly give $d\sigma/d\Omega$ to a good approximation.

The $\bar{n}$ angular distribution obtained from the reaction $\bar{p}C\to\bar{n}X$ is shown in Fig. 2a as a function of the $\bar{n}$ laboratory angle. For comparison, the $\bar{n}$ angular distribution from the reaction $\bar{p}p\to\bar{n}n$ at the same momentum is shown in Fig. 2b. The error bars in these figures show statistical errors only. The total systematic error amounts to ±8.5%. We note the similarity of the two angular distributions. This fact strongly suggests that the $\bar{n}$ production on the carbon nucleus is, as expected, a quasi-free process corresponding to the elementary process $\bar{p}p\to\bar{n}n$.

Figure 3 shows the ratio $d\sigma/d\Omega(\bar{p}C\to\bar{n}X)$ to $d\sigma/d\Omega(\bar{p}p\to\bar{n}n)$ as a function of the $\bar{n}$ laboratory angle. Since both $\bar{p}C\to\bar{n}X$ and $\bar{p}p\to\bar{n}n$ reactions were measured with the same apparatus, most of the systematic errors are expected to cancel out for this ratio. Therefore, only the statistical errors are considered for this ratio.

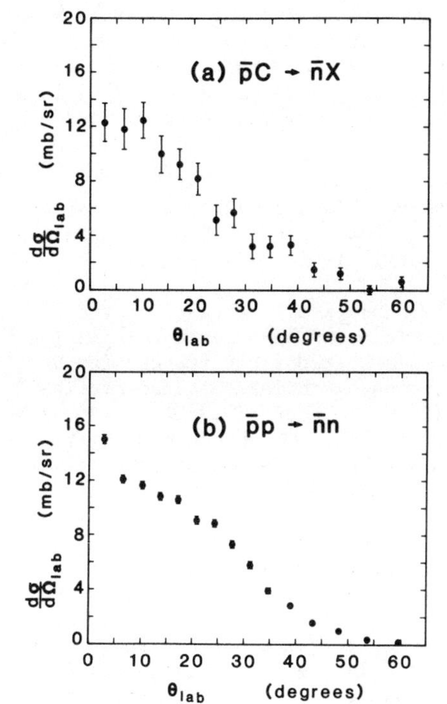

Fig. 2. (a) The $\bar{n}$ angular distribution for the reaction $\bar{p}C\to\bar{n}X$ and (b) the same for the elementary process $\bar{p}p\to\bar{n}n$, both at 590 MeV/c as a function of the $\bar{n}$ laboratory angle.

By integrating the differential cross sections up to 60°, we obtain the ratio of the $\bar{n}$ production cross sections $\sigma(\bar{p}C\to\bar{n}X)/\sigma(\bar{p}p\to\bar{n}n) = 0.86\pm0.05$. This means that the $\bar{n}$ production cross section per bound proton in the carbon nucleus is 0.14±0.01 times that for a free proton.

Theoretical formulation of incoherent particle production from nuclei at high energies is given by Kölbig and Margolis (1968) on the basis of a Glauber-type multiple-scattering theory (Glauber 1959). The incoherent differential cross section is then given as

$$\left(\frac{d\sigma}{d\Omega}\right)_{\text{incoherent}} = \left(\frac{d\sigma}{d\Omega}\right)_{\text{free}} N(A;\sigma_1,\sigma_2) + \cdots\cdots , \qquad (1)$$

where

$$N(A;\sigma_1,\sigma_2) = \frac{1}{\sigma_2-\sigma_1}\int [\, e^{-\sigma_1 T(\vec{b})} - e^{-\sigma_2 T(\vec{b})}\,]\, d^2b \qquad (2)$$

$$T(\vec{b}) = A\int_{-\infty}^{\infty} \rho(\vec{b},z)\, dz$$

and $(d\sigma/d\Omega)_{\text{free}}$ is the differential cross section for the two-body elementary process, σ_1 and σ_2 are the total cross sections for the incoming and produced particles on a nucleon, and $\rho(\vec{r})$ is the nucleon density distribution of the nucleus normalized to unity. The first term on the right-hand side of Eq. (1) accounts for the particle production with no secondary interactions before and/or after production. The rest of the terms give corrections due to elastic scattering of the incoming and produced particles with the nucleons in the nucleus.

The similarity of the observed $\bar{n}$ angular distributions in both $\bar{p}C \to \bar{n}X$ and $\bar{p}p \to \bar{n}n$, however, suggests that the first term is sufficient to explain our data, provided that proper normalization is taken into account. For this purpose, we replaced σ_1 and σ_2 in Eq. (2) by the annihilation cross sections σ_{A1} and σ_{A2} to obtain $d\sigma/d\Omega(\bar{p}C \to \bar{n}X)$ as

$$\frac{d\sigma}{d\Omega}(\bar{p}C \to \bar{n}X) = \frac{d\sigma}{d\Omega}(\bar{p}p \to \bar{n}n) \cdot \frac{1}{2} N(A; \sigma_{A1}, \sigma_{A2}) \quad , \qquad (3)$$

where a factor 1/2 accounts for the fact that only the bound protons contribute to this reaction.

For numerical calculation, the $\bar{p}$-nucleon and $\bar{n}$-nucleon annihilation cross sections are represented by the $\bar{p}p$ annihilation cross section (Lowenstein et al 1981), $\sigma_A = 36.5 + 28.9/p$ mb. The nucleon density distribution is taken to be $\rho(\vec{r}) = \{1 + \exp[(r-R)/a]\}^{-1}$ with $R = r_0 A^{1/3}$ and $a = 0.52$ fm. The value of r_0 was changed over a physically reasonable range, $r_0 = 1.0 \sim 1.2$ fm. The momentum of the produced $\bar{n}$ was assumed to be given by the two-body kinematics of the free process $\bar{p}p \to \bar{n}n$. The curve in Fig. 3 shows the calculated result with $r_0 = 1.1$ fm. For $r_0 = 1.0$ and 1.2 fm, the curve is shifted by -8% and +8%, respectively.

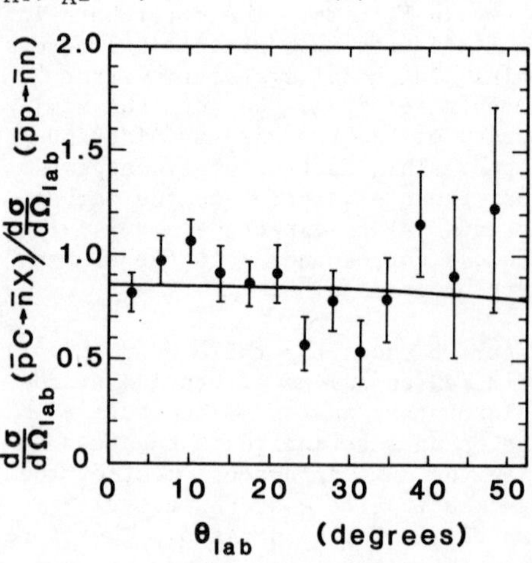

Fig. 3. The ratio of the differential cross section of the reaction $\bar{p}C \to \bar{n}X$ to that of the elementary process $\bar{p}p \to \bar{n}n$. The curve shows the calculated result using Eq. (3) for $r_0 = 1.1$ fm.

4. Conclusions

We have reported the first observation of the $\bar{n}$ production from the $\bar{p}C$ interaction. The shape of the $\bar{n}$ angular distribution is similar to that of the corresponding elementary process $\bar{p}p \to \bar{n}n$. The ratio $\sigma(\bar{p}C \to \bar{n}X)/\sigma(\bar{p}p \to \bar{n}n)$ is found to be 0.86 ± 0.05 at the incident beam momentum of 590 MeV/c. This ratio is well explained by a simple formalism of the incoherent $\bar{n}$ production preceded and followed by the absorption of the antinucleons by the bound nucleons in the nucleus.

References

Garreta D et al 1984 Phys. Lett. 135B 266

Glauber R J 1959 Lectures in Theoretical Physics (New York:Interscience) vol.1 p 315

Kölbig K S and Margolis B 1968 Nucl. Phys. B 6 85

Lowenstein D I et al 1981 Phys. Rev. D 23 2788

Nakamura K et al 1984a Phys. Rev. Lett. 52 731

Nakamura K et al 1984b Phys. Rev. Lett. to be published

Nakamura K and Tanimori T 1984 these Proceedings

Inst. Phys. Conf. Ser. No. 73: Section 8
Paper presented at VII Eur. Symp. Antiproton Interactions, Durham 1984

Model analysis of the p̄p annihilation process at low energies

S Saito, Y Hattori, J H Kim and T Yamagata

Department of Physics, Tokyo Metropolitan University, Setagaya-ku, Tokyo, Japan 158

Abstract. Assuming that, in the annihilation process of an antiproton with a proton, quarks and antiquarks are in a thermal equilibrium state before they recombine into mesons, we calculate the quark (antiquark) distributions, from which meson distributions are derived. The parameters in the model are fixed so that the pion multiplicity and inclusive pion momentum distributions at p_{lab} = 700 Mev/c agree with the experiment. Comparisons with the experimental data are done on the energy dependence of the average pion multiplicity and meson production rates.

1. Formalism

The single particle potential $V(\vec{r})$ is given as

$$V(\vec{r}) = \tfrac{1}{2}m\omega^2 r^2 + G_{LS}\hbar\omega\vec{L}\cdot\vec{S}$$

and the residual interaction $V'(\vec{r}_{ij})$ between a quark-antiquark pair of the same quantum numbers is introduced as

$$V'(\vec{r}_{ij}) = G_\alpha r_{ij}^\alpha + G_S\{\frac{8\pi}{3}\vec{S}_i\cdot\vec{S}_j\delta(\vec{r}_{ij}) + \frac{1}{r_{ij}^3}[3(\vec{S}_i\cdot\hat{r}_{ij})(\vec{S}_j\cdot\hat{r}_{ij}) - \vec{S}_i\cdot\vec{S}_j]\}$$

$$\alpha = 1,\ 2$$

The quark density in each level specified by the quantum numbers n, l, j (radial quantum number, orbital angular momentum and total angular momentum quantum numbers respectively) is determined by minimizing the free energy of the system. Since the number of quarks is not constant due to the possible pair creation of quarks from the vacuum, it is convenient to use the thermal Hartree-Fock-Bogoliubov(THFB) method(Goodman 1981, Tanabe et al 1981). The zero temperature corresponds to the vacuum and the chemical potential should be identified with the quark mass m.

Using the data of quark distribution, we calculate the multiplicities of mesons produced through recombination of quarks and antiquarks. Here we assume that the recombination takes place only between a quark and antiquark pair of the same quantum number, but of the opposite color. The number of produced mesons with definite spin, parity, isospin and G-parity (J^P, I^G) is determined by the statistical weights calculated from the spin decomposition. All the mesons, except pions, will decay into either two or three pions depending on their sign of G-parity. The mesons with (J^P, I^G) = ($0^-,1^-$), ($1^-,1^+$), ($1^-,0^-$), ($2^+,0^+$) will be identified with π, ρ, ω, f-mesons, respectively, and the rest of them are to produce background (or phase space) pions.

2. Analysis

The THFB program has been performed at p_{lab} = 700 MeV/c, the only energy which can supply sufficient number of experimental data(Hamatsu et al 1977), to determine the parameters in our model. For simplicity the energy levels with quantum numbers which do not satisfy $2n + \ell < 3$ are truncated, and we have found the following results:

(i) The potential strength $\hbar\omega$ decreases as the quark mass m increases for the fixed values of the temperature and other parameters (G_{LS}, G_S, G_α). In particular, $\hbar\omega$ - m relation is almost linear and independent of the temperature and other parameters. The experimental value of the inclusive negative pion (1.63 ± 0.3) is reproduced at T = 117 MeV when m is 250 MeV, the same temperature obtained in our previous study of antiproton-proton elastic scattering(Kimura and Saito 1981). To obtain the same multiplicity of negative pion, the temperature must be increased as the quark mass becomes small. For instance, we find T = 136 MeV corresponding to m = 10 MeV.

(ii) A large value of the L-S force is inevitable to get a relatively large number of f-mesons. This can be understood from the fact that a large L-S force lowers the level (n, l, j) = (0, 2, 3/2), from which spin 2 particles can be created, whereas little change results for the multiplicities of particles with spin 0 and 1. Our prediction of the f-meson production rate is still lower than the experimental value(Hamatsu et al 1977) by a factor of 2 or 3. We should recall, however, that the experimental data is not very well established due to the large background compared to the f-meson peak.

(iii) The effects of the residual interactions were not so large as to fix the parameters G_α and G_S within the errors of the experimental data available at present. Nevertheless we can see some tendency of their effects which can be summarized as follows: Both decreasing G_α and increasing G_S decrease the population of quarks. This might sound rather contradictory because a strong attraction of the confining force would lower the levels. This is not the case, however, since the vacuum will absorb some parts of quark pairs as the attractive force becomes strong.

(iv) Momentum distribution of quarks and inclusive pions

From the information of the quark density on the energy levels, the momentum distribution and the distribution in the configuration space will be calculated in a straightforward way. Under the assumption that the intermediate mesons decay into either two or three pions depending on their G-parity, which we have adopted previously, we calculate the momentum distribution of inclusive pions. For simplicity we ignore the spin of the meson and use the two and three body decay functions at the rest frame. Their motion around the center of mass will be taken into account by broadening the energy levels of the quarks in terms of the Breit-Wigner formula in which the width is approximated by the average kinetic energy of quarks. We find that the set of parameters m = 140 MeV, $\hbar\omega$ = 152.5 MeV, G_{LS} = 0.9 can reproduce the momentum distribution of inclusive pions as shown in Fig. 1. Using the same values of the parameters the meson multiplicities are given in Table 1.

(v) Energy dependence

Once fixed the parameters in our model at p_{lab} = 700 MeV/c, only the temperature varies as the total energy of the system changes. Various quantities are calculated as functions of energy. Among them only the inclusive pion multiplicity data(Baltay et al 1966, Amaldi et al 1966) is available to be compared with our prediction. Again their agreement is quite satisfactory as seen in Fig. 2.

3. Conclusion

Our model predicts various low energy inclusive phenomena of antiproton-proton annihilation processes. Available experimental data at preset are not sufficient to fix the parameters in the model, which describe the character of the force between quarks and antiquarks. Therefore it is desirable to have more information about the inclusive phenomena at low energies in the future experiments.

References

Amaldi U et al 1966 Nuovo Cimento XLVIA 171
Baltay et al 1966 Phys. Rev. 145 1103
Goodman L 1981 Nucl. Phys. A352 30
Hamatsu R et al 1977 Nucl. Phys. B123 189
Kimura M and Saito S 1981 Nucl. Phys. B178 477
Tanabe K, Sugawara-Tanabe K and Mang H J 1981 Nucl. Phys. A357 20

Table 1

	Experiment	Theory
n_{π^-}	1.63 ± 0.03	1.631
n_{ρ^0}	0.26 ± 0.01	0.243
n_{ω}	0.22 ± 0.01	0.243
n_f	0.12 ± 0.01	0.042
$n_{\text{all mesons}}$	3.27 ± 0.13	3.55
n_{quark}	--	2.2

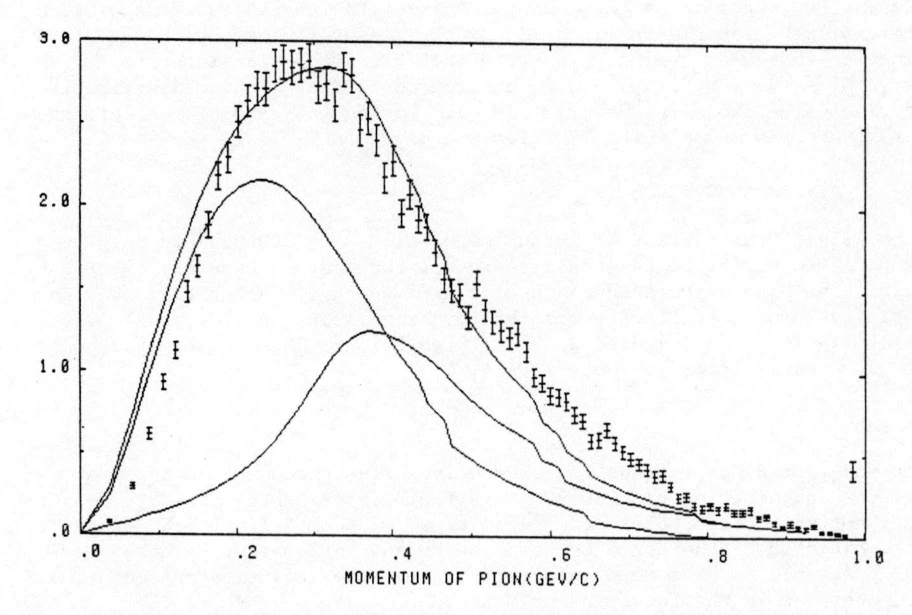

Fig. 1 Inclusive momentum distribution of negative pions at 700 MeV/c. Solid line represents the best fit of our model. Curve A corresponds to two body pion decays, curve B to three body decays and curve C to the sum.

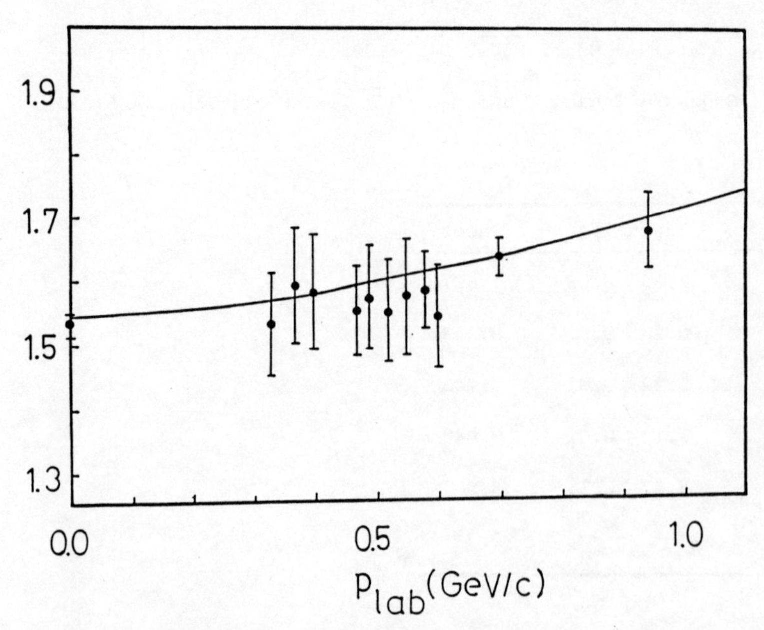

Fig. 2 Energy dependence of the multiplicity of the negative pions. Solid line shows our prediction at $G_{LS} = 0.9$, $G_1 = 0$ and $G_S = 0$.

Measurement of the real-to-imaginary ratios of the $\bar{p}p$ and $\bar{p}n$ forward amplitudes between 367 and 764 MeV/*c*

T Takeda†, J Chiba†, H Fujii†, T Fujii†, H Ikeda§, H Iwasaki†$, T Kageyama†$, S Kuribayashi†, K Nakamura†$, T Sumiyoshi†$ and Y Takada¶

†Department of Physics, University of Tokyo, Tokyo 113, Japan
§National Laboratory for High Energy Physics (KEK), Oho, Ibaraki 305, Japan
¶Institute of Applied Physics, University of Tsukuba, Sakura, Ibaraki 305, Japan

Abstract. Preliminary results are presented for the real-to-imaginary ratios of the $\bar{p}p$ and $\bar{p}n$ forward-scattering amplitudes in the incident momentum range between 367 and 764 MeV/c.

1. Introduction

Measurement of the real-to-imaginary ratio, ρ, of the $\bar{N}N$ forward scattering amplitude at low momenta provides a means to investigate the analytic structure of the $\bar{N}N$ scattering amplitude in the unphysical region, because it is sensitive to the contribution from the high-mass region of the unphysical cut of the $\bar{N}N$ amplitude via the forward dispersion relation. Recent development of $\bar{N}N$ potential models also makes the study of ρ at low momenta very interesting, particularly in relation to the spin-dependent effects (Lacombe et al 1983, Timmers 1984).

Previously, $\rho(\bar{p}p)$ was measured down to the incident $\bar{p}$ momentum of $\sim$350 MeV/c (Iwasaki et al 1981, 1983, Cresti et al 1983). However, no measurement has been reported for $\rho(\bar{p}n)$. We present here preliminary results of a measurement of $\rho(\bar{p}p)$ with high statistical accuracy and also preliminary results of the first measurement of $\rho(\bar{p}n)$, both in the momentum range between 367 and 764 MeV/c. These results, however, have been obtained without taking the spin-dependent effects into account. An alternative analysis with the spin-dependent effects taken into account is under way.

2. Experiment

The experiment was performed in a low-momentum separated beam line, K3, at KEK. The experimental layout is shown in Fig. 1. The incident $\bar{p}$'s were identified by using the pulse heights of scintillation counters C1-C3 and the TOF between C1 and C3. Four multiwire proportional chambers (MWPC1-MWPC4) were located in front of a target to determine the beam direction. The rms angular resolution for the incident track was 0.3°. The target assembly consisted of three Mylar cells, 10 cm in diameter: a 217-mm long liquid-hydrogen cell, an empty cell having the same structure and a 115-mm long liquid-deuterium cell. The scattered particles including the transmitted beam were measured by a scintillation counter C4, scintillator hodo-

$Present address: National Laboratory for High Energy Physics (KEK), Oho, Ibaraki 305, Japan.

scope H2 and MWPC5-MWPC7. The rms resolution for the scattering angle was 0.5°. Further details of the experiment are described elsewhere (Sumiyoshi et al 1982).

3. Results

To obtain $\rho(\bar{p}p)$, we have fitted the following expression to the measured differential cross section data:

$$\frac{d\sigma}{d\Omega} = \left| f_C(\vec{q}) + f_N(\vec{q}) \right|^2$$

$$f_N(\vec{q}) = \frac{k\sigma_T}{4\pi}(\rho+i)\exp(-\frac{B}{2}q^2) ,$$

where q is the momentum transfer in the center-of-momentum system, k is the incident momentum and f_C is the $\bar{p}p$ Coulomb amplitude. The total cross section σ_T, the slope parameter B and the real-to-imaginary ratio ρ are the parameters to be determined experimentally.

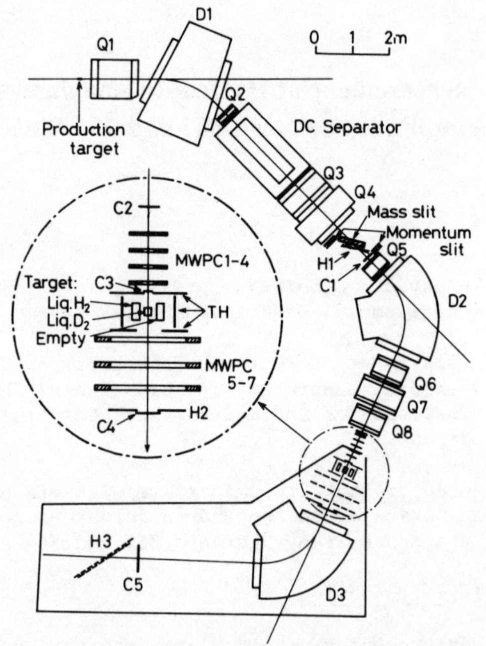

Fig. 1. Experimental arrangement. D1-D3, bending magnets; Q1-Q8, quadrupole magnets; C1-C5, scintillation counters; H1-H3 and TH, scintillator hodoscopes; and MWPC1-MWPC7, multiwire proportional chambers.

For the $\bar{p}n$ process, a similar fitting has been applied to the $\bar{p}d$ differential cross section data. To construct the $\bar{p}d$ amplitude from the $\bar{p}p$ and $\bar{p}n$ amplitudes, we used the Glauber approximation (Franco and Glauber 1966, Franco 1971). In this approximation, the $\bar{p}d$ differential cross section is given by

$$\frac{d\sigma}{d\Omega} = \int \exp(i\vec{q}\cdot\vec{r}) \, |\phi(\vec{r})|^2 \, |F(\vec{q},\vec{r})|^2 \, d^3r ,$$

where $\phi(\vec{r})$ is the deuteron wave function and $F(\vec{q},\vec{r})$ is the Glauber amplitude expressed by the single-particle amplitudes parametrized as

$$f_j(q) = \frac{k\sigma_j}{4\pi}(\rho_j+i)\exp(-\frac{B_j}{2}q^2) \qquad j = \bar{p}p, \bar{p}n \quad .$$

The values of σ_T were determined by extrapolating our differential cross section data from the angular region where the Coulomb effect is negligibly small to zero degree. For the $\bar{p}p$ process, the results are consistent with the previous data (Carroll et al 1974, Hamilton et al 1980, Sumiyoshi et al 1982, Nakamura et al 1984). For the $\bar{p}d$ process, the existind data (Carroll et al 1974, Hamilton et al 1980) are systematically different from each other, and our total cross section results lie in between these data. In fitting the differential cross section data, we used our own results for the values of σ_T. The values of B and ρ were determined by the fit.

The obtained values of $\rho(\bar{p}p)$ are shown in Fig. 2 together with the other data (Iwasaki et al 1983, Cresti et al 1983, Kaseno et al 1976), and those of $\rho(\bar{p}n)$ are shown in Fig. 3. The errors shown in these figures are statistical only. (For $\rho(\bar{p}p)$, the size of the error bars for the present experiment is smaller than the data points plotted in Fig. 2.) The sources of the systematic errors are dominated by the uncertainties in σ_T and B.

For $\rho(\bar{p}n)$, the uncertainty in $\rho(\bar{p}p)$ is an additional source of error. We estimate that the systematic errors due to these uncertainties are ±0.1 for $\rho(\bar{p}p)$ and ±0.2 for $\rho(\bar{p}n)$.

4. Discussion

The $\bar{p}p$ real-to-imaginary ratio is close to zero at low momenta and seems to slowly increase with incident momentum. This behavior is generally consistent with the previous experiments (Iwasaki et al 1983, Cresti et al 1983) except that the present results show less momentum dependence. However, this discrepancy is well within the quoted systematic errors. The values of $\rho(\bar{p}n)$ is positive at low momenta even if the systematic error is taken into consideration. At higher momenta, $\rho(\bar{p}n)$ seems to approach zero.

The momentum dependence of $\rho(\bar{p}p)$ is consistent with the predictions of potential models. The solid and dashed curves in Fig. 2 are the theoretical predictions of typical potential models (Bryan and Phillips 1968, Lacombe et al 1983). For $\rho(\bar{p}n)$, on the other hand, the observed momentum dependence is different from the theoretical predictions of potential models (Timmers 1984, Lacombe 1984), although these models well describe the momentum dependence of $\rho(\bar{p}p)$. It should be noted, however, that both potential models (Timmers 1984, Lacombe 1984) take the spin-dependent effects into consideration. Therefore, it may not be appropriate to directly compare their predictions with the present results which were derived without taking account of the spin-dependent effects.

A dispersion-relation analysis (Grein 1977) shows that a pole-like term is necessary in the $\bar{p}p$ forward amplitude near threshold to describe $\rho(\bar{p}p)$. If this is also the case for the $\bar{p}n$ amplitude, the behavior of $\rho(\bar{p}n)$ is similar to that of $\rho(\bar{p}p)$, i.e., negative at low momenta and close to zero at higher momenta. On the other hand, if a pole-like term does not exist in

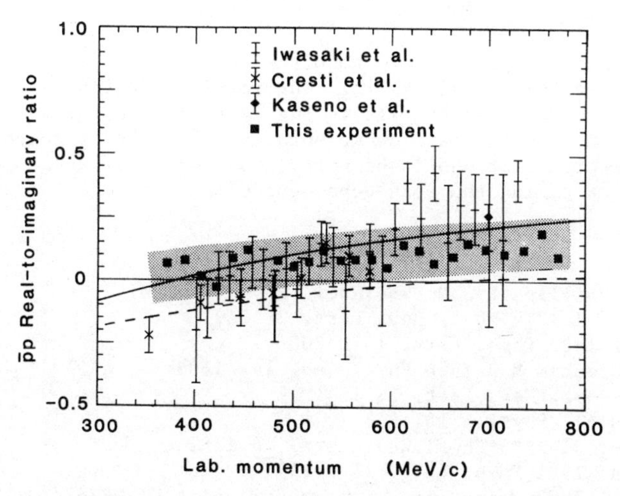

Fig. 2. Preliminary results for the real-to-imaginary ratio of the $\bar{p}p$ forward-scattering amplitude. The shaded region indicates the magnitude of possible systematic errors. The solid curve shows the prediction of the Bryan-Phillips non-static potential model (Bryan and Phillips 1968). The dashed curve shows the prediction of the Paris-potential model (Lacombe et al 1983).

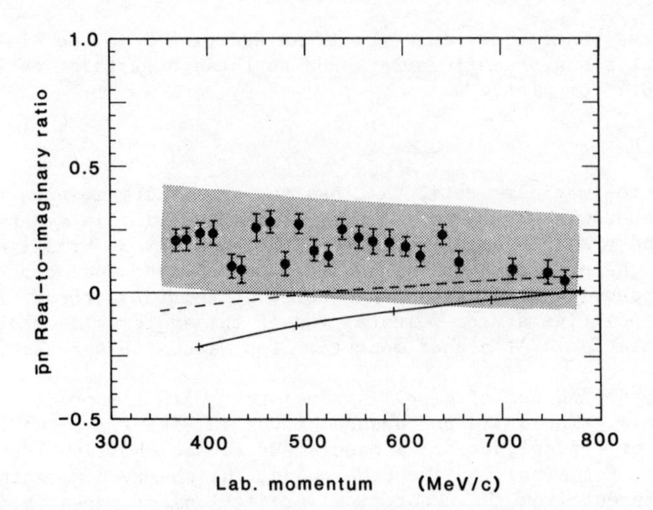

Fig. 3. Preliminary results for the real-to-imaginary ratio of the $\bar{p}n$ forward-scattering amplitude. The shaded region indicates the magnitude of possible systematic errors. The solid curve shows the prediction of the Nijmegen-potential model (Timmers 1984). The dashed curve shows the prediction of the Paris-potential model (Lacombe 1984).

the $\bar{p}n$ amplitude near threshold, $\rho(\bar{p}n)$ has a large positive value at momenta less than 200 MeV/c and approaches zero at higher momenta. Therefore, our results suggest that a pole-like term near threshold does not exist in the $\bar{p}n$ forward amplitude, or that it is very small if at all.

5. Conclusions

We have presented the preliminary results for $\rho(\bar{p}p)$ and $\rho(\bar{p}n)$ in the momentum range between 367 and 764 MeV/c. The results for $\rho(\bar{p}p)$ are consistent with the previous measurements within the systematic errors. For the $\bar{p}n$ process, $\rho(\bar{p}n)$ is positive at low momenta and seems to approach zero with increasing momentum. It should be noted, however, that these results were obtained without taking the spin-dependent effects into consideration.

References

Bryan R A and Phillips R J N 1968 Nucl. Phys. B 5 201
Carroll A S et al 1974 Phys. Rev. Lett. 32 247
Cresti M et al 1983 Phys. Lett. 132B 209
Franco V and Glauber R J 1966 Phys. Rev. 142 1195
Franco V 1971 Phys. Rev. Lett. 27 1541
Grein W 1977 Nucl. Phys. B 131 255
Hamilton R P et al 1980 Phys. Rev. Lett. 44 1182
Iwasaki H et al 1981 Phys. Lett. 103B 247
Iwasaki H et al 1983 University of Tokyo preprint UT-HE-83/13
Kaseno H et al 1976 Phys. Lett. 61B 203
Lacombe M et al 1983 Phys. Lett. 124B 443
Lacombe M 1984 private communication
Nakamura K et al 1984 Phys. Rev. D 29 349
Sumiyoshi T et al 1982 Phys. Rev. Lett. 49 628
Timmers P H 1984 private communication

Author Index

Adiels L, *297*
Aerts A M, *491*
Ahmad S, *131*, *287*
Åkesson T, *69*
Amsler C, *131*, *287*
Andersson B, *447*
Ando F, *495*
Angelopoulos A, *305*
Apostolakis A, *305*
Armenteros R, *131*, *287*
Armstrong T, *305*, *323*
Ashford V, *205*
Aslanides E, *187*
ASTERIX collaboration, *131*, *287*
Auld E G, *131*, *287*
Axen D, *131*, *287*

Backenstoss G, *297*
Baglin C, *339*
Bailey D, *131*, *287*
Baird S, *339*
Baker C A, *137*
Balestra F, *251*
Bardin G, *143*
Barger V, *375*
Barlag S, *131*, *287*
Barnes P D, *201*
Bassalleck B, *305*
Bassompierre G, *339*
Batty C J, *137*
Batusov Yu A, *251*
Beard C I, *147*
Beer G, *131*, *287*
Bendiscioli G, *251*
Benulis C, *305*
Berger E L, *349*
Bergström I, *297*
Berrada M, *187*
Besold R, *201*
Biard J, *305*
Bing O, *187*
Birien P, *187*, *201*
Birsa R, *147*
Bizot J C, *131*, *287*
Bocquet J P, *187*
Bol K D, *195*
Bonner B E, *201*
Borreani G, *339*
Bos K, *147*
Bradamante F, *147*
Breunlich W, *201*
Bridges D, *315*
Brient J C, *339*
de Brion J P, *143*
Broll C, *339*
Brom J M, *339*
Brown H, *315*
Brückner W, *157*
Bruge G, *187*
Büche G, *175*, *181*, *507*
Buenerd M, *195*
Bugg D V, *147*
Bugge L, *339*
Buran T, *339*
Burgun G, *143*
Burq J P, *339*
Bussa M P, *251*
Bussière A, *339*
Busso L, *251*
Buzzo A, *339*

Calabrese R, *143*
Capon G, *143*
Caria M, *131*, *287*
Carius S, *175*, *297*
Carlin R, *143*
Cester R, *339*
Chanel M, *119*
Charalambous S, *175*, *297*
Chardalas M, *175*, *181*, *507*
Chaumeaux A, *187*
Chauvin J, *195*
Chemarin M, *339*
Chevallier M, *339*
Chiba J, *523*
Chu C, *323*
Clark S A, *137*
Clement J, *323*
Clough A S, *147*
Clover M R, *195*
Comyn M, *131*, *287*
Cooper M, *297*
Cresti M, *259*
Cugnon J, *499*

Daftari I, *315*
Dahme W, *131*, *287*
Dalla Torre-Colautti S, *147*
Dalpiaz P, *143*
Dalpiaz P F, *143*
Damgaard P H, *349*
Daniel H, *181*, *507*

Davies J D, *137*
Debbe R, *205*, *315*
Dedoussis S, *175*, *181*, *507*
Degli-Agosti S, *147*
Delcourt B, *131*, *287*
Denes P, *305*
Derks W, *503*
Derre J, *143*
DeVries R M, *195*
DiBitonto D, *355*
DiGiacomo N J, *195*
Döbbeling H, *157*
Donnachie A, *433*
Doser M, *131*, *287*
Dosselli U, *143*
Dousse J C, *195*
Drake D M, *187*
Duch K D, *131*, *287*
Duclos J, *143*
Dutty W, *201*

Edgington G A, *147*
von Egidy T, *175*, *181*, *507*
Eisenstein R A, *201*
Elinon C, *323*
Erdmann K, *131*, *287*
Erell A, *187*
Ericsson G, *201*
Escoubes B, *339*
Eyrich W, *201*

Falomkin I V, *251*
Faure J L, *143*
Fay J, *339*
Fayard L, *81*
Feld F, *131*, *287*
Ferrero L, *251*
Ferroni S, *339*
Fickinger W, *205*, *315*
Filippini V, *251*
Findeisen C, *175*, *297*
Frankenberg R, *201*
Franklin G, *201*
Fransson K, *297*
Franz J, *201*
Fujii H, *523*
Fujii T, *515*, *523*
Fumagalli G, *251*
Furic M, *323*

Garreta D, *187*
Gasparini F, *143*
Gastaldi U, *131*, *287*
Geich-Gimbel Ch, *421*
Gervino G, *251*
Giorgi M, *147*
Girard C, *339*
Gorringe T P, *137*
Gracco V, *339*
Graf N, *305*
Grasso A, *251*
Gray L, *315*
Green A M, *213*
Guaraldo C, *251*
Guillaud J P, *339*
deGuzman A, *315*

Hadjifotiadou D, *297*
Hall J R, *147*
Hamman N, *201*
Hancock A D, *175*, *181*, *507*
von Harrach D, *157*
Hartman C, *323*
Hartmann F J, *175*, *181*, *507*
Hattori Y, *519*
Hauth J, *175*
Heel M, *131*, *287*
Heer E, *147*
Hertzog D, *201*
Hess R, *147*
Hicks A, *323*
Hill R, *305*
Hofmann A, *201*
Howard B, *131*, *287*
Howard R, *131*, *287*
Huet M, *143*
Hungerford E, *323*

Ikeda H, *523*
Ille B, *339*
Iwasaki H, *523*

Jaffe R L, *1*
Janouin S, *187*
Jeanjean J, *131*, *287*
Johansson T, *201*

Kageyama T, *515*, *523*
Kalinowsky H, *131*, *287*
Kalogeropoulos T, *315*
Kane G L, *471*
Kanert W, *175*, *181*, *507*
Kapustinsky J S, *195*
Karl G, *281*
Kayser F, *131*, *287*
Kerek A, *297*
Kilian K, *201*
Kilvington A I, *137*
Kim J H, *519*
Kirsebom K, *339*
Kishimoto T, *323*
Klempt E, *131*, *287*

Kluyver J C, *147*
Kneis H, *157*
Koch H, *175*, *181*, *507*
Kochowski C, *143*
Köhler Th, *175*, *181*, *507*
Komninos N, *305*
Kreissl A, *175*, *181*, *507*
Kruk J, *323*
Kunne R A, *147*
Kuribayashi S, *523*

Lafarge D, *143*
Lambert M, *339*
Landua R, *131*, *287*
Lassila K E, *511*
Lebrun D, *195*
Lechanoine-Leluc C, *147*
Lefèvre P, *119*
Legrand D, *187*
Leistam L, *339*
Lemaire M C, *187*
Leon M, *181*, *507*
Lewis R A, *305*, *323*
Lichtenstadt J, *187*
Limentani S, *143*
Lingeman E W A, *137*
Linssen L, *147*
Litchfield P J, *263*
Lochstet W, *323*
Lodi Rizzini E, *251*
Loucatos S, *37*
Lowe J, *137*
Lowenstein D, *323*
Lundby A, *339*

McFarlane W K, *305*
McGaughey P L, *195*
Macri M, *339*
Maggiora A, *251*
Maher C, *201*
Majewski S, *157*
Mandelkern M, *305*
Marchetto F, *339*
Marciano C, *251*
Marel G, *143*
Marino R, *205*, *315*
Marshall G, *131*, *287*
Martin A, *147*
Martin P, *195*
Mattera L, *339*
Mayer B, *187*
Mayes B, *323*
Meneguzzo A, *143*
Menichetti E, *339*
Möhl D, *119*
Moir J, *137*
Monnand E, *187*
Moser E, *181*, *507*
Moss B, *323*
Mouellic B, *339*
Mougey J, *187*
Müller R, *201*
Mutchler G, *323*

Nakamura K, *329*, *515*, *523*
Nelson J M, *137*
Ng A, *511*
Nguyen H, *131*, *287*
Nicholas M, *175*, *181*, *507*
Nicklas G, *201*
Nicolescu B, *411*
Nilsson A, *175*, *181*, *507*
Nomachi M, *157*

Oh B Y, *305*
Onel Y, *147*
Ortner H, *201*

Pain J, *187*
Panzieri D, *251*
Papaelias P, *305*
Papastefanou K, *297*
Pastrone N, *339*
Paul S, *157*
Pauli E, *143*
Pavlopoulos P, *297*
Pawlek P, *201*
Peaslee D, *315*
Peccei R D, *51*
Peebles P J E, *243*
Peng J C, *187*
Penzo A, *147*
Perrin P, *187*
Peruzzo L, *259*
Petridou Ch, *315*
Petrillo L, *339*
Petrucci F, *143*
Pia M G, *339*
Pinsky L, *323*
Piragino G, *251*
Playfer S M, *137*, *305*
Polikanov S M, *201*
Pontecorvo G B, *251*
Posocco M, *143*
Poth H, *175*, *181*, *507*
Poulet M, *339*
Povh B, *157*
Pozzo A, *339*
Press J L, *305*
Prevot N, *131*, *287*
Pyle G J, *137*

Raich U, *175*, *181*, *507*
Rana N C, *235*
Ransome R D, *157*
Rapin D, *147*
Ray R, *305*
Reidy J J, *175*, *181*, *507*
Repond J, *175*, *297*
Rimondi F, *403*
Rinaudo G, *339*
Robertson L, *131*, *287*
Robinson D K, *205*, *315*
Rohmann D, *175*, *181*, *507*
Rössle E, *201*
Rotondi A, *251*
Rozaki H, *305*

Sabev C, *131*, *287*
Sai F, *515*
Sainio M E, *205*
Saito S, *519*
Sakamoto S, *137*, *515*
Sakelliou L, *305*
Sakitt M, *205*
Santroni A, *339*
Sapozhnikov M G, *251*
Sartori G, *259*
Sato S, *515*
Savriè M, *143*
Schaefer U, *131*, *287*
Schiavon P, *147*
Schledermann H, *201*
Schmidt G, *181*, *507*
Schmitt H, *201*
Schneider R, *131*, *287*
Schreiber O, *131*, *287*
Schuhl A, *143*
Schultz D, *305*
Schultz J, *305*
Severi M, *339*
Shapiro I S, *165*
Shibata T-A, *157*
Simon D J, *119*
Simone G, *143*
Skelly J F, *205*
Skjevling G, *339*
Smith G A, *305*, *323*
Smith G R, *195*
Sondheim W E, *195*
Soulliere M, *305*
Spyropoulou-Stassinaki M, *305*
Squier G T A, *137*
Straumann U, *131*, *287*
Stugu B, *339*
Suffert M, *175*, *181*, *507*
Sumiyoshi T, *523*
Sunier J W, *195*
de Swart J J, *503*

Takada Y, *515*, *523*
Takahashi T, *515*
Takeda T, *515*, *523*
Tang L, *323*
Tanimori T, *329*, *515*
Tauscher L, *175*, *297*
Tecchio L, *143*
Timmers P H, *503*
Tomasini F, *339*
Tosello F, *251*
Treichel M, *157*
Tröster D, *297*
Truöl P, *131*, *287*
Tsokos K, *349*
Tuominiemi J, *19*
Tzanakos G, *315*

UA1 collaboration, *19*, *355*
UA2 collaboration, *37*, *81*
UA5 collaboration, *421*, *465*

Valbusa U, *339*
Vandermeulen J, *499*
Vascon M, *251*
Venugopal R, *315*
Villari A, *147*
Voci C, *143*
Von Lintig R D, *305*

Walcher Th, *157*
Ward D R, *465*
Webber B R, *99*
Welsh R E, *137*
Wharton W, *201*
White B L, *131*, *287*
Whitmore J, *305*
Williams C, *297*
Winter R G, *137*
von Witsch W, *323*
Wodrich W R, *131*, *287*
Woldt P, *201*
Wolfe D M, *305*
Wolfendale A W, *235*

Xue Y, *323*

Yamagata T, *519*
Yamamoto S S, *515*
Yavin A I, *187*

Zanella G, *251*
Zenoni A, *251*
Ziegler M, *131*, *287*
Zioutas K, *297*